Palgrave Foundations

A series of introductory texts across a wide range of subject areas to meet the needs of today's lecturers and students

Foundations texts provide complete yet concise coverage of core topics and skills based on detailed research of course requirements suitable for both independent study and class use – *the firm foundations for future study.*

Published

History of English Literature
Biology
Chemistry
Contemporary Europe
Economics
Economics for Business
Modern British History
Nineteenth-Century Britain
Physics
Politics

Forthcoming

British Politics
Marketing
Maths for Science and Engineering
Modern European History
Sociology

Physics

Second edition

JIM BREITHAUPT

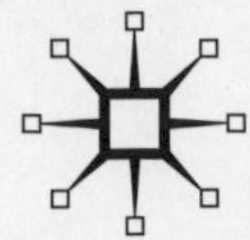

First edition 1999
Second edition 2003

Published by
PALGRAVE MACMILLAN
Houndmills, Basingstoke, Hampshire RG21 6XS
and 175 Fifth Avenue, New York, N.Y. 10010
Companies and representatives throughout the world

PALGRAVE MACMILLAN is the global academic imprint of the Palgrave Macmillan division of St. Martin's Press, LLC and of Palgrave Macmillan Ltd. Macmillan® is a registered trademark in the United States, United Kingdom and other countries. Palgrave is a registered trademark in the European Union and other countries.

ISBN 1–4039–0055–8 paperback

A catalogue record for this book is available from the British Library.

This book is printed on paper suitable for recycling and made from fully managed and sustained forest sources.

Typeset by Footnote Graphics, Warminster, Wilts

10 9 8 7 6 5 4 3 2 1
12 11 10 09 08 07 06 05 04 03

Printed in China

Contents

PART 4

Electricity

PART 5

Fields

Preface

This book provides a first course in physics for those studying access or foundation programmes in Physics at university or college, and for non-specialists on degree courses in Biological sciences, Chemistry and Engineering for whom physics is a subsidiary subject in their first year at university. The book is also suitable for trainee science teachers and medical students who need to develop a solid background in physics.

No previous experience of the subject is assumed as the book starts from 'scratch'. Guidance on how the book may be used is provided opposite. The introductory section on 'Units and Measurements' is designed to equip students with essential knowledge and skills in order to gain confidence and make rapid progress at the outset of the course. Each unit commences with a list of objectives and is divided into topics. Each topic introduces mathematical skills as appropriate and ends with questions to ensure the topic is assimilated before progressing to the next. A list of the 65 experiments described in the book is provided on pp. 434–435. Each unit ends with a summary of the key points and a set of revision questions. The book contains over 400 questions with complete solutions as well as a further 112 questions with numerical answers only.

The second edition has been revised and reformatted with a website as a new feature that may be found at http://www.palgrave.com/foundations/breithaupt. The website contains case studies on applications of physics and mathematical exercises to develop the essential mathematical skills featured in the book. In addition, the spreadsheet modelling features referred to in the book can be accessed through the website, and some proofs and experiments have been moved to the website. This has allowed clearer page layouts and the addition of some new material on U-values.

The units of the book are grouped into parts, with each part being a well-defined area of Physics. The first four parts (Waves, Materials, Mechanics and Electricity) each provide an independent route into the subject, thus providing flexibility for lecturers as well as for students. The part on Fields and the part on Atomic and Nuclear Physics build on topics studied in the first four parts. The final part on Further Physics provides an opportunity for selected in-depth studies appropriate to the needs of each student, either in support of specialist studies in another academic discipline or in preparation for further specialist studies in Physics or another academic discipline.

The final part of the book includes an appendix on the use of spreadsheets in physics, an appendix listing all the experiments described in the book and an appendix on where mathematical skills are introduced in the book. In addition, a summary of equations and a list of useful data is provided. Complete solutions to all revision questions in the book are provided in this part. Key terms, definitions and equations are highlighted in bold in the text and a glossary of key terms is also provided for reference and revision purposes.

The features described above are intended to enable the book to be used by the student in class, in the laboratory, in 'tutorial' workshops and for private study. The book should provide flexible support for teachers of foundation courses in Physics, used as appropriate with students according to their backgrounds and specific needs. Whether you are a student or a teacher, I hope the book serves your needs.

Jim Breithaupt

To the Student: How to use this book

- This book is designed to introduce you to Physics and to give you confidence to use the subject or to progress to further studies in the subject.
- The book is designed so that you can work on the subject with the minimum of help.
- Make sure you master the introduction 'Units and measurements' at the outset.
- Take note of the advice offered at the start of each part.
- If possible, study each unit topic-by-topic in sequence.
- Use the glossary of key definitions when a term from another topic holds up your progress.
- Do the questions at the end of each topic and at the end of each unit. When you complete each question, check your solution against the solution provided in the answers section. Questions with numerical answers only are also provided for further practice.
- Move on to the next topic only when you are satisfied that you have mastered the previous topic.
- Establish where your physics course links to your other courses and develop these links as you progress through your physics course.
- Use the location guide to mathematical skills (see p. 436) if necessary if a difficult mathematical skill reappears.
- Keep a folder of your laboratory reports and retain your own solutions for revision purposes.
- Use the website http://www.palgrave.com/foundations/breithaupt as appropriate. The mathematical exercises are intended to be used when practice is needed in each mathematical skill. The spreadsheet section provides rapid access to software for modelling dynamics and decay processes referred to in the book. Case studies are provided to illustrate applications of basic principles. Some proofs and experiments from the 1st edition are now only on the website and these are for reference as appropriate.

Acknowledgements

The production of this book has involved a considerable number of people and I wish to thank all who have contributed to the task of converting the manuscript into the book. In particular I would like to thank Frances Arnold and Suzannah Burywood of the publishers, Liz Jones and Footnote Graphics for their particular expertise, advice and courtesy. I am indebted to my wife and my daughters for their cheerful support and encouragement throughout.

Grateful acknowledgement is made to the following for permission to use copyright material:

AEA Technology plc p. 321; Associated Press Ltd p. 419; Professor Castle, School of Mechanical and Materials Engineering, Surrey University p. 69; Greg Evans International pp. 84, 144; f8 Imaging pp. 4, 5, 21, 31, 67, 97, 156, 165, 180, 219, 260, 383; Oxford University Press for a photograph taken by R. Morrison, from W. Llowarch *Ripple Tank Studies of Wave Motion*, 1961, p. 17; Steve Redwood p. 220; Dr B E Richardson, Department of Physics and Astronomy, Cardiff University p. 30; Ann Ronan at Image Select pp. 1, 238, 305, 308; Science Photo Library pp. 11, 12, 28, 57, 68, 130, 166, 352; Alan Thomas pp. 108, 173, 198; The Welcome Trust p. 63.

Every effort has been made to trace all the copyright-holders, but if any have been inadvertently overlooked the publisher will be pleased to make the necessary arrangements at the first opportunity.

Units and Measurements

Philosophy and practice in science

Experiments play a vital part in the development of all branches of science. Over five hundred years ago, Nicolas Copernicus's theory that the Earth and the planets orbit the Sun was rejected by the Church because it displaced the Earth from the centre of the Universe. Copernicus had no experimental evidence to support his theory, so there was no reason to change the theory established in Ancient Greece. Many years after Copernicus, Galileo used the newly invented telescope to observe the motion of the moons of Jupiter and he realised that the planets orbit the Sun just like the moons of Jupiter orbit Jupiter. Galileo was arrested when he published his evidence for the Copernican view of the Universe but his evidence could not be denied as it was based on observations.

Scientists now believe the Universe originated in a massive explosion, referred to as the Big Bang. The evidence is based on accurate and reliable measurements by astronomers. In the mid-twentieth century, some astronomers theorised that the Universe was in a steady state, i.e. the same now as it has always been. The steady state theory was rejected in favour of the Big Bang Theory when astronomers discovered the presence of low-intensity microwave radiation in all directions in space – radiation that originated in the Big Bang which is still travelling through space. The controversy was settled by experimental evidence which supports the Big Bang theory. Nevertheless, experimental evidence in support of a theory is not the same as proof, and just one experiment is all that is needed to disprove a theory. The more experimental tests that are found to support a theory, the stronger that theory becomes.

Fig. 1 Galileo, the father of the Scientific Age.

The S.I. system of units

Different units for the same quantity can be very misleading. For example, energy supplied by gas is measured in therms, whereas energy supplied by electricity is in kilowatt hours. If gas and electricity suppliers used the same unit, the consumer would be able to compare costs more effectively. Scientists have agreed on a single system of units to avoid unnecessary effort and time converting between different units of the same quantity. This system, the **Système International,** is based on a defined unit for each of five physical quantities, as listed in Table 1. Units of all other quantities are derived from the S.I. base units.

Table 1 S.I base units

Physical quantity	Unit
Mass	kilogram (kg)
Length	metre (m)
Time	second (s)
Electric current	ampere (A)
Temperature	kelvin (K)

Numbers and units

Derived units

The following examples show how the units of all other physical quantities are derived from the base units.

(a) The unit of area is the square metre, abbreviated to m^2.

(b) The unit of volume is the cubic metre, abbreviated to m^3.

(c) Density is defined as mass per unit volume. Its unit is the kilogram per cubic metre, abbreviated to kg m^{-3}. The negative sign in a unit means 'per' (e.g. per m^3).

(d) The unit of speed is the metre per second, abbreviated to m s^{-1}.

Note

The cubic centimetre (cm^3) and the gram (g) are used widely outside science and are therefore allowed exceptions to the system of prefixes illustrated here.

Standard form

Numerical values smaller than 0.001 and greater than 1000 are usually expressed in **standard form** for convenience. The numerical value is written as a number between 1 and 10, multiplied by an appropriate power of ten corresponding to how many places the decimal point needs to be moved until the number is between 1 and 10. If the decimal point is moved to the right, the power of 10 will have a negative value. Engineers usually express numerical values in **ternary form** where the power of ten is a multiple of 3 (i.e. 10^3, 10^6, etc.)

Numerical prefixes

These are used to avoid unwieldy numerical values. The most common numerical prefixes are:

nano-	10^{-9}	(n)
micro-	10^{-6}	(μ)
milli-	10^{-3}	(m)
kilo-	10^{3}	(k)
mega-	10^{6}	(M)
giga-	10^{9}	(G)

Examples

(a) The mean radius of the Earth is 6 360 000 m. The decimal point after the final zero needs to be moved six places to the left to arrive at 6.36. Therefore, in standard form, the Earth's radius is 6.36×10^6 m.

(b) The speed of light in free space is 300 000 000 m s^{-1}. This is written as 3.0×10^8 m s^{-1} in standard form.

(c) The wavelength of yellow light from a sodium lamp is 0.000 000 59 m or 5.9×10^{-7} m in standard form. Note the negative power of ten because the decimal point is moved to the right to arrive at 5.9.

Significant figures

If a numerical value is expressed in standard notation, the number of digits in the decimal number is the number of significant figures of the numerical value. For example, in (c) above, the wavelength of yellow light is given to 2 significant figures whereas in (a), the radius of the Earth is expressed to 3 significant figures. In (b) above, the speed of light is expressed to 2 significant figures as 3.0×10^8; a value of 3×10^8 is expressed to one significant figure only.

Calculators give numerical answers to as many significant figures as there are digits on the display. This can be misleading, since the answer to a calculation should always be to the same number of significant figures as the data supplied. Most data for problem-solving exercises are usually given to no more than 3 significant figures. A numerical answer should therefore be rounded up or rounded down to match the number of significant figures of the data supplied. The general rule is to round up if the last digit is 5 or more; round down otherwise. For example, a value of 6.36 rounded to 2 significant figures would become 6.4, whereas a value of 6.33 would become 6.3.

Reminder

Mass: 1000 g = 1 kg

Length:

1000 mm = 100 cm = 1 m

Area:

10^6 mm^2 = 10^4 cm^2 = 1 m^2

Volume:

10^9 mm^3 = 10^6 cm^3 = 1 m^3

Questions U1

1. Copy and complete the following conversions.
   (a)(i) 500 mm = _____ m, **(ii)** 3.2 m = _____ cm,
   (iii) 9560 cm = _____ m.
   (b)(i) 0.45 kg = _____ g, **(ii)** 1997 g = _____ kg,
   (iii) 54000 kg = _____ × 10 g.
2. Write the following values in standard form to 3 significant figures.
   (i) 150 million kilometres in metres,
   (ii) 365 days in seconds,
   (iii) 630 nanometres in metres,
   (iv) 25.78 micrograms in kilograms,
   (v) 1502 metres in millimetres,
   (vi) 1.245 micrometres in metres.

Symbols and equations in physics

Physical quantities are usually represented by a standard symbol. For example, time is usually represented by the symbol t. Greek symbols are sometimes used (e.g. the Greek letter ρ (pronounced rho) is used for density). The symbol represents a numerical value and its unit.

Equations and formulae in physics are usually presented in symbolic form. For example, the word equation

$$\text{density} = \frac{\text{mass}}{\text{volume}}$$

can be written $\rho = \dfrac{m}{V}$

Note that the S.I. unit of density is kilogram per cubic metre, usually abbreviated as kg m^{-3}.

To use an equation:

- write down the equation and the values of the known quantities
- rearrange the equation so the unknown quantity is the subject of the equation

- substitute the known values into the rearranged equation. In simple equations, include the units.
- calculate the unknown quantity and write down its numerical value and unit.

Worked example

Calculate the mass of a steel block of volume $0.15\ m^3$ and density $7900\ kg\ m^{-3}$.

Solution

$\rho = \frac{m}{V}$ $\quad V = 0.15\ m^3 \quad \rho = 7900\ kg\ m^{-3}$

Rearranging the equation gives $m = \rho V = 7900\ kg\ m^{-3} \times 0.15\ m^3 = 1185\ kg$.

Worked example

The area, A, of a circle of diameter, d, is given by the formula $A = \frac{\pi d^2}{4}$.

Calculate the diameter of a circular plate of area $0.10\ m^2$.

Solution

Rearrange the equation to give $d^2 = \frac{4A}{\pi}$

Substituting $A = 0.10\ m^2$ gives $d^2 = \frac{4 \times 0.10}{\pi} = 0.127\ m^2$

Hence $d = \sqrt{0.127} = 0.36\ m$ (to 2 sig. figs).

Note

Electronic calculators save time, provided you don't press the wrong button. Calculator layouts and usage may differ from one type to another but the essential principles of operation remain unchanged. A scientific calculator is needed for physics at this level.

Making measurements in physics

You will meet a wide range of instruments for making measurements in your physics course. Instruments such as stopwatches, metre rules and top pan balances are easier to use than others, such as oscilloscopes and Geiger counters. Many instruments, such as frequency meters and electrical multimeters, have digital read-outs instead of scales. However, accurate and reliable measurements from any instrument require practice and care. Here are some hints and tips on how to use the more common instruments.

- The **top-pan balance** is used to measure mass. It should be level and in a draught-free environment. Its read-out can easily be checked using a set of standard masses. Most top-pan balances have more than one range. To avoid overloading a balance, always use the highest range first and then select the most appropriate range for the object under test. A 'tare' button on the balance enables the read-out to be set to zero with a beaker on the pan. If the beaker is removed, filled and then replaced, the read-out gives the mass of its contents directly. The precision of a balance (eg. to 0.01 g) is given by the least significant figure on the read-out.

Fig. 2 A top-pan balance.

- **Vernier calipers** are used to measure the length of an object to within 0.1 mm, for lengths up to about 200 mm. They may be used to measure the external diameter or the internal diameter of a pipe. Fig. 3 shows a vernier scale. Each interval of the slider scale is 0.9 mm. This is so that the position of the zero of the slider scale beyond the nearest millimetre mark on the main scale can be determined by seeing where a mark on the slider scale is aligned with a millimetre mark on the main scale.
- A **micrometer** is used to measure lengths to within 0.01 mm, up to about 50 mm. Fig. 4 shows a micrometer. Each complete turn of the barrel alters the gap by 0.5 mm exactly. There are usually 50 equal divisions on the barrel scale so that each division corresponds to 0.01 mm. If used to measure the diameter of a wire, several readings should be made at different positions and an average value calculated.
- A **stopwatch** is used to measure time intervals and may be checked against the 'speaking clock' on any phone system. Human error in starting and stopping a stopwatch can be minimised by counting down.

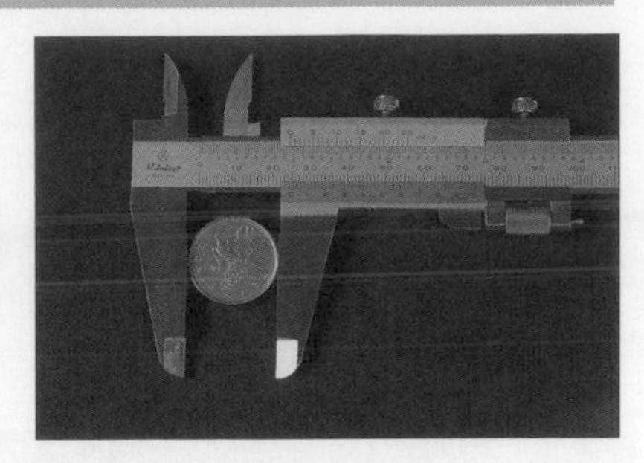

Fig. 3 Using vernier calipers.

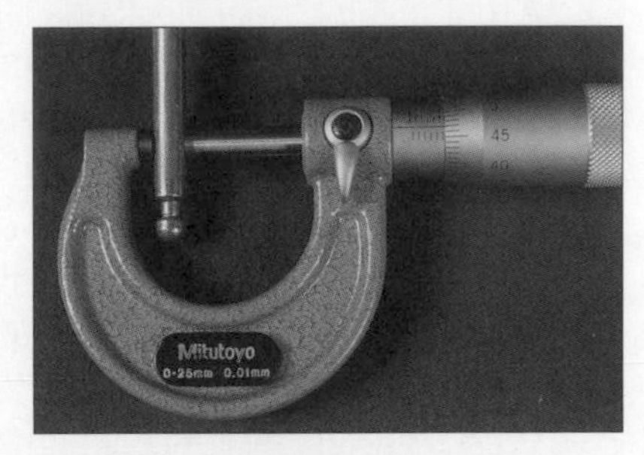

Fig. 4 Using a micrometer.

Density tests

1. Regular solids

Measure the mass and dimensions of rectangular and cylindrical blocks of different materials. Calculate the volume of each object using the appropriate formula from Fig. 5. Use your results to calculate the density of each object, using the formula

$$\text{density} = \frac{\text{mass}}{\text{volume}}$$

If possible, use a data book of density values to identify the material of each object. Record your measurements and calculations.

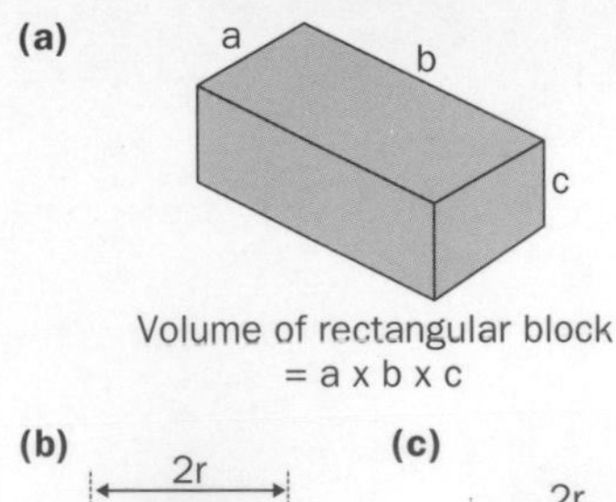

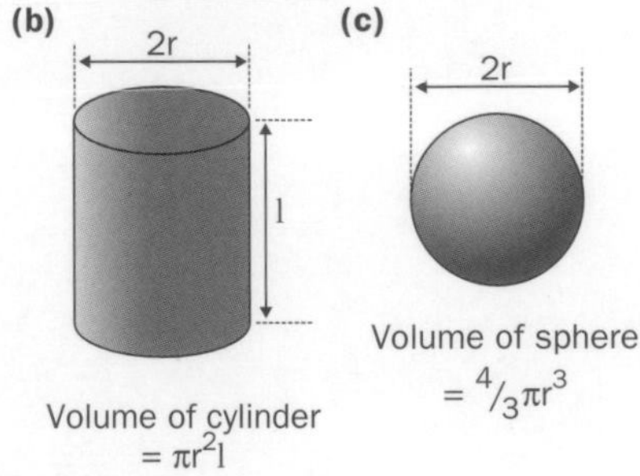

Fig. 5 Volume formulae.

2. Liquids

Measure the mass of a given quantity of liquid using a suitable beaker and a top-pan balance. Also, measure the volume of the liquid using a suitable measuring cylinder, as in Fig. 6a. Hence calculate the density of the liquid.

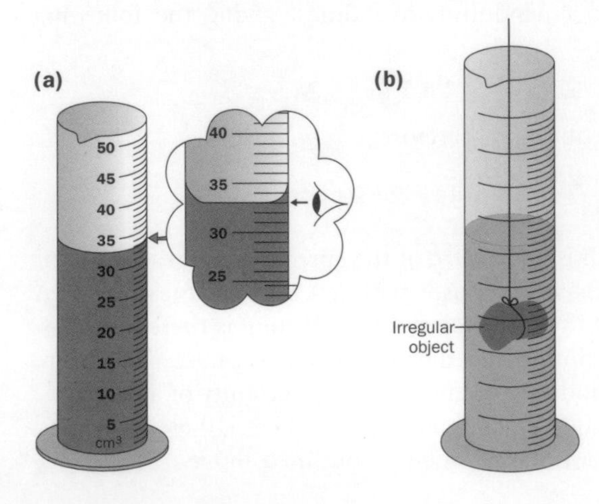

Fig. 6 Using a measuring cylinder **(a)** to measure the volume of a liquid, **(b)** to measure the volume of an irregular object.

Note

Give your density values in kg m^{-3}. Use the conversion factors on p. 3 to convert from grams to kilograms, etc. Measuring cylinders are calibrated in cm^3. To calculate a volume in m^3, divide by the volume in cm^3 by 10^6. Note that 1 g cm^{-3} = 1000 kg m^{-3}.

3. Irregular objects

Measure the mass of the object. Then suspend the object from a thread and immerse it in water in a measuring cylinder, as in Figure 6(b). The rise of the water level is equal to the volume of the object. Hence calculate the density of the object.

Questions U2

1. A rectangular steel block has dimensions 120 mm × 80 mm × 25 mm. Its mass is 1900 g. Calculate **(a)** its mass in kilograms, **(b)** its volume in m^3, **(c)** its density in kg m^{-3}.
2. A solid metal cylinder of diameter 32 mm and length 22 mm has a mass of 48 g. Calculate **(a)** its mass in kilograms, **(b)** its volume in m^3, **(c)** its density in kg m^{-3}.
3. An empty can of mass 160 g was filled to a depth of 120 mm with water. The total mass of the can and contents was 1.120 kg. The density of water is 1000 kg m^{-3}.

   (a) Calculate **(i)** the mass of water in the can, **(ii)** the volume of water in the can, **(iii)** the internal diameter of the can.

   (b) A glass object was then immersed in the can, causing the water level to rise by 5 mm. The mass of the tin and its contents was re-measured and found to be 1.225 kg. Calculate **(i)** the mass of the glass object in kilograms, **(ii)** its volume in m^3, **(iii)** its density in kg m^{-3}.

Uncertainty and accuracy

In the density tests on p. 5, how confident can you be of your results? Statements such as 'I'm very confident about my results because I was very careful' carry little weight in science. Measurements need to be checked by repeating the readings for each measurement and using them as outlined below to calculate the following:

1. The **mean value,** which is obtained by adding the readings and dividing by the number of readings. For example, suppose a stopwatch is used to time 20 swings of a pendulum five times, giving the following measurements.

   25.2 s 24.7 s 25.1 s 24.7 s 24.9 s

   The mean value of the timing is therefore

   $$\frac{25.2 + 24.7 + 25.1 + 24.7 + 24.9}{5} = 24.9 \text{ s.}$$

2. The **uncertainty**, which is a measure of the spread of the readings. For example, the timings above are mostly within 0.2 s of the mean value. A reasonable estimate of the uncertainty in the timing is therefore 0.2 s. The mean value of the timing may then be written as 24.9 ± 0.2 s. More precise statistical methods for obtaining the uncertainty of a measurement are used by scientists and engineers. At this level, there is no need to go further than the 'estimating' method outlined above.

Note

The probable error of a measurement is an alternative indicator of the spread of a set of readings.

Repeated readings of a measurement with a sufficiently precise instrument will, in general, give a spread of readings. The readings will vary due to **errors** of measurements which may be human errors (e.g. reaction time when stopping and starting a stopwatch) or errors due to varying conditions. If there is no pattern to the variation in a set of readings, the errors are said to be **random**. If, however, some sort of pattern or trend is evident in successive measurements, the errors are said to be **systematic**.

If repeated readings of a measurement are the same, the instrument could be replaced by a more precise instrument. If this is not possible, the precision of the instrument is a reasonable estimate of the uncertainty of the measurement. For example, suppose the read-out on a top-pan balance shows 35.6 g. The measurement could lie between 35.55 and 35.65 g. The uncertainty in the measurement is therefore 0.05 g. Since there is also a similar uncertainty for the zero reading, the uncertainty in the mass is therefore 0.1g. The value of the mass is therefore written as 35.6 ± 0.1 g.

The **overall uncertainty** of a quantity calculated from several measurements can be estimated by using the upper and lower limits of each measurement. For example, suppose the width of a rectangular sheet of metal lies between 20.8 and 21.2 mm and its length lies between 26.8 and 27.2 mm. The area of the sheet therefore lies between 557 mm² (= 20.8mm × 26.8mm) and 577 mm² (=21.2 mm × 27.2 mm). The area is therefore 567 ± 10 mm².

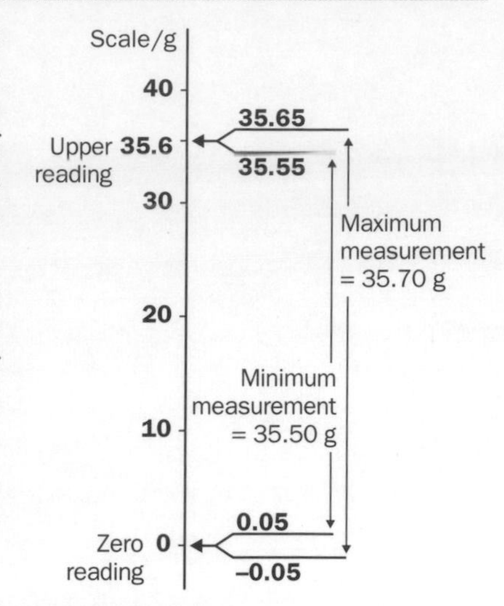

Fig. 7 Probable errors.

Exercise

Revisit the density tests you carried out and estimate the uncertainty in your measurements. Hence determine the percentage uncertainty for each result.

See www.palgrave.com/foundations/breithaupt for more about uncertainties in calculations.

Worked example

A metal cube has dimensions 20 mm × 20 mm × 20 mm, each dimension accurate to 0.2 mm. Its mass is 64.2 ± 0.2 g. Calculate (a) its volume, (b) the uncertainty in its volume, (c) its density, (d) the uncertainty in its density.

Solution

(a) Volume $= 20 \times 20 \times 20\ \text{mm}^3 = 8000\ \text{mm}^3 = 8.0 \times 10^{-6}\ \text{m}^3$.

(b) Maximum volume $= 20.2 \times 20.2 \times 20.2\ \text{mm}^3 = 8240\ \text{mm}^3$.

Minimum volume $= 19.8 \times 19.8 \times 19.8\ \text{mm}^3 = 7760\ \text{mm}^3$.

The uncertainty in the volume is therefore $240\ \text{mm}^3 = 2.4 \times 10^{-7}\ \text{m}^3$.

(c) Density $= \dfrac{\text{mass}}{\text{volume}} = \dfrac{64.2 \times 10^{-3}\ \text{kg}}{8.0 \times 10^{-6}\ \text{m}^3} = 8025\ \text{kg m}^{-3}$.

(d) The maximum value of the density

$$= \frac{\text{maximum mass}}{\text{minimum volume}} = \frac{64.4 \times 10^{-3}\ \text{kg}}{7760 \times 10^{-9}\ \text{m}^3} = 8300\ \text{kg m}^{-3}$$

The minimum value of density

$$= \frac{\text{minimum mass}}{\text{maximum volume}} = \frac{64.0 \times 10^{-3}\ \text{kg}}{8240 \times 10^{-9}\ \text{m}^3} = 7770\ \text{kg m}^{-3}.$$

The value of the density is therefore $8030 \pm 270\ \text{kg m}^{-3}$ to 3 significant figures.
The uncertainty of $270\ \text{kg m}^{-3}$ is approximately 3% of the measured density.
The percentage uncertainty or 'error' in the measured density is therefore 3%.

Summary

- **The S.I. system** of scientific units is based on the metre, the kilogram, the second, the ampere (electric current) and the kelvin (temperature). All other units are derived from these five base units.
- **Prefixes** commonly used in science are nano- (n), micro- (μ), milli- (m), kilo- (k), mega- (M) and giga-(G), representing 10^{-9}, 10^{-6},10^{-3},10^{3}, 10^{6} and 10^{9} respectively.
- **Commonly used instruments** in physics include vernier calipers and the micrometer (lengths), the top-pan balance (mass), the stopwatch (time intervals) and the measuring cylinder (volume).
- The **uncertainty** of a measurement gives the range within which the measurement lies.
- **Area and volume formulae**
  - The area of a circle of radius r $= \pi r^2$
  - The surface area of a sphere of radius r $= 4\pi r^2$
  - The volume of a rectangular block = length × width × height
  - The volume of a cylinder of radius, r, and height, H $= \pi r^2 H$
  - The volume of a sphere of radius r $= \frac{4}{3}\pi r^3$

Revision Questions

1. Which of the following instruments, A – E, would you use to measure
(a) the volume of a liquid, **(b)** the thickness of a coin, **(c)** the mass of a beaker, **(d)** the diameter of a wire?
A vernier calipers **B** top-pan balance **C** measuring cylinder
D mm rule **E** micrometer

2. Fig. 8 shows a top-pan balance with an empty beaker on. The balance read zero before the beaker was placed on the pan.

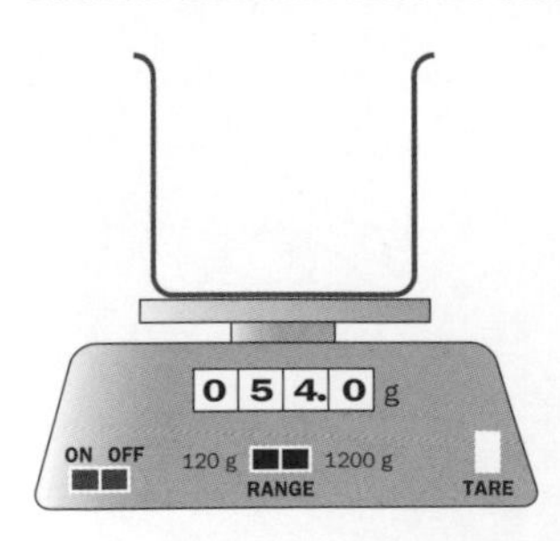

Fig. 8

(a) (i) What is the mass of the beaker? **(ii)** How accurate is the read-out?
(b) The beaker was removed and partly filled with liquid. When the beaker was replaced on the pan, the read-out displayed 157.4 g.
(i) Calculate the mass of liquid in the beaker. **(ii)** Write down the uncertainty in the mass of liquid.
(c) The liquid was poured into the measuring cylinder shown in Fig. 9.
(i) What is the volume of the liquid in the measuring cylinder in Fig. 9? **(ii)** Estimate the uncertainty in this volume measurement.
(d)(i) Calculate the density of the liquid. **(ii)** Estimate the percentage uncertainty in this value of density.

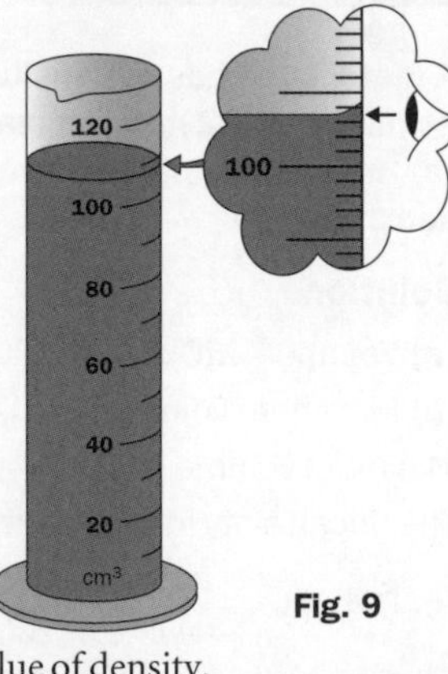

Fig. 9

3. A rectangular metal plate of thickness 2.5 mm is 100 mm long and 65 mm wide. Its mass is 120 g. Calculate **(a)** the distance around its edge in metres, **(b)** the area of one of its flat faces in m^2, **(c)** its volume in m^3, **(d)** its density in kg m^{-3}.

4. A solid aluminium rod of length 1.2 m has a uniform diameter of 52 mm. The density of aluminium is 2700 kg m^{-3}. Calculate **(a)** its volume in m^3, **(b)** its mass in kg.

5. (a) Calculate the volume of a steel sphere of diameter 24 mm. **(b)** Calculate the diameter of a steel sphere of volume 0.60 cm^3.

Waves

Unit 1. Properties of Waves
Unit 2. Sound
Unit 3. Optics
Unit 4. Electromagnetic Waves

Part 1 introduces you to the basic principles of wave motion and shows you how these basic principles are used in a wide range of applications including sound and sight, imaging techniques, communications and energy transfer. The section provides plenty of opportunities for practical work and it offers a gradual introduction to the use of mathematics in Physics.

Properties of Waves

Objectives

After working through this unit, you should be able to:

- ▶ describe different types of waves and their uses
- ▶ state the meaning of the amplitude, wavelength, frequency and speed of a wave
- ▶ explain what is meant by phase difference
- ▶ relate the speed to the frequency and the wavelength of a wave
- ▶ describe reflection, refraction and diffraction as properties of all waves
- ▶ describe what is meant by a transverse wave and give examples
- ▶ describe what is meant by a longitudinal wave and give examples
- ▶ describe polarisation as a property of a transverse wave

Contents

1.1 Types of waves

Waves are used to transfer energy or to carry information. For example, microwaves are used for heating food in a microwave oven or for carrying TV or phone channels around the world via satellite links. Here is a list of different types of waves you will meet in this section. Some of the uses of each type of wave are also listed.

- **Sound waves** are vibrations in a medium that travel through the medium. For example, when you speak, your vocal cords make the surrounding air vibrate and the vibrations spread out. The speed of sound through air at room temperature is about 340 m s^{-1}. In general, sound waves travel faster through solids and liquids than through air. Sound waves above the range of the human ear are called **ultrasonics**. Ultrasonic imaging systems are used in hospitals for pre-natal scans.

- **Electromagnetic waves** are vibrating electric and magnetic fields that travel through space without the need for a medium. In essence, vibrations of the electric field generate vibrations of the magnetic field in

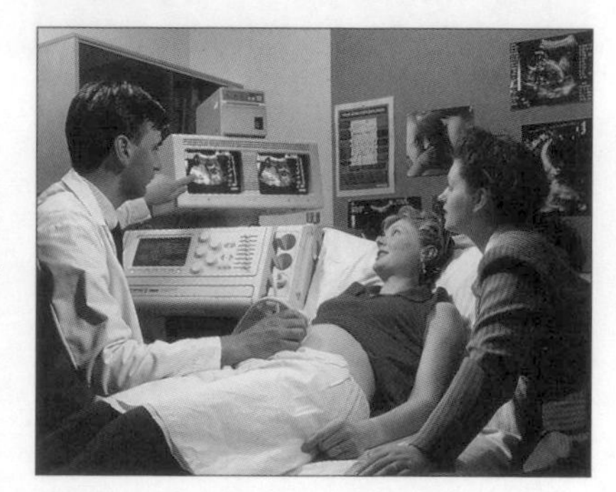

Fig. 1.1 Ultrasound scanning is a diagnostic technique which uses high-frequency sound waves. Here, a woman in early pregnancy receives an ultrasound scan of her abdomen to monitor foetal development. The doctor moves a hand-held transducer over the woman's lubricated skin, while two-dimensional images of the foetus can be seen on the screens.

the surrounding space which generates vibrations of the electric field further away, etc. The spectrum of electromagnetic waves is divided according to wavelength, as indicated in Table 1.1. Note that all electromagnetic waves travel at the same speed of 300000 km s^{-1} through space.

- **Seismic waves** are vibrations of the Earth, usually created by earthquakes (Fig. 1.2a). By recording seismic waves from earthquakes, the internal structure of the Earth has been determined.
- **Water waves** are useful for studying the general properties of waves. The ripple tank on p. 14 shows how the general properties of waves can be investigated. Electrical power from off-shore floating wave machines could make a significant contribution to national power supplies.

Table 1.1 Spectrum of electromagnetic waves

Type of wave	Radio waves	Microwaves	Infra-red	Visible	Ultraviolet	X-radiation	Gamma radiation
Wavelength	>0.1 m	0.1 m–0.1 mm	0.1 mm–700 nm	700 nm–400 nm	400 nm–1 nm	<1 nm	<0.001 nm
Applications	radio and TV	radar, heating	heating, security sensors	the eye, photography	UV ink driers, sun lamps	radiography	gamma therapy, sterilisers

(a)

(b)

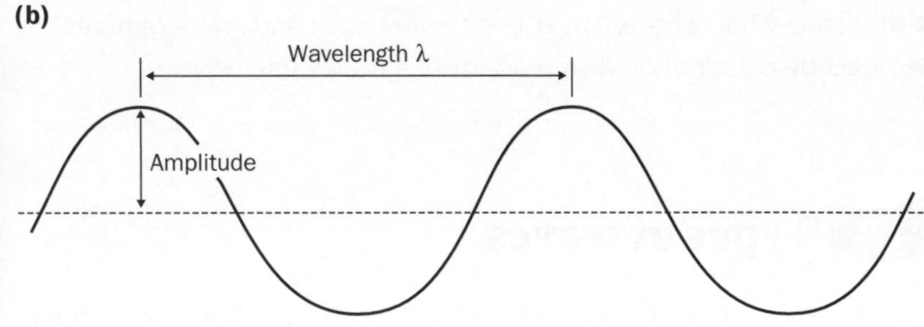

Fig. 1.2 **(a)** damage caused to houses in the Marina district of San Francisco during the Iowa Prieta earthquake of October 17th 1989, **(b)** snapshot of a travelling wave.

1.2 Measuring waves

Fig. 1.2b shows a snapshot of a wave on a rope travelling from left to right. The peaks and crests propagate (i.e. travel) to the right. Suppose the rope was invisible except for a spot at one point. How would you describe the motion of that spot?

The spot moves up and down along a line at right angles to the direction of propagation. If this seems odd, watch a small object floating on water when some waves pass it. The object bobs up and down.

- **The amplitude** of a wave is the maximum distance of a point from its equilibrium position. In Fig. 1.2b, it is the height of a wave from the middle.

- **The wavelength** of a wave is the distance from one peak on a wave to the next peak. The symbol for wavelength is the Greek letter λ (pronounced 'lambda').
- **The frequency** of a wave is the number of complete cycles of oscillation of a given point per second. This is the same as the number of wavelengths that pass a point in one second. The symbol for frequency is f, or the Greek letter ν (pronounced 'new'). The unit of frequency is the hertz (symbol Hz), equal to 1 cycle per second.

Note

The time period, T, of a wave is the time taken for one complete cycle to pass a point. For a wave of frequency f, $T = 1/f$.

Wave speed

Fig. 1.3 shows the wave in Fig. 1.2b again and then one cycle later, at time T from its initial position. Each peak has moved a distance of one wavelength in time T.

$$\text{The speed of propagation, } v = \frac{\text{distance moved}}{\text{time taken}}$$

$$= \frac{\lambda}{T} = \frac{\lambda}{(1/f)} = f\lambda$$

$$\mathbf{v = f\lambda}$$

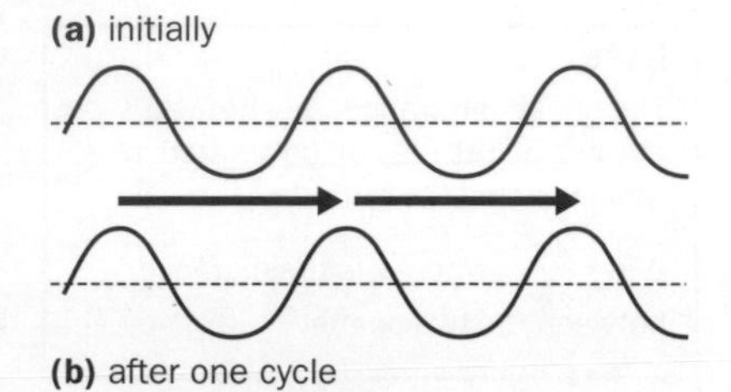

Fig. 1.3 One cycle later.

Worked example 1.1

The speed of light in air is 3.0×10^8 m s^{-1}. Calculate the frequency of light waves of wavelength 590 nm.

Solution

$v = 3.0 \times 10^8 \text{ m s}^{-1} \qquad \lambda = 590 \text{ nm} = 5.90 \times 10^{-7} \text{ m}$

$v = f\lambda \therefore f = \frac{v}{\lambda} = \frac{3.0 \times 10^8 \text{ m s}^{-1}}{5.90 \times 10^{-7} \text{ m}} = 5.1 \times 10^{14} \text{ Hz.}$

Phase difference

In Fig. 1.2b, any two points separated by a distance equal to a whole number of wavelengths move up and down together. They vibrate **in phase** which means that at any instant they are at the same distance and direction from equilibrium and are moving in the same direction.

Consider the wave in Fig. 1.4. Which points are in phase with point O? The **phase difference** between any two points is the fraction of a complete cycle between the two points. The symbol for phase difference is Δϕ (pronounced 'delta phi'). Phase difference is usually expressed either in:

- **degrees**, on a scale where 360° corresponds to a complete cycle, or
- **radians**, which is an angle scale in which 2π radians = 360°.

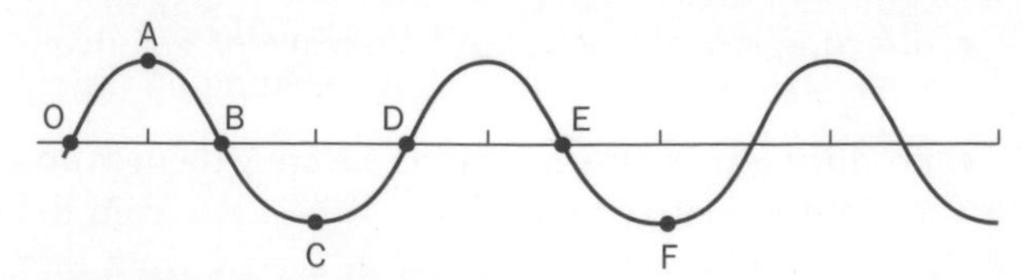

Fig. 1.4 Phase difference.

Table 1.2 Phase differences relating to Fig. 1.4

Point	Phase difference relative to O in degrees	in radians
A	90	π/2
B	180	π
C	270	3π/2
D	360	2π
E	180	π

Questions 1.2

1. Use a millimetre rule to measure **(a)** the amplitude, **(b)** the wavelength of the wave in Fig. 1.2b.
2. In Fig. 1.2b, the frequency of the wave was 5.0 Hz. Calculate **(a)** the time period of the wave, **(b)** the speed of propagation of the wave.
3. The speed of sound in air at room temperature is 340 m s^{-1}. Calculate **(a)** the frequency of sound waves in air of wavelength 0.10 m, **(b)** the wavelength of sound waves in air of frequency 3000 Hz.
4. The speed of radio waves in air is 3.0×10^8 m s^{-1}. Calculate **(a)** the frequency of radio waves of wavelength 1500 m in air, **(b)** the wavelength of radio waves in air of frequency 1.2 MHz.
5. In Fig. 1.4, what is the phase difference in **(a)** degrees, **(b)** radians between

   (i) A and F, **(ii)** B and D, **(iii)** C and F, **(iv)** O and F?

Note

The phase difference, $\Delta\phi$, between any two points may be calculated in radians using the formula $\Delta\phi = \frac{2\pi x}{\lambda}$, where x is the distance between the two points.

1.3 Properties of waves

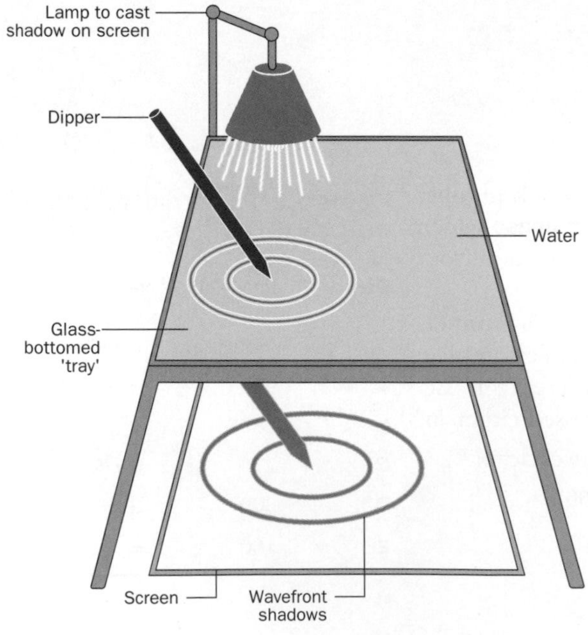

Using a ripple tank

Fig. 1.5 shows how to set up and use a ripple tank. This can be used to investigate reflection, refraction and diffraction of waves as outlined below. Note that each wavefront joins points that vibrate in phase with one another. The wavefront shadows on the screen can be thought of as wavecrests.

Fig. 1.5 A ripple tank.

Reflection

Investigate the reflection of straight and circular waves from different-shaped surfaces. For example, Fig. 1.6 shows how to generate and direct straight waves at a plane (i.e. straight) reflector. If possible, try this and other tests such as:

- directing straight waves at a concave or a convex reflector
- producing circular waves and reflecting them from straight and curved reflectors.

Compare your observations with the patterns described below.

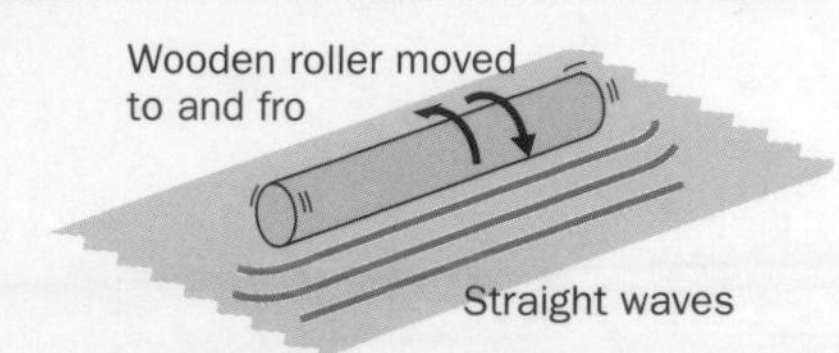

Fig. 1.6 Producing straight waves.

Straight waves reflected by a straight reflector

The angles between the reflector and the incident wavefronts and between the reflector and the reflected wavefronts are the same. The same effect is observed when a ray of light is reflected off a mirror.

Straight waves reflected by a concave reflector

The reflected waves converge at a point which is referred to as the **focal point** of the reflector. The detector of a satellite dish must be positioned at the focal point to pick up the strongest signal.

Circular waves reflected by a straight reflector

The waves reflect as if they originated at the same distance behind the reflector as the point source is in front. This effect is the same as when you look in a plane mirror and observe your image at the same distance behind the mirror as you are in front.

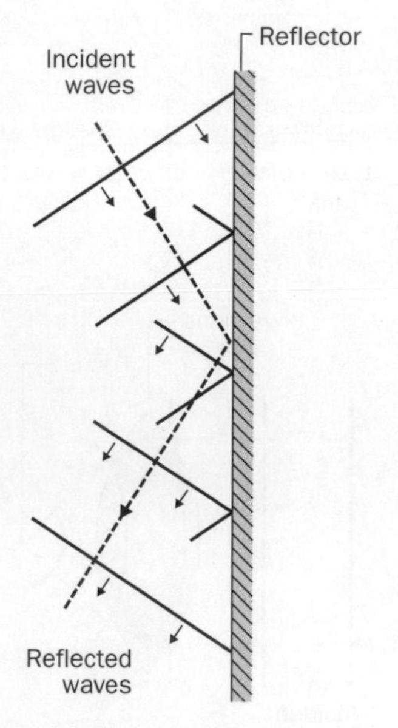

Fig. 1.7 Straight waves reflected by a straight reflector.

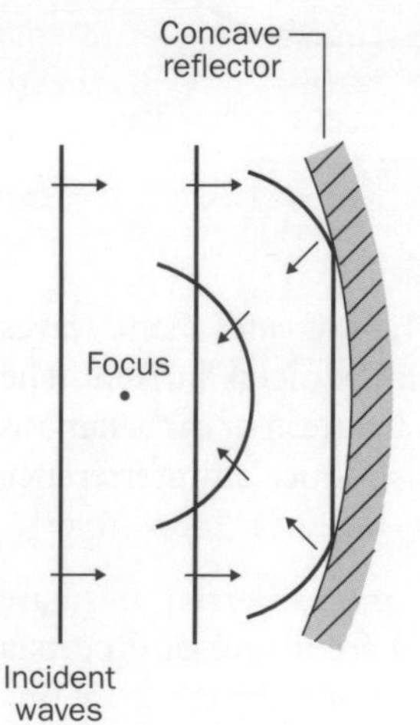

Fig. 1.8 Straight waves reflected by a concave reflector.

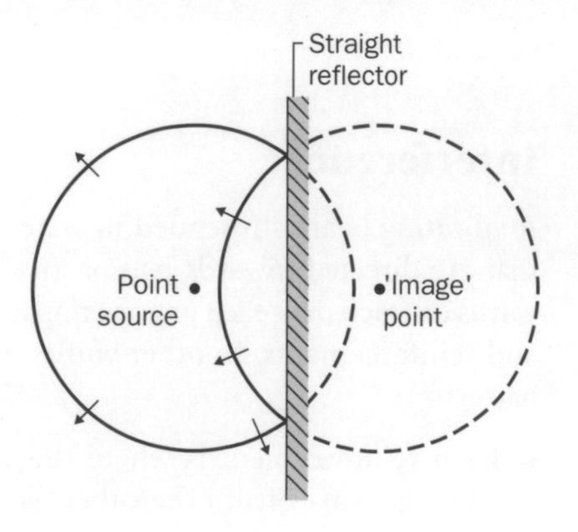

Fig. 1.9 Circular waves reflected by a straight reflector.

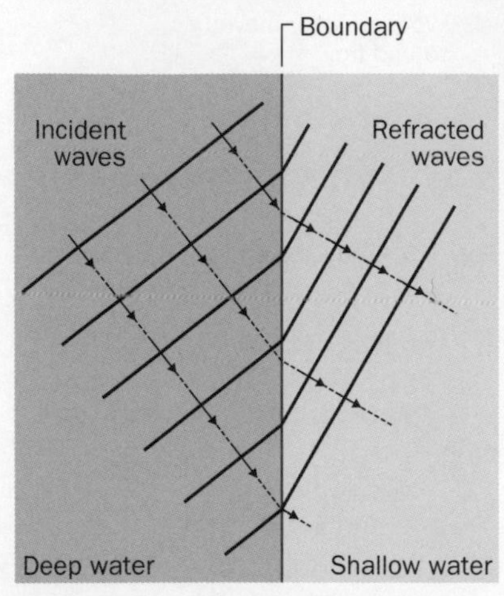

Fig. 1.10 Refraction of water waves in a ripple tank.

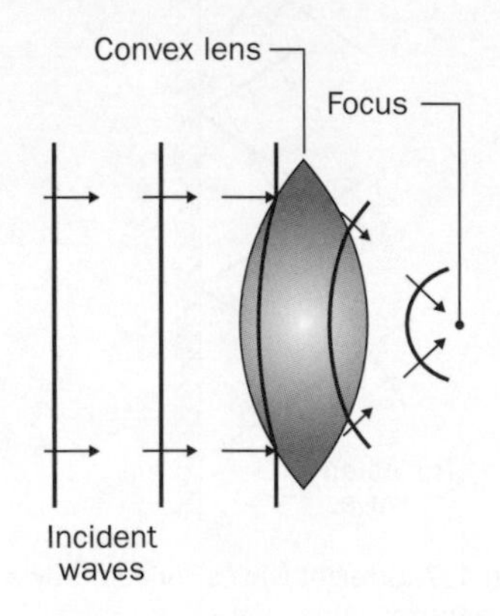

Fig. 1.11 Action of a lens.

Refraction

Water waves change direction when they travel across a boundary between deep and shallow water at a non-zero angle to the boundary. This change of direction is called **refraction**. It occurs because water waves travel faster in deep water than in shallow water. Refraction also takes place when a light ray travels from one transparent medium to another. See p. 39. Fig. 1.11 also shows how refraction explains the formation of an image by a convex lens.

Diffraction

Waves spread out when they pass through a gap or around the edge of an obstacle. This effect is known as **diffraction**. Fig. 1.13 shows plane waves passing through a narrow gap. The narrower the gap, the greater the amount of diffraction. Diffraction of light, on passing through an aperture, limits the detail of an image seen through a microscope or a telescope. The reduced depth of field in a high-speed photograph is caused by diffraction. This is because the aperture of a camera must be as wide as possible to enable a high-speed photograph to capture as much light as possible. Light from background objects is diffracted less if the aperture is widened, forming images in focus in front of the film.

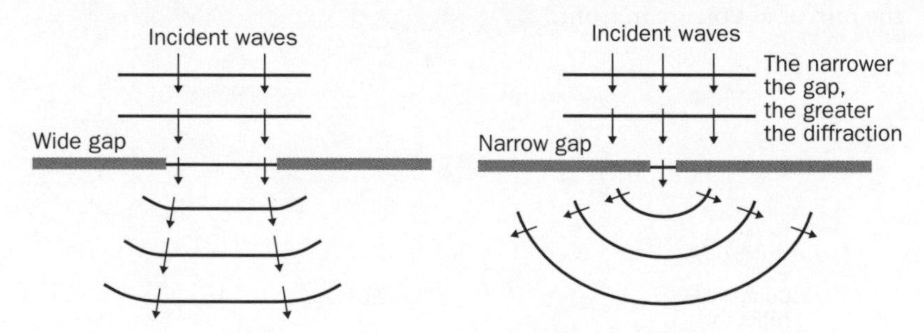

Fig. 1.12 Diffraction at a wide gap. **Fig. 1.13** Diffraction at a narrow gap.

Interference

A vibrating beam suspended in water, as in Fig. 1.14, creates plane waves that are directed towards two narrow gaps created by metal barriers. The waves diffracted by each gap overlap and produce a pattern of cancellations and reinforcements. In other words, the two sets produce an interference pattern.

- Each reinforcement is where the waves from one gap arrive in phase with the waves from the other gap (e.g. a crest from one gap meets a crest from the other gap to produce a supercrest or a trough from one gap meets a trough from the other gap to produce a supertrough).
- Each cancellation is where the waves from one gap arrive exactly out of phase with the waves from the other gap (e.g. a crest from one gap meets a trough from the other gap).

Note

The waves from one gap pass through the waves from the other gap. Each crest or trough that meets another crest or trough produces a cancellation or a reinforcement for an instant.

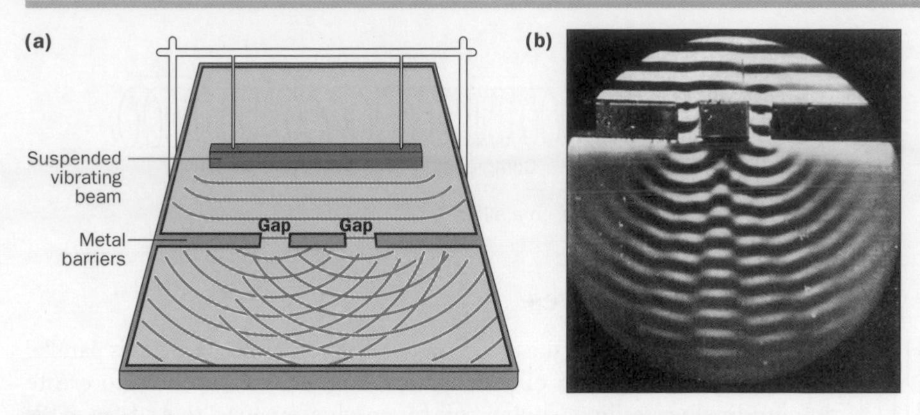

Fig. 1.14 Interference of water waves in a ripple tank **(a)** ripple tank arrangement **(b)** photograph showing the interference of water waves passing through two narrow, closely spaced slits.

The wave nature of light was deduced from the fact that light from a point source passing through two narrow, closely spaced slits produces an interference pattern consisting of alternate bright and dark fringes. Each dark fringe is where light from one slit cancels light from the other slit. As explained in Unit 3, this effect is used to deduce the wavelength of light.

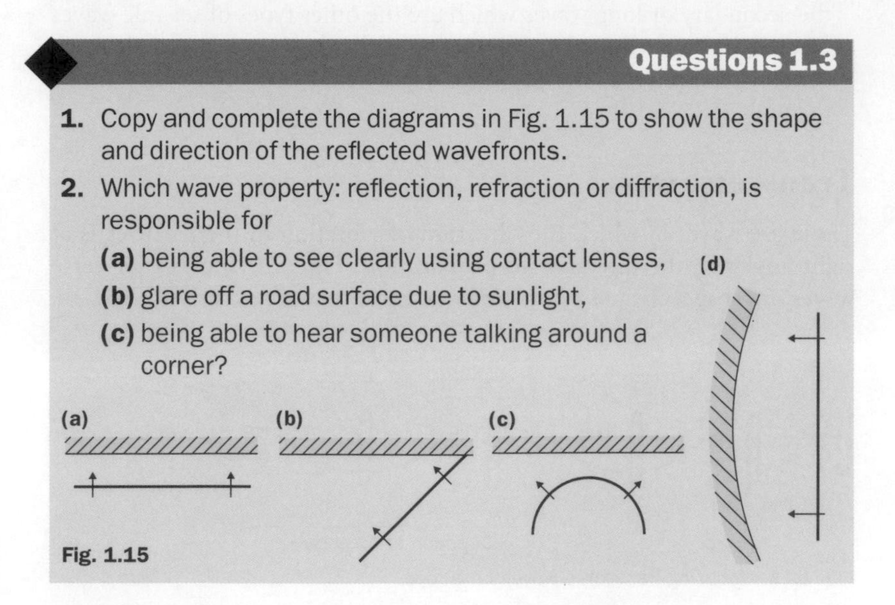

Questions 1.3

1. Copy and complete the diagrams in Fig. 1.15 to show the shape and direction of the reflected wavefronts.
2. Which wave property: reflection, refraction or diffraction, is responsible for
   (a) being able to see clearly using contact lenses,
   (b) glare off a road surface due to sunlight,
   (c) being able to hear someone talking around a corner?

Fig. 1.15

1.4 Longitudinal and transverse waves

When waves passes through a medium, the particles of the medium vibrate. The frequency of vibration of the particles is the same as the frequency of the waves. Most types of waves may be classified as either **longitudinal** or **transverse**, depending on whether the direction of vibration of the particles is along or perpendicular to the direction of propagation of the waves.

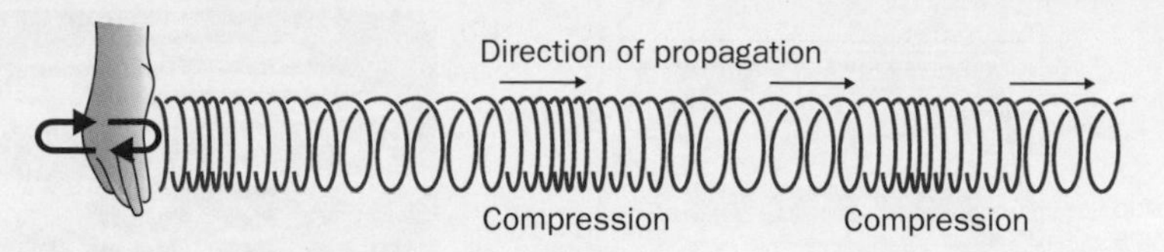

Fig. 1.16 Longitudinal waves on a slinky.

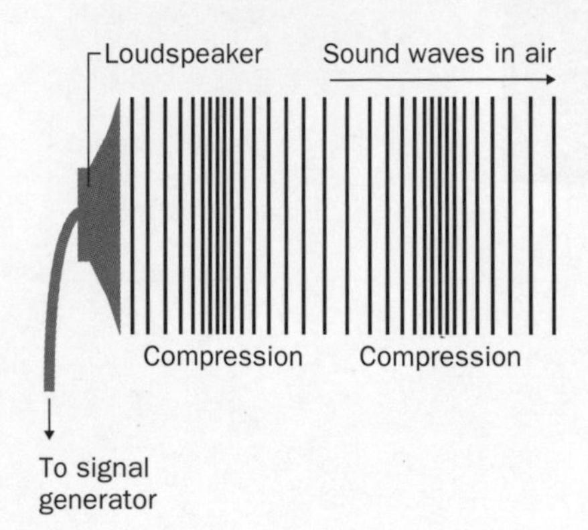

Fig. 1.17 Using a loudspeaker to create sound.

Longitudinal waves

These are waves in which the direction of vibration of the particles is parallel (i.e. along) the direction of propagation. Fig. 1.16 shows how to create longitudinal waves in a 'slinky' coil by moving one end to and fro. This generates a series of compressions which travel along the slinky.

- **Sound waves** are longitudinal waves. A loudspeaker supplied with alternating current creates sound waves because the diaphragm of the loudspeaker is forced to move to and fro. The diaphragm compresses the surrounding air in front of it as it moves forwards, then it moves back before creating another compression. Effectively, the air vibrates to and fro as the sound waves pass through it.
- **Primary seismic waves** are longitudinal waves. They travel faster than the secondary or long waves which are the other types of seismic waves created in an earthquake. Primary seismic waves can travel through solids and liquids as they push and pull on the medium they travel through.

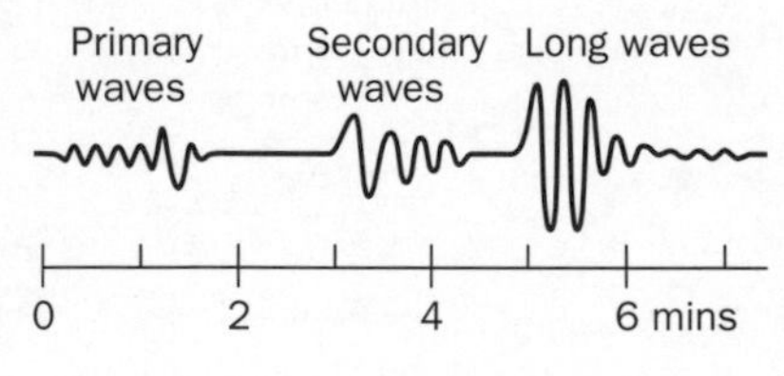

Fig. 1.18 Seismic waves.

Transverse waves

These are waves in which the direction of vibration of the particles is at right angles to the direction of propagation. Fig. 1.19 shows transverse waves on a rope, created by moving one end of the rope from side to side.

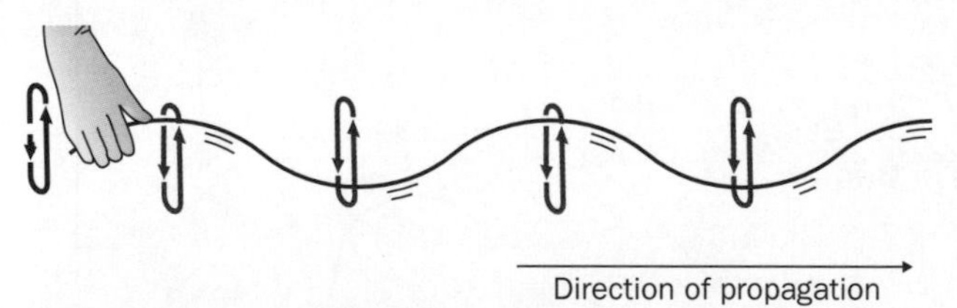

Fig. 1.19 Producing transverse waves on a rope.

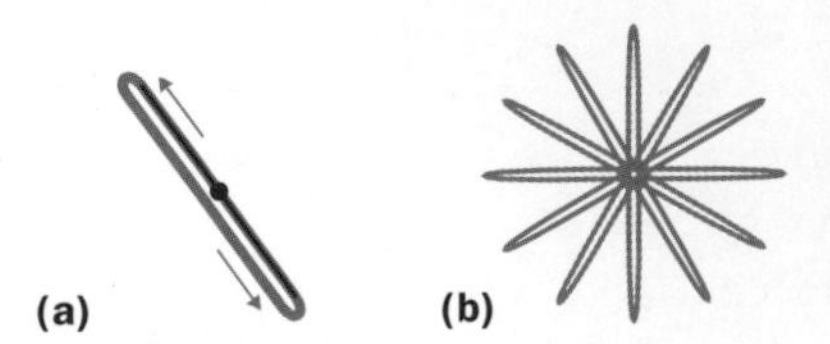

Fig. 1.20 **(a)** polarised and **(b)** unpolarised waves seen end-on.

Transverse waves are said to be **polarised** if the direction of vibration remains in the same plane, as in Fig. 1.19. If the direction of vibration changes, the waves are said to be **unpolarised**. Seen end-on, polarised waves would appear as a single line as in Fig. 1.20 whereas unpolarised waves would not.

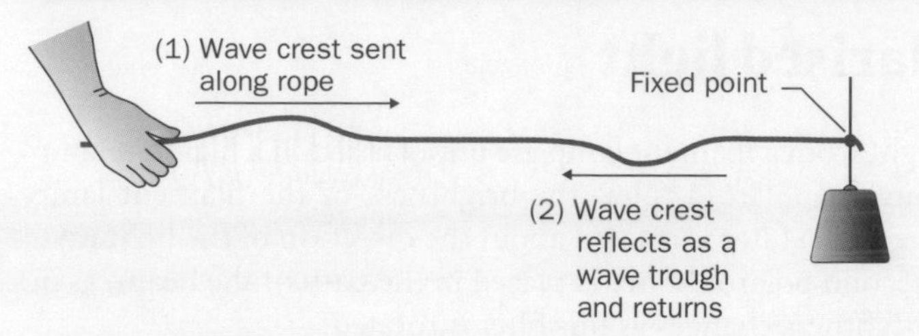

Fig. 1.21 Reflection of waves on a string.

Notes

1. Waves on water are neither transverse nor longitudinal. Observe a cork on water as waves pass it. The cork bobs up and down and also moves to and fro. Its motion takes it round and round in a vertical circle.

2. Seismic waves comprise three types: primary, secondary and surface (or long) waves. Surface waves travel more slowly than primary or secondary waves but they cause much more damage. They are like waves on water because they shake the ground from side to side as well as making it move to and fro.

- **Electromagnetic waves** are transverse waves. The electric field and the magnetic field of an electromagnetic wave vibrate at right angles to the direction of propagation and at right angles to each other. Because they are transverse in nature, electromagnetic waves can be polarised or unpolarised. The plane of polarisation of an electromagnetic wave is defined as the plane of vibration of its electric field. Radio waves from a transmitter aerial are polarised, which is why the aerial of a radio receiver must be aligned correctly to pick up maximum signal strength. Light can be polarised using polaroid filters, as described on p. 20.

- **Secondary seismic waves** are transverse waves. They travel more slowly than primary seismic waves but faster that surface long waves. Secondary seismic waves shake the material they travel through from side to side. Transverse waves cannot pass through a liquid because liquid molecules slide past each other.

- **Waves on a string or wire** are transverse in nature. If a string is fixed at one end, waves travelling towards that end reflect back along the string, reversing their phase on reflection.

See www.palgrave.com/foundations/breithaupt for experiments on wave properties using microwaves.

Questions 1.4

1. **(a)** How would you make transverse waves on a 'slinky' coil?

   (b) State whether each of the following type of wave is longitudinal or transverse:

   (i) ultrasonic waves, **(ii)** microwaves, **(iii)** X-rays.

2. Use your knowledge of longitudinal and transverse waves to explain

   (a) how sound waves make the ear drum vibrate

   (b) why the picture on a portable TV fades if the aerial is turned through 90°.

1.5 Polarised light

Sunlight and light from a filament lamp are unpolarised. If a filament lamp is viewed through a polaroid filter, the brightness of the filament lamp does not change as the filter is rotated about the direction of the light rays. However, if a second polaroid filter is placed in the path of the beam, as in Fig. 1.22, the brightness changes as this filter is rotated.

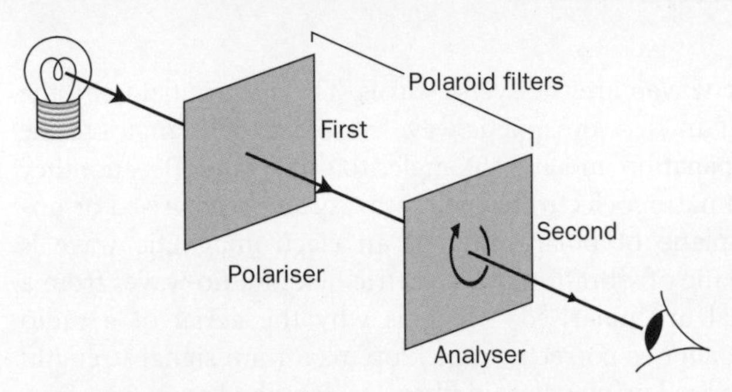

Fig. 1.22 Using polaroid filters.

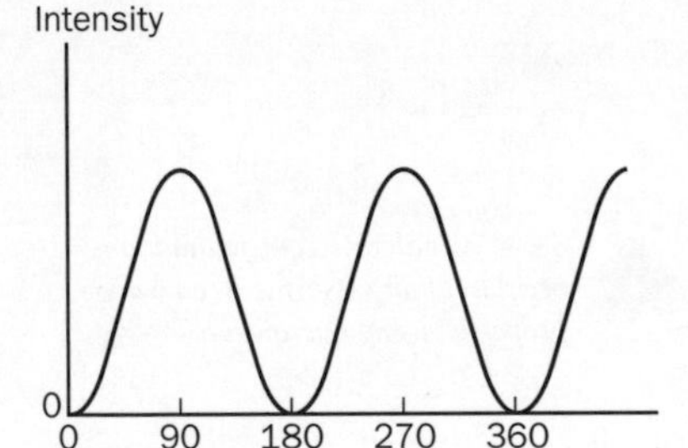

Fig. 1.23 Variation of transmitted intensity with angle of rotation.

The filament lamp produces unpolarised light, which means the plane of vibration of the light waves along a light ray changes at random, as outlined on p. 18. In Fig. 1.22:

- the first polaroid filter, referred to as the 'polariser', allows light waves through only if the plane of polarisation of the waves is in a certain direction. The light transmitted by the first filter is therefore polarised. This occurs because the molecules in the filter are aligned with each other and they only transmit light waves that are polarised in the same direction as the molecules are aligned
- the second filter, referred to as the 'analyser', cuts out all the light from the first filter if the two filters are aligned at 90° to each other. Fig. 1.23 shows how the intensity of light changes as the second filter is rotated from the 'crossed' position. The intensity rises then falls to zero again as a result of rotating the second filter by exactly one-half turn. The two filters are then once again aligned at 90° to each other.

Fig. 1.24 shows a simple analogy of the action of two polaroid filters on an unpolarised light ray. Unpolarised waves on a rope are polarised by the first 'letterbox'. The waves only fully pass through the second 'letterbox' if it is aligned with the first. If the two 'letterboxes' are at 90° to each other, the waves are unable to pass through the second letterbox.

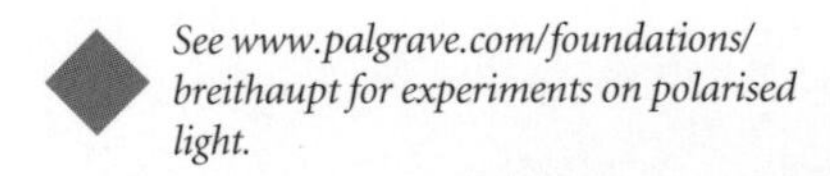

See www.palgrave.com/foundations/breithaupt for experiments on polarised light.

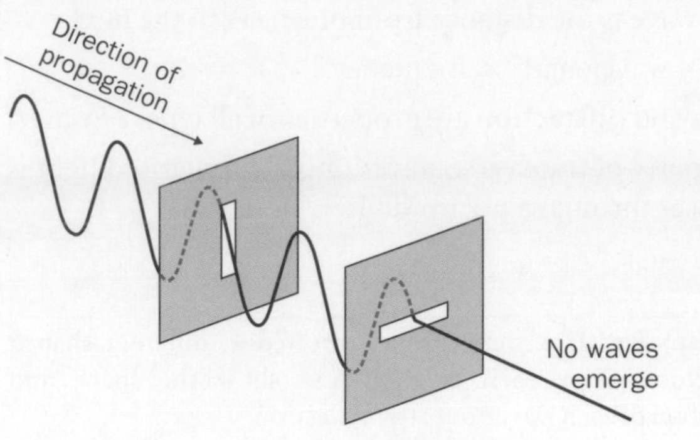

Fig. 1.24 A polarisation analogy.

Polarisation applications

- **Polaroid sunglasses** cut out glare due to reflected sunlight. This is because light is polarised (partly or completely, depending on the angle of incidence of the light) when it reflects from water. Someone underwater in an outdoor swimming pool can be seen more easily through polaroid sunglasses. This is because the polaroid lenses cut out sunlight reflected from the water surface.

Fig. 1.25 Using polaroid sunglasses.

- A **calculator display** makes use of the polarisation of light. The display consists of 'pixels', each comprising a liquid crystal sandwiched between two polaroid filters. The molecules of the liquid crystal are long, spiral molecules that can rotate the plane of polarisation of light when they line up with each other. With no voltage applied, light is reflected from the cell by a mirror under the pixel. When a voltage is applied between the two filters, the molecules line up and rotate the plane of polarisation of light passing through the cell by exactly 90°, so no light can pass through the cell.

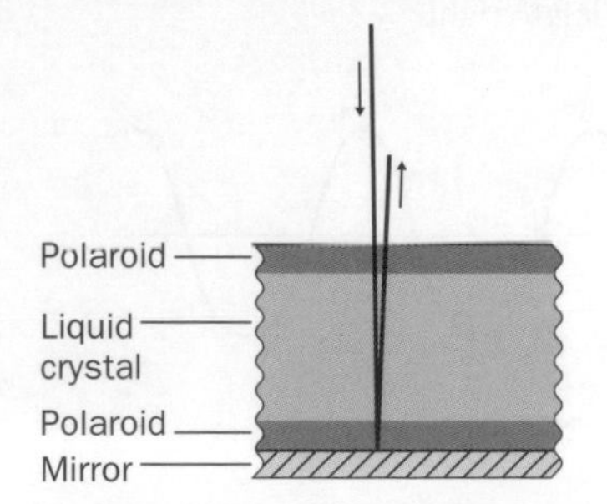

Fig. 1.26 Inside an LCD display.

Summary

- ◆ **Transverse waves** vibrate at 90° to the direction of propagation. Examples of transverse waves include electromagnetic waves, secondary seismic waves and waves on strings and wires.
- ◆ **Longitudinal waves** vibrate along the direction of propagation. Examples of longitudinal waves include sound waves and primary seismic waves.
- ◆ The **amplitude** of a wave is the displacement of a point from equilibrium to peak displacement.
- ◆ The **phase difference** between any two points at distance x apart $= 2\pi x / \lambda$.
- ◆ The **frequency** of a wave is the number of complete cycles that pass a point each second.

- The **wavelength** of a wave is the distance from one peak to the next.
- The **speed** of a wave = wavelength × frequency.
- **Reflection, refraction** and **diffraction** are properties of all types of waves.
- **Polarisation** is a property of transverse waves only. Unpolarised light is polarised when it passes through a polaroid filter.

Revision questions

1.1. (a) List the main parts of the electromagnetic spectrum of waves, in order of increasing wavelength.
(b) State with which part of the electromagnetic spectrum each of the following wavelengths is associated.
(i) 10 m **(ii)** 600 nm **(iii)** 1 mm **(iv)** 10^{-14} m

1.2. (a) Describe the difference between a transverse wave and a longitudinal wave.
(b) State which of the following types of waves are transverse and which are longitudinal
sound waves ultrasonics microwaves light
primary seismic waves

1.3. Fig. 1.27 shows a snapshot of a transverse wave travelling from left to right.

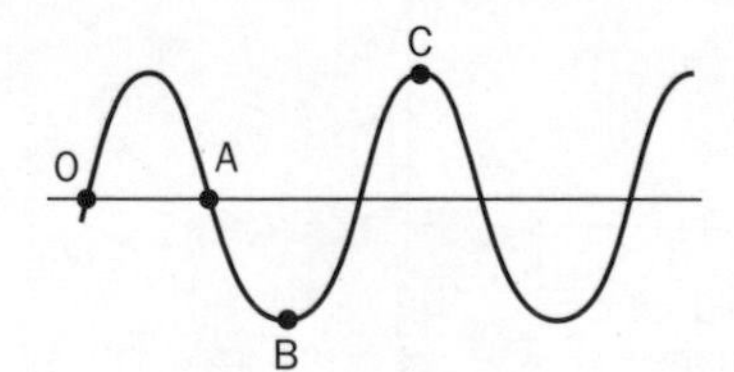

Fig. 1.27

(a) Use a millimetre rule to measure **(i)** the amplitude, **(ii)** the wavelength of the wave.
(b) Calculate the phase difference in radians between point O and **(i)** point A, **(ii)** point B, **(iii)** point C.
(c) The frequency of the waves is 0.5 Hz. What is the displacement of
(i) point O 0.5 s later, **(ii)** point O 3.0 s later, **(iii)** point A 1.0 s later, **(iv)** point A 2.5 s later?

1.4. (a) The speed of light in air is 3.0×10^8 m s^{-1}. Calculate :
(i) the frequency of light waves of wavelength 500 nm
(ii) the wavelength of light waves of frequency 5.0×10^{14} Hz.
(b) The speed of sound in air at room temperature is 340 m s^{-1}. Calculate:
(i) the frequency of sound waves of wavelength 0.05 m
(ii) the wavelength of sound waves of frequency 12 kHz.

1.5. (a) Fig. 1.28 shows waves directed at different shaped reflectors. Copy each diagram and show the shape and direction of each wavefront after reflection.

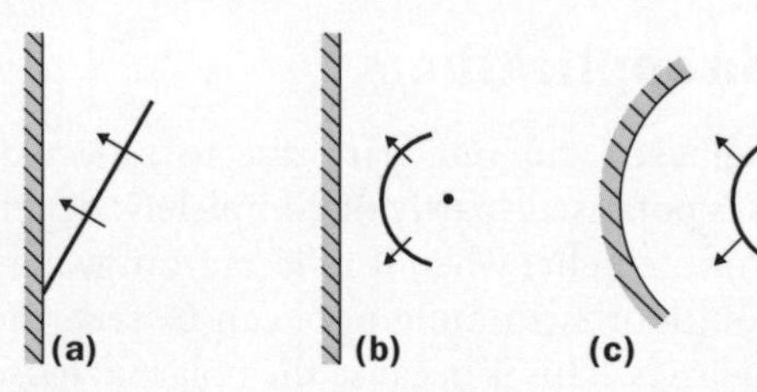

Fig. 1.28

(b) Explain why a satellite dish is **(i)** concave in shape, **(ii)** fitted with a detector above the dish surface.
(c) Explain why water waves do not reflect from a sloping beach.

1.6. (a) Fig. 1.29 shows straight waves about to pass across a boundary between deep and shallow water. The waves travel faster in deep water than in shallow water.

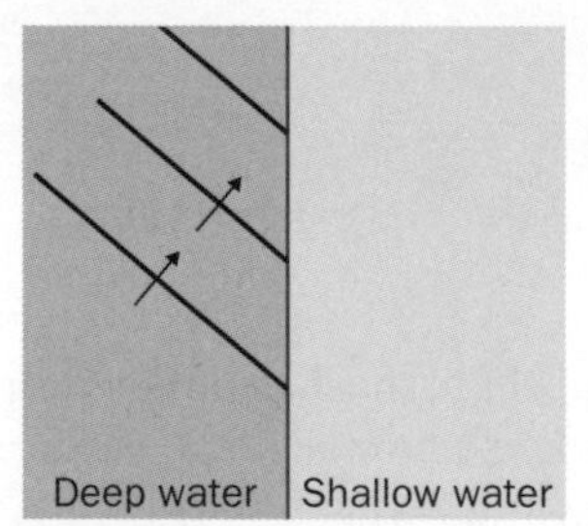

Fig. 1.29

(i) Copy the diagram and indicate the shape and direction of the waves in the shallow water.
(ii) What can you say about the wavelength and direction of the waves in the shallow water compared with the deep water?
(b) Explain why waves running onto a beach are usually parallel to the beach.

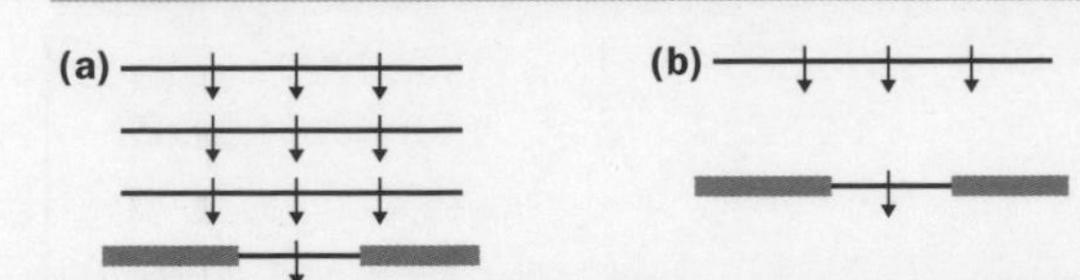

Fig. 1.30

1.7. (a) Copy and complete the two diagrams in Fig. 1.30 showing the waves after passing through each gap.

(b) Explain why a large-diameter satellite dish is more difficult to align than a small-diameter dish.

1.8. (a) Describe how you would use a polaroid filter to find out if light from a particular light source is polarised.

(b) Light from a filament lamp is observed through two polaroid filters.

(i) Explain why the filament lamp cannot be seen when one filter is aligned in a certain direction relative to the other filter.

(ii) Describe what is observed when the polaroid nearer the observer is rotated through 360° from the position in (i).

(iii) How would the observation differ if the other filter had been turned through 360° instead?

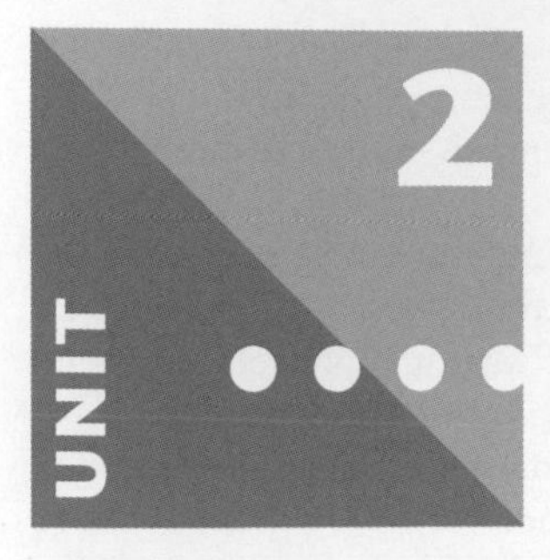

Sound

Contents

Objectives

After working through this unit, you should be able to:

- ▶ describe the nature of sound waves and measure the speed of sound in air
- ▶ describe the main properties of sound, including absorption, reflection, refraction and diffraction
- ▶ describe the structure of the ear and recall the main characteristics of human hearing
- ▶ describe the main properties and uses of ultrasonics
- ▶ relate the fundamental frequency of a vibrating wire or resonance tube to the length of the wire or tube
- ▶ explain the relationship between the overtones and the fundamental frequency for a vibrating wire or resonance tube

Displacement, +, 0, −, R, C, R, C, R, Position

Fig. 2.1 Transverse representation.

> **Note**
> Fig. 2.1 is a more convenient way to represent a sound wave and will be used from now on. However, don't forget that sound waves are longitudinal and diagrams like Fig. 2.1 are for convenience.

2.1 The nature and properties of sound

Producing sound

An object vibrating in air creates sound waves in the surrounding air because the surface of the vibrating object pushes and pulls on the air. Sound waves are longitudinal, as explained on p. 17, and compression waves spread out from the object. A sound wave may therefore be thought of as a series of compressions (where the air is at higher pressure than normal) and rarefactions (where the air is at lower pressure than normal). The density of the air is higher than normal at a compression and lower than normal at a rarefaction.

Fig. 1.17 represents how the density varies along the line of propagation at an instant of time. The areas of maximum compression and rarefaction correspond to zero displacement of the layers of air, as represented by the transverse wave diagram in Fig. 2.1.

Measuring sound waves

1. Generate sound waves at constant frequency using a small loudspeaker connected to a signal generator. Use a microphone connected to an oscilloscope to display the waveform of the sound waves from the loudspeaker on the oscilloscope screen, as in Fig. 2.2, which shows a waveform called a 'sine wave'. This is a waveform of sound waves of a single frequency only. Use the oscilloscope to measure the sound frequency, as shown in Fig. 2.2.
2. Display the sound waveforms of different sources of sound (e.g. tuning fork, whistle, musical instruments, voice) on the oscilloscope. Each sound source except noise has a characteristic waveform. The tuning fork produces a sine wave but the other waveforms are more complicated. Noise is unpleasant to listen to because the waveform varies at random, not rhythmically.

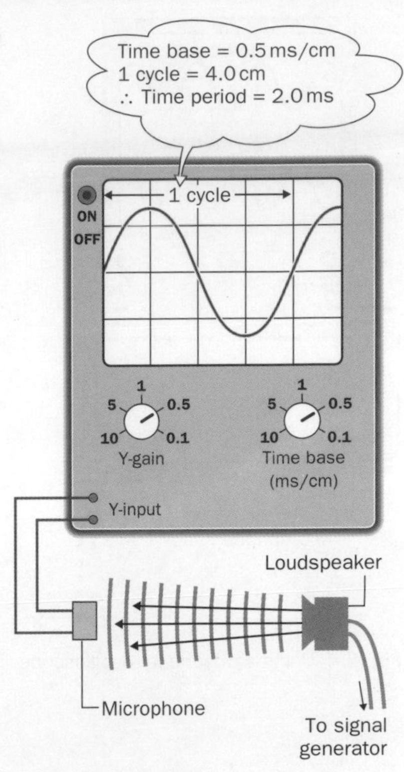

Fig. 2.2 Using an oscilloscope to measure the frequency of sound waves.

Echoes

Sound waves reflect off hard surfaces. Clap your hands in an empty sports hall and you may hear an echo of the clap due to sound waves reflected from the walls, if the wall surface is smooth. If the surface is rough, the reflected sound waves are broken up and the original waveform is lost.

Echo sounders are used for depth tests on board ships. Ultrasonic pulses are directed from a transmitter (called a transducer) on board the ship to the sea bed. The pulses reflected from the sea bed are detected by the transducer. The time, t, between transmitting and detecting a pulse is measured and used to calculate the distance to the sea bed. Since the time taken for a pulse to travel from the detector to the sea bed is ½t,

the distance to sea bed, s, = speed × time ÷ 2, = $\upsilon t \div 2$

where υ = the speed of sound in sea water.

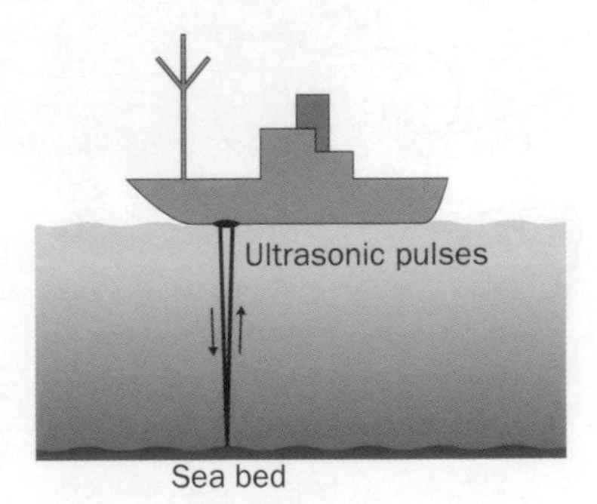

Fig. 2.3 Depth testing.

The speed of sound

Sound travels faster in solids and liquids than in gases. The speed of sound in a gas is faster, the lighter the gas molecules are and the higher the temperature. Fig. 2.4 shows how to measure the speed of sound in air in a laboratory.

- The signal from the signal generator is displayed on one beam of the oscilloscope. The other beam is used to display the microphone signal which is delayed because of the time taken by sound to travel from the loudspeaker to the microphone. As a result, the microphone trace is out of phase with the loudspeaker trace.
- If the distance between the microphone and the loudspeaker is increased by exactly one wavelength, the microphone trace moves through exactly one cycle relative to the other trace. This can be used to measure the wavelength of the sound waves. The frequency can be measured from the signal generator. The speed of sound is then calculated from speed = wavelength × frequency.

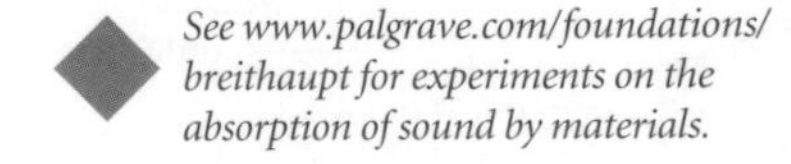

See www.palgrave.com/foundations/breithaupt for experiments on the absorption of sound by materials.

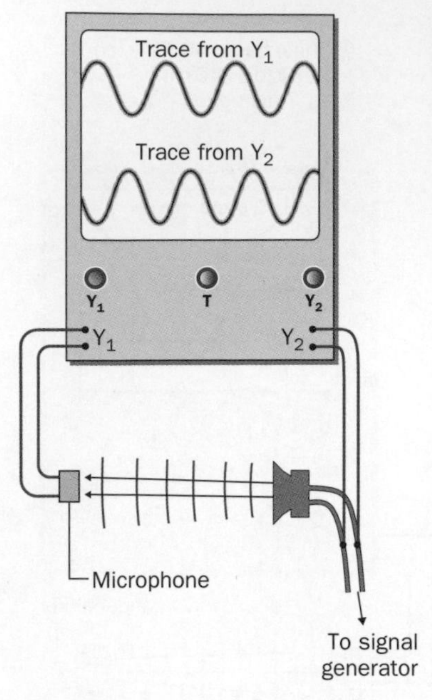

Fig. 2.4 Using a dual beam oscilloscope.

Questions 2.1

1. Fig. 2.5 shows a sound waveform displayed on an oscilloscope screen.

(a) Calculate the frequency of the sound waves if the time base control was set at 5 ms per cm.

(b) Copy the display and make further sketches to show how the waveform changes if **(i)** the loudness of the sound is increased, **(ii)** the frequency of the sound is reduced.

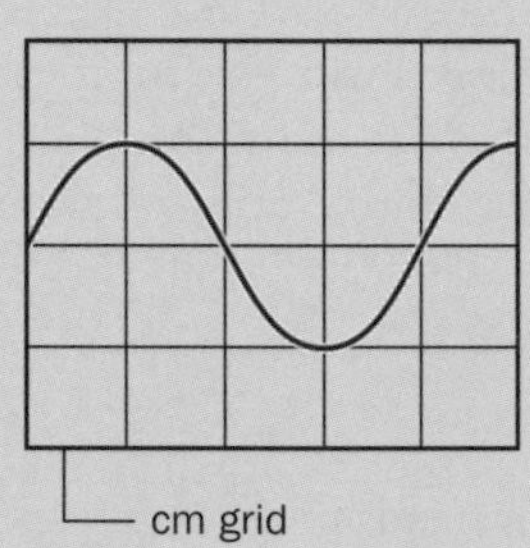

Fig. 2.5

Fig. 2.6

2. The speed of sound in water is 1500 m s^{-1}. Fig. 2.6 shows pulses from a depth-finder on an oscilloscope screen. The pulses are produced at a rate of 2 per second.

(a) Use Fig. 2.6 to estimate the time taken for each pulse to travel from the ship to the sea bed.

(b) Calculate the depth of the sea bed at this position.

2.2 The human ear

The faintest sound that the normal human ear can detect is about 1 million million times quieter than the loudest sound it can withstand without damage. The ear is a remarkable organ but it needs to be protected from extremely loud sounds.

Fig. 2.7 shows a cross-section through the human ear. Sound waves make the tympanic membrane (ear drum) vibrate and these vibrations are transmitted to the oval window via the bones of the middle ear. These bones amplify the force of the vibrations and filter out unwanted noise (e.g. physiological sounds). They also protect the inner ear by switching to a less sensitive mode of vibration if the sound becomes excessive. The vibrations are transmitted through the fluid of the cochlea in the inner ear where they are detected by very sensitive 'hair cells' attached to the basilar membrane. These are nerve cells and when stimulated they send electrical signals via the auditory nerve to the brain.

(a)

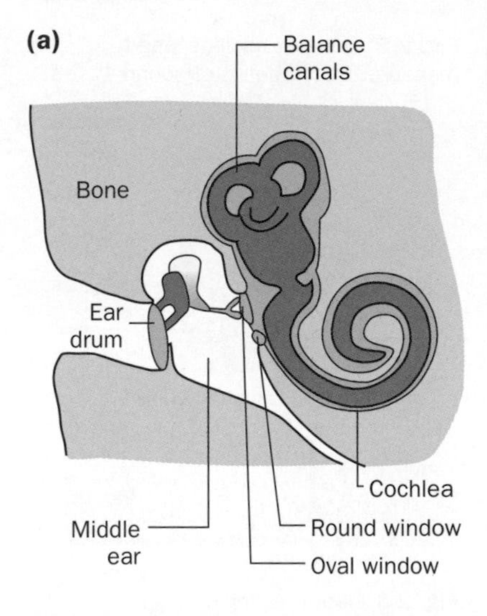

(b)

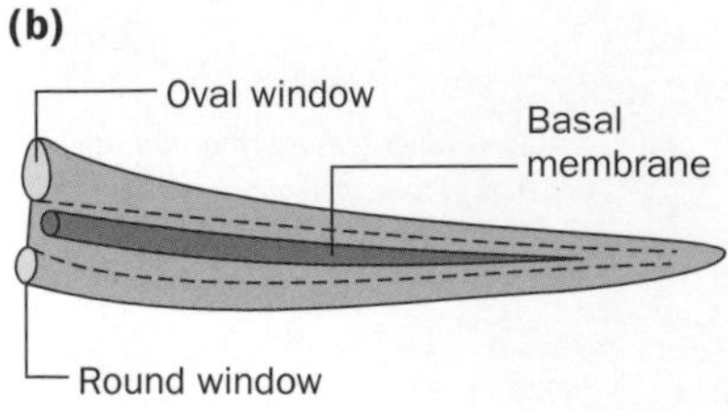

Fig. 2.7 The human ear **(a)** cross-section, **(b)** cochlea 'unrolled'.

Notes

1. The muscles that hold the bones of the middle ear together slacken when the ear is subjected to excessive sound. This prevents all the sound energy transmitted by the ear drum from arriving at the oval window. When someone leaves a very noisy event, the muscles take a little while to become taut again. This is why a person leaving a noisy nightclub has temporary difficulty with hearing. Frequent over-exposure results in permanent hearing loss, as the muscles weaken permanently and the bones wear away.
2. Frequency discrimination is due to the hair cells on the basilar membrane. The pattern of stimulation of the hair cells along the basilar membrane depends on the sound frequency and it is thought that the nerve impulses are triggered by different patterns.

The decibel scale

This is a 'times ten' scale for loudness, similar to the Richter scale for earthquakes in which every further point corresponds to ten times as much energy released. However, on the decibel (dB) scale, an increase of 10 decibels (or 1 'bel') corresponds to a tenfold increase in sound energy. Fig. 2.8 shows decibel levels of some everyday sounds.

- Zero dB is defined as the faintest possible sound that can be heard.
- Every 10 dB increase in loudness corresponds to a tenfold increase in sound energy.

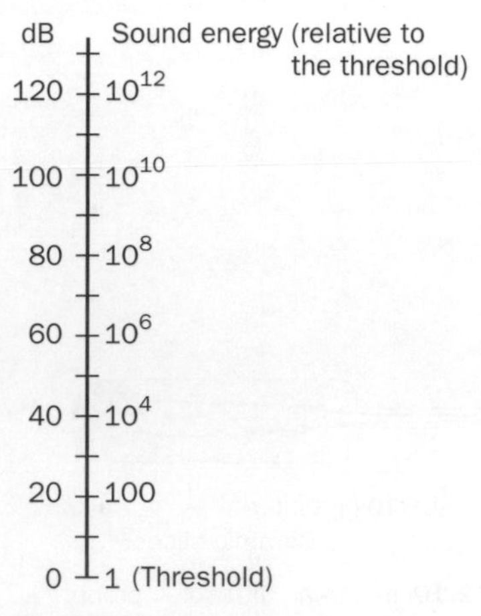

Fig. 2.8 The decibel scale.

Hearing tests

1. Use the apparatus in Fig. 2.2 to determine the frequency range of your ears. Take care, as loud sounds can damage your ears. The upper frequency limit, about 18 kHz for a normal young adult, decreases with age. Fig. 2.9 shows the frequency response curve of the ear. At what frequency are your ears most sensitive?
2. Use a decibel meter to measure the loudness of different sounds. At its most sensitive, it responds to the faintest sound detectable. At its least sensitive, it scarcely responds unless the sound is very, very loud.

Questions 2.2

1. **(a)** What are the functions of the bones of the middle ear?
   (b) Compare a recording of your own voice with what you hear when you speak. Describe and account for the differences.
   (c) A student walks from a room where the loudness level is 40 dB to a hall where the loudness level is 70 dB.
   (i) What is the increase in loudness in decibels?
   (ii) How much more sound energy is the student subjected to in the hall in comparison with the room?
2. For the frequency response curve in Fig. 2.9:
   (a) What frequency is the young person most sensitive to?
   (b) What is the hearing loss in dB of the older person at the most sensitive frequency?
   (c) How does the hearing of the older person compare with that of the younger person?

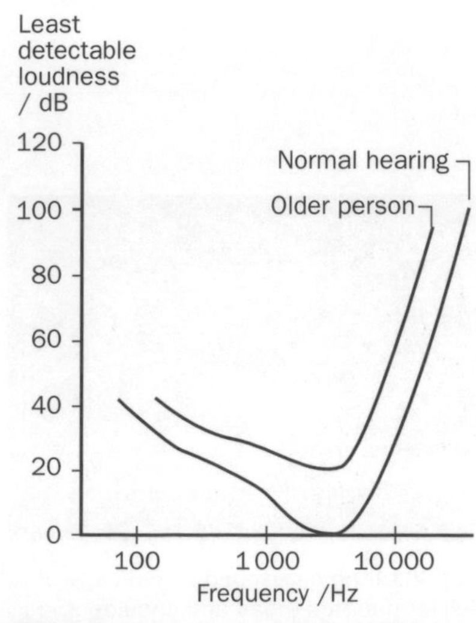

Fig. 2.9 Frequency response of the ear.

2.3 Ultrasonics

Ultrasonics are sound waves of frequencies too high for the human ear to detect (i.e. above about 18 kHz).

- Ultrasonic scanners are used in hospitals to produce images of babies in the womb. X-rays would damage the baby because they ionise matter which they pass through, whereas ultrasonics are non-ionising and are therefore used instead of X-rays. Ultrasonic devices are also used for the detection of cracks inside metals and for depth location at sea.
- The power of ultrasonics is used to clean street lights by immersing the lighting unit in a tank of water and using ultrasonic waves to dislodge the dust particles from the surfaces of the unit. In hospitals, ultrasonic power is used to pulverise kidney stones, thus avoiding the need for surgical removal.

Producing and detecting ultrasonic waves

An **ultrasonic probe**, sometimes referred to as a transducer, is designed to produce and detect ultrasonic pulses. It contains a thin slice of a certain ceramic material that vibrates strongly when an alternating voltage is applied between its two faces. The applied frequency is matched to the natural frequency of vibration of the slice so it **resonates** when the alternating voltage is applied. This is the same principle as pushing a child on a swing; if the frequency of the pushes matches the natural frequency of the swing, the amplitude of oscillation of the swing becomes very large.

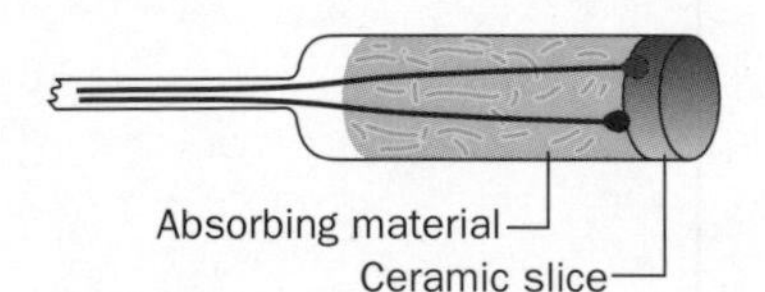

Fig. 2.10 Inside an ultrasonic probe.

Probes in ultrasonic medical scanners produce ultrasonics at a frequency of about 1.5 MHz. The speed of ultrasonic waves in the ceramic used is about 3.8 km s^{-1}. Hence the wavelength of ultrasonics in the crystal is about 2.5 mm (= speed / frequency = 3800 m s^{-1} / 1.5×10^6 Hz). For resonance, the thickness of the slice needs to be half a wavelength, corresponding to ultrasonics travelling from one surface and back in one complete cycle.

In an **ultrasonic scanner**, the pulses are produced by repeatedly switching the alternating voltage on for a brief interval. Between transmission of successive pulses, the probe is automatically connected to a circuit that detects the alternating signal produced when reflected ultrasonic pulses hit the slice. Ultrasonic waves travel at a speed of about 1000 m s^{-1} in body tissue and therefore a pulse would travel a typical distance of 1 m through the body (across and back) in about 1 ms. Internal boundaries in the body (e.g. bone-tissue boundaries, boundaries between different types of tissue) reflect the pulses.

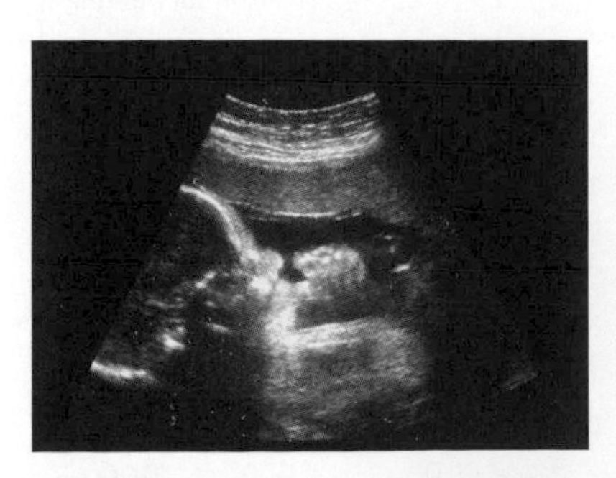
Fig. 2.11 An ultrasound scan image of a full-term foetus (head and shoulders) sucking it's thumb.

- In the A-scan system, the reflected pulses are displayed on an oscilloscope screen and the time taken for each pulse to travel from the probe and back again can be measured. The depth of the boundary causing a reflected pulse can then be calculated in the same way as for depth finding.
- In the B-scan system, an image is built up on the display screen as the probe is moved over the body surface. Sensors attached to the probe

determine the direction of the display trace and each reflected pulse makes the trace brighter.

Notes

1. The greater the difference in the density either side of an interface, the stronger the reflected pulse. The probe is applied to the skin via a paste. This is necessary to prevent strong reflection occuring at the air–skin boundary, otherwise the pulses would not enter the body.
2. The ultrasonic frequency needs to be high, otherwise the pulses would spread out from the probe due to diffraction. At 1.5 MHz, the ultrasonic wavelength in the body is about 2.5 mm, which is much smaller than the probe width, therefore minimising diffraction.

Questions 2.3

1. **(a)** In an ultrasonic scanning system, why is it essential that the ultrasonic pulses should be very brief in comparison with the rate at which the pulses are produced?

   (b) Calculate the time taken for a 1.5 MHz transducer to produce 10 cycles of an ultrasonic wave.

   (c) The time between successive 1.5 MHz pulses of a scanner system is 1 ms. Calculate the ratio of the width of a pulse consisting of 10 cycles to the time interval between successive pulses.
2. **(a)** An ultrasonic transducer in a cleaning unit emits an ultrasonic beam at a frequency of 40 kHz from a flat transducer of width 1 cm. Calculate the wavelength of 40 kHz ultrasonic waves in water. The speed of sound in water is 1500 m s^{-1}.

   (b) Diffraction is not significant if the width of the source of waves is much greater than the wavelength. Discuss whether or not diffraction is **(i)** significant, **(ii)** desirable in an ultrasonic cleaning unit.

2.4 Vibrating strings

Musical instruments produce rhythmical sounds provided they are played correctly. The notes produced by a string or a wind instrument are characteristic of the type of instrument. Any waveform of a musical note can be analysed to find the frequencies from which it is composed and the relative loudness of each frequency present.

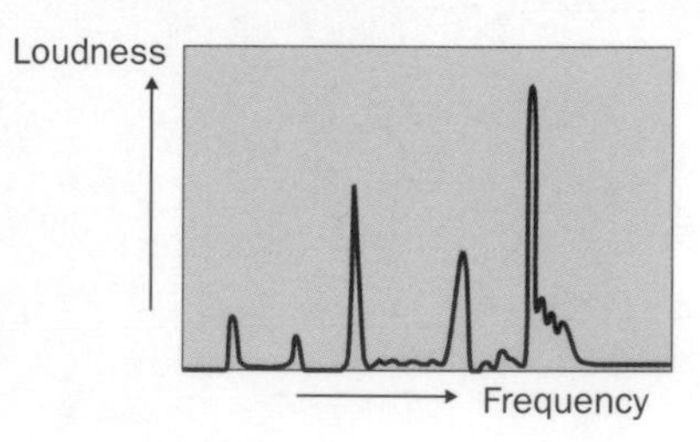

Fig. 2.12 The frequency spectrum of a musical note.

Stationary waves on a string

A stationary wave pattern in which the wave peaks and troughs do not travel along the string can be set up on a string or wire under tension, using an arrangement as shown in Fig. 2.13. The mechanical oscillator is driven from a signal generator and it vibrates with a small amplitude, sending waves along the string. These waves reflect at the fixed end of the string and they return to the oscillator in phase with the waves being produced at that instant. Just like pushing a child on a swing at the right instant, the amplitude of vibration builds up as the reflected waves add to the waves being produced to create a stationary wave pattern.

The fundamental frequency

The simplest pattern is called the **fundamental** mode of vibration of the string, as shown in Fig. 2.13. The amplitude

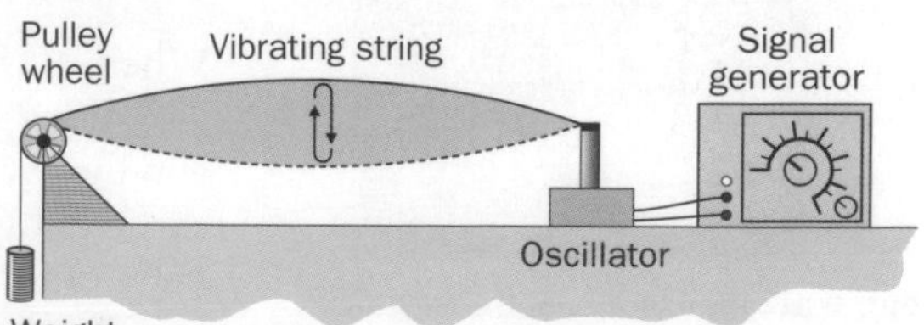

Fig. 2.13 Stationary waves on a string.

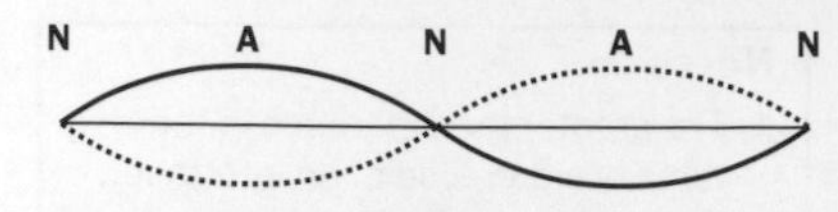

(a) 1st overtone: frequency $f_1 = 2f_0$
frequency $\lambda_1 = 2L/2$

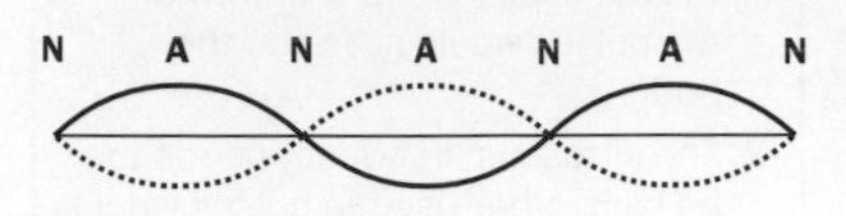

(b) 2nd overtone: frequency $f_2 = 3f_0$
frequency $\lambda_2 = 2L/3$

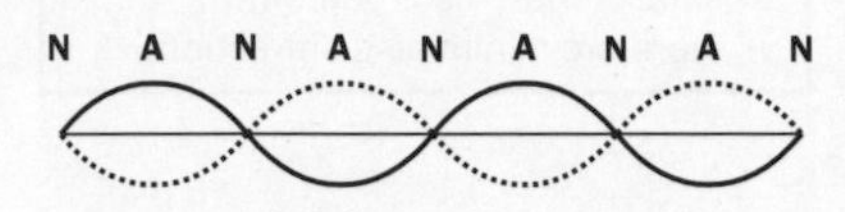

(c) 3rd overtone: frequency $f_3 = 4f_0$
frequency $\lambda_3 = 2L/4$

N = node **A** = antinode
F_0 = fundamental frequency

Fig. 2.14 Stationary wave patterns.

of vibration is zero at the ends and maximum in the middle of the string. This pattern occurs when the oscillator goes through exactly one cycle in the time taken for a wave to travel from the oscillator to the fixed end and back. In other words, the time for one cycle, T, = 2L ÷ υ, where L is the string length and υ is the speed of the waves on the string. Since the fundamental frequency, f_o, = 1 ÷ T, then

$$f_o = \frac{\upsilon}{2L}$$

Overtones

Further patterns called **overtones** are produced at frequencies $2f_o$, $3f_o$, $4f_o$, etc, corresponding to the oscillator going through 2 or 3 or 4, etc. cycles in the time taken for a wave to travel the length of the string and back. In general, a stationary wave pattern occurs if $2L \div \upsilon = mT$, where m is equal to a whole number. Fig. 2.14 shows the pattern of vibration for the fundamental and the overtones.

Note that:

1. The frequency, f, $= 1/T = m\upsilon/2L = mf_o$.
2. The wavelength, λ, $= \upsilon/f = 2L/m$. In other words, the string length, L, $= m\lambda/2$ (i.e. a whole number of half wavelengths).
3. The fundamental frequency increases with tension in the string and decreases with length.

Nodes and antinodes

- The points of no displacement, referred to as **nodes**, are fixed in position.
- The distance between adjacent nodes is exactly one-half wavelength.
- The points of maximum displacement, referred to as **antinodes**, are exactly midway between the nodes. No energy transfer occurs because the antinodes do not move along the string.
- Two points between adjacent nodes, or separated by an even number of nodes, vibrate in phase.
- Two points separated by an odd number of nodes vibrate out of phase by 180°.
- Each stationary wave pattern is due to the reflected waves combining with the waves from the oscillator. In other words, a stationary wave pattern is produced when two sets of waves of the same frequency and amplitude pass through each other.

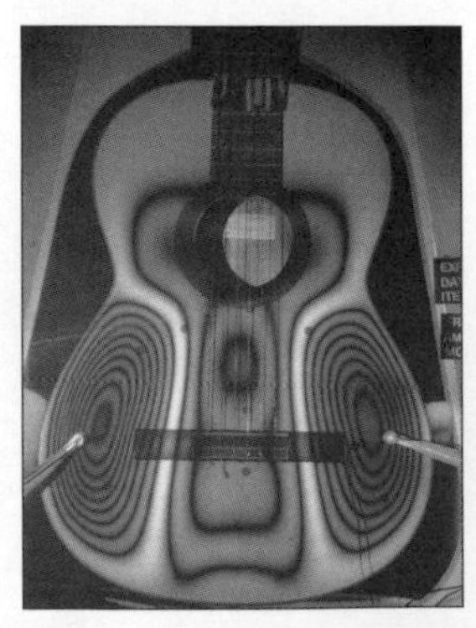

Fig. 2.15 Photograph of the fringe patterns formed by light reflecting from a vibrating guitar

Sound waves from a string instrument

When a wire or string of a musical instrument produces sound, its vibrations are usually a combination of its fundamental mode of vibration and the overtones. The loudness of each frequency component depends on exactly how the string is made to vibrate. The vibrating string makes the body of the instrument vibrate, creating sound waves in the surrounding air. The shape of the body affects the quality of the note produced since it causes some frequencies to be suppressed and enhances others.

Questions 2.4

1. Fig. 2.16 shows a wire under tension which is vibrating at its fundamental frequency of 120 Hz.

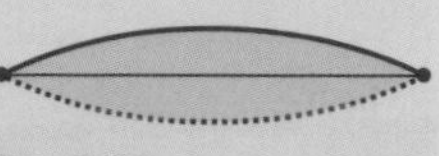

Fig. 2.16

   (a) Sketch the pattern of vibration of the wire when it vibrates at its second overtone frequency of 360 Hz.

   (b) What would be the fundamental frequency of this wire if its length is **(i)** doubled, **(ii)** halved, without change of tension?

2. **(a)** Sketch the pattern of vibration of a wire vibrating at twice its fundamental frequency.

   (b) Describe **(i)** how the amplitude of vibration varies with position along the wire, **(ii)** how the phase difference between any two points on the wire varies with distance apart.

Fig. 2.17 The saxophone, a wind instrument, creates sound as a result of resonance in tubes.

2.5 Acoustic resonance

Wind instruments create sound as a result of resonance in tubes. For example, the note from a trumpet is due to vibrations of the air created by the lips at the mouthpiece, causing the air in the tube of the trumpet to resonate. An organ pipe resonates with sound when air is blown in at one end over a sharp edge, creating eddy currents that make the air in the pipe resonate. In a reed instrument such as a clarinet, the vibrations of a reed in the mouthpiece cause the air in the tube to vibrate and resonate with sound.

The general principle of resonance was outlined in the previous section. Think about a child on a swing again. The swing has a natural frequency of oscillation; if a force is applied to it periodically, the amplitude of swing becomes very large if the forcing frequency matches the natural frequency of the swing, because the periodic force is applied at the same point in each cycle of the swing, causing the amplitude to build up. The same is true when acoustic resonance occurs. The air column has its own natural frequencies of vibration. If the air is made to vibrate at one of these frequencies, the amplitude of vibration builds up and the sound becomes very loud.

Resonance in a pipe closed at one end

Fig. 2.18 shows how this can be demonstrated using a small loudspeaker connected to a signal generator. Stationary wave patterns are established in the air column at resonance, as a result of sound waves from the loudspeaker reinforcing sound waves reflected from the closed end. In every case, there is a node at the closed end of the pipe and an antinode at the speaker, a short distance beyond the open end of the pipe. The lowest frequency at which the pipe resonates is its **fundamental frequency**, f_o. Further resonances occur at frequencies $3f_o$, $5f_o$, $7f_o$, etc.

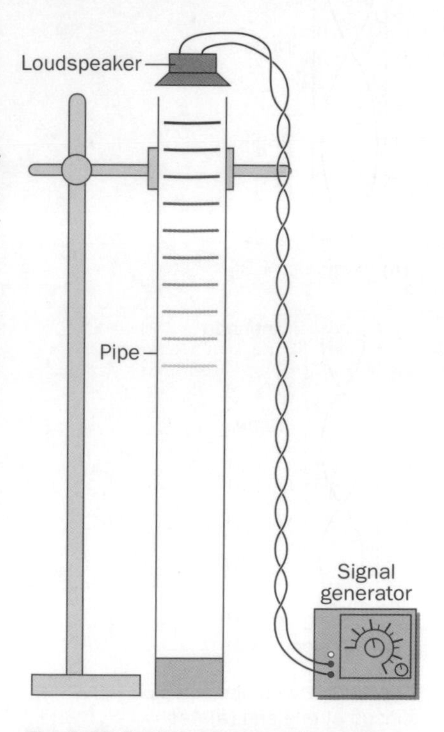

Fig. 2.18 Testing resonance in a pipe.

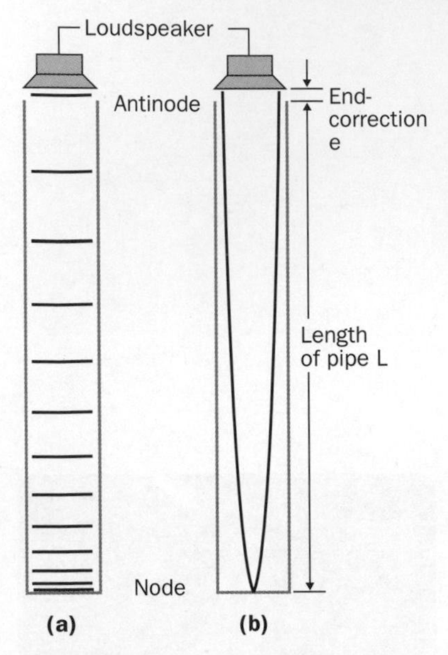

Fig. 2.19 At the fundamental frequency **(a)** fundamental vibrations at frequency f_0, **(b)** transverse representation.

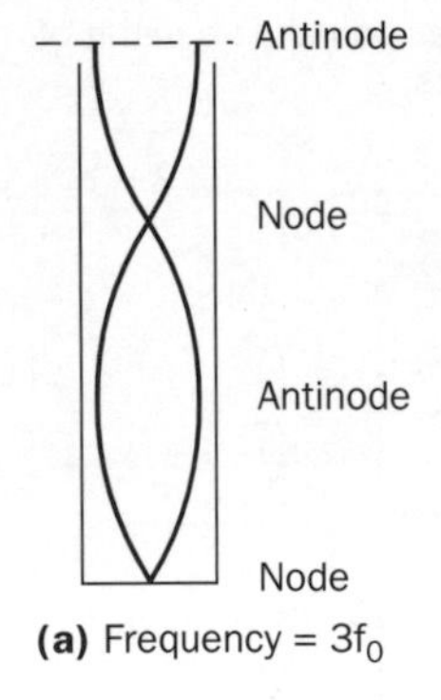

(a) Frequency = $3f_0$

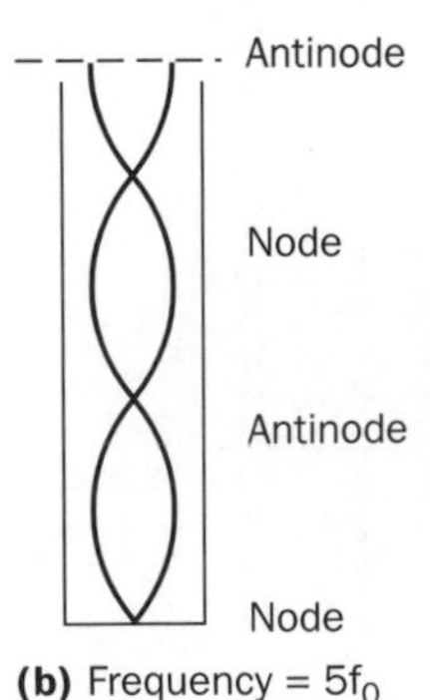

(b) Frequency = $5f_0$

Fig. 2.20 Overtones in a pipe closed at one end **(a)** 1st overtone, **(b)** 2nd overtone.

1. At the **fundamental frequency, f_o**, the stationary wave pattern corresponds to a distance of $\frac{1}{4}\lambda$ (the distance from a node to the nearest antinode) between the speaker and the closed end of the pipe. In other words:

   $$L + e = \tfrac{1}{4}\lambda$$

   where e (called the 'end correction') is the distance between the speaker and the open end of the pipe, υ is the speed of sound in the pipe and λ_o is the fundamental wavelength.

   Hence $f_0 = \dfrac{\upsilon}{\lambda_0} = \dfrac{\upsilon}{4(L+e)}$

2. **Overtones** at frequencies $3f_o$, $5f_o$, $7f_o$, etc. are due to stationary waves in which there are an odd number of quarter wavelengths between the speaker and the closed end, corresponding to a node at the closed end and an antinode at the speaker. In other words:
   - $L + e = \frac{3}{4}\lambda_1$ for the first overtone, where λ_1 = 1st overtone wavelength,
     so frequency $= \dfrac{\upsilon}{\lambda_1} = \dfrac{3\upsilon}{4(L+e)} = 3f_0$
   - $L + e = \frac{5}{4}\lambda_2$ for the second overtone, where λ_2 = 2nd overtone wavelength,
     so frequency $= \dfrac{\upsilon}{\lambda_2} = \dfrac{5\upsilon}{4(L+e)} = 5f_0$
   - $L + e = \frac{7}{4}\lambda_3$ for the third overtone, where λ_3 = 3rd overtone wavelength,
     so frequency $= \dfrac{\upsilon}{\lambda_3} = \dfrac{7\upsilon}{4(L+e)} = 7f_0$

Worked example 2.1

A pipe of length 0.845 m, closed at one end, is known to resonate at frequencies of 100 Hz and 300 Hz. The speed of sound in the pipe is 340m s^{-1}.

(a) Calculate the fundamental frequency of this pipe.

(b) Calculate (i) the wavelength of sound in the pipe at its fundamental frequency, (ii) the end-correction.

Solution

(a) The resonant frequencies fit the pattern 1 ; 3 ; 5 ; etc. The fundamental frequency is therefore 100 Hz.

(b)(i) Fundamental wavelength, $\lambda_o = \dfrac{\upsilon}{f_o} = \dfrac{340}{100} = 3.40$ m

(ii) End-correction, $e = \frac{1}{4}\lambda_o - L = \dfrac{3.40}{4} - 0.845 = 0.005$ m.

Resonance in a pipe open at both ends

Fig. 2.21 shows a loudspeaker used to make an open pipe resonate with sound. This occurs because sound from the loudspeaker partially reflects when it reaches the other end. The result is that the reflected waves reinforce the waves from the speaker to set up a stationary wave pattern in the pipe. Each pattern has an antinode at either end.

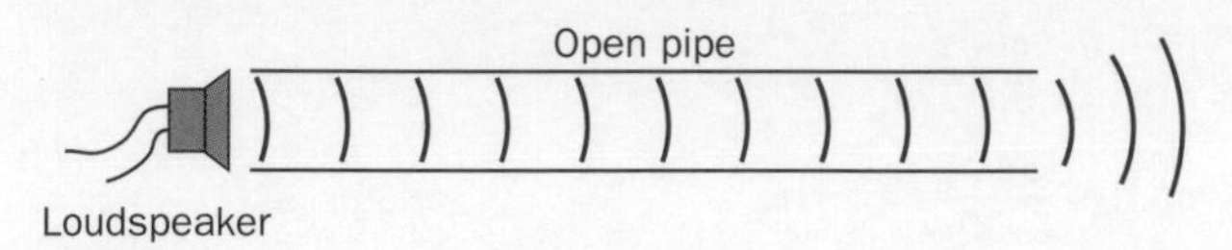

Fig. 2.21 Resonance in an open-ended pipe.

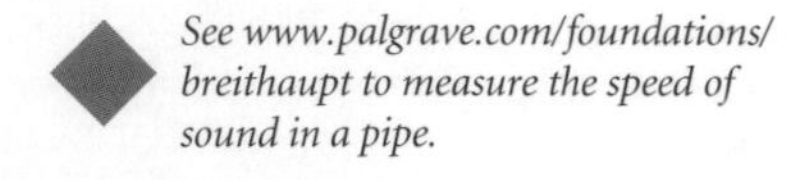

See www.palgrave.com/foundations/breithaupt to measure the speed of sound in a pipe.

1. At the fundamental frequency, f_o, the lowest frequency for resonance, the length of the air column is exactly equal to one-half wavelength. This is the distance between adjacent antinodes. The condition for this resonance is therefore $\frac{1}{2}\lambda_o = (L + 2e)$, since there is an end-correction at either end, w§here λ_o is the fundamental wavelength.

Hence $f_o = \dfrac{v}{\lambda_o} = \dfrac{v}{2(L + 2e)}$

2. Overtones occur at frequencies $2f_o$, $3f_o$, $4f_o$, etc. In general, resonance occurs if there are a whole number of half wavelengths between the antinodes at the ends. In other words

- $L + 2e = \frac{2}{2}\lambda_1$ for the first overtone, where λ_1 = 1st overtone wavelength,

  so frequency $= \dfrac{v}{\lambda_1} = \dfrac{v}{(L + 2e)} = 2f_o$

- $L + 2e = \frac{3}{2}\lambda_2$ for the second overtone, where λ_2 = 2nd overtone wavelength,

  so frequency $= \dfrac{v}{\lambda_2} = \dfrac{3v}{2(L + 2e)} = 3f_o$

- $L + 2e = \frac{4}{2}\lambda_3$ for the third overtone, where λ_3 = 3rd overtone wavelength,

  so frequency $= \dfrac{v}{\lambda_3} = \dfrac{4v}{2(L + 2e)} = 4f_o$

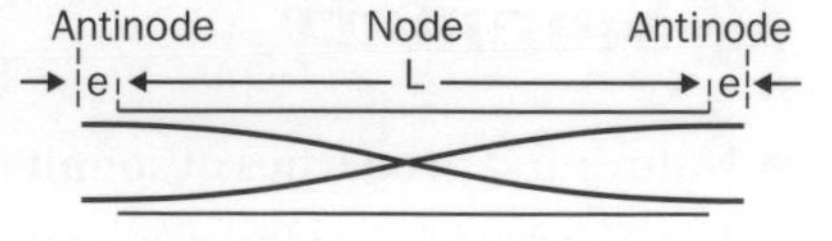

Fig. 2.22 At the fundamental frequency.

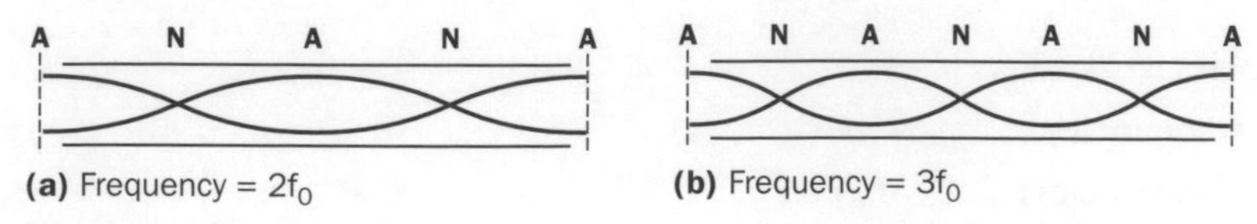

(a) Frequency = $2f_o$ **(b)** Frequency = $3f_o$

Fig. 2.23 Overtones in a pipe open at both ends **(a)** 1st overtone, **(b)** 2nd overtone.

Worked example 2.2

Calculate the fundamental frequency of sound in an open-ended pipe of length 1.40 m. Neglect end-corrections. The speed of sound in the pipe is 340 m s^{-1}.

Solution

Ignoring end-corrections, the length of the pipe is one-quarter of the fundamental wavelength, λ_o.

Hence $\lambda_o = 4 \times 1.40$ m $= 5.60$ m.

$\therefore$ Using $f_o = \frac{v}{\lambda_o}$ gives $f_o = \frac{340}{5.60} = 61$ Hz.

Questions 2.5

1. A pipe of length 0.815 m is closed at one end. The lowest frequency at which it resonates is 104 Hz.

(a) Calculate the frequency of the first overtone.

(b) The speed of sound in the pipe is 340 m s^{-1}. Calculate
(i) the wavelength of sound of frequency 104 Hz in the pipe,
(ii) the end-correction.

2. A car silencer box is open at both ends and has a length of 0.60 m. The speed of sound in the pipe is 340 m s^{-1}. Estimate **(i)** the wavelength of sound in the box when it resonates at its fundamental frequency, **(ii)** the fundamental frequency.

Summary

◆ **Nature and properties of sound waves**

1. Sound waves are longitudinal.
2. Soft materials absorb sound waves; smooth hard surfaces cause echoes.
3. Sound travels faster in solids and liquids than in gases.

◆ **Ultrasonics:** sound waves of frequency above the range of the human ear.

◆ **The human ear** is most sensitive at a frequency of about 3000 Hz.

◆ **The decibel scale (dB)** is a 'times ten' scale used to measure loudness.

◆ **Stationary waves**

1. on a vibrating wire occur at frequencies f_o, $2f_o$, $3f_o$, etc.
2. in a pipe closed at one end occur at frequencies f_o, $3f_o$, $5f_o$, etc.
3. in a pipe open at both ends occur at frequencies f_o, $2f_o$, $3f_o$, etc.

◆ **Comparison of travelling waves and stationary waves**

	Amplitude	**Phase difference between any two points**
1. Travelling wave	same at all positions	increases steadily with distance
2. Stationary wave	max at antinodes; zero at nodes	180° × no. of nodes between the two points

Revision questions

2.1. Fig. 2.24 shows an oscilloscope screen displaying the waveform of sound from a loudspeaker connected to a signal generator.
(a) (i) Measure the time period of the waveform, **(ii)** Calculate the frequency of the sound.
(b) The speed of sound in air at room temperature is 340 m s^{-1}. Calculate the wavelength of the sound waves in (a).

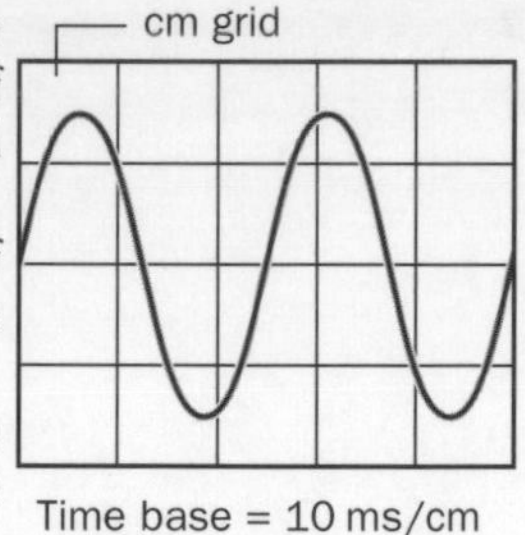

Fig. 2.24

(c) Sketch the trace on the screen if the frequency is halved without changing the time base setting.

2.2. (a) (i) Which of the materials A–D listed below is most effective at absorbing sound?
A cardboard   B brick   C water   D cushion material
(ii) Which of the above materials is most effective at reflecting sound?
(b) (i) A ship at sea in a fog is near some cliffs. The captain sounds a short blast of a siren and hears an echo 1.2 s later. How far is the ship from the cliffs? Assume the speed of sound as for Q2.1(b).
(ii) How could the captain determine if the ship is moving towards or away from the cliffs?

2.3. (a) (i) State the upper frequency limit of hearing of a young person with normal hearing.
(ii) Give one reason why the upper frequency limit could be lower?
(b) What is the increase in sound energy in a room if the loudness increases from **(i)** 40 dB to 50 dB, **(ii)** 50 dB to 60 dB, **(iii)** 60 dB to 100 dB?
(c) A person with poor hearing has a hearing loss of 20 dB at the most sensitive frequency. How many motor bikes, each producing the same noise, would need to be present for the same effect as one motor bike for normal hearing?

2.4. (a) In an ultrasonic scanner used on a patient, several strong reflected pulses are received by the transducer for every pulse emitted.
(i) Why is there more than one reflected pulse for each emitted pulse?
(ii) Give two reasons why the reflected pulses from each emitted pulse vary in strength.
(b) In an ultrasonic transducer,
(i) why is it essental that the frequency of the alternating voltage applied to the ceramic slice is the same as the natural frequency of the slice?
(ii) why does the ultrasonic frequency need to be of the order of MHz instead of about 40 kHz?

2.5. A guitar string is tuned to the same frequency as a tuning fork of frequency 256 Hz by altering the tension of the string.
(a) How would the frequency of the note produced by the guitar string change if **(i)** the string is tightened further, **(ii)** the string is shortened without change of tension?
(b) The note from the guitar string is recorded and analysed using a computer. Fig. 2.25 shows how the frequency spectrum of the note. Explain **(i)** the presence of notes of frequency 256 Hz and 768 Hz, **(ii)** the absence of a note of frequency 512 Hz.

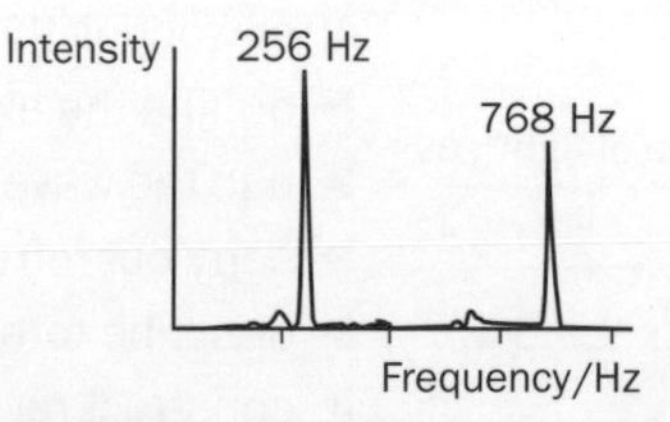

Fig. 2.25

2.6. The fundamental frequency of vibration of a string is inversely proportional to its length. A steel wire of length 0.80 m has a fundamental frequency of vibration of 384 Hz.
(a) Calculate its fundamental frequency if its length is changed to **(i)** 0.40 m, **(ii)** 1.20 m.
(b) What change of length would be needed to raise the frequency by 5% from 384 Hz?

2.7. A open-ended pipe of length 1.10 m resonates with sound at a frequency of 150 Hz. The speed of sound in the pipe is 340 m s^{-1}.
(a) (i) Calculate the wavelength of sound waves in the pipe of frequency 150 Hz.
(ii) Sketch the stationary wave pattern in the pipe when it resonates at 150 Hz.
(iii) Calculate the end-correction of the pipe.
(b) What would be the next highest frequency at which this pipe would resonate?
(c) What would be the fundamental frequency of the pipe in (a) if it was to be closed at one end?

2.8. The length of the pipes of a certain organ range from 40 mm to 4 m. The speed of sound in the pipes is 340 m s^{-1}.
(a) If each pipe is open at both ends, calculate the range of the fundamental frequencies of the organ.
(b) Calculate the first three overtone frequencies of the shortest pipe.

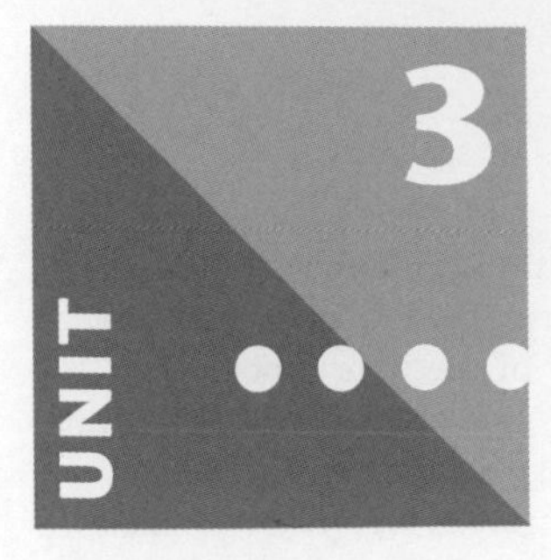

Optics

Contents

Objectives

After working through this unit, you should be able to:

- ▶ outline the experimental evidence for the wave theory of light
- ▶ use the wave theory of light to explain reflection and refraction
- ▶ carry out refractive index calculations
- ▶ describe total internal reflection and its application to optical fibres
- ▶ construct ray diagrams to explain the formation of images by mirrors, lenses, microscopes and telescopes
- ▶ outline the factors that govern the quality of the image formed by an optical instrument
- ▶ carry out calculations using the lens formula

3.1 Interference and the nature of light

It isn't obvious that light consists of waves. Over three centuries ago, Isaac Newton put forward the theory that light consisted of tiny particles (he referred to these as 'corpuscles') which were thought to bounce off mirrors and speed up on entering glass or water. About the same time, another less famous scientist called Christiaan Huygens thought that light was composed of waves. His theory also could explain reflection and it predicted that light would slow down on entering glass or water. There was no experimental evidence at the time to resolve the question over the speed of light in glass or water. Since Newton's theory of gravitation and his laws of motion were accepted, the corpuscular theory of light held sway and Huygen's theory was rejected for over a century.

The first experimental evidence for Huygen's theory was produced by Thomas Young, who made a detailed study of light passing through two closely spaced slits. Young observed a pattern of bright and dark fringes which he could only explain using the idea that light waves from one slit interfered with light waves from the other slit. See Fig. 1.14 for a general description of interference.

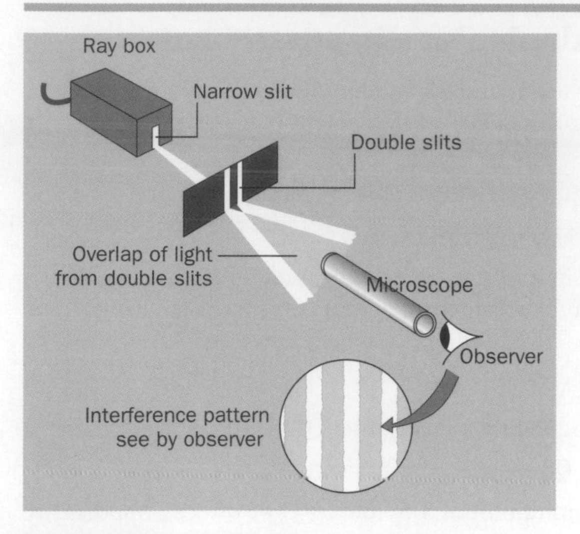

Fig. 3.1 Interference of light.

Interference of light

A parallel beam of light is directed at two closely spaced slits in an opaque plate. Light is diffracted on passing through each slit. With the aid of a microscope, an interference pattern of bright and dark fringes is observed where light from one slit overlaps light from the other slit. According to Newton's theory of light, there should be just two bright bands, one opposite each slit. In fact, the number of bright and dark fringes depends on how closely spaced and how wide the slits are. With care, more than six or so dark fringes are easily visible. Huygen's wave theory explains the pattern of fringes. In addition, it can be combined with the measurements to enable the wavelength of light to be worked out. This is very small; over 2000 wavelengths can fit into 1 mm.

As explained on p. 16, each dark fringe occurs where light from one slit cancels light from the other slit. Each bright fringe occurs where light from one slit reinforces light from the other slit. The two slits emit waves with a constant phase difference because part of each wavefront from the point light source passes through each slit. The two slits are said to be **coherent** sources of waves because they emit waves with a constant phase difference. In general, at any point, P, where the waves overlap, the waves from one slit travel a different distance to that point in comparison with the waves from the other slit. The phase difference therefore changes with the position of P.

1. At any point where a bright fringe is formed, the waves from the two slits arrive in phase and therefore reinforce each other.
2. Where a dark fringe is formed, the waves arrive exactly out of phase and therefore cancel each other out.

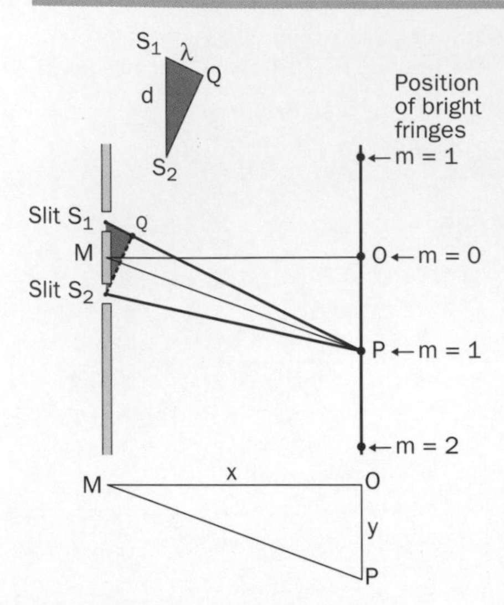

Fig. 3.2 Double slits theory.

The theory of the double slits experiment

Fig. 3.2 shows two slits, S_1 and S_2, and an interference pattern formed along a line parallel to S_1S_2. For reinforcement (i.e. a bright fringe) at a point R,

$$S_1R - S_2R = m\lambda, \text{ where m is a whole number}$$

1. m = 0 gives the central bright fringe at O.
2. m = 1 gives the first bright fringe either side of the central bright fringe; hence for reinforcement at point P, the first bright fringe from the centre

$$S_1P - S_2P = \lambda$$

3. Let Q be the point on line S_1P such that $QP = S_2P$; hence

$$S_1P - S_2P = QS_1 = \lambda$$

Let M be the midpoint of S_1S_2 and let O be directly opposite M on the screen. Since triangles MOP and S_1S_2Q are almost the same shape (i.e. similar), then

$$\frac{\text{distance } QS_1}{\text{distance } S_1S_2} \text{ is almost equal to } \frac{\text{distance OP}}{\text{distance OM}}$$

$$\therefore \frac{\lambda}{d} = \frac{y}{X}$$ **where d = the slit spacing S_1S_2**
y = fringe spacing OP
X = slit-screen distance OM

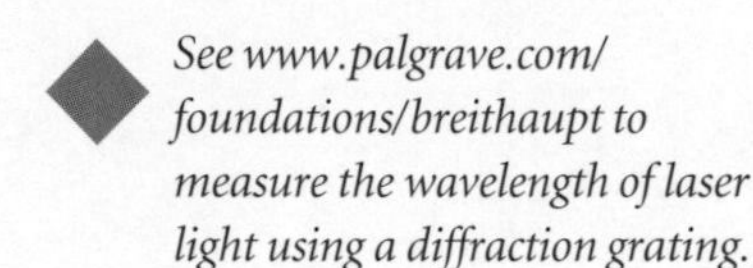

See www.palgrave.com/foundations/breithaupt to measure the wavelength of laser light using a diffraction grating.

Measurement of the wavelength of light using double slits

Use the arrangement in Fig. 3.1 to measure the fringe spacing. A pin in focus, positioned in front of the microscope, will give the plane of the fringes. The microscope can also be used to measure the slit spacing. A metre rule is used to measure X. Coloured filters may be used to measure the wavelength of light of each colour. Fig. 3.3 shows how the wavelength of light varies with colour. Use the above formula to calculate λ.

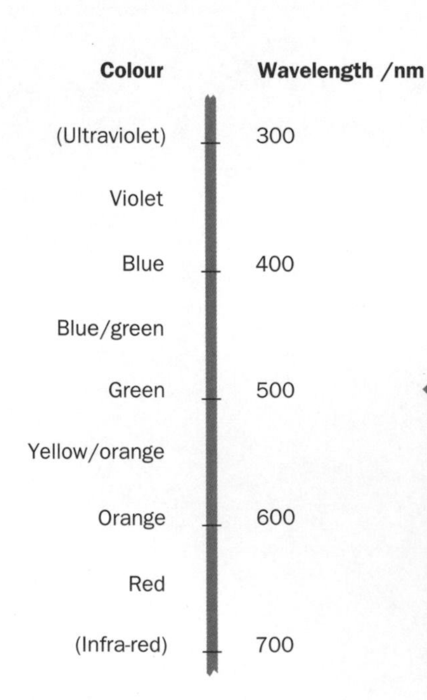

Fig. 3.3 Wavelength and colour.

Worked example 3.1

A laser beam is directed at two closely spaced slits 0.4 mm apart. The fringe pattern is observed on a screen placed 1.50 m from the slits. The fringe spacing was measured at 2.4 mm. Calculate the wavelength of the light used.

Solution

d = 0.4 mm, X = 1.50 m, y = 2.4 mm

Rearranging $\frac{\lambda}{d} = \frac{y}{X}$, gives $\lambda = \frac{yd}{X} = \frac{2.4 \times 10^{-3} \times 0.4 \times 10^{-3}}{1.50}$

$= 6.4 \times 10^{-7}$ m = 640 nm.

ometry

t-angled triangle shown
ngle θ (pronounced
he sine function is
quation

$\frac{\text{...site}}{\text{...nuse}} = \frac{BC}{AB}$

sine of an angle, key the
egrees on to the display of
lator then press the 'sin'

he angle for a given value of
function (i.e. the inverse of
e funtion), key the value on to
splay of your calculator then
the buttons marked 'inv' then

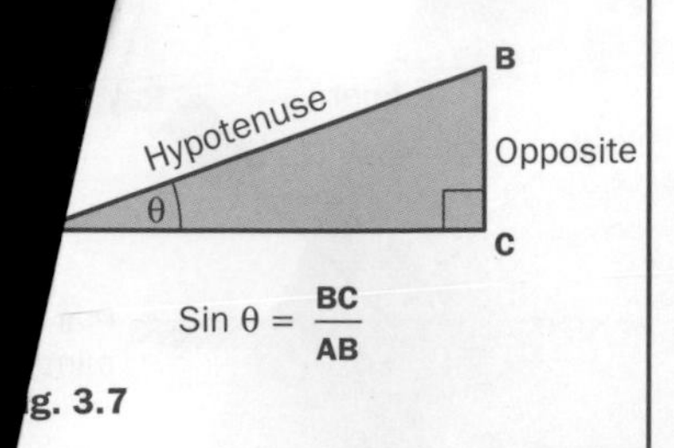

$\text{Sin } \theta = \frac{BC}{AB}$

g. 3.7

Using the wave theory to explain refraction

Light travels more slowly in a transparent substance than in air. Fig. 3.9 shows what happens to plane waves on passing across the boundary between air and a transparent substance.

1. Consider triangle WXY; angle WXY = the angle of incidence, i.

WY = the wavelength in air, λ_o (opposite angle i)

$$\therefore XY = \frac{\lambda_o}{\sin i} \left(\text{since } \sin i = \frac{WY}{XY}\right)$$

2. Consider triangle XYZ; angle XYZ = angle of refraction, r.

XZ = wavelength in the substance, λ

$$\therefore XY = \frac{\lambda}{\sin r} \left(\text{since } \sin r = \frac{XZ}{XY}\right)$$

Hence

$$\frac{\lambda}{\sin r} = \frac{\lambda_o}{\sin i}$$

Rearranging this equation gives

$$\frac{\sin i}{\sin r} = \frac{\lambda_o}{\lambda}$$

It follows that the refractive index of the substance, $n = \frac{\lambda_o}{\lambda}$

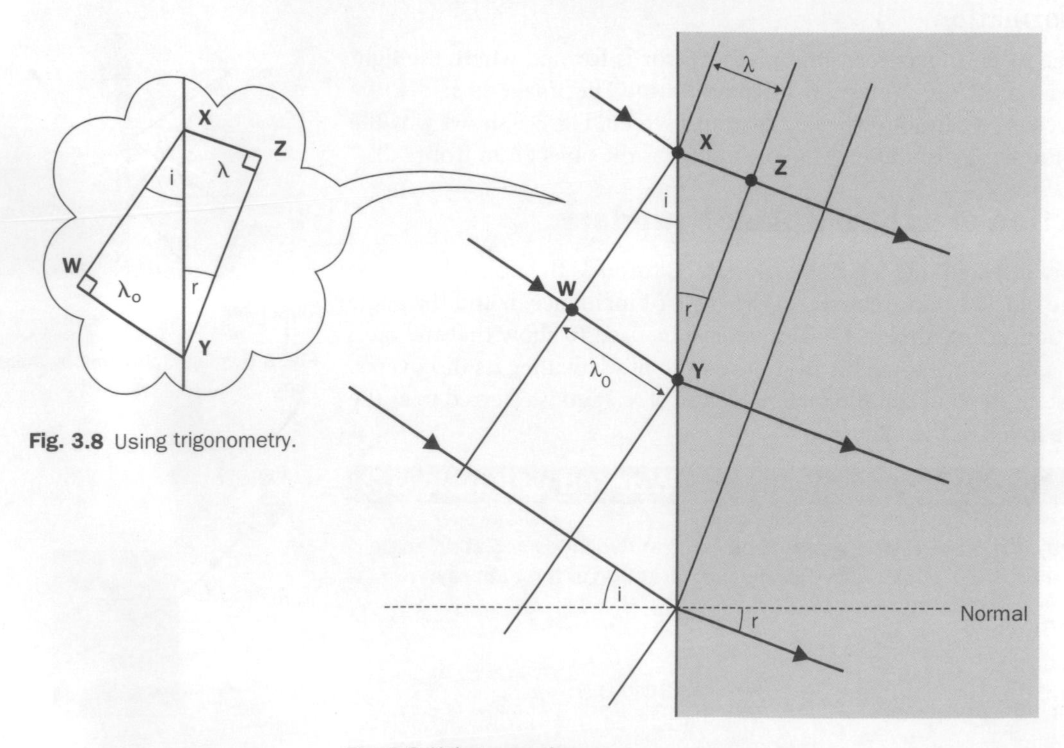

Fig. 3.8 Using trigonometry.

Fig. 3.9 Using wave theory.

1. A double slits experiment was set up using two slits at a spacing o observed at a distance of 0.80 m from the slits.

   (a) The fringe spacing was 0.90 mm. **(i)** Calculate the wavelength of was the light?

   (b) The light source was replaced by a different light source producing li Calculate the fringe spacing produced by this light source.

2. Describe and explain how the fringe pattern would change in Q1 if **(a)** one **(b)** the two slits were replaced by a pair of slits which were **(i)** the same wid **(ii)** wider but at the same spacing.

3.2 Reflection and refraction of light

Reflection at a plane mirror

The law of reflection

Fig. 3.4 shows a ray of light reflecting off a plane mirror. The ripple tank experiment on p. 14 shows plane waves reflecting off a straight reflector. This is what happens with light except the scale is much smaller. Note that the incident ray is at the same angle to the normal as the reflected ray. This is the **law of reflection**.

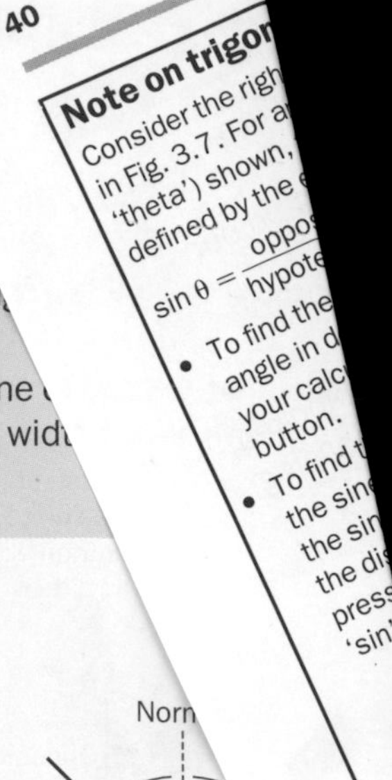

Fig. 3.4 Reflection at a plan

Image formation

The image of an object seen in a plane mirror is formed where the light rays, after reflection, appear to originate from. The image is said to be **virtual** because it cannot be projected onto a screen. Fig. 3.5 shows why the image is the same distance behind the mirror as the object is in front.

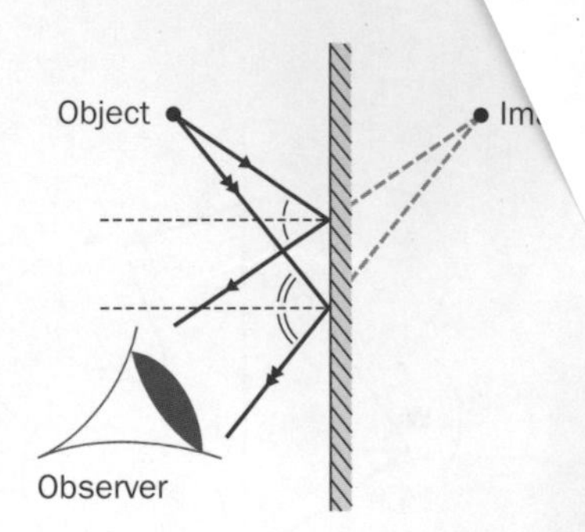

Fig. 3.5 Image formation by a plane mirror.

Refraction of light at a plane boundary

A light ray directed into a glass block refracts towards the normal where it enters the block. Measurements of the angle of incidence, i, and the angle of refraction, r, for different values of i can be used to show that the ratio sin i/sin r is a constant. This is known as Snell's law after its discoverer. The constant depends on the material of the block and is referred to as the **refractive index** of the material.

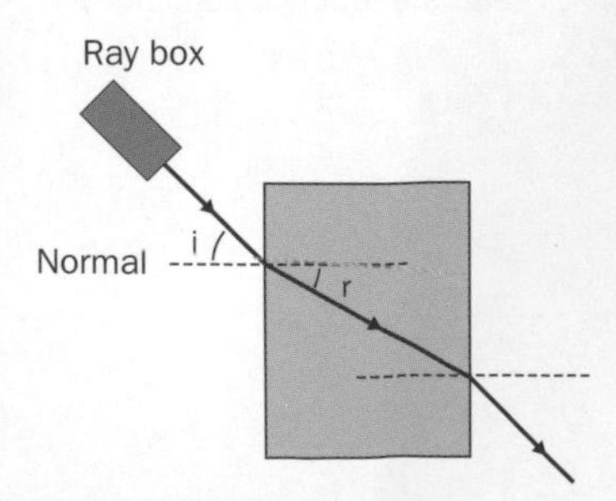

Fig. 3.6 Refraction of light.

Worked example 3.2

A light ray is directed into a glass block of refractive index 1.5 at an angle of incidence of 30°. Calculate the angle of refraction of the light ray.

Solution

$i = 30°, n = 1.5$

Rearranging $\frac{\sin i}{\sin r} = n$, gives $\sin r = \frac{\sin i}{n} = \frac{\sin 30}{1.5} = \frac{0.5}{1.5} = 0.33 \therefore r = 19.3°$

Since wave speed = frequency × wavelength and the frequency does not change on passing from air into the transparent substance, the refractive index is also equal to

$$\frac{\text{the speed of light in air}}{\text{the speed of light in the substance}}$$

Note

If the light ray travels from the substance into air, it bends away from the normal and the ratio

$$\frac{\sin i}{\sin r} = \frac{1}{n}\left(= \frac{\lambda}{\lambda_o}\right)$$

Worked example 3.3

The refractive index of glass is 1.5. Calculate the speed of light in glass. The speed of light in air is 3.0×10^8 m s^{-1}.

Solution

$$\frac{\text{speed of light in air}}{\text{speed of light in glass}} = 1.5$$

Rearranging this equation gives

$$\text{speed of light in glass} = \frac{\text{speed of light in air}}{1.5} = 2.0 \times 10^8 \text{ m s}^{-1}.$$

Total internal reflection

When a light ray passes from a transparent substance into air, the light ray refracts away from the normal at the boundary. Suppose the angle of incidence is increased until the refracted ray emerges along the boundary. This angle of incidence is called the **critical angle.** If the angle of incidence exceeds the critical angle, the light ray undergoes **total internal reflection** at the boundary.

The critical angle, c, can be calculated if the refractive index, n, is known by using Snell's law and the fact that the angle of refraction, r, is 90° when the angle of incidence is equal to the critical angle. Since angle i is in the transparent substance,

$$\frac{\sin i}{\sin r} = \frac{\sin c}{\sin 90} = \frac{1}{n}$$

As $\sin 90 = 1$, then $\sin c = \frac{1}{n}$

Note

Light undergoes total internal reflection at a boundary between two transparent substances, provided:

- the refractive index of the incident medium, $n_i > n_r$, the refractive index of the refracted medium
- the angle of incidence > the critical angle for the interface which is given by $\sin c = \frac{n_r}{n_i}$.

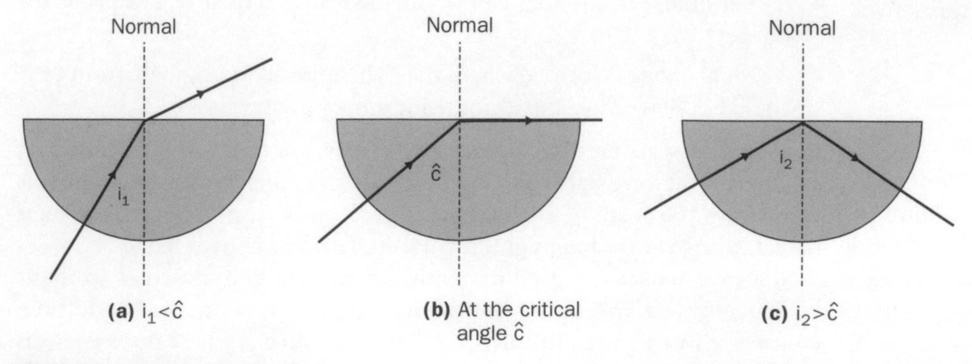

Fig. 3.10 Total internal reflection.

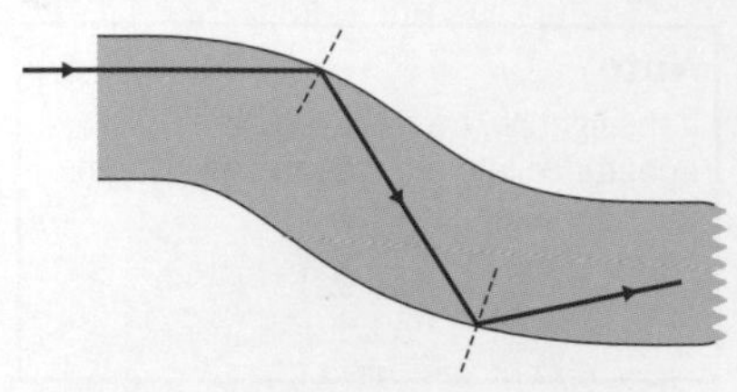
Fig. 3.11 An optical fibre.

See www.palgrave.com/foundations/breithaupt for the medical endoscope.

Optical fibres

These are used in medicine, to see inside the body and to guide light pulses in digital communication systems. Fig. 3.11 shows a light ray guided by a thin optical fibre. The ray undergoes total internal reflection each time it is incident on the fibre–air boundary, provided the fibre is not too bent. The fibre is usually surrounded by a transparent cladding of lower refractive index than the core. Because light is reflected at the core–cladding boundary, the cladding ensures no light can cross from one fibre to another at points of contact.

Worked example 3.4

A light ray is directed normally into a glass prism of refractive index 1.5, as shown in Fig. 3.12.

(a) Calculate the critical angle of the glass

(b) Copy the diagram and complete the path of the light ray through the prism.

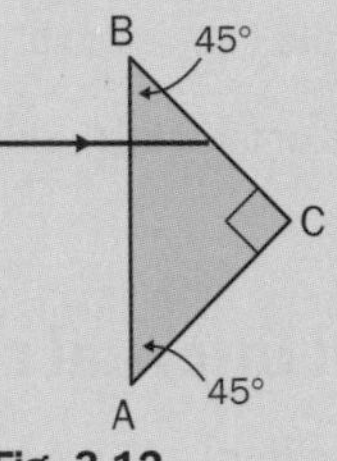

Fig. 3.12

Solution

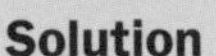

(a) $\sin c = \frac{1}{n} = \frac{1}{1.5} = 0.67 \quad \therefore c = 42°$

(b) The light ray passes into the prism normally (i.e. along the normal) through face AB. The angle of incidence at face BC is 45° which exceeds the critical angle. Therefore the ray undergoes total internal reflection at BC and is incident on face AC at 45°. The ray again undergoes total internal reflection at face AC, hence it passes out through AB normally, without refraction.

Questions 3.2

1. (a) Copy and complete the diagram in Fig. 3.13 to show how an image of a point object is formed in a plane mirror.

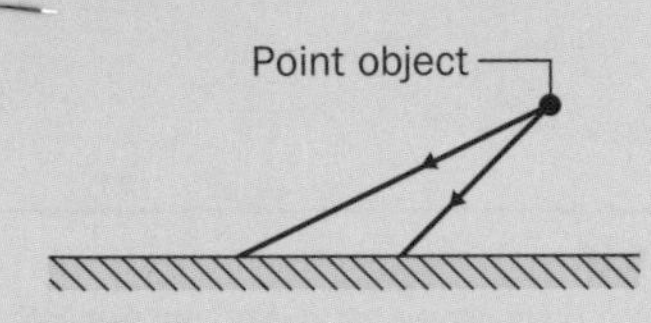

Fig. 3.13

(b) A person stands 1.5 m in front of a vertical plane mirror. How far away is the person from her image?

2. An optical fibre consists of a transparent core of refractive index 1.5, surrounded by a transparent layer of cladding of refractive index 1.2.

(a) With the aid of a diagram, explain the purpose of the cladding.

(b) Calculate the critical angle at the core–cladding boundary.

(c) Discuss whether or not a larger refractive index for the cladding is desirable.

3.3 Lenses

Lenses are used to form real or virtual images.

- A **real image** is one that can be formed on a screen (e.g. a projector image).
- A **virtual image** is formed where the light appears to originate from (e.g. an object viewed in a mirror or using a magnifying glass).

Convex lenses are used in cameras, projectors, microscopes and telescopes and to correct long sight. The eye contains a flexible convex lens which is sometimes too weak to enable near objects to be seen clearly. This sight defect, referred to as long sight, is corrected using a convex lens.

Concave lenses are used in combination with convex lenses to make high-quality lens systems for optical instruments and cameras. In addition, concave lenses are used to correct short sight which is where the eye lens is too strong to see distant objects clearly.

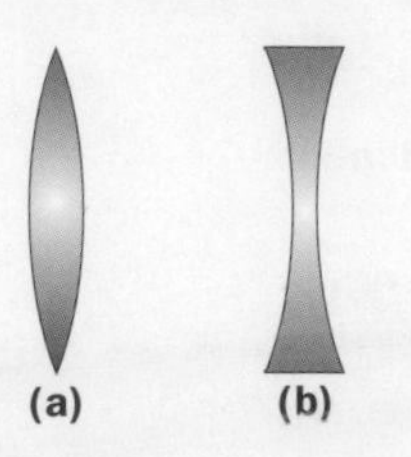

Fig. 3.14 Lens shapes
(a) convex, (b) concave.

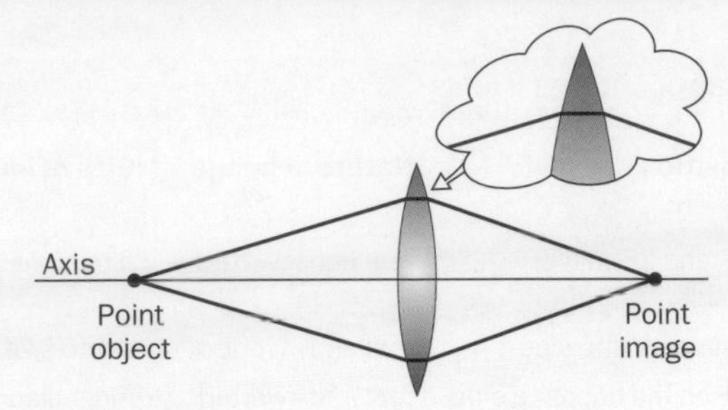

Fig. 3.15 Image formation by a convex lens.

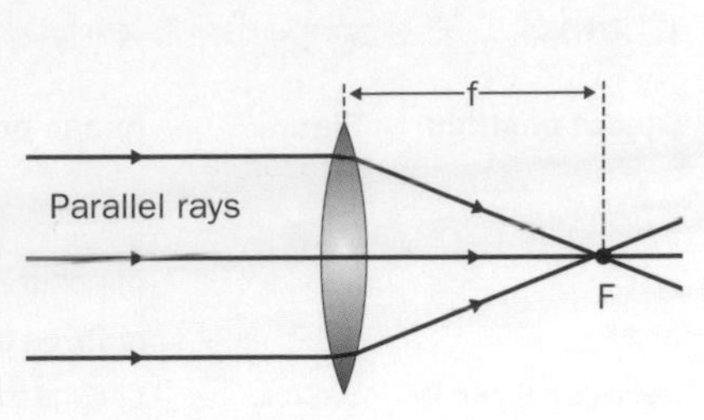

Fig. 3.16 Focal length.

The convex lens

A convex lens forms a real image by making light from a point object converge to form a point image. Fig. 3.15 shows how this is achieved.

Focal length, f

This is the distance from the lens to the **focal point**, **F**, of the lens, which is the point where a beam of light parallel to the lens axis is brought to a focus.

Ray diagrams

The position and size of the image formed by a convex lens can be determined by constructing a ray diagram. Fig. 3.17 shows how to do this. The image formed in Fig. 3.17 is real, inverted and magnified (i.e. larger than the object).

The position, nature and size of the image depends on the position of the object in relation to the focal length of the lens, as shown in Figs 3.17, 18, 19 and 20 and summarised in Table 3.1.

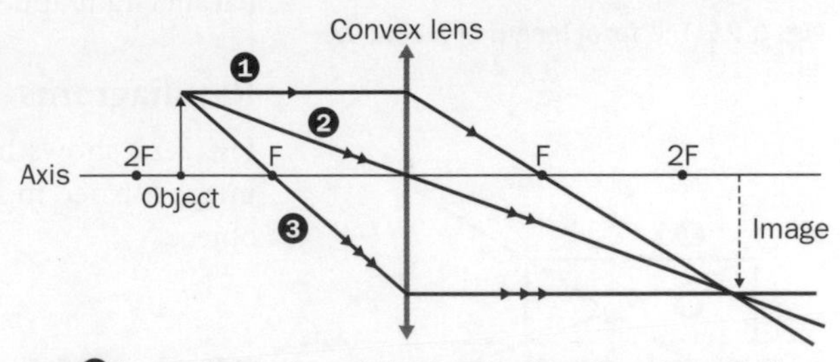

Ray ❶ is parallel to the axis before the lens then through F

Ray ❷ is through the lens centre without refraction

Ray ❸ is through F before the lens then parallel to the axis

Fig. 3.17 Constructing a ray diagram for a convex lens.

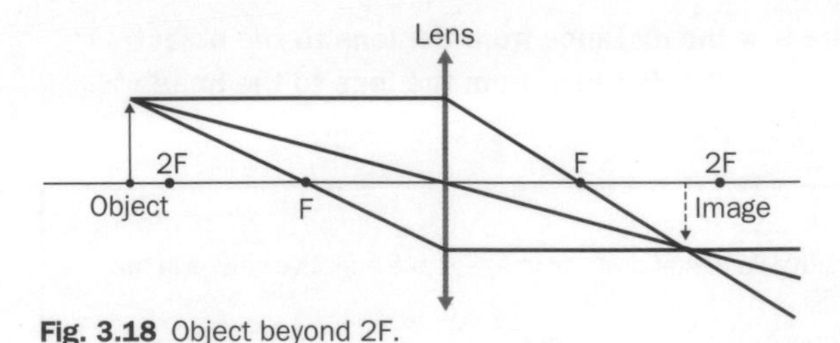

Fig. 3.18 Object beyond 2F.

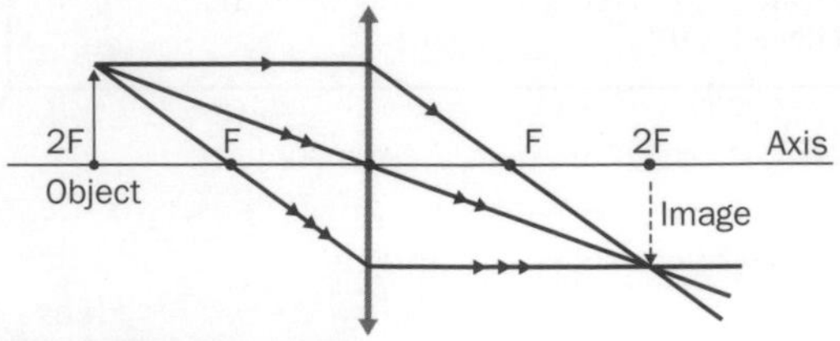

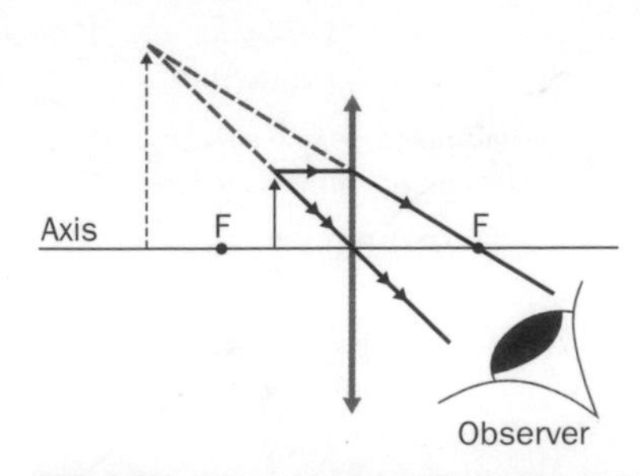

Fig. 3.20 Object between F and the lens.

Fig. 3.19 Object at 2F.

Table 3.1 The convex lens and its applications

Object position	Figure reference	Image position	Nature of image	Size of image	Application
beyond 2F	3.18	between F and 2F on the opposite side	real and inverted	diminished	camera, eye
at 2F	3.19	at 2F on the opposite side	real and inverted	same size	inverter
between F and 2F	3.17	beyond 2F on the opposite side	real and inverted	magnified	projector
at F		no image formed			to produce a parallel beam
inside F	3.20	further away on the same side	virtual and upright	magnified	magnifying glass

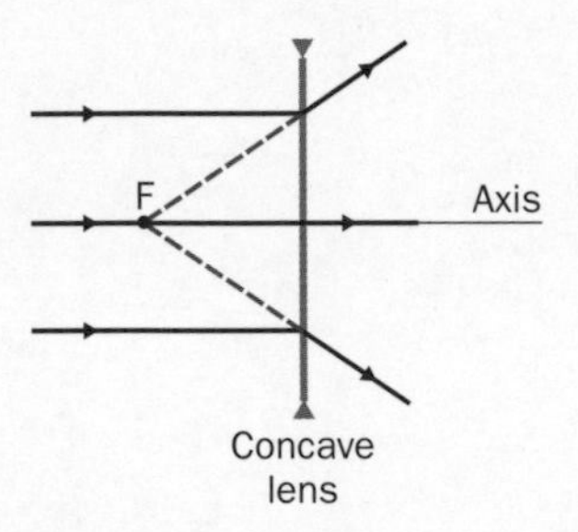

Fig. 3.21 The focal length of a concave lens.

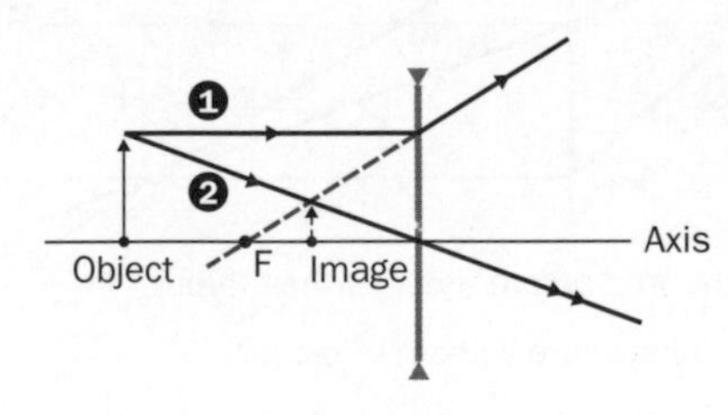

Ray ❶ is parallel to the axis before the lens, then diverges as if from F

Ray ❷ is through the centre of the lens without refraction

Fig. 3.22 A ray diagram for a concave lens.

See palgrave.com/foundations/breithaupt for a proof of the lens formula.

The concave lens

A concave lens makes light from a point object diverge more or converge less.

Focal length, f

Fig. 3.21 shows the effect of a concave lens on a parallel beam of light. The focal length is the distance from the lens to the point where a beam of parallel light appears to originate.

Ray diagrams

Fig. 3.22 shows how to construct a ray diagram for a concave lens. The image formed in Fig. 3.22 is virtual, upright and closer to the lens than the object.

The lens formula

$$\frac{1}{u} + \frac{1}{v} = \frac{1}{f}$$ **where u = the distance from the lens to the object**
v = the distance from the lens to the image

Notes

1. Real images take positive values and virtual images take negative values in the lens formula.
2. The focal length of a convex lens is given a positive value. The focal length of a concave lens is given a negative value.
3. Linear magnification $= \frac{\text{image height}}{\text{object height}} = \frac{v}{u}$

Worked example 3.5

A convex lens of focal length 0.10 m is used to form a magnified image of an object of height 5 mm placed at a distance of 0.08 m from the lens. Calculate the position, nature and size of the image.

Solution

$u = 0.08$ m, $f = 0.10$ m

Using the lens formula gives $\frac{1}{0.08} + \frac{1}{v} = \frac{1}{0.10}$

$$12.5 + \frac{1}{v} = 10$$

$$\therefore \quad \frac{1}{v} = 10 - 12.5 = -2.5$$

$$v = -\frac{1}{2.5} = -0.40 \text{ m}$$

The image is virtual and its position is 0.40 m from the lens on the same side as the object. The linear magnification is (–)0.40 ÷ 0.08 = 5. The image height is therefore 25 mm (= 5 × 5 mm).

Determination of the focal length of a convex lens

The focal length of a convex lens can be determined by the plane mirror method, as follows.

Using illuminated cross-wires as the object, a plane mirror is used to reflect light from the lens back towards the object. The position of the object is adjusted to where a clear image of the cross-wires can be seen alongside the cross-wires. The object is then at the focal point of the lens since the beam from the lens to the mirror is exactly parallel and at 90° to the mirror. See Fig. 3.24.

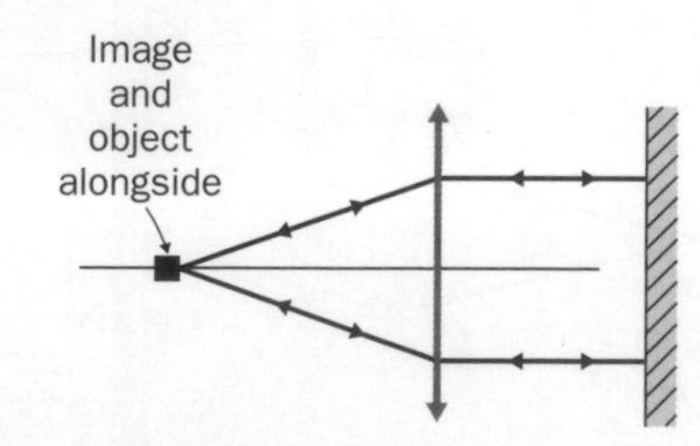

Fig. 3.24 The plane mirror method.

Questions 3.3

1. **(a)** By means of a ray diagram, determine the position and nature of the image formed by a convex lens of focal length 0.20 m if the object distance is **(i)** 0.50 m, **(ii)** 0.25 m, **(iii)** 0.15 m.

   (b) Use the lens formula to calculate the image distance and the linear magnification for each object position in (a).

2. **(a)** With the aid of a diagram, explain how a convex lens may be used as a magnifying glass.

   (b) With the aid of a diagram, explain why the image of an object viewed through a concave lens appears smaller and closer than the object.

See www.palgrave.com/foundations/breithaupt to measure the focal length of a convex lens by measuring the image distance for different object distances.

3.4 Optical instruments

The quality of the image formed by an optical instrument depends on the quality of the optical components (e.g. lenses) in the instrument. The two main effects that cause image defects are:

- **spherical aberration**, where the outer rays from a point object are focused to a position different from the inner rays. The shape of the lens or mirror surface determines the extent of this effect.
- **chromatic aberration**, a lens defect due to the splitting of white light when it is refracted. The effect can be minimised using convex and concave lenses of different refractive indices.

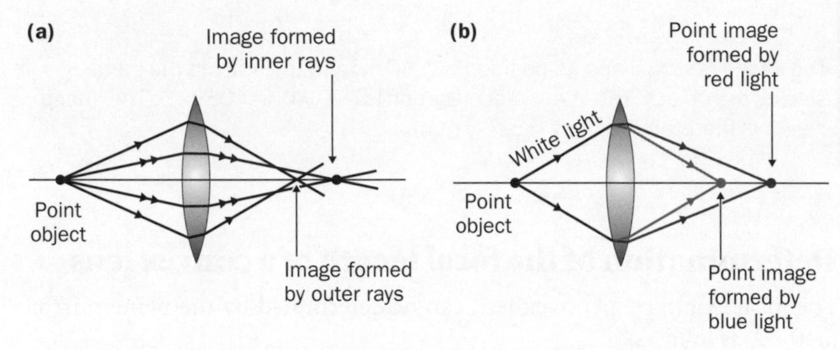

Fig. 3.25 Lens aberrations **(a)** spherical aberration, **(b)** chromatic aberration.

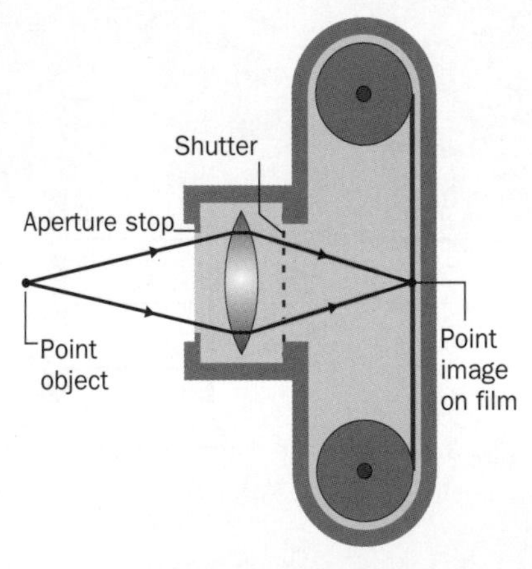

Fig. 3.26 The camera.

The camera

The convex lens forms a real image on the film inside the camera. In a digital camera, the film is replaced by a CCD (charge-coupled device).

- The lens position is adjusted for different object distances. The nearer the object, the greater the distance from the lens to the film.
- The aperture width controls the amount of light entering the camera. A wide aperture will produce minimum diffraction to give very clear images but little depth of focus.
- The shutter must open and close rapidly to photograph a fast-moving object. In such a situation, the aperture width must be as wide as possible to let as much light as possible into the camera. As a result, the depth of focus is much reduced in an 'action' photograph.

The optical microscope

The objective lens is used to form a magnified, real image of the object. The eyepiece lens is used to view this 'intermediate' image, enabling a large, magnified, virtual image to be seen by the observer looking into the eyepiece. In normal adjustment, the position of the final image is at the observer's **near point** of vision.

- The image of a point object is smudged due to diffraction of light as it passes through the objective lens aperture. More image detail can be seen using blue light rather than ordinary white light. Because blue light has a smaller wavelength than all the other components of white light (except violet), there is less diffraction using blue light.
- Cross-wires (or a transparent scale) are seen in focus with the final image, provided the cross-wires are located at the same position as the intermediate image.

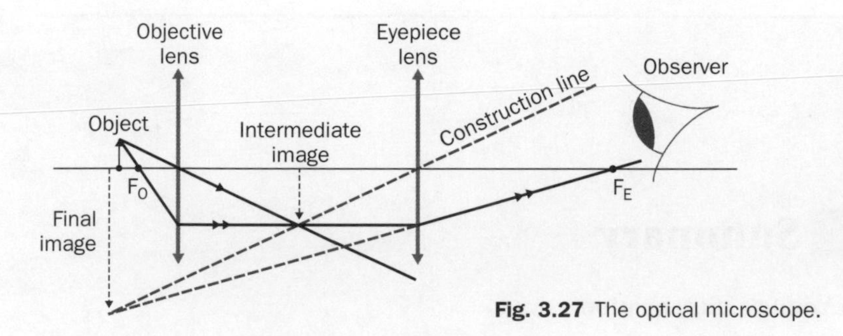

Fig. 3.27 The optical microscope.

The refracting telescope

The objective lens forms an intermediate, real image of a distant object. This image is magnified using the eyepiece so the observer sees a final virtual image at infinity. The best position for the eye is at the 'eye-ring', as shown in Fig. 3.28.

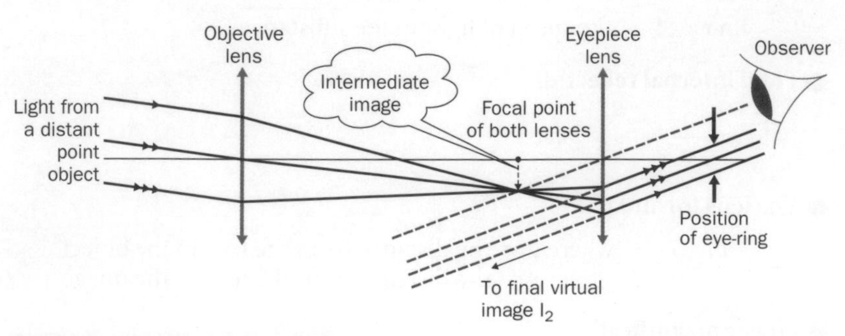

Fig. 3.28 The refracting telescope.

- The magnifying power of a telescope is measured by how many times larger the final image appears in comparison with the object. For example, a telescope that makes the Moon appear 10 times wider than if viewed directly has a magnifying power of $\times 10$.
- The wider the objective, the more light is collected by the telescope. This is why stars that cannot be seen directly are visible using a telescope. In addition, diffraction is less and more detail can be seen, although beyond a width of about 10 cm, atmospheric smudging limits the detail that can be seen.

Note

It can be shown that

$$\text{the magnifying power of a telescope} = \frac{\text{the focal length of the objective}}{\text{the focal length of the eyepiece}}$$

Summary

◆ **Young's slits equation**

$$\frac{\lambda}{d} = \frac{y}{X}$$ where d = the slit spacing S_1S_2
y = fringe spacing OP
X = slit-screen distance OM

◆ **The law of reflection**

The incident ray is at the same angle to the normal as the reflected ray.

◆ **Snell's law of refraction**

The refractive index of a substance, n

$$= \frac{\sin i}{\sin r} = \frac{\lambda_o}{l} = \frac{\text{the speed of light in air}}{\text{the speed of light in the substance}}$$

◆ **Total internal reflection**

$$\sin c = \frac{1}{n}$$

◆ **The lens formula**

$$\frac{1}{u} + \frac{1}{v} = \frac{1}{f}$$ where u = the distance from the lens to the object
v = the distance from the lens to the image

◆ **Linear magnification**

$$= \frac{\text{image height}}{\text{object height}} = \frac{v}{u}$$

Revision questions

3.1. An interference pattern was observed using two closely spaced slits to view a narrow source of light.
(a) With the aid of a diagram, explain why the pattern consists of alternate bright and dark fringes.
(b) The pattern was observed on a white screen at a distance of 0.90 m from the slits. The slit spacing was 0.50 mm.
(i) The wavelength of the light observed was 590 nm. Calculate the spacing between adjacent bright fringes.
(ii) The light source was replaced by a different light source. The bright fringes were observed at a spacing of 0.80 mm. Calculate the wavelength of this light source.

3.2. (a) State an approximate value for the wavelength of **(i)** red light, **(ii)** blue light.
(b) A beam of yellow light of wavelength 590 nm enters a glass lens of refractive index 1.55.
(i) Calculate the wavelength of this light in the glass.
(ii) What change occurs in the speed of this light beam when it travels from air to glass?

3.3. (a) When a ray of light passes from air into glass, does it refract towards or away from the normal?
(b) Calculate the angle of refraction of a ray of light that is incident on a plane surface of a glass block of refractive index 1.50 at an angle of incidence of **(i)** 25°, **(ii)** 50°.
(c) Fig. 3.29 shows a ray of light that enters a semi-circular glass block of refractive index 1.50 through the curved face.
(i) For an angle of incidence of 40°, calculate the angle of refraction of the light where it leaves the block.

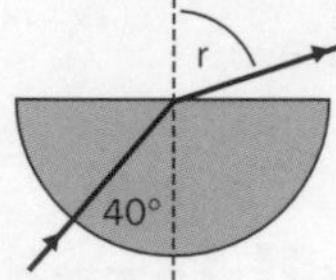

Fig. 3.29

(ii) Calculate the critical angle of the block.
(iii) Sketch the path of a light ray in Fig. 3.29 that is incident on the flat face at an angle of incidence of 60°.

3.4. (a) Construct a ray diagram to determine the nature, position and magnification of the image formed by a convex lens of focal length 0.15 m when the object distance from the lens is **(i)** 1.00 m, **(ii)** 0.20 m, **(iii)** 0.10 m.
(b) Use the lens formula to calculate the image distance and the linear magnification for each object distance in (a) above.
(c) Calculate the image distance for an object of height 12 mm at a distance of 0.20 m from a concave lens of focal length 0.30 m, and state the nature and size of the image.

3.5. (a) With the aid of a diagram, explain the function of the convex lens in a camera.
(b) A certain camera is used to take a photograph of a distant scene. What adjustment should be made to the camera to photograph someone indoors? Explain your answer.

3.6. (a) Draw a ray diagram to show the formation of a virtual image by an optical microscope.
(b) Explain why more detail is visible in a microscope image viewed in blue light.
(c) Give one reason why an image observed through a certain microscope is distorted at the edges.

3.7. (a) With the aid of a suitable diagram, explain the function of the lenses in a refracting telescope.
(b) What are the advantages and disadvantages of using a wide objective lens in a telescope?

Electromagnetic Waves

Contents

Objectives

After working through this unit, you should be able to:

- ▶ explain the splitting of white light by a prism and explain the colour of an object
- ▶ describe and explain continuous and line spectra
- ▶ describe how to measure optical spectra using a spectrometer and diffraction grating
- ▶ state similarities and differences between different parts of the electromagnetic spectrum
- ▶ describe the production and common uses of different types of electromagnetic waves

4.1 The visible spectrum

Using a prism to produce a visible spectrum

White light is split into the colours of the spectrum when it is refracted. Fig. 4.1 shows the splitting of a beam of white light by a prism. The beam is split into colours on entering the prism. The splitting of white light into colours is called **dispersion.** On leaving the prism, the colours are separated further.

The reason why the beam is split into the colours of the spectrum is because the speed of light in glass depends on wavelength and hence on

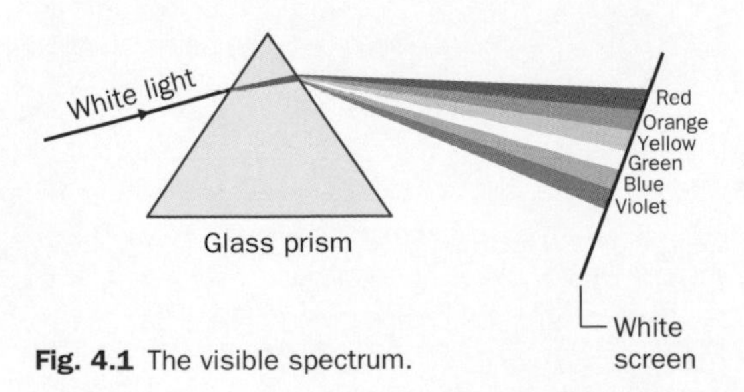

Fig. 4.1 The visible spectrum.

colour. Consequently, the refractive index of glass depends on colour, being greatest for violet and blue and least for red light. A prism bends blue light more than red light (blue bends best!) and therefore the red part of a spectrum produced by a prism is always the part which is least refracted.

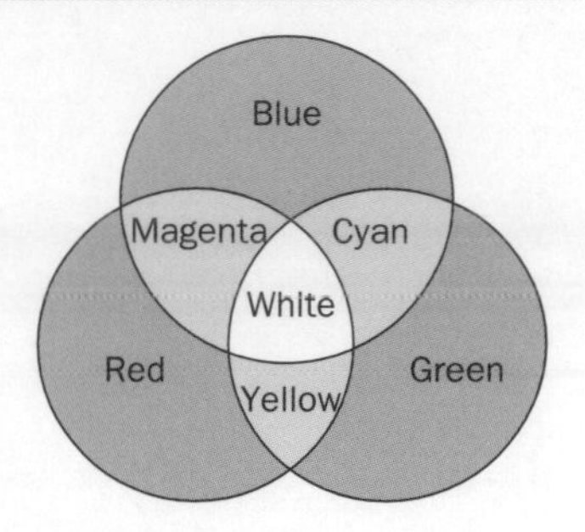

Fig. 4.2 Primary colours overlapping.

Primary and secondary colours

The primary colours of light are red, green and blue. Fig. 4.2 shows the effect of these colours overlapping on a white screen.

- Where all three colours overlap, white light is observed.
- Where two primary colours overlap, a secondary colour is observed. For example, yellow is a secondary colour because it is produced where red and green overlap.
- Adding a secondary colour to the other primary colour produces white light. The secondary colour and the primary colour are said to be complementary. For example, yellow is complementary to blue since blue and yellow overlap to produce white.

Questions 4.1

1. **(a)** Sketch a diagram to show how a glass prism can be used to split a narrow beam of white light into the colours of the spectrum.

   (b) Describe and explain the appearance of a white light spectrum displayed on a blue screen.

2. **(a)** State the three primary colours and the secondary colour that complements each primary colour.

   (b) What colour would you expect a shirt that is yellow in normal light to appear in **(i)** red light, **(ii)** blue light?

4.2 Types of spectra

Continuous spectra

The white light spectrum is an example of a continuous spectrum because the colours change continuously with change of position across the spectrum, from deep red to violet, corresponding to a continuous range of wavelengths. The Sun, the stars and filament lamps all produce light with a continuous range of wavelengths and therefore give light spectra which are continuous.

The temperature of the light source determines the distribution of light energy emitted in each part of the spectrum.

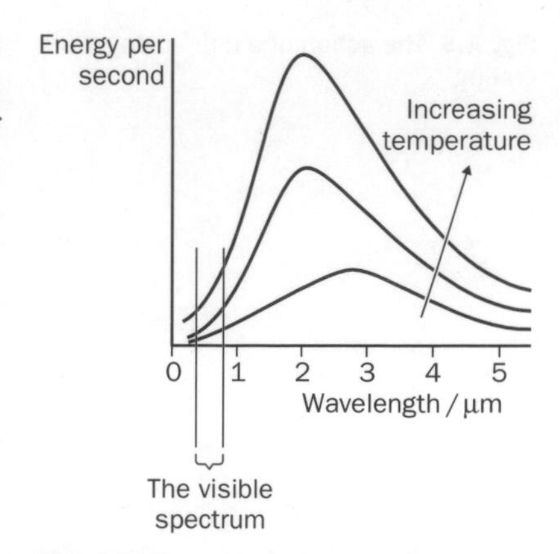

Fig. 4.3 The energy distribution across a continuous spectrum.

1. Observe a low voltage filament light bulb when the current through it is gradually increased from zero to the current for normal brightness. The current heats the filament enough to make the filament emit light. The filament glows dull-red at first, then it becomes orange-red and then yellow. This is because the higher its temperature, the smaller the wavelength at which it emits most strongly.

2. A carbon arc lamp glows white because its temperature is so high that it emits blue light as well as the other colours of the spectrum. If its temperature is raised even more, it glows blue-white as it emits mostly in the blue part of the spectrum.

3. The above observations can be applied when comparing the colour of two stars. For example, a blue star is hotter than a red star. The Sun is a yellow star and its surface is therefore hotter than the surface of a red star.

Line emission spectra

Fig. 4.4 Line emission spectra.

The spectrum of light from a discharge tube (e.g. a neon tube) or from a vapour lamp (e.g. a sodium street lamp) is composed of narrow vertical lines, each line being a different colour. This is because the light source emits light of certain wavelengths only. When a narrow beam of light from such a light source is split by a prism, each colour of light present appears as a discrete line in the spectrum. The pattern of lines makes up the line emission spectrum of the light source. The wavelength of each line can be measured using a **diffraction grating** as explained below. If the light source is a single element (e.g. neon in a neon tube), the line spectrum is characteristic of the element and can be used to identify the element. If the light source is due to two or more elements, the line spectrum can be used to identify the elements present. This is how astronomers identify the elements present in stars.

The theory of the diffraction grating

The diffraction grating is used to measure the wavelength of light of any colour. It consists of many closely spaced parallel slits on a transparent slide. When a parallel beam of light of a single colour is directed normally at the grating, the effect of the slits is to diffract the beam into certain well-defined directions only, as shown in Fig. 4.5. The action of a diffraction grating is due to light waves diffracting as they pass through the slits and the diffracted waves reinforcing each other in certain directions only.

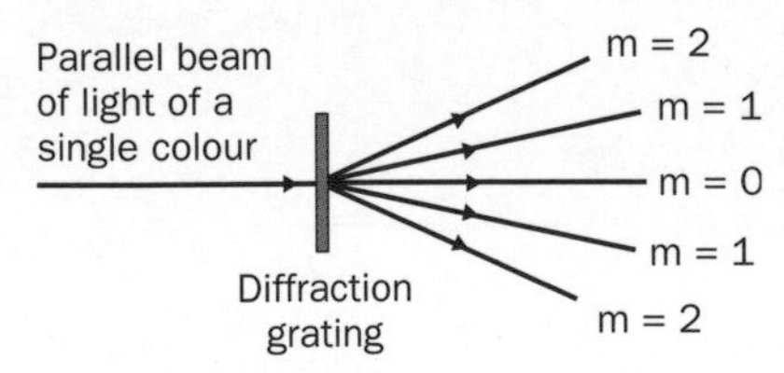

Fig. 4.5 The action of a diffraction grating.

The diffracted beams are numbered in order from the centre. Fig. 4.6 illustrates the formation of the diffracted wavefronts.

1. To form the first order wavefronts, light from one slit travels a distance of one wavelength before it is in phase with light emerging through the next slit. As shown by Fig. 4.6, this extra distance is equal to $d \sin \theta_1$ where d is the spacing between adjacent slits and θ_1 is the angle of diffraction. Hence the angle of diffraction of the first order is given by the equation,

   $$d \sin \theta_1 = \lambda$$

2. To form the second order wavefronts, light from one slit travels an extra distance of two wavelengths to be in phase with light emerging through the next slit. Using the same reasoning as above therefore gives

   $$d \sin \theta_2 = 2\lambda$$

   for the angle of diffraction, θ_2, of the second order beam.

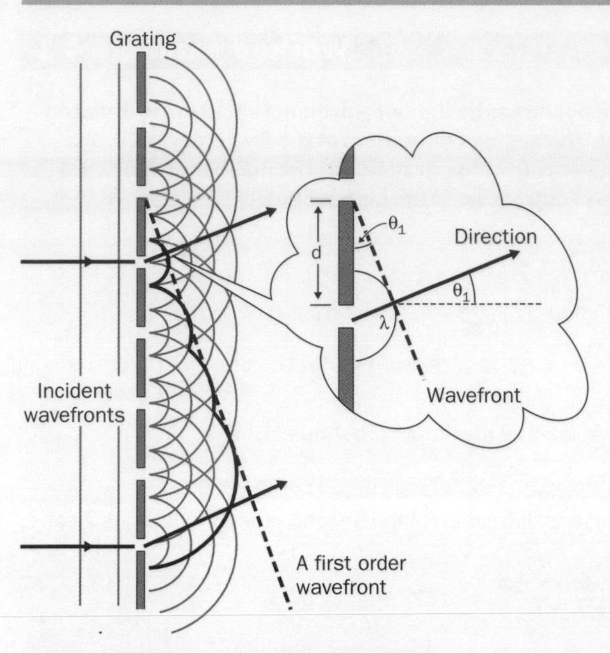

Fig. 4.6 The theory of the diffraction grating.

3. In general, the mth order wavefront is due to light from any slit travelling an extra distance of m wavelengths to be in phase with light emerging through the next slit. The angle of diffraction of the mth order wavefront, θ_m, is therefore given by the equation, $d \sin \theta_m = m\lambda$.

$$d \sin \theta_m = m\lambda$$

θ_m = angle of diffraction of mth order
d = slit spacing (centre-to-centre)
λ = wavelength of light

Notes

1. The number of lines per unit width on the grating, N, = 1/d. Gratings are usually specified in terms of N rather than d. For example, for a grating that has 300 lines per millimetre, $N = 300\ mm^{-1}$ and $d = 1/300 = 3.33 \times 10^{-4}$ mm, since d = 1/N.

2. The maximum number of orders for light of a certain wavelength, λ, incident normally on a grating of spacing, d, is determined by the condition that the maximum value of sin θ is 1 (which is when θ = 90°). To determine the maximum order number, substitute 1 for $\sin \theta_m$ in the diffraction grating equation to give m = d/λ. The maximum value of m is therefore d/λ rounded down to the nearest whole number.

Measurement of the wavelength of light using a laser

A laser produces a narrow, parallel beam of light which is highly monochromatic (i.e. a single colour corresponding to a very narrow wavelength range). Use laser safety goggles and do **NOT** look along the direct beam or any reflected beam. The beam from the laser is directed normally

Worked example 4.1

A parallel beam of monochromatic light of wavelength 640 nm is directed normally at a grating. The angle of diffraction of the first order is 22.6°. Calculate (a) the slit spacing of the grating, (b) the number of diffracted beams, (c) the angle of diffraction of the highest order.

Solution

(a) $\lambda = 640$ nm, $m = 1$, $\theta_m = 22.6°$

Using $d \sin \theta_m = m\lambda$ gives $d \sin 22.6 = 1 \times 640 \times 10^{-9}$

$$\therefore d = 1 \times \frac{640 \times 10^{-9}}{\sin 22.6} = 1.67 \times 10^{-6} \text{ metres}$$

(b) $\frac{d}{\lambda} = \frac{1.67 \times 10^{-6}}{640 \times 10^{-9}} = 2.62$ $\therefore$ maximum order number = 2

There are two diffracted beams either side of the central beam.

(c) For $m = 2$, the angle of diffraction of the second order beam, θ_2, is given by $d \sin \theta_2 = 2\lambda$

$$\therefore \sin \theta_2 = \frac{2\lambda}{d} = \frac{2 \times 640 \times 10^{-9}}{1.67 \times 10^{-6}} = 0.776$$

$$\therefore \theta_2 = 50.9°$$

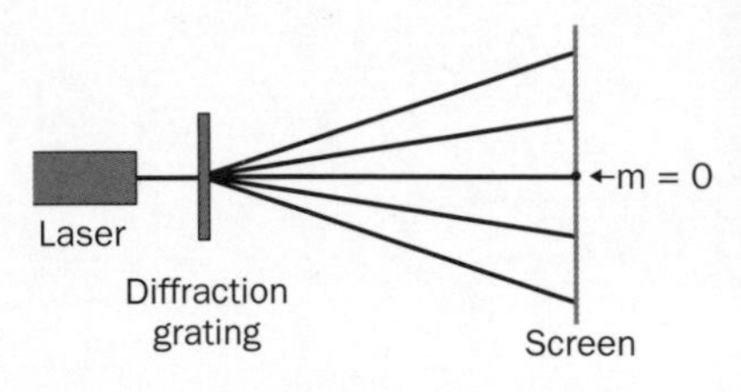

Fig. 4.7 Using a laser.

at a large, white screen so a spot of laser light is observed on the screen. The diffraction grating is then positioned between the laser and the screen, parallel to the screen, as in Fig. 4.7. The diffracted orders are seen as a horizontal row of light spots on the screen.

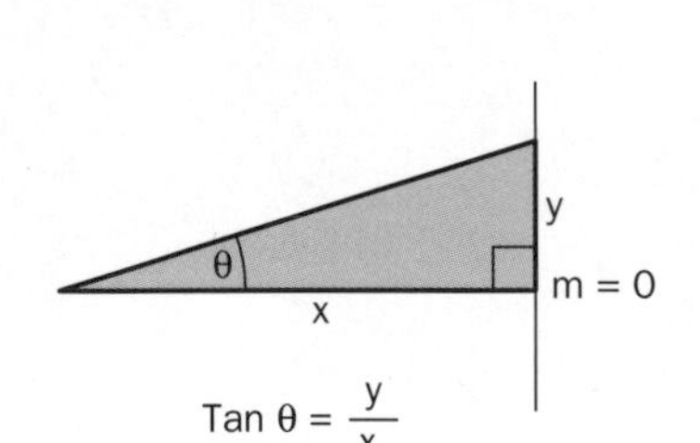

Fig. 4.8 Calculating the angle of diffraction.

Measurements and theory:

1. The perpendicular distance, X, from the grating to the screen is measured and the distance, y, of each spot from the central order spot is measured. The angle of diffraction of each order can be calculated using the trigonometric formula $\tan \theta = \frac{y}{X}$. See Fig. 4.8.
2. The diffraction grating equation is used to measure the wavelength of the laser light. Each order should give the same value of wavelength, although errors of measurement may give some variation, in which case the average value should be calculated.

Note

The reason why you must never look along a laser beam is that the eye lens would focus it to a tiny spot on the retina, concentrated sufficiently to damage the retina permanently.

Using a spectrometer

This is used to measure accurately the wavelengths of light in a spectrum. The light source is used to illuminate a narrow slit at the focal point of the collimator lens so a parallel beam of light emerges from the collimator tube. The diffraction grating is placed on the spectrometer turntable at 90° to the beam. The telescope is used to locate each diffracted beam, observed as a narrow line which is an image of the slit. Each wavelength of light present gives a narrow line of a particular colour for each order number. Thus each order gives a pattern of spectral lines on each side of the central line. The scale attached to the telescope is used to measure the position of each line. From the colour and scale reading of each observed line, the angle of

See www.palgrave.com/foundations/breithaupt for more about using a spectrometer.

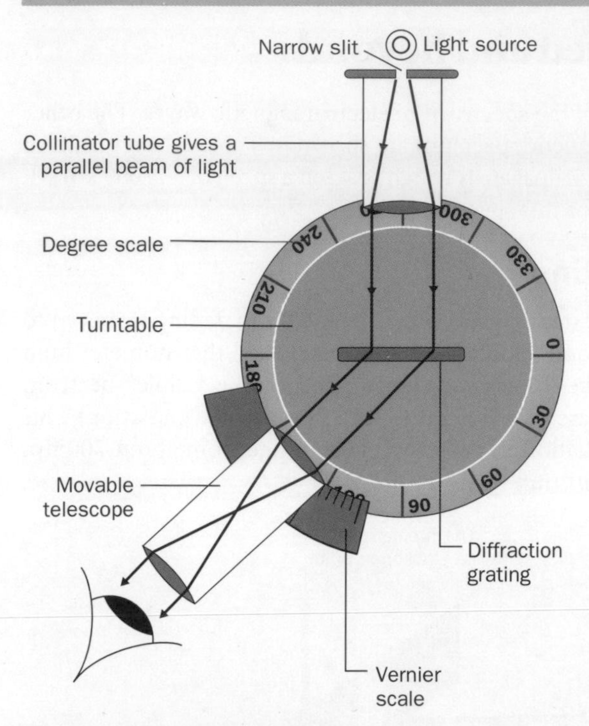

Fig. 4.9 Using a spectrometer.

diffraction and the order number can be worked out and the wavelength of each line calculated.

> **Note**
> The degree scale of a spectrometer is usually sub-divided into minutes, where 60 minutes (symbol ´) equals 1°.

Questions 4.2

1. The spectrum of light from a sodium vapour lamp was observed and measured using a spectrometer and a diffraction grating which had 600 lines per millimetre. The position of the telescope was adjusted until the crosswires were centred on the zero order slit image. The scale reading was then 101° 22´.

   (a) The telescope was then used to locate and measure the diffracted orders. A first order yellow line was observed at 122° 06´. Calculate the wavelength of this yellow line.

   (b) Calculate the angle of diffraction of each further order of this wavelength on the same side as the observed first order line.

2. A spectrometer fitted with a diffraction grating was used to observe the spectrum of light from a certain discharge lamp known to emit red light at a wavelength of 629 nm.

   (a) A first order red line was observed at an angle of diffraction of 10° 52´. Calculate the number of lines per millimetre of the grating.

   (b) Calculate the maximum number of diffracted orders to be expected at 629 nm.

4.3 Infra-red and beyond

Light is just one part of the spectrum of electromagnetic waves. The other parts of the electromagnetic spectrum are shown in Table 1.1 on p. 12. All electromagnetic waves propagate through space at a speed of 300 000 km s^{-1}. The symbol **c** is used for this speed.

In space, no object can travel as fast as electromagnetic waves. The speed of light in a medium is less than in free space and is equal to c/n where n is the refractive index of the medium.

Infra-red radiation

The visible spectrum extends to no more than 700 nm, fading from red to deep red to black beyond about 700 nm. A blackened thermometer bulb placed just beyond the longer wavelength end of the visible spectrum should show an increase of temperature due to infra-red radiation. This is electromagnetic radiation in the wavelength range from about 700 nm (= 0.7 μm) to about 0.1 mm (= 100 μm).

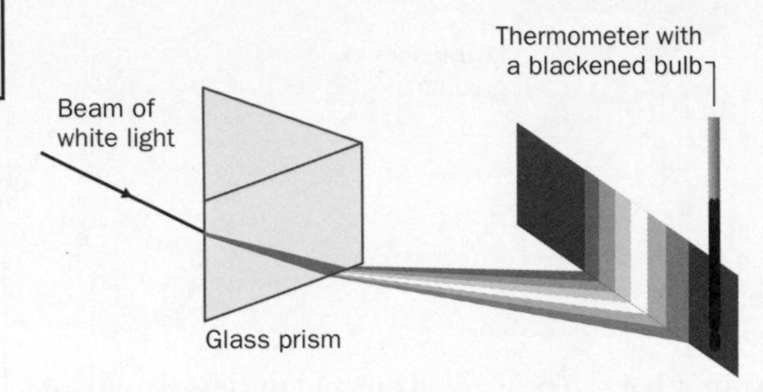

Fig. 4.10 Detecting infra-red radiation.

Producing infra-red radiation

Infra-red radiation is emitted by every object. The hotter an object is, the more infra-red radiation it emits from its surface. If an object is hot enough to glow, its surface emits visible, as well as infra-red radiation. The colder an object is, the less infra-red radiation it emits and the longer the wavelength of the radiation. If you enter a refrigeration room, you soon feel cold because you radiate away much more infra-red radiation to the walls of the room than you get back from the walls. As a result, you lose internal energy.

Absorbing infra-red radiation

The amount of infra-red radiation absorbed by a silvered surface is much less than for a black surface. Also, a surface that is smooth and shiny absorbs much less than a matt, rough surface. A good absorber of infra-red radiation is a good emitter; a poor absorber is a poor emitter. This is why a

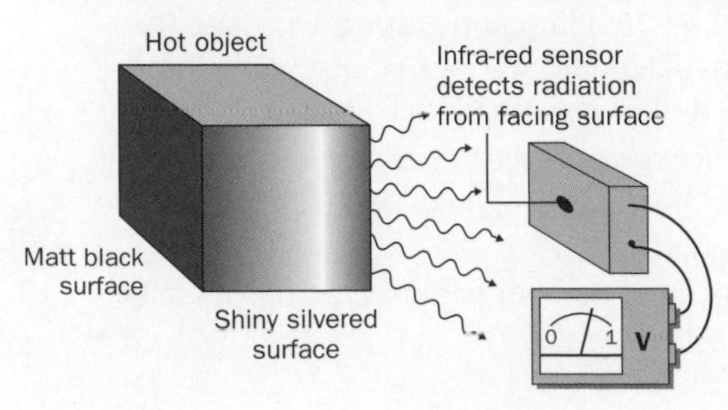

Fig. 4.11 Testing different surfaces.

hot object wrapped in shiny aluminium foil retains internal energy longer than if unwrapped.

Detecting infra-red radiation

Infra-red radiation can be detected using electronic sensors, as in an alarm system. A night-vision camera used for surveillance purposes forms an infra-red image because it is fitted with an array of electronic sensors that detects infra-red radiation. Such cameras are used to 'see' people and animals in darkness, provided the surroundings are at a different temperature to the surface temperature of the bodies. Special infra-red sensitive photographic film can also be used to take infra-red photographs. Glass absorbs infra-red radiation beyond about 2 μm, and other transparent substances such as quartz or certain liquids are used to make lenses and prisms capable of transmitting infra-red radiation beyond about 2 μm.

Using infra-red radiation

1. Infra-red radiation of wavelength about 1μm is used to carry digital signals along optical fibres. The frequency of such radiation is of the order of 3×10^{14} Hz, which is about 10000 times higher than microwave frequencies. Because the wavelength is so small, much more data can be carried in the form of digital signals in comparison with microwaves or radio waves. For example, a microwave link to or from a satellite can carry several TV channels. However, a fibre optic link can carry many more TV channels.
2. Infra-red radiation is used for heating purposes. For example, a halogen hob of an electric cooker contains a high-intensity compact bulb that is positioned between a reflector and the transparent plate of the hob. When the hob is switched on, infra-red radiation from the bulb is absorbed by a saucepan on the hob, after passing through the transparent plate. Infra-red heaters are used for industrial drying, for example to dry car bodies after spraying with paint.

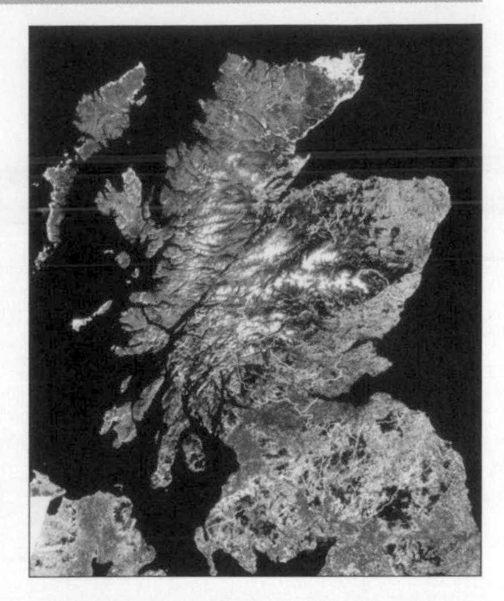

Fig. 4.12 An infra-red image of Scotland, taken from a satellite.

Questions 4.3a

1. Satellites fitted with infra-red cameras produce pictures which show populated areas, rural areas, rivers and coastlines. Why is it possible for such a camera to tell the difference between **(a)** sea and land, **(b)** a rural area and an urban area?
2. Explain why your hand would feel **(a)** warm if you placed it near a hot plate, **(b)** cold if you placed it near a beaker of ice.

Microwave radiation

Microwave radiation is electromagnetic radiation in the wavelength range from about 1 mm to about 0.1 m. The atmosphere does not absorb microwaves and so they are used for 'line of sight' communications. They are reflected by metals and partly absorbed by non-metals.

Producing and detecting microwaves

The magnetron valve is a vacuum tube which is designed to produce high-intensity microwave radiation. A microwave oven is fitted with a magnetron valve that produces microwaves at a frequency of 2.45 GHz. Electromagnetic waves of this frequency penetrate food and are absorbed by water molecules, causing internal heating of the food. Low power microwaves are also produced by a different type of valve, known as a klystron valve, and by solid-state devices.

Microwaves can be detected using a point-contact diode in parallel with a sensitive meter or an amplifier.

Using microwaves

1. Microwaves are used for heating food and for drying non-conducting materials such as cotton. In a microwave oven, microwaves penetrate the food and heat it up. The food must be in a non-conducting dish, otherwise dangerous voltages are generated in the dish. Also, the food is rotated on a turntable because stationary waves and nodes are set up in the oven by microwaves reflected from different sides of the oven. No microwave power is delivered to food at a node, so the food is rotated in the oven to heat all parts.
2. Microwaves are used for communications because they pass through the atmosphere without significant absorption and they can be directed in beams with much less diffraction than radio waves. They are used to carry digital signals from ground stations to satellites and vice versa, and from one part of the country to another via transmitter and receiver dishes on towers and hilltops. Microwave frequencies are of the order of 10 GHz, sufficiently high to carry much more information in digital form than a copper cable or on radio waves.

Questions 4.3b

1. What properties of microwaves makes them suitable for **(a)** heating food rapidly, **(b)** carrying digital signals to and from a satellite?
2. In a microwave oven, why is it essential that **(a)** the food rotates on a turntable, **(b)** there is a safety switch so no microwaves are produced if the door is open?

Radio waves

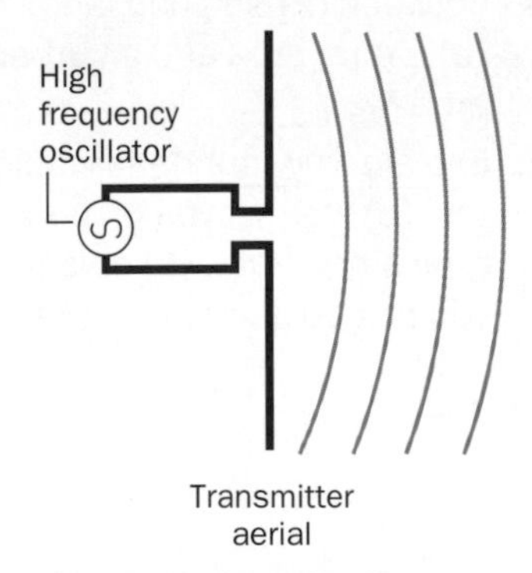

Fig. 4.14 Producing radio waves.

Radio waves are electromagnetic waves of wavelengths about 0.1 m and longer. Radio waves are used for radio and TV broadcasting, mobile phone communications and local emergency channels. The radio spectrum is sub-divided into bands, as shown in Fig. 4.13.

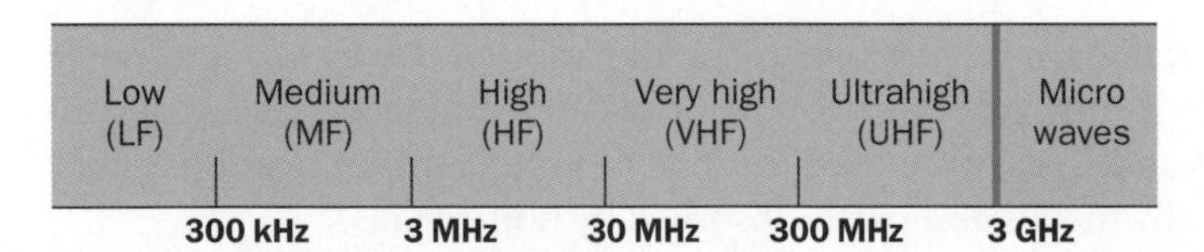

Fig. 4.13 The radio spectrum.

Producing and detecting radio waves

Radio waves are generated when a high-frequency, alternating voltage is applied to a suitable transmitter aerial. The electrons in the aerial are forced to and fro, which makes them generate radio waves in the surrounding space. The radio-wave frequency is equal to the frequency of the alternating voltage. When radio waves pass across a receiving aerial, they force the electrons in the aerial to move to and fro along the aerial, thus generating an alternating voltage which can then be detected.

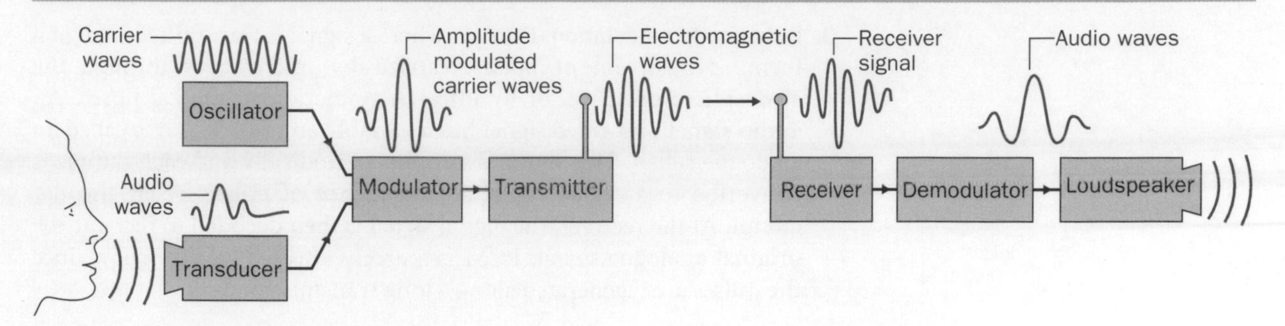

Fig. 4.15 Amplitude modulation.

Using radio waves

Radio waves are used to carry information through the atmosphere and through space. There are three main methods of carrying information using radio waves:

1. **Amplitude modulation (AM)** is where an audio or TV signal modulates the amplitude of the carrier wave. Long wave and medium wave radio channels use AM. A radio channel needs a 'bandwidth' of 4 kHz to carry most of the frequencies in an audio signal. A TV channel has a bandwidth of 8 MHz and is therefore carried by radio waves in the UHF band of frequencies, from 400 to 900 MHz.
2. **Frequency modulation (FM)** is where the frequency of the carrier wave is modulated by the audio signal. A greater bandwidth is necessary, so carrier frequencies need to be in the VHF band, higher than for AM. FM broadcasts suffer less from distortion than AM.

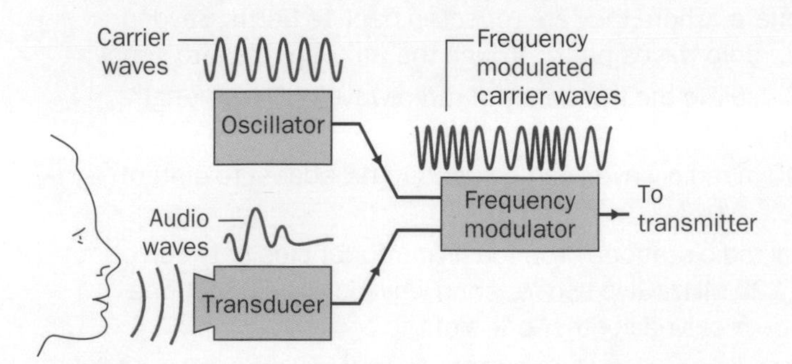

Fig. 4.16 Frequency modulation.

3. **Pulse code modulation** (**PCM**) is where a signal is transmitted in digital form as a stream of pulses, carried by microwaves through the atmosphere or space, or by infra-red radiation in optical fibres. An audio signal or a video signal has a variable amplitude and is called an analogue signal. This signal is sampled periodically and each sample is converted to a digital signal as a sequence of pulses by an encoder circuit. At the receiver, the digital signal is then decoded to recreate the original analogue signal. PCM is scarcely affected by distortion since the pulses are regenerated along a long transmission link.

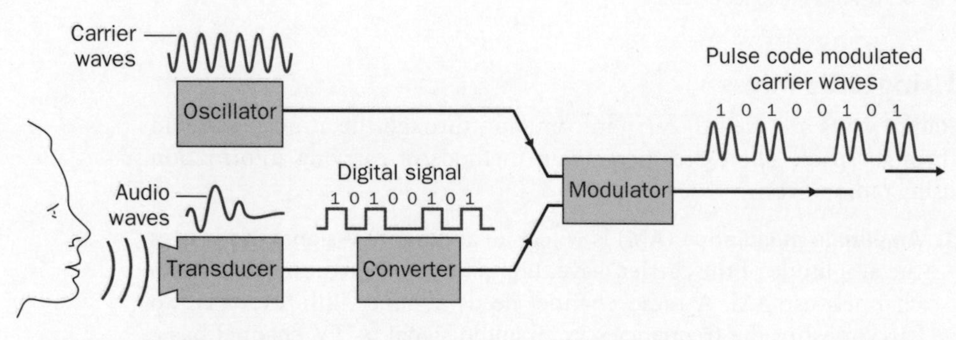

Fig. 4.17 Pulse code modulation.

Questions 4.3c

1. Radio waves of frequencies up to about 300 kHz follow the Earth's curvature. Radio waves from about 300 kHz to 30 MHz travel along the line of sight up to the ionosphere in the upper atmosphere, where they are reflected back to Earth. Beyond 30 MHz, radio waves pass through the atmosphere into space.

   (a) (i) Calculate the frequency of radio waves of wavelength 1500 m.

   (ii) 1500 m radio waves can be used to broadcast to distant countries. Why?

   (b) Local radio stations broadcast at frequencies between about 80 and 120 MHz. Give two reasons why local radio stations cannot be received in other parts of the country.

2. **(a)** The picture quality of a portable TV is affected by altering the orientation of the aerial. What property of electromagnetic waves is responsible for this?

   (b) A satellite TV dish points directly to a satellite in orbit in space, high above the Equator. Such dishes are larger and more difficult to align in Northern Europe than in Southern Europe. Why?

4.4 Ultraviolet radiation, X-radiation and gamma radiation

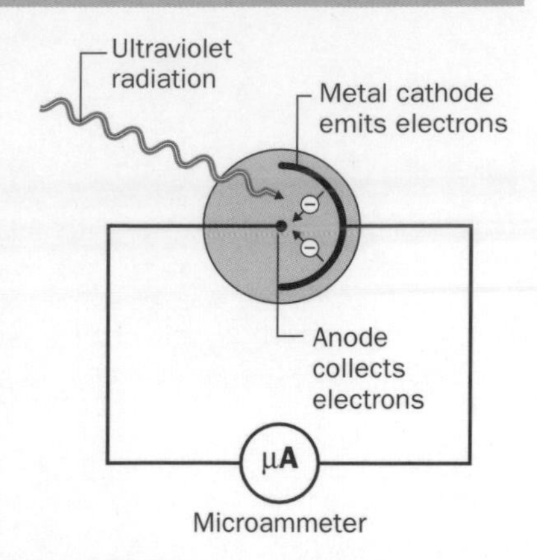

Fig. 4.18 A vacuum photocell.

Ultraviolet radiation

Ultraviolet radiation is electromagnetic radiation beyond the violet end of the visible spectrum from wavelengths of about 400 nm to about 1 nm. Ultraviolet radiation is harmful to the eyes and its presence in sunlight causes sunburn and skin cancer.

Producing ultraviolet radiation

The Sun is a natural source of ultraviolet radiation. The ozone layer in the Earth's upper atmosphere filters out ultraviolet radiation from the Sun, but the depletion of this layer due to certain chemicals (e.g. CFCs) released into the atmosphere is a cause of continuing concern. People who stay outdoors in summer are advised to use appropriate skin creams to protect the skin from ultraviolet radiation. Sun lamps and certain types of discharge tubes also produce ultraviolet radiation.

Detecting ultraviolet radiation

Fluorescent dyes glow when exposed to ultraviolet radiation. The molecules in these dyes absorb ultraviolet radiation and emit visible light. Washing powders that contain such dyes makes clothing glow in ultraviolet radiation and give a 'whiter than white' effect in sunlight.

Photographic film is sensitive to ultraviolet radiation down to about 200 nm. Photocells are also used to detect ultraviolet radiation. In a vacuum photocell connected to a microammeter, ultraviolet radiation causes electrons to be emitted from the metal cathode, which creates a current through the microammeter.

Using ultraviolet radiation

1. Security marker pens, used to mark equipment and banknotes, contain invisible ink which glows in ultraviolet radiation. Fluorescent pens can produce outstanding displays in ultraviolet radiation.
2. High-power ultraviolet lamps are used to rapidly dry printing ink. The paper is not warmed by the radiation, which is absorbed only by the ink.

Questions 4.4a

1. Describe how you could detect whether or not ultraviolet radiation is present beyond the violet part of the visible spectrum.
2. Explain **(a)** why a white shirt glows under ultraviolet light, **(b)** how a skin cream designed to protect the skin from ultraviolet radiation works.

X-rays and gamma radiation

Electromagnetic radiation of wavelength less than about 1 nm is either X-radiation or gamma (γ) radiation depending on whether it is produced by an X-ray tube or by radioactive decay. X and γ-radiation both knock electrons off atoms and therefore create ions. Ionising radiation is harmful because it damages living cells. See p. 326 for more information.

1. In an X-ray tube, electrons are accelerated through a very high voltage onto a metal target, where they are suddenly stopped. As a result, X-rays are emitted by the electrons at the point of impact. Fig. 4.19 shows the construction of an X-ray rube.

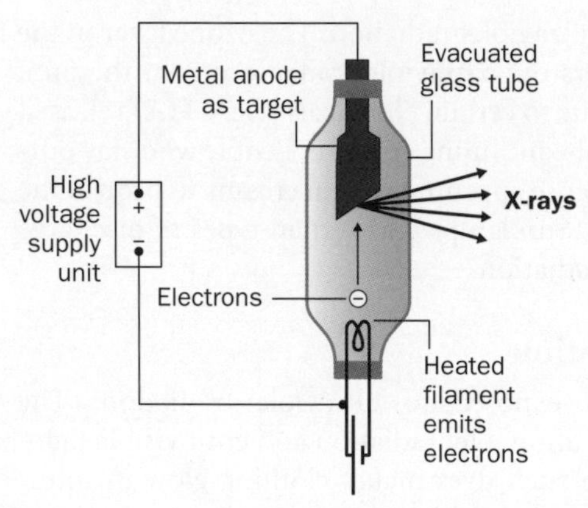

Fig. 4.19 Inside an X-ray tube.

2. Unstable nuclei become stable or less unstable as a result of emitting one of three types of radiation known as alpha, or beta, or gamma radiation. Alpha radiation is stopped by paper or thin metal foil, whereas beta radiation requires several millimetres of lead to be stopped. Gamma radiation is even more penetrating and is only stopped by several centimetres of lead. See p. 323 for more about radioactivity.

X-rays in medicine

X-rays are used to take photographs of internal organs and bones. This is possible because X-rays pass through body tissue but not through bone. When an X-ray photograph is taken of a broken bone in a limb, X-rays are directed through the limb at a light-proof photographic plate or film behind the limb. The X-rays form a shadow of the bone on the film since it absorbs the X-rays whereas the surrounding tissue does not. If the bone is fractured, the fracture can be seen on the X-ray photograph.

Notes

1. Low energy X-rays are removed from the beam by placing a metal plate between the tube and the patient. Such X-rays would be absorbed by body tissue, causing unnecessary cell damage.
2. To take an X-ray photograph of organs such as the stomach, the organ is filled with a contrast medium which absorbs X-rays. For example, a barium drink is given to a patient who is to have a stomach X-ray.
3. Personnel in an X-ray unit must not be exposed to X-rays. The X-ray tube is surrounded by a thick lead shield with a window that allows a beam of X-rays through.

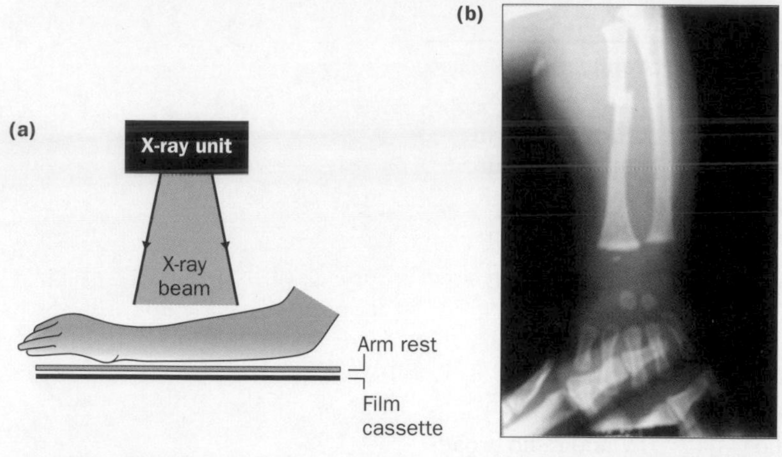

Fig. 4.20 Taking an X-ray picture **(a)** arrangement, **(b)** an X-ray photograph showing a fractured radius in the forearm of an infant.

Questions 4.4b

1. When an X-ray picture is taken, why is it important that
   (a) the X-rays should originate from a small spot on the metal target,
   (b) the photographic film is in a light-proof cover?
2. **(a)** Calculate the frequency of gamma radiation of wavelength 1×10^{-14} m. The speed of electromagnetic waves in space is 3.0×10^{8} m s^{-1}.
   (b) (i) Why is gamma radiation harmful?
   (ii) Why is gamma radiation used to destroy tumours?

Summary

◆ **Dispersion is the splitting of white light into the colours of the spectrum**

◆ **Primary (and complementary secondary) colours** are red (cyan), green (magenta), blue (yellow).

◆ **The wavelength range of light** varies from about 400 nm for blue light to about 650 nm for red light.

◆ **Spectra**

(a) A continuous spectrum contains all the colours of the spectrum.

(b) A line spectrum consists of well-defined vertical lines of different colours.

◆ **The diffraction grating equation**

$d \sin \theta_m = m \lambda$

θ_m = angle of diffraction of mth order
d = slit spacing (centre-to-centre)
λ = wavelength of light

◆ **Beyond the visible spectrum**

	Wavelength range	Source	Applications
X- and gamma radiation	less than 1 nm	X-ray tube, unstable nuclei	X-ray pictures in medicine, gamma therapy
ultraviolet radiation	1–400 nm approx.	UV lamps, the Sun	security markers
infra-red radiation	0.7–100 μm approx.	hot objects	heating, night-viewing
microwave radiation	0.1–100 mm approx.	magnetron valve	heating, communications
radio waves	more than 100 mm	radio transmitters	TV and radio broad-casting, mobile phone links

Revision questions

4.1. With the aid of a diagram, describe how you would
(a) split a narrow beam of white light into the colours of the spectrum,
(b) demonstrate that the critical angle for red light in glass is more than the critical angle for blue light.

4.2. Describe and explain the colour of
(a) a yellow shirt in blue light,
(b) a red poster carrying blue print in **(i)** blue light, **(ii)** yellow light.

4.3. A laser beam was directed normally at a diffraction grating with 300 lines per millimetre. Diffracted beams were observed on a white screen at a distance of 1.50 m from the grating.
(a) The first order beams were located on the screen at a distance of 58.0 cm apart as in Fig. 4.21. Calculate **(i)** the angle of diffraction of the first order beams, **(ii)** the wavelength of the laser light.

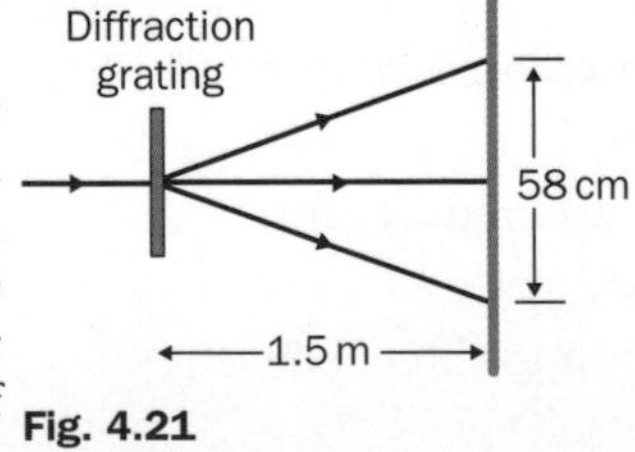

Fig. 4.21

(b) Calculate the maximum number of diffracted orders from this grating with this light.

4.4. A spectrometer fitted with a diffraction grating is used to measure the wavelength of a prominent blue line in the line spectrum of an element. In addition, the angle of diffraction of a first order prominent yellow line at 590 nm was measured. The telescope scale readings were as follows:

1st order blue line; left of centre = 95° 39′ right of centre = 125° 33′
1st order yellow line; left of centre = 89° 52′ right of centre = 131° 20′
(a) Calculate **(i)** the angle of diffraction for each line, **(ii)** the number of lines per millimetre of the grating, **(iii)** the wavelength of the blue light.
(b) Calculate the angle of diffraction for each further order of **(i)** the blue light, **(ii)** the yellow light.

4.5. **(a)** Which part of the electromagnetic spectrum includes electromagnetic waves of wavelength **(i)** 100 m **(ii)** 0.01 mm **(iii)** 100 nm?
(b) Calculate the frequency of **(i)** radio waves of wavelength 1000 m in air, **(ii)** microwaves of wavelength 3 cm in air, **(iii)** light of wavelength 0.60 μm in air. The speed of electromagnetic waves in air is 3.0×10^8 m s^{-1}.

4.6. What property or properties makes each of the following types of electromagnetic radiation suitable for the application listed?
(a) Infra-red radiation; night surveillance cameras
(b) Ultraviolet radiation; security marker pens
(c) Microwaves; satellite links
(d) X-rays; X-ray imaging
(e) Gamma rays; gamma therapy
(f) Radio waves in the UHF range; mobile phone channels.

Properties of Materials

Part 2 considers the mechanical and thermal properties of materials, and relates these properties to materials at the molecular level. It provides plenty of opportunities for practical work on the measurement of the properties of materials and, as with Part 1, it offers a gradual introduction to the use of mathematics in Physics. Applications of the properties of materials are also developed, including the strength of solid materials, control of heat transfer and measurement of pressure.

Matter and Molecules

Objectives

After working through this unit, you should be able to:

- ▶ describe distinctive properties of each of the three states of matter
- ▶ use the kinetic theory of matter to explain the main properties of each state of matter
- ▶ classify substances as elements, compounds or mixtures
- ▶ understand the nature and composition of atoms and molecules
- ▶ define and use the Avogadro constant to calculate atomic masses and sizes
- ▶ relate the different types of bonds between atoms and molecules to the distinctive properties of the three states of matter

Contents

5.1 States of matter

At a definite temperature, most substances are in one of three states of matter: solid, liquid or gas. The physical characteristics that distinguish the three states of matter are:

- **shape** – a characteristic of a solid substance since a liquid takes the shape of its container, and a gas in a closed container fills its container completely or spreads out into the atmosphere if the container is open. The shape of a solid is due to its inability to flow. Its shape can be changed by applying forces to distort it. Elasticity is the ability of a solid to regain its shape after forces applied to it have been removed. Liquids and gases are fluids because they flow and therefore have no characteristic shape.
- **surfaces** – present only in solids and liquids. A coloured gas released into the atmosphere spreads out and soon becomes invisible as its concentration decreases. The surface of a solid is a definite boundary between the solid and the liquid or gas that surrounds it. There is little or no transfer of matter between a solid and the substance which surrounds it. The open surface of a liquid is a boundary between the

Fig. 5.1 Solid, liquid and gas.

liquid and the atmosphere; this is well-defined, although transfer of matter across the boundary can and does take place (e.g. water vapour from hot water).

Fig. 5.2 Electron micrograph of common table salt, showing roughly cubic grains with rounded edges (presumably rounded during production as an aid to pouring).

Shape and surface tests

1. Place a drop of a saturated solution of common salt on a warm glass slide and observe the growth of crystals in the liquid using a microscope. Repeat the test with a saturated solution of copper sulphate. Crystals of common salt differ in shape to crystals of copper sulphate.

Models of crystals made using polystyrene spheres show how different shapes can be produced according to how the spheres are arranged. Crystal shapes depend on how the atoms and molecules of the crystal are arranged.

2. Observe some other crystalline substances, including metals if possible, using a microscope if necessary. A metal is composed of tiny crystals called **grains**. The size and nature of the grains in a metal determine its strength.

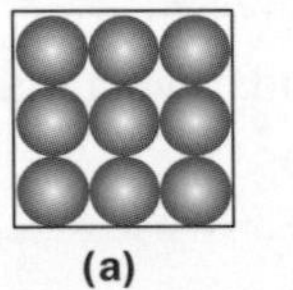

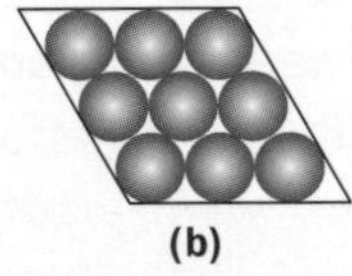

Fig. 5.3 Crystal models **(a)** cubic packing, **(b)** close packing.

Fig. 5.4 Micrograph of a thin section of common brass showing the distinct grain structure of this metal alloy.

3. Heat a glass rod until it melts and you will find it becomes very flexible. Use safety spectacles. When removed from the source of heat, glass quickly solidifies in whatever shape it has at that point. Glass is an example of an **amorphous** solid, which is a solid without a characteristic shape and no long-range order in the arrangement of its particles.

4. Place some coloured crystals in a beaker of water and observe the crystals as they gradually **dissolve.** Gradually, the crystals disappear as the particles they are formed from move away into the solution. This is an example of **diffusion**, which is where one substance spreads out into another. Another example is diffusion of a coloured gas in a sealed gas jar into an adjoining gas jar of air when the seal is removed, as in Fig. 5.5.

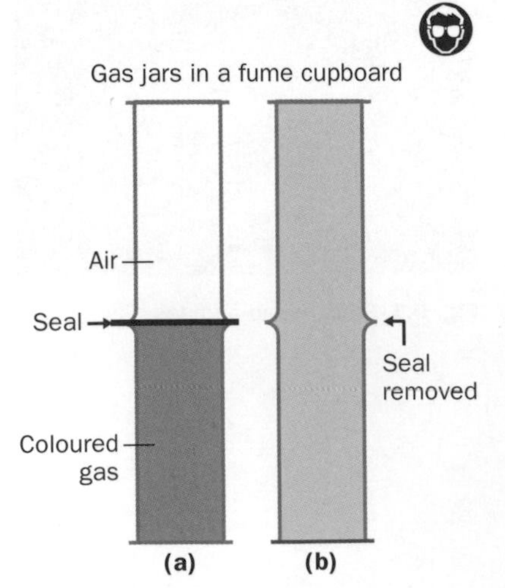

Fig. 5.5 Diffusion of a gas in air **(a)** before and **(b)** after the seal is removed.

Changes of state

Pure substances change from one state to another at a well-defined temperature. For example, pure ice melts at 0°C and water boils at 100°C at atmospheric pressure.

- The melting point of a pure substance is the temperature at which it changes from a solid to a liquid when heated, or from a liquid to a solid when cooled.
- The boiling point of a pure substance is the temperature at which it changes from the liquid to the gaseous state as a result of boiling. This is not the same as vaporisation, which occurs from the liquid surface at any temperature.

The energy needed to change the state of a substance is referred to as **latent heat**. Fig. 5.6 shows the possible changes of state of a pure substance.

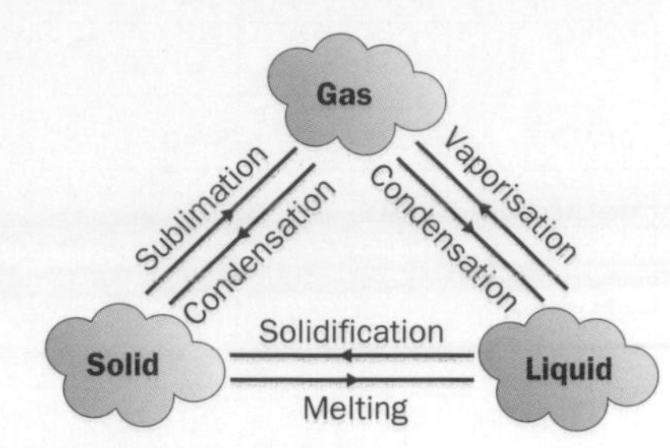

Fig. 5.6 Change of state.

Questions 5.1

1. **(a)** Is jelly a solid or a liquid? Give a reason for your answer.
   (b) Is an aerosol spray liquid or gas? Give a reason for your answer.
2. Describe the changes of state when **(a)** ice cubes in a beaker are heated until steam is produced, **(b)** wet clothing on a washing line dries out.

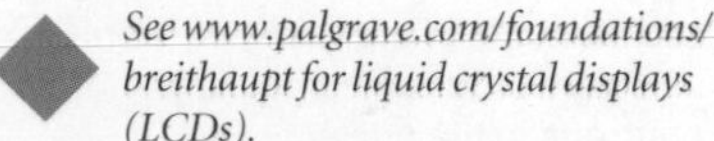
See www.palgrave.com/foundations/breithaupt for liquid crystal displays (LCDs).

5.2 Elements and compounds

Make a list of all the substances within sight. Even without moving, your list is likely to be very long, yet all substances are made from no more than about 90 elementary substances referred to as the **chemical elements**. These cannot be broken down into other substances. There are 92 naturally-occuring elements and a few more artifical elements. Each element has its own symbol (e.g. C for carbon, Pb for lead).

A **compound** is a substance composed of two or more elements in fixed proportions. For example, water is composed of the elements hydrogen and oxygen in proportions 1 : 8 by mass.

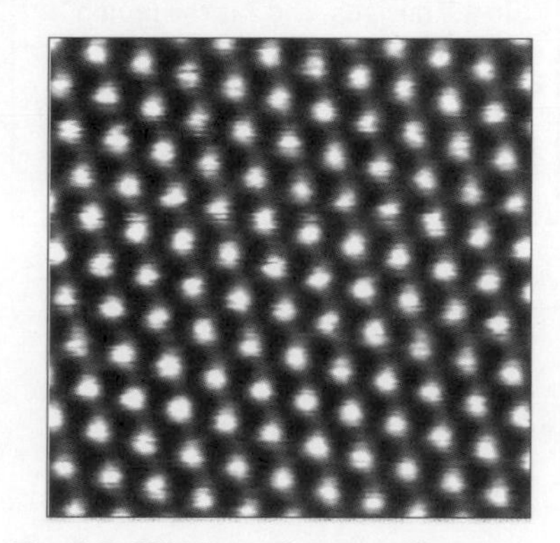
Fig. 5.7 STM (scanning tunnelling microscope) image of carbon atoms.

Atoms and elements

An **atom** is the smallest particle of an element which is characteristic of the element. The atoms of a given element are identical to each other but they differ from the atoms of every other element. The atomic theory of matter was developed almost two centuries ago. Now, using high-power atomic microscopes, it is possible to 'see' the patterns of atoms on the surface of metals.

Inside every atom, there is a positively-charged **nucleus** surrounded by negatively charged particles called **electrons**. The electron is held in orbit round the nucleus by the force of attraction due to its opposite charge. The magnitude of the charge on an electron is represented by the symbol e.

The lightest known atom is the hydrogen atom, which has a single positively charged particle called a **proton** as its nucleus. The proton is

	Mass	Charge
proton	1u	+e
neutron	1u	0
electron	$\frac{1}{1850}$ u	−e

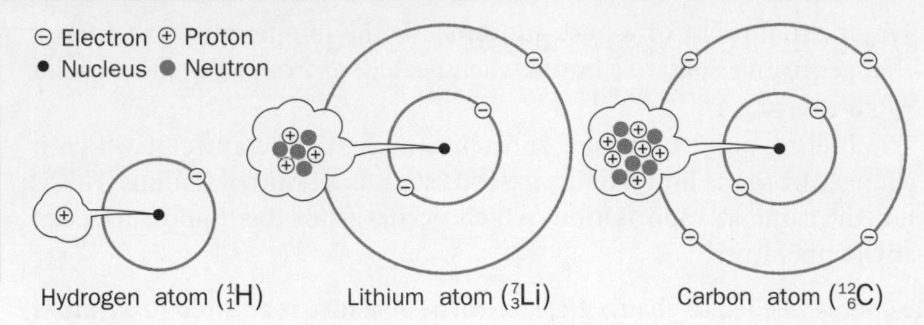

Fig. 5.8 Inside the atom.

Notes

1. The **atomic number**, Z, of an element is the number of protons in the nucleus of an atom of the element. All the atoms of a given element contain the same number of protons in each nucleus. This is sometimes called the **proton number** of the atom. Atoms of the same element can have different numbers of neutrons in the nuclei.
2. **Isotopes** are atoms of the same element with different numbers of neutrons. For example, deuterium is an isotope of hydrogen in which each nucleus consists of a proton and a neutron. An isotope is represented by the symbol ^A_ZX, where X is the chemical symbol of the element, A is the number of neutrons and protons in the nucleus, and Z is the proton number.
3. **One unit of atomic mass** (1 u) is defined as one-twelfth of the mass of a carbon-12 atom, equal to 1.66×10^{-27} kg.

approximately 1850 times heavier than the electron. Its charge is exactly equal and opposite to that of the electron. An uncharged atom contains the same number of electrons and protons. The nuclei of all other types of atoms contain uncharged particles called **neutrons** as well as protons. The mass of a neutron is about the same as that of a proton.

Questions 5.2a

1. Determine the number of protons and the number of neutrons in one atom of each of the isotopes below.

   (a) ^3_1H **(b)** $^{16}_8O$ **(c)** $^{23}_{11}Na$ **(d)** $^{206}_{82}Pb$ **(e)** $^{238}_{92}U$
2. Write down the symbol for

   (a) the isotope of helium (He), which contains 2 neutrons and 2 protons,

   (b) the isotope of carbon (C), which contains 6 protons and 6 neutrons.

Molecules and compounds

A **molecule** is a combination of atoms joined together. For example, an oxygen molecule consists of two oxygen atoms joined together. The chemical symbol for this molecule is therefore written O_2. Another example is the carbon dioxide molecule, which consists of two oxygen atoms joined to a carbon atom. The chemical symbol for a carbon dioxide molecule is CO_2.

A **compound** consists of identical molecules, each with the same atomic composition. This is why a compound is composed of elements in **fixed** proportions. For example, carbon dioxide is a compound consisting of carbon dioxide molecules. For every atom of carbon, there are two atoms of oxygen.

The proportion of each element present in a compound can be determined from the molecular formula if the atomic masses are known. For example, the atomic mass of carbon is approximately 12 u and the atomic mass of oxygen is approximately 16 u. Therefore, the mass of a carbon dioxide molecule is approximately 44 u [$= 12 + (2 \times 16)$] and the proportion of carbon to oxygen by mass is 0.375 ($= 12 \div 32$). Molecules composed of chains of atoms can be very long.

Estimating the size of an oil molecule

A tiny oil drop placed on a clean water surface spreads out into a circular patch just one molecule thick. A V-shaped thin wire is dipped into the oil and shaken so just one droplet remains on the wire. The diameter, d, of this droplet can be estimated by using a magnifying glass and a millimetre scale. The water surface is then sprinkled with very light powder after being cleaned. When the droplet is placed on the water surface, the oil spreads into a circular patch, pushing the powder away. The diameter, D, of the patch can then be measured.

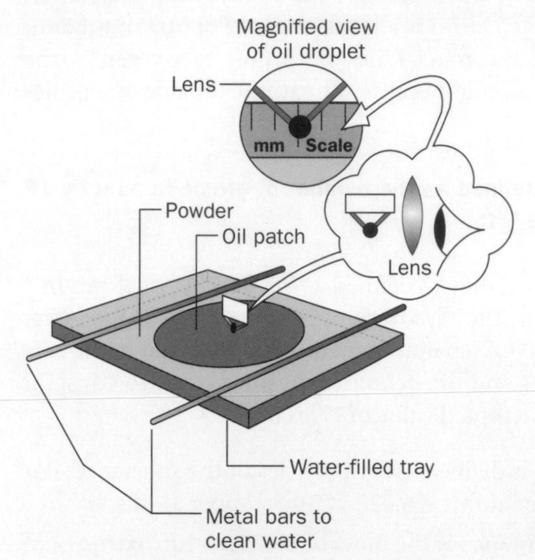

Fig. 5.9 Estimating the size of an oil molecule.

See www.palgrave.com/foundations/breithaupt for more about this experiment.

The thickness, t, of the patch can be estimated by assuming the volume of the droplet ($= 4\pi r^3 \div 3$, where its radius, $r = d \div 2$) equals the volume of the patch ($= \pi D^2 t \div 4$).

Worked example 5.1

An oil droplet of diameter 1.0 mm spreads out to form a circular patch of diameter 650 mm when placed on a clean water surface. Use this information to estimate the length of an oil molecule.

Solution

Volume of oil patch $= \dfrac{\pi D^2 t}{4} = \dfrac{\pi (0.65)^2 t}{4}$, where t is the thickness of the patch

Volume of oil droplet $= \dfrac{4\pi r^3}{3} = \dfrac{4\pi (0.5 \times 10^{-3})^3}{3}$

$$\therefore \frac{\pi (0.65)^2 t}{4} = \frac{4\pi (0.5 \times 10^{-3})^3}{3}$$

Estimate of size of oil molecule $= t = \dfrac{4\pi (0.5 \times 10^{-3})^3 \times 4}{3\pi (0.65)^2} = 1.6 \times 10^{-9}$ m

Questions 5.2b

1. Use the molecular formula given to calculate the mass of one molecule of each of the following compounds or elements in atomic mass units. The relative atomic mass of each element is listed below in atomic mass units.

   (a) water H_2O, **(b)** carbon monoxide CO, **(c)** methane CH_4, **(d)** ammonia NH_3, **(e)** copper oxide CuO.

   (H = 1, C = 12, N = 14, O = 16, Cu = 64)

2. An oil droplet of diameter 0.8 mm spreads out on water to form a circular patch of diameter 350 mm. Estimate the size of an oil molecule from this data.

5.3 The Avogadro constant, N_A

When atoms combine to form molecules, they do so only in fixed ratios. For example, every carbon dioxide molecule is composed of one carbon atom and two oxygen atoms. The mass ratio of carbon to oxygen in a single molecule is always 12 : 32 because the ratio of the mass of a carbon atom to the mass of an oxygen atom is always 12 : 16. For any number of carbon dioxide molecules, the mass ratio of carbon to oxygen is the same as that in a single molecule because all carbon dioxide molecules are alike.

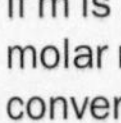

Note
If m is in atomic mass units, then the molar mass in grams = m because the conversion factor from atomic mass units to grams is $1 \div N_A$. For example, the mass of 1 atom of oxygen = 16 u = $16 \div N_A$ grams. Hence the molar mass of N_A oxygen atoms = 16 grams.

The Avogadro constant is defined as the number of atoms in exactly 12 grams of the carbon isotope $^{12}_{6}C$.

Originally, the Avogadro constant was defined as the number of atoms in 1 gram of hydrogen. However, the discovery of isotopes of hydrogen (e.g. deuterium) and the difficulty of separating them, eventually led to the carbon-12 scale of atomic mass and the definition of the Avogadro constant in terms of carbon-12. The accepted value of N_A is 6.022×10^{23}.

- One **mole** of a substance is defined as N_A particles of the substance. For example, 1 mole of copper atoms is 6.022×10^{23} copper atoms.
- The **molar mass** of a substance is the mass of 6.022×10^{23} particles of the substance. By definition, the molar mass of carbon-12 is exactly 12 grams. Since 1 atom of carbon-12 has a mass of 12 u because 1 u is defined as 1/12th of the mass of a $^{12}_{6}C$ atom, then 1 u = 1.66×10^{-24} grams (= $1 \div N_A$) or 1.66×10^{-27} kg. For a substance consisting of particles of mass m, its molar mass is equal to mN_A.

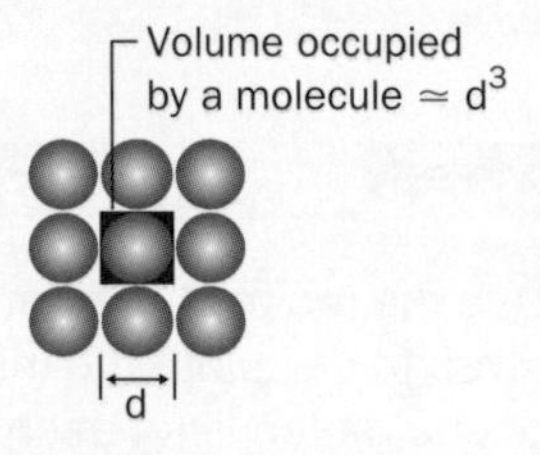

Fig. 5.10 Estimate of atomic size.

Density and atomic size

The atoms in a solid element are in direct contact with each other.

- The mass of 1 atom of an element = $M \div N_A$, where M is its molar mass
- The volume of 1 atom of an element =

$$\frac{\text{mass of 1 atom}}{\text{density } \rho} = \frac{M}{N_A \rho}$$

Assuming each atom occupies a cube of volume d^3, where d is the diameter of an atom, then the diameter of an atom can be estimated from the equation

$$d^3 = \frac{M}{N_A \rho}$$

In practice, atoms are packed together in a number of different ways, depending on the type of atom. Nevertheless, the above formula gives a reasonable estimate of atomic size.

Note
To calculate d, the cube root of $M \div N_A\rho$ must be calculated. This is written $\left(\frac{M}{N_A \rho}\right)^{1/3}$

Use the button marked y^x on your calculator as follows:

Step 1 Work out $M \div N_A \rho$ and keep the answer on your calculator display.

Step 2 Press the y^x button and key in $(1 \div 3)$; Press = and the calculator will then display the value of $(M \div N_A \rho)^{1/3}$

Worked example 5.2

Copper has a density of 8000 kg m^{-3} and an atomic mass of 64.
(a) Calculate the mass of a copper atom in kg, (b) Estimate the diameter of a copper atom.
The Avogadro constant, $N_A = 6.02 \times 10^{23}$

Solution

(a) 64 g of copper contains N_A copper atoms.

$$\therefore \text{the mass of 1 copper atom} = \frac{64\text{ g}}{6.02 \times 10^{23}} = 1.1 \times 10^{-22}\text{ grams} = 1.1 \times 10^{-25}\text{ kg}$$

(b) Volume of a copper atom

$$= \frac{\text{mass of 1 copper atom}}{\text{density}} = \frac{1.1 \times 10^{-25}}{8000} = 1.32 \times 10^{-29}\text{ m}^3$$

Diameter of atom, $d = (1.32 \times 10^{-29})^{1/3} = 2.4 \times 10^{-10}\text{ m} = 0.24\text{ nm}$

Questions 5.3

1. **(a)** Calculate the number of atoms in 1 kg of **(i)** carbon, **(ii)** uranium.
   (b) Calculate the number of molecules in 1 kg of **(i)** carbon dioxide (CO_2), **(ii)** methane (CH_4).
   (Relative atomic masses: H = 1, C = 12, O = 16, U = 238.)
2. The density of aluminium is 2700 kg m^{-3} and its relative atomic mass is 27.
   (a) Calculate the mass of an aluminum atom.
   (b) Estimate the diameter of an aluminum atom.
   The Avogadro constant, $N_A = 6.02 \times 10^{23}$.

5.4 Intermolecular forces

Electrons in atoms

The electrons in an uncharged isolated atom surround the nucleus in 'shells', each shell corresponding to a fixed energy level and capable of holding a certain number of electrons. The innermost shell can hold two electrons, the next shell eight electrons, the third shell eight electrons, the next shell ten electrons. The number of electrons in each shell was first worked out from the rows and columns of the **Periodic Table**. This is a table of the elements, listed in order of increasing atomic mass in such a way that elements with similar chemical properties are in the same vertical column. See p. 74.

Examples

1. The uncharged carbon atom has six electrons, two in its innermost shell and four in the next shell.
2. The uncharged oxygen atom has eight electrons, two in its innermost shell and six in the next shell.
3. The uncharged sodium atom has 11 electrons, two in its innermost shell, eight in the next shell and a single electron in the third shell.
4. The uncharged chlorine atom has 17 electrons, two in its innermost shell, eight in the second shell and seven in its third shell.

Bonds

Atoms interact to form molecules, and molecules interact to form solids and liquids. The different types of forces that hold atoms and molecules together depend on the electron structure of the atoms. The underlying principle is that an atom has minimum energy when its electron shells are complete. Each type of bond achieves this in a different way.

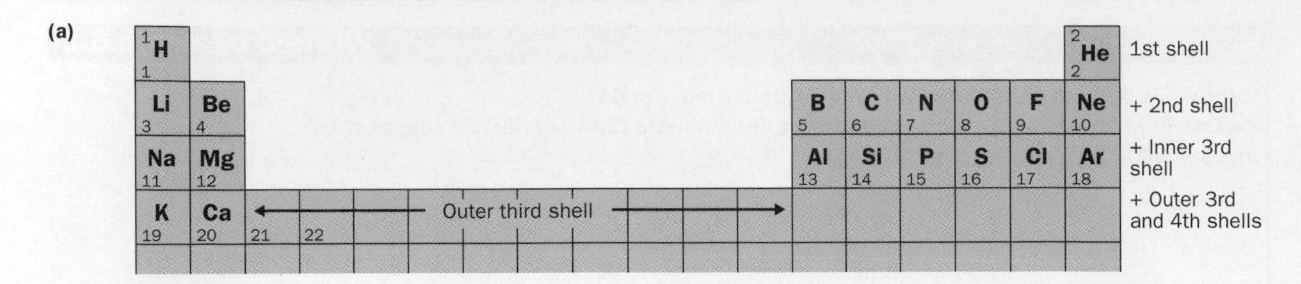

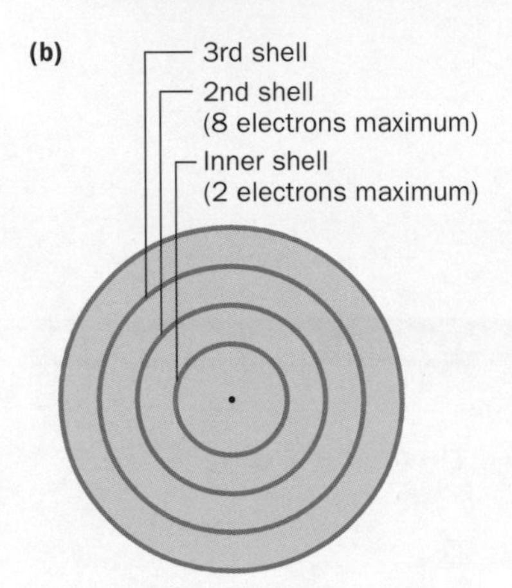

Fig. 5.11 (a) the Periodic Table and **(b)** electron shells.

An **ionic bond** is formed when one atom gives up one or more electrons to another atom. An **ion** is a charged atom, formed as a result of an uncharged atom gaining or losing one or more electrons. For example, sodium chloride consists of positive sodium ions and negative chlorine ions. Each sodium atom has lost its outer electron and each chlorine atom has gained an electron in its outer shell. In this way, both types of atoms have full electron shells.

Sodium chloride crystals are cubic in shape because the positive ions and negative ions arrange themselves in a cubic lattice, each ion held in place by the ionic bonds between it and the neighbouring, oppositely charged ions, as shown in Fig. 5.13.

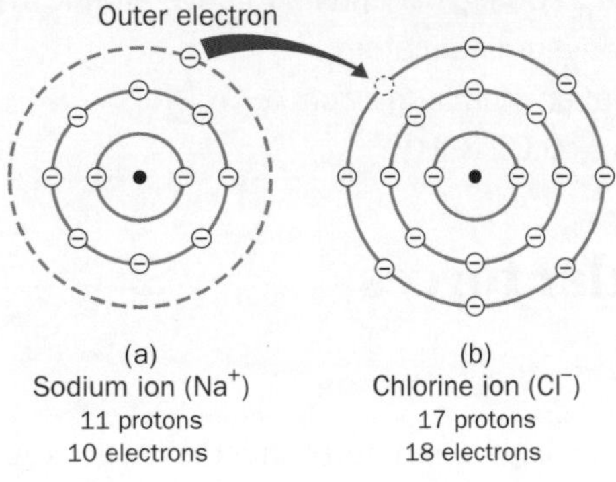

Fig. 5.12 The ionic bond.

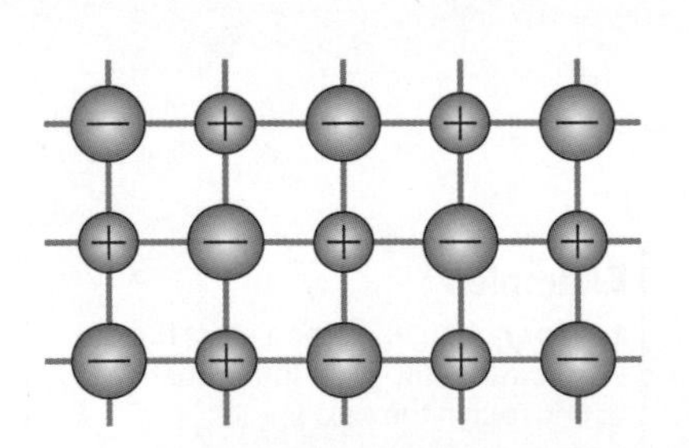

Fig. 5.13 A cubic lattice.

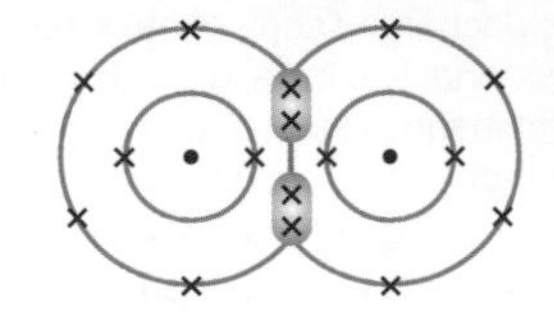

Fig. 5.14 Covalent bonds.

A **covalent bond** is formed as a result of two atoms with incomplete electron shells sharing electrons to achieve complete shells. Each bond consists of two electrons, one donated by each atom. For example, an oxygen molecule consists of two oxygen atoms held together by two covalent bonds, as shown in Fig. 5.14. Each oxygen atom in the molecule has eight electrons in its outer shell.

Metallic bonding is where atoms lose their outer electrons to form a structure of positive ions surrounded by a 'sea' of electrons that holds the structure together. The electrons that break free are referred to as **conduction** electrons because they are responsible for electrical conduction when a voltage is applied. This type of bonding exists only in metals because uncharged metal atoms possess outer-shell electrons.

Molecular bonds, sometimes called van der Waals bonds, are weak attractive bonds between uncharged atoms and molecules. Two uncharged molecules attract each other because the electrons of each molecule are attracted by the nuclei of the atoms of the other molecule. Uncharged molecules in a liquid are prevented from leaving the liquid by these bonds, which have a range of no more than a few molecular diameters. In comparison, the ionic bond has a much greater range.

In general, the force between two atoms or molecules or ions attracts the two particles together but prevents them from becoming too close. At very short-range, the electron shells of one particle repel the electron shells of the other particle. The variation of force with separation is shown in Fig. 5.17.

- The **equilibrium separation**, r_o, is the point where attraction is balanced by repulsion
- The **bond energy**, ε, is the energy needed to separate two molecules from equilibrium separation to infinity.

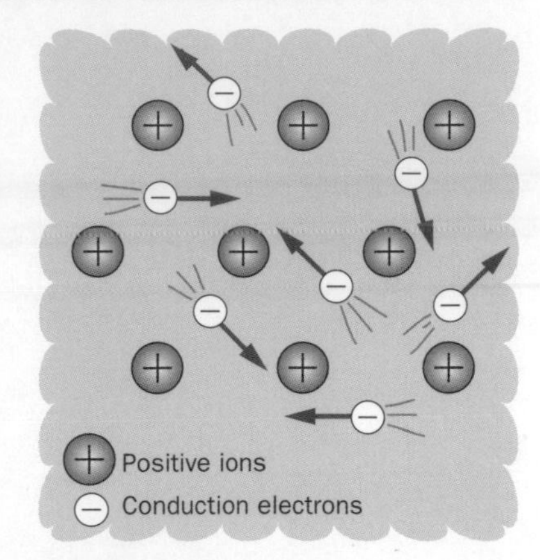

Fig. 5.15 Metallic bonds.

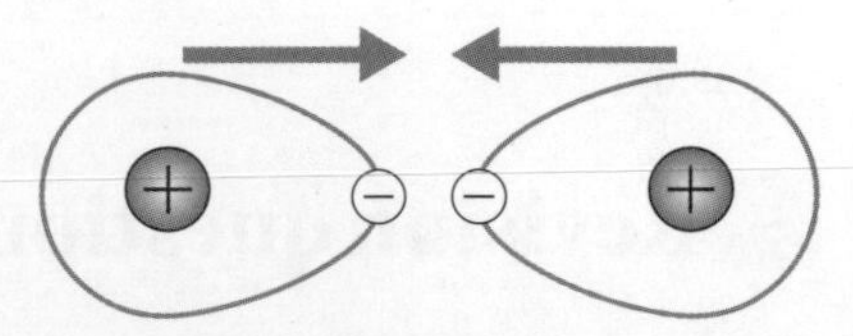
Fig. 5.16 Molecular bonds.

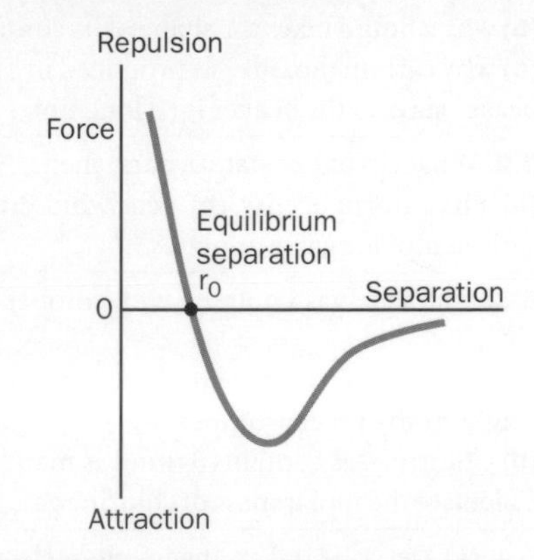

Fig. 5.17 Force versus separation for two molecules.

Molecules and materials

Solids and liquids do not diffuse into the atmosphere because of the presence of attractive forces between the molecules of the substance. The molecules are in contact with each other, locked together in a solid and free to move about in a liquid.

In a gas, intermolecular forces are either very weak or non-existent, so gas molecules can move away from each other. The molecules in a gas move about at random, independently of each other, coming into contact only when they collide and bounce off each other.

Molecules composed of more than one atom are held together by covalent bonds. These bonds are localised and in general do not affect the interaction between molecules.

Questions 5.4

1. **(a)** State the type of bond that exists between **(i)** the atoms of a carbon dioxide molecule, **(ii)** the atoms in solid copper, **(iii)** the atoms in a sodium chloride crystal, **(iv)** water molecules, **(v)** air molecules.

   (b) Describe the arrangement of electrons in **(i)** a carbon atom, **(ii)** a carbon dioxide molecule.
2. State whether or not the molecules in each of the following materials **(i)** are in contact, **(ii)** move about at random:

   (a) air **(b)** water **(c)** wood **(d)** ice.

Summary

- **Solids** possess shape and surfaces; are composed of molecules locked together and in contact.
  Liquids possess surfaces but no shape; are composed of molecules in random motion and in contact.
  Gases possess no surfaces and no shape; are composed of molecules in random motion not in contact.
- **Atoms** consist of protons (mass = 1u, charge = +1) and neutrons (mass = 1u, charge = 0) in the nucleus, surrounded by electrons (mass = 0.005 u approx, charge = −1).
  1 atomic mass unit (u) = 1/12th of the mass of a carbon-12 atom.
  The Avogadro constant, N_A, is the number of atoms in 12 g of carbon-12.
  1 mole of a substance = N_A particles of the substance.
- **Types of bonds:** ionic, covalent, metallic, molecular.

Revision questions

The Avogadro constant, $N_A = 6.02 \times 10^{23}$

5.1. Use your knowledge of atoms and molecules to explain
(a) why a solid possesses its own shape,
(b) why a liquid takes the shape of its container,
(c) why carbon dioxide gas produced in a chemical reaction in a beaker stays in the beaker for a long time.

5.2. What change of state occurs when
(a) a hail storm occurs, **(b)** a car windscreen mists up, **(c)** solid carbon dioxide gas evaporates.

5.3. Chlorine gas contains two isotopes of chlorine, $^{37}_{17}Cl$ and $^{35}_{17}Cl$.
(a) How many protons and how many neutrons are there in a single atom of each isotope?
(b) Chlorine gas contains 3 times as many $^{35}_{17}Cl$ as the other type. Calculate the molar mass of chlorine gas.

5.4. (a) Describe the arrangement of electrons in an uncharged atom of **(i)** carbon, **(ii)** hydrogen.
(b) The molecular formula of methane is CH_4.
(i) What is the molar mass of methane?
(ii) Calculate the mass, in kg, of 1 molecule of methane.
(iii) Describe the arrangement of electrons in a molecule of methane.

5.5. (a) $^{235}_{92}U$ is an unstable isotope of uranium. How many protons and how many neutrons are present in the nucleus of an atom of this isotope?
(b) Calculate the number of atoms present in 1 gram of this isotope.

5.6. The density of water is 1000 kg m^{-3} and it has a molar mass of 0.018 kg.
(a) Calculate the mass of a single water molecule.
(b) Estimate the size of a water molecule.

5.7. The density of lead is 11340 kg m^{-3} and its relative atomic mass is 207.
(a) Calculate the mass of a lead atom.
(b) Estimate the diameter of a lead atom.

5.8. (a) Air is approximately 80% nitrogen (molar mass 28 g) and 20% oxygen (molar mass 32 g). Show that the molar mass of air is 29 g.
(b) The density of air at atmospheric pressure and room temperature is about 1.2 kg m^{-3}.
(i) Estimate the average distance between air molecules.
(ii) Comment on this answer in the light of the knowledge that the diameter of an air molecule is about 0.3 nm.

Thermal Properties of Materials

Objectives

After working through this unit, you should be able to:

- ▶ convert temperature in °C to kelvins and vice versa
- ▶ calculate the thermal expansion of a solid for a given temperature rise
- ▶ measure the specific heat capacity of different substances
- ▶ measure the specific latent heat of fusion and of vaporisation of different substances
- ▶ carry out specific heat capacity calculations and specific latent heat calculations
- ▶ describe conduction, convection and radiation as heat transfer mechanisms
- ▶ carry out thermal conduction and radiation calculations

Contents

6.1 Thermal expansion

Temperature and heat

Heat is energy transferred due to a temperature difference. When a substance is heated, its atoms and molecules gain energy. When the temperature rises, the molecules vibrate faster in a solid, or move about more rapidly in a liquid or a gas. When a solid is heated at its melting point or a liquid is heated at its boiling point, the energy supplied is used by the molecules to break free from each other.

Temperature is the degree of hotness of an object. To define a scale of temperature, numerical values are assigned to reproducible and reliable degrees of hotness.

- The **Celsius scale** (in °C) is defined by:

  1. Ice point (0°C) which is the temperature of pure melting ice.

  2. Steam point (100°C) which is the temperature of steam at atmospheric pressure.

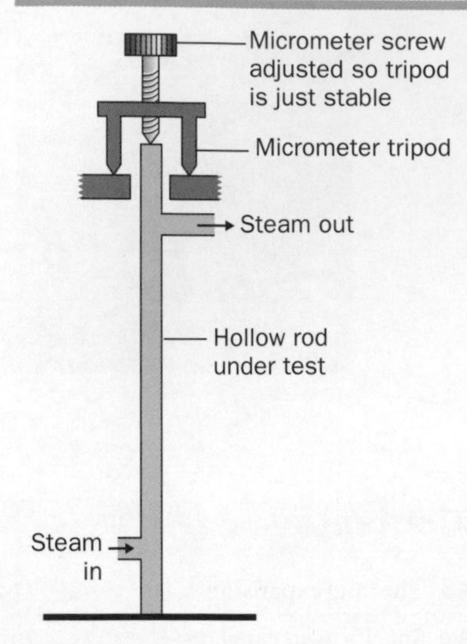

Fig. 6.1 Thermal expansion.

- The **absolute scale** in kelvins (K) is defined by:
  1. Absolute zero (0 K) which is the lowest possible temperature, equal to −273°C (see p. 375).
  2. The triple point of water (273 K) which is the point at which ice, water and water vapour are in thermal equilibrium. This is a more reliable standard than ice point which varies slightly with pressure. The temperature of the triple point is assigned a numerical value of 273 K so that the difference between the two scales for any given degree of hotness is 273 degrees.

temperature in kelvins = temperature in °C + 273

The subject of thermometry is developed further in Unit 25.

Thermal expansion

A solid expands when it is heated and contracts when it cools. Fig. 6.1 shows how this can be demonstrated and measured. Heating a solid makes the atoms in the solid vibrate more and causes the average separation between atoms to increase. The change of length, ΔL, due to thermal expansion of a length of a solid, L, is

1. proportional to its length L,
2. proportional to the increase of temperature ΔT,
3. dependent on the material.

The coefficient of thermal expansion, α, of a solid rod is defined as

$$\frac{\text{change of length } \Delta L}{\text{length } L \times \text{temperature rise } \Delta T}$$

Table 6.1 gives values of α for some different materials. The unit of α is K^{-1} (per kelvin). The equation below is used to calculate the change of length, ΔL, given the length, temperature change and coefficient of thermal expansion.

$$\Delta L = \alpha L \Delta T$$

Table 6.1 Coefficient of thermal expansion values

Material	Coefficient of thermal expansion K^{-1}
steel	1.1×10^{-5}
brass	1.9×10^{-5}
aluminium	2.6×10^{-5}
invar*	0.1×10^{-5}

* nickel–steel alloy

Worked example 6.1

Calculate the change of length of a steel girder of length 45.0 m, when its temperature rises from 5°C to 30°C. The coefficient of thermal expansion of steel is 1.1×10^{-5} K^{-1}.

Solution

$L = 45.0$ m, $\Delta T = (30 + 273) - (5 + 273) = 25$ K, $\alpha = 1.1 \times 10^{-5}$ K^{-1}.

$\Delta L = \alpha L \Delta T = 1.1 \times 10^{-5} \times 45.0 \times 25 = 1.2 \times 10^{-2}$ m $= 1.2$ cm.

Applications of thermal expansion

1. The **bimetallic strip** consists of a strip of brass bonded to a strip of steel. The strip bends towards the steel when its temperature rises. This

happens because brass expands more than steel for a given rise in temperature. The bimetallic strip is used as a temperature-controlled switch in alarms and heating appliances.

2. **Expansion gaps** in bridges, railway tracks and buildings are necessary to allow for thermal expansion in summer. These gaps are usually filled with a soft material like rubber, to prevent chunks of rock falling in.

3. A **radiator thermostat** contains oil in a sealed container. A rod attached to a valve protrudes into the oil. When the oil heats up, it expands and forces the rod partly out to open the valve. The valve is closed when the engine is cold and opens when the engine has warmed up.

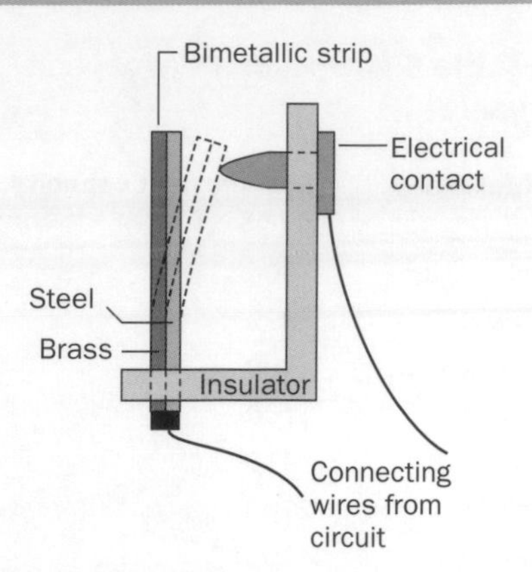

Fig. 6.2 A bimetallic strip.

Questions 6.1

1. With the aid of a diagram, explain how a bimetallic strip can be used to switch an electric heater off when the temperature rises too much.
2. Use the data in Table 6.1 to calculate the change of length of
   (a) a brass rod of length 0.80 m when its temperature rises from 15°C to 20°C,
   (b) an aluminium bar of length 1.50 m when its temperature falls from 20°C to −5°C.

6.2 Specific heat capacity

Energy and power

The **power** of an electrical appliance is a measure of the rate at which the appliance uses electrical energy. The unit of power is the **watt** (W). This is used for any situation where energy is being transferred. For example, a 1000 watt electric heater uses ten times as much energy in a given time as a 100 W light bulb.

The electrical energy used by an electrical appliance is expressed in kilowatt hours (kW h). Thus a 2 kW electric heater which runs for 4 hours uses 8 kW h of electrical energy or 8 'units'. If the unit cost of electricity is 6 p, the cost of running this electric heater for 4 hours is therefore 48 p.

The scientific unit of energy is the **joule** (J) which is equal to the electrical energy used by a 1 watt electrical appliance in 1 second. In other words, 1 watt is equal to a rate of energy transfer of 1 J s^{-1}. Prove for yourself that 1 kW h of electrical energy is 3.6 MJ.

$$\text{power (in W)} = \frac{\text{energy transferred (in J)}}{\text{time taken (in seconds)}}$$

The electrical energy supplied to a low voltage heater may be measured using a **joulemeter**, as shown in Fig. 6.3. The joulemeter may be used to determine the power of an electric heater by measuring the energy transferred in a given time, then using the equation above.

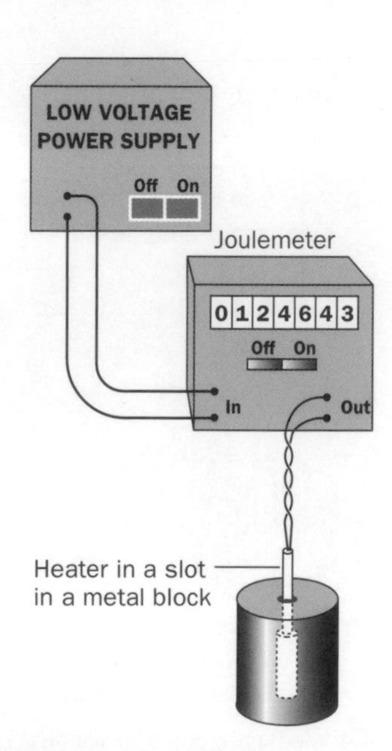

Fig. 6.3 Using a joulemeter.

Table 6.2 Specific heat capacity values

Material	Specific heat capacity / $J\,kg^{-1}\,K^{-1}$
aluminium	900
brass	370
copper	390
glass	700
steel	470
oil	2100
water	4200
concrete	850

Note

The symbol Q is used to denote the quantity of heat transfer.

Heating things up

Some materials heat up more easily than others. For example, in summer, a car in sunlight becomes much hotter than a large swimming pool. The temperature rise of an object depends on:

- the energy supplied to the object; the more energy that is supplied, the greater the temperature rise
- the mass of each substance in the object; the greater its mass, the smaller its temperature rise
- the nature of the substance.

The **specific heat capacity**, **c**, of a substance is the energy required to raise the temperature of unit mass of the substance by unit rise of temperature. The unit of c is $J\,kg^{-1}\,K^{-1}$. The value of c for some common substances is shown in Table 6.2. For example, the specific heat capacity of water is $4200\,J\,kg^{-1}\,K^{-1}$, which means that 4200 J of energy must be supplied to 1 kg of water to raise its temperature by 1 K.

In general, to heat mass, m, of a substance from temperature T_1 to temperature T_2, the energy needed, Q, can be written as

$$Q = mc(T_2 - T_1)$$

where c is the specific heat capacity of the substance.

Worked example 6.2

A 3 kW electric kettle contains 1.5 kg of water at 15°C.

(a) Calculate the energy needed to raise the temperature of this amount of water to 100°C.

(b) How long does it take for a 3 kW heater to supply this amount of energy?

(c) The time taken to heat this amount of water from 15°C to 100°C in the kettle is longer. Why?

Solution

(a) Energy needed $= mc(T_2 - T_1) = 1.5 \times 4200 \times (100 - 15) = 5.4 \times 10^5$ J

(b) Rearranging the power equation gives

$$\text{time taken} = \frac{\text{energy}}{\text{power}} = \frac{5.4 \times 10^5}{3000} = 180 \text{ seconds}$$

(c) Heat is lost from the kettle; the kettle uses some energy as the kettle body becomes hot.

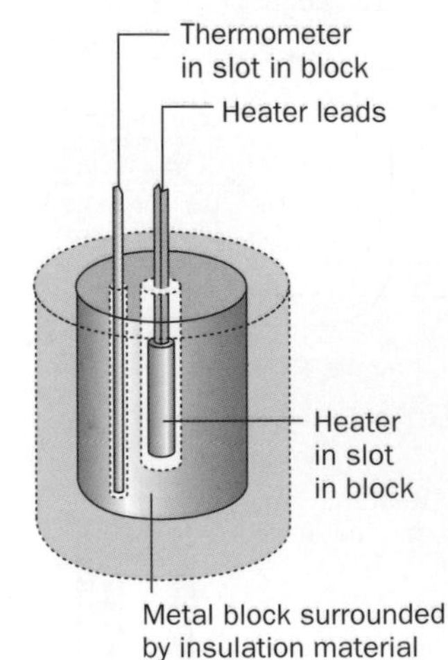

Fig. 6.4 Measuring the specific heat capacity of a metal block.

Measuring specific heat capacity

1. **For a metal**, an insulated metal block is heated using a low voltage heater inserted into a slot in the block. A thermometer is inserted in a different slot. The mass of the block alone is measured using a top pan balance. The block is then arranged as in Fig. 6.4. Its initial temperature is measured and the joulemeter is used to measure the energy supplied

in a given time. The highest temperature is measured when the heater is switched off. The specific heat capacity is calculated from

$$\frac{\text{energy supplied}}{\text{mass} \times \text{temperature rise}}$$

2. **For a liquid**, a low voltage heater is used to heat a measured mass of the liquid in an insulated metal calorimeter, as in Fig. 6.5. The mass of the calorimeter must be measured separately. The joulemeter is used to measure the energy supplied. A thermometer is used to measure the initial and highest temperature of the liquid.

   Assuming no heat loss occurs, the energy supplied to the liquid = electrical energy supplied − energy supplied to heat the calorimeter.

$$\therefore m_l c_l (T_2 - T_1) = E - m_{cal} c_{cal} (T_2 - T_1)$$

where m_l = mass of liquid, m_{cal} = mass of the calorimeter, c_l = specific heat capacity of the liquid, c_{cal} = specific heat capacity of the calorimeter, T_2 = highest temperature, T_1 = initial temperature, E = electrical energy supplied.

Using the above equation, the specific heat capacity of the liquid can therefore be calculated. The above method assumes no heat loss, which can be tested by measuring the temperature of the liquid after the heater is switched off.

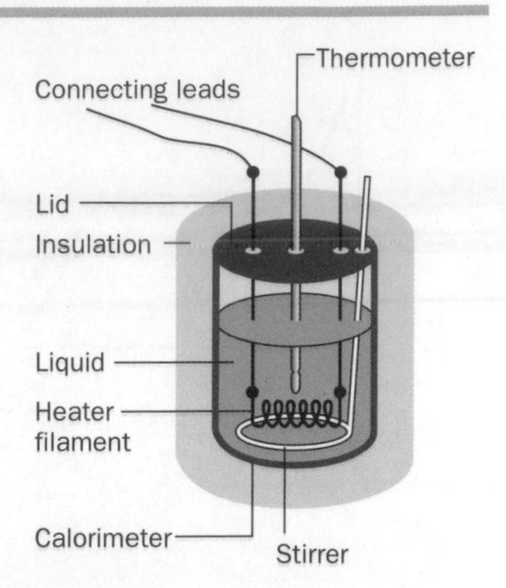

Fig. 6.5 Measuring the specific heat capacity of a liquid.

Questions 6.2

1. Use the data in Table 6.2 to calculate the energy needed to raise the temperature of
   (a) 5.0 kg of oil from 10°C to 50°C,
   (b) a concrete block of mass 1500 kg from 5°C to 35°C,
   (c) an aluminium tank of mass 15 kg containing 95 kg of water from 20°C to 80°C.
2. The following measurements were made in an experiment to determine the specific heat capacity of a liquid by heating the liquid in an insulated calorimeter:

Mass of empty copper calorimeter = 55 g
Mass of calorimeter + liquid = 143 g
Initial temperature of the liquid = 15°C
Final temperature of liquid = 62°C
Energy supplied = 9340 J

Use the above data and the information in Table 6.2 to calculate the specific heat capacity of the liquid.

Worked example 6.3

An insulated copper water tank of mass 27 kg contains 160 kg of water.

(a) Calculate the energy needed to heat the water and the tank from 15°C to 45°C. The specific heat capacity of copper is 390 J kg^{-1} K^{-1} and of water is 4200 J kg^{-1} K^{-1}.

(b) The water in the tank is heated by a 3.0 kW electric heater. Calculate the time taken by the heater to supply the energy in (a).

Solution

(a) Energy needed to heat the copper tank = $27 \times 390 \times (45-15) = 3.2 \times 10^5$ J = 0.32 MJ

Energy needed to heat the water = $160 \times 4200 \times (45-15) = 2.02 \times 10^7$ J = 20.2 MJ

Total energy needed = 20.2 + 0.3 = 20.5 MJ

(b) Time taken = $\frac{\text{energy needed}}{\text{power}} = \frac{20.5 \times 10^6}{3000} = 6800$ s.

6.3 Specific latent heat

Melting and solidification

If a pure solid is heated, its temperature rises until it melts. At its melting point, no further change of temperature occurs even though the solid continues to be heated. The energy supplied at the melting point is called

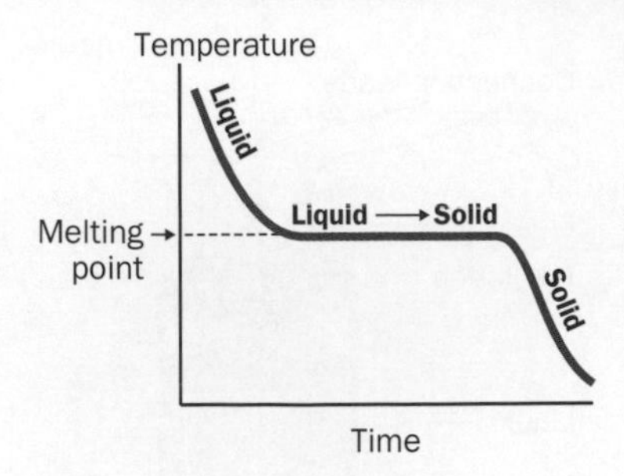

Fig. 6.6 At the melting point.

latent heat because it does not cause a rise of temperature. The reason is that the energy is used to enable the molecules of the solid to break the bonds that hold them in a rigid structure. When a liquid is cooled sufficiently, it solidifies at its melting point. Energy is released when this occurs as the molecules bind each other together in a rigid structure.

The **specific latent heat of fusion** of a substance, l, is the energy required to melt unit mass of substance without change of temperature. The unit of specific latent heat is J kg^{-1}.

1. The quantity of heat supplied to melt mass, m, of a substance can be calculated from the equation

$$Q = ml.$$

2. The same equation is used to calculate the energy removed when mass m solidifies.

Worked example 6.4

Calculate the energy needed to melt 120 g of ice at 0°C and heat the melted ice to 40°C. The specific latent heat of ice is 340 000 J kg^{-1} and the specific heat capacity of water is 4200 J kg^{-1} K^{-1}.

Solution

Energy needed to melt the ice = $0.12 \times 340\,000 = 41\,000$ J

Energy needed to heat the melted ice = $0.12 \times 4200 \times (40 - 0) = 20\,200$ J

Total energy needed = $41\,000 + 20\,200 = 61\,200$ J

Measuring the specific latent heat of fusion of ice

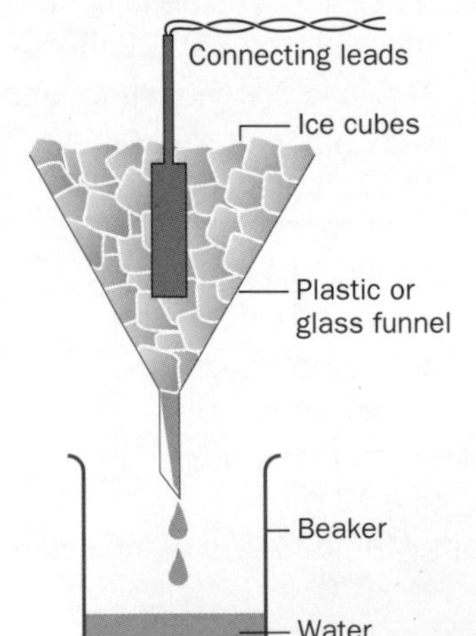

Fig. 6.7 Measuring the specific latent heat of fusion of ice.

A low voltage heater is surrounded by ice cubes packed into a large funnel, as in Fig. 6.7.

1. Without the heater switched on, the mass of melted ice, m_0, from the funnel is measured in a certain time.
2. The measurement is repeated with the heater on, connected to the power supply via a joulemeter.

If m_1 represents the mass of melted ice collected with the heater on, the mass of ice melted due to the heater is $(m_1 - m_0)$. The specific latent heat of ice can therefore be calculated by dividing the energy supplied by $(m_1 - m_0)$.

Vaporisation and condensation

When a liquid is heated, its temperature rises until it reaches its boiling point. Energy supplied at the boiling point enables the molecules throughout the liquid to break free from each other and enter the vapour state. Because no change of temperature occurs at the boiling point, the energy supplied is referred to as **latent heat**. This energy is released when a vapour condenses as the molecules form bonds.

The **specific latent heat of vaporisation** of a substance, ***l***, is the energy required to vaporise unit mass of liquid without change of temperature. The unit of specific latent heat is J kg^{-1}.

1. The quantity of heat supplied to turn mass, m, of a liquid into a vapour or vice versa can be calculated from the equation $\mathbf{Q = m}\boldsymbol{l}$, where l is the specific latent heat of vaporisation. For example, the specific latent heat of vaporisation of water is 2.3 MJ kg^{-1}. Therefore, the energy needed to boil away 0.5 kg of water is 1.15 MJ (= 0.5 × 2.3 MJ).

Worked example 6.5

A 2.5 kW electric kettle contains 2.0 kg of water.

(a) Calculate the energy needed to boil away 1.5 kg of water.

(b) (i) Estimate the time taken for this amount of water in the kettle to boil away.

(ii) What assumption have you made in your estimate?

The specific latent heat of vaporisation of water is 2.3 MJ kg^{-1}.

Solution

(a) Energy needed = 1.5 × 2.3 MJ = 3.45 MJ

(b)(i) Time taken $= \dfrac{\text{energy needed}}{\text{power}} = \dfrac{3.45 \times 10^6}{2500} = 1380$ s

(ii) There is no heat loss from the kettle.

Measuring the specific latent heat of vaporisation of a liquid

The liquid is heated to its boiling point in the apparatus shown. At its boiling point, the vapour from the liquid surrounds the liquid container and enters the vertical condenser tube where it condenses and is collected in a beaker at the bottom of the tube. An empty beaker of known mass is used to collect the condensed liquid for a certain time. The joulemeter is used to measure the energy supplied in this time. The specific latent heat of vaporisation is calculated from energy supplied ÷ mass of liquid collected.

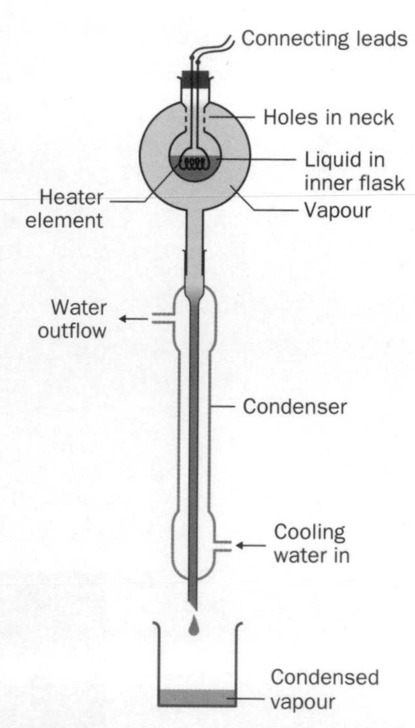

Fig. 6.8 Measuring the specific heat of vaporisation of a liquid.

Questions 6.3

1. Calculate the energy needed to turn 0.20 kg of ice at −10°C into steam at 100°C. The specific latent heat of ice is 340 000 J kg^{-1}, the specific latent heat of steam is 2.3 MJ kg^{-1}, the specific heat capacity of ice is 2100 J kg^{-1} K^{-1} and the specific heat capacity of water is 4200 J kg^{-1} K^{-1}.
2. A jet of steam is directed into a polythene beaker containing 120 g of water at 20°C. The steam heats the water to 52° C. Assuming no heat loss from the beaker, calculate

   (a) the energy needed to heat the water from 20°C to 52°C,

   (b) the mass of steam that condensed in the beaker.

   The specific latent heat of steam is 2.3 MJ kg^{-1} and the specific heat capacity of water is 4200 J kg^{-1} K^{-1}.

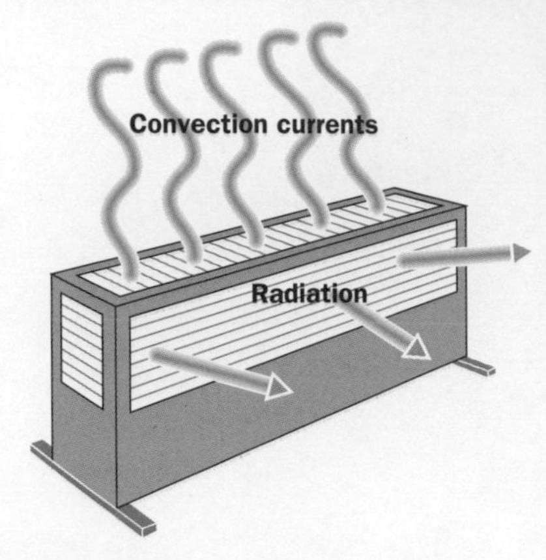

Fig. 6.9 Home heating.

6.4 Heat transfer

Heat is energy transfer due to a temperature difference. The three methods by which heat transfer occurs are thermal conduction, convection and radiation. Conduction takes place in solids, liquids and gases, convection takes place in liquids and gases only. For example, heat from a hot radiator reaches other parts of a room due to thermal conduction through the radiator panel which heats the air and causes convection currents, and also radiates infra-red radiation throughout the room.

Convection

When a fluid is heated it becomes less dense and it rises because it is less dense than the cold fluid. In a closed space, the fluid circulates as the hot fluid rises and cold fluid is drawn in to the source of heating, where it heats up and rises. This process is known as **convection**. Some examples are listed below.

1. Hot gases from the flames of a gas fire rise, drawing air into the gas fire. The products of combustion include carbon monoxide, which is lethal if allowed to build up. For this reason, gas fires must be well-ventilated so that the products of combustion escape into the atmosphere and fresh air is drawn into the room where the heater is.
2. A hot air balloon rises because a gas burner heats the air in the balloon and makes it less dense than the surrounding air. Without repeated bursts of heating, the hot gas in the balloon would cool and the balloon would sink as the air in it became more dense.
3. Water in a hot water tank is usually drawn off near the top of the tank.

(a)

(b)

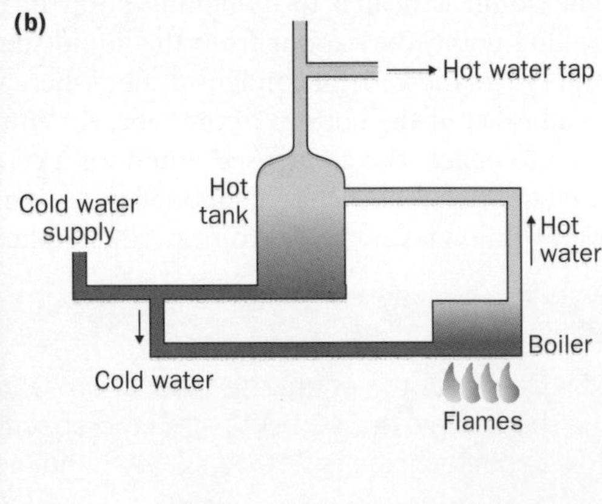

Fig. 6.10 Convection examples **(a)** a hot air balloon, **(b)** a hot water system.

The hot water from the boiler enters the tank and rises to the top. If the feed pipe for the hot water was near the bottom of the tank, it would draw cold water, unless all the water in the tank was hot.

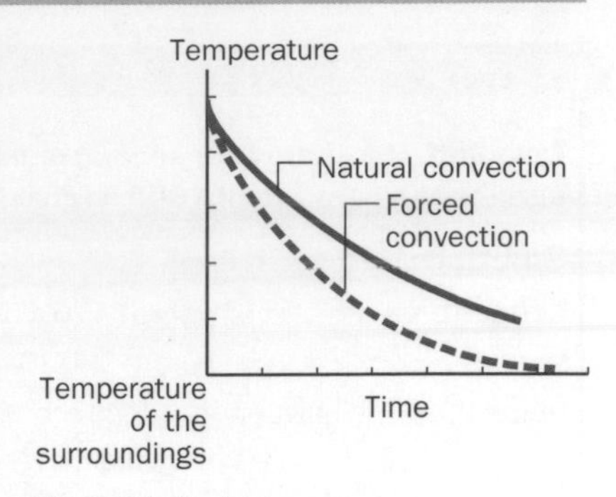

Fig. 6.11 Cooling curves.

Investigation of convection in air

1. **Natural convection** Measure the temperature of a hot object at regular intervals as it cools naturally in air. The measurements may be plotted as a temperature versus time graph, as in Fig. 6.11. The gradient of this graph at any point is the rate of temperature loss of the object at that point. The results show that the rate of temperature loss decreases as the excess temperature of the object above its surroundings decreases.
2. **Forced convection** Repeat the measurements in the presence of an airstream from a hair dryer or fan. The temperature decreases more rapidly than without the fan.

See www.palgrave.com/foundations/breithaupt for more information about this investigation.

Radiation

Every object emits electromagnetic radiation due to its temperature. This radiation is known as **thermal radiation** and consists mostly of infra-red radiation, although it includes visible radiation if the temperature is sufficiently high.

Thermal radiation is absorbed most effectively by matt black surfaces and least effectively by shiny, silvered surfaces. A **black body** is a body that absorbs all the radiation incident on its surface. For example, a small hole in the surface of a hollow object would act as a black body as any radiation entering the hole would be completely absorbed by the surface of the cavity. The Sun and the stars may be considered as black bodies since any radiation incident on the surface would be absorbed.

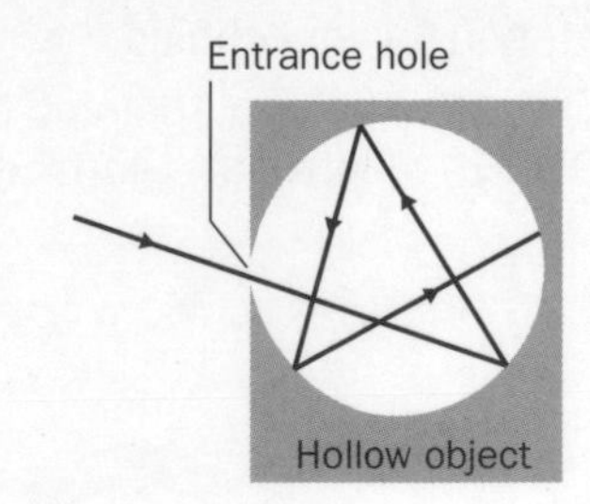

Fig. 6.12 A black body.

A surface that is a good absorber of thermal radiation is also a good emitter. The radiation from a black body is referred to as **black body radiation**. The energy per second radiated from a surface depends on

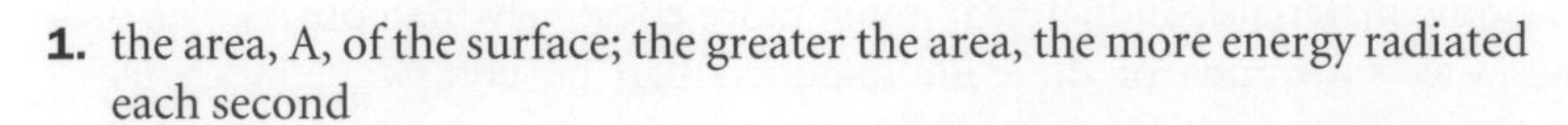

1. the area, A, of the surface; the greater the area, the more energy radiated each second
2. the surface temperature, T; the hotter an object's surface is, the more energy it radiates per second
3. the nature of the surface; the surface of a black body radiates more energy per second than any other surface of the same area and at the same temperature.

Stefan's law of radiation states that the energy per second, W, radiated from a surface of area A at absolute temperature T is given by the equation

$$W = e \sigma A T^4$$

where σ is a constant known as the Stefan constant, and e, the emissivity of the surface, is the ratio of the energy per second radiated from the surface to energy per second radiated by a black body surface of equal area at the same temperature. The value of σ is 5.67×10^{-8} W m^{-2} K^{-4}.

Note

Fig. 4.3 shows how the energy radiated per second varies with wavelength for different surface temperatures. The wavelength, λ_o, at which each curve peaks decreases as the absolute temperature, T, rises in accordance with the relationship

$\lambda_o T = 0.0029$ **K m**.

This relationship is known as **Wien's law of radiation**.

Worked example 6.6

Calculate the energy per second radiated from a glowing fuse wire of length 20 mm and diameter 0.3 mm at a surface temperature of 1600 K, given its surface emissivity is 0.2.

Solution

Surface area $A = 2\pi r L$, where $r = 0.15$ mm $= 1.5 \times 10^{-4}$ m, and $L = 20$ mm $= 0.020$ m

Hence $A = 2\pi \times 1.5 \times 10^{-4} \times 0.020 = 1.89 \times 10^{-5}$ m^2

Using Stefan's Law, energy per second radiated from the surface $= e\sigma AT^4$

$= 0.2 \times 5.67 \times 10^{-8} \times 1.89 \times 10^{-5}$ m$^2 \times (1600)^4 = 1.40$ J s^{-1}

Questions 6.4

1. Fig. 6.13 shows a cross-section of a vacuum flask designed to keep a liquid hot.
   (a) Why is the glass vessel in the flask silvered?
   (b) Why is there a vacuum between the inner and outer surfaces of the vessel?
   (c) Why is a lid essential?
   (d) What would be a suitable material to surround the vessel? Give a reason for your choice.
2. Estimate the heat radiated per second from a hot plate of diameter 0.10 m which is at a surface temperature of 1200 K.

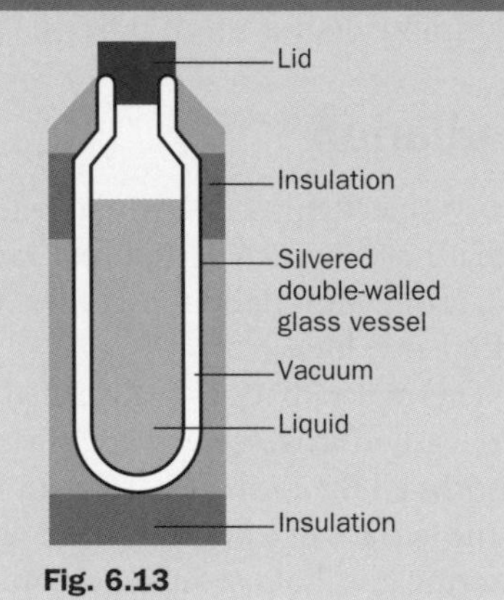

Fig. 6.13

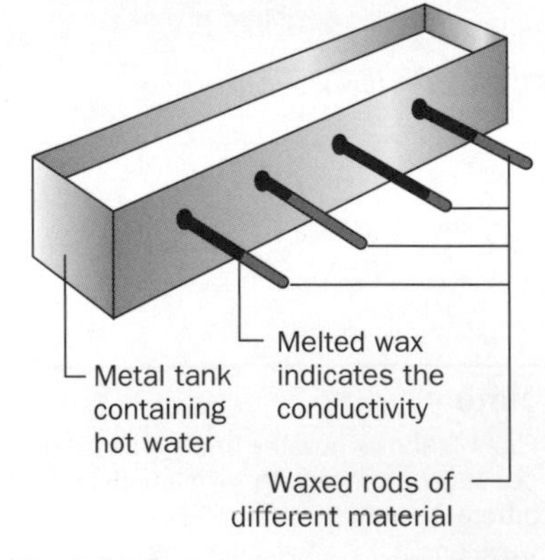

Fig. 6.14 Comparing thermal conduction.

6.5 More about thermal conduction

Some materials conduct heat much more effectively than others. Fig. 6.14 shows how rods of different materials can be compared in terms of thermal conductivity.

Metals are better conductors of heat than non-metals because of the presence of conduction electrons in metals. When a metal is heated, the electrons gain energy and move about faster, transferring energy to atoms and electrons elsewhere.

Non-metals do not contain conduction electrons and therefore do not conduct as well as metals. Thermal conduction in a non-metal takes place as a result of vibrations of the atoms spreading throughout the non-metal. Heating a non-metal makes the atoms at the point of heating vibrate more and these vibrations cause atoms to vibrate in other parts of the non-metal. This process does occur in a metal but much less than energy transfer due to conduction electrons.

Consider the heat flow along a conductor of uniform cross-sectional area, A, and of length, L, which has one end at fixed temperature, T_1, and

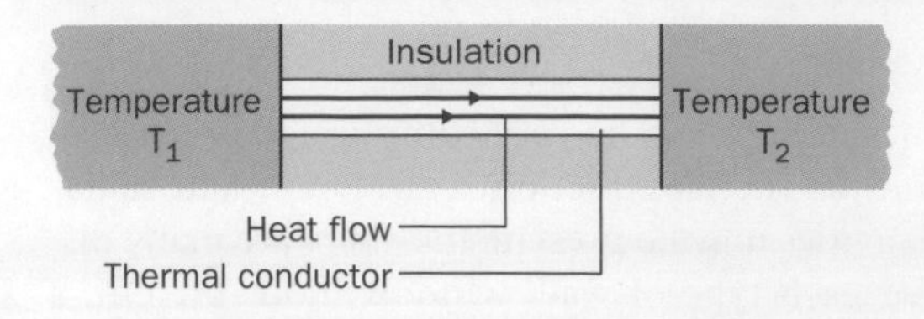

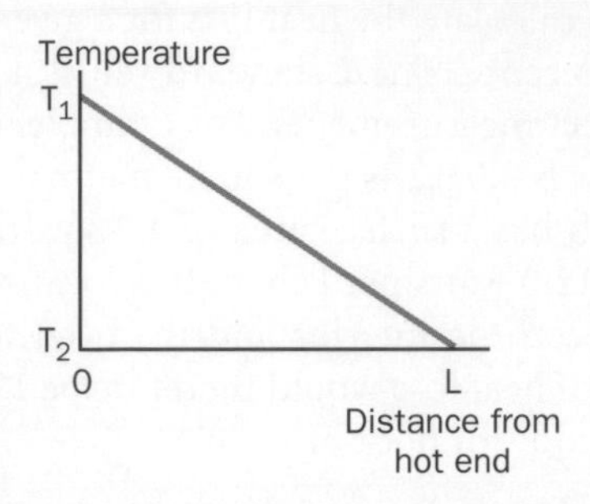

Fig. 6.15 Temperature versus position along an insulated thermal conductor.

the other end at a lower fixed temperature, T_2. Provided the conductor is in an insulated jacket so there is no heat loss from its sides, the temperature falls uniformly along its length from T_1 to T_2, as shown in Fig. 6.15. This shows that the temperature gradient along the bar, the temperature drop per unit length, is constant.

The energy per second conducted along the bar (i.e. the heat flow), Q/t is

1. proportional to the area of cross section, A
2. proportional to the temperature gradient $\frac{(T_1 - T_2)}{L}$
3. dependent on the thermal conductivity of the material.

Hence the heat flow

$$\frac{Q}{t} = kA\frac{(T_1 - T_2)}{L}$$

where k is the **coefficient of thermal conductivity** of the material. The unit of k is W m^{-1} K^{-1}. Table 6.3 gives values of k for different materials.

Table 6.3 Thermal conductivity values

Material	Thermal conductivity /W m^{-1} K^{-1}
aluminium	210.0
copper	390.0
glass	0.7
cardboard	0.2
felt	0.04

Worked example 6.7

Calculate the heat loss per second through a rectangular 6 mm glass pane of dimensions 1.20 m × 0.60 m, for a temperature difference of 15 K. The thermal conductivity of glass is 0.72 W m^{-1} K^{-1}.

Solution

$$\frac{Q}{t} = kA\frac{(T_1 - T_2)}{L} = \frac{0.72 \times (1.20 \times 0.60) \times 15}{6.0 \times 10^{-3}} = 1.3 \times 10^3\ \text{W}$$

See www.palgrave.com/foundations/breithaupt for an experiment to measure the thermal conductivity of copper.

Questions 6.5

1. Calculate the heat flow per second per square metre through a single-brick wall of thickness 120 mm, when the temperature on one side is 20°C and the temperature on the other side is 5°C.

   The thermal conductivity of brick is 0.40 $W\ m^{-1}\ K^{-1}$.

2. An aluminium pan of diameter 120 mm is used to boil water on a gas cooker. In 2 minutes, 0.10 kg of water boils away.

   (a) Calculate the energy per second needed to boil water away at this rate.

   (b) Estimate the temperature of the underside of the base of the pan.

   The specific latent heat of steam is 2.3 $MJ\ kg^{-1}$ and the thermal conductivity of aluminium is 210 $W\ m^{-1}\ K^{-1}$.

3. A portable box-shaped building has four windows, each 1.0 m high and 1.5 m wide. Use the U-value in Table 6.4 to calculate:

   (a) the total window area, **(b)** the total heat loss per second through the windows for **(i)** single-glazed windows, **(ii)** double-glazed windows, when the temperature difference is 10°C.

   (b) the cost per day of maintaining the temperature difference in (a) using electrical heating at a cost of 1 p per MJ for each type of window.

U-values

Heat loss calculations from buildings can be very complicated because all three heat transfer mechanisms (i.e. conduction, convection and radiation) are usually involved, cavities may be present and there are many different types of surfaces. For each type of wall, window, floor or roof, a standard value for the heat loss per second per square metre for a 1 kelvin temperature difference is used to calculate the heat loss for a given surface area and a given temperature difference. These standard values, known as **U-values**, are determined by direct measurements. For example, if the U-value for a certain type of window is 3.0 watts per square metre per kelvin, then a window of this type which has a surface area of 4.0 square metres would transmit heat at a rate of 12.0 watts per kelvin. In a location where the temperature difference between the interior and the exterior of the building was 10 kelvins, the rate of heat loss would therefore be 120 watts. In general terms, for a wall or window or floor or roof:

the heat loss per second (in watts) = U-value × surface area (in square metres) × temperature difference (in kelvins or °C)

where **the U-value** (of the wall or window or roof or floor) is defined as **the heat loss per second per square metre for a 1 kelvin temperature difference.** U-values are often expressed in watts per square metre per °C (abbreviated $W\ m^{-2}\ {}^{\circ}C^{-1}$) rather than watts per square metre per kelvin (abbreviated $W\ m^{-2}\ K^{-1}$). The numerical value is the same whichever of the above two units is used because a temperature difference of 1 kelvin is the same as a temperature difference of 1°C.

The total heat loss per second from a building can be calculated for a given temperature difference if the U-value for every different type of external wall, window, roof, floor and door in the building is known and the relevant areas are measured. Table 6.4 lists some typical U-values.

Table 6.4 Typical U-values

Type of surface	U-value / $W\ m^{-2}\ {}^{\circ}C^{-1}$
Cavity brick wall without cavity wall insulation	1.6
Cavity brick wall with insulation	0.6
Tiled roof with no loft insulation	1.9
Tiled roof with loft insulation	0.6
Single-glazed window	4.3
Double-glazed window	3.2
Floor	0.5

Worked example 6.8

A house has a cavity brick wall of total area 280 m^2 with no cavity insulation, double glazed windows of total area 18 m^2, a floor of area 110 m^2 and an insulated roof of area 130 m^2.

Use the data in Table 6.4 to calculate the total rate of loss of heat from the house on a day when the temperature outdoors is 15°C higher than the temperature indoors.

Solution

Heat loss per second from the walls = U-value × surface area × temperature difference
= 0.6 × 280 × 15 = 2520 W

Heat loss per second from the windows = 3.2 × 18 × 15 = 864 W

Heat loss per second from the roof = 0.6 × 130 × 15 = 1170 W

Heat loss per second from the floor = 0.5 × 110 × 15 = 825 W

Total heat loss per second = 2520 + 864 + 1170 + 825 = 5400 W (to 2 significant figures)

Summary

- **Temperature** in kelvins (K) = temperature in °C + 273.
- **Thermal expansion**
  Change of length = $\alpha L \Delta T$, where L = length, ΔT = temperature change, α = coefficient of thermal expansion.
- **Energy** in joules = power in watts × time taken in seconds.
- **Specific heat capacity**
  Energy transferred = $mc(T_2 - T_1)$ where m = mass, c = specific heat capacity, T_2 = final temperature, T_1 = initial temperature.
- **Specific latent heat**
  Energy needed or released = ml, where m = mass, l = specific latent heat of fusion or of vaporisation.
- **Heat transfer** is due to conduction, convection and radiation.
- **Stefan's law of radiation**
  Power radiated, $W = e\sigma AT^4$, where σ = the Stefan constant, e = the emissivity of the surface, A = surface area, T = absolute temperature.
- **Surfaces**
  A good absorber is a good emitter; a poor absorber is a poor emitter. A black body is a body that absorbs all the radiation incident on it.
- **Thermal conduction**
  Metals are the best conductors of heat because of the presence of conduction electrons.

  Heat transfer per second, $\frac{Q}{t} = \frac{kA(T_1 - T_2)}{L}$

  where A = area of cross section, $T_1 - T_2$ is the temperature difference across length L.
- **U-value** (of the wall or window or roof or floor) is defined as the heat loss per second per square metre for a 1 kelvin temperature difference.
  heat loss per second = U-value × surface area × temperature difference.

Revision questions

6.1. (a) Convert each of the following temperatures into kelvins: **(i)** 100°C, **(ii)** −30°C.
(b) Convert each of the following temperatures into °C: **(i)** 100 K **(ii)** 1000 K.

6.2. (a) (i) Explain why expansion gaps are necessary in between adjacent concrete spans of road on a bridge.
(ii) When a steel tyre is fitted on a locomotive wheel, the tyre is heated, placed on the wheel and then allowed to cool. Explain why this procedure is necessary to ensure it fits tightly on the wheel.
(b) Calculate the thermal expansion of a steel cable of length 120 m when its temperature changes from –5°C to 35°C. The coefficient of thermal expansion of steel is $1.1 \times 10^{-5} K^{-1}$.

6.3. In an experiment to determine the specific heat capacity of a piece of metal of mass 235 g, the metal object is suspended in steam at 100°C then transferred to an insulated copper calorimeter of mass 65 g containing 185 g of water at 15°C. As a result, the temperature of the water increases to 27°C. Calculate
(a) the energy needed to raise the temperature of **(i)** the water, **(ii)** the copper calorimeter from 15°C to 27°C,
(b) the specific heat capacity of the metal.
The specific heat capacity of copper is 390 J $kg^{-1}K^{-1}$ and the specific heat capacity of water is 4200 J $kg^{-1}K^{-1}$.

6.4. A copper calorimeter of mass 80 g containing 120 g of water at 16°C was placed in a refrigerator. The water in the calorimeter took 35 minutes to freeze solid. Calculate
(a) the energy removed from the calorimeter and the water when the water **(i)** cools to 0°C, **(ii)** freezes,
(b) the rate at which energy was removed. (Use the specific heat capacity values in question 6.3, specific latent heat of ice = 340 kJ kg^{-1}.)

6.5. When an object at temperature T_1 is in an enclosure at a lower temperature T_0, the net rate of heat transfer due to radiation = $e\sigma AT_1^4 - e\sigma AT_0^4$, where A is the surface area of the body and e is its emissivity.
(a) Calculate the net heat transfer due to radiation from a radiator of surface area 1.6 m^2 when its surface temperature is 45°C, in a room at a temperature of 25°C. The surface emissivity of the radiator is 0.60.
The Stefan constant, $\sigma = 5.67 \times 10^{-8}$ W m^{-2} K^{-4}.
(b) Calculate the heat loss per second per square metre from the ground on a clear night when the ground temperature is 10°C and the air temperature is 0°C. Assume the surface emissivity of the ground is 1.

6.6. The insulation jacket of a copper tank has a surface area of 0.95 m^2 and is 15 mm thick. When the outer surface of the copper tank was 54°C, the temperature on the outside of the insulation jacket was 26°C.
(a) Calculate **(i)** the temperature gradient across the insulation jacket, **(ii)** the heat per second conducted through the insulation.
(b) The base of the tank rested on a cardboard pad of thickness 10 mm. The underside of the cardboard was at 18°C and the upper side was at 22°C. Calculate the heat per second conducted through the cardboard pad. The area of the pad in contact with the tank was 0.12 m^2. The thermal conductivity values for the insulation material and for cardboard are 0.037 and 0.2 W $m^{-1}K^{-1}$, respectively.

6.7. A householder proposes to fit loft insulation in her home which has a total ceiling area of 80 m^2. This insulation reduces the U-value of the roof from 2.0 W m^{-2} °C^{-1} to 0.5 W m^2 °C^{-1}. Using the U-values in Table 6.4, calculate for a temperature difference of 20°C, the heat loss per second (**a**) without loft insulation, (**b**) with loft insualtion.

6.8. A mobile home of length 9.0 m, width 3.0 m and height 2.5 m has a flat roof and single glazed windows of total area 4.0 m^2.
(a) Calculate **(i)** the total non-window area of the sides and roof, **(ii)** the total window area, **(iii)** the floor area.
(b) (i) Calculate the heat loss per second when the external temperature is 5°C lower than the internal temperature. U-values; window = 4.3 W m^{-2} °C^{-1}, sides and roof = 2.5 W m^{-2} °C^{-1}, floor = 2.0 W m^{-2} °C^{-1}.
(ii) Calculate the cost of maintaining the temperature in (b)(i) for 1 week using heating costing 2 p per MJ.

Strength of Solids

Objectives

After working through this unit, you should be able to:

- ▶ calibrate and use a spring to measure force
- ▶ describe the effect of forces on different materials
- ▶ appreciate the meaning of strength, stiffness, elasticity, hardness, toughness and brittleness
- ▶ calculate the breaking stress of a wire
- ▶ measure and carry out calculations on the stress, strain and Young modulus of a material
- ▶ describe the general features of the stress versus strain relationship for different materials
- ▶ relate the features of a stress versus strain curve to the structure of the material

Contents

7.1 Measuring force

Mass and weight

The **weight** of an object is proportional to its mass. Mass is the amount of matter an object has and is measured in kilograms. Weight is a force and is therefore measured in **newtons** (**N**), as explained on p. 148. The weight of a 1 kg mass is 9.8 N on the surface of the Earth. This relationship can be used to calculate the weight of any object, given its mass in kilograms. The weight, W, of an object of mass, m, is calculated from the equation:

$$W = mg$$

where g = the force of gravity per unit mass on the object.

The value of g decreases with height above the Earth and is 9.8 N kg^{-1} on the surface of the Earth.

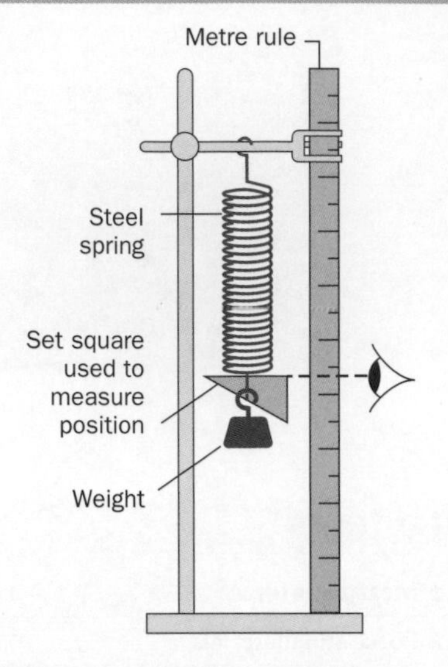

Fig. 7.1 Investigating the extension of a spring.

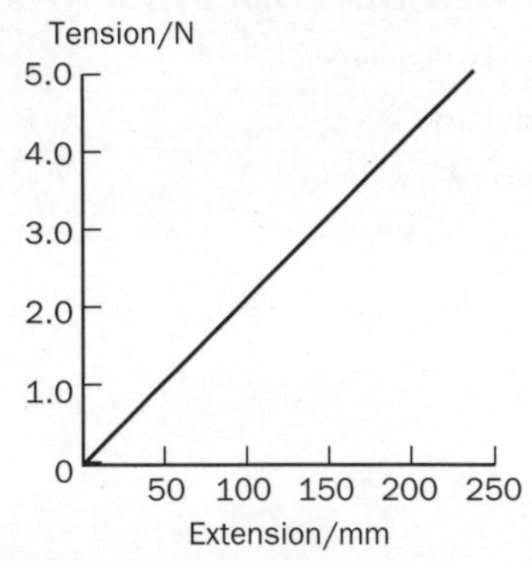

Fig. 7.2 A graph of tension versus extension.

Using a spring balance as a force meter

A steel spring suspended freely from a fixed point extends when a weight is hung on its lower end. Its change of length from its inital unstretched length is referred to as its **extension**. When the spring is stretched beyond its unstretched length, it is said to be under **tension**. When it supports a weight at rest as in Fig. 7.1, the tension in the spring is equal and opposite to the weight.

The arrangement in Fig. 7.1 can be used to measure the extension of a spring for different weights. The measurements may be plotted as a graph of tension on the vertical axis against extension on the horizontal axis.

> **Safety note**
> Always wear impact-resistant safety spectacles when carrying out stretching tests in case the weight drops off suddenly and the spring flies up.

Hooke's Law

For a steel spring, the measurements define a straight line that passes through the origin, as in Fig. 7.2. The weight of an object can be determined by measuring the extension of the spring when it supports the object, then reading the tension off the graph to give the object's weight.

The graph shows that the tension is proportional to the extension because the line is straight and it passes directly through the origin. This relationship, known as **Hooke's Law**, may be expressed as an equation

$$T = ke$$

where T = tension (in N), e = the extension (in m) and k is referred to as the spring constant (in N m^{-1}).

> **Note**
> Rearranging the equation for Hooke's Law gives $k = T \div e$, which is the gradient of the line in Fig. 7.2 because the line passes through the origin. Show that k = 25 N m^{-1} for the spring that was used to generate the graph.

Questions 7.1

1. A steel spring was suspended vertically from a fixed point. Without any weight hung from its lower end, its length was 300 mm.
   (a) When a 0.40 kg mass was suspended from its lower end, it stretched to a length of 420 mm. Calculate **(i)** the weight of a mass of 0.40 kg, **(ii)** the extension of the spring, **(iii)** the spring constant.
   (b) The 0.40 kg mass was replaced by an object of unknown mass which caused the spring to stretch to a length of 390 mm. Calculate **(i)** the weight of the object, **(ii)** the mass of the object.
2. The following measurements were obtained in an experiment to determine the spring constant of a steel spring:

Weight/N	0.0	2.0	4.0	6.0	8.0
Spring length/cm	25.0	30.2	35.0	40.2	44.8

   (a) Make a table to show the extension for different values of tension.
   (b) **(i)** Plot a graph of tension on the vertical axis against extension on the horizontal axis.
   (ii) Calculate the spring constant of the spring.

7.2 Force and materials

Material properties

The effect of forces on solid materials varies widely from one material to another. For example, a biscuit cracks suddenly when subjected to excessive force; in comparison, a bar of toffee will bend if subjected to a slowly increasing force. The same material will, however, behave like a biscuit if subjected to an impact force of short duration.

- **Strength** is a measure of how much force is needed to break an object. Suitable heat treatment of steel objects makes them stronger. Some objects become weaker when subjected to repeated bending.
- **Stiffness** is a measure of the difficulty of stretching or bending an object. For example, a steel strip is much stiffer than a rubber strip. The spring constant of a spring is a measure of its stiffness. Strength and stiffness depend on the dimensions of the material as well as its nature.
- **Elasticity** is the property of a material that enables it to regain its shape after being distorted. For example, a rubber band regains its original length when it is released after being stretched. The **elastic limit** of an object is the limit to which it can be distorted and still be able to regain its shape. A strip of polythene does not regain its shape after being stretched because its elastic limit is very small. An object stretched beyond its elastic limit is said to exhibit **plastic behaviour**.
- **Hardness** is the resistance of the surface of a material to scratches and dents. The hardness of two surfaces may be compared by dropping a metal punch onto each surface from the same height and comparing the diameters of the dents created.

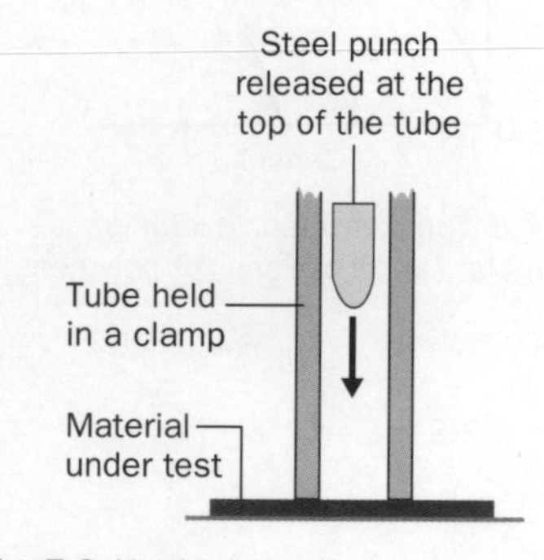

Fig. 7.3 Hardness testing.

Safety note

Always wear impact-resistant safety spectacles when carrying out hardness tests in case bits fly up from the material under test.

- **Toughness** is the ability of a material to withstand impact forces without cracks developing, which would weaken the material when a material is distorted. For example, the sole of a sports shoe must be very tough to withstand repeated flexing and impacts. Once a crack develops in the sole of a sports shoe, it widens and the sole splits. In contrast, a **brittle** material snaps suddenly without 'giving' when subjected to excessive force.

Investigating stiffness and elasticity

The apparatus in Fig. 7.1 may be modified as shown in Fig. 7.4 to investigate how easily a strip of material stretches and whether or not it regains its shape easily. The results for these tests are plotted with tension on the vertical axis and extension on the horizontal axis, as in Fig. 7.5.

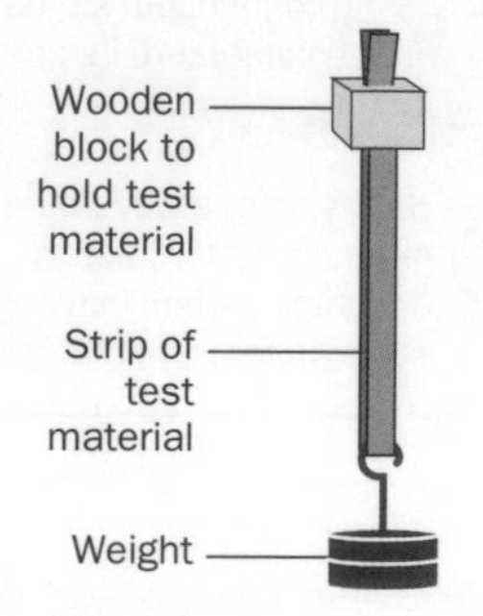

Fig. 7.4 Testing materials.

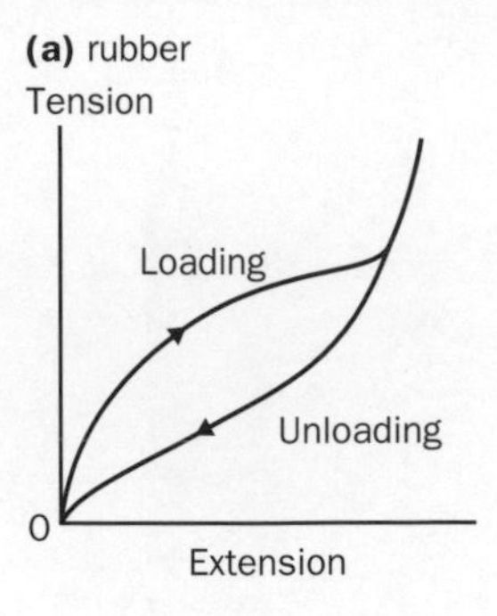

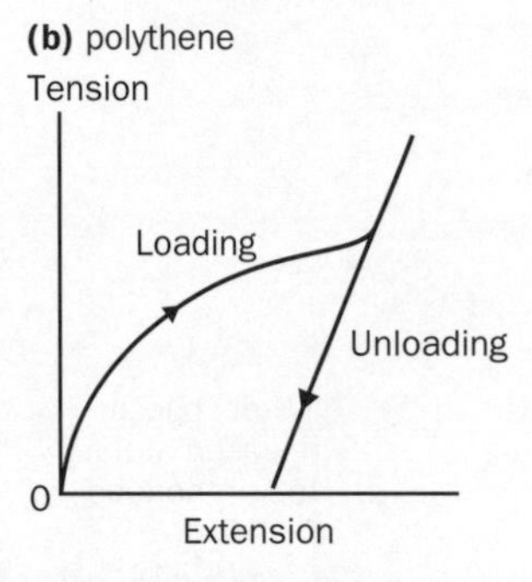

Fig. 7.5 Tension versus extension graphs for **(a)** rubber and **(b)** polythene.

Safety note

Always wear impact-resistant safety spectacles when carrying out stretching tests in case the weight drops off and the material under test flies up.

1. A rubber band stretches easily then becomes much stiffer. The extension is not proportional to the weight. It regains its original length when all the weights are removed. If the weight is reduced in steps, its length for a given weight is greater than when it was loaded with that weight. This effect is referred to as **mechanical hysteresis** because the extension 'lags' behind the weight. The area of the loop is a measure of the energy used when the material is deformed and allowed to regain its shape.

2. A polythene strip also stretches easily then becomes much stiffer. Its extension is not proportional to the weight. However, unlike rubber, polythene does not regain its original length, as its elastic limit is very small and it undergoes plastic behaviour when it is extended noticeably.

Polymers

Rubber and polythene are examples of polymers, which are materials composed of long 'chain' molecules. Rubber is a natural polymer, unlike polythene. A polymer is formed when much smaller identical molecules called 'monomers' link together, forming covalent bonds as they join end to end like the links in a chain. In rubber and polythene, the polymer molecules become tangled up. When rubber or polythene are stretched, the molecules are straightened and the material becomes difficult to extend further. Rubber molecules tend to curl when released, which is why a rubber band regains its original length when released.

7.3 Stress and strain

Breaking stress

The force needed to break a wire depends on the diameter of the wire and the material of the wire. Four times as much force is needed to break a wire of diameter 1mm than is needed to break a wire of diameter 0.5 mm, made of the same material. This is because the area of cross-section of the thicker wire is four times as much as the area of cross-section of the thinner wire.

The **breaking stress** of a wire is defined as the breaking force per unit area of cross-section.

If F is the force required to break a wire of cross-sectional area A,

$$\textbf{the breaking stress, } \sigma_b = \frac{F}{A}$$

Notes

1. The unit of stress is the pascal (Pa), equal to $1\ \text{N m}^{-2}$. The symbol for stress is σ (pronounced 'sigma').
2. For a wire of diameter d, its cross-sectional area $A = \frac{\pi d^2}{4}$
3.  Safety note: always wear impact-resistant safety spectacles when carrying out strength tests on materials.

Questions 7.2

1. Describe the ideal mechanical properties of **(a)** a biscuit, **(b)** a running shoe, **(c)** a dustbin.
2. Discuss whether or not a rubber band would make a suitable force meter instead of a steel spring.

Worked example 7.1

Calculate the force needed to break a nylon wire of diameter 0.80 mm. The breaking stress of nylon is 5.0×10^7 Pa.

Solution

Area of cross-section, $A = \pi (0.8 \times 10^{-3})^2 / 4 = 5.0 \times 10^{-7}$ m²

Rearranging the equation for breaking stress gives $F = \sigma_b A$

$= 5.0 \times 10^7 \times 5.0 \times 10^{-7} = 25$ N

Measuring stress and strain

Stress is defined as the force per unit area acting perpendicular to the area. The unit of stress, the pascal (Pa) is equal to 1 N m^{-2}. For a wire of cross-sectional area A, under tension T, the stress, σ, is calculated from the equation

$$\sigma = \frac{T}{A}$$

For a wire suspended vertically from a fixed point, supporting a weight W, the tension is equal to the weight.

Strain is defined as the change of length per unit length. Since strain is a ratio, there is no unit for strain. For a wire of unstretched length L, which extends to a length L + e, where e is its extension, the strain is calculated from the equation

$$\text{strain} = \frac{e}{L}$$

The relationship between stress and strain for a wire can be investigated using the arrangement in Fig. 7.8. The control wire supports a micrometer

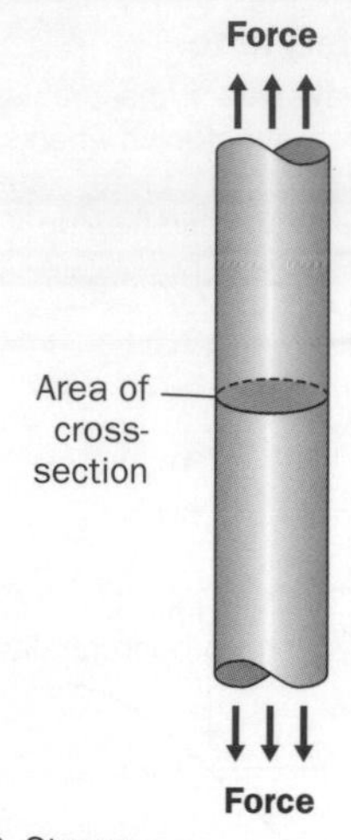

Fig. 7.6 Stress.

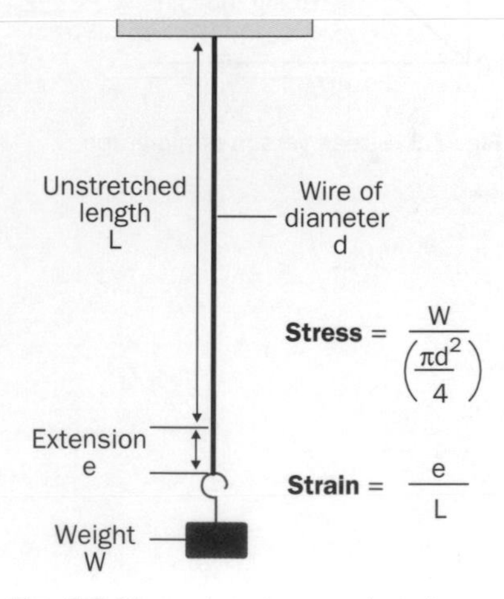

Fig. 7.7 Measuring stress and strain.

See www.palgrave.com/foundations/breithaupt for more about measuring stress and strain.

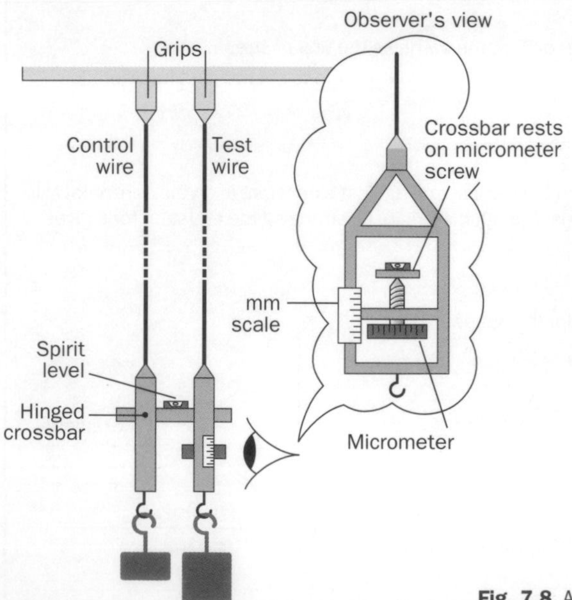

Fig. 7.8 A wire under test.

Safety note

Always wear impact-resistant safety spectacles when carrying out stretching tests in case the wire snaps and flies up.

which is used to measure the change of length of the test wire when the test wire is loaded and unloaded. Each time a measurement is made, the micrometer is adjusted so the spirit level is horizontal before the micrometer reading is taken.

1. The two wires are loaded sufficiently to ensure each wire is straight. The micrometer is adjusted as explained above and its reading noted.
2. The initial length, L, of the test wire is then measured using a metre rule with a millimetre scale. The diameter of the wire is also measured using a micrometer at several points along the wire to give an average value, d, for the diameter of the wire.
3. The test wire is then loaded with a known weight, W, its micrometer is adjusted as above and its reading noted. The procedure is repeated as W is increased in steps and then decreased in steps to zero.
4. The extension, e, from the initial length, L, is calculated from the test wire micrometer readings. The strain is then calculated for each reading by dividing the extension by the inital length. The area of cross-section, A, of the wire is calculated using $A = \pi d^2 \div 4$ and the stress ($= W \div A$) calculated for each reading.
5. The results may be plotted as a graph of stress on the vertical axis against strain on the horizontal axis, as shown by Fig. 7.9. The graph shows that the stress is proportional to the strain. This relationship holds up to a limit referred to as *the limit of proportionality*. Provided the limit of proportionality is not exceeded, the value of stress ÷ strain is constant. This value is known as the **Young modulus of elasticity** of the material.

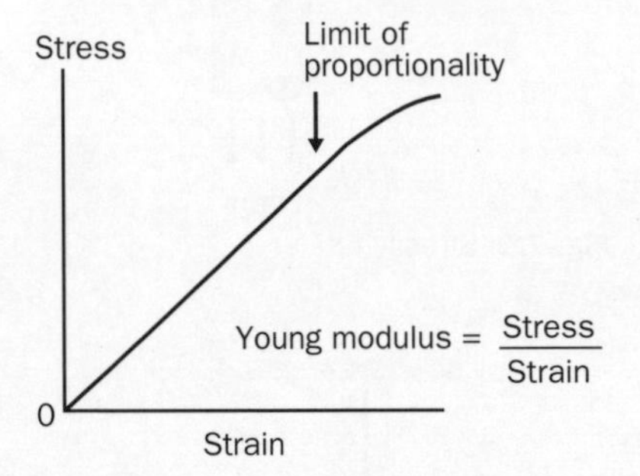

Fig. 7.9 Stress versus strain graph.

$$\textbf{Young modulus of elasticity, } E = \frac{\text{stress}}{\text{strain}}$$

Notes

1. The unit of E is the pascal (Pa), the same as the unit of stress.
2. For a wire of unstretched length, L, and area of cross-section, A, under tension, T,

$$E = \frac{\text{stress}}{\text{strain}} = \frac{T/A}{e/L} = \frac{TL}{Ae}$$

3. A graph of tension, T, on the vertical axis against extension, e, on the horizontal axis also gives a straight line through the origin. Rearranging the equation for E gives

$$T = \frac{AE}{L}e$$

which is the equation for this graph.

The gradient of the line $= \frac{AE}{L}$

Hence $E = \text{gradient (of T v. e line)} \times \frac{L}{A}$

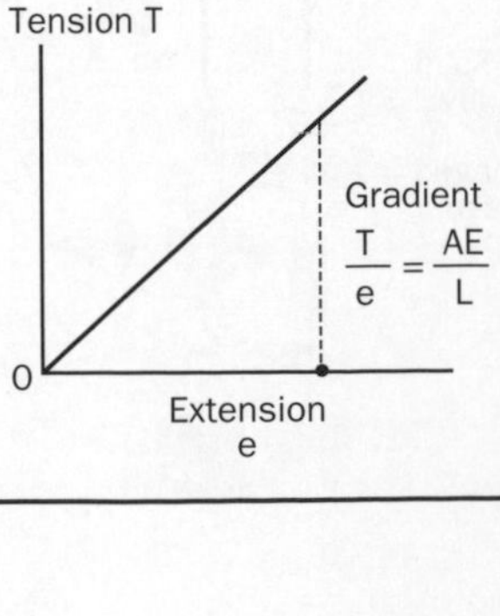

Tension T

Gradient $\frac{T}{e} = \frac{AE}{L}$

0

Extension e

Fig. 7.10 Tension versus extension for a steel wire.

Worked example 7.2

A vertical steel wire of diameter 0.35 mm and unstretched length 1.22 m extends by a distance of 3.2 mm when it is loaded with a 50 N weight. Calculate (a) (i) the stress and (ii) the strain in the wire, (b) the Young modulus of steel.

Solution

(a) (i) Area of cross-section, $A = \pi(0.35 \times 10^{-3})^2 \div 4 = 9.62 \times 10^{-8}\ m^2$

$$\therefore \text{Stress} = \frac{T}{A} = \frac{50}{9.62 \times 10^{-8}} = 5.2 \times 10^{8}\ Pa$$

(ii) $\text{Strain} = \frac{e}{L} = \frac{3.2 \times 10^{-3}}{1.22} = 2.6 \times 10^{-3}$

(b) Young modulus $E = \frac{\text{stress}}{\text{strain}} = 2.0 \times 10^{11}\ Pa$

Questions 7.3

1. A 2.37 m steel wire of diameter 0.38 mm is suspended vertically from a fixed point and used to support a weight of 115 N at its lower end.

(a) The Young modulus of steel is 2.0×10^{11} Pa. Calculate **(i)** the stress in the wire, **(ii)** the extension of the wire.

(b) The breaking stress of steel is 1.1×10^{9} Pa. Calculate the maximum weight this wire can support when vertical without snapping.

2. A wire of length 1.28 m and of diameter 0.28 mm was stretched by increasing its tension in steps. The extension of the wire was measured at each step and tabulated as below.

Tension/N	0.0	5.0	10.0	15.0	20.0	25.0
Extension/mm	0.0	0.8	1.6	2.4	3.4	4.6

(a) Plot a graph of tension on the vertical axis against extension on the horizontal axis.

(b)(i) Estimate the limit of proportionality of the wire.

(ii) Calculate the Young modulus of elasticity of the wire.

7.4 Stress versus strain curves

A stress–strain curve is characteristic of the material and does not depend on the dimensions of the sample tested. Fig. 7.11 shows a stress–strain curve for a steel wire, with stress on the vertical axis against strain on the horizontal axis. Its features are therefore due to the nature and properties of the material and are determined by the structure of the material.

This type of curve is obtained by using a 'tensometer' which stretches the material and measures the force as it is stretched.

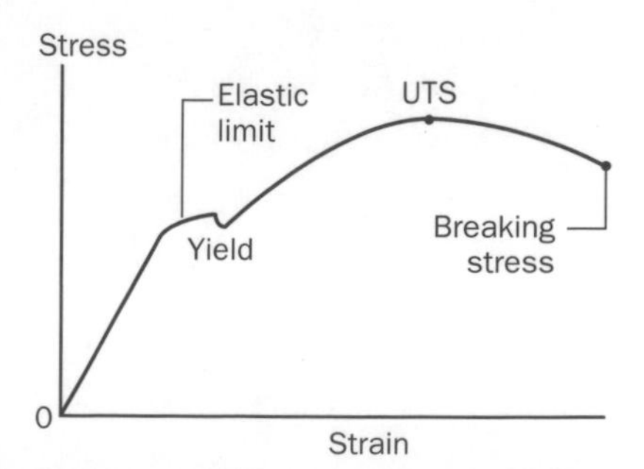

Fig. 7.11 Stress versus strain for steel.

- **Proportionality:** the stress is proportional to the strain up to the 'limit of proportionality'. The gradient of this straight section is the Young modulus of the material. See p. 96.
- **The elastic limit:** this is the limit beyond which the material becomes permanently stretched.

Fig. 7.12 A tensometer in use.

- **Plastic behaviour:** beyond the elastic limit, the material does not regain its original length and its behaviour is described as plastic.
- **Yield:** the sample suddenly 'gives' a little at a point referred to as the upper yield point. The stress drops slightly to the lower yield point before rising again as the sample is stretched.
- **The ultimate tensile stress** (UTS) of the material is the point where the sample is strongest. The material is unable to withstand stress beyond its UTS.
- **The breaking stress:** this is the stress at the breaking point of the sample. Stretching a wire beyond its UTS causes it to become narrower at its weakest point where internal cavities form. The effective cross-sectional area is reduced at this point which increases the stress locally, reducing the area of cross-section even more until the wire snaps.

Solid materials and their structure

Solid materials are classified in terms of internal structure in three main categories.

Crystalline solids

The atoms in a crystal are arranged in a regular pattern to give the crystal a distinctive shape. A metal is composed of many tiny crystals called **grains**, each with the same atomic arrangement but aligned at random to each other.

Amorphous solids

The atoms in an amorphous solid are locked together at random. Glass is an example of an amorphous solid.

Polymers

A polymer is composed of long molecules usually tangled together. Polymers that set permanently and cannot be remoulded by heating contain **cross-links** between the molecules.

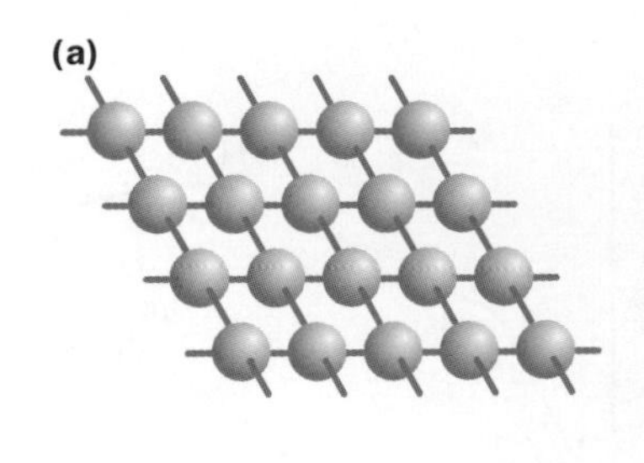

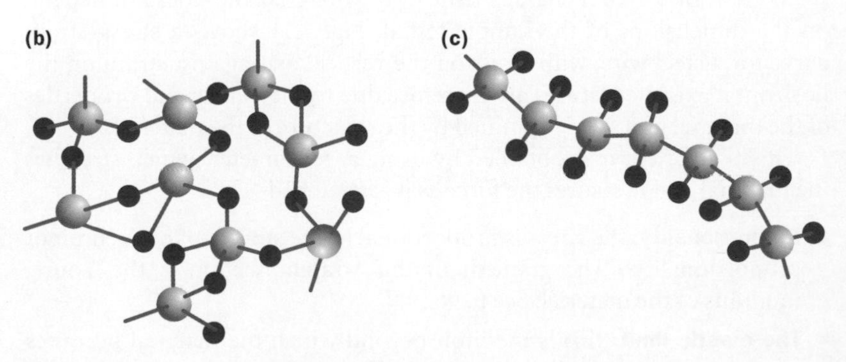

Fig. 7.13 Solid materials (a) crystalline, (b) amorphous, (c) polymer.

Molecular interpretation of stress versus strain curves

The general features of the force between two atoms in a solid are explained on p. 74. Fig. 7.14 shows how the force changes with distance apart. The equilibrium separation corresponds to the average separation between the atoms in a solid.

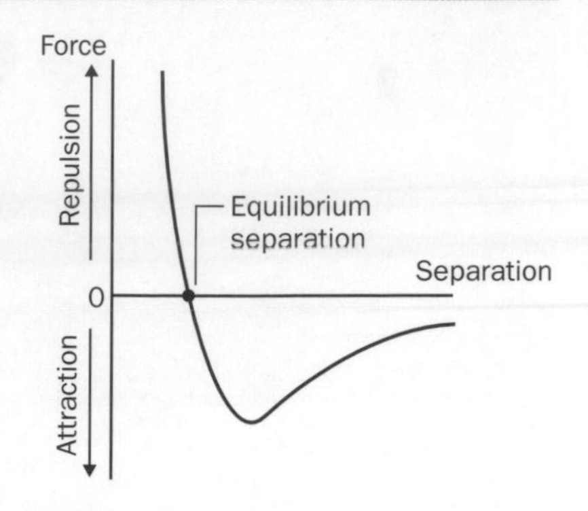

Fig. 7.14 Force versus separation for two atoms.

- **Proportionality:** this is because the graph of force versus separation is straight near to the equilibrium separation. Therefore, the force is proportional to the change of separation from equilibrium. This is why the tension in a wire is proportional to its extension.

- **Limit of proportionality:** this is because the force versus separation graph departs from a straight line curve as the separation is increased more and more.

- **Plastic behaviour:** this is because the atoms are made to slip past each other to take up new positions in the structure. This is easily achieved in a polymer such as PVC, but not in an elastic band due to the presence of strong cross-links which prevent the molecules slipping past each other. In a metal, plastic behaviour occurs when planes of atoms are forced to slip past each other. However, because a metal consists of countless grains, the tendency of one grain to slip is opposed by the surrounding grains, so a metal is much stronger than a single crystal. In addition, the presence of **interstitial atoms** strengthens a metal by preventing crystal planes slipping past each other.

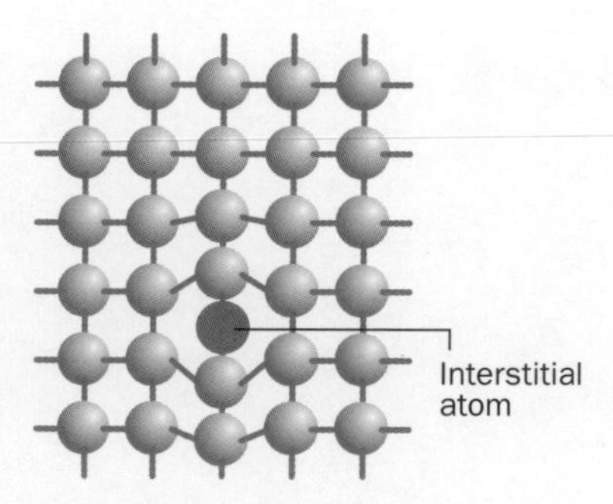

Fig. 7.15 Interstitial atoms.

- **Ultimate tensile strength:**

  (a) Polymer molecules line up with each other when a polymer is stretched as much as possible. To break the polymer, the applied force must be sufficient to break each molecule.

  (b) In an amorphous solid under increasing stress, tiny cracks appear at the surface and become progressively wider and deeper as the stress becomes concentrated at each crack. Amorphous solids such as glass are brittle because there are no internal boundaries, as in a metal, to prevent surface cracks from propagating into the material.

  (c) In a metal, the presence of **dislocations** affect its strength. A dislocation is where a plane of atoms is squeezed out by adjacent planes. Little stress is needed to make a dislocation move through a grain and reach the grain boundary. The grain size affects the strength of the metal because there are more grain boundaries to stop the movement of dislocations if the grains are small. Heat treatment affects grain size, which is why steel is made brittle if it is heated strongly then rapidly cooled by plunging it into water.

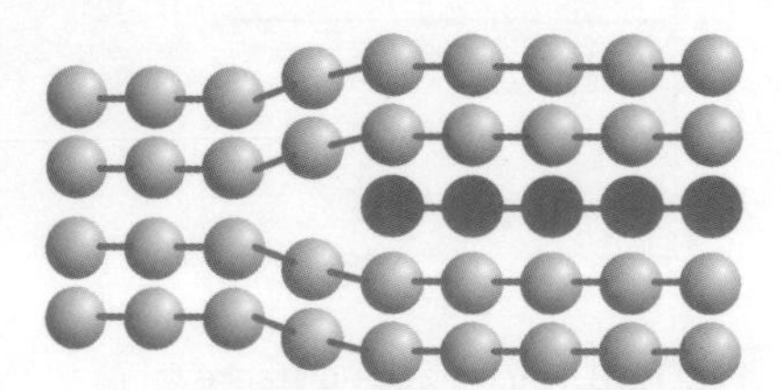

Fig. 7.16 Dislocations.

Questions 7.4

1. **(a) (i)** Sketch a curve to show how the force between two atoms in a solid varies with separation.
   (ii) Use your curve to explain why the extension of a wire is proportional to its tension, up to a limit.
   (b) Explain the difference between elastic and plastic behaviour of a solid.

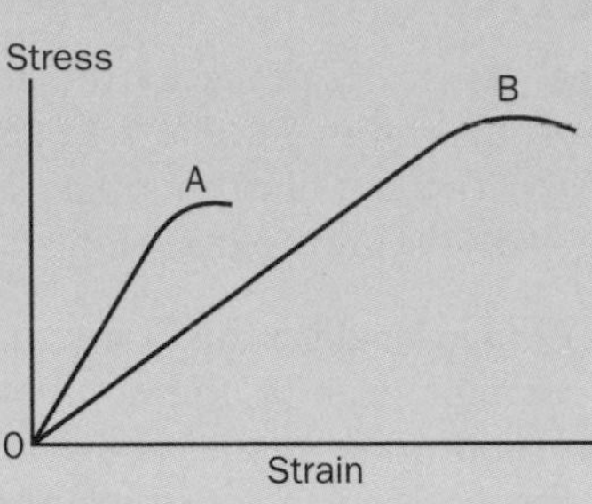

Fig. 7.17

2. Fig. 7.17 shows the stress–strain curves for two different materials, A and B, on the graph. Which material is **(a)** stronger **(b)** stiffer? In each case, give a reason for your answer.

7.5 Elastic energy

Force and energy

Energy is the capacity of a body to do work. Work is done when a force moves its point of application in the direction in which it acts. Lift an object and you have transferred energy from your muscles to the object. Stretch a spring and you have transferred energy from your muscles to the spring. Energy is transferred when work is done by a force moving its point of application in the direction of the force.

1 Joule of energy is transferred when a force of 1 N moves its point of application through a distance of 1 m in the direction of the force.

Work done by a force = force × distance moved in the direction of the force
(in joules) (in newtons) (in metres)

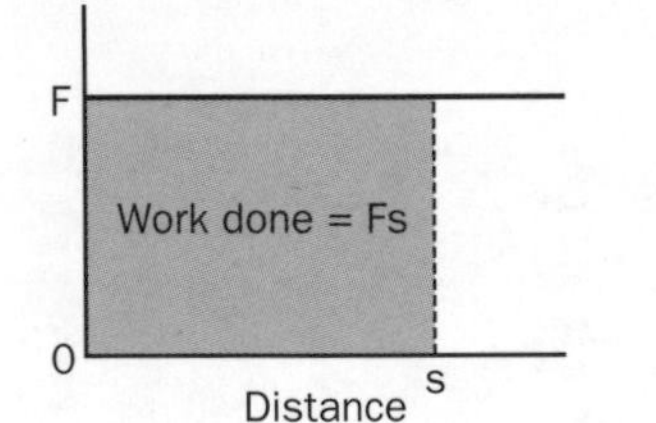

Fig. 7.18 Force versus distance for a constant force.

Force versus distance graphs

1. For a constant force, F, acting over distance, s, the work done = Fs. Fig. 7.18 shows the force versus distance graph for a constant force. The work done is represented by the area under the force line on the graph.
2. For a spring, the force increases with extension, as in Fig. 7.19. To extend the spring when under tension, F′, by a small amount, δs, the work done = F′δs.This is represented on Fig. 7.19 by the strip of width δs under the line.

∴ the total work done to stretch the spring from zero to extension e
= total area under the line = $\frac{1}{2}$ Fe

where F is the force needed to extend the spring to extension e. The area under the line is a triangle of height, F, and base length, e. The area of this triangle = $\frac{1}{2}$ base × height = $\frac{1}{2}$Fe.

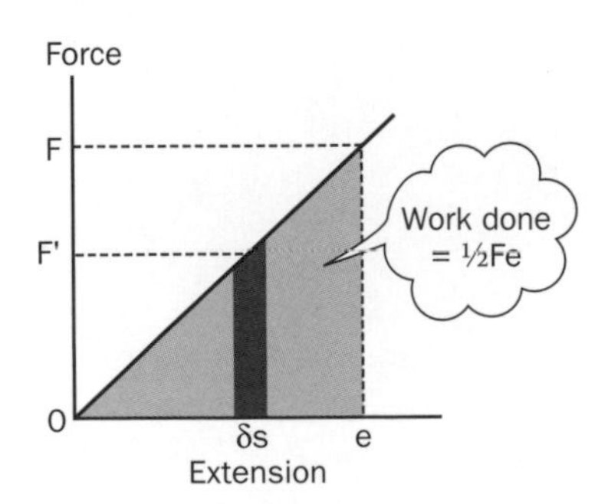

Fig. 7.19 Force versus extension for a spring.

As the work done is stored in the spring as elastic energy, the energy stored in a stretched spring = $\frac{1}{2}Fe$.

Since F = ke, where k is the spring constant of the spring, the energy stored = $\frac{1}{2}Fe = \frac{1}{2}ke^2$

Energy stored in a stretched spring = $\frac{1}{2}ke^2$

where e = extension and k= spring constant.

Energy stored in a stretched wire

The tension, T, in a stretched wire is related to the extension, e, of the wire in accordance with the equation

$$T = \frac{AE}{L}e$$

where L is the unstretched length of the wire, A is its area of cross-section and E is the Young modulus of the material of the wire.

∴ the energy stored in the wire at extension e = $\frac{1}{2}Te = \frac{\frac{1}{2}(AE)}{L}e^2$

Note

Since the volume of the wire =AL, the energy stored per unit volume

$= \frac{\frac{1}{2}Te}{AL} = \frac{1}{2}$ stress × strain

Worked example 7.3

A 1.25 m steel wire of diameter 0.30 mm is suspended vertically from a fixed point and used to support an 85 N weight. Calculate (a) the extension of the wire, (b) the energy stored in the wire, (c) the energy stored per unit volume in the wire. The Young modulus of steel is 2.0 × 10^{11} Pa.

Solution

(a) Rearranging

$$E = \frac{TL}{Ae} \text{ gives } e = \frac{TL}{AE} = \frac{85 \times 1.25}{\frac{1}{4}\pi(0.30 \times 10^{-3})^2 \times 2.0 \times 10^{11}} = 7.5 \times 10^{-3}\text{ m}$$

(b) Energy stored = $\frac{1}{2}Te = \frac{1}{2} \times 85 \times 7.5 \times 10^{-3} = 0.32$ J

(c) Volume of wire = AL = $\frac{1}{4}\pi(0.30 \times 10^{-3})^2 \times 1.25 = 8.8 \times 10^{-8}\text{ m}^3$

$$\text{Energy stored per unit volume} = \frac{0.32\text{ J}}{8.8 \times 10^{-8}\text{ m}^3} = 3.6 \times 10^6\text{ J m}^{-3}$$

Questions 7.5

1. The unstretched length of a steel spring is 320 mm. When it was suspended vertically from a fixed point and used to support a weight of 5.0 N at rest, its length extended to 515 mm. Calculate
   (a) **(i)** the extension of the spring, **(ii)** the spring constant.
   (b) the energy stored in the spring.
2. A nylon guitar string of diameter 0.50 mm was stretched from an unstretched length of 750 mm to a length of 785 mm as a result of turning the tension key.
   (a) Calculate **(i)** the extension, **(ii)** the tension in the string at a length of 785 mm.
   (b) Calculate the energy stored per unit volume in the string at this length.
   The Young modulus of nylon = 3.0 × 10^9 Pa.

Summary

- **Weight** in newtons = mass × g.
- **Hooke's Law:** For a steel spring, tension, T = ke, where e = extension, k = spring constant.
- **Material properties:** Strength and stiffness depend on the dimensions and on the material of an object. Elasticity, hardness and toughness are material properties (i.e. do not depend on dimensions).
- **Stress** is defined as the force per unit area acting perpendicular to the area.
- **Strain** is defined as the change of length per unit length.
- **The Young modulus of elasticity,** E, of a material is defined as stress ÷ strain.
- **The area under a force versus extension graph** represents the work done.
- **The energy stored in a stretched spring** = $\frac{1}{2}ke^2$, where e = extension, k = spring constant.
- **The energy stored in a stretched wire** = $\dfrac{\frac{1}{2}(AE)e^2}{L}$

Revision questions

Where necessary, assume $g = 9.8\ \text{N kg}^{-1}$.

7.1. A steel spring was suspended vertically from a fixed point. Its length when unloaded was 500 mm.
(a) When it was loaded with a 6.0 N weight, it extended to a length of 845 mm when at rest. Calculate **(i)** the spring constant, **(ii)** the energy stored in the spring at this extension.
(b) The 6.0 N weight was replaced with an unknown weight that made the spring stretch to a length of 728 mm. Calculate **(i)** the weight of the unknown object, **(ii)** the mass of the unknown object.

7.2. (a) Fig. 7.20 is a graph of extension versus weight for two different materials, A and B, of the same length, thickness and width. Which material is **(i)** strongest? **(ii)** stiffest?

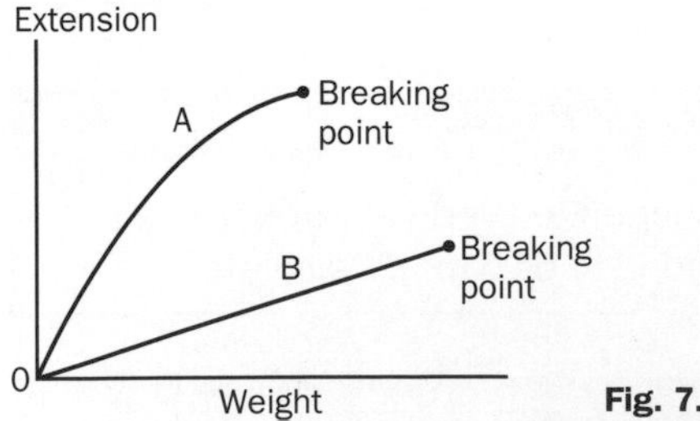

Fig. 7.20

(b) (i) Sketch curves on the same axes to show how the tension of **(i)** an elastic band, **(ii)** a strip of polythene varies with extension.
(c) Use your graphs to explain the difference between plastic and elastic behaviour.

7.3. A metal wire of diameter 0.20 mm was suspended vertically from a fixed point and used to support a 5.0 N weight which straightened it out. Its length was then measured at 1.393 m. The weight supported was then increased to 65 N and its length increased to 1.414 m. Calculate **(a)** the extension of the wire, **(b)** the Young modulus of the wire, **(c)** the energy stored in the wire.

7.4. The steel lifting cable of a crane has a diameter of 30 mm and a maximum length of 55 m. The maximum stress in the cable is not allowed to exceed 2.0×10^8 Pa.
(a) Calculate the maximum weight the cable is allowed to support.
(b) Calculate **(i)** the extension of the cable when it supports this maximum weight, **(ii)** the energy stored per unit volume in the cable when extended by this maximum weight. The Young modulus of steel is 2.0×10^{11} Pa.

7.5. The jaws of a vice open by 0.5 mm for each full turn on the handle. A copper cylinder of diameter 10 mm and length 60 mm is placed in the vice which is then tightened up by half a turn. The Young modulus for copper is 1.3×10^{11} Pa. Calculate **(a)** the reduction of length of the copper, **(b)** the force needed to compress it by this amount.

7.6. A passenger lift of weight 1800 N is supported by eight steel cables, each of diameter 5.0 mm and 65 m in length. The maximum allowed stress in the cables is 1.0×10^8 Pa. The Young modulus of steel is 2.0×10^{11} Pa. Calculate
(a) the maximum weight of passengers allowed in this lift,
(b) the extension of each lift cable when it supports this weight.

Pressure

Objectives

After working through this unit, you should be able to:

- carry out calculations relating to pressure, force and area
- describe the principles and applications of hydraulics
- calculate the pressure due to a liquid column
- describe how pressure is measured, including atmospheric pressure
- explain the principle of flotation in terms of pressure

Contents

8.1 Pressure and force

Snow shoes are designed with a large contact area so that the wearer does not sink into snow when walking across it. The weight of the wearer is effectively spread out over as large an area as possible. The opposite reason explains why a sharp knife cuts more easily than a blunt knife. The area of the knife in contact with the material it cuts into is less for a sharp knife than for a blunt knife. As a result, the pressure of the knife on the material is greater for a sharp knife than for a blunt knife.

The pressure, p, of a force, F, acting normally (i.e. at right angles) on a surface is defined by the equation

$$p = \frac{F}{A}$$

where A is the area over which the force acts. The unit of pressure is the pascal (Pa), equal to $1\ \mathrm{N\ m^{-2}}$.

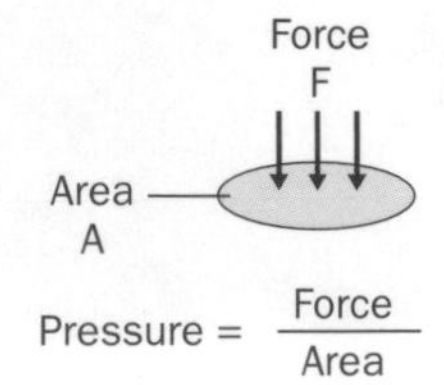

Fig. 8.1 Pressure and force.

> **Note**
> An area of $1\ \mathrm{m^2} = 10\ 000\ \mathrm{cm^2}$ $(= 100\ \mathrm{cm} \times 100\ \mathrm{cm}) = 10^6\ \mathrm{mm^2}$ $(= 1000\ \mathrm{mm} \times 1000\ \mathrm{mm})$.

Worked example 8.1

A brick of weight 55 N has dimensions 100 mm × 150 mm × 250 mm. Calculate the pressure of the brick on the ground when it is (a) end-on, (b) side-on, (c) face-on.

Solution

(a) $A = 0.100 \times 0.150 = 1.5 \times 10^{-2}\ \mathrm{m^2}$. $p = \frac{F}{A} = \frac{55}{1.5 \times 10^{-2}} = 3.7 \times 10^3\ \mathrm{Pa}$

Worked example 8.1 continued

(b) $A = 0.100 \times 0.250 = 2.5 \times 10^{-2}\ m^2$.

$$p = \frac{F}{A} = \frac{55}{2.5 \times 10^{-2}} = 2.2 \times 10^3\ Pa$$

(c) $A = 0.150 \times 0.250 = 3.75 \times 10^{-2}\ m^2$.

$$p = \frac{F}{A} = \frac{55}{3.75 \times 10^{-2}} = 1.5 \times 10^3\ Pa$$

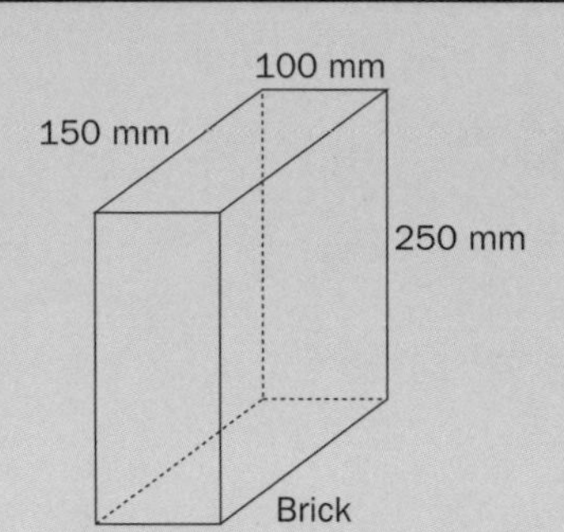

Fig. 8.2

Questions 8.1

1. Explain each of the following:
   (a) A sharp needle is easier than a blunt needle to push through a cloth.
   (b) A farm tractor is fitted with wide tyres to prevent it sinking into earth.
2. **(a)** A person of weight 750 N stands upright with both feet on the floor. The contact area of each foot with the floor is $8.5 \times 10^{-3}\ m^2$. Calculate the pressure on the floor.
   (b) The four tyres of a car, of total weight 12000 N, are at a pressure of 250 kPa. Calculate the area of each tyre in contact with the ground.

8.2 Hydraulics

A hydraulic machine is designed to make use of the fact that a liquid is almost incompressible. When a force is applied to a liquid in a sealed system, the liquid is put under pressure. This pressure is transmitted throughout the liquid.

- A **hydraulic press** is designed to squeeze an object using a much smaller force. Fig. 8.3 shows the principle of operation of a hydraulic press. The applied force, F_1, acts on a piston of area A_1, creating pressure

$$p = \frac{F_1}{A_1}$$

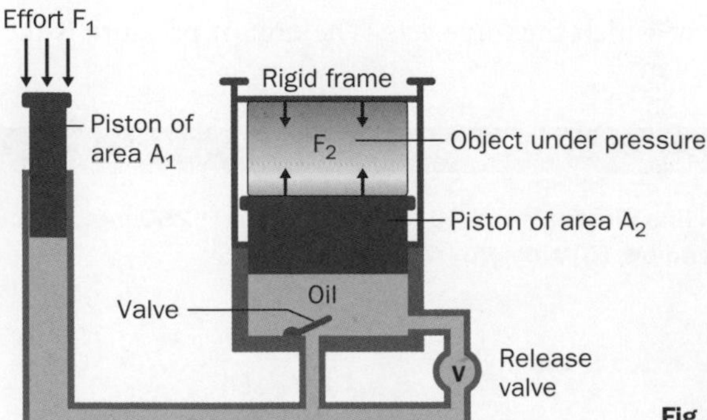

Fig. 8.3 The hydraulic press.

in the oil in the cylinder. This pressure is transmitted to the wide 'ram' piston to compress the object in the frame.

The force, F_2, acting on the wide piston

$$= pA_2 = \frac{F_1A_2}{A_1}$$

Since A_2 is much greater than A_1, then force F_2 is much greater than force F_1. In other words, the applied force is multiplied by a factor equal to the area ratio $A_2 \div A_1$. Air must be absent from the system, otherwise the pressure is used to compress the air instead of being transmitted through the oil. When the release valve is closed, the inlet valve prevents the oil from leaving the wide cylinder. Opening the release valve relieves the pressure in the wide cylinder, allowing the object to be removed.

- **Hydraulic brakes** use pressure created when the driver applies a force to the foot pedal. This creates pressure on the brake fluid in the master cylinder. This pressure is transmitted to each of the slave cylinders, where it forces the disc pads to grip the wheel disc. Fig. 8.4 shows the idea. No air should be allowed into the system otherwise the force of the brake pedal is used to compress the air instead of creating pressure in the system, making the brakes ineffective.
- A **hydraulic lift** is designed to lift an object using a much smaller force. The lift platform is supported by four vertical legs, each attached to a piston. Compressed air is used to force oil into the cylinders to raise the pistons and the platform. Fig. 8.5 shows how this is achieved.

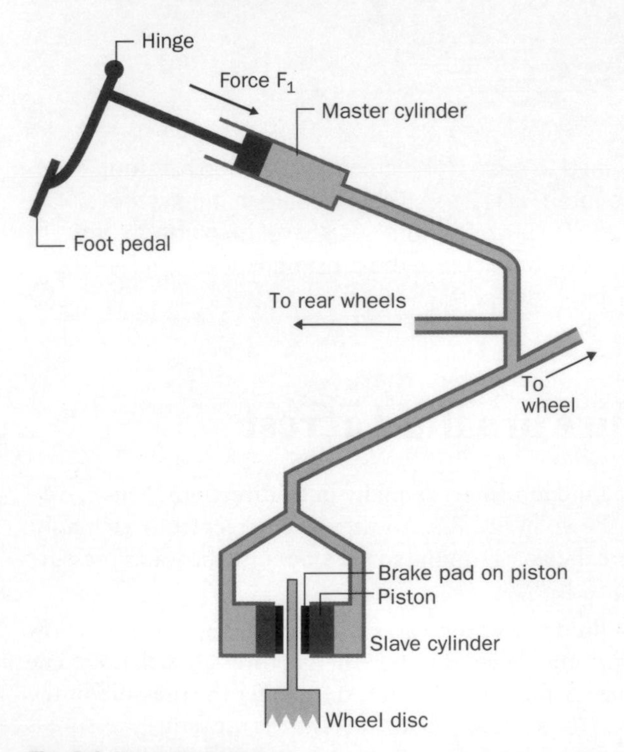

Fig. 8.4 Hydraulic brakes.

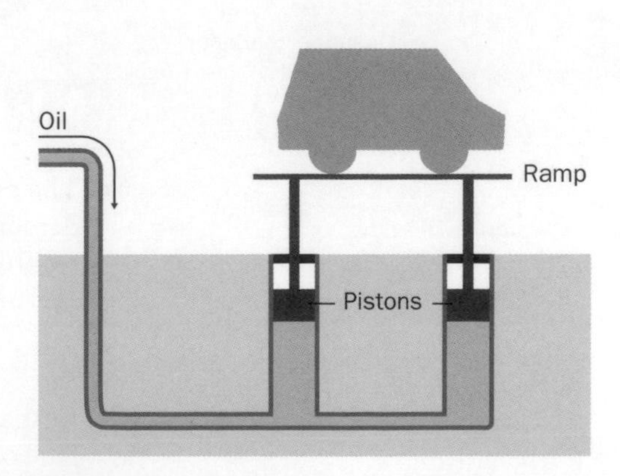

Fig. 8.5 The hydraulic lift.

Worked example 8.2

A hydraulic press has a narrow cylinder of diameter 2.0 cm and a wide 'ram' cylinder of diameter 30 cm. A force of 120 N is applied to the narrow cylinder. Calculate
(a) the pressure created in the system, (b) the force on the 'ram' piston.

Solution

(a) Area of the narrow cylinder, $A_1 = \dfrac{\pi(2.0 \times 10^{-2})^2}{4} = 3.1 \times 10^{-4}\ \text{m}^2$

$\therefore p = \dfrac{F}{A_1} = \dfrac{120}{3.1 \times 10^{-4}} = 3.8 \times 10^5\ \text{Pa}$

(b) Force on the ram piston, $F_2 = pA_2$

where $A_2 = \dfrac{\pi(30.0 \times 10^{-2})^2}{4} = 7.1 \times 10^{-2}\ \text{m}^2$

$\therefore F_2 = 3.8 \times 10^5 \times 7.1 \times 10^{-2} = 2.7 \times 10^4\ \text{N}.$

Questions 8.2

1. Fig. 8.6 shows the internal construction of a hydraulic car jack. Explain why a small effort applied to the car jack is capable of raising a much greater load.

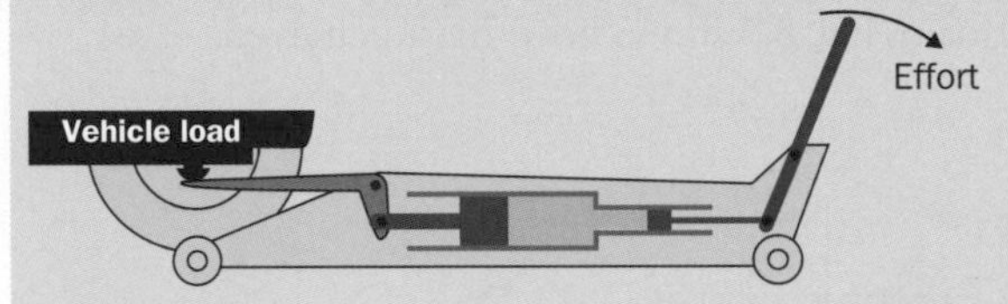

Fig. 8.6

2. A hydraulic lift used to raise a vehicle in a workshop has four pistons, each of area 0.012 m^2. The pressure in the system must not exceed 500 kPa. The total weight of the platform and the pistons is 2500 N. Calculate the maximum additional weight that the platform can support.

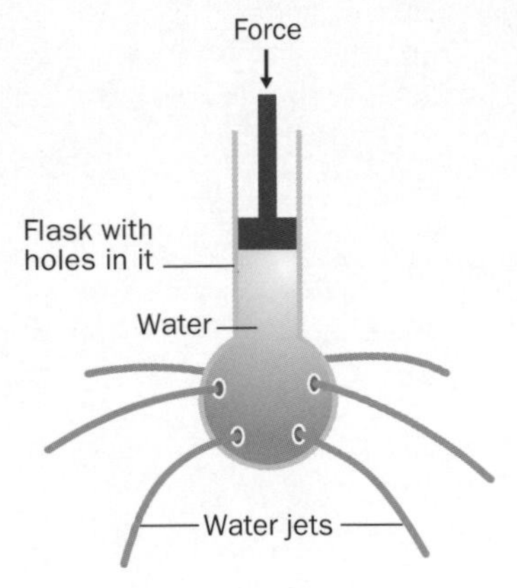

Fig. 8.7 Pressure acts equally in all directions.

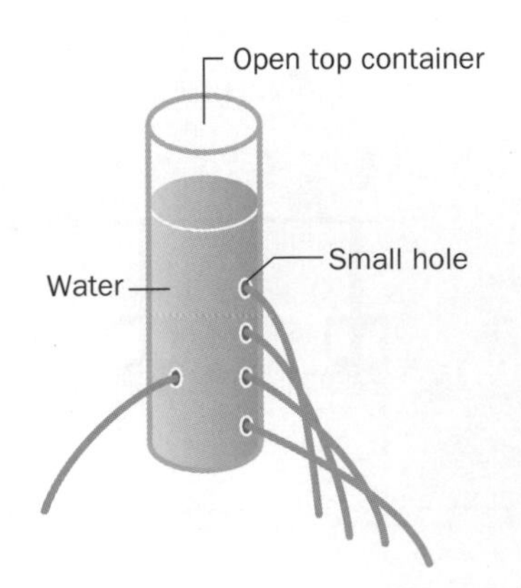

Fig. 8.8 Pressure increases with depth.

8.3 Pressure in a fluid at rest

- **The pressure in a fluid at rest acts equally in all directions.** This can be demonstrated as shown in Fig. 8.7. A water jet emerges from each hole, regardless of where the holc is on the surface, because the water pressure acts in all directions.
- **The pressure in a fluid at rest increases with depth.** Fig. 8.8 shows how this can be demonstrated. A jet of water emerges through each hole. The further the hole below the water surface, the greater the pressure of the jet. For two holes at the same depth, the pressure is the same. Pressure is unchanged along the same level in a fluid at rest.

Pressure of a liquid column

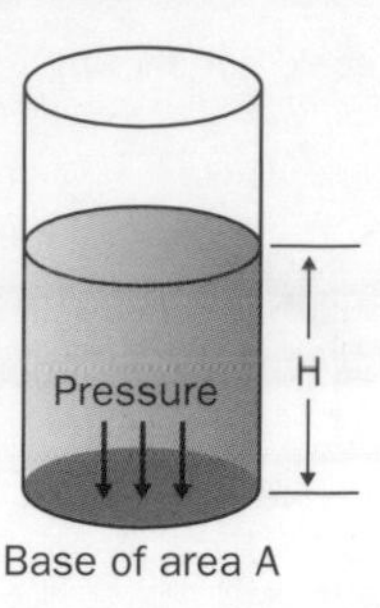

Fig. 8.9 Pressure due to a liquid column.

The pressure, p, at the base of a column of liquid of height H, is given by the equation

$$p = H\rho g$$

where ρ = the liquid density. To prove this equation, consider the liquid column in Fig. 8.9. The volume of liquid in the cylinder = HA, where A is the cross-sectional area of the cylinder.

$\therefore$ the mass of liquid in the cylinder = volume × density = $HA\rho$
the weight of liquid in the cylinder = mass × g = $HA\rho g$

$$\text{the pressure on the base} = \frac{\text{weight of liquid}}{\text{base area}} = \frac{HA\rho g}{A} = H\rho g$$

Worked example 8.3

(a) Calculate the pressure in water at a depth of 10.0 m.

(b) What would be the height of a column of mercury with the same pressure at its base as at a depth of 10.0 m in water?

Density of water = 1000 kg m^{-3}, density of mercury = 13 600 kg m^{-3}

Solution

(a) $p = H\rho g = 10.0 \times 1000 \times 9.8 = 9.8 \times 10^4$ Pa

(b) Rearrange $p = H\rho g$ to give $H = \frac{p}{\rho g} = \frac{9.8 \times 10^4}{13600 \times 9.8} = 0.74$ m

Questions 8.3

1. Calculate the pressure at a depth of 20.0 m in sea water of density 1030 kg m^{-3}.
2. The mean value of atmospheric pressure at sea level is 101 kPa. Calculate the depth of sea water that would give a pressure equal to 101 kPa.

8.4 Measurement of pressure

Blood pressure, gas pressure and tyre pressure are just three examples of the many situations where pressure measurements need to be made.

The U-tube manometer

This is used to measure the **excess pressure** of gas above atmospheric pressure. The height difference, H, between the liquid surfaces in each arm of the U-tube is measured. The gas pressure to be measured acts on the liquid surface in the closed arm. This is equal to the pressure on the same level in the open arm, which is due to the column of liquid of height H and the atmosphere.

$\therefore$ the gas pressure = $H\rho g$ + atmospheric pressure, where ρ is the density of the manometer liquid.
$\therefore$ the 'excess' pressure of the gas above atmospheric pressure = $H\rho g$

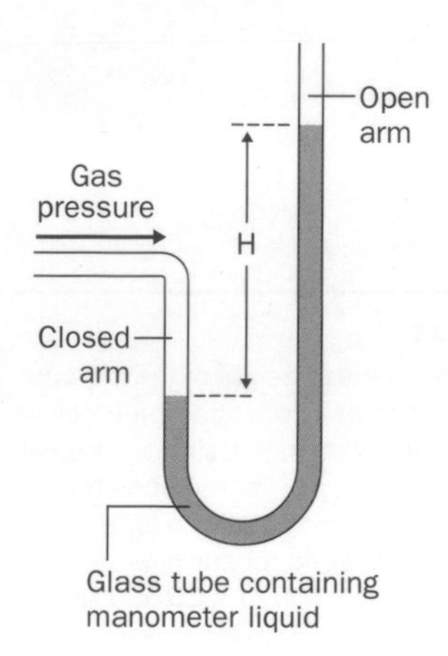

Fig. 8.10 The U-tube manometer.

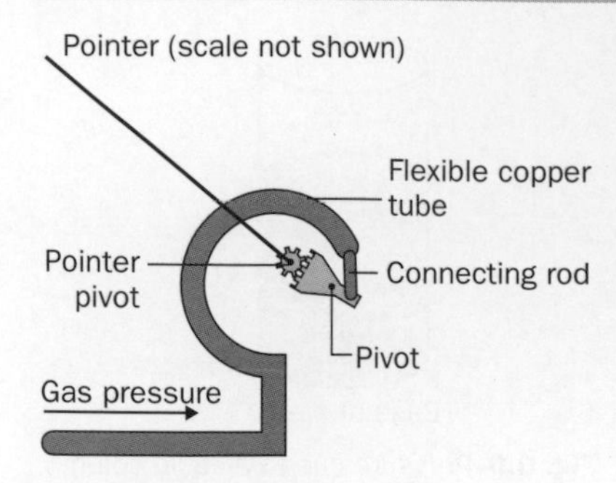

Fig. 8.11 The Bourdon gauge.

The Bourdon gauge

This is more robust than the U-tube manometer. The pressure of the gas forces the copper tube to unbend slightly, which makes the pointer move across the scale. The scale is usually calibrated to measure the absolute pressure of the gas. To avoid confusion, the word 'absolute' is introduced here as the sum of the excess pressure and atmospheric pressure.

The electronic pressure gauge

An electronic pressure gauge contains a piezoelectric pressure sensor that generates a voltage when pressure is applied to it. The sensor is part of the inner surface of a sealed hollow container which the pressure is applied to. If there is a vacuum in the container, an electronic pressure gauge is usually calibrated to give readings of the absolute pressure of the fluid.

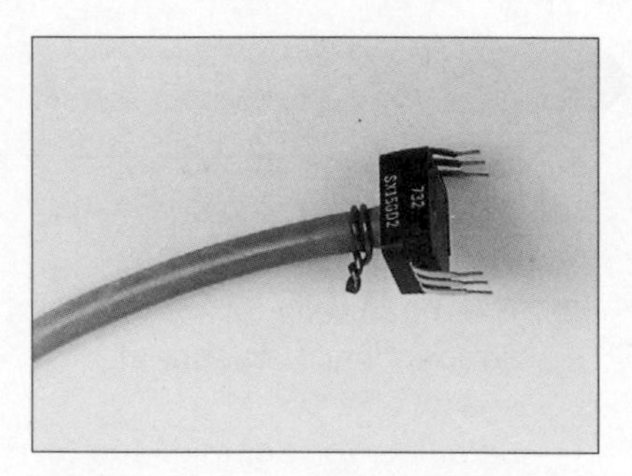

Fig. 8.12 An electronic pressure gauge.

The barometer

Barometers measure atmospheric pressure. At sea level, atmospheric pressure varies from day to day between about 100 and 102 kPa.

- The **Fortin barometer** is an inverted glass tube with its lower end immersed in a mercury reservoir. The space at the top of the tube is a vacuum and exerts no pressure on the mercury column in the tube. The pressure at the base of the mercury column, at the same level as the open surface of mercury in the reservoir, is the same as atmospheric pressure. Hence the pressure of the mercury column is equal to atmospheric pressure. Atmospheric pressure is calculated using the equation

  $$p = H\rho g$$

  where ρ is the density of mercury and H is the height of the mercury column.
- The **aneroid barometer** is more robust than the Fortin barometer. A partially evacuated hollow box with flexible sides is acted upon by

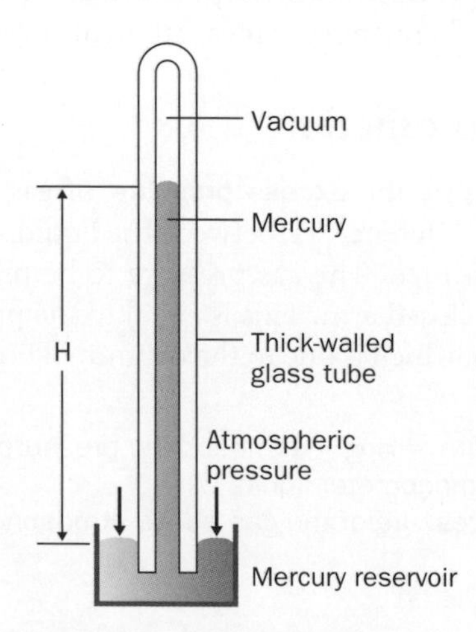

Fig. 8.13 The Fortin barometer.

Notes

1. The average height of the mercury column of a Fortin barometer at sea level is 760 mm. Using $p = H\rho g$ and 13600 kg m^{-3} for the density of mercury gives 101 kPa for the mean value of atmospheric pressure at sea level ($= 0.760$ m $\times$ 13600 kg m^{-3} $\times$ 9.8 N kg^{-1}). This value is referred to as 'standard' pressure.
2. Pressure is sometimes quoted in 'mm of mercury' (mm Hg) instead of pascals. The formula $p = H\rho g$ is used to convert from mm of mercury to pascals.

atmospheric pressure. This operates a pointer via a lever system. When atmospheric pressure changes, the outer face of the box moves and makes the pointer move across the scale of the instrument.

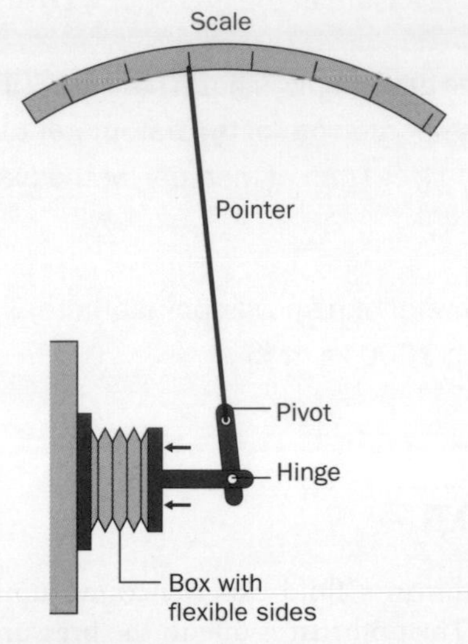

Fig. 8.14 The aneroid barometer.

The sphygmomanometer

A sphygmomanometer is used to measure blood pressure. An inflatable cuff is connected to a manometer and fitted round the arm of patient just above the elbow. A stethoscope is placed over an artery in the lower arm. The cuff is inflated manually to restrict the blood flow in the upper arm. The cuff pressure is gradually released and the manometer reading is noted when the blood flow can be heard again. This pressure is referred to as the **systolic pressure**. The flow of blood becomes quieter when turbulence ceases at a lower pressure, known as the **diastolic pressure**. Blood pressure is usually measured and quoted in 'mm of mercury', even where the mercury manometer is replaced by an electronic pressure gauge.

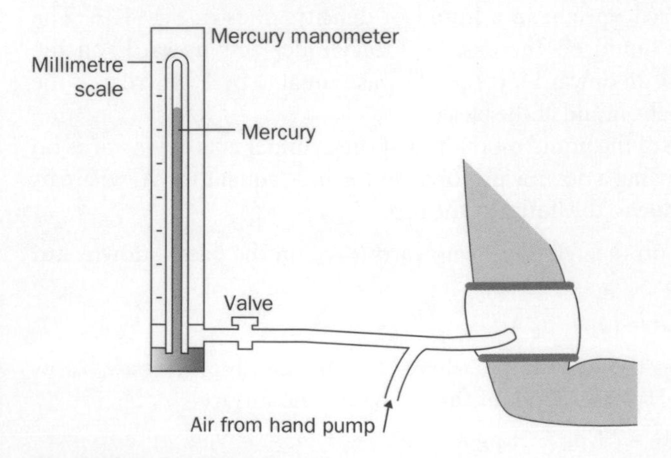

Fig. 8.15 The sphygmomanometer.

Questions 8.4

1. (a) A U-tube manometer containing water is used to measure the pressure of the domestic gas supply, giving a reading of 25 cm. Calculate the excess pressure of the gas supply **(i)** in kPa, **(ii)** as a percentage of standard pressure (101 kPa).

(b) Why is it important that the pressure of the domestic gas supply is not **(i)** too low, **(ii)** too high?

2. (a) In a Fortin barometer, explain why the mercury does not drop out of the barometer tube.

(b) The blood pressure of a healthy young adult is normally 80 mm of mercury for the diastolic pressure and 120 mm of mercury for the systolic pressure.

(i) Calculate each of these pressures in kPa.

(ii) Why is mercury rather than water used in the manometer of a sphygmomanometer?

(density of water = 1000 kg m^{-3}, density of mercury = 13600 kg m^{-3})

8.5 Flotation

An object in equilibrium in a fluid experiences an upthrust equal and opposite to its weight. The upthrust is due to the pressure of the fluid on the object.

1. If the upthrust is greater than the weight, the object will rise (e.g. a ball released under water).
2. If the upthrust is equal to the weight, the object will float (e.g. a boat).
3. If the upthrust is less than the weight, the object will sink (e.g. an overloaded raft).

Archimedes' Principle

The upthrust on an object in a fluid is equal to the weight of fluid displaced by the object.

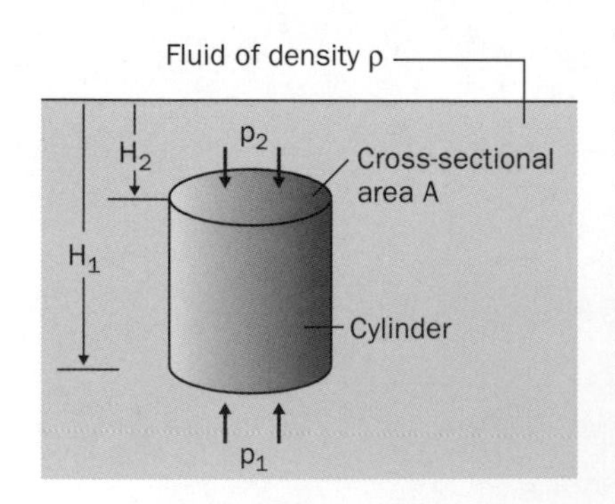

Fig. 8.16 Archimedes principle.

To prove this statement, consider a uniform cylinder of cross-sectional area, A, immersed upright in a liquid of density, ρ, as in Fig. 8.16. The pressure of the liquid on the base of the cylinder acts upwards on the cylinder, causing an upward force on the base equal to p_1 A, where p_1 is the pressure due to the liquid at the base.

The pressure of the liquid on the top of the cylinder acts downwards on the cylinder, causing a downward force on the base equal to p_2 A, where p_2 is the pressure due to the liquid at the top.

∴ the upthrust on the cylinder = upward force on the base – downward force on the top

$$= p_1A - p_2A = (p_1 - p_2)A$$

Since $p_1 = H_1\rho g$ and $p_2 = H_2\rho g$, where H_1 is the depth of the base below the surface and H_2 is the depth of the top below the surface,

∴ upthrust $U = (H_1\rho g - H_2\rho g)A, = (H_1 - H_2)A\rho g, = V\rho g$
since V, the volume of the object $= (H_1 - H_2)A$

The mass of liquid displaced by the object, m, = $V\rho$

hence $V\rho g = mg$ = weight of liquid displaced

∴ upthrust, U = weight of liquid displaced

Floating and sinking

A rowing boat becomes lower and lower in the water as more and more people get into it. It displaces more and more water as it floats lower in the water. The upthrust on it therefore increases, enabling the extra weight to be supported. However, if the water level reaches the upper edge of the hull, no further upthrust is possible because the boat cannot displace any more water. Any additional weight added to the boat at this stage would sink it.

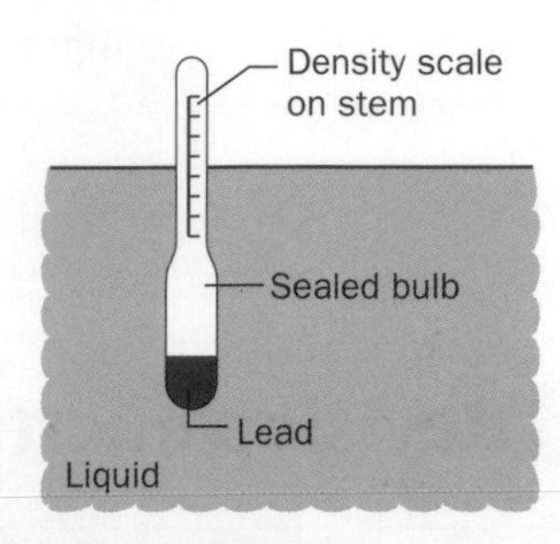

Fig. 8.17 The hydrometer.

The hydrometer

This is used to measure the density of a liquid. It consists of a weighted bulb of air with a stem which is vertical when the hydrometer is floating at rest. The scale is linear provided the stem has a uniform diameter. The scale may be calibrated using liquids of known density.

Worked example 8.4

An empty, flat-bottomed iron barge of weight 450 kN floats with the base of its hull 1.5 m below the water line. Calculate

(a) the area of the base of the boat, (b) the maximum weight of cargo it can carry if it can float no more than 2.0 m lower in the water.

The density of water = 1000 kg m^{-3}

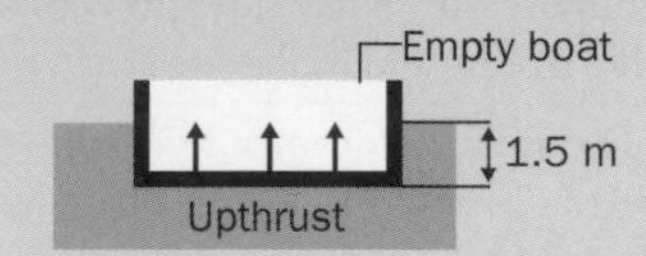

Fig. 8.18

Solution

(a) Upthrust = weight of boat = 450 kN ∴ Weight of water displaced = 450 kN.

Since weight of water displaced = $HA\rho g$, where H is the depth of the boat in the water and A is the base area,

$$HA\rho g = 450\,000 \therefore A = \frac{450\,000}{H\rho g} = \frac{450\,000}{1.5 \times 1000 \times 9.8} = 31\text{ m}^2$$

(b) At 2.0 m lower, the additional depth $\Delta H = 2.0$ m

∴ the additional weight of water displaced = $\Delta HA\rho g = 2.0 \times 31 \times 1000 \times 9.8 = 6.0 \times 10^5$ N

Questions 8.5

1. A person of weight 700 N floats on an air bed of dimensions 2.0 m × 0.60 m in a swimming pool. Calculate **(a)** the pressure on the underside of the air bed, **(b)** the depth of the air bed in the water. The density of water = 1000 kg m^{-3}.
2. A flat-bottomed ferry boat floats at rest in sea water of density 1050 kg m^{-3} with the underside of its hull 1.20 m below the water line. When it carries its maximum permitted load of 40 000 N, it floats 0.05 m lower in the water. Calculate **(a)** the weight of the ferry boat when unloaded, **(b)** the depth of the hull below the water line in lake water of density 1000 kg m^{-3} when fully loaded.

Summary

- **Pressure** = force per unit area acting perpendicular to the area.
- **Equation for pressure** $p = \frac{F}{A}$
- **Pressure of a liquid column** of height H = Hρg, where ρ is the density of the liquid.
- **Pressure-measuring devices:**
  the U-tube manometer, the Bourdon gauge, the electronic pressure gauge, the Fortin barometer, the aneroid barometer (for measuring atmospheric pressure), the sphygmomanometer (for measuring blood pressure).
- **The upthrust** on an object in a fluid is equal to the weight of fluid displaced.

Revision questions

8.1. A rectangular paving stone has dimensions 450 mm × 300 mm × 25 mm and has a density 2600 kg m^{-3}. Calculate
(a) the weight of the stone,
(b) the pressure of the stone on the ground when it rests **(i)** horizontally on its flat side, **(ii)** on its shortest edge with this edge horizontal and its flat surface in a vertical plane.

8.2. The ram piston of a hydraulic press is in a cylinder of diameter 45 cm and is capable of applying a force of 120 kN.
(a) Calculate the pressure in the cylinder when the ram piston exerts maximum force.
(b) The effort is applied to a narrow piston of diameter 2.5 cm. Calculate the effort necessary to make the ram piston exert a force of 120 kN.

8.3. (a) Atmospheric pressure at sea level is 101 kPa on average. The density of air at sea level is 1.2 kg m^{-3}. Calculate the height of an atmosphere of constant density equal to 1.2 kg m^{-3}.
(b) The atmosphere extends to a height of more than 100 km above the Earth. What does your calculation in (a) tell you about the atmosphere?

8.4. (a) The pressure of a person's lungs can be measured by blowing into one side of a long, water-filled manometer, as in Fig. 8.19. In a test with an adult, the height difference in the water levels was about 2.0 m. Calculate the lung pressure, in pascals, of this person. The density of water = 1000 kg m^{-3}.
(b) Why is it very difficult to breathe through a long hollow tube under water at a depth below about 2.0 m?

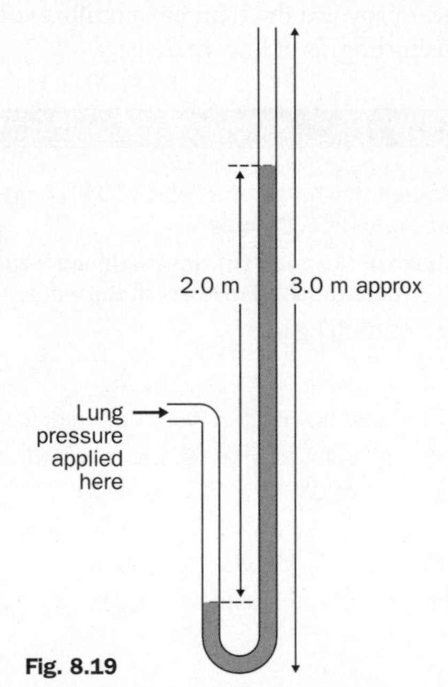

Fig. 8.19

8.5. (a) With the aid of a labelled diagram, describe the construction of a Fortin barometer and explain how it is used to measure atmospheric pressure.
(b) When the blood pressure of a person is measured, the cuff should normally be wrapped round the upper arm above the elbow and level with the heart. How would the blood pressure readings be affected if the arm were raised so the cuff was significantly above the heart?

8.6. (a) Explain why a submerged submarine rises to the surface when compressed air is used to force water from its ballast tanks.
(b) In an experiment to test a loaded test tube as a hydrometer, the test tube was floated upright in water so 40 mm of its length was above water and 60 mm of its length below water. It was then placed in a liquid of unknown density and its length above water was measured at 32 mm. Calculate the density of the liquid.

Mechanics

Part 3 introduces you to key principles of mechanics, including equilibrium, the link between force and motion and the link between work, energy and power. Experiments are provided to reinforce understanding and to link with applications such as friction and efficiency. The mathematical level in this part increases from the basic skills developed in Parts 1 and 2 to a level which should enable you to cope with the mathematical demands in all other parts of this book.

Forces in Equilibrium

Objectives

After working through this unit, you should be able to:

- ▶ represent a force geometrically or numerically
- ▶ combine two or more forces
- ▶ resolve a force into perpendicular components
- ▶ describe the condition for a point object to be in equilibrium
- ▶ draw a free-body diagram of the forces acting on a body
- ▶ state and apply the principle of moments to a body in equilibrium
- ▶ understand and use the concept of centre of gravity

Contents

9.1 Force as a vector

Vectors and scalars

An air traffic controller needs to know the position, height, speed and direction of motion of an aircraft in order to monitor its flight path. Each piece of information is important, including the direction of motion. The position and height of an aircraft gives its **displacement**, which is its distance and direction from the air traffic controller. Its speed and direction of motion define its **velocity**.

Displacement and velocity are examples of **vector** quantities because they have a direction as well as a magnitude. Distance and speed are examples of **scalar** quantities because they have a magnitude only. Any vector may be represented by an arrow of length in proportion to the magnitude of the vector and in the appropriate direction.

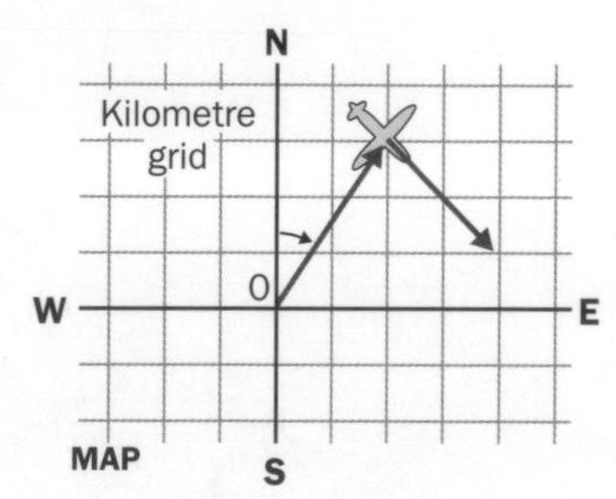

Velocity = 60 ms^{-1} south-east
Displacement from 0 = 3.5 km 36°E of due North (height not shown)

Fig. 9.1 Vectors and scalars.

- A **vector** is defined as any physical quantity that has a direction. Examples include weight, force, displacement (i.e. distance in a given direction), velocity, acceleration and momentum.
- A **scalar** is any physical quantity that has no direction. Examples include mass, energy and speed.

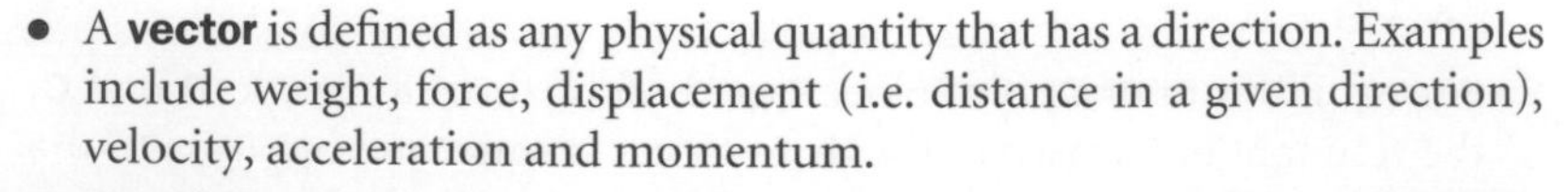

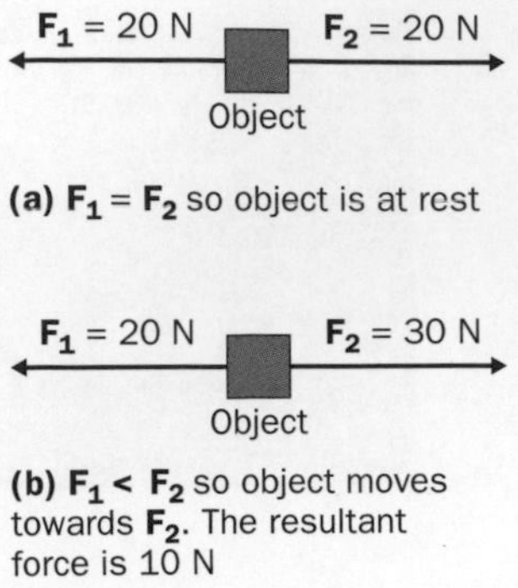

F_1 = 20 N Object F_2 = 20 N

(a) $\mathbf{F_1} = \mathbf{F_2}$ so object is at rest

F_1 = 20 N Object F_2 = 30 N

(b) $\mathbf{F_1} < \mathbf{F_2}$ so object moves towards $\mathbf{F_2}$. The resultant force is 10 N

Fig. 9.2 Forces in opposite directions.

Force as a vector

Force is measured in units called **newtons** (N).The weight of a mass of 1 kg is approximately 10 N on the surface of the Earth. An object acted on by two forces will be at rest if the two forces are equal and opposite, as in Fig. 9.2a. However, if one of the forces is larger than the other, their combined effect is the difference between the two forces, as in Fig. 9.2b.

Each force in Fig. 9.2 can be represented by a vector (i.e. an arrow of length proportional to the force in the appropriate direction). From now on in this book, a bold symbol will denote a vector. For example, in Fig. 9.2b, $\mathbf{F_1}$ denotes force vector F_1 and $\mathbf{F_2}$ denotes force vector F_2.

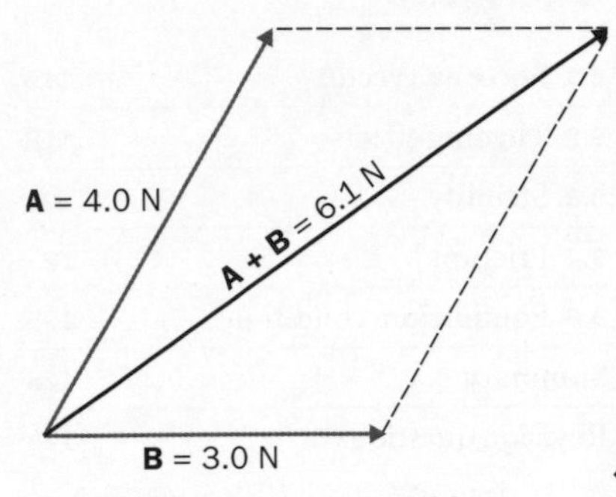

Fig. 9.3 The parallelogram of forces.

The parallelogram of forces

Fig. 9.3 shows how to work out the effect of two forces, **A** and **B**, acting on a point object but not along the same line. Each force is represented by a vector, such that the two vectors form adjacent sides of a parallelogram. The combined effect of the two forces, the **resultant**, is the diagonal of the parallelogram between the two vectors. The resultant is therefore given by force vector **B** added onto force vector **A**. Fig. 9.4 shows how to balance out the resultant of **A** and **B** with an equal and opposite force, **C**.

Questions 9.1a

1. A point object is acted on by a force of 5.0 N acting due East and a force of 12.0 N. Use the parallelogram method to determine the magnitude and direction of the resultant force when the 12.0 N force acts **(a)** due East, **(b)** due West, **(c)** due North, **(d)** 60° North of due East, **(e)** 60° North of due West.
2. A point object is acted on by forces of 6.0 N and 8.0 N. Determine the magnitude of the resultant of these two forces if the angle between them is **(a)** 90°, **(b)** 45°.

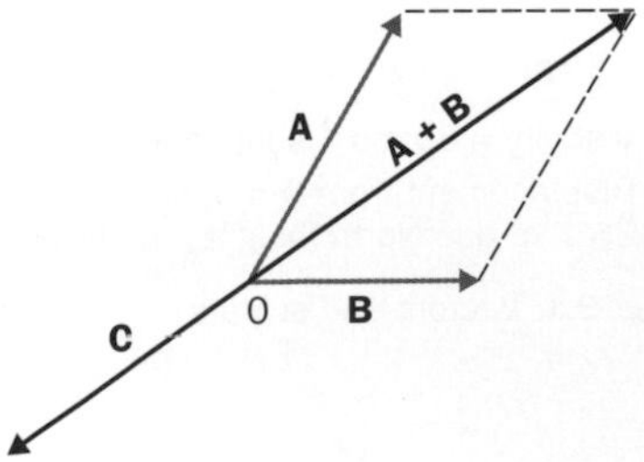

A A + B C 0 B

Fig. 9.4 Equilibrium.

The condition for equilibrium of a point object

In general, a point object acted on by two or more forces is in equilibrium (i.e. at rest) if the resultant of the forces is zero.

1. For equilibrium of a point object acted on only by two forces, **A** and **B**, the two forces must be equal and opposite to each other. The resultant of the two forces must be zero.

   $$\mathbf{A} + \mathbf{B} = 0$$

2. For equilibrium of a point object acted on by three forces, **A**, **B** and **C**, the resultant of any two of the three forces is equal and opposite to the third force. For example, $\mathbf{A} + \mathbf{B} = -\mathbf{C}$. Rearranging this equation gives

   $$\mathbf{A} + \mathbf{B} + \mathbf{C} = 0$$

The three force vectors add together to give a zero resultant if the object is in equilibrium. This can be represented as a triangle, as in Fig. 9.5.

3. In general, a point object is in equilibrium if the forces acting on it form a vector diagram which is a closed polygon. This rule is called the **closed polygon** rule.

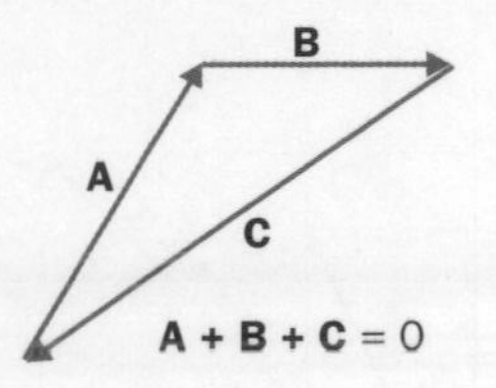

Fig. 9.5 Equilibrium of a point object.

Resolving a force into perpendicular components

This is a mathematical technique which enables equilibrium situations to be analysed without scale diagrams. Sketch diagrams are nevertheless useful to visualise situations.

Consider a force, **A**, acting at point, O, of an x–y coordinate system, as in Fig. 9.6a. The magnitude of **A** is represented by A and the direction of **A** is at angle θ to the x-axis. This force may be resolved into two perpendicular components

1. $A_x = A \cos \theta$ along the x-axis.
2. $A_y = A \sin \theta$ along the y-axis.

Force A may be written in terms of its components in the form

$$\mathbf{A} = A\cos\theta\mathbf{i} + A\sin\theta\mathbf{j}$$

where i and j are vectors of unit length along the x-axis and the y-axis respectively.

Using Pythagoras' theorem gives

$$A = \sqrt{(A_x^2 + A_y^2)}$$

and using the rules of trigonometry gives

$$\tan\theta = \frac{A_y}{A_x}$$

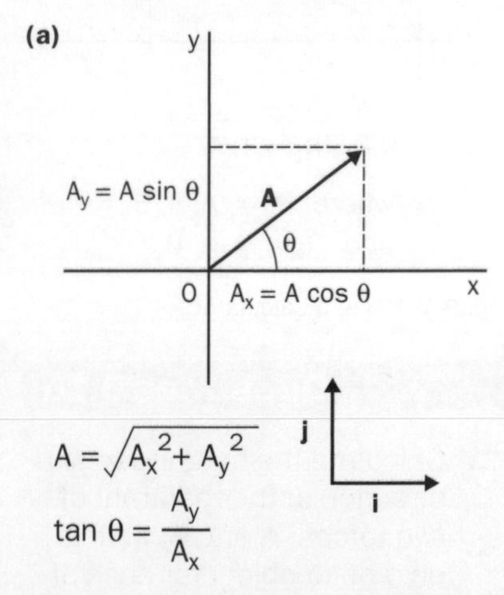

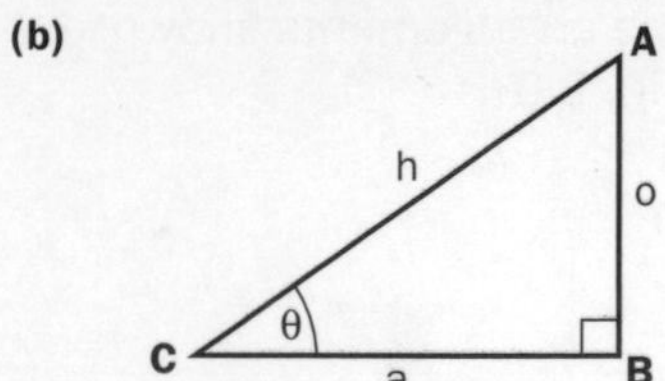

Fig. 9.6 Resolving a force.

Notes

1. Pythagoras' theorem states that for the right-angled triangle ABC in Fig. 9.6b, $AB^2 + BC^2 = AC^2$
2. The following trigonometric functions are defined from this triangle as follows,

$$\sin\theta = \frac{AB}{AC}\left(=\frac{o}{h}\right) \quad \cos\theta = \frac{BC}{AC}\left(=\frac{a}{h}\right) \quad \tan\theta = \frac{AB}{BC}\left(=\frac{o}{a}\right)$$

where o is the opposite side and a is the adjacent side to θ, and h is the hypotenuse.

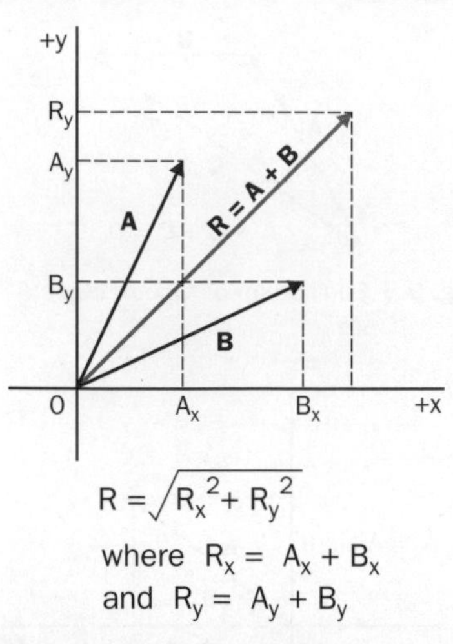

$$R = \sqrt{R_x^2 + R_y^2}$$

where $R_x = A_x + B_x$
and $R_y = A_y + B_y$

Fig. 9.7 Using a calculator.

How to calculate the resultant of two or more forces

Consider a point object, O, acted on by two or more forces, **A**, **B**, **C**, etc, as shown in Fig. 9.7. The resultant of these forces can be calculated by following these steps:

1. Resolve each force into perpendicular components along the x-axis and the y-axis.
2. For each axis, add the components, taking account of + or − directions, to give the component of the resultant along each axis.
3. Use Pythagoras' equation to calculate the magnitude of the resultant, and use the trigonometry equations to calculate the direction of the resultant.

Questions 9.1b

1. Calculate the magnitude and direction of the resultant of two forces, **A** and **B**, acting on a point object for each of the arrangements shown in Fig. 9.9.

(a)

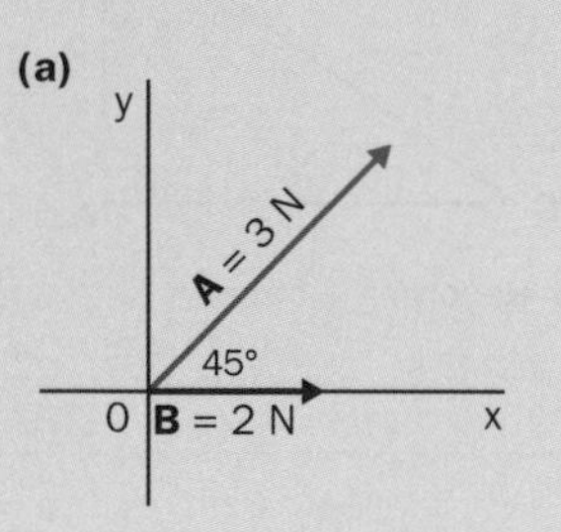

(b)

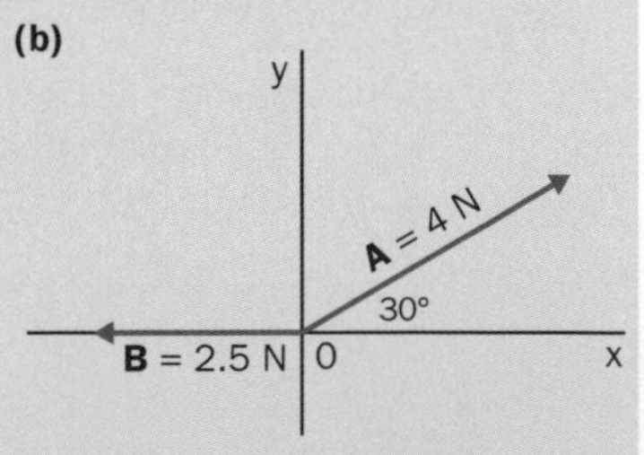

Fig. 9.9

2. For each arrangement shown in Fig. 9.9, state the magnitude and direction of a third force, **C**, acting on O which would cancel out the resultant of **A** and **B**.

Worked example 9.1

A point object, O, is acted on by three forces, A, B and C, as shown in Fig. 9.8. Calculate the magnitude and direction of the resultant of these three forces.

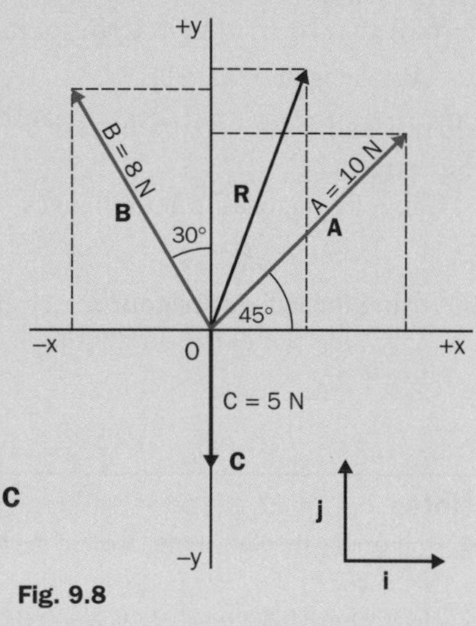

Fig. 9.8

Solution

Resolving the three forces into components along the i-axis and the j-axis gives:

$\mathbf{A} = 10 \cos 45\mathbf{i} + 10 \sin 45\mathbf{j}$

$\mathbf{B} = -8 \sin 30\mathbf{i} + 8 \cos 30\mathbf{j}$

$\mathbf{C} = -5\mathbf{j}$

Therefore the resultant, **R**, = **A** + **B** + **C**

$= (10 \cos 45 - 8 \sin 30)\,\mathbf{i}$

$+ (10 \sin 45 + 8 \cos 30 - 5)\,\mathbf{j}$

thus $R = 3.1\,\mathbf{i} + 9.0\,\mathbf{j}$

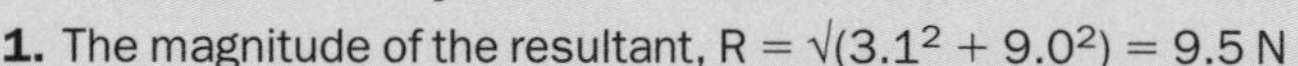

1. The magnitude of the resultant, $R = \sqrt{(3.1^2 + 9.0^2)} = 9.5$ N
2. The angle between the resultant and the x-axis, θ, is given by

$$\tan\theta = \frac{R_y}{R_x} = \frac{9.0}{3.1} = 2.90$$

Hence θ = 71°

9.2 Turning effects

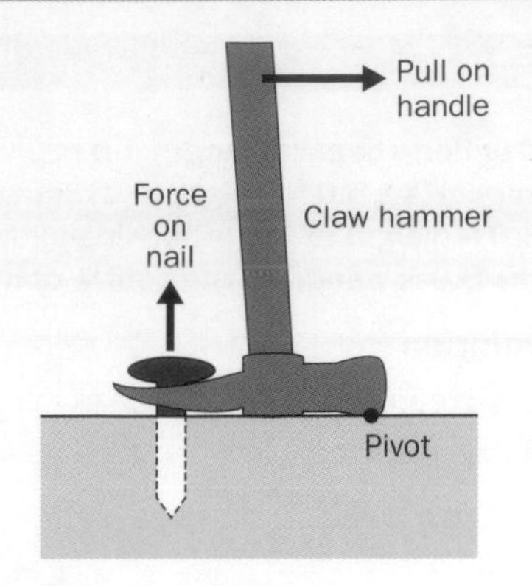

Fig. 9.10 Applying leverage.

When you need to use a spanner to unscrew a nut, the longer the handle of the spanner, the easier the task will be. The same applies to a claw hammer if ever you need to remove a nail from a piece of wood. The longer the handle, the greater the leverage about the turning point.

The moment of a force about a point is defined as the magnitude of the force × the perpendicular distance from its line of action to the point.

The unit of a moment is the newton metre (Nm). The term **torque** is used to describe the moment of a force about an axis.

For example, if a force of 40 N is applied to a spanner at a perpendicular distance of 0.20 m from the turning point, the moment of the force about that point is equal to 8.0 Nm (= 40 N × 0.20 m). The same moment could be achieved by applying a force of 80 N at a perpendicular distance of 0.1 m from the pivot (= 80 N × 0.1 m). Now you can see why a lesser force gives the same leverage if it acts at a greater distance. Prove for yourself that a force of 16 N acting at a perpendicular distance of 0.5 m from the turning point gives the same moment as 40 N at 0.20 m.

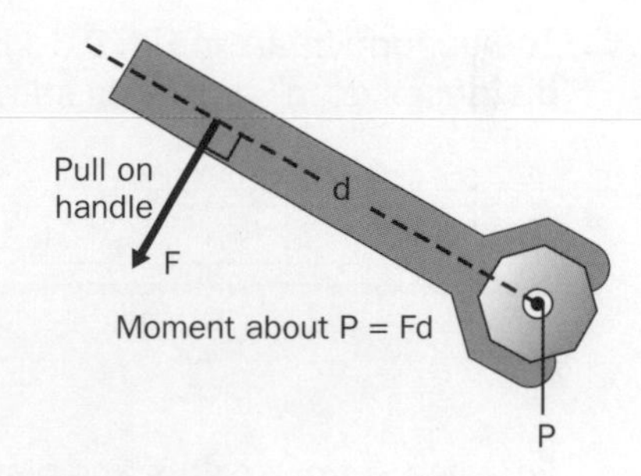

Fig. 9.11 Moment = force x perpendicular distance.

The principle of moments

Any object that is not a point object can be turned when one or more forces act on it. For example, a door turns on its hinges when it is given a push. The term *body* is used for any object that is not a point object.

- A body acted on by only one force cannot be in equilibrium and will turn if the force does not act through its centre of gravity.
- A body acted on by two or more forces may or may not turn, depending on the position and direction of the forces acting on it. For example, a balanced see-saw carrying a child either side of the pivot will become unbalanced if one of the children climbs off. For balance, a child of weight 300 N would need to be nearer the pivot than a child of weight 250 N. The condition for a balanced see-saw is that the moment of the child on one side about the pivot should equal the moment of the other child about the same pivot. For example, if the 250 N child is 1.20 m from the pivot (moment = 250 × 1.20 = 300 Nm), the other child would need to be 1.0 m from the pivot (moment = 300 N × 1.0 m) for balance.

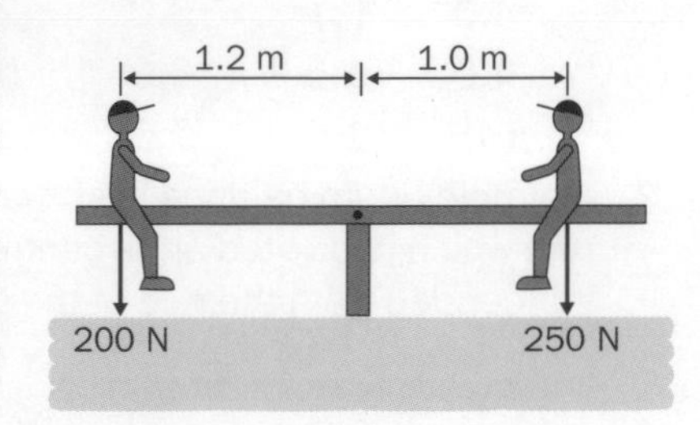

Fig. 9.12 A balanced see-saw.

The balanced see-saw is an example of a more general rule, the Principle of Moments, which applies to any body in equilibrium.

The Principle of Moments states that for a body in equilibrium, the sum of the clockwise moments about any point is equal to the sum of the anticlockwise moments.

Worked example 9.2

A uniform beam of length 4.0 m pivoted horizontally about its centre, supports a 5.0 N weight at a distance of 1.5 m from the centre, and an 6.0 N weight at the other side of the centre. Calculate the distance from the 6.0 N weight to the centre of the beam.

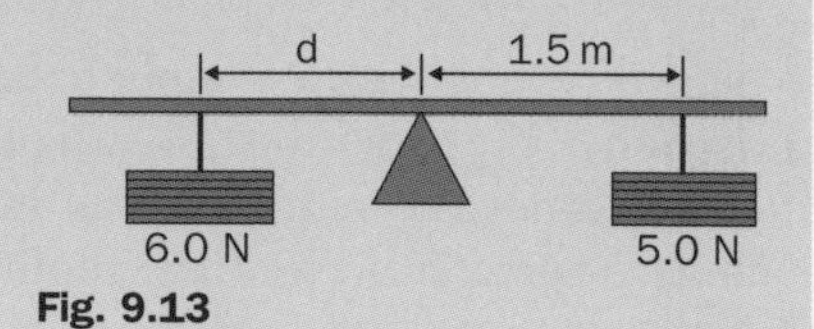

Fig. 9.13

Solution

Let d represent the distance from the centre to the 6.0 N weight.

Applying the principle of moments about the centre gives $6.0\,d = 5.0 \times 1.5$

Hence $d = \dfrac{5.0 \times 1.5}{6.0} = 1.25$ m

Questions 9.2

1. The uniform beam in Fig. 9.14 is pivoted at its centre and supports three weights, W_1, W_2 and W_3, at distances d_1, d_2 and d_3 as shown.

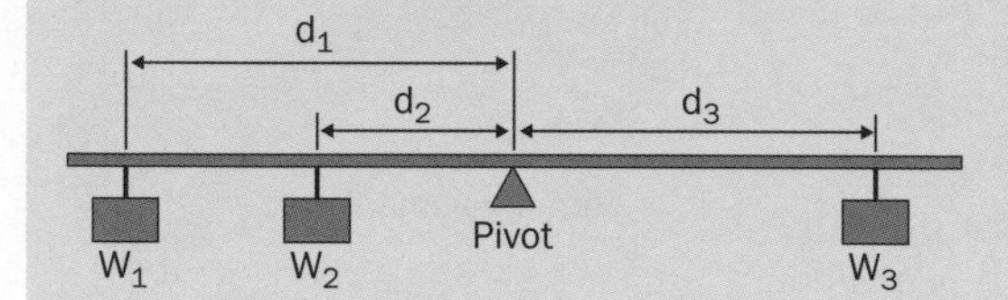

Fig. 9.14

For each set of weights and distances in the Table, calculate the value of the missing quantity necessary to maintain the beam in equilibrium.

W_1/N	d_1/m	W_2/N	d_2/m	W_3/ N	d_3/m
6.0	2.0	4.0	1.0	8.0	**(a)**
4.0	**(b)**	3.0	2.0	8.0	1.5
4.0	1.5	6.0	**(c)**	4.5	2.0
(d)	1.0	2.0	2.0	4.0	1.5

2. A simple beam balance is shown in Fig. 9.15. The scale pan is suspended from one end of a metre rule which is pivoted at its centre. The rule is balanced by adjusting the position of a 2.0 N weight suspended from the rule at the other side of the pivot to the scale pan.

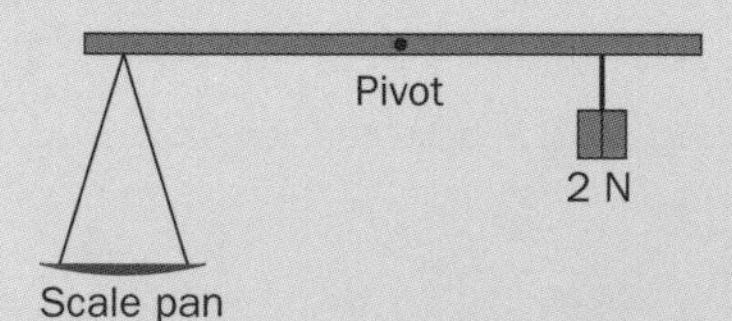

Fig. 9.15

(a) When the scale pan is empty, the 2.0 N weight must be positioned 150 mm from the centre of the rule to keep the rule horizontal. Calculate the weight of the scale pan.

(b) An object, X, of unknown weight is placed on the scale pan and the 2.0 N weight is moved to a distance of 360 mm from the centre to keep the rule horizontal. Calculate the weight of X.

9.3 Stability

Mass and weight

If you want to lose weight, go to the Moon where the force of gravity is one sixth of the Earth's gravity. Unfortunately, you won't become slimmer because your mass will be unchanged.

- The **mass** of an object is the amount of matter it possesses. The unit of mass is the kilogram (kg), defined by means of a standard kilogram at BIPM (Bureau International des Poids et Mesures), Paris.
- The **weight** of an object is the force of gravity acting on it. The unit of weight is the newton (N).
- The **force of gravity per unit mass**, *g*, at the surface of the Earth, is 9.8 N kg^{-1}. The value of *g* changes slightly with position on the Earth's surface, ranging from 9.78 N kg^{-1} at the equator to 9.81 N kg^{-1} at the poles. The cause of this variation is due partly to the shape of the Earth not being exactly spherical and partly to the effect of the Earth's rotation. Strictly, this latter effect is not due to the force of gravity but arises due to the circular motion of an object at the equator. The force of gravity on an object on the Moon, at 1.6 N kg^{-1}, is much smaller than on an object on the Earth because the Moon is smaller in size than the Earth.

For an object of mass m, its weight, W, can be calculated from the equation

$$W = mg$$

Stable and unstable equilibrium

At a bowling alley, have you noticed how easily a pin falls over? Each pin is designed to be top-heavy with a small base. Fig. 9.16 shows a bowling pin in comparison with 'Wobbly Walter', a bottom-heavy child's toy that rights itself when knocked over.

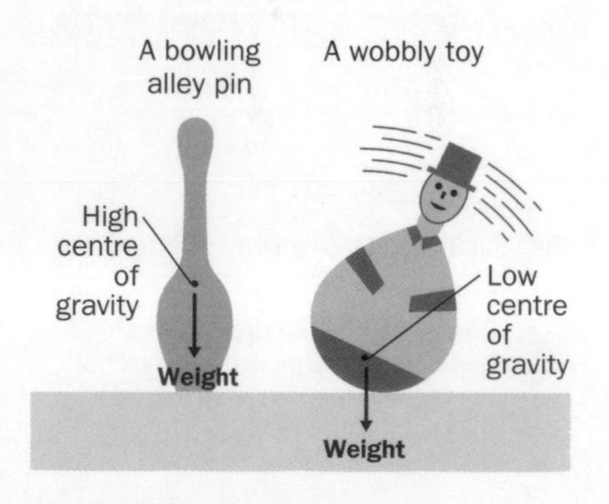

Fig. 9.16 Equilibrium.

- The child's toy is an example of an object in **stable equilibrium** – it returns to equilibrium when knocked over because its centre of gravity is always over its base.
- The standing bowling pin is an example of an object in **unstable equilibrium** – it moves away from equilibrium when released from a tilted position.
- The bowling pin lying on its side is an example of an object in **neutral equilibrium** – it stays wherever it is moved to, neither returning to or moving away from where it was.

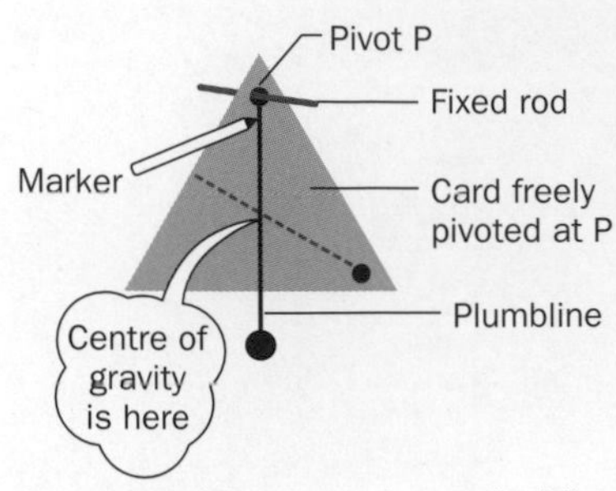

Use a plumbline to mark a vertical line through the pivot on the card. Repeat for a different pivot. The centre of gravity is where the lines intersect.

Fig. 9.17 Locating the centre of gravity of a flat card.

Centre of gravity

The centre of gravity of an object is the point where its weight can be considered to act.

- An object freely suspended from a point and then released will come to rest with its centre of gravity directly beneath the point. Fig. 9.17 shows how to find the centre of gravity of a flat object.
- A tall object tilted too far will topple over if released. This happens if it is tilted beyond the point where its centre of gravity is directly above the point about which it turns, as in Fig. 9.18. If released beyond this position, the moment of the weight about the point of turning causes the object to overbalance.

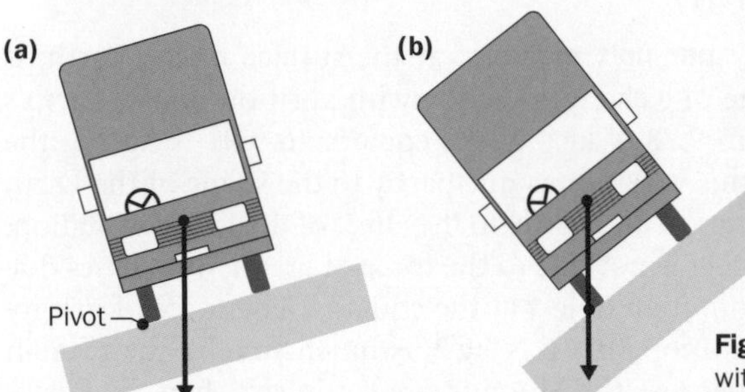

Fig. 9.18 Stability **(a)** tilting without toppling, **(b)** on the point of toppling.

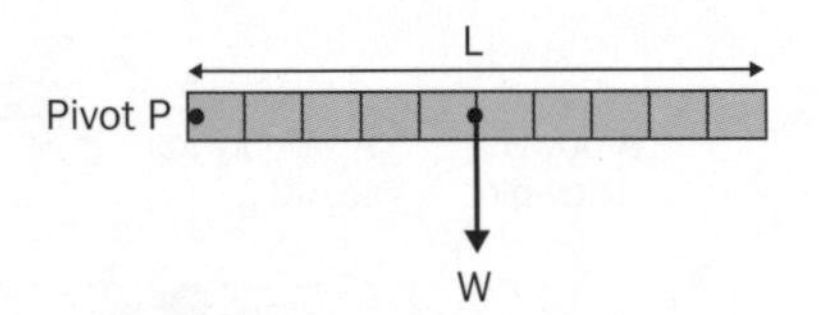

Fig. 9.19 Centre of gravity of a rule.

See www.palgrave.com/foundations/breithaupt for a proof that the centre of gravity of a uniform beam is at its midpoint.

- The moment of an object due to its weight, W, about any point = Wd, where d is the perpendicular distance from the point to the vertical line through its centre of gravity. For example, Fig. 9.19 shows a rule pivoted at one end, held horizontal by a vertical string attached to the other end. The moment of the rule's weight about the pivot can be proved to be equal to WL/2, where L is its length. Therefore its centre of gravity is at a distance L/2 from the pivot.
- An object may be balanced by a single vertical force applied at its centre of gravity. For example, it ought to be possible to support a plate on the end of a vertical rod, provided the plate is placed with its centre on the end of the rod. Another example is when someone carries a ladder in a horizontal position; the task is much easier if the ladder is supported at its centre of gravity, because the ladder is balanced at this point.

Worked example 9.3

A road gate consists of a 5.0 m uniform steel tube of weight 120 N which has a 480 N counterweight fixed to one end, as shown in Fig. 9.20. It is pivoted about a horizontal axis which passes through the beam 0.55 m from the centre of the counterweight.

(a) Calculate the position of the centre of gravity of the road gate.

(b) Calculate the magnitude and direction of the force that must be applied to the counterweight to keep the beam horizontal.

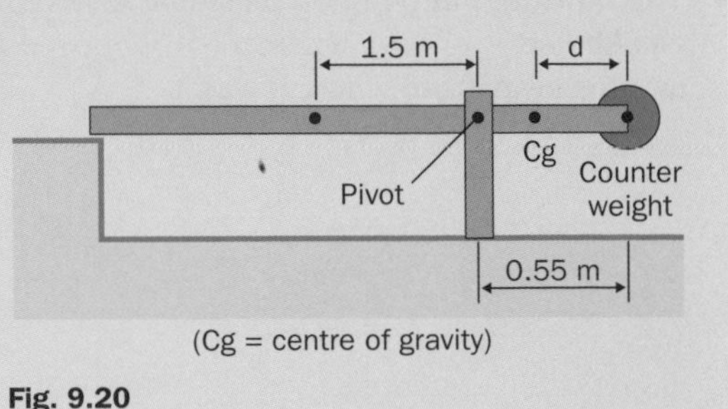

Fig. 9.20

Worked example 9.3 continued

Solution

(a) Let d = the distance from the centre of gravity C_g of the road gate to the counterweight. The distance from the centre of gravity of the counterweight to the centre of gravity of the tube = 2.5 m. The moment of the tube about the counterweight = 120 N × 2.5 m = 300 N m anticlockwise. This is the same as the moment of the whole gate about the counterweight = (480 + 120) d anticlockwise.

Hence 600 d = 300 Nm $\therefore$ d = 300 ÷ 600 = 0.5 m.

(b) Because C_g is between the counterweight and the pivot, the road gate will turn clockwise if no additional force is applied to it. Let F = the upward vertical force on the counterweight needed to keep the road gate horizontal.

Applying the principle of moments about the pivot, the moment of the road gate about the pivot
= (480 + 120) × 0.05 = 30 N m clockwise.

The moment of force F about the pivot = 0.55 F anticlockwise.

Hence 0.55 F = 30

$\therefore$ F = 30 ÷ 0.55 = 55 N

9.4 Friction

Friction exists between any two solid surfaces that slide over one another. Walking and running would be very difficult without friction (e.g. on ice). However, friction causes undesirable wear and heating in machines and engines. Friction always acts on a surface in the opposite direction to the movement of the surface. Oil between two solid surfaces reduces the friction because it keeps them apart and reduces the points of direct contact.

Measuring friction

The force of friction between two flat surfaces can be measured by:

- measuring the force needed to pull a block across a fixed horizontal surface. One of the surfaces to be tested is fixed to the underside of the

Fig. 9.23 Forces on a running shoe.

Questions 9.3

1. A placard consists of a uniform pole of length 4.0 m and of weight 30 N which has a square board of weight 20 N fixed to one end, as shown in Fig. 9.21. Calculate the distance from the other end of the pole to the centre of gravity of the placard.

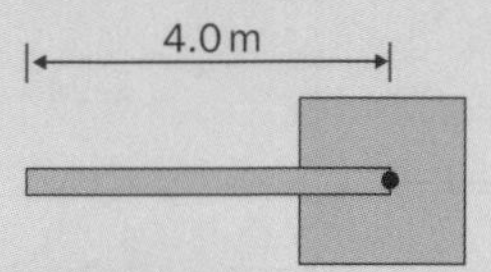

Fig. 9.21

2. An advertising sign consists of a uniform board as in Fig. 9.22. The total mass of the board is 15 kg.

(a) Calculate the weight of the rectangular section of the board.

(b) Calculate the weight of the square section of the board.

(c) Calculate how high above the base the centre of gravity of the whole board is.

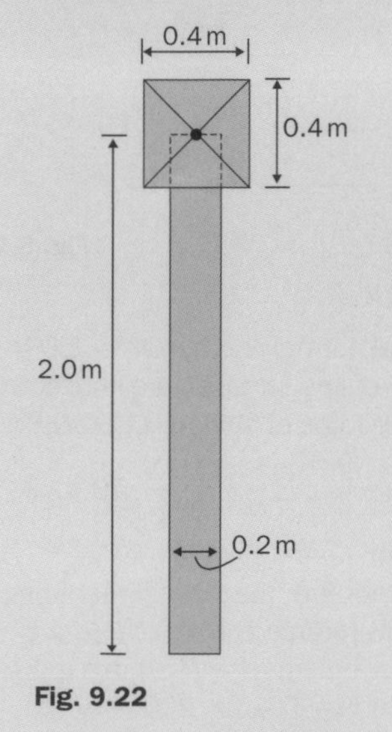

Fig. 9.22

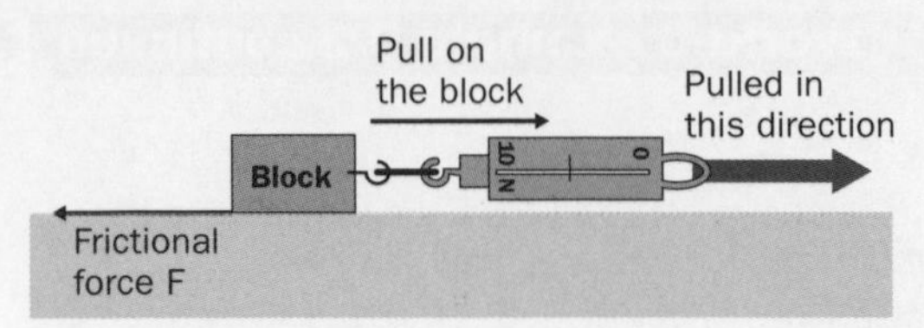

Fig. 9.24 Measuring friction.

block. The other surface is the fixed surface. If the applied force is gradually increased from zero using a spring balance, as in Fig. 9.24, the block does not slide until the applied force reaches a certain value. Once the block starts to move, the applied force becomes slightly less. At the point of sliding, the force needed to make the block move is equal to the maximum possible frictional force between the two surfaces, referred to as the limiting value of the frictional force

- using an inclined plane as in Fig. 9.25. The angle of the incline is increased until the block on the plane is at the point of sliding down the incline. The limiting value of the frictional force is equal to the component of the block's weight, **W sin** θ, acting down the incline.

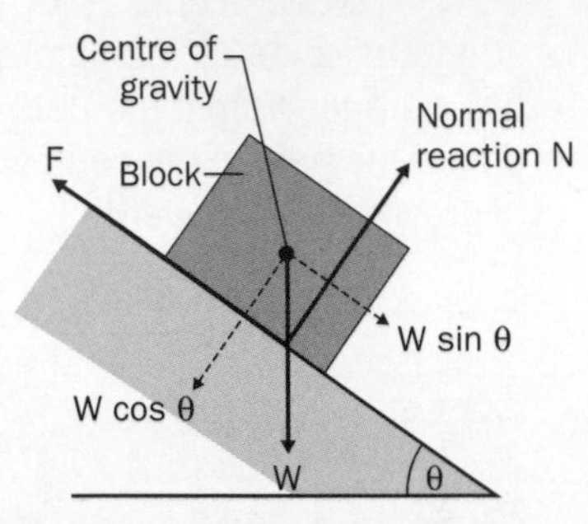

Fig. 9.25 Using an inclined plane.

The coefficient of static friction, μ

The arrangement in Fig. 9.24 may be used to prove that the limiting value of the frictional force, F, is proportional to the weight of the block. Since the weight of the block on a horizontal surface is equal and opposite to the normal reaction, N, of the surface on the block, the frictional force F is proportional to N. In other words, the ratio F / N is a constant for the two surfaces. This ratio is referred to as the coefficient of static friction, μ.

$$\mu = \frac{F}{N}$$

In Fig. 9.25, the normal reaction on the block is equal to Wcos θ, the component of the block's weight normal to the surface. If the plane is made steeper, the block will slip if its component of weight parallel to the slope, W sin θ, exceeds the frictional force. At the point of sliding, the frictional force is equal to W sin θ. Therefore the coefficient of static friction, μ

$$= \frac{F}{N} = \frac{W \sin\theta}{W \cos\theta} = \tan\theta, \quad \text{since } \frac{\sin\theta}{\cos\theta} = \tan\theta$$

Investigating friction

Design and carry out an experiment to measure the coefficient of static friction between two surfaces both without and with oil present.

Free-body force diagrams

This is a diagram that just shows a body with the forces acting on it, without showing the other bodies that cause these forces. Force diagrams can become very complicated where the forces on more than one body are

Worked example 9.4

A cupboard of weight 350 N is pushed at steady speed across a horizontal floor, by a force of 150 N applied horizontally near its base. Calculate (a) the coefficient of static friction between the two surfaces that slide over each other, (b) the force needed if the weight of the cupboard is reduced to 200 N by removing some of its contents.

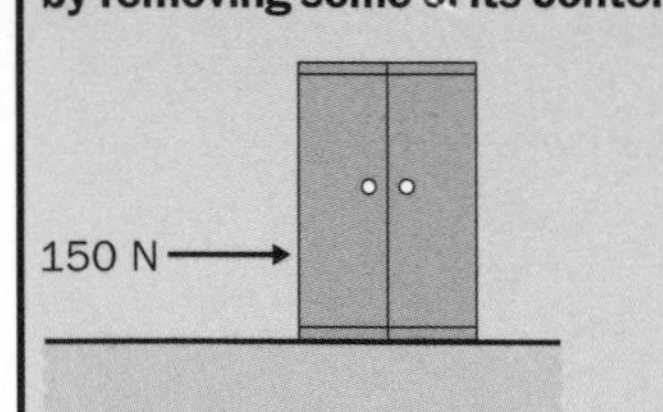

Fig. 9.26

Solution

(a) The normal reaction N = 350 N since the surfaces are horizontal. The force of friction is 350 N.

Hence $\mu = \frac{F}{N} = \frac{150}{350} = 0.43$

(b) With less weight in the cupboard, the normal reaction is now reduced to 200 N. Hence the force needed to move it, F $= \mu N = 0.43 \times 200 = 86$ N.

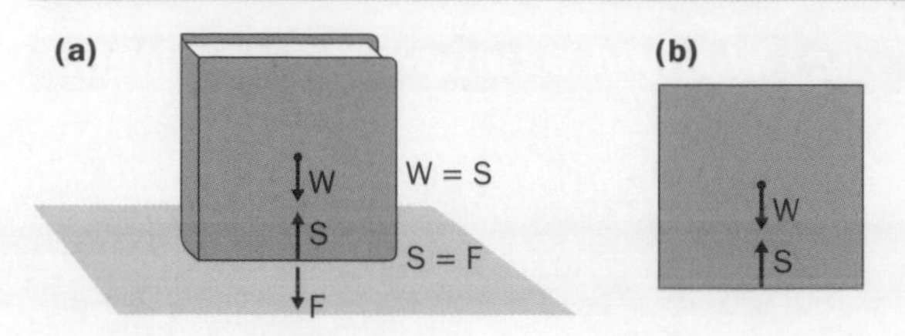

W = weight of book
S = support force on the book from the table
F = force of book on the table

Fig. 9.27 Force diagrams **(a)** forces on the book and table, **(b)** free-body force diagram.

shown. For example, Fig. 9.27a shows the forces acting on a book and on the table on which it rests. The free-body diagram of the book in Fig. 9.27b is much easier to understand.

9.5 Equilibrium conditions

For any body to be in equilibrium:

- the force vectors must form a closed polygon
- the principle of moments must hold.

If the lines of action of the forces lie in the same plane, the first condition can be established by resolving all the forces into two perpendicular directions in that plane. In each direction, the force components one way should be equal to the force components the other way.

Worked example 9.5

A uniform shelf of weight 20 N and of width 0.60 m is hinged horizontally to a vertical wall by a small bracket and a metal stay at the centre of the shelf, as shown in Fig. 9.30. The stay makes an angle of 50° with the wall.

(a) Show that the moment of the shelf about the hinge is 6.0 Nm.

(b) Calculate the tension in the stay.

(c) Calculate the magnitude and direction of the force of the hinge on the shelf.

50°
Hinge H
0.6 m

Fig. 9.30

Solution

(a) The centre of gravity of the shelf is 0.30 m from the hinge.
$\therefore$ the moment of the shelf about the hinge = 20 N × 0.30 m = 6.0 Nm.

(b) Fig. 9.31 shows a free-body diagram of the forces acting on the shelf. Since the stay is at 40° to the shelf, the perpendicular distance from the hinge to the stay, d = 0.60 sin 40 = 0.39 m.

Questions 9.4

1. A block of weight 12 N rests on an inclined plane. The angle of the plane to the horizontal, θ, is increased until the block slips at $\theta = 58°$.

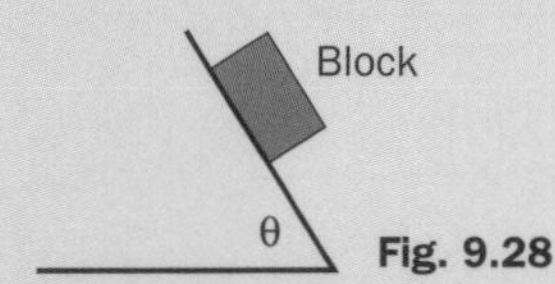

Fig. 9.28

(a) Show that the limiting value of the frictional force is 10.2 N.

(b) Calculate the normal component of the reaction force on the block due to the slope.

(c) Calculate the coefficient of static friction, μ, between the block and the incline.

2. A ladder of weight 220 N and of length 8.0 m rests with its lower end on a horizontal concrete surface and its upper end against a smooth wall. Calculate the maximum angle between the ladder and the wall at which no slippage can occur, for a coefficient of friction between the ladder and the floor of 0.4. (Hint: take moments about the lower end of the ladder to calculate the normal reaction on the ladder due to the wall.)

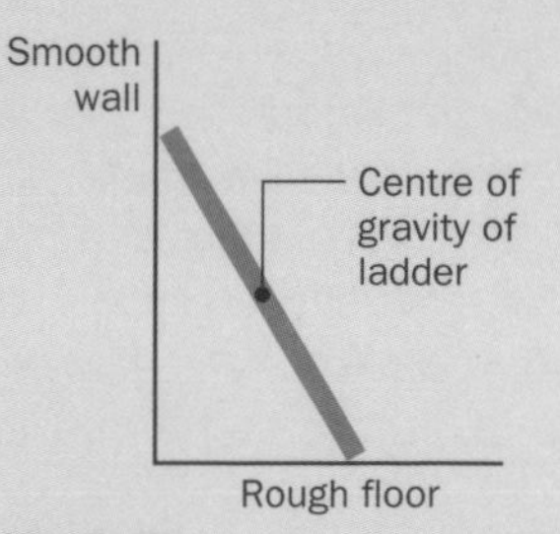

Fig. 9.29

Worked example 9.5 continued

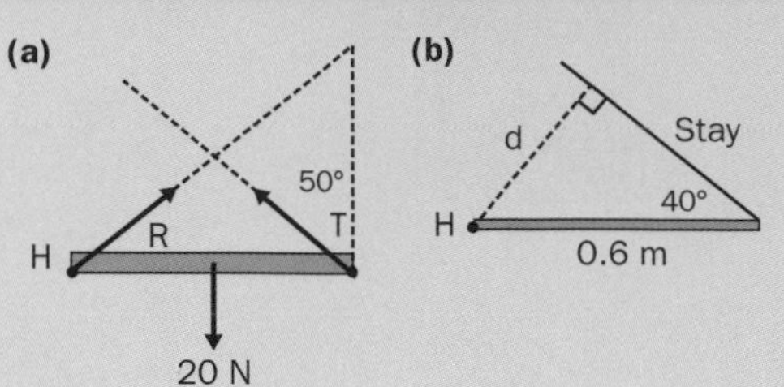

Fig. 9.31

Taking moments about the hinge:

the total clockwise moment = 6.0 N m (due to the shelf's weight)

the total anticlockwise moment = T d, where T is the tension in the stay.

Hence T d = 6.0

$$\therefore T = \frac{6.0}{d} = \frac{6.0}{0.39} = 15.6 \text{ N}$$

(c) The three forces acting on the shelf form a closed polygon, as in Fig. 9.32. The magnitude and direction of the force of the hinge on the shelf, R, can therefore be determined by a scale diagram.

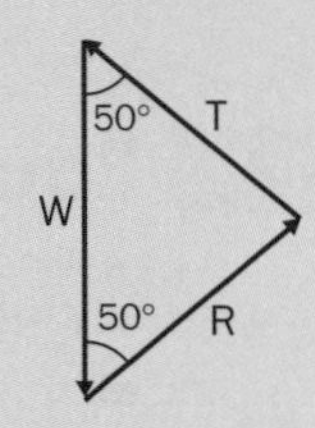

Fig. 9.32

Alternatively, the forces can be resolved into vertical and horizontal components, as in Table 9.1. The vertical and horizontal component of R can then be determined, to give its magnitude and direction.

Table 9.1 Vertical and horizontal components of R

	Vertical components	Horizontal components
W	20 N down	0.0
T	15.6 cos 50 up	15.6 sin 50 leftwards
R	R_v (assume up)	R_h (rightwards)

From the Table, $R_v = 20 - 15.6 \cos 50 = 10.0$ N upwards

$R_h = 15.6 \sin 50 = 11.9$ N rightwards

Hence $R = \sqrt{(R_v^2 + R_h^2)} = \sqrt{(10.0^2 + 11.9^2)} = 15.6$ N

The angle between the line of action of R and the wall, θ, is given by

$$\tan\theta = \frac{R_h}{R_v} = \frac{11.9}{10.0}$$

Hence θ = 50°.

Worked example 9.6

A plank of weight 160 N and of length 5.0 m rests horizontally on two pillars. One pillar, X, supports the plank at one end and the other pillar, Y, supports it 1.0 m from the other end.

(a) Sketch a free-body diagram for this arrangement.

(b) Calculate the support force from each pillar on the plank.

Solution

(a) See Fig. 9.33.

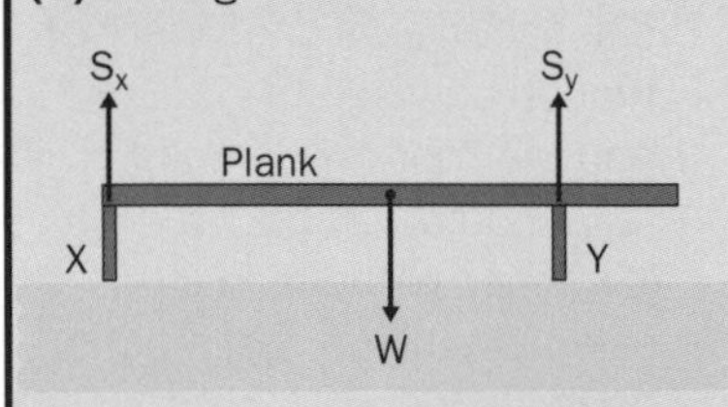

Fig. 9.33

(b) Let S_x and S_y represent the support forces at X and Y.

$\therefore S_x + S_y = W$, where W = 160 N

Taking moments about X;

$S_y \times 4.0 = W \times 2.5$,

hence $S_y = 160 \times \frac{2.5}{4.0} = 100$ N

Hence $S_x = 160 - 100 = 60$ N

Couples

A **couple** consists of two equal and opposite forces, acting parallel to each other but not along the same straight line. For a pair of equal and opposite forces, F, at perpendicular distance, d, apart,

the moment of the couple = Fd about any point

This can be proved by choosing an arbitrary point, P, along a perpendicular line between the two forces at distance d_1 from one of the forces.

The moment of this force about P = Fd_1.

The moment of the other force about P = $F(d - d_1)$ in the same sense of rotation.

Therefore, the total moment = $Fd_1 + F(d - d_1)$ = Fd which is independent of the position of P.

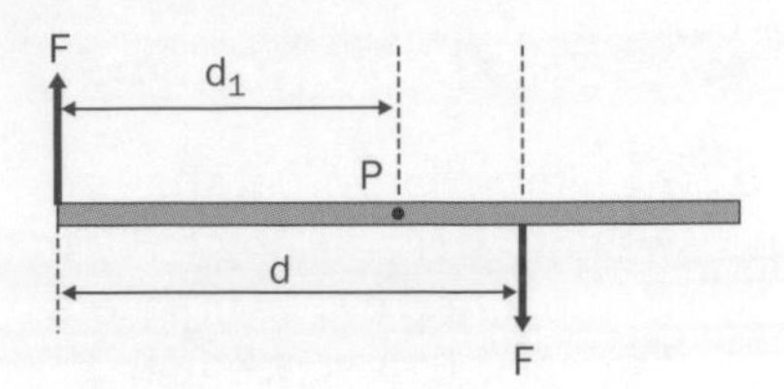

Fig. 9.34 A couple.

Simplifying free-body force diagrams

A single force, F, acting on a body at any point has the same effect as an equal force acting through the centre of gravity and a couple of moment Fd, where d is the perpendicular distance from the centre of gravity to the line of action of the force. Fig. 9.35 shows why this is so. This rule means that all the forces acting on a body can be reduced to a single resultant force acting through the centre of gravity **and** a resultant couple. It also means that for a body to be in equilibrium, the resultant force on it must be zero **and** there must be no overall couple.

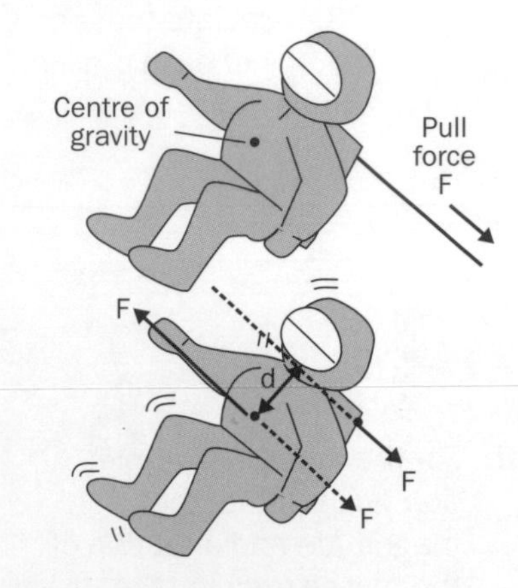

Fig. 9.35 Forces in space.

Worked example 9.7

A caravan of total weight 3200 N is towed at steady speed along a horizontal road. The centre of gravity of the caravan is 1.0 m in front of the wheels at a height of 1.5 m above the ground. The perpendicular distance between the wheel axis and the vertical line through the point of attachment of the tow bar is 3.0 m. The towing vehicle pulls on the caravan with a force of 2.2 kN, as shown in Fig. 9.36.

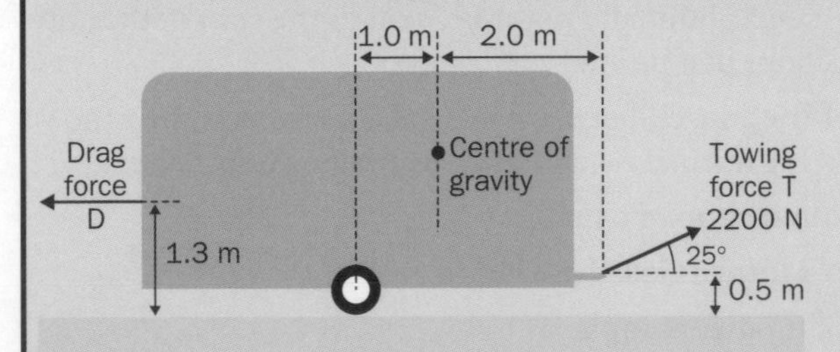

Fig. 9.36 Pulling a caravan.

(a) Sketch the free-body force diagram showing the forces acting on the caravan. Assume the drag force acts horizontally above the road.

(b) (i) By resolving the forces vertically and horizontally, consider the vertical components and show that the normal reaction of the road on the road wheels is equal to 2.3 kN.

(ii) The coefficient of friction between the road and the tyres is 0.4. Calculate the frictional force on the road wheels and hence calculate the drag force on the caravan.

(c) Use the principle of moments to show that the line of action of the drag force is 1.3 m above the road.

Solution

(a) See Fig. 9.37.

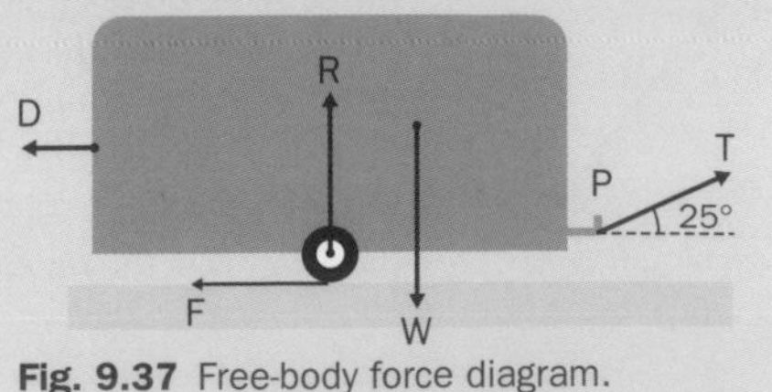

Fig. 9.37 Free-body force diagram.

(b) (i) Vertically:

$R + 2200 \sin 25 = 3200$

Hence $R = 3200 - 2200 \sin 25 = 2270$ N

Questions 9.5

1. A uniform horizontal shelf of weight 25 N and length 3.0 m is supported by two brackets fixed at 0.5 m from either end. The shelf supports a can of paint of weight 40 N at 1.0 m from one end, as in Fig. 9.38. Calculate the support force of each bracket on the shelf.

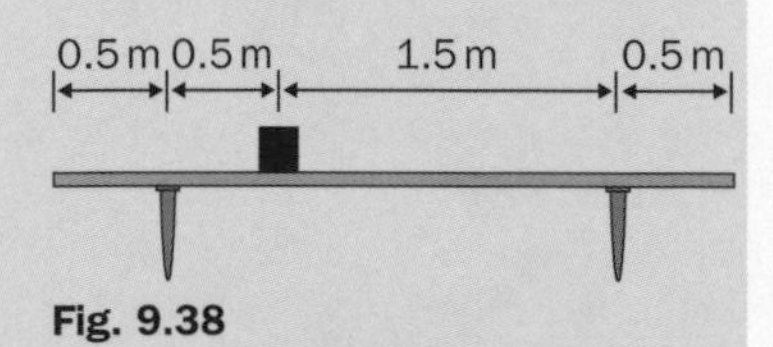

Fig. 9.38

2. For the worked example above, calculate the magnitude and direction of the towing force on the caravan if the drag force is reduced by 50% at the same height.

Fig. 9.39

Worked example 9.7 continued

(ii) $F = \mu R = 0.4 \times 2270 = 908$ N

Hence the drag force, $D = 2200 \cos 25 - 908 = 1086$ N.

(c) Apply the principle of moments about the point where the tow bar is attached to the caravan, which is 0.5 m above the road and 3.0 m in front of the wheels, as in Fig. 9.37.

Total clockwise moment $= 3.0R + 0.5F = 7264$ N.

Total anticlockwise moment $= 2.0W + (h - 0.5)D = 5857 + 1086h$, where h is the height of the drag force above the road.

Hence $5857 + 1086h = 7264 \therefore h = (7264 - 5857) \div 1086 = 1.3$ m.

Summary

◆ **The components of a force, F, parallel and perpendicular to a line at** angle θ, are F cos θ parallel to the line and F sin θ perpendicular to the line.

◆ **The moment of a force, F, about a point = Fd,** where d is the perpendicular distance from the point to the line of action of the force.

◆ **The principle of moments** states that for a body in equilibrium, the sum of the clockwise moments about any point is equal to the sum of the anticlockwise moments about that point.

◆ **The coefficient of friction,** μ, between two solid surfaces is the ratio of the frictional force to the normal reaction between the two surfaces.

◆ **Conditions for equilibrium of a body**

1. The force vectors form a closed polygon.
2. The principle of moments applies.

Revision questions

9.1. A concrete paving stone of dimensions 50 mm × 500 mm × 800 mm rests end-on against a smooth wall, as shown in Fig. 9.40.

(a) The density of the concrete is 2700 kg m^{-3}. Calculate **(i)** the mass, **(ii)** the weight of the paving stone.

(b) The paving stone is at angle θ to the wall.

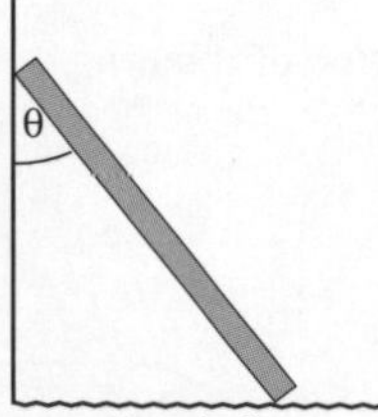

Fig. 9.40

(i) Sketch a free-body diagram showing the forces acting on the paving stone, including the frictional force and the normal reaction at the point of contact on the floor.

(ii) Why is the normal reaction at the point of contact on the floor equal and opposite to the weight?

(iii) The coefficient of friction between the paving stone and the floor is 0.40. Calculate the frictional force due to the floor on the paving stone.

(iv) Show that the maximum value of θ for no slip is 39°.

9.2. Fig. 9.41 shows a rubbish skip and its contents being raised from a driveway.

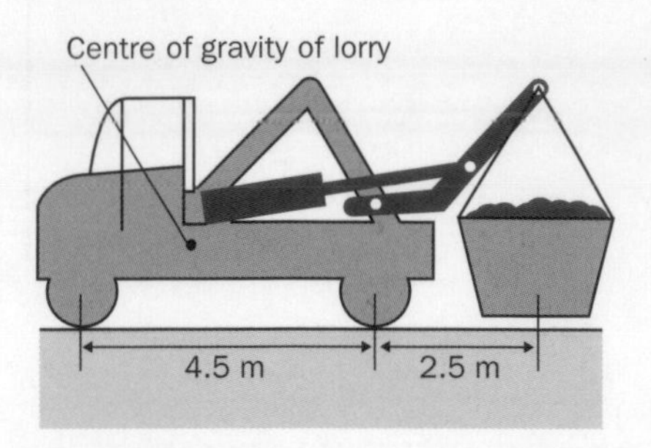

Fig. 9.41

(a) Describe how the distribution of the total weight of the skip and the lorry on the front and rear wheels changes as the skip is lifted off the roadway onto the back of the lorry.
(b) The weight of the lorry without the skip is 120 kN and its centre of gravity is 3.2 m from the back axle. The front axle is 4.5 m from the back axle. For a total weight of 10 kN for the skip and its contents, calculate the reaction of the ground on each pair of wheels just as the lorry raises the skip off the ground. In this position, the centre of gravity of the skip is 2.5 m from the back axle.

9.3. Fig. 9.42 shows a diagram of a model arm in which a 500 mm rule, pivoted at its 10 mm mark, represents the lower arm pivoted at the elbow joint. The spring balance, attached to the rule at its 50 mm mark, represents the biceps muscle.

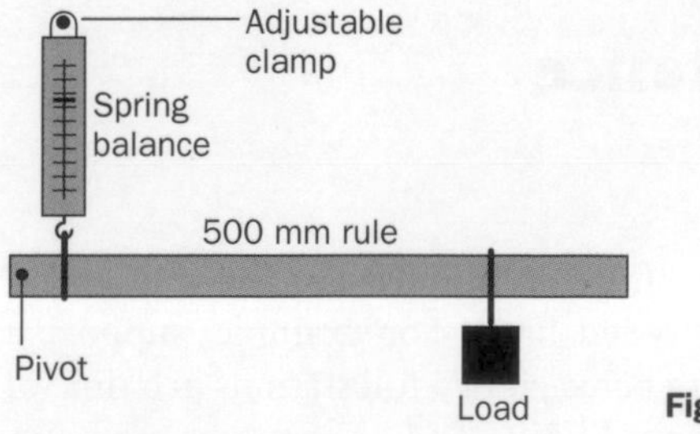

Fig. 9.42

(a) With no load on the rule, the spring balance reads 6 N after being adjusted so it is vertical and the rule is horizontal.
(i) Show that the weight of the rule is 1.0 N.
(ii) Calculate the force of the pivot on the rule.
(b) A weight of 2.0 N is suspended from the rule at its 410 mm mark. The spring balance is readjusted until it is vertical and the rule is horizontal. Calculate the reading on the spring balance now.

9.4. A uniform beam of length 8.0 m and of weight 2400 N rests horizontally on pillars at either end. A pulley hoist of weight 80 N is suspended from the beam at 2.0 m from one end. The hoist is to be used to lift a crate of mass 50 kg, as shown in Fig. 9.43.

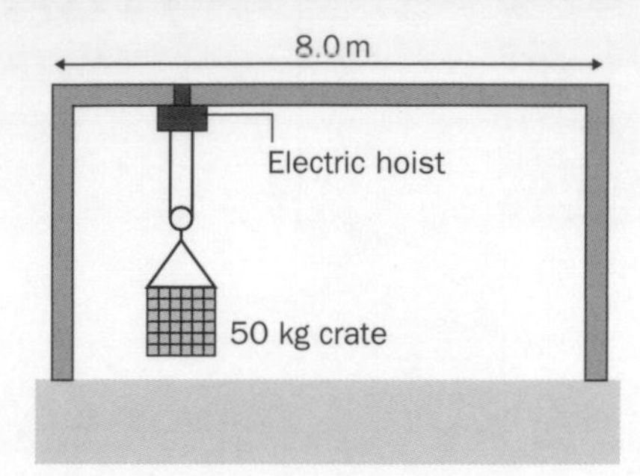

Fig. 9.43

(a) Calculate the downward pull on the beam due to the hoist when the crate is being raised.
(b) Calculate the support forces on the beam due to the pillars at either end when the downward force in (a) is exerted on the beam.

9.5. A piano of weight 1.5 kN rests on four small wheels at each corner. The centre of gravity of the piano is midway along its length at a perpendicular distance of 0.25 m from its rear panel and at a height of 0.70 m above the floor, as shown in Fig. 9.44.
(a) The base of the piano is 0.40 m wide. Calculate the maximum horizontal force that could be applied at the back edge of the piano, 1.5 m above the floor, to make it tilt about its front wheels.
(b) A horizontal force of 300 N is necessary to move the piano. Calculate the maximum height at which this force can be applied without tilting the piano about its front wheels.

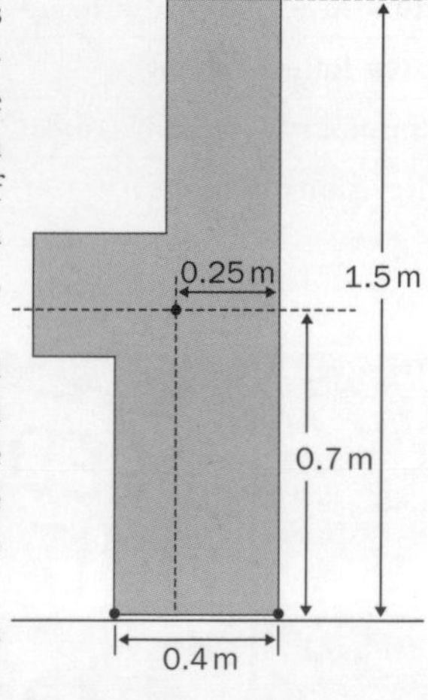

Fig. 9.44

Dynamics

Contents

Objectives

After working through this unit, you should be able to:

- ▶ state the difference between displacement and distance and between speed and velocity
- ▶ plot a displacement versus time graph and use it to find the velocity of an object and the distance it moves
- ▶ plot a velocity versus time graph and use it to find the acceleration of an object
- ▶ explain what is meant by uniform speed and uniform acceleration
- ▶ carry out calculations using the equations for uniform acceleration

Fig. 10.1 This speed limit sign in the USA instructs drivers that their speed should not exceed 35 mph.

10.1 Speed and distance

On the road

On motorways in certain countries, fines are imposed at toll stations on motorists who have exceeded the speed limit. For example, suppose a motorist travels a distance of 150 km between two toll stations in a time of 75 minutes on a motorway with a speed limit of 110 kilometres per hour. The motorist's average speed is 2 kilometres per minute (= 150 kilometres /75 minutes) which is equal to 120 kilometres per hour (= 2 km per minute × 60 minutes) which is over the speed limit. Ignoring speed limits can be costly as well as dangerous.

Conversion factors

The S.I. unit of speed is the metre per second (m s^{-1}). Since 1 kilometre equals 1000 metres and there are 3600 seconds in 1 hour, then a speed of 100 km h^{-1} is equal to 28 m s^{-1} (= 100 × 1000 m / 3600 s). Note that 18 km h^{-1} = 5 m s^{-1} exactly.

Speed limits and speedometer readings are usually stated either in kilometres per hour (km h^{-1}) or in miles per hour (mph). The speed limit on

UK motorways is 70 mph which is equal to 112 km h^{-1}, since 1 mile = 1.6 km. Prove for yourself that 112 km h^{-1} is equal to 31 m s^{-1}.

The speed equation

Speed is defined as distance travelled per unit time. For an object that travels distance, s, in time, t, at **constant speed,** υ,

$$\upsilon = \frac{s}{t}$$

This equation can be rearranged to give

$$s = \upsilon t, \text{ or } t = \frac{s}{\upsilon}$$

If the speed of an object changes, the equation $\upsilon = s/t$ gives the **average speed** over time t.

Worked example 10.1

A motorist on a motorway journey of length 50 km travels the first 30 km in exactly 20 minutes and completes the rest of the journey at a constant speed of 100 km h^{-1}.

(a) Calculate the average speed of the motorist in metres per second over the first 30 km of the journey.

(b) (i) Calculate the time taken, in seconds, for the second part of the journey.

(ii) Sketch a graph to show how the motorist's speed varied with time for the whole journey.

(iii) Calculate the average speed in metres per second for the whole journey.

Solution

(a) Average speed $= \frac{\text{distance}}{\text{time}} = \frac{30\,000}{20 \times 60} = 25 \text{ m s}^{-1}$

(b) (i) 100 km h^{-1} converts to 28 m s^{-1} as explained earlier.

$\therefore \text{time} = \frac{\text{distance}}{\text{speed}} = \frac{20\,000 \text{ m}}{28 \text{ m s}^{-1}} = 714 \text{ s}$

(ii) See Fig. 10.2.

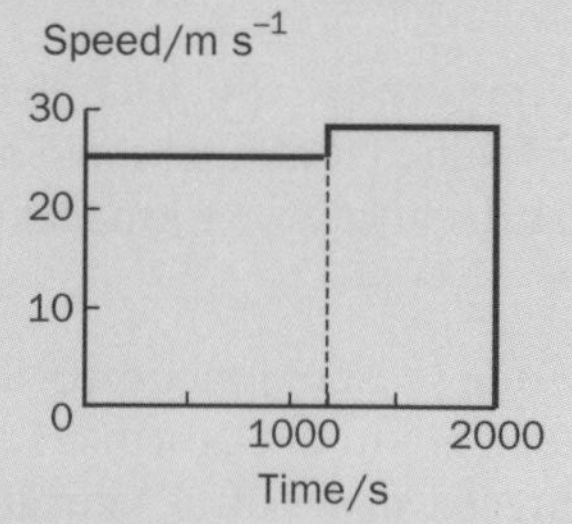

Fig. 10.2

(iii) Average speed $= \frac{\text{distance}}{\text{time}} = \frac{50\,000 \text{ m}}{(1200 + 710) \text{ s}} = 26 \text{ m s}^{-1}$

Questions 10.1

The road map in Fig. 10.3 shows the progress of a motorist from joining a motorway to leaving it 2 hours later. The total distance travelled by the motorist on the motorway was 180 km, including a 20 km section of roadworks which took 30 minutes to pass through.

1. The motorist travelled at a steady speed of 110 km h^{-1} for the first hour. How far did the motorist travel in this time?

2. The motorist took 30 minutes to travel 20 km through the roadworks. What was the motorist's average speed in km h^{-1} in this section?

3. **(a)** How far was the last section of the journey after the roadworks?

(b) How long did this last section take to complete?

(c) What was the average speed of the motorist in this final section?

4. **(a)** Sketch a graph to show how the motorist's speed changed with time.

(b) What was the average speed for the whole journey in **(i)** km h^{-1}, **(ii)** m s^{-1}?

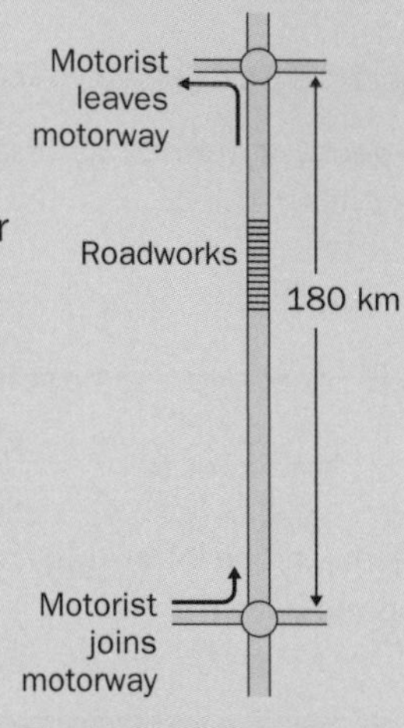

Fig. 10.3 On the road.

10.2 Speed and velocity

Vectors and scalars

A vehicle travelling due South on a motorway at a speed of 100 km h^{-1} does not have the same velocity as a vehicle travelling at the same speed due North on the opposite carriageway. Velocity is speed in a given direction and is therefore a vector quantity. The idea of a vector was introduced in Unit 9. A **vector** is a physical quantity which has direction as well as magnitude. Displacement and velocity are examples of vectors. A **scalar** is a physical quantity that has magnitude only. Distance and speed are examples of scalars.

- **Displacement** is distance travelled in a given direction.
- **Velocity** is speed in a given direction.

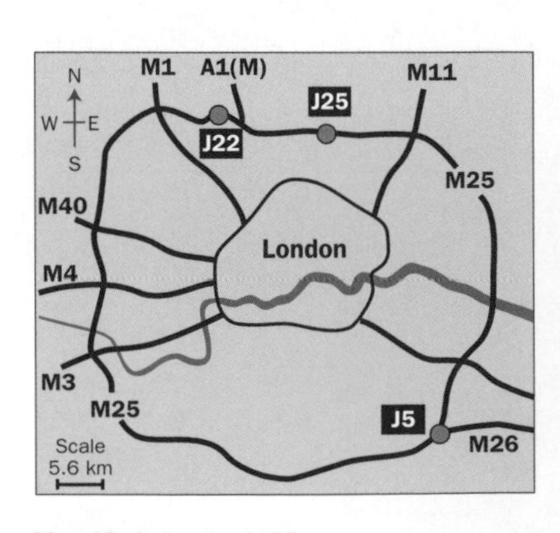

Fig. 10.4 On the M25.

On the M25

Fig. 10.4 shows the M25, London's orbital motorway, which stretches 196 km around the capital. Suppose a motorist joined it at junction 5 and travelled anticlockwise to junction 25, a total distance of 64 km, taking 40 minutes for the journey.

1. Junction 25 is 44 km due North and 15 km due West of junction 5. The motorist's displacement on exit was therefore 44 km due North, 15 km due West from the point of entry. This is a direct distance of 46.4 km at an angle of 19° West of due North. Fig. 10.5 shows how to use Pythagoras' theorem and the rules of trigonometry to work out this distance and angle.

2. For the whole journey, the average speed

$$= \frac{\text{distance}}{\text{time}} = \frac{64 \text{ km}}{(40 \div 60) \text{ hr}} = 96 \text{ km h}^{-1}$$

3. The motorist's direction changed gradually during the journey, initially travelling North at Junction 5 and ending the journey travelling westwards at junction 25. The velocity of the motorist's vehicle therefore changed, even if the speed remained the same throughout.

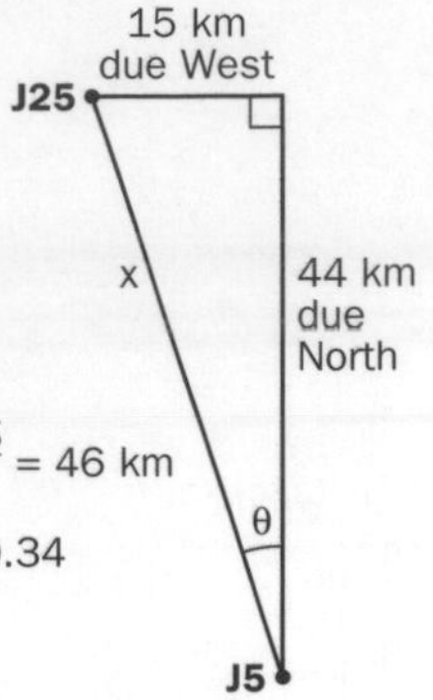

J25, 15 km due West, x, 44 km due North, θ, J5

$x = \sqrt{15^2 + 44^2} = 46$ km

$\tan\theta = \frac{15}{44} = 0.34$

$\theta = 19°$

Fig. 10.5 Using trigonometry.

Question 10.2

1. A motorist joins the M25 at junction 22. He travels a distance of 84 km clockwise before leaving the motorway 70 minutes later at junction 5, which is 48 km due South and 30 km due East of junction 22.

(a) Calculate the average speed of the motorist on the motorway **(i)** in km h^{-1}, **(ii)** in m s^{-1}.

(b) Calculate the displacement from junction 22 to junction 5.

(c) What direction was the motorist travelling on **(i)** joining the motorway, **(ii)** leaving the motorway?

10.3 Acceleration

Uniform acceleration along a straight line

Consider an aircraft that accelerates from rest on a runway, taking 40 s to reach its take-off speed of 120 m s^{-1}. Fig. 10.6 shows its speed increases 3 m s^{-1} each second during take-off. The graph shows that the speed increases steadily since the gradient of the line is constant. This is an example of **uniform acceleration** which is where the speed changes at a constant rate. The acceleration of the aircraft is 3 m s^{-2}.

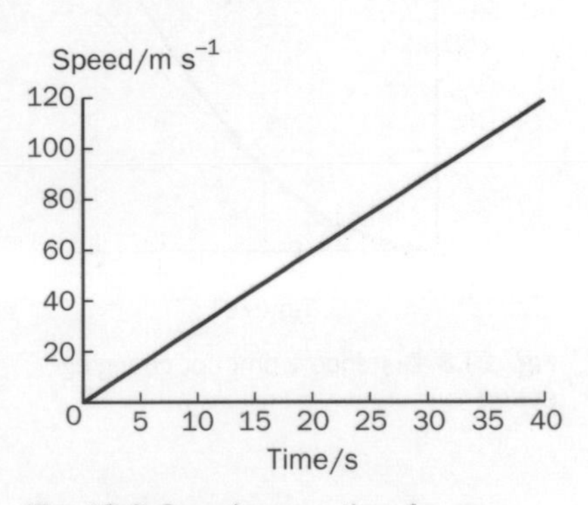

Fig. 10.6 Speed versus time for an aircraft taking off.

Acceleration is defined as change of velocity per second.

The unit of acceleration is the metre per second per second, abbreviated as m s^{-2} or m/s^2. Deceleration, sometimes described as negative acceleration, is where an object slows down.

Consider an object that accelerates uniformly from initial speed, u, to speed, v, in time, t, without a change of direction. Its change of speed is $(v - u)$, therefore its acceleration, a, can be calculated from the equation below.

$$a = \frac{(v - u)}{t}$$

The above equation may be rearranged as follows to calculate v given values for u, a and t.

$at = (v - u)$

hence

$v = u + at$

Worked example 10.2

A car is capable of accelerating from rest to a speed of 20 m s^{-1} in 4.0 s.
(i) Calculate its acceleration.
(ii) Calculate its speed after a further 2.0 seconds if the acceleration does not change.

Solution

(i) Acceleration, $a = \frac{(20 - 0)}{4.0} = 5.0\ \text{m s}^{-2}$

(ii) For $a = 5.0\ \text{m s}^{-2}$, $u = 0$, $t = 4.0 + 2.0 = 6.0\ \text{s}$
$\therefore v = u + at = 0 + (5.0 \times 6.0) = 30\ \text{m s}^{-1}$

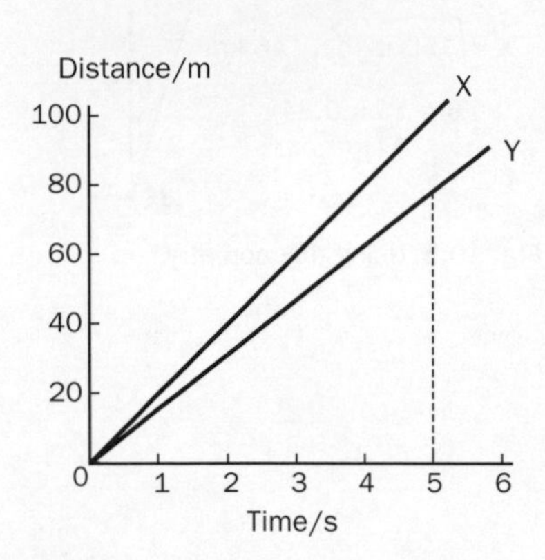

Fig. 10.7 Distance v time for constant speed

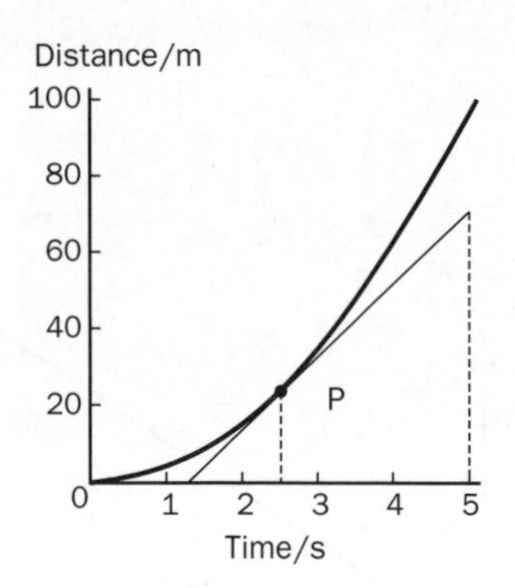

Fig. 10.8 Distance v time for changing speed

Using graphs to represent motion

1. Distance versus time graphs

For **constant speed**, the distance travelled increases steadily with time. Fig. 10.7 shows distance v. time graphs on the same axes for two vehicles moving at different speeds. The speed of X is greater than the speed of Y because X travels a greater distance than Y in the same time.

The speed of X or Y can be determined from the **gradient** of the appropriate line. In Fig. 10.7, the height and the base of the gradient triangle for Y represent the distance moved (= 75 m) and the corresponding time taken (= 5.0 s) The gradient or slope, the height divided by the base, therefore represents the speed which is 15 m s^{-1} (= 75 m / 5 s). Show for yourself that the speed of X is 20 m s^{-1}.

For **changing speed**, the distance travelled per second varies. Fig. 10.8 shows how the distance changes with time for an object moving in a straight line with constant acceleration. The gradient becomes steeper as its speed increases.

To find the speed at any given point, the gradient of the tangent to the curve is measured at that point. The tangent at a point is a straight line that touches the curve without cutting through the curve at that point. Show for yourself that in Fig. 10.8 the speed at point P is 20 m s^{-1}.

The gradient of a distance versus time graph represents the speed.

2. Speed versus time graphs

Consider an object that accelerates uniformly from initial speed, u, to speed v, in time, t, without a change of direction. Its motion is represented by the speed–time graph in Fig. 10.9.

Its **acceleration, a** $= \frac{\text{change of velocity}}{\text{time taken}} = \frac{(v - u)}{t}$

The gradient of a speed versus time graph represents the acceleration.

In Fig. 10.9, the gradient of the line is $\frac{(v - u)}{t}$

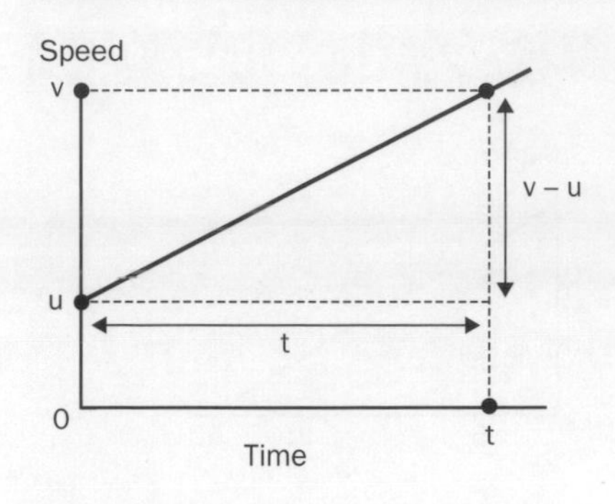

Gradient of line = $\frac{v-u}{t}$ = acceleration

Area under line = $\frac{(u+v)t}{2}$ = distance moved

Fig. 10.9 Uniform acceleration.

since $(v - u)$, the change of speed, is represented by the height of the gradient triangle and t, the time taken, represents the base.

The **distance travelled**, **s** = average speed × time taken

$$= \frac{(u+v)}{2}t, \quad \text{since the average speed is } \frac{(u+v)}{2}$$

The area under the line of a speed versus time graph represents the distance travelled.

In Fig. 10.9, the area under the line is $\frac{(u+v)}{2}t$

This is because the shape under the line (a trapezium) has the same area as a rectangle of the same width (t) and of height $\frac{(u+v)}{2}$

which represents motion at a constant speed of $\frac{(u+v)}{2}$ for time t.

See www.palgrave.com/foundations/breithaupt for information about investigating motion.

Worked example 10.3

A vehicle accelerates from rest to a speed of 15 m s^{-1} in 5.0 s at uniform acceleration.

(a) Sketch a graph to show how its speed changes with time.

(b) Calculate (i) the acceleration of the vehicle, (ii) the distance travelled from rest in 5 seconds.

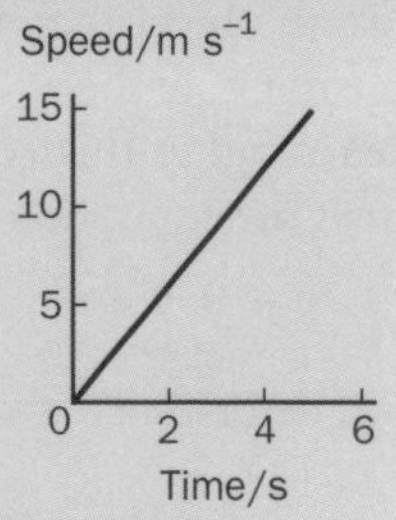

Solution

(a) See Fig. 10.10.

Speed/m s^{-1} (0, 5, 10, 15) vs Time/s (0, 2, 4, 6)

Fig. 10.10

(b) (i) Acceleration = gradient of speed v. time graph = $\frac{(15-0)}{5.0}$ = 3.0 m s^{-2}

(ii) Distance travelled = area under the speed v. time graph from 0 to 5 s

$= \frac{1}{2}$ height × base $= \frac{1}{2} \times 15 \times 5 = 37.5$ m

Questions 10.3

1. A vehicle travelling at 30 m s^{-1} leaves a motorway at a slip road and comes to rest 20 s later at traffic lights at the end of the slip road. Fig. 10.11 shows how its speed decreases with time.
   (a) Show that the acceleration of the vehicle is –1.5 m s^{-2}.
   (b) Calculate the distance travelled along the slip road by the vehicle.
2. A train left a station and accelerated uniformly from rest for 30 s to a speed of 12 m s^{-1}. It then travelled at this speed for 100 s before decelerating uniformly to rest in 20 s. Fig. 10.12 shows how its speed changed with time.
   (a) Calculate its acceleration and the distance it travelled in the first 30 s.
   (b) (i) Explain why its acceleration was zero during the next 100 s.
   (ii) Show that the distance it travelled in this time was 1200 m.
   (c) Calculate its acceleration and distance travelled in the last 20 s.
   (d) Calculate the total distance it travelled and show that its average speed for the whole journey was 10 m s^{-1}.

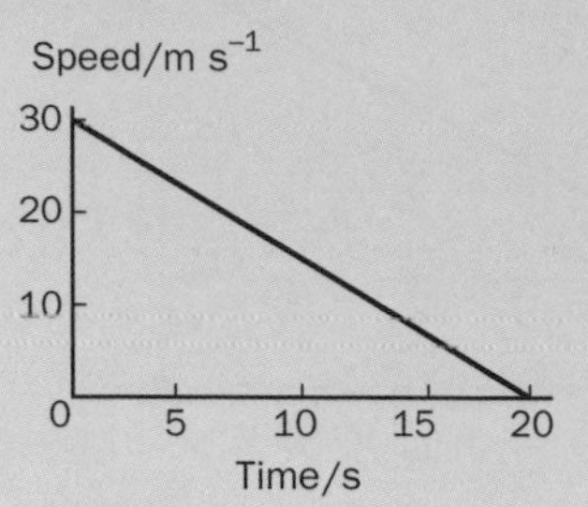

Fig. 10.11

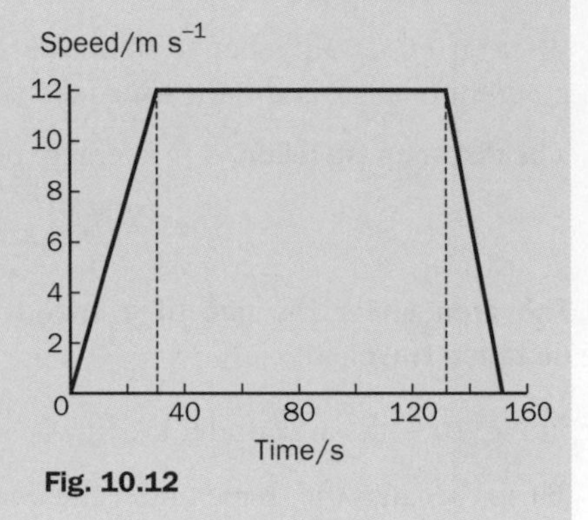

Fig. 10.12

10.4 The equations for uniform acceleration

For an object that accelerates uniformly from initial speed, u, to speed, v, in time, t, without a change of direction, its acceleration, $a = (v - u) \div t$. Rearranging this equation gives $at = (v - u)$ and making v the subject gives the equation in the form

$$\mathbf{v = u + at} \qquad \textbf{(equation 1)}$$

its distance travelled, $\mathbf{s = \frac{(u + v)}{2} t}$ **(equation 2)**

Substituting 'u + at' for v in equation 2 gives an equation for s without v present, as follows:

$$s = \frac{(u + (u + at))}{2} t = \frac{(2u + at)}{2} t = \frac{(2ut + at^2)}{2} = ut + \tfrac{1}{2} at^2$$

$$\mathbf{s = ut + \tfrac{1}{2} at^2} \qquad \textbf{(equation 3)}$$

Equations 1 and 2 may be combined to eliminate t as follows

$(v - u) = at$ from equation 1

$(v + u) = \frac{2s}{t}$ from equation 2

Multiplying the two equations together gives:

$$(v - u)(v + u) = at\frac{2s}{t}$$

which gives $v^2 + uv - uv - u^2 = 2as$
hence $v^2 - u^2 = 2as$
or $\mathbf{v^2 = u^2 + 2as}$ **(equation 4)**

Using equations

Symbols such as u, v, a, s and t are used instead of words to represent physical quantities. When an equation is used to calculate an unknown quantity, all the other quantities must be known. To solve for the unknown quantity:

1. Write the equation down and list the quantities with known values.
2. If one or more of the known quantities is zero, the equation can be simplified by substituting zero for these quantities.
3. Rearrange the equation in symbolic form to make the unknown quantity the subject of the equation. You may prefer to substitute the known values into the equation before rearranging the equation. If the numerical values are unwieldy, this approach can lead to errors.
4. Write the rearranged equation again with the known quantities substituted in place of the symbols.
5. Finally, calculate the unknown quantity, writing down its magnitude and the unit as the answer.
6. Note that a negative value for acceleration represents a deceleration.

Worked example 10.4

In a crash test, a car moving at a speed of 20 m s^{-1} is driven automatically at a concrete wall. The impact causes the front part of the car to crumple, shortening the vehicle by a distance of 0.60 m. Calculate (a) the time taken for the impact, (b) the deceleration of the car on impact.

Solution

Initial speed $u = 20\ \text{m s}^{-1}$, Final speed $v = 0$.

Distance moved during the impact, $s = 0.60$ m.

(a) To find the time taken, t, choose the equation which contains s, u, v and t.

$$s = \frac{(u + v)}{2}t = \frac{ut}{2} \quad \text{since } v = 0$$

Rearranging this equation gives $t = \frac{2s}{u} = \frac{2 \times 0.60}{20} = 0.06$ s

(b) To find the deceleration, choose an equation with acceleration, a,

$$a = \frac{(v - u)}{t} = \frac{0 - 20}{0.06} = -330\ \text{m s}^{-2}$$

Notes

1. The acceleration is negative which means it slows down.
2. The acceleration could have been calculated using $v^2 = u^2 + 2as$. Since $v = 0$, this equation reduces to $0 = u^2 + 2as$. This rearranges as $a = -u^2/2s = -20^2/2 \times 0.6 = -330\ \text{m s}^{-2}$.

Questions 10.4

1. An aircraft takes off from a standstill position on an aircraft carrier with the assistance of a catapult mechanism. It reaches a speed of 85 m s^{-1} in a distance of 210 m from rest. Calculate **(a)** the time taken for the launch, **(b)** the acceleration during take-off.
2. A bullet moving at a speed of 115 m s^{-1} hits a tree and embeds itself to a depth of 45 mm. Calculate **(a)** the time taken for the impact, **(b)** the deceleration of the bullet.
3. A cricket ball moving at a speed of 28 m s^{-1} is caught with one hand by a cricketer. The cricketer's hand moves back through a distance of 1.2 m in stopping the ball. Calculate the deceleration of the cricket ball.

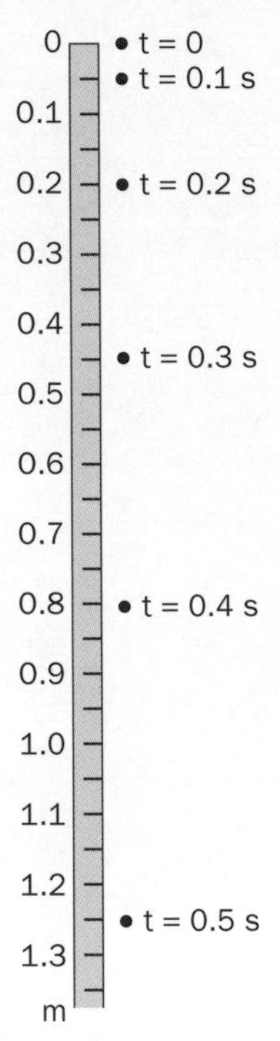

Fig. 10.13 Free fall.

10.5 Acceleration due to gravity

Free fall

A freely falling object is acted on by the force of gravity only. Fig. 10.13 represents a video sequence of a sphere falling freely. The video camera recorded the object's position against a vertical scale every tenth of a second after it was released. Table 10.1 below gives the distance fallen from its initial position every 0.10 s after release.

Table 10.1 An object in free fall

Time, t, after release/s	0.000	0.100	0.200	0.300	0.400	0.500
Distance fallen, s, after release/m	0.000	0.050	0.200	0.450	0.800	1.250

If the acceleration is constant, the distance fallen, s, in time t, is given by

$$s = ut + \tfrac{1}{2}at^2$$

Since the object was released from rest, its initial speed, u = 0. Hence the above equation reduces to

$$\mathbf{s = \tfrac{1}{2}at^2}$$

A graph of s on the vertical axis against t^2 on the horizontal axis should therefore give a straight line of gradient $\frac{1}{2}$ a, as shown in Fig. 10.14. From this graph, it can be concluded that

1. the acceleration is constant because the graph is a straight line
2. the acceleration is equal to 9.8 m s^{-2} since the gradient ($= \frac{1}{2}$a) = 4.9 m s^{-2}.

The acceleration due to gravity of an object falling freely is constant, equal to 9.8 m s^{-2}.

As explained on p. 150, the symbol *g* is used to denote this acceleration. The value of *g* varies slightly over the Earth's surface from 9.81 m s^{-2} at the poles to 9.78 m s^{-2} at the Equator. The value of 9.8 m s^{-2} for *g* is an average value over the Earth's surface.

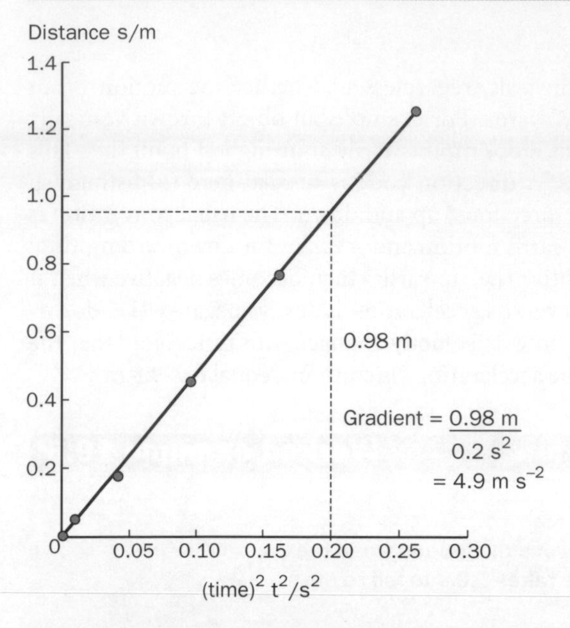

s/m	0	0.05	0.20	0.45	0.80	1.25
t/s	0	0.10	0.20	0.30	0.40	0.50
t/s^2	0	0.01	0.04	0.09	0.16	0.25

Fig. 10.14 Distance versus time^2.

Maths workshop; straight line graphs

Fig. 10.15 shows a straight line drawn on a graph. The general equation for this line is $\mathbf{y = mx + c}$, where m is the gradient and c is the intercept on the y-axis.

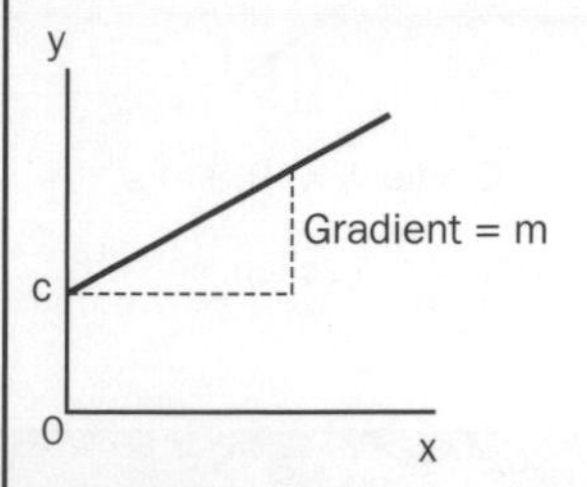

Fig. 10.15 $y = mx + c$.

Compare the equation $s = \frac{1}{2}at^2$ with $y = mx + c$. Plotting s on the y-axis and t^2 on the x-axis therefore gives a straight line passing through the origin ($c = 0$) with a gradient $\frac{1}{2}a$.

Experiment to measure *g* using the falling ball method

Use an electronic timer to time a steel ball falling from rest through a measured distance. Repeat the measurement several times to give an average time. Repeat the procedure for different distances. Plot a graph of distance (on the vertical axis) against time^2 as Fig. 10.14. Measure the gradient of the graph and determine *g*.

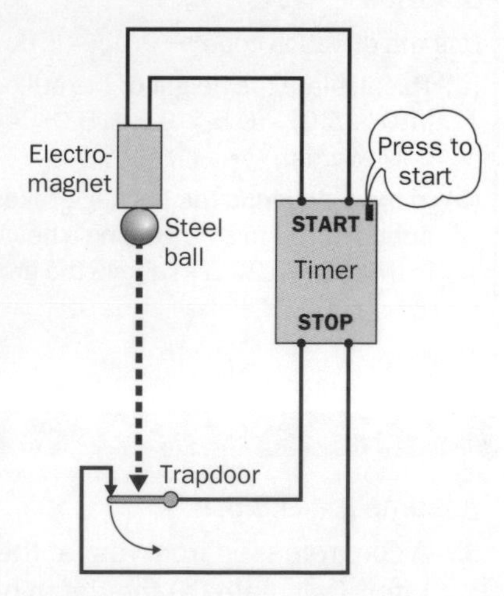

Fig. 10.16 The falling ball method to determine g

Worked example 10.5

Assume $g = 9.8$ m s^{-2}

An object is released from rest at the top of a tower of height 47 m and falls to the ground. Calculate (a) the time it takes to fall to the ground, (b) its speed just before impact at the ground.

Solution

Initial speed, $u = 0$, distance fallen, $s = 47$ m, acceleration, $a = 9.8$ m s^{-2}

(a) To find the time taken, t, use $s = ut + \frac{1}{2}at^2$

Substituting $u = 0$ gives $s = \frac{1}{2}at^2$, hence $t^2 = \frac{2s}{a} = \frac{2 \times 47}{9.8} = 9.6$

Hence $t = \sqrt{9.6} = 3.1$ s.

(b) To find the speed, v, just before impact, use $v = u + at = 0 + (9.8 \times 3.1)$ $= 30$ m s^{-1}

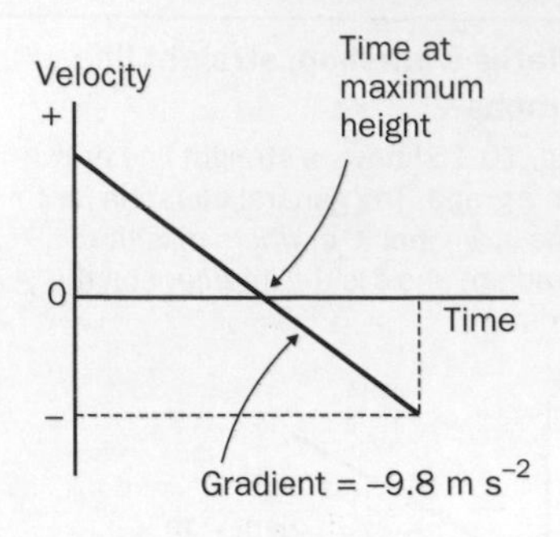

Fig. 10.18 Vertical projection.

Up and down

Gravity always acts downwards, regardless of whether the motion of an object is upwards or downwards. For example, an object thrown vertically up into the air slows down, stops momentarily at its highest point then falls back towards the ground. A direction code is needed here to distinguish between the two possible directions, up and down. The usual convention is to assign + values for upwards motion and – values for downward motion. Its velocity is initially positive (i.e. upwards) then becomes negative when it moves downwards. However, its acceleration is always negative (i.e. downwards). Fig. 10.18 shows how its **velocity** changes with time. Note that the gradient of the line (i.e. the acceleration) is constant, equal to – 9.8 m s^2.

Worked example 10.6

Assume $g = 9.8$ m s^{-2}

A package is released from a hot air balloon at a certain height above the ground when the balloon is ascending at a steady speed of 4.0 m s^{-1}. The package takes 5.0 s to fall to the ground. Calculate the height of the balloon

(a) when the package is released,

(b) when the package hits the ground.

Solution

Use the direction code '+ is up; – is down'; $u = +4.0$ m s^{-1}, $a = -9.8$ m s^{-2}, $t = 5.0$ s

(a) To calculate the height of the balloon when the package is released, use $s = ut + \frac{1}{2}at^2 = (4.0 \times 5.0) - (0.5 \times 9.8 \times 5.0^2) = -102.5$ m. The minus sign means the package has moved downwards.

(b) In the 5 seconds the package takes to fall to the ground, the balloon moves up at a steady speed of 4.0 m s^{-1}, gaining a height of 20.0 m (= 4.0 m $s^{-1} \times 5.0$ s) in this time. The balloon is therefore 122.5 m above the ground when the package hits the ground.

Fig. 10.19

Questions 10.5a

Assume $g = 9.8$ m s^{-2}

1. A coin released from rest at the top of a well strikes the bottom of the well 1.8 s later. Calculate **(a)** the depth of the well, **(b)** the speed of the coin just before impact.
2. A package is released from a hot air balloon at a height of 36 m above the ground when the balloon is descending at a steady speed of 4.0 m s^{-1}. Calculate **(a)** the speed of the package just before impact, **(b)** the time taken by the package to fall to the ground.

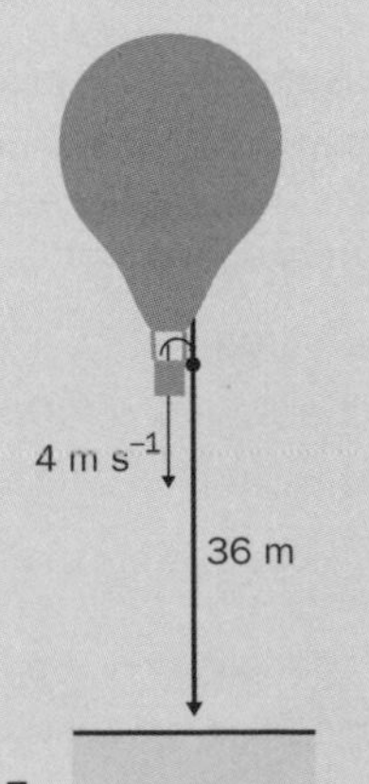

Fig. 10.17

Questions 10.5b

Assume $g = 9.8 \text{ m s}^{-2}$

1. A ball is thrown vertically into the air and takes 4.4 s to return to the thrower. Calculate **(a)** the time it takes to reach maximum height, **(b)** its speed of projection, **(c)** its maximum height, **(d)** its speed on returning to the thrower.
2. A helicopter is ascending vertically at a steady speed of 15 m s^{-1} when a package is dropped from it at a height of 90 m above the ground. Calculate **(a)** the speed of impact of the package at the ground, **(b)** the time taken for the package to fall to the ground.

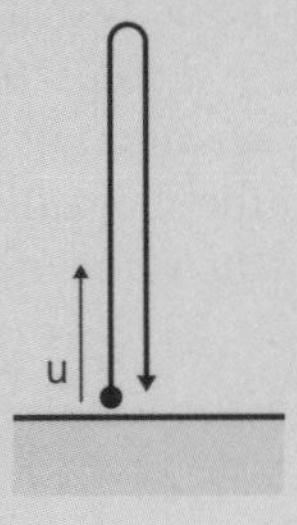

Fig. 10.20

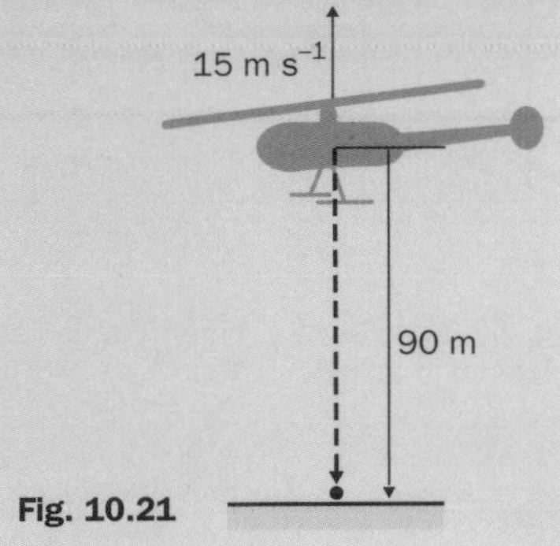

Fig. 10.21

10.6 Rates of change

Velocity

The **velocity** of an object is its rate of change of displacement with time. This may be written in the form

$$v = \frac{ds}{dt}$$

where v is the velocity, s is the displacement and d/dt means 'rate of change'. Fig. 10.22 shows how the displacement of an object released in water changes with time as the object falls. The gradient of any displacement–time curve is the velocity. In Fig. 10.22, the velocity becomes constant due to the resistance of the water. This velocity is referred to as the **terminal velocity** of the ball. To find the velocity at any point on the curve, a tangent to the curve is drawn and the gradient of the tangent is measured.

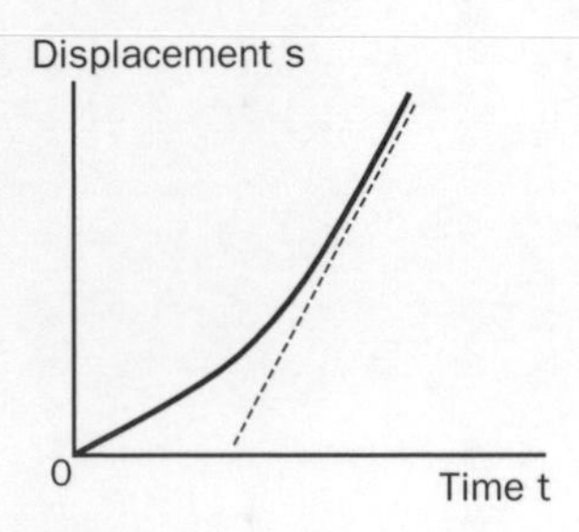

Fig. 10.22 Displacement v. time for an object released in a fluid.

Acceleration

The **acceleration** of an object is its rate of change of velocity. This may be written in the form

$$a = \frac{dv}{dt}$$

where a is the acceleration, v is the velocity and d/dt means 'rate of change'. Fig. 10.23 shows how the velocity of an object released in water increases with time. The gradient of this curve is the acceleration, which can be determined by measuring the gradient of the tangent to the curve. In Fig. 10.23, the acceleration becomes zero as the object reaches its terminal velocity. See Fig. 12.10, p. 161.

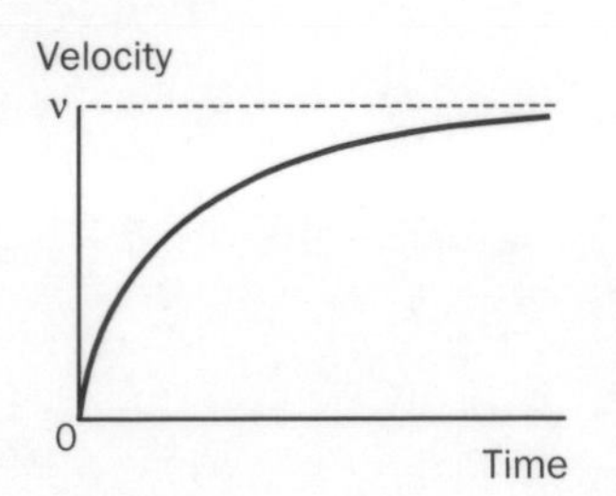

Fig. 10.23 Velocity v. time for an object released iin a fluid.

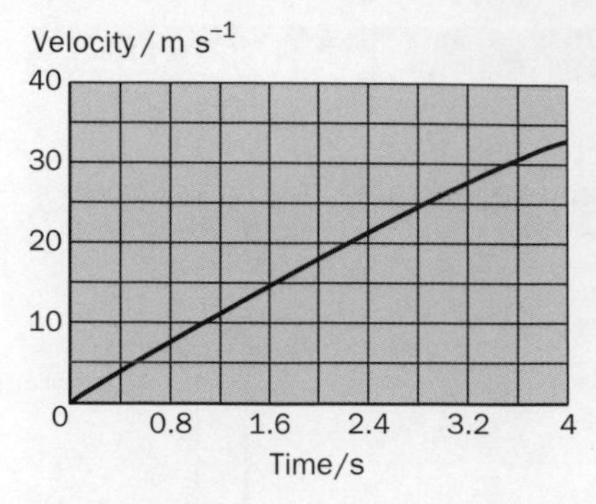

Fig. 10.24 Velocity v. time for an object released in a fluid.

See www.palgrave.com/foundations/breithaupt to model motion in a straight line using a spreadsheet.

A spreadsheet exercise

Let the acceleration of an object at an instant = a. In a short time interval, δt, its change of velocity, δυ = aδt.

Let υ = its velocity at the start of the interval, then its velocity at the end of the interval = υ + δυ; this can then used as the velocity at the start of the next time interval. This sequence can be used repeatedly to calculate the velocity of an object if its acceleration is known.

Table 10.2 shows how the velocity of an object released in a fluid changes; its acceleration a= g − kυ, where k is a constant determined by the shape of the object and the resistance to flow of the fluid. In Table 10.2, the value of k= 0.1 and δt = 0.2 seconds.

Table 10.2 Using a spreadsheet

t	υ	a = g − 0.1υ	δυ = a δt	υ + δυ
0.000	0.000	9.800	1.960	1.960
0.200	1.960	9.600	1.920	3.880
0.400	3.880	9.410	1.880	5.760
0.600	5.760	9.220	1.840	7.610
0.800	7.610	9.040	1.810	9.420
1.000	9.420	8.860	1.770	11.190

The results are plotted on a graph shown in Fig. 10.24, in which the calculations are continued to t = 4.0 s. The velocity eventually becomes constant because the fluid resistance eventually counteracts the force of gravity.

The values in the Table can be calculated using a spreadsheet, as explained on p. 141. The instructions can be used to calculate the motion of any object, given its acceleration in terms of its displacement and velocity. The data from a spreadsheet can be imported into a chart. Fig. 10.24 is a graph of velocity against time imported from a spreadsheet used to calculate the data in the table above.

Questions 10.6

Assume $g = 9.8$ m s^{-2}

1. Use the dynamics equations to calculate the displacement and velocity of a ball projected vertically upwards with an initial velocity of 24.5 m s^{-1}, every 0.5 s after projection for 5.0 s.
2. Plot graphs of the results in Q1 to show how the displacement and velocity vary with time.
3. Repeat the procedure using a spreadsheet and display the results in charts to show how the displacement and position vary with time.

Summary

- **Displacement** is distance in a given direction.
- **Speed** is rate of change of distance.
- **Velocity** is rate of change of displacement.
- **Acceleration** is rate of change of velocity.
- **Equations for uniform acceleration:**

  $v = u + at \quad s = \frac{1}{2}(u + v)t \quad s = ut + \frac{1}{2}at^2 \quad v^2 = u^2 + 2as$

- **Graphs**
  1. **Displacement v. time:** gradient = velocity
  2. **Velocity v. time:** gradient = acceleration, area under line = displacement.

Revision questions

Assume $g = 9.8$ m s^2

10.1. (a) A car initially at rest, accelerates at 1.5 m s^{-2} for 20 s, then travels at constant speed for 90 s, then decelerates to rest in a further 40 s. Its direction does not change.
(a) Calculate its maximum speed.
(b) Sketch a graph to show how its speed changes from its initial position to where it stops again.
(c) Calculate its deceleration.
(d) Calculate **(i)** the distance moved in each of the three parts of its journey, **(ii)** its average speed for the whole journey.

10.2. A cyclist accelerates uniformly from rest to a speed of 6.5 m s^{-1} in 30 s, then brakes to rest in a distance of 80 m.
(a) Calculate the acceleration and distance moved in the first 30 s.
(b) Calculate the deceleration and how long the brakes are applied before the cyclist stops.
(c) (i) Sketch a speed v. time graph. **(ii)** Calculate the average speed of the cyclist.

10.3. A tennis ball is thrown vertically upwards at an initial speed of 15 m s^{-1}.
(a) Calculate its maximum height and the time it takes to reach this height.
(b) (i) Calculate its velocity when it is 5.0 m above the ground moving downwards. **(ii)** How long does it take to reach this position?

10.4. A parachutist is descending vertically at a steady speed of 3.2 m s^{-1} when she releases an object from a height of 100 m above the ground.
(a) Calculate the speed of the object just before impact.
(b) How long does it take the object to fall freely to the ground?
(c) How far is the parachutist above the ground when the object strikes the ground?

10.5. A rocket is launched vertically and accelerates uniformly at 8.0 m s^{-2} for 40 s. Its engine then cuts out and it eventually returns to the ground.
(a) Calculate **(i)** its speed and height when the engine cuts out, **(ii)** its maximum height, **(iii)** its velocity just before impact, **(iv)** its time of flight.
(b) Sketch a velocity v. time graph for the whole flight of the rocket.

Force and Motion

Contents

Objectives

After working through this unit, you should be able to:

- ▶ calculate the motion of a body acted upon by a resultant force
- ▶ explain the link between mass and weight
- ▶ relate the resultant force on a body to its change of momentum
- ▶ explain the thrust of a jet or a rocket
- ▶ use the principle of conservation of momentum to analyse collisions and explosions

Fig. 11.1 On ice there is virtually no friction.

11.1 Newton's laws of motion

The nature of force

Friction is almost entirely absent on ice. A moving vehicle on an icy road is very difficult to control because there is hardly any grip between its wheels and the road surface. At a bend, a vehicle on ice would continue in a straight line because the icy surface is unable to exert a force on the wheels.

In the absence of any resultant force acting on it, an object remains at rest or continues to move at constant velocity. This statement is known as **Newton's 1st law of motion**. Essentially, this law defines a force as anything that can change the velocity of an object.

- An ice hockey puck moving on ice keeps moving without change of direction or change of speed because there is negligible frictional force between the puck and the ice surface. No force means no change of motion.
- When a cupboard is pushed across a floor, a force must be applied to keep the cupboard moving. This is because the hidden force of friction balances out the applied force, so there is no resultant force on the cupboard and it keeps moving at constant velocity.

Push force

Friction

Fig. 11.2 The hidden force of friction.

Force and acceleration

To investigate the link between force and acceleration, we need to eliminate friction, then time the motion of an object when a constant force is

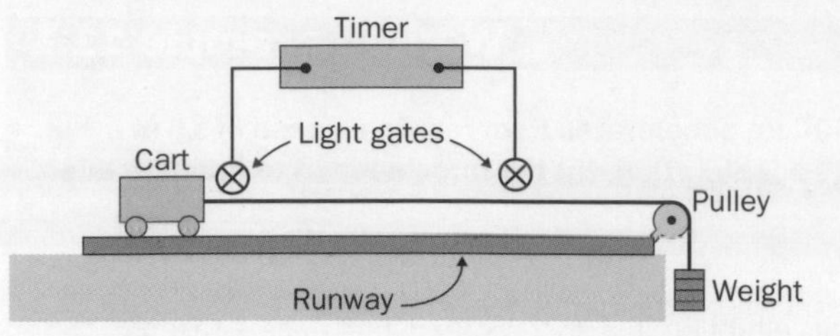

Fig. 11.3 Force and acceleration.

1. A cart with friction-free bearings is pulled along a horizontal track by a string which hangs over a pulley and supports a falling weight.
2. Different forces are applied using different weights, transferred to or from the cart to keep the total mass constant.
3. A pair of light gates is used to time how long the cart takes to travel from rest at the first light gate through a measured distance.
4. The acceleration, a, is calculated using $a = 2s \div t^2$ (from '$s = ut + \frac{1}{2}at^2$' with $u = 0$) where s is the distance between the light gates and t is the time taken for the cart to travel through this distance.
5. Other methods of measuring acceleration may be used. For example, a motion sensor linked to a computer as in Fig. 10.10 can be used to record and display the change of speed with time.

Notes

1. The equation $F = ma$ can be rearranged to give

$a = \frac{F}{m}$ to find a, given values of F and m

or $m = \frac{F}{a}$ to find m, given values of F and a.

2. If several forces act on an object and F is the resultant of these forces, the acceleration, a, of the object is given by

$$a = \frac{F}{m}$$

3. If the resultant force F is zero, the acceleration of the object is zero in accordance with Newton's 1st Law.

applied to it. Fig. 11.3 shows one way this can be achieved. Table 11.1 gives some typical results from such an experiment.

The results show that the force is proportional to the mass × acceleration.

Force, $F \propto$ mass, $m \times$ acceleration, a

This may be written as an equation

$$\mathbf{F = ma}$$

where the unit of force, the newton (N), is defined as the amount of force that would give a mass of 1 kg an acceleration of 1 m s^{-2}.

The equation $F = ma$ is **Newton's second law** for constant mass. You will meet the general form of Newton's second law of motion in Section 11.2.

Table 11.1 Typical results

Force/N	1.0	2.0	3.0	1.0	2.0	3.0
Mass/kg	0.5	0.5	0.5	1.0	1.0	1.0
Acceleration/ms^{-2}	2.0	4.0	6.0	1.0	2.0	3.0
Mass × acceleration/kg ms^{-2}	1.0	2.0	3.0	1.0	2.0	3.0

Worked example 11.1

A car of total mass 600 kg accelerates from rest to a speed of 15 m s^{-1} in 12 s. Calculate (a) its acceleration, (b) the force needed to produce this acceleration.

Solution

(a) Data to calculate acceleration, a: Initial speed, u = 0, time taken, t = 12 s, final velocity, v = 15 m s^{-1}

Using '$v = u + at$' gives $a = \frac{(v - u)}{t} = \frac{(15 - 0)}{12} = 1.25$ m s^{-2}

(b) $F = ma = 600$ kg × 1.25 m s^{-2} = 750 N

Weight and weightlessness

A freely falling object descends at constant acceleration, *g*. This acceleration is due to the force of gravity on the object (i.e. its weight). Therefore, using '$F = ma$' gives its weight as its mass × *g*.

Weight, W = mg, for an object of mass m

An object in free fall is unsupported, but it is not weightless because the force of gravity continues to act on it. An astronaut 'floating' in an orbiting space vehicle is unsupported but not weightless. The force of gravity (i.e. the astronaut's weight) is necessary to keep the astronaut in orbit, otherwise the astronaut would move off at a tangent.

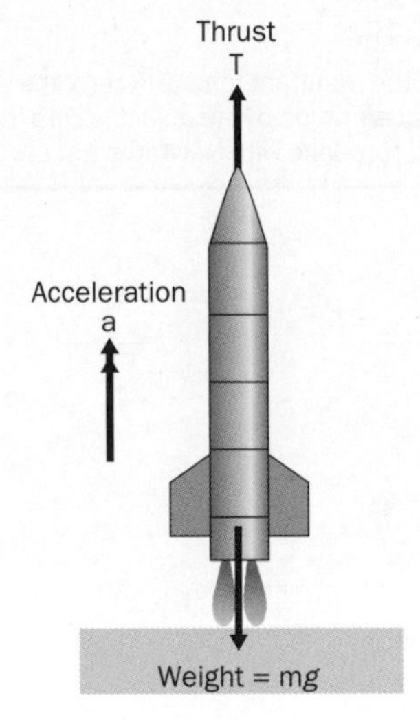

Fig. 11.4 Blast-off.

On the launch pad

For a rocket of mass, m, to take off vertically from its launch pad, the thrust, T, from its engines must exceed its weight, mg. The resultant force is T − mg. Using '$F = ma$' therefore gives

$T - mg = ma$, where a is the acceleration of the rocket.

Worked example 11.2

$g = 9.8$ m s^{-2}

A rocket of mass 5000 kg is powered by rocket engines capable of developing a total thrust of 69 000 N. Calculate (a) the resultant force on lift-off, (b) the acceleration at lift-off.

Solution

(a) Resultant force = thrust − weight = 69 000 − (5 000 × 9.8) = 20 000 N

(b) $\text{Acceleration} = \frac{\text{resultant force}}{\text{mass}} = \frac{20\,000}{5000} = 4.0$ m s^{-2}

Lift problems

When a descending lift stops, the passengers in the lift feel a greater push than normal from the floor of the lift as it slows down. This extra push is necessary to stop the passengers. The lift cable is under extra tension as the lift stops.

Consider a lift of mass, m, supported by a cable as shown.

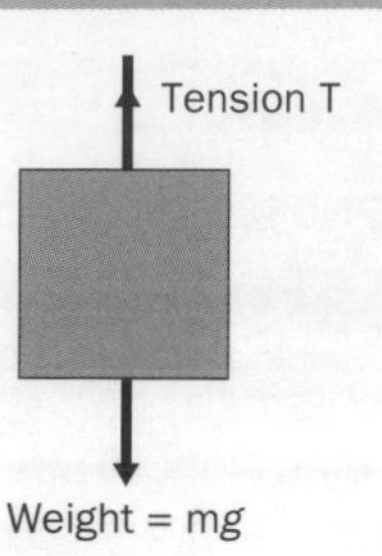

Fig. 11.5 Lift problems.

1. When the lift is at rest or moving at constant speed, the tension, T, in the cable is equal and opposite to the weight of the lift, i.e. $T = mg$.
2. When the lift is accelerating or decelerating, the tension is not equal to the weight. The difference between the tension and the weight is the resultant force and is therefore equal to the mass $\times$ the acceleration, a, of the lift. In general terms

   $T - mg = ma$

   where upward quantities are assigned + values and downward quantities are assigned − values.

(a) Lift accelerating as it moves upwards:

$T > mg$, since $a > 0$ (as acceleration is directed up, in the same direction as its velocity).

Worked example 11.3

$g = 9.8\ \text{m s}^{-2}$

A lift of mass 450 kg and a maximum load capacity of 320 kg is supported by a steel cable. Calculate the tension in the cable when

(a) it ascends at a constant speed,

(b) it ascends with an acceleration of 2.0 m s^{-2},

(c) it descends and decelerates at 2.0 m s^{-2},

(d) it ascends with a deceleration of 2.0 m s^{-2},

(e) it descends with an acceleration of 2.0 m s^{-2}.

Solution

(a) $a = 0$ ∴ tension in the lift cable, $T = mg = (450 + 320) \times 9.8 = 7550$ N

(b) $a = +2.0\ \text{m s}^{-2}$ ∴ $T - mg = ma$, where T is the tension

Hence $T = mg + ma = (770 \times 9.8) + (770 \times 2.0) = 9090$ N

(c) The velocity is downwards and the deceleration is upwards.

∴ $a = +2.0\ \text{m s}^{-2}$ (+ for upwards which is the opposite sign to the velocity)

Tension, T, is given by $T - mg = ma$

∴ $T = mg + ma = 770 \times 9.8 + 770 \times 2.0 = 9090$ N

(d) The velocity is upwards and the deceleration is downwards ∴ $a = -2.0\ \text{m s}^{-2}$

∴ Tension, T, is given by $T - mg = ma$

∴ $T = mg + ma = (770 \times 9.8) + (770 \times -2.0) = 6010$ N

(e) The velocity is downwards and the acceleration is downwards

∴ $a = -2.0\ \text{m s}^{-2}$

∴ Tension, T, is given by $T - mg = ma$

∴ $T = mg + ma = (770 \times 9.8) + (770 \times -2.0) = 6010$ N

Questions 11.1

1. A 1200 kg vehicle travelling at a speed of 25 m s^{-1} brakes to a halt in 8.0 s. Calculate **(a)** its deceleration, **(b)** the force needed to produce this deceleration.
2. A rocket of mass 8000 kg lifts off with an initial acceleration of 6.0 m s^{-2}. Calculate **(a)** its weight at lift-off, **(b)** the thrust from its engines.

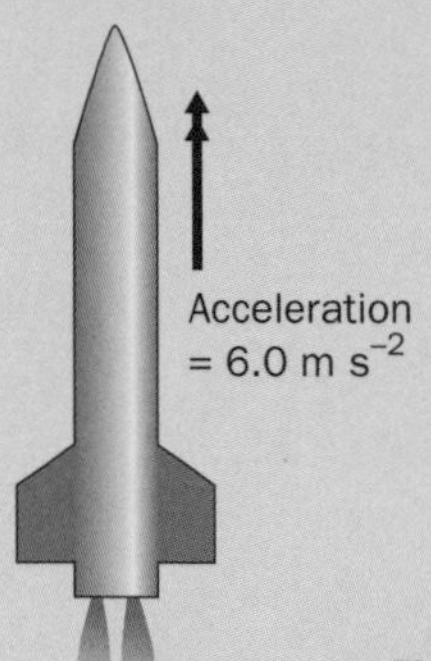

Fig. 11.6

3. A lift of total mass 1500 kg is supported by a steel cable. Calculate the cable tension when the lift
   (a) moves at constant speed,
   (b) moves upwards with an acceleration of 1.2 m s^{-2},
   (c) slows down with a deceleration of 1.2 m s^{-2} as it moves downwards.

(b) Lift accelerating as it moves downwards:
$T < mg$, since $a < 0$ (as acceleration is directed down, in the same direction as the velocity).

(c) Lift decelerating as it moves upwards:
$T < mg$, since $a < 0$ (as acceleration is directed down, in the opposite direction to its velocity).

(d) Lift decelerating as it moves downwards:
$T > mg$, since $a > 0$ (as acceleration is directed up, in the opposite direction to the velocity).

11.2 Force and momentum

A moving body possesses momentum due to its motion. If ever you have collided with someone running at top speed, you will know about momentum!

The momentum of a moving body is defined as its mass × its velocity.

The unit of momentum is the kilogram metre per second (kg m s^{-1}). Momentum is a vector quantity in the same direction as the velocity of the body. For a body of mass, m, moving at velocity, υ, its momentum is given by the equation

momentum = mυ

Consider a body of constant mass acted on by a constant force, F, which increases its speed from u to υ in time t without a change of direction. Its acceleration, a, can be written as

$$a = \frac{(\upsilon - u)}{t}$$

$$\text{hence force, } F = ma = \frac{m(\upsilon - u)}{t}$$

$$= \frac{m\upsilon - mu}{t}$$

$$= \frac{\text{change of momentum}}{\text{time taken}}$$

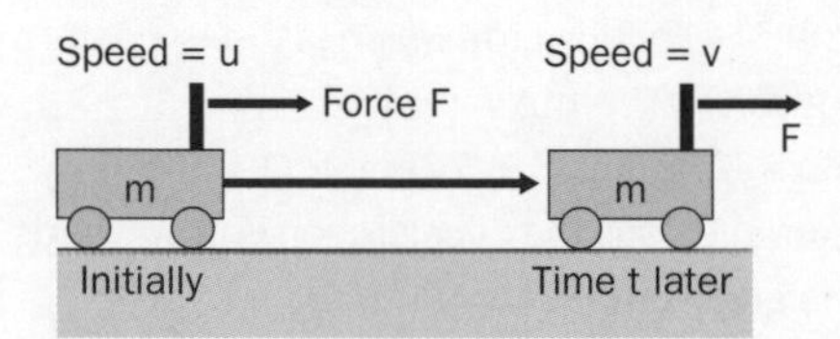

Fig. 11.7 Force and motion.

Worked example 11.4

A railway locomotive of mass 30 000 kg, moving at a speed of 8.5 m s^{-1}, was stopped on a level track in a time of 15.0 s when its brakes were applied.

Calculate (a) its change of momentum, (b) the braking force.

Solution

(a) Change of momentum = final momentum − initial momentum = 0 − (30 000 × 8.5) = −255 000 kg m s^{-1}

(b) $\text{Force} = \dfrac{\text{change of momentum}}{\text{time taken}} = \dfrac{-255\,000}{15} = -17\,000\text{ N}$

(−indicates a retarding force)

Fig. 11.8

Worked example 11.5

A rubber ball of mass 0.12 kg, moving at a speed of 25 m s^{-1}, rebounds perpendicularly from a smooth vertical wall without loss of speed. The time of contact between the wall and the ball was 0.20 ms. Calculate (a) the change of momentum of the ball due to the impact, (b) the force of impact.

Solution

(a) Initial momentum = 0.12 × 25 = 3.0 kg m s^{-1}

Final momentum = − 0.12 × 25 = − 3.0 kg m s^{-1} (− for away from the wall; + for towards the wall)

∴ Change of momentum = final momentum − initial momentum =(−3.0) − (3.0) = −6.0 kg m s^{-1}

(b) $\text{Force of impact} = \dfrac{\text{change of momentum}}{\text{impact time}} = \dfrac{-6.0}{0.20 \times 10^{-3}} = -30\,000\text{ N}$

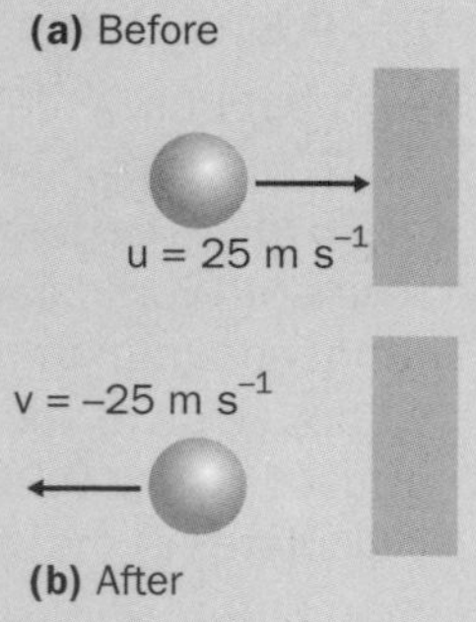

(a) Before

u = 25 m s^{-1}

v = −25 m s^{-1}

(b) After

Fig. 11.9

Note

The minus sign means that the force on the ball is directed away from the wall.

Newton's second law of motion

The equation

$$\text{force} = \frac{\text{change of momentum}}{\text{time taken}}$$

is an expression of Newton's second law of motion.

The rate of change of momentum of an object is proportional to the resultant force on it.

In more general terms, Newton's second law may be written as

$$\mathbf{F} = \frac{\mathbf{d}}{\mathbf{dt}}(\mathbf{m}v)$$

where $\dfrac{d}{dt}$ means 'rate of change' (see p. 141)

1. For constant mass,

$$F = \frac{d}{dt}(mv) = m\frac{dv}{dt} = ma$$

since acceleration, a = rate of change of velocity $\frac{dv}{dt}$

2. For a constant rate of change of mass, $\frac{dm}{dt}$

(e.g. a rocket or a jet engine),

$$F = \frac{d}{dt}(mv) = v\frac{dm}{dt}$$

where v is the relative speed at which mass is lost or gained.

Questions 11.2

$g = 9.8\ m\ s^{-2}$

1. A rubber ball of mass 0.120 kg, moving at a speed of 25 m s^{-1} perpendicular to a smooth vertical wall, rebounds from the wall without loss of speed in an impact lasting 0.004 s. Calculate **(a)** the change of momentum of the ball, **(b)** the impact force.
2. A rocket of mass 2000 kg is launched vertically with an initial acceleration of 6.5 m s^{-2}. Calculate **(a)** the weight of the rocket on the launch pad, **(b)** the thrust of its engines.

Worked example 11.6

A rocket's engines generate a thrust of 127 kN by expelling hot gas at a speed of 1500 m s^{-1}.

(a) Calculate the rate at which mass is expelled from the rocket's engines.

Fig. 11.10

(b) The rocket carries 7500 kg of fuel. Calculate how long it would take to burn this amount of fuel.

Solution

(a) Use $F = v\frac{dm}{dt}$, hence $127\,000 = 1500\frac{dm}{dt}$ to give $\frac{dm}{dt} = \frac{127\,000}{1500}$

$= 85\ kg\ s^{-1}$

(b) Since 85 kg of fuel is burned each second, the time taken to burn 7500 kg

$= \frac{7500}{85} = 88\ s$

11.3 Conservation of momentum

An impact puzzle

In an impact between a heavy lorry and a car, which vehicle experiences the greatest impact force? The car suffers most damage so you might think it experiences the greatest force. In fact, the two vehicles experience equal and opposite forces. The impact damage to the car is greater though because it is lighter. Whenever two bodies interact, they exert equal and opposite forces on one another. This is known as **Newton's third law of motion.**

When two bodies interact, they exert equal and opposite forces on one another.

Some further examples of Newton's third law at work are discussed below.

- When a chair is made to slide across a rough floor, the floor exerts a force on the chair in the opposite direction to its motion. The chair exerts an equal force on the floor in the direction the chair is moving.
- Someone leaning on a wall pushes on the wall with a force equal and opposite to the force of the wall on that person.

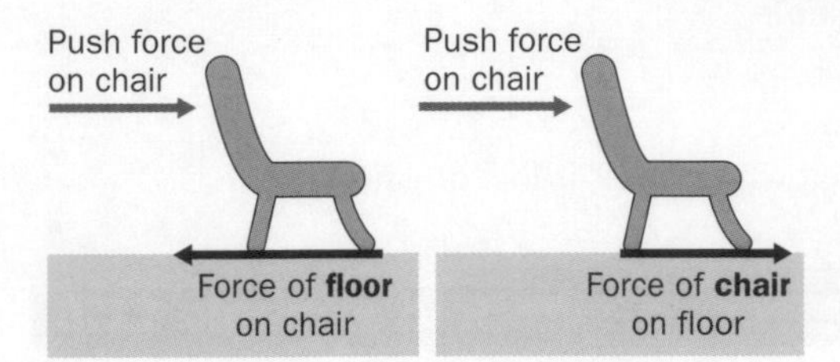

Fig. 11.11 Equal and opposite forces.

Collisions

The **impulse** of a force is defined as the force × the time for which it acts. From Newton's second law,

$$\text{force} = \frac{\text{change of momentum}}{\text{time taken}}$$

Hence the impulse of a force = force × time = change of momentum. Therefore, when two bodies X and Y interact, the impulse of X on Y is equal and opposite to the impulse of Y on X since the force of X on Y is equal and opposite to the force of Y on X. Since impulse equals change of momentum, the change of momentum of X is equal and opposite to the change of momentum of Y when X and Y interact.

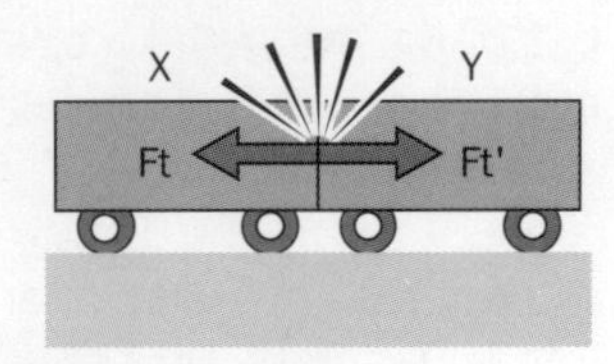

Fig. 11.12 Equal and opposite impulses.

In a collision between two bodies, momentum is transferred from one body to the other. Therefore, the momentum lost by one body is equal and opposite to the momentum gain of the other body. The total momentum is therefore unchanged, because one object loses the same amount of momentum as the other object gains. This is an example of the principle of conservation of momentum.

The principle of conservation of momentum states that when two or more bodies interact, the total momentum is conserved (i.e. not changed), provided no external force acts on the bodies.

Consider a body of mass, m_1, moving along a straight line at an initial speed, u_1, which collides with another body of mass, m_2, moving along the same line at speed, u_2. The two bodies move apart after the collision at speeds, v_1 and v_2, along the same straight line as before.

The total initial momentum $= m_1u_1 + m_2u_2$
The total final momentum $= m_1v_1 + m_2v_2$

Using the principle of conservation of momentum therefore gives

the total final momentum = the total initial momentum

$$(m_1v_1 + m_2v_2) = (m_1u_1 + m_2u_2)$$

(a)

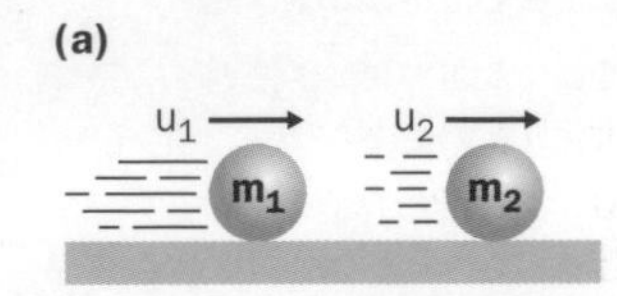

(b)

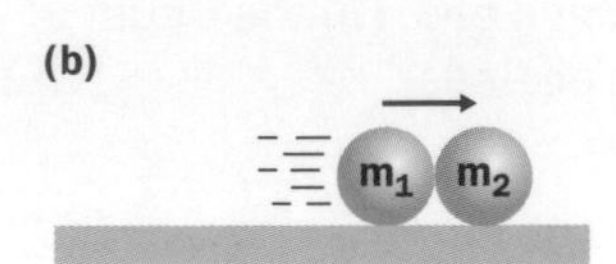

(c)

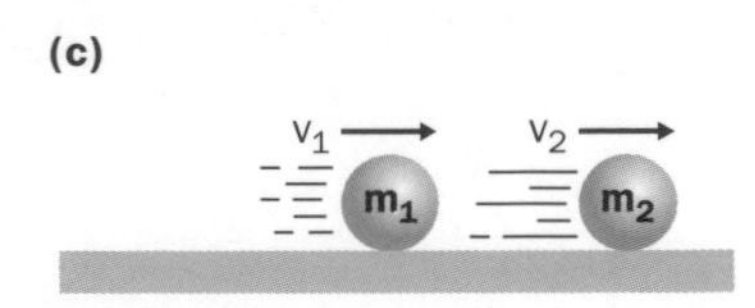

Fig. 11.13 Conservation of momentum **(a)** before impact, **(b)** on impact, **(c)** after impact.

Experiment to verify the principle of conservation of momentum

Fig. 11.14 shows a moving dynamics cart, A, on a horizontal track about to collide with a stationary cart, B. The two carts stick together on impact. Use a motion sensor to measure the speed of the dynamics cart before and after impact, on a horizontal runway.

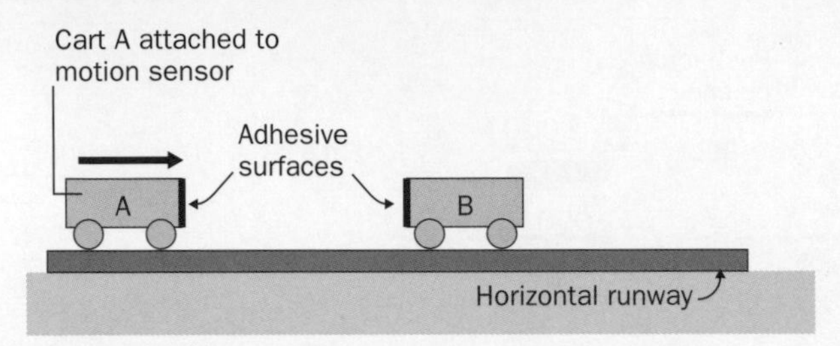

Fig. 11.14 Testing conservation of momentum.

The total momentum before impact = mass of A × speed of A before impact.

The total momentum after impact = (mass of A + mass of B) × speed of A after impact.

The measurements should demonstrate that the total momentum is unchanged by the impact.

Worked example 11.7

A vehice of mass 600 kg, moving at a speed of 30 m s^{-1}, collides with a stationary vehicle of mass 900 kg. The two vehicles lock together on impact. Calculate (a) the speed of the two vehicles immediately after the impact, (b) the force of the impact, if the duration of the collision was 0.15 s, (c) the acceleration of each vehicle during the impact.

Fig. 11.15

Solution

(a) Using the principle of conservation of momentum,

$(m_1 v_1 + m_2 v_2) = (m_1 u_1 + m_2 u_2)$

$600\,V + 900\,V = (600 \times 30) + (900 \times 0)$

where V is the speed of the two bodies locked together immediately after the impact.

Hence $1500\,V = 18\,000$

$V = \dfrac{18\,000}{1500} = 12\text{ m s}^{-1}$

Worked example 11.7 (continued)

(b) To calculate the force on vehicle 1,

use $F = \frac{m_1v_1 - m_1u_1}{t} = \frac{(600 \times 12) - (600 \times 30)}{0.15} = \frac{7200 - 18000}{0.15}$

$= -72\,000$ N

An equal and opposite force is exerted on vehicle 2.

(c) Acceleration $= \frac{\text{force}}{\text{mass}}$

For vehicle 1, acceleration $= \frac{-72\,000\text{ N}}{600\text{ kg}} = -120\text{ m s}^{-2}$ (i.e. deceleration)

For vehicle 2, acceleration $= \frac{72\,000\text{ N}}{900\text{ kg}} = 80\text{ m s}^{-2}$

Explosions

In an explosion, bits fly in all directions, each bit taking momentum away. Assuming the object that exploded was initially at rest, the total initial momentum must have been zero. The total final momentum is therefore also zero since the total momentum does not change.

To understand why the total final momentum is zero, consider an explosion where two bodies, X and Y, initially at rest, fly apart. They push each other away with equal and opposite forces for the same length of time. Therefore they exert equal and opposite impulses on each other. Since impulse is change of momentum, they therefore gain equal and opposite amounts of momentum. The gain of momentum of X is equal and opposite to the gain of momentum of Y because they move in opposite directions. The two objects recoil from each other with equal and opposite amounts of momentum.

Momentum is a vector quantity so if one direction is defined as positive and the other direction negative, one object carries away positive momentum and the other object carries away an equal amount of negative momentum.

Momentum of X after explosion $= m_x\upsilon_x$ where m_x is the mass of X and υ_x is the speed of X immediately after the explosion

Momentum of Y after the explosion $= -\,m_y\upsilon_y$ (− for the opposite direction to X) where m_y is the mass of Y and υ_y is the speed of Y immediately after the explosion

$\therefore$ Total momentum after the explosion $= m_x\upsilon_x - m_y\upsilon_y$

Also, the total initial momentum = 0 since the objects were initially at rest

$\therefore$ according to the principle of conservation of momentum,
$m_x\upsilon_x - m_y\upsilon_y = 0$
$\therefore m_x\upsilon_x = m_y\upsilon_y$

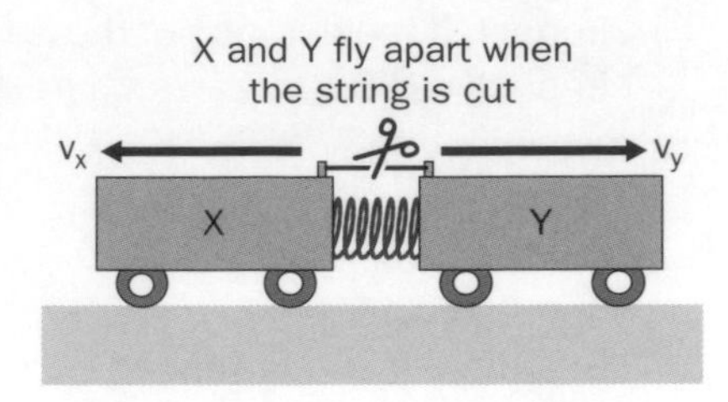

Fig. 11.16 A controlled explosion.

See www.palgrave.com/foundations/breithaupt for an experiment to compare masses using the principle of conservation of momentum.

Worked example 11.8

A student of mass 60 kg, standing on a stationary skateboard of mass 3.0 kg, jumps off the skateboard at a speed of $0.5\ m\ s^{-1}$. Calculate the speed of recoil of the skateboard.

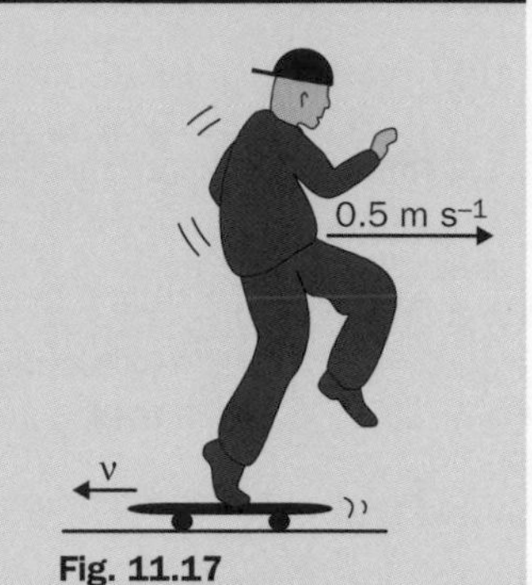

Fig. 11.17

Solution

Momentum of student immediately after jumping off the skateboard

$= 60 \times 0.5 = 30\ kg\ m\ s^{-1}$

Momentum of skateboard = 3.0 v, where v is the speed of the skateboard.

$3.0\,v = 30 \therefore v = 10\ m\ s^{-1}$

Questions 11.3

1. A vehicle of mass 800 kg, moving at a speed of $25\ m\ s^{-1}$, collides with a second vehicle of mass 1600 kg which is initially at rest. The impact reduces the speed of the first vehicle to $5\ m\ s^{-1}$ without change of direction. Calculate **(a)** the loss of momentum of the first vehicle, **(b)** the speed of the second vehicle immediately after the impact, **(c)** the force of the impact, if the duration of the impact was 0.2 s.
2. An artillery gun of total mass 500 kg is designed to fire a shell of mass 4.0 kg at a speed of $115\ m\ s^{-1}$. Calculate the recoil speed of the gun.

Summary

◆ **Momentum** = mass × velocity.

◆ **Newton's laws of motion:**

1st law An object remains at rest or in uniform motion unless acted on by a resultant force.

2nd law The resultant force on an object is proportional to the rate of change of momentum of the object.

3rd law When two objects interact, they exert equal and opposite forces on each other.

◆ **The principle of conservation of momentum** states that the total momentum of any system of bodies is constant, provided no external resultant force acts on the system.

◆ **Equation for Newton's second law**

In general, $F = \frac{d}{dt}(mv)$

For constant mass, $F = m\frac{dv}{dt} = ma$

This may be written as

$F = m\frac{(v - u)}{t} = \frac{mv - mu}{t}$ for constant acceleration.

For constant rate of change of mass, $F = v\frac{dm}{dt}$

Revision questions

$g = 9.8 \text{ m s}^{-2}$

11.1. (a) A vehicle of total mass 1500 kg accelerates from rest to a speed of 12 m s^{-1} in 10 s. Calculate **(i)** the acceleration of the vehicle, **(ii)** the force required to cause this acceleration.
(b) A lift of total mass 850 kg is suspended from four identical steel cables. Calculate the tension in each cable when the lift

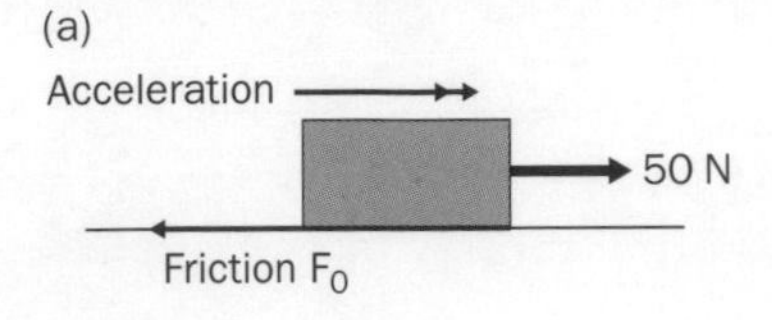

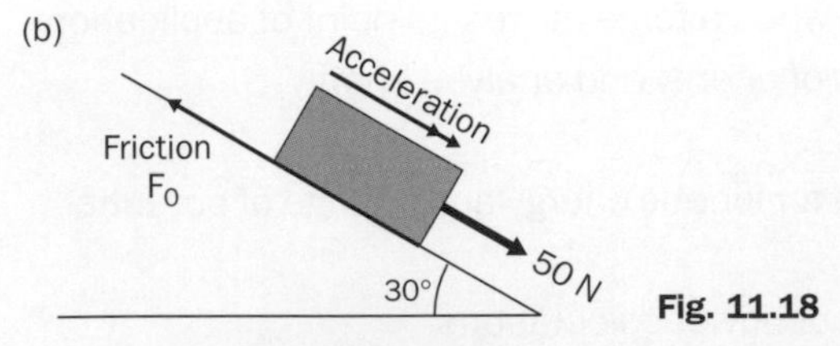

Fig. 11.18

(i) descends at steady speed, **(ii)** descends with a deceleration of 0.5 m s^{-2}.
(c) A rectangular block of mass 12 kg on a rough surface, is acted on by a force of 50 N parallel to the surface. The coefficient of friction between the block and the surface is 0.4. Calculate the acceleration of the block if the surface is **(i)** horizontal, **(ii)** inclined at 30° to the horizontal such that the force acts down the slope.

11.2. An air-rifle pellet of mass 0.001 kg, moving at a speed of 100 m s^{-1}, strikes a tree and embeds itself to a depth of 50 mm in the wood. Calculate **(a)** the loss of momentum of the pellet due to the impact, **(b)** the duration of the impact, **(c)** the force of impact.

11.3. A rocket of total mass 1200 kg accelerates directly upwards off the launch pad with a constant thrust from its engines of 16 kN.
(a) Calculate its weight and show that its acceleration is 3.5 m s^{-2}.
(b) Calculate the rate of loss of mass of fuel from the rocket if the exhaust gases are ejected at a speed of 1200 m s^{-1}.
(c) Calculate how long its fuel would last if its initial mass of fuel was 800 kg.

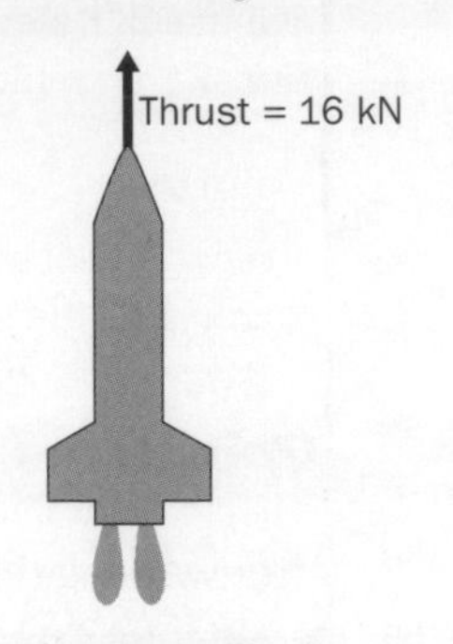

Fig. 11.19

11.4. A railway wagon of mass 1000 kg, moving along a horizontal track at a speed of 3.0 m s^{-1}, collided with a wagon of mass 2000 kg and coupled to it. Calculate the speed of the two wagons immediately after the impact if the second wagon was initially
(a) at rest,
(b) moving at a speed of 2.0 m s^{-1} in the same direction as the 1000 kg wagon was moving,
(c) moving at a speed of 2.0 m s^{-1} in the opposite direction to the direction the 1000 kg wagon was moving in before the impact.

11.5. In an experiment to measure the mass of an unknown object, the unknown object is fixed to a dynamics cart of mass 0.50 kg. A spring-loaded cart of mass 0.60 kg is then placed in contact with the first cart on a horizontal track. When the spring is released, the two carts recoil from each other due to the action of the spring. The spring-loaded cart moves away at a speed of 0.35 m s^{-1} and the other cart moves in the opposite direction at a speed of 0.22 m s^{-1}. Calculate the mass of the unknown object.

Energy and Power

Contents

Objectives

After working through this unit, you should be able to:

- ▶ calculate the work done when a force moves its point of application
- ▶ describe different forms of energy and analyse energy transformations
- ▶ derive and use formulae for kinetic energy and change of potential energy
- ▶ define power and carry out power calculations
- ▶ relate the power of an engine to force and speed
- ▶ explain what is meant by useful energy,waste energy and efficiency when energy changes occur
- ▶ carry out efficiency calculations and explain the main causes of inefficiency
- ▶ describe sources of useful energy

12.1 Work and energy

The meaning of work

Working out in a gym usually involves lifting weights or moving about vigorously. A weightlifter who raises a 400 N weight by 1.0 m does twice as much work as someone who raises a 200 N weight by 1.0 m or who raises a 400 N weight by 0.5 m. On an exercise bike, you do work by pedalling to make an axle rotate against friction. The force of friction can be altered by adjusting the 'brake system' pressing on the axle.

Work is done by a force when the force moves its point of application in the direction of the force. For example, work is done by the force of gravity when an object falls. However, if an object is moved horizontally, no work is done by the force of gravity on the object because there is no vertical movement of the object. On an exercise bike, work is done by the force exerted on the pedals but only when the pedals are moving.

Fig. 12.1 Doing work on a cardio-vascular machine.

The work done by a force is defined as the force × the distance moved by the point of application of the force in the direction of the force.

The unit of work is the **joule** (J), defined as the work done when a force of one newton moves its point of application through one metre in the direction of the force.

Work done, W = Fd

where F is the force and d is the distance moved by the point of application of the force in the direction of the force.

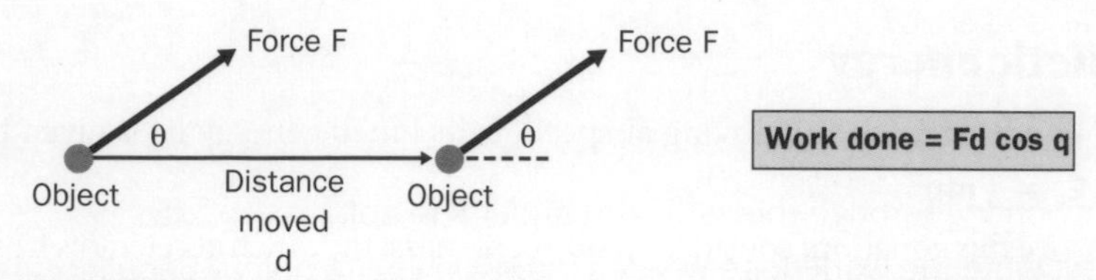

Fig. 12.2 Work and force.

Where the direction of movement is at angle θ to the direction of the force, the component of force in the direction of movement is F cos θ. Hence the work done = Fd cos θ. Note that if θ = 90°, the work done is zero, since cos θ = 90° and the force is at right angles to the displacement.

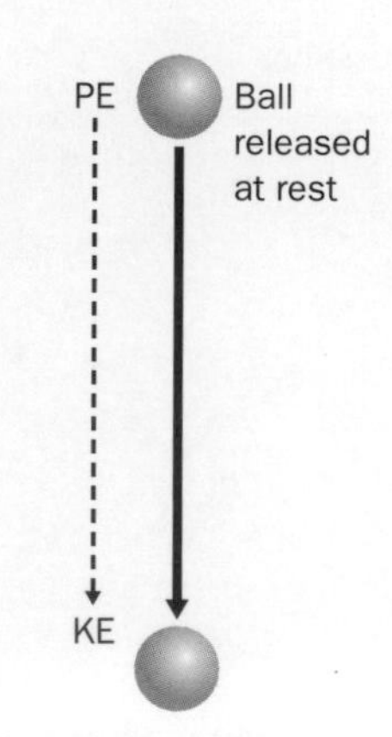

Fig. 12.3 A falling ball.

Energy

- **Energy is defined as the capacity to do work.** The unit of energy is the joule. An object can lose or gain energy, either as a result of work done by a force or through heat transfer. For example, if 100 J of work is done by a force acting on an object and 40 J of heat transfer from the object occurs, the net energy gain by the object is 60 J.
- **Energy exists in different forms**, including kinetic energy (i.e. energy of a moving object due to its motion), potential energy (i.e. energy due to position or stored energy), chemical energy, nuclear energy, electrical energy, light and sound.
- **Energy can be transformed from any one form into other forms.** Whenever energy changes from one form into other forms, the total amount of energy is unchanged. This general rule is known as the **principle of conservation of energy**.

Figs 12.3 and 12.4 show two examples of energy transformations. In Fig. 12.3, the ball loses potential energy as it falls and gains kinetic energy. Because air resistance is negligible, the gain of kinetic energy is equal to the loss of potential energy.

In Fig. 12.4, the weight gains potential energy as it rises as a result of electrical energy from the battery enabling the motor to do work. Not all the electrical energy from the battery is converted to potential energy, as heat is produced in the wires due to their resistance; friction at the bearings of the motor also generates heat.

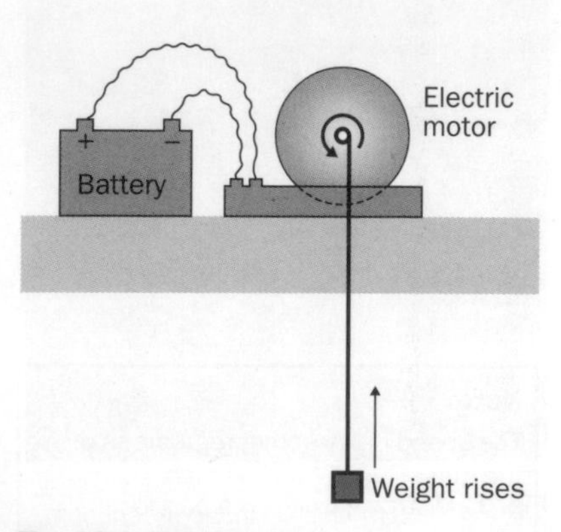

Fig. 12.4 Raising a load.

Gravitational potential energy

To raise a body, a force must be applied which is equal and opposite to its weight. The work done by the applied force is therefore equal to the weight of the body × its height gain. This is equal to the gain of potential energy of the body. (Note: strictly, potential energy is gained by the body and the Earth as a result of separating them.) Since the weight of an object equals mg, where m is its mass,

the change of potential energy, $\Delta E_p = mgh$

where h is its height gain or loss.

Kinetic energy

For a body of mass, m, moving at speed, v, its kinetic energy E_k is given by

$$E_k = \tfrac{1}{2} mv^2$$

To prove this equation, consider an object of mass, m, which accelerates from rest to speed, v, in time, t, due to a constant force, F, acting on the object.

1. The distance it moves, $s = \frac{1}{2}(u + v)t = \dfrac{vt}{2}$
2. The force acting on it, $F = ma = \dfrac{m(v - u)}{t} = \dfrac{mv}{t}$

Hence the work done by the force = Fs

$$= \frac{mv}{t} \times \frac{vt}{2} = \tfrac{1}{2} mv^2$$

which is therefore the kinetic energy of the body.

The equation above is not valid at speeds approaching the speed of light, c (= 300 000 km s^{-1}) and Einstein's famous equation, $E = mc^2$, is used in such circumstances.

Worked example 12.1

On a roller coaster, a passenger train of total mass 2500 kg descends from rest at the top of a steep incline 75 m high. Calculate (a) the loss of potential energy as a result of descending the incline, (b) (i) its kinetic energy, (ii) its speed at the bottom of the incline.

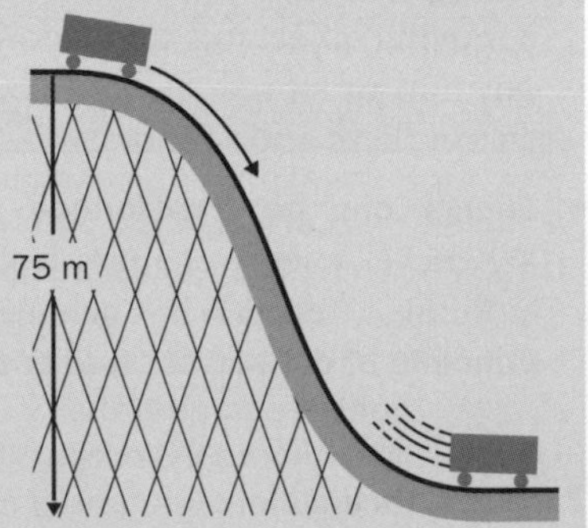

75 m

Fig. 12.5

Solution

(a) Loss of p.e. = mgh = 2500 × 9.8 × 75 = 1.84 MJ

(b) (i) Gain of k.e. = loss of p.e. provided air resistance and friction are negligible. Hence its k.e. at the bottom = 1.84 MJ

(ii) $\frac{1}{2} mv^2 = 1.84 \times 10^6$ J, hence $\frac{1}{2} \times 2500 \times v^2 = 1.84 \times 10^6$

$\therefore v^2 = 2 \times 1.84 \times 10^6 \div 2500 = 1470$

Hence $v = 38$ m s^{-1}

Note

The speed is the same, regardless of the mass, since $\frac{1}{2} mv^2 = mgh$

$\therefore v = \sqrt{(2gh)}$ which is independent of m.

Questions 12.1

$g = 9.8\ m\ s^{-2}$

1. **(a)** A ball of mass 0.12 kg is thrown vertically up at a speed of 22 m s^{-1}. Calculate **(i)** its initial kinetic energy, **(ii)** its kinetic energy at maximum height, **(iii)** its gain of potential energy on reaching maximum height, **(iv)** its maximum height gain.

   (b) A bobsleigh of total mass 150 kg accelerates down a 1500 m incline from rest, descending a total height of 200 m and reaching a speed of 55 m s^{-1} at the foot of the incline. Calculate **(i)** the loss of potential energy of the bobsleigh, **(ii)** its gain of kinetic energy, **(iii)** the work done against friction and air resistance, **(iv)** the average force resisting the motion of the bobsleigh.

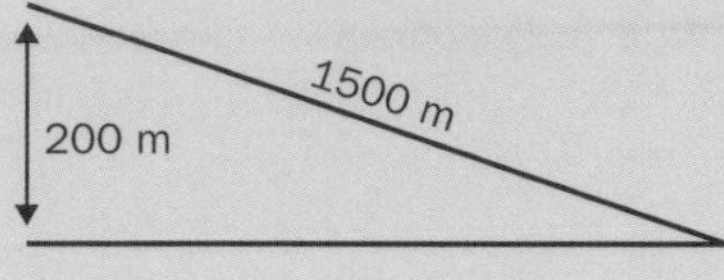

Fig. 12.6

12.2 Power and energy

In an aerobics class, a student of weight 540 N steps repeatedly on and off a box of height 0.5 m. For each step, 270 J (= 540 N × 0.5 m) of work is done. If the student takes 1.8 s for each complete step, then she does 150 J of work each second (= 270 J ÷ 1.8 s). This is her power output. In the weightlifting section, another student lifts a weight of 300 N through a height of 0.80 m in a time of 3.0 s. The weightlifter does 240 J (= 300 N × 0.80 m) of work in this process and has a power output of 80 J s^{-1}.

Power is defined as rate of transfer of energy.

The unit of power is the **watt**, equal to a rate of energy transfer of 1 J s^{-1}. Note that 1 kilowatt = 1000 watts.

For energy transfer, E, in time t,

the average power, $P = \dfrac{E}{t}$

Worked example 12.2

A student of weight 450 N climbed 4.2 m up a vertical rope in 15 s. Calculate (a) the gain of potential energy, (b) the power of the student's muscles.

Solution

(a) Gain of p.e. = 450 N × 4.2 m = 1890 J

(b) Power = $\dfrac{\text{p.e. gain}}{\text{time taken}} = \dfrac{1890}{15\ s}$ J = 126 W

Worked example 12.3

A vehicle of total weight 20 000 N travels up an incline for 100 s at a steady speed of 8.0 m s^{-1}, gaining a height of 120 m. (a) Calculate the gain of potential energy of the vehicle. (b) Estimate the output power of its engine.

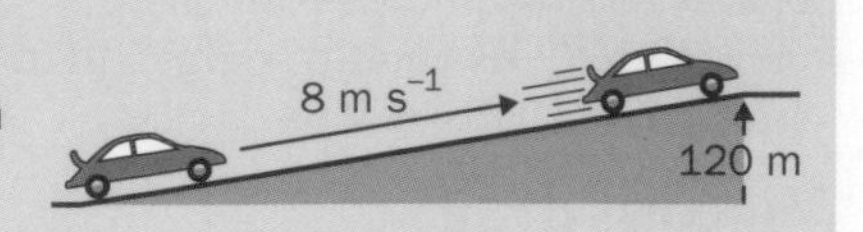

Fig. 12.7

Solution

(a) Gain of p.e. = mgh = 20 000 × 120 = 2.4 MJ

(b) Assume all the work done by the engine increases the p.e.

$$\therefore \text{output power} = \frac{\text{work done by the engine}}{\text{time taken}} = \frac{2.4\ \text{MJ}}{100\ \text{s}} = 24\,000\ \text{W}$$

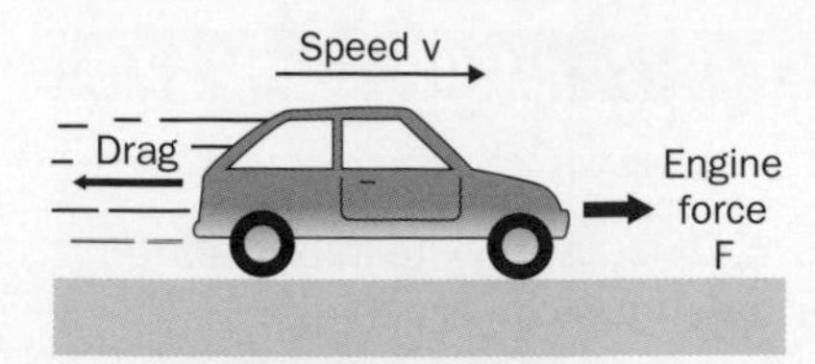

Fig. 12.8 Engine power.

Engine power and speed

Consider a vehicle of mass, m, initially at rest, driven by an engine that delivers a constant driving force, F. The initial acceleration of the vehicle is therefore F ÷ m. Because the drag force, D, on it due to air resistance increases with speed, the resultant force becomes less and less as the vehicle speeds up. Its acceleration therefore becomes less and less and it reaches a maximum speed when the drag force is equal and opposite to the driving force. In general terms

$$\text{acceleration, } a = \frac{F - D}{m}$$

- The top speed is reached when D = F, corresponding to a = 0 in the above equation.
- The initial acceleration = F ÷ m, corresponding to $\upsilon = 0$ in the above equation.
- The output power from the engine, P_E = work done per second = engine force × distance moved per second.

  $$P_E = Fv$$

- The difference between the work done by the engine and the gain of kinetic energy is the energy wasted due to resistive forces. At top speed, all the work done by the engine is wasted.

Drag factors

The drag force on a moving vehicle depends on the speed of the vehicle, its shape and on the area, A, of the vehicle as seen from the front. A moving vehicle creates turbulence in the surrounding air except at very low speed. Investigations have shown that where turbulence is created, the drag force, D, is given by the following equation

$$D = C_D \times \tfrac{1}{2} A\rho\upsilon^2$$

where υ is the vehicle speed, ρ is the fluid density and C_D is the drag coefficient of the vehicle. The term $\frac{1}{2}A\rho\upsilon^2$ is the force due to pressure on a flat plate of area, A, moving at speed, υ, hence C_D represents the shape of the vehicle. Fig. 12.9 shows how the drag coefficient varies with speed for different vehicles.

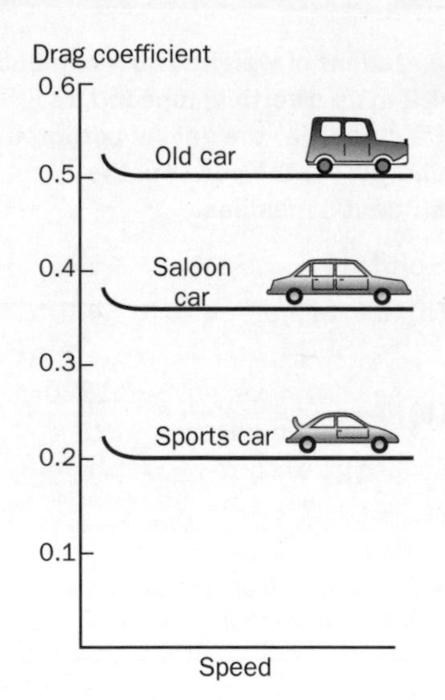

Fig. 12.9 Shape and drag.

Viscous drag

Internal friction in a fluid causes a drag force on surfaces which move relative to the fluid. Such friction is due to the **viscosity** of the fluid, which is a measure of how easily it can flow. When viscous effects are dominant and turbulence is negligible, the drag force is proportional to the speed according to the equation

$$D = k\upsilon$$

where k is a constant that depends on the object and on the viscosity of the fluid it moves through. For example, a ball released at rest in water reaches terminal speed when the drag force becomes equal and opposite to its weight. Its acceleration, a, decreases in accordance with the equation

$$a = \frac{F - k\upsilon}{m}$$

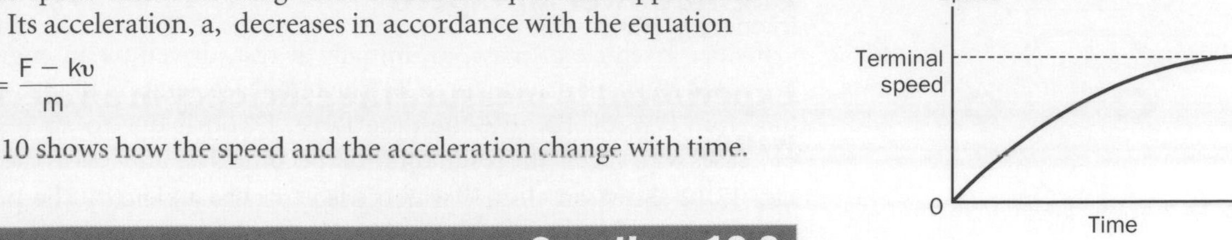

Fig. 12.10 shows how the speed and the acceleration change with time.

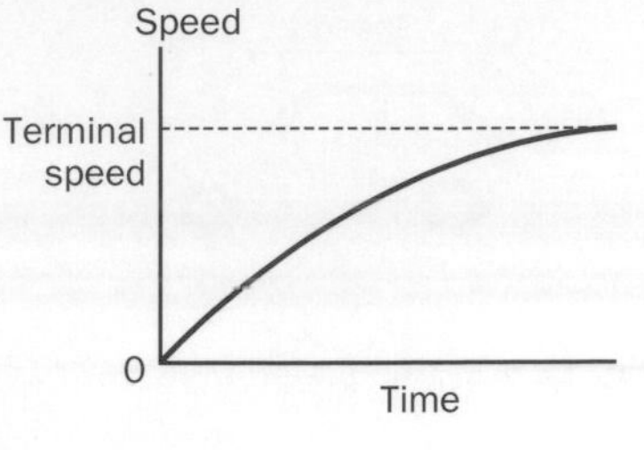

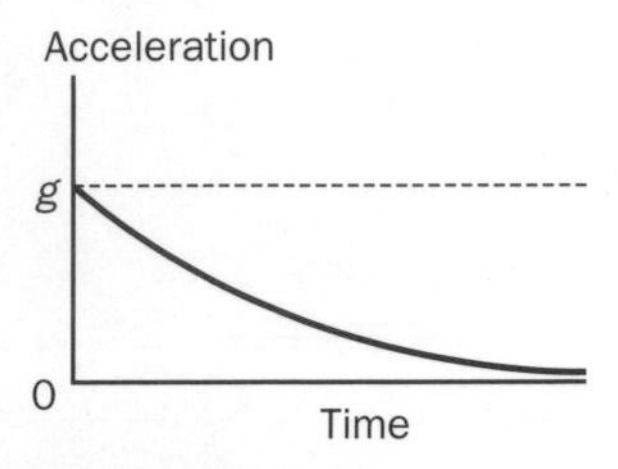

Fig. 12.10 Terminal speed.

Questions 12.2

$g = 9.8\ \mathrm{m\ s^{-2}}$

1. **(a)** A sprinter of mass 70 kg is capable of reaching a speed of $10\ \mathrm{m\ s^{-1}}$ from rest in 3.5 s. Calculate **(i)** the kinetic energy gain of the sprinter, **(ii)** the average power output of the sprinter's leg muscles.
   (b) In a keep-fit session, a person of weight 55 kg repeatedly steps on and off a box of height 0.40 m at a rate of 20 steps per minute. Calculate **(i)** the person's gain of potential energy as a result of stepping onto the box, **(ii)** the average power output of the person's leg muscles.
2. A locomotive engine of mass 20 000 kg, which has an output power of 175 kW, pulls carriages of total mass 40 000 kg at a maximum speed of $55\ \mathrm{m\ s^{-1}}$ along a level track. Calculate **(a)** the output force of the engine, **(b)** the total resistive force opposing the motion, **(c)** the average force needed to stop the train in a distance of 1000 m if the engine is switched off and the brakes are applied.

See www.palgrave.com/foundations/breithaupt to investigate motion using a spreadsheet.

Note

The equation $a = \frac{F - k\upsilon}{m}$ can be written

$a = \alpha + b\upsilon$, where $\alpha = \frac{F}{m}$ and $b = \frac{-k}{m}$

The general dynamics spreadsheet can be used to solve this equation numerically. The results may then be displayed as a graph. See p. 432.

12.3 Efficiency

Useful energy

In Fig. 12.12, an electric motor connected to a battery is used to raise a weight. Chemical energy from the battery is converted to potential energy of the weight. Some of the chemical energy is wasted due to friction in the motor and resistance heating in the connecting wires. The chemical energy converted to potential energy is referred to as **useful energy**, the energy needed for the task of raising the weight.

Friction in a machine causes energy to be wasted. The energy supplied to the machine to enable a given task to be carried out must be greater than the useful energy needed because energy is wasted due to friction, increasing the energy of the surroundings.

The efficiency of a machine is defined as

$$\frac{\text{the useful power delivered by the machine (i.e. output power)}}{\text{the power supplied to the machine (i.e. input power)}}$$

This is the same as the ratio

$$\frac{\text{the useful energy delivered by the machine}}{\text{the total energy supplied to the machine in the same time}}$$

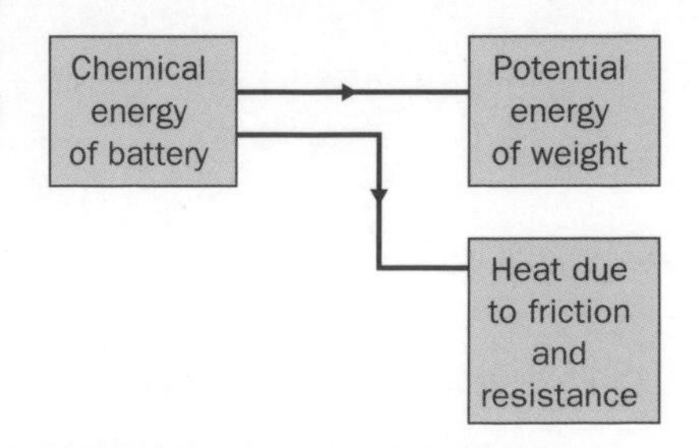

Fig. 12.11 An energy flow diagram.

Efficiency values sometimes are quoted as a percentage (i.e. the above fraction × 100).

Experiment to measure the efficiency of an electric winch

Fig. 12.12 shows an electric winch used to raise a weight. The potential energy gained by the weight can be calculated from its mass, m, and its height gain, h. The winch motor is connected to a suitable power supply via a joulemeter. The joulemeter is used to measure the energy supplied to the motor when the motor is used to raise a known weight by a measured height. The efficiency can be calculated by dividing the p.e. gain (= mgh) by the energy supplied via the joulemeter.

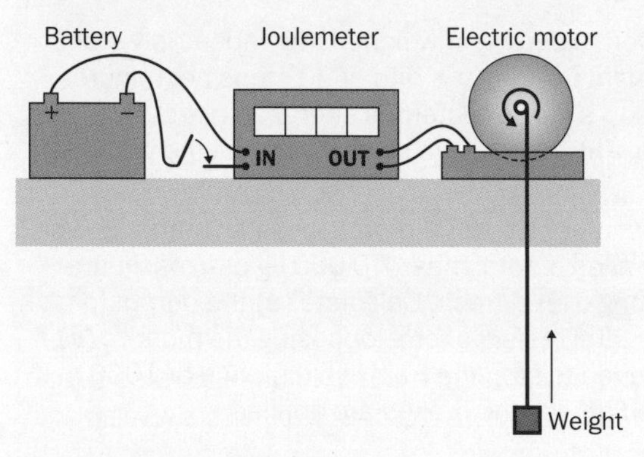

Fig. 12.12 Measuring efficiency.

Sample results

Mass = 0.50 kg, height gain = 1.2 m

Increase of joulemeter reading = 8.2 J

$$\text{Efficiency} = \frac{\text{p.e. gain}}{\text{electrical energy supplied}} = \frac{0.50 \times 9.8 \times 1.2}{8.2} = \frac{5.9}{8.2} = 0.71$$

The problem with energy

Energy tends to spread out and become less useful when it changes from one form into other forms. For example, friction between the moving parts of a machine causes heating and reduces the efficiency of the machine. Energy is wasted due to friction and this energy spreads out, becoming less useful as a result. Even in processes where energy is concentrated (for example, charging a car battery), energy must be used to make the process happen and some of this energy will be wasted (for example, the charging current will heat the circuit wires when a car battery is charged). Another example is when energy is stored in an elastic object. In theory, all the stored energy in a perfectly elastic object can be used usefully. For example, a perfectly elastic ball released above a horizontal surface would repeatedly rebound to the same height. In practice, the ball would lose height after each successive bounce and eventually it would stop bouncing. Part of the reason for this is that such a ball is not perfectly elastic. Its initial potential energy is gradually converted to internal energy of the ball and its surroundings. In general, when energy changes from one form into other forms, the total energy is conserved but it spreads out and becomes less useful. This is why fuel reserves must be used as efficiently as possible.

Worked example 12.4

A 300 W electric winch takes 15 s to raise a crate of total weight 540 N through a height of 1.6 m. Calculate (a) the energy supplied to the winch in this time, (b) the potential energy gain of the crate, (c) the energy wasted when the crate is raised 1.5 m, (d) the efficiency of the electric winch.

Solution

(a) Energy = power × time = $300 \times 15 = 4500$ J

(b) P.e. gain = $540 \times 1.6 = 864$ J

(c) Energy wasted = $4500 - 864 = 3636$ J

(d) Efficiency = $\frac{864}{4500} \times 100\% = 19\%$

Questions 12.3

1. A 300 W electric motor attached to a winch is used to raise a load of total weight 320 N through a height of 9.5 m in 25 s. Calculate **(a)** the electrical energy supplied to the motor in this time, **(b)** the useful energy supplied by the motor, **(c)** the energy wasted, **(d)** the efficiency of the motor.
2. A load of 150 N is raised with the aid of a pulley system using an effort of 40 N. The effort must move through a distance of 4 m to raise the load by 1 m. Calculate **(a)** the p.e. gain of the load when it is raised by 2.5 m, **(b)** the work done by the effort to raise the load by 2.5 m, **(c)** the efficiency of the pulley system.

12.4 Energy resources

Fuel supply and demand

Each person in a developed country uses about 5000 J of energy each second on average, corresponding to about 150 thousand million joules per year. Fossil fuels such as wood, coal, oil and gas release no more than about 50 MJ per kilogram when burned. Prove for yourself that at least 3000 kg of fuel each year is needed to supply your energy needs, before taking account of inefficiencies and energy needed to extract and transport fuel to where it is needed. With these factors included, you probably need about 10 000 kg of fuel each year to maintain your lifestyle. The world's fuel reserves are finite and unlikely to last more than 200 years or so at the

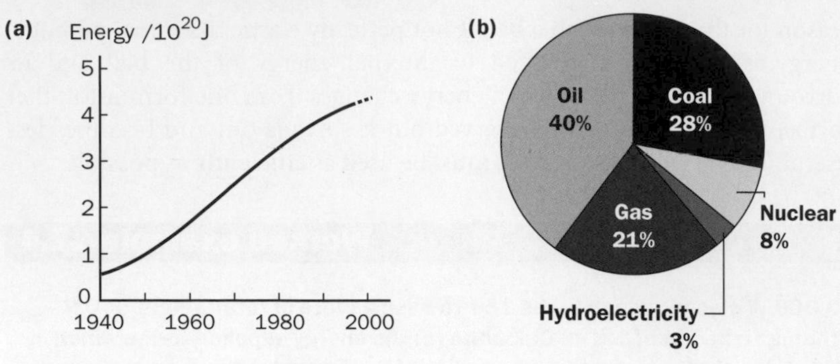

Fig. 12.13 World energy demand.

present rate of consumption. In comparison, people in remote areas in third-world countries use far less fuel, mostly wood for cooking. As living standards rise in these countries, energy consumption will rise and fuel reserves will be used more quickly.

Fig. 12.13a shows how the world demand for energy has risen over the past 70 years or so. In 1990, world demand exceeded 3×10^{20} J and could well double by 2020. Present reserves of fossil fuels have been estimated as equivalent to about 2 million million tonnes of coal, corresponding to about 5×10^{22} J or about 100 years of supplies at an energy consumption rate of 5×10^{20} J per year.

Table 12.1 shows world fuel usage (1990) and fuel reserves for different types of fuels. Present reserves of oil and gas are unlikely to last much beyond 2050. The fact that coal reserves are likely to last much longer has interesting, long-term implications. The present generation of nuclear power stations uses only the U-235 content of uranium which consists of more than 99% U-238 and less than 1% U-235. At the present rate of use, world uranium reserves are likely to last no more than about 100 years. The fast-breeder nuclear reactor uses plutonium as its fuel. This is an artificial element which is produced from U-238 in thermal reactors. The world's uranium reserves could therefore last many centuries if fast-breeder reactors replaced present thermal reactors.

Table 12.1 Fuel reserves in millions of tonnes of coal equivalent (Mtce)

	World reserves	World use per year
Oil	211 000	4800
Gas	199 000	2800
Coal	1 032 000	3300
Nuclear	54 000[a]	900
	2 670 000[b]	–
Hydro	renewable	320

[a] Thermal
[b] Fast Breeder

Fuel and the environment

Fossil fuels

Fossil fuels produce so-called 'greenhouse' gases, such as carbon dioxide, when burned. These gases are thought to be responsible for atmospheric warming in recent decades which could cause the polar ice caps to melt. Low-level countries bordering the oceans would be threatened if the ice caps melted. In addition, burning coal releases sulphur dioxide into the atmosphere which then dissolves in rain droplets to produce acid rain. This can have a severe effect on plant growth far away from the coal-burning sites.

Nuclear fuel

Nuclear fuel releases about a hundred thousand times as much energy per kilogram as fossil fuel. Your annual energy needs would be met by about

100 kg of uranium. In a nuclear reactor, energy is released when nuclei of U-235 are split as a result of neutron bombardment. Each fission (i.e. split) of a U-235 nucleus releases two or three neutrons, which then proceed to split other U-235 nuclei. This chain reaction is controlled in a nuclear reactor by the presence of neutron-absorbing control rods which ensure only one neutron per fission goes on to produce further fission. See p. 349.

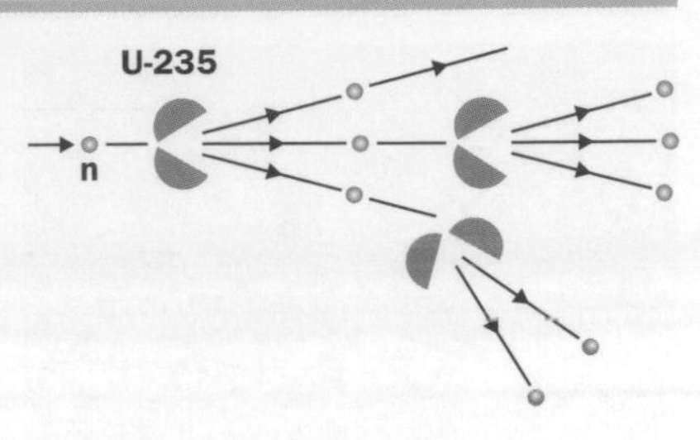

Fig. 12.14 A chain reaction.

The fission products in a fuel rod are extremely radioactive. In addition, the spend fuel rods contain highly radioactive plutonium produced as a result of U-238 nuclei absorbing neutrons. After removal from the reactor, the spent fuel rods are first placed in cooling ponds for several months until they are cool enough to be processed. The unused uranium and plutonium is removed and stored for possible re-use at this stage. The fission products are highly radioactive for many years and are stored in sealed containers in safe, underground conditions.

Renewable energy resources

The Sun radiates energy at a rate of about 4×10^{26} J every second. In one year, the Earth receives about 5×10^{24} J of energy from the Sun. World energy demands would be met if just 1% of 1% of the energy received by the Earth from the Sun could be used. Solar heating panels and solar cells use sunlight directly. Solar panels are an example of a renewable energy resource, which is an energy resource that does not use fuel. Other renewable energy resources such as hydroelectric power, wind power and wave power use solar energy indirectly. The Sun heats the atmosphere and the oceans in the equatorial regions, causing circulation currents that generate weather systems responsible for waves and winds.

Solar panels

At the Earth above the atmosphere, a surface of area 1 m^2, facing the Sun directly, would receive 1400 J every second. Much less energy is received at sea level outside the equatorial regions. Even so, a solar heating panel of area 1 m^2 facing South, in Northern Europe or North America, could absorb 500 J per second from the Sun.

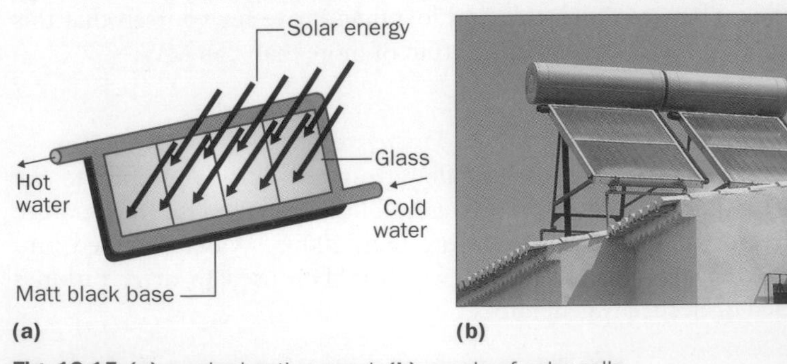

Fig. 12.15 **(a)** a solar heating panel, **(b)** panels of solar cells.

Fig. 12.16 Aerogenerators on a wind farm. Wind power creates no emissions and is reasonably efficient.

Solar cells in panels are used to provide power to satellites in orbit above the atmosphere. At present, these cells are not very efficient, although they have been developed to the extent that they can provide sufficient power to specially designed vehicles such as 'Solar challenger'.

Hydroelectric power

Hydroelectric power could make a significant contribution to energy supplies in hilly areas. A lot of rainfall is necessary. Prove for yourself that 1 cm of rain (density 1000 kg m^{-3}) over an area of 1 km^2 ($= 10^6$ m^2) running downhill by a vertical distance of 500 m would release 5×10^{10} J of potential energy. At present, hydroelectric power stations worldwide meet no more than about 3% of world energy needs.

Wind power

Wind power, generated by means of aerogenerators, makes a small but increasing contribution to national energy needs in some countries. The blades of an aerogenerator are forced to rotate by the wind, turning an electricity generator in the aerogenerator tower. A wind farm of aero-generators, each capable of producing up to 1 MW of electrical power, could meet the needs of a small town.

Wave power

See www.palgrave.com/foundations/breithaupt for more about energy resources.

Wave power is generated by floating devices when waves make one part of the device move relative to the other parts. Scientists reckon that more than 50 kW of power can be obtained per metre length of wavefront in certain coastal areas.

Tidal power

Tidal power is due to the rise and fall of ocean tides. This occurs as a result of the gravitational pull of the Moon on the Earth's oceans. A tidal barrier is designed to trap each high tide, releasing it gradually through turbines that drive electricity generators. Sites such as the Severn Estuary in Britain are particularly suitable, as the tide is heightened by the shape of the estuary. Sea water over an area of 10 km by 10 km, dropping in height by a 5 m drop, would release 12.5 million million joules of potential energy over 6 hours between high tide and low tide. Prove for yourself that this corresponds to an average power output of more than 250 MW.

Geothermal power

Geothermal power is due to heat flow from the Earth's interior to the surface. In some parts of the Earth, this heat flow has raised the temperature of underground rock basins to more than 200°C. Water pumped into these rock basins produces steam, which is then used to drive turbines connected to electricity generators.

Questions 12.4

1. **(a)** Solar cells at a certain site are capable of generating 500 W of electrical power per square metre. What would be the area of a panel of these cells capable of generating 5 kW of electrical power?
   (b) A solar heating panel is capable of heating water flowing through it at a rate of 0.012 kg s^{-1}, from 15°C to 30°C. Calculate the rate at which solar energy is absorbed by this panel. The specific heat capacity of water is 4200 J kg^{-1} K^{-1}.
2. The power, P, generated by an aerogenerator, depends on the wind speed, v, in accordance with the equation $P = \frac{1}{2}\rho A v^3$, where A is the area swept out by the blades and ρ is the density of air. Calculate the power from an aerogenerator which has blades of length 15 m when the wind speed is **(a)** 5 m s^{-1}, **(b)** 15 m s^{-1}. The density of air is 1.2 kg m^{-3}.

Summary

- **Work** = force × distance moved in the direction of the force.
- **Energy** is the capacity of a body to do work.
- **Power** = rate of transfer of energy.
- **The principle of conservation of energy** states that when energy changes from one form to other forms, the total amount of energy is unchanged.
- **Efficiency** $= \frac{\text{output power}}{\text{input power}} = \frac{\text{useful energy delivered}}{\text{total energy supplied}}$
- **Equations**
  Kinetic energy $= \frac{1}{2}mv^2$
  Change of potential energy = mgh
  Power, $P = \frac{\text{Energy transferred}}{\text{time taken}}$
- **Energy resources**
  Fuels: fossil fuels (coal, oil, gas), nuclear fuel (uranium-235 and plutonium).
  Renewable energy resources: solar heating panels, solar cells, hydroelectricity, wind power, wave power, tidal power, geothermal power.

Revision questions

12.1. A luggage trolley and luggage cases, total mass 120 kg, is pushed up a ramp of length 40 m which is inclined at an angle of 5°, in 45 seconds.

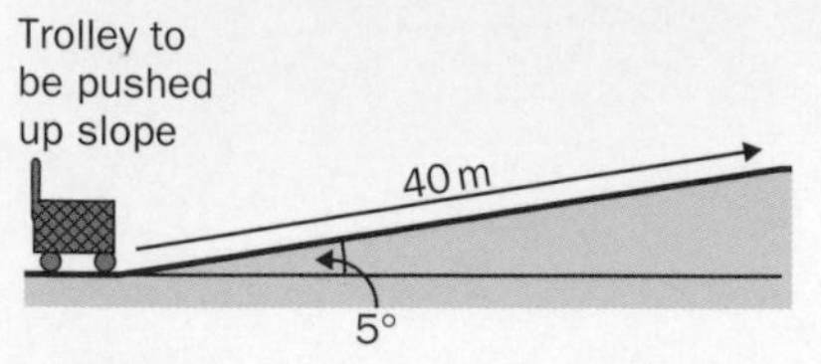

Fig. 12.17

(a) Calculate **(i)** the height gain of the trolley, **(ii)** the gain of potential energy of the trolley and the luggage.
(b) **(i)** Calculate the minimum force needed to push the trolley up the ramp. **(ii)** Explain why more force is needed in practice.
(c) Calculate the minimum amount of work done each second.

12.2. A car of total mass 600 kg travelled at a constant speed of 21 m s^{-1} up a steady 1 in 20 incline.
(a) Calculate **(i)** the height gain of the car each second, **(ii)** its potential energy gain each second.
(b) The output power of the car engine when travelling up the incline was 25 kW. Calculate **(i)** the work done against resistive forces each second, **(ii)** the resistive force.

12.3. A pile driver consisting of an iron block of mass 4000 kg is dropped from rest, through a height of 3.0 m, onto the top end of a vertical steel girder of mass 6000 kg which has its lower end resting on the ground.

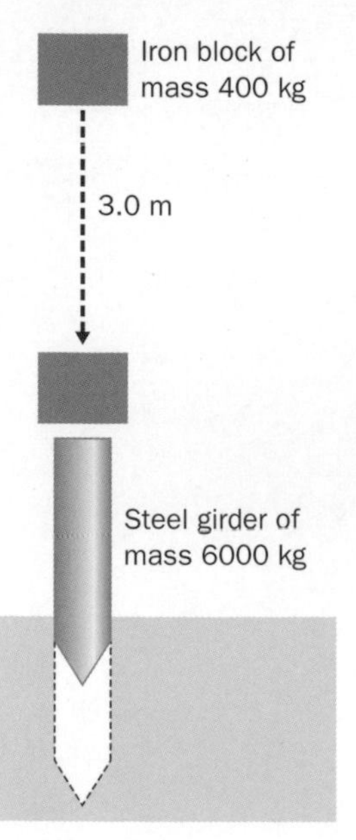

Fig. 12.18

(a) Calculate the speed of the pile driver just before impact with the girder.
(b) Use the conservation of momentum to calculate the speed of the pile driver and the girder just after impact, assuming the pile driver does not rebound from the girder.
(c) As a result of the impact, the girder penetrates 1.5 m into the ground. Calculate the force of friction on the girder due to the ground.

12.4. A steel ball of mass 0.25 kg hanging from the end of a thread at rest, is in contact with a steel ball of mass 0.15 kg which also hangs from a thread. The heavier ball is raised through a height of 0.10 m with the thread taut, by drawing it aside. It is then released and strikes the lighter ball, causing the lighter ball to gain a height of 0.05 m.

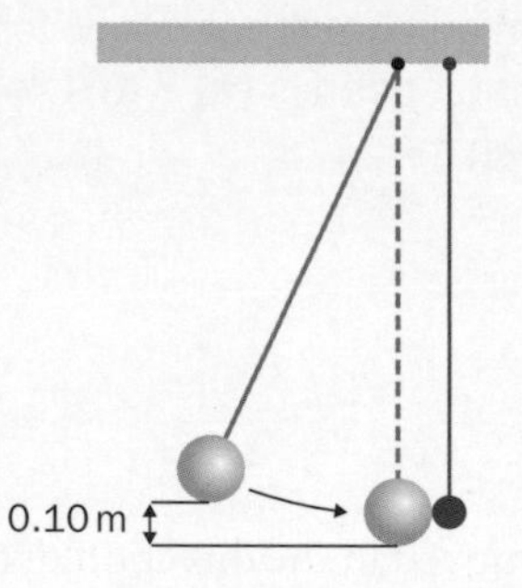

Fig. 12.19

(a) Calculate **(i)** the gain of potential energy of the heavier ball as a result of drawing it to one side, **(ii)** the speed of the heavier ball just before impact.
(b) Calculate **(i)** the gain of potential energy of the lighter ball as a result of gaining a height of 0.05 m, **(ii)** the speed of the lighter ball just after impact.
(c) **(i)** Use the principle of conservation of momentum to calculate the velocity of the heavier ball just after the impact. **(ii)** Hence calculate the height gain of the heavier ball as a result of the impact.

12.5. A vehicle of mass 1200 kg, travelling at a speed of 30 m s^{-1}, collides with a stationary car of mass 800 kg. The two vehicles lock together on impact.
(a) Calculate the speed of the two vehicles immediately after impact.
(b) Calculate the loss of kinetic energy due to the impact.
(c) The impact causes the 1200 kg vehicle to slow down in a time of 0.12 s. Calculate the force of the impact.

Electricity

Where would you be without electricity? We take electricity for granted – until there's a power cut. Electricity in every home is perhaps the greatest development of the 20th century. In Part 4, we consider the nature of electricity, how it is controlled and stored in electric circuits and how it is measured. The principles of electronics and its applications are also considered. Mathematical skills from earlier parts can be consolidated in this part, through circuit calculations, before moving on to the more demanding later parts.

Introduction to Electricity

Objectives

After working through this unit, you should be able to:

- ▶ explain the nature and cause of static electricity
- ▶ explain the nature of conductors, insulators and semiconductors
- ▶ measure an electric current and relate charge and current
- ▶ define and measure potential difference, and relate electrical power to current and potential difference
- ▶ describe the characteristics of common electrical components
- ▶ define resistance and calculate the resistance of a combination of resistors

Contents

13.1 Static electricity

Electric charge

Certain insulating materials, such as polythene and perspex, are easily charged with static electricity by rubbing with a dry cloth. Test a plastic ruler by rubbing it with a dry cloth, then see if it can pick up small bits of paper or if it generates a tiny spark when you hold its charged surface near a finger. Static electricity was first discovered in Ancient Greece when it was found that amber, a naturally occurring substance, produced sparks when rubbed. The word 'electricity' is derived from the Greek word for amber.

When investigations using many different materials were carried out over 200 years ago, scientists discovered that there are two types of electric charge. They discovered that objects carrying the same type of charge always repel each other, whereas objects carrying different charges always attract each other. This simple rule explains many effects of static electricity.

Like charges repel; unlike charges attract.

The two types of charge cancel each other out. For example, if a charged object is placed in an insulated metal can, the can also becomes charged. If a second charged object carrying the other type of charge is also placed in the can, the can becomes uncharged. One type of charge, the charge on glass rubbed with silk, was therefore labelled 'positive' and the other type 'negative'.

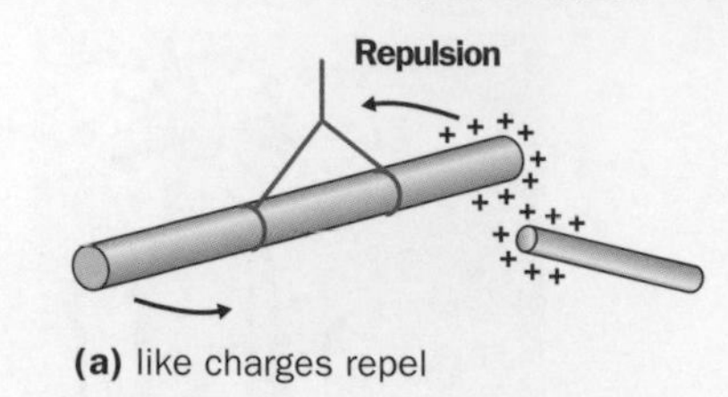

(a) like charges repel

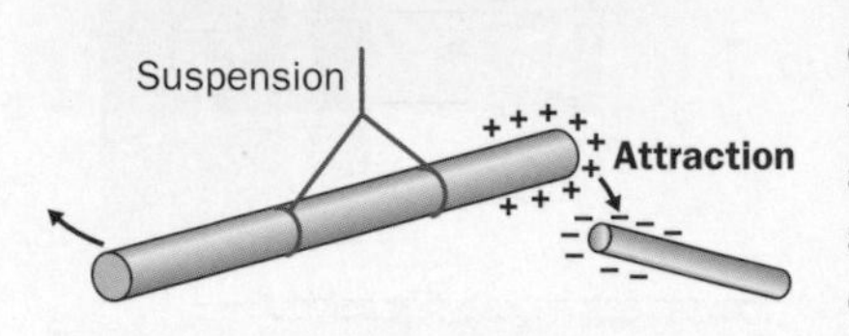

(b) unlike charges attract

Fig. 13.1 Attraction and repulsion.

Inside the atom

The atom is the smallest particle of an element that is characteristic of the element. Every atom contains a positively charged nucleus, where most of the atom's mass is located. The nucleus is surrounded by electrons which are much lighter, negatively charged particles. The electrons are held in the atom by the force of attraction of the oppositely charged nucleus. They occupy certain orbits called 'shells' round the nucleus. Each shell can hold a certain number of electrons, which fill the shells from the innermost outwards. In an uncharged atom, the total negative charge of the electrons is equal and opposite to the positive charge of the nucleus. Fig. 13.2 represents an atom of the metallic element, sodium.

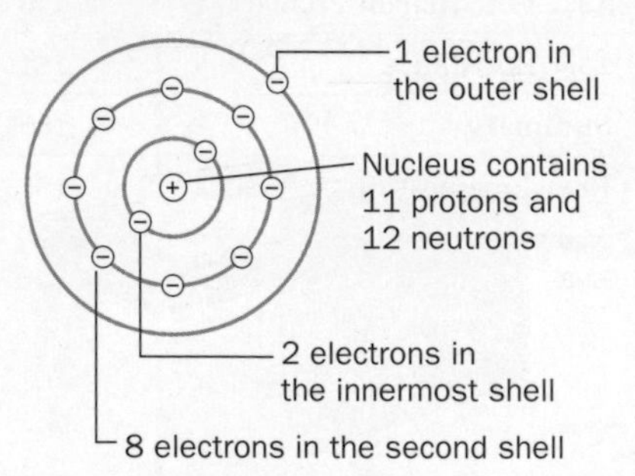

Fig. 13.2 Inside the atom.

Explaining static electricity

Rubbing a perspex rod with a dry cloth causes the rod to become positively charged. The reason why this occurs is because the friction between the rod and the cloth detaches electrons from the atoms of the rod and they transfer to the cloth. Rubbing a polythene rod with a dry cloth causes the rod to become negatively charged. The reason why this occurs is because the rod gains electrons from the dry cloth.

Experiment to investigate static electricity

1. **The gold leaf electroscope** is designed to detect static electricity. Charge placed on the cap spreads to the stem and the gold leaf of the electroscope. Because the stem and leaf acquire the same charge, the leaf is repelled by the stem. If the cap is touched by hand, the leaf returns to its original position because the charge on the electroscope leaks to earth through the body. Test different insulating materials to find out how easily each material can be charged by friction.
2. **To determine the type of charge on a charged object**, charge the electroscope with a known charge (e.g. negative charge from a charged polythene rod). If the leaf rises further when an unknown charged object is held above the cap, the unknown object must carry the same type of charge as the known charge.
3. **Charging by induction** An insulated metal object can be charged without direct contact with a charged body. Fig. 13.4 shows how this occurs. In (a), a negative rod held near an uncharged insulated conductor forces electrons to the other side of the conductor. Touching the conductor briefly allows these electrons to flow to earth. When the rod is removed, the conductor is left with a positive charge. In (b), a positive rod held near the conductor attracts electrons in the conductor. When the cap is touched briefly, electrons flow from earth onto the conductor. When the rod is removed, the conductor is left negatively charged.

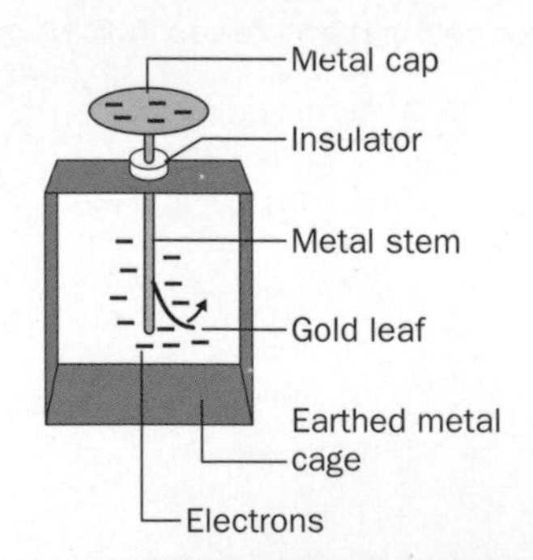

Fig. 13.3 The gold leaf electroscope.

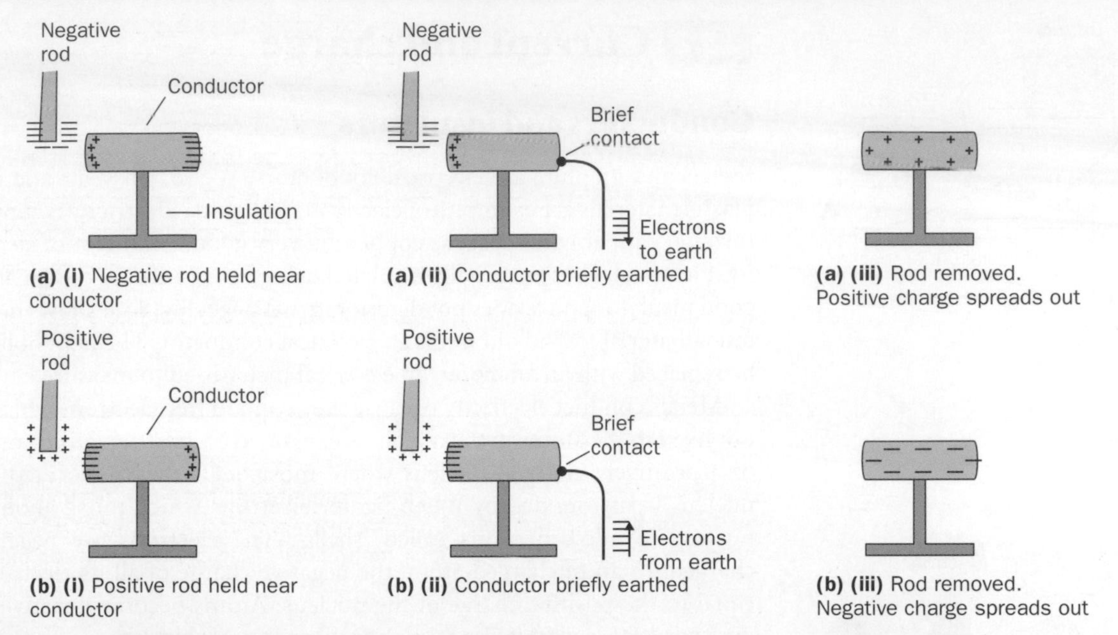

Fig. 13.4 Charging by induction **(a)** positively **(b)** negatively.

Static hazards

Sparks generated by static electricity can cause flammable vapours and powders to explode. Antistatic flooring in an operating theatre is necessary to prevent any static electricity building up, as anaesthetic gases are flammable. Oil tanks and pipes are earthed to prevent static electricity building up. Earthing an object allows electrons to transfer between it and the ground, which prevents static electricity building up.

Fig. 13.5 A conducting strip fitted to a vehicle prevents the build-up of static electricity.

Questions 13.1

1. **(a)** Explain in terms of electrons why a polythene rod becomes negatively charged when it is rubbed with a dry cloth.

   (b) A charged polythene rod is held near one end of a charged rod that is freely suspended in a horizontal position. The suspended rod is repelled by the polythene rod. What type of charge does the suspended rod carry?

   (c) Given a polythene rod, a perspex rod, a dry cloth and a gold leaf electroscope, describe how you would find out the type of charge carried by a rod of unknown material that can be charged by rubbing with the dry cloth.

2. **(a)** Integrated circuits can be damaged by static electricity. Describe how the metal pins of a microcomputer chip could become accidently charged without direct contact with a charged body.

   (b) When oil is delivered from an oil tanker to a storage tank, why is it necessary to earth the tanker and the delivery pipe?

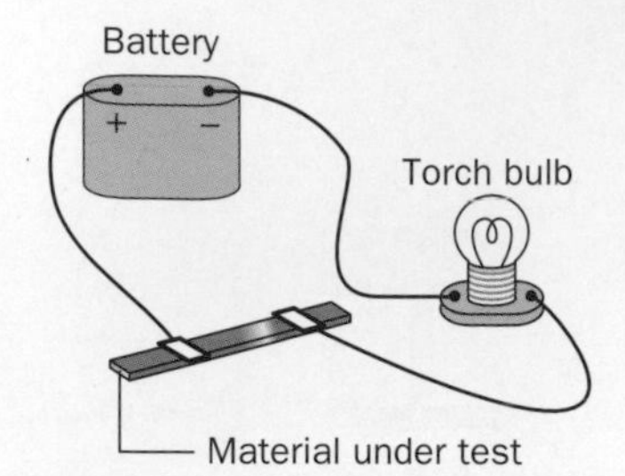

Fig. 13.6 Testing conductors and insulators.

13.2 Current and charge

Conductors and insulators

Metals and graphite are electrical conductors. Wood, glass, air and most plastic materials are examples of electrical insulators. Electricity is supplied through copper wires because copper is a very good conductor of electricity. Electrical wires are usually insulated using PVC, since PVC is flexible, a good insulator and it does not deteriorate with age. Fig. 13.6 shows how to test a material to find out if it is an electrical conductor. The light bulb can be replaced with an **ammeter**, an electrical meter used to measure current.

Metals conduct electricity because they contain free electrons which are not fixed to the atoms in the metal. As explained on p. 70, an atom consists of a positively charged nucleus where most of its mass is located. The nucleus is surrounded by much lighter electrons which move about the nucleus in allowed orbits called 'shells'. The electrons are negatively charged. In an uncharged atom, the negative charge of all its electrons is equal to the positive charge of the nucleus. Atoms become negatively or positively charged ions as a result of gain or loss of electrons.

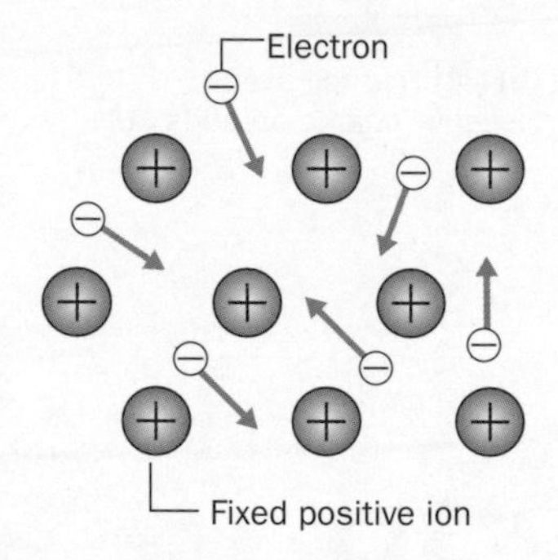

Fig. 13.7 Electrical conduction.

- In a **metal**, the atoms lose their outermost electrons to form a lattice of positive ions. The outermost electrons move about at random between the ions, acting like a glue which holds the ions in place. These electrons are referred to as **conduction** (or free) electrons and they are responsible for the conduction of electricity by the metal. In an electric circuit, the electric current in a metal is the flow of charge due to free electrons being attracted towards the positive end of the metal.
- In an **electrical insulator**, all the electrons are fixed rigidly to the atoms and cannot move throughout the material. As explained on p. 74, the bonds that hold atoms together in non-metallic substances are due to the transfer or sharing of electrons between the atoms. These electrons are referred to as **valence** electrons. If an insulator is connected in an electric circuit, no current is possible in any part of the circuit because the electrons in the insulator cannot move towards the positive end of the insulator.
- **Semiconductors** are materials that are insulators only at a temperature of absolute zero. All the electrons in a semiconductor at absolute zero are rigidly held by individual atoms in the structure. Above absolute zero, a semiconductor conducts electricity because some of the electrons break free from the atoms. The higher the temperature of a semiconductor, the more easily it conducts electricity. This is because more and more electrons break free from individual atoms as the temperature of the semiconductor is raised. Semiconductors, such as the elements silicon and germanium, are used to make integrated circuits used in microcomputers. Compound semiconductors, such as gallium arsenide (GaAs), are used to make opto-electronic devices such as laser diodes.

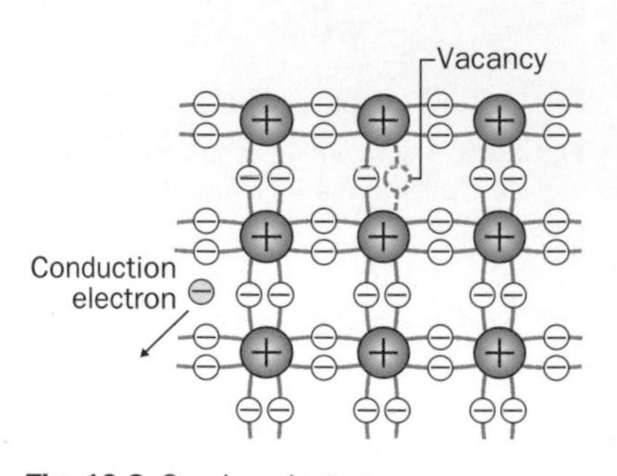

Fig. 13.8 Semiconductors.

A simple electric circuit

An electric current is a flow of charge. When a torch bulb is connected to a battery, the torch bulb lights because electrons flow around the circuit, transferring energy from the battery to the bulb. Work is done on the electrons inside the battery and they are then forced from the negative pole of the battery, through the bulb and back to the battery via the positive pole. The electrons repeatedly collide with the atoms of the bulb filament as they are forced through the filament, transferring their energy to the atoms as a result.

Electricity flows one way only in the electric circuit shown in Fig. 13.6. This is an example of a direct current circuit. Later, you will meet alternating current circuits, in which the current direction repeatedly reverses. The one-way nature of a direct current was first discovered by André Ampère over two centuries ago. Before Ampère's discovery, it was thought that positive electricity from the positive pole of a battery reacted in the filament with negative electricity from the negative pole to produce heat and light. Ampère discovered that an electric current along a wire created a magnetic field around the wire. He did this by showing that the electric current caused the needle of a nearby plotting compass to deflect. More significantly, he showed that reversing the connections of the wire to the battery made the needle deflect in the opposite direction. This can only happen if current flows one way only around the circuit.

Ampère did not know whether an electric current was a flow of positive charge or a flow of negative charge. He thought it was a flow of positive charge, so he marked the direction of the current from 'positive to negative', and this convention became the established rule. A century later, it was discovered that electrons which are responsible for current flow in metals are negatively charged. Nevertheless, the 'positive to negative' rule for the current direction in a direct circuit is still the standard convention. The importance of Ampère's work was recognised by naming the S.I. unit of electric current as the **ampere** (**A**). This is defined in terms of the magnetic effect of an electric current, as explained on p. 263.

Electric current effects

1. The **heating effect** of an electric current can be demonstrated using the arrangement in Fig. 13.9. The current is increased by adjusting the variable resistor. As the current increases, the thin strand of wire heats up and glows red then orange before it melts. This is how a fuse works. If the current becomes too large, the fuse wire melts and the circuit is then broken. Take care not to touch the wire when it is hot. The element of an electric heater is a wire that heats up when an electric current passes through it.

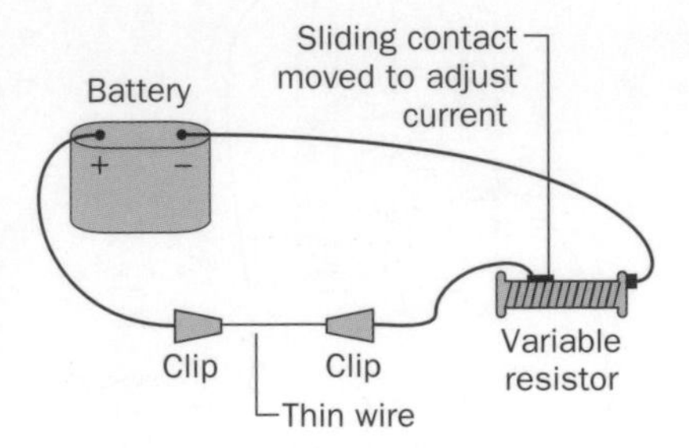

Fig. 13.9 The heating effect of an electric current.

2. The **magnetic effect** of an electric current can be demonstrated using a plotting compass, or by winding an insulated wire around a nail as shown in Fig. 13.10a. When connected to a suitable battery, the magnetic effect of the current magnetises the nail, which can then attract small iron or steel objects such as paper clips. The nail loses its magnetism

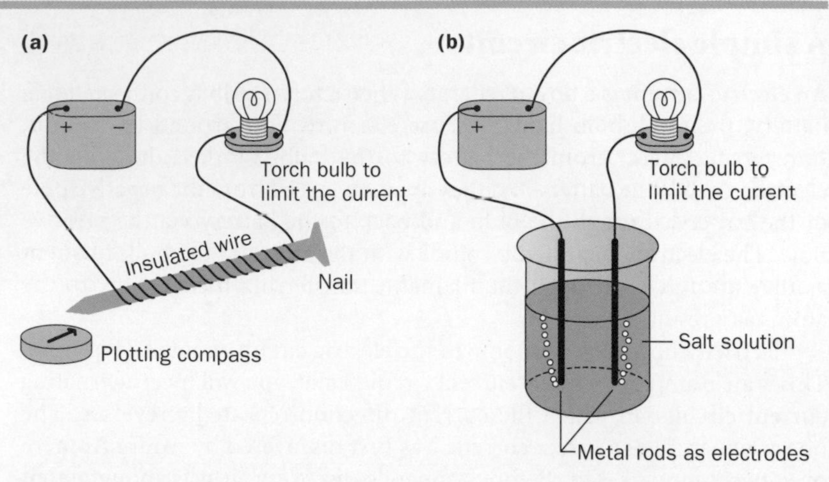

Fig. 13.10 More effects of an electric current **(a)** magnetic effect, **(b)** chemical effect.

when the current is switched off. An electromagnet used to move vehicles in a scrapyard is no more than a coil of insulated wire wrapped round an iron core.

3. The **chemical effect** of an electric current can be demonstrated by passing an electric current through a beaker of salt water. The two metal rods in the solution connected to the circuit are called **electrodes**. Gas is evolved at the electrodes, demonstrating that the solution is decomposed by the electric current. This is an example of **electrolysis.**

Current and charge

In the circuit in Fig. 13.6, the amount of electric charge passing through the torch bulb is proportional to the number of electrons that pass through it. Each electron carries the same tiny amount of negative charge. The current through the torch bulb is a measure of the charge or the number of electrons that pass through the bulb each second.

The link between current and charge is demonstrated in Fig. 13.11 where electrons are transferred from the negative to the positive plate, causing a reading to register on the meter. Each time the ball touches the negative plate, it gains electrons to become negative. It is then attracted to the positive plate where it releases electrons to become positively charged due to a shortage of electrons. It is then attracted back to the negative plate to continue the cycle, transferring the same amount of negative charge each time it moves from the negative to the positive plate. If the plates are moved closer together, the ball shuttles back and forth between the two plates at a faster rate, causing the current to increase. The current is therefore a measure of the rate of transfer of charge.

One coulomb of charge is defined as the charge transferred in one second when the current is exactly one ampere.

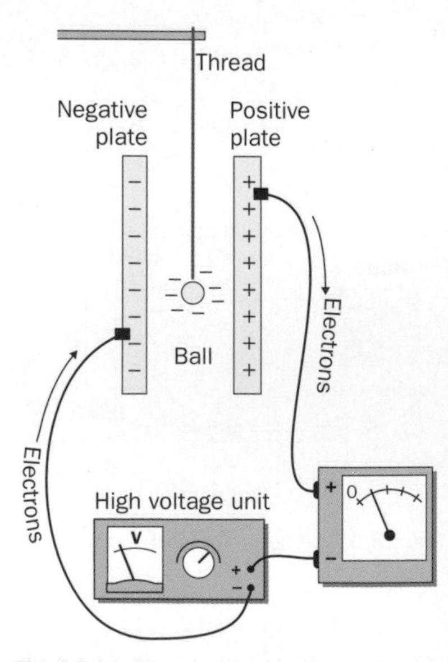

Fig. 13.11 The shuttling ball experiment.

In more general terms, the charge, Q, transferred in coulombs (abbreviation C) is given by the equation

$Q = It$

where I = current in amperes, t = the time in seconds.

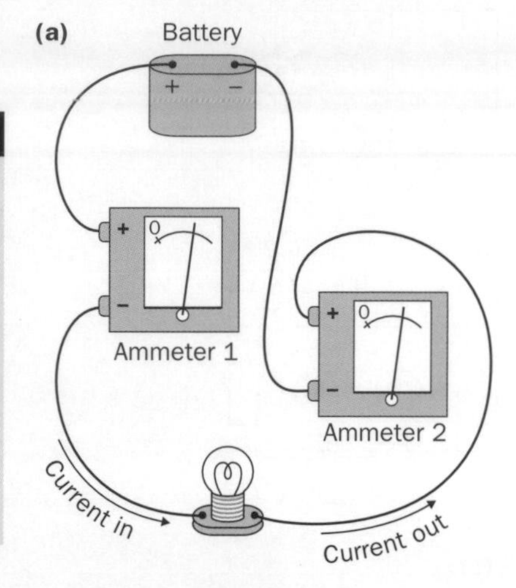

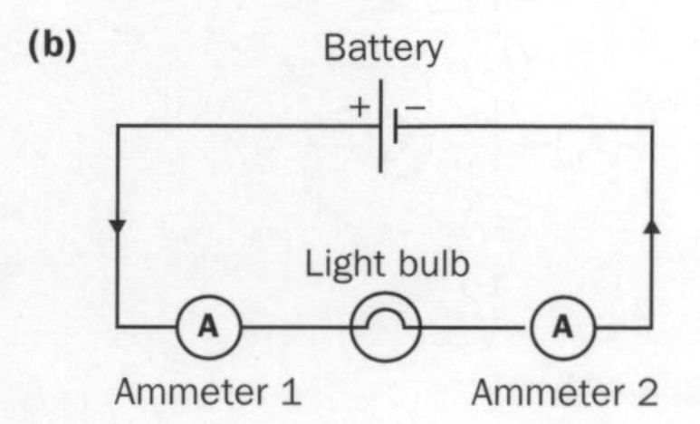

Fig. 13.12 Current entering and leaving a component **(a)** circuit connections, **(b)** circuit diagram.

Worked example 13.1

(a) Calculate the charge that passes a point in a circuit in one minute when the current is 0.5 A.

(b) Calculate the current that would transfer, in 10 s, the same amount of charge as a current of 0.5 A in 1 min.

Solution

(a) Charge $Q = It = 0.5\ \text{A} \times 60\ \text{s} = 30\ \text{C}$

(b) Rearrange $Q = It$ to give $I = \frac{Q}{t} = \frac{30\ \text{C}}{10\ \text{s}} = 3.0\ \text{A}$.

Current rules

1. **Current is measured using an ammeter.** To measure the current in a wire, an ammeter must be connected in series with the wire. An ideal ammeter has no electrical resistance so it does not affect the current to be measured.
2. **The current leaving a component is equal to the current entering a component.** In any component through which a steady current passes, the number of electrons leaving the component is equal to the number of electrons entering the component in the same time. In other words, the current out of a component is equal to the current into the component.
3. **Components in series pass the same current.** Fig. 13.13 shows two torch bulbs **in series** connected to a battery. An electron from the battery passes through one bulb then through the other bulb before returning to the battery. The number of electrons per second passing through one bulb is the same as through the other bulb. In other words, the same current passes through each bulb.
4. **At a junction in a circuit, the total current entering the junction is equal to the total current leaving the junction.** Fig. 13.14 shows a circuit in which two torch bulbs, X and Y, are connected to a battery. Each bulb can be switched on or off without affecting the other one. The two bulbs are said to be **in parallel** with each other. An electron from the battery passes either through X or through Y before returning to the battery. Consequently, the flow of electrons from the battery divides at junction Q and recombines at junction P. Measurements show that the reading of the ammeter in series with the battery is equal to the sum of the readings of the ammeters in series with the torch bulbs.

∴ current from the battery = current through X + current through Y

More generally, the total current entering a junction is equal to the total current leaving the junction.

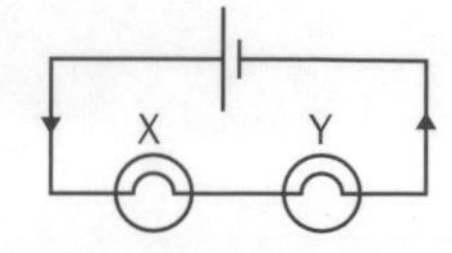

Fig. 13.13 Components in series.

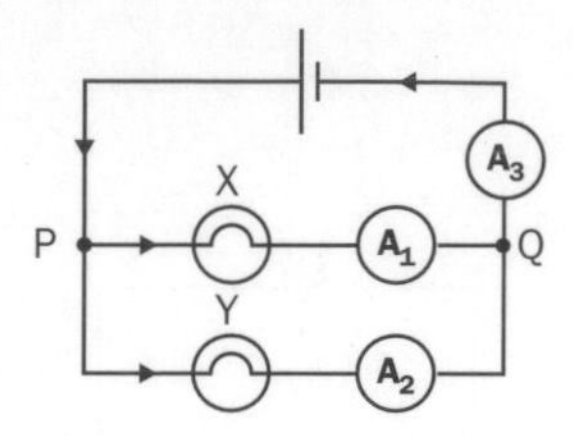

Fig. 13.14 Components in parallel.

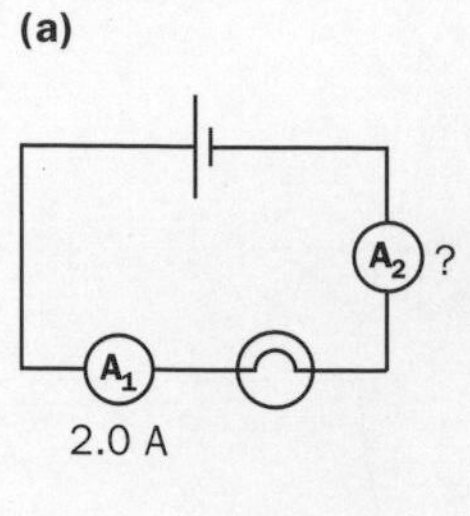

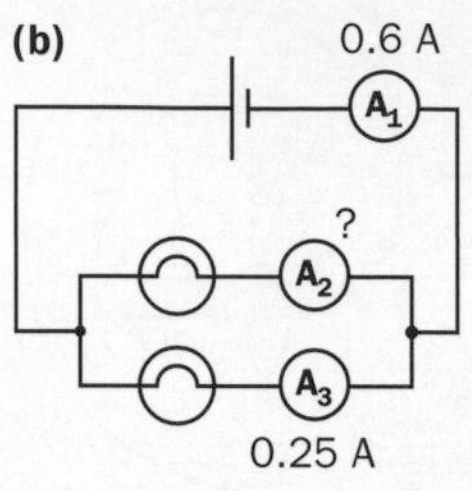

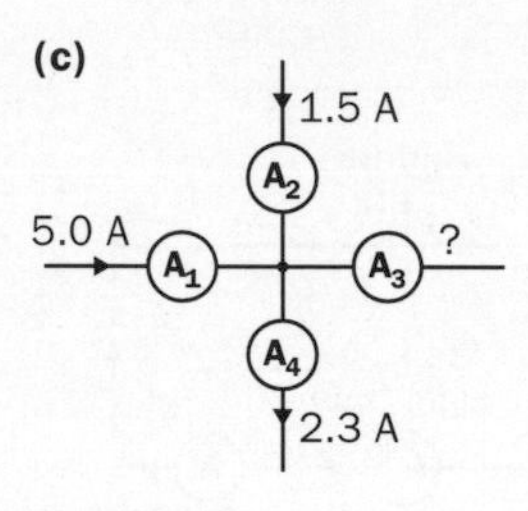

Fig. 13.15

Questions 13.2

1. **(a)** Which of the following materials are not electrical conductors?

   air aluminium glass germanium graphite

   (b) Copy and complete the Table below by inserting the missing values.

Current/A	**(i)**	0.60	15.00	**(iv)**
Charge/C	15.00	**(ii)**	45.00	25.00
Time/s	3.00	0.50	**(iii)**	10.00

2. Calculate the missing current in each of the circuits in Fig. 13.15.

13.3 Potential difference

Electrical energy and potential difference

In Fig. 13.9, an electric current passes through the torch bulb filament only if the filament is part of a complete circuit and the circuit includes a battery. The battery converts chemical energy into electrical energy and is therefore a source of electrical energy. The torch bulb converts electrical energy into heat and light and is therefore a sink of electrical energy.

The potential difference between two points in an electrical circuit is the electrical energy delivered per unit charge when charge flows from one point to the other.

The unit of potential difference is the **volt** (V) which is equal to one joule per coulomb. For example, when a 6.0 V battery is connected to a 6.0 V torch bulb, every coulomb of charge from the battery delivers 6.0 J of electrical energy to the bulb. Potential difference is often referred to as **voltage.**

$$V = \frac{E}{Q}$$

where E = energy delivered and Q = charge.

Voltage rules

1. **A voltmeter is used to measure the potential difference between two points in a circuit.** An ideal voltmeter has infinite resistance so it does not draw any current from the circuit to which it is connected. Fig. 13.16 shows a voltmeter connected across a torch bulb which is in series with a variable resistor. The voltmeter is connected in parallel with the bulb so it measures the potential difference across the bulb only. Charge from the battery delivers some of its electrical energy to the bulb and some to the variable resistor.

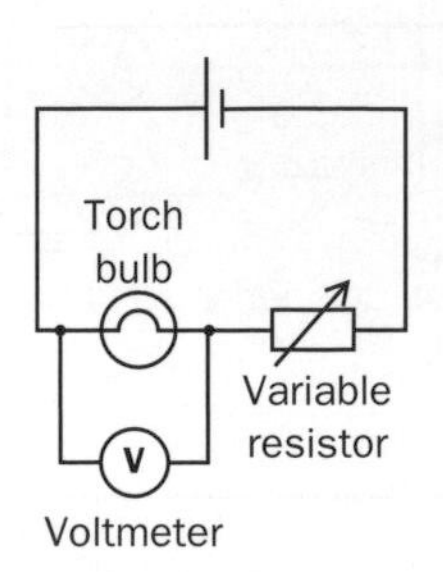

Fig. 13.16 Using a voltmeter.

2. **For two or more components in series, the total voltage is equal to the sum of the individual voltages.** In Fig. 13.17, the two light bulbs are in series and therefore pass the same current. Each coulomb of charge passing through bulb A also passes through bulb B. The potential difference across each light bulb is the electrical energy delivered per coulomb of charge.

   Suppose the p.d across light bulb A is 3 V and the p.d. across light bulb B is 9 V. Each coulomb of charge delivers 3 J of electrical energy to A and 9 J to B. Therefore the total electrical energy delivered per coulomb is 12 J (= 9 J + 3 J). Hence the total p.d. is 12 V across A and B.

   $$V = V_A + V_B$$

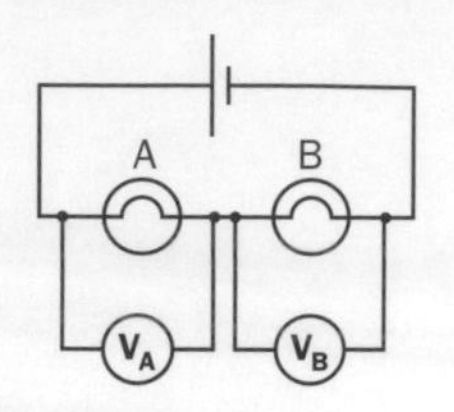

Fig. 13.17 Adding voltages.

3. **For components in parallel, the voltage is the same.** In Fig. 13.18, bulbs X and Y are in parallel with each other and in series with bulb Z and a battery. Each coulomb of charge from the battery flows around the circuit through Z then through X or Y. Suppose the p.d. across Z is 9 V and the battery p.d. is 12 V. Each coulomb of charge therefore delivers 12 J to the light bulbs as a result of delivering 9 J to Z and 3 J to either X or Y. Thus the p.d across either X or Y is 3 V.

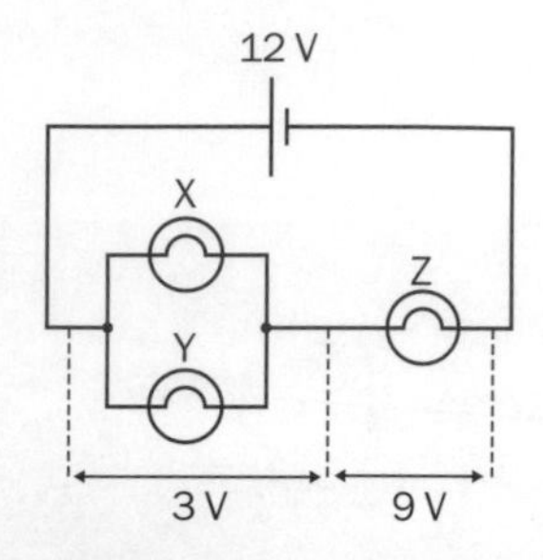

Fig. 13.18 Parallel components have the same voltage.

See www.palgrave.com/foundations/breithaupt for an experiment to calibrate a voltmeter.

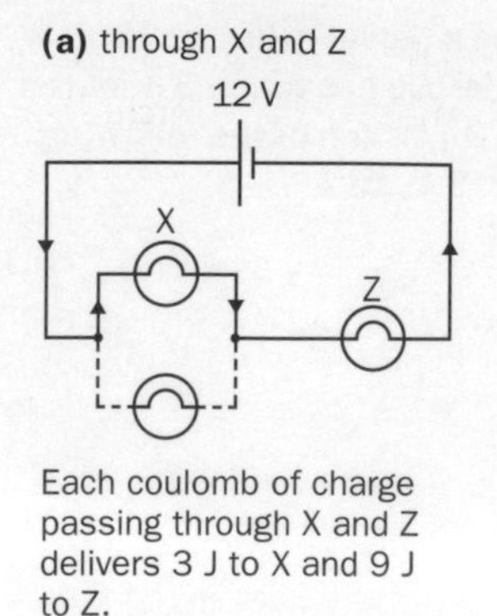

(b) through Y and Z

12 V

Z

Y

Each coulomb of charge passing through Y and Z delivers 3 J to Y and 9 J to Z.

Fig. 13.19 Two routes for charge flow.

Electrical power

As explained on p. 159, power is the rate of transfer of energy. Thus a heating element with a rating of 1000 W dissipates heat at a rate of 1000 J s^{-1}. If the potential difference across the torch bulb in Fig. 13.16 is represented by V, then the energy, E, delivered to the bulb as a result of charge Q passing through the bulb is QV. In time t, the charge flow, Q = It, where I is the current. Hence the energy delivered is given by the equation

$$E = QV = ItV$$

Since the power, P, supplied to the torch bulb is the energy delivered per second,

$$P = \frac{E}{t} = \frac{ItV}{t} = IV$$

Energy delivered in time t, E = ItV
Power supplied, P = IV

where I = current, V = potential difference.

Fuse ratings

A fuse wire is designed to melt if excessive current passes through it. The current rating of a fuse is the maximum current the fuse is capable of conducting without melting. The current rating of a fuse intended to protect an appliance from excessive current can be calculated from the power and potential difference (i.e. voltage) using the equation $I = P \div V$.

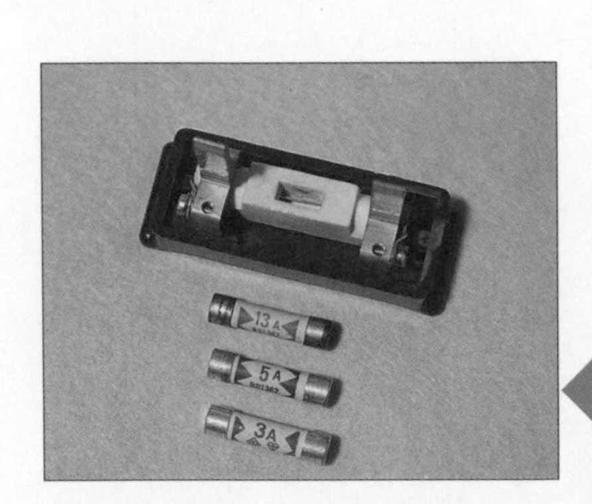

Fig. 13.20 Domestic fuses. The top fuse is from a domestic fuse box and the three smaller fuses below fit into 3-pin plugs to protect individual appliances.

Worked example 13.2

An electric kettle operates at 240 V and has a power rating of 3000 W. (a) Calculate (i) the energy supplied to the kettle in exactly 5 minutes, (ii) the current through the kettle element. (b) Which of the following fuses is suitable for this appliance: 1 A, 3 A, 5 A, 13 A?

Solution

(a) (i) Energy supplied = power × time = 3000 W × 300 s = 9.0×10^5 J s^{-1}

(ii) Rearrange P = IV to give $I = \frac{P}{V} = \frac{3000}{240} = 12.5$ A

(b) The 13 A fuse.

Notes

1. SI units must be used throughout. Current is always in amperes, potential difference in volts, time in seconds, energy in joules and power in watts.
2. Prefixes commonly used for numerical values include mega (M) for 10^6, kilo (k) for 10^3, milli (m) for 10^{-3} and micro (μ) for 10^{-6}.
3. Domestic electricity meters record the electrical energy used in kilowatt hours (kW h). 1 kW h is the electrical energy supplied at a constant rate of 1 kW for 1 hour. Prove for yourself that 1 kW h = 3.6 MJ.
4. The power equation can be rearranged as
$$I = \frac{P}{V}$$
to find I, given values for P and V, or as
$$V = \frac{P}{I}$$
to find V, given values for P and I.

Questions 13.3

1. A 12 V, 36 W electric heater is connected to a 12 V battery. Calculate **(a)** the current, **(b)** the charge flow through the heater in 10 minutes, **(c)** the energy delivered to the heater in this time.
2. In a test to determine the power and current supplied to a 230 V electric oven, all the appliances in the household were switched off and the electricity meter reading noted. The electric oven was then switched on for exactly 30 minutes and the meter reading was found to have increased by 2.8 kW h. Calculate **(a)** the electrical energy supplied in joules, **(b)** the charge flow in this time, **(c)** the current.

13.4 Resistance

Symbols and diagrams

You need to learn the standard symbols used for each type of component in circuit diagrams. Major problems could arise if a motorist who needs to change a fuse in a vehicle lighting circuit confuses the symbols for a fuse and an electric heater. The standard symbols for common electrical components are shown in Fig. 13.21. In addition to recognising these symbols, you need to know the general characteristics of each type of component.

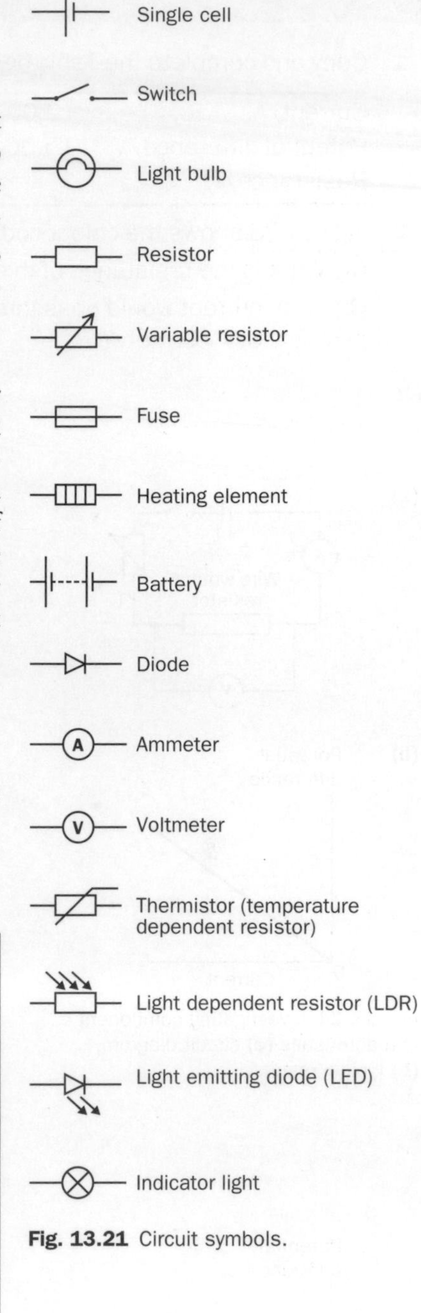

Fig. 13.21 Circuit symbols.

Resistors

The current in a circuit depends on the voltage of the power supply or battery and on the resistance of the circuit components. The electrical resistance of a conductor is due to the opposition by the atoms of the conductor to the flow of electrons through the conductor. This is not unlike a central-heating system, where the flow of water depends on the width of the pipes and radiators as well as the pump which forces water around the system. The pipes and radiators resist the flow of water. The narrower the pipes, the greater the resistance to flow. A resistor is designed to limit the current in a circuit by resisting the flow of charge through it.

The **resistance** R of a component is defined by the equation

$$R = \frac{V}{I}$$

where V = the potential difference across the component, and I = the current through the component.

The unit of resistance is the **ohm** (symbol Ω pronounced 'omega') which is equal to 1 volt per ampere.

Notes

1. The above equation can be rearranged to $V = IR$ to calculate V from values of I and R or to $I = V \div R$ to calculate I from values of V and R.
2. The power dissipated in a resistor, $P = IV = I^2R$ since $V = IR$.
3. Resistors are colour-coded to show the value of resistance. Fig. 13.22 shows the resistor colour code.

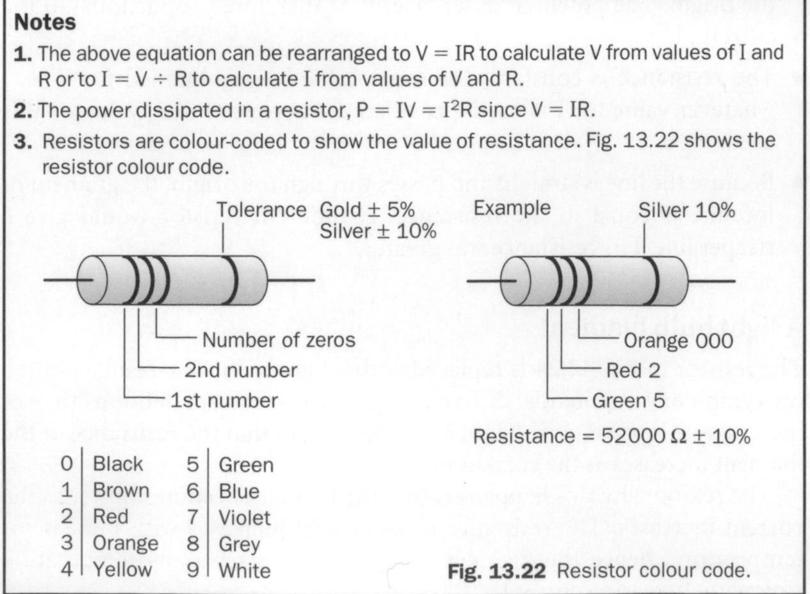

Fig. 13.22 Resistor colour code.

Questions 13.4a

1. Copy and complete the Table below.

Current/A	2.00	0.30	?	2.5×10^{-3}	?
Potential difference/V	15.00	?	1.50	6.00	5.0×10^{4}
Resistance/Ω	?	6.8×10^{2}	2.2 k	?	100 M

2. Fig. 13.23 shows the colour-coding of a certain resistor in a circuit.

(a) What is the resistance of this resistor?

(b) What current would pass through this resistor if it was connected across a 6.0 V battery?

Red
Violet
Orange

Fig. 13.23

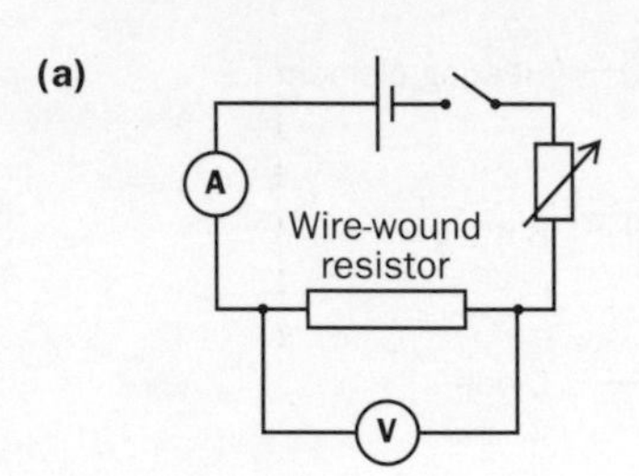

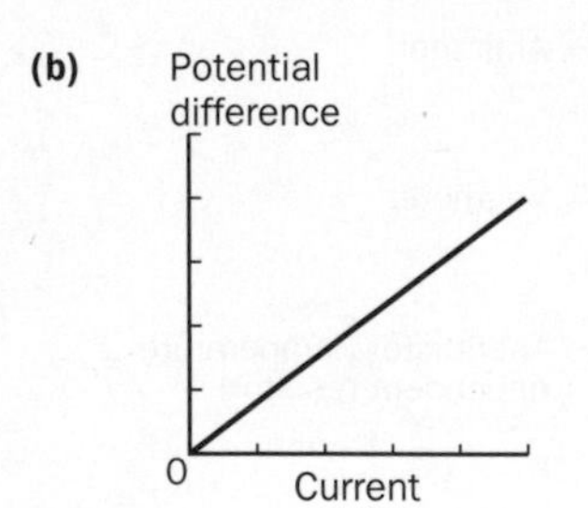

Fig. 13.24 Investigating component characteristics **(a)** circuit diagram, **(b)** typical results.

Experiment to investigate the characteristics of circuit components

A wire-wound resistor

Fig. 13.24a shows how the current through a wire-wound resistor can be measured for different potential differences applied across the resistor. The variable resistor is adjusted to alter the current. The measurements may be plotted on a graph of y = potential difference against x = current, as in Fig. 13.24b.

- The results plotted on the graph define a straight line passing through the origin. The potential difference, V, is therefore proportional to the current, I.
- The resistance is constant, the same value being obtained for V ÷ I whatever value for I is used, provided the corresponding value of V is used.
- Because the line is straight and passes through the origin, the gradient of the line is equal to the resistance. A different resistor would give a steeper line if its resistance was greater.

A light bulb filament

The resistor in Fig. 13.24 is replaced with a light bulb. The results plotted on a graph of y = potential difference against x = current define a curve of increasing steepness, as in Fig. 13.25. This shows that the resistance of the filament increases as the current increases.

The reason why this happens is that the filament becomes hotter as the current increases. The resistance of any metal increases with increase of temperature, hence the filament resistance increases as its temperature rises with increasing current.

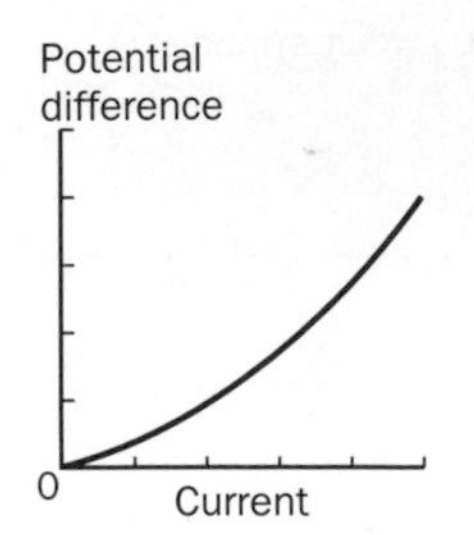

Fig. 13.25 Potential difference versus current for a light bulb filament.

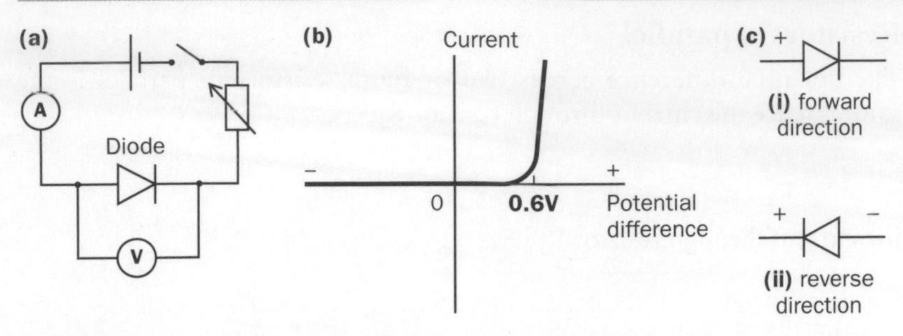

Fig. 13.26 Testing a diode **(a)** circuit diagram, **(b)** typical results, **(c)** diode directions.

A silicon diode

A diode conducts electricity in one direction only, referred to as the 'forward' direction. Fig. 13.26a shows a diode connected into a circuit in the forward direction. Reversing the diode in the circuit enables it to be tested in the reverse direction.

The results are usually plotted on a graph of y = current against x = potential difference, as in Fig. 13.26b. This graph shows that

- the forward resistance of the diode is very low when it conducts
- the diode has a very high resistance in the reverse direction.

In addition, a silicon diode only conducts if the applied potential difference exceeds a threshold of about 0.6 V. The current increases sharply with little change of potential difference beyond about 0.6 V.

Light-emitting diodes (LEDs) emit light when they conduct. For this reason, they are used as indicators in electronic circuits. The threshold voltage for an LED is higher than for an ordinary diode.

If the current through the diode is allowed to become too great, the diode overheats and stops conducting permanently. For this reason, a silicon diode in a circuit is always connected in series with a resistor, which therefore limits the current passing through the diode.

Worked example 13.3

A light-emitting diode with a maximum current rating of 40 mA and a threshold voltage of 2.0 V is to be used in a circuit where the p.d. is 5.0 V, as in Fig. 13.27. Calculate the resistance of a suitable resistor that should be connected in series with the LED.

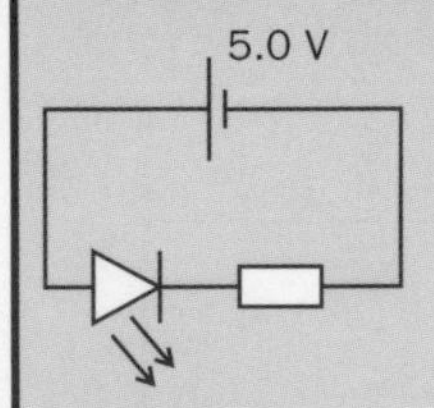

Fig. 13.27

Solution

The p.d. across the resistor
= 5.0 − 2.0 = 3.0 V

Current = 0.04 A

∴ resistance of resistor,

$$R = \frac{V}{I} = \frac{3.0\ \text{V}}{0.04\ \text{A}} = 75\ \Omega$$

Resistor combination rules

Resistors in series

The current, I, is the same through each resistor.
Hence

the p.d. across resistor R_1, $V_1 = IR_1$
the p.d. across resistor R_2, $V_2 = IR_2$
the p.d. across resistor R_3, $V_3 = IR_3$, etc.
Since the total p.d. $V = V_1 + V_2 + V_3 +$, etc.

$$\text{the total resistance, } R = \frac{V}{I} = \frac{V_1 + V_2 + V_3 +, \text{etc.}}{I}$$

$$\therefore R = \frac{V_1}{I} + \frac{V_2}{I} + \frac{V_3}{I} +, \text{etc.} = R_1 + R_2 + R_3 +, \text{etc.}$$

$\mathbf{R = R_1 + R_2 + R_3 +, etc.}$ for resistors in series.

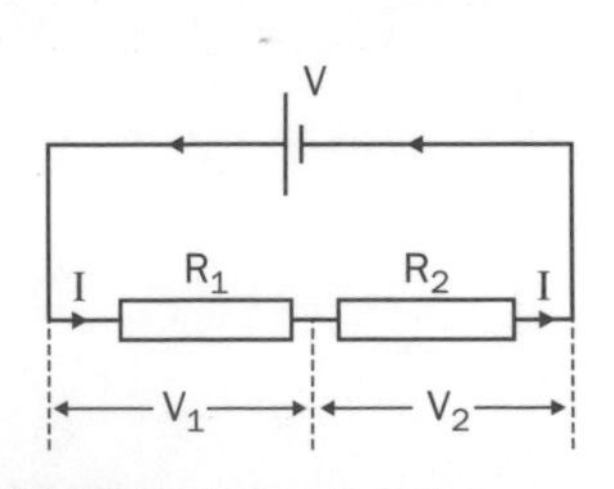

Fig. 13.28 Resistors in series.

Worked example 13.4

A 12 Ω resistor is connected in series with a 6 Ω resistor and a 9 V battery. Sketch the circuit diagram and calculate (a) the resistance of the two resistors in series, (b) the current and p.d. for each resistor.

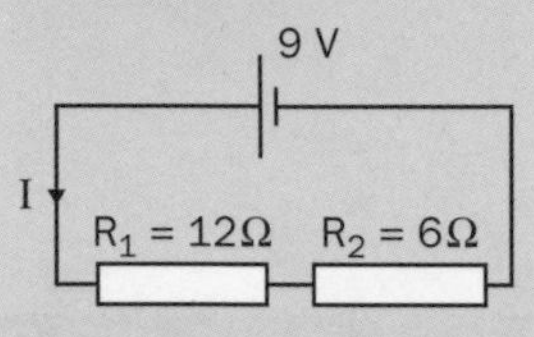

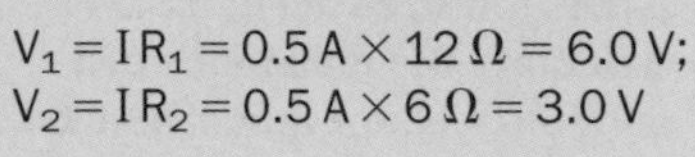

Fig.13.29

Solution

(a) $R = 12 + 6 = 18\,\Omega$

(b) $I = \frac{V}{R} = \frac{9\,V}{18\,\Omega} = 0.5$ A for both resistors

$V_1 = IR_1 = 0.5\,A \times 12\,\Omega = 6.0\,V$; $V_2 = IR_2 = 0.5\,A \times 6\,\Omega = 3.0\,V$

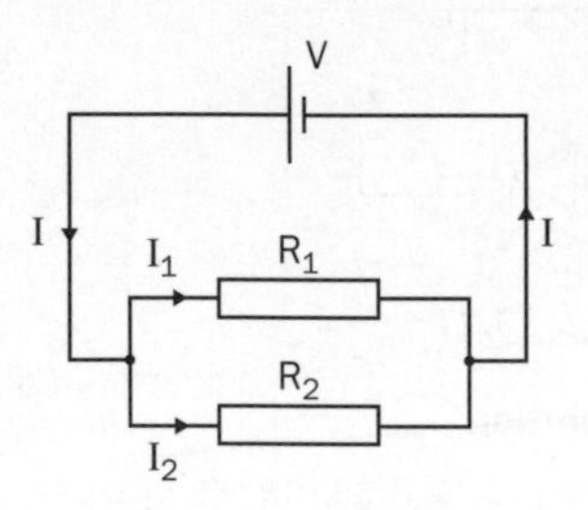

Fig. 13.30 Resistors in parallel.

Resistors in parallel

The potential difference across two or more resistors in parallel is the same. Hence the current through resistor R_1,

$$I_1 = \frac{V}{R_1}$$

the current through resistor R_2,

$$I_2 = \frac{V}{R_2}$$

the current through resistor R_3,

$$I_3 = \frac{V}{R_3}, \text{ etc}$$

Since the total current through the combination

$$I = I_1 + I_2 + I_3 +, \text{etc.}$$

and the total resistance

$$R = \frac{V}{I}$$

then $\frac{1}{R} = \frac{I}{V} = \frac{I_1 + I_2 + I_3 +, \text{etc.}}{V}$

hence $\frac{1}{R} = \frac{I_1}{V} + \frac{I_2}{V} + \frac{I_3}{V} +, \text{etc.} = \frac{1}{R_1} + \frac{1}{R_2} + \frac{1}{R_3} +, \text{etc.}$

$$\frac{1}{R} = \frac{1}{R_1} + \frac{1}{R_2} + \frac{1}{R_3} +, \text{etc.}$$ **for resistors in parallel.**

Worked example 13.5

A 5 Ω resistor is connected in parallel with a 20 Ω resistor and the combination is connected to a 12 V battery. Sketch the circuit diagram and calculate (a) the resistance of the two resistors, (b) the current from the battery, (c) the current through each resistor, (d) the power dissipated in each resistor.

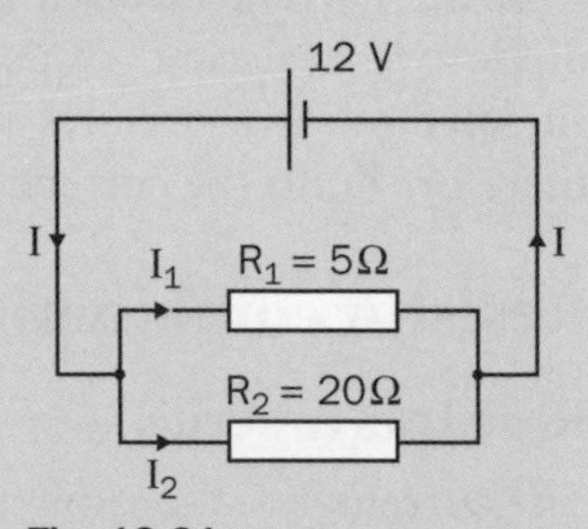

Fig. 13.31

Solution

The circuit diagram is shown in Fig. 13.31.

(a) Using $\frac{1}{R} = \frac{1}{R_1} + \frac{1}{R_2}$, $\frac{1}{R} = \frac{1}{5} + \frac{1}{20} = 0.20 + 0.05 = 0.25$

$\therefore R = \frac{1}{0.25} = 4.0\,\Omega$

(b) Using $I = \frac{V}{R}$ gives $I = \frac{12}{4.0} = 3.0$ A

(c) $I_1 = \frac{V}{R_1} = \frac{12}{5} = 2.4$ A; $I_2 = \frac{V}{R_2} = \frac{12}{20} = 0.6$ A

(d) Power, $P = IV$ $\therefore P_1 = 2.4 \times 12 = 28.8$ W; $P_2 = 0.6 \times 12 = 7.2$ W

Summary

◆ **Law of force between charged objects**

Like charges repel; unlike charges attract.

◆ **Equations**

Charge, $Q = It$, where I = current, t = time

Potential difference, $V = \frac{E}{Q}$

where E = electrical energy transferred by charge Q

Power, $P = IV$

Resistance, $R = \frac{V}{I}$

Resistor combination rules $R = R_1 + R_2$ for two resistors in series

$\frac{1}{R} = \frac{1}{R_1} + \frac{1}{R_2}$ for two resistors in parallel

◆ **Units**

Current in amperes (A), charge in coulombs (C), p.d. in volts (V), power in watts (W), resistance in ohms (Ω).

◆ **Circuit rules**

1. **components in series:** same current; total p.d. = sum of individual p.ds.
2. **components in parallel:** same p.d.; total current = sum of individual currents.

Questions 13.4b

1. **(a)** A 3 Ω resistor is connected in series with a 6 Ω resistor and a 9 V battery. Sketch the circuit diagram and calculate **(i)** the resistance of the two resistors in series, **(ii)** the current through each resistor, **(iii)** the p.d. across and power dissipated in each resistor.

   (b) The two resistors in (a) are connected in parallel to the same battery. Sketch the circuit diagram and calculate **(i)** the resistance of the two resistors in parallel, **(ii)** the current through each resistor, **(iii)** the p.d. across and power dissipated in each resistor.

2. **(a)** Sketch the possible resistance combinations of a 2 Ω resistor, a 3 Ω resistor and 6 Ω resistor.

   (b) Calculate the resistance of each combination in (a).

Revision questions

13.1. Explain the following observations:

(a) If held briefly on the ceiling, a balloon will stick to the ceiling if it has been rubbed beforehand.

(b) A gold leaf electroscope is charged by direct contact with a perspex rod. The leaf of the electroscope rises further when a plastic ruler is rubbed, then held over the cap of the electroscope.

(c) A person sitting on a plastic chair receives an electric shock when she stands up and touches a door handle.

13.2. (a) Explain why the resistance of a semiconductor falls when it is heated.

(b) Explain why the resistance of a metal wire rises when it is heated.

13.3. A 1.5 V, 0.25 A torch bulb is connected to a 1.5 V battery for exactly 10 minutes.

(a) Calculate **(i)** the charge passing through the bulb in this time, **(ii)** the power supplied to the bulb, **(iii)** the electrical energy delivered to the bulb in 10 minutes.

(b) The battery is capable of lighting the bulb for 1 hour. Calculate **(i)** the charge passing through the bulb in 1 hour, **(ii)** the total electrical energy the battery can deliver.

13.4. Fig. 13.32 shows two 3 V, 0.1 A torch bulbs, X and Y, in parallel with each other. The combination is in series with a variable resistor and a 4.5 V battery. The variable resistor is adjusted until the two bulbs light normally.

(a) Calculate **(i)** the current from the battery, **(ii)** the p.d. across the variable resistor, **(iii)** the power supplied to each bulb, **(iv)** the power supplied by the battery to the circuit.

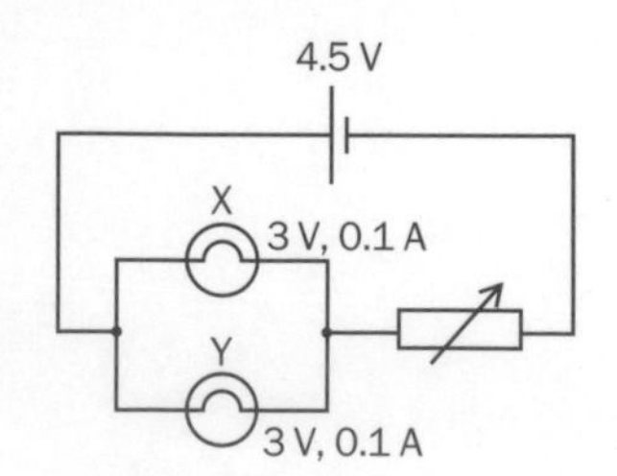

Fig. 13.32

(b) Account for the difference between the power supplied by the battery and the power supplied to the two bulbs.

13.5. (a) A hair drier is rated at 230 V, 750 W maximum. Calculate **(i)** the maximum current, **(ii)** the number of units of electricity it would use in 30 minutes.
(b) Which of the following fuse ratings would be suitable for this hairdrier:
1 A   3 A   5 A   13 A

13.6. In an experiment to measure the resistance of a heating element, the following measurements were made:

Current / A	0.00	0.25	0.38	0.48	0.57	0.66
Potential difference / V	0.00	2.00	4.00	6.00	8.00	10.00

(a) Plot a graph of y = potential difference against x = current.
(b) Use your graph to determine the resistance of the resistor at **(i)** 5.0 V, **(ii)** 10.0 V.
(c) Explain why the resistance changed as the potential difference was increased.

13.7. A 2.0 A diode with a threshold voltage of 0.6 V is to be connected in its forward direction in series with a 9.0 V battery, and a resistor chosen to limit the current to 2.0 A.
(a) Sketch the circuit diagram.
(b) Calculate **(i)** the potential difference across the resistor, **(ii)** the resistance of the resistor.

13.8. A 6 Ω resistor and a 4 Ω resistor are connected in series with each other and a 6.0 V battery.
(a) Sketch the circuit diagram.
(b) Calculate **(i)** the resistance of the two resistors in series, **(ii)** the battery current, **(iii)** the p.d. across each resistor, **(iv)** the power dissipated in each resistor.

13.9. A 6 Ω resistor and a 12 Ω resistor are connected in parallel with each other and a 12 V battery.
(a) Sketch the circuit diagram.
(b) Calculate **(i)** the resistance of the two resistors in parallel, **(ii)** the battery current, **(iii)** the current through each resistor, **(iv)** the power dissipated in each resistor.

13.10. In the circuit in Q13.8, a 3 Ω resistor is connected in parallel with the 6 Ω resistor.
(a) Sketch this circuit diagram.
(b) Calculate **(i)** the total resistance of the combination, **(ii)** the battery current, **(iii)** the p.d. across each resistor, **(iv)** the current through each resistor, **(v)** the power dissipated in each resistor.

Electric Circuits

Objectives

After working through this unit, you should be able to:

- ▶ define and measure the e.m.f. and internal resistance of a cell
- ▶ describe a potential divider and use a potentiometer to compare the e.m.f.s of two cells
- ▶ use a Wheatstone bridge to measure resistance
- ▶ define and measure resistivity and conductivity
- ▶ explain how to extend the range of an ammeter or a voltmeter
- ▶ use a multimeter to make electrical measurements

Contents

14.1 Cells and batteries

Inside a cell

Cells and batteries are used whenever a portable source of electrical power is needed. In addition, certain cells are used to calibrate voltmeters and to check other electrical measurements. One of the first batteries was made by Volta in 1800, who stacked discs of zinc and silver alternately in a pile, interleaved with paper soaked in brine. He discovered that the battery produced an electric current through a wire connected between the ends of the pile. He also found that an insulated metal object was charged positively when connected to the silver disc at the end of the pile, with the lead disc at the other end earthed. When the connections were reversed, the object charged negatively.

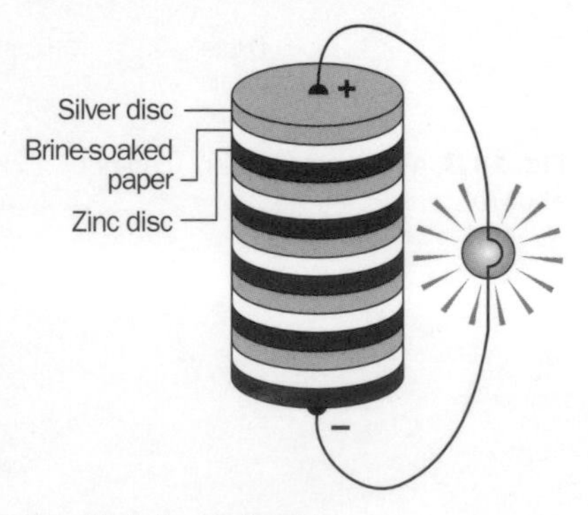

Fig. 14.1 A voltaic pile.

Primary cells are not rechargeable. Once a primary cell has run down, it must be replaced by a fresh cell. Torch batteries and non-rechargeable calculator batteries consist of one or more primary cells in series.

Fig. 14.2 shows a simple primary cell consisting of a copper plate and a zinc plate as electrodes, with dilute sulphuric acid as the electrolyte. Zinc is more reactive than copper in dilute sulphuric acid. Zinc atoms leave the plate and go into the solution as positive zinc ions, making the zinc plate negative. Hydrogen ions in the acid are attracted to the copper plate,

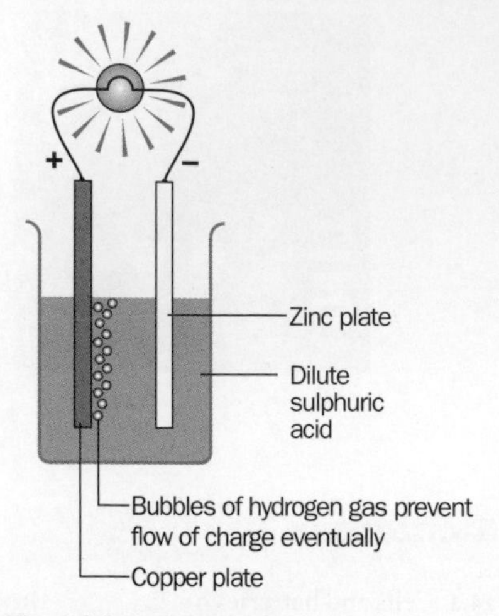

Fig. 14.2 A simple cell.

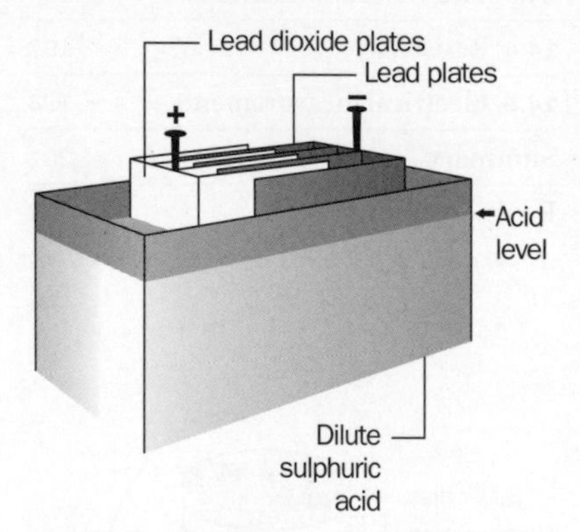

Fig. 14.3 A lead-acid battery (cap not shown).

Table 14.1 Types of cells

Type of cell	EMF / V	Rechargeable
alkaline cell	1.5 V	No
mercury cell	1.35 V	No
nickel–cadmium cell	1.2 V	Yes
lead-acid cell	2.0 V	Yes

making the copper plate positive. In a complete circuit, electrons would be forced around the circuit, leaving the cell via the zinc plate and returning via the copper plate.

In a torch battery or a calculator battery, the electrolyte is a paste and the metal case of the cell is one of the electrodes.

Secondary cells are rechargeable. Most car batteries consist of six 'lead-acid' cells in series. Each cell consists of a set of lead plates as one electrode, interleaved with a set of lead dioxide plates as the other electrode, as shown in Fig. 14.3. Lead from the lead plates dissolves in the dilute sulphuric acid, each lead atom from the plate going into solution as a positive ion, leaving behind electrons on the lead plate. Hydrogen ions attracted to the lead dioxide plate combine with oxygen atoms in the lead dioxide to form water molecules. The electrolyte becomes weaker, therefore, as the lead dissolves more and more.

To recharge the cell, a battery charger is used to force electrons into the cell via the lead plate. The lead ions in the solution are attracted back onto the lead plates, lead dioxide forms on the other plate and the strength of the acid is increased.

Electromotive force

A cell forces electrons around a complete circuit, from negative to positive. The electrons are supplied with electrical energy in the cell as a result of chemical reactions inside the cell. They are then forced out of the cell at the negative terminal, provided the cell is part of a complete circuit. As the electrons pass through each component in the circuit, the electrons do work in the component, converting their electrical energy into other forms of energy.

The electromotive force (e.m.f.) of a cell is the electrical energy per unit charge created in the cell.

The unit of e.m.f. is the volt, equal to 1 joule per coulomb. Note that the volt is also the unit of potential difference. However, potential difference is defined as electrical energy delivered per unit charge, whereas e.m.f. is electrical energy created per unit charge. Thus a source of e.m.f. is a source of electrical energy, whereas a potential difference is a 'sink' of electrical energy.

The e.m.f. of a cell depends on the substance used as the electrolyte and on the materials which the electrodes are made from. Table 14.1 gives the e.m.f.s of some different types of cells.

When a cell is in a circuit with current passing through the cell, energy is wasted in the cell because of the cell's **internal resistance**. The electrons use electrical energy to move through the cell against the resistance of the electrolyte and the electrodes. The electrical energy per coulomb delivered to the circuit components is less than the e.m.f. of the cell because electrical energy is used inside the cell. The potential difference across the cell terminals (i.e. the electrical energy delivered to the circuit components) is therefore less than the e.m.f. of the cell. The difference between the e.m.f. and the cell p.d. is in effect 'lost voltage' inside the cell. The internal resistance of the cell is the lost voltage per unit current.

Electrical energy per coulomb created in the cell (the cell's e.m.f.)

= electric energy per coulomb delivered to the circuit (cell p.d.)
+ electrical energy per coulomb used inside the cell (lost voltage)

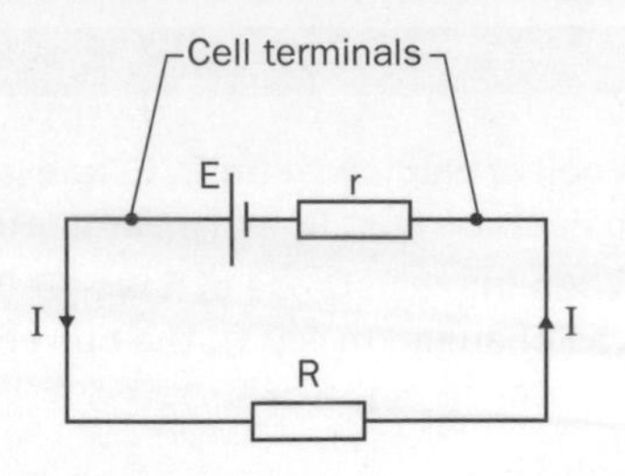

Fig. 14.4 Internal resistance.

In the circuit shown in Fig. 14.4, a cell of e.m.f. E, and internal resistance, r, is connected to a resistor, R. Hence

$$E = IR + Ir$$

where I is the current, the term IR is the cell p.d. and the term Ir is the lost voltage.

Worked example 14.1

A cell of e.m.f. 1.5 V and internal resistance 0.5 Ω is connected to a 4.5 Ω resistor. Calculate (a) the total resistance of the circuit, (b) the current through the 4.5 Ω resistor, (c) the potential difference across the cell terminals.

Solution

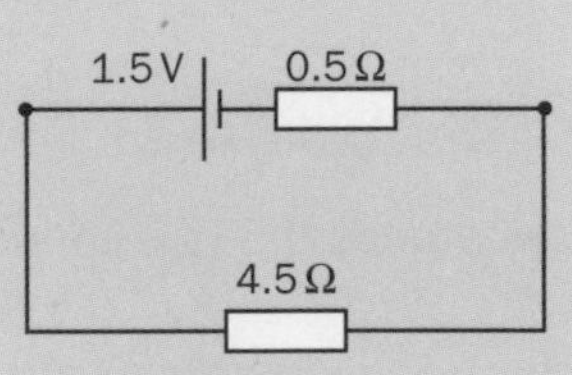

Fig. 14.5

(a) Total resistance = 4.5 + 0.5 = 5.0 Ω since the internal resistance and the external resistance are in series.

(b) Using $I = \frac{V}{R}$ gives $I = \frac{1.5}{5.0} = 0.3$ A

(c) The cell p.d. is the same as the p.d. across the 4.5 Ω resistor = 0.3 A × 4.5 Ω = 1.35 V

Note

The lost voltage = 0.3 A × 0.5 Ω = 0.15 V, which is the difference between the e.m.f. and the cell p.d.

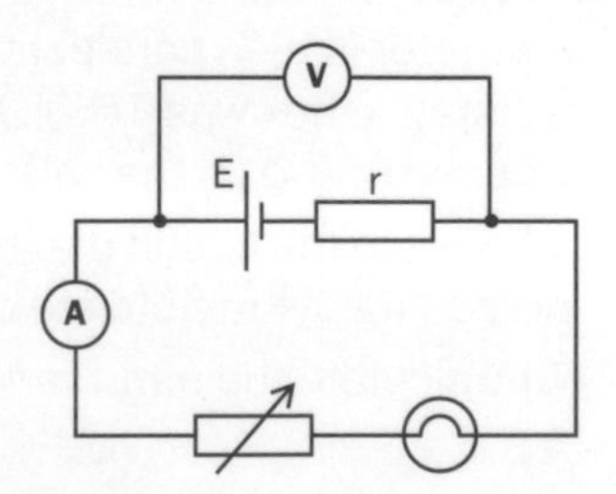

Fig. 14.6 Testing a cell.

Experiment to measure the e.m.f. and the internal resistance of a cell

Use the circuit in Fig. 14.6 to measure the cell p.d. for different currents delivered by the cell. Use the variable resistor to alter the current. Note the torch bulb is present to limit the current from the cell. Tabulate your measurements and use them to plot a graph of the cell p.d. on the vertical axis against the current on the horizontal axis. Use the theory below and your graph to calculate the e.m.f. of the cell and its internal resistance.

- **Theory** The cell p.d. V = IR = E − Ir where E is the cell e.m.f., r is the cell's internal resistance and I is the current. The above equation is the equation for a straight line, y = mx + c, provided V is plotted on the y-axis and I on the x-axis. Thus the gradient, m = −r and the y-intercept, c = E as in Fig. 14.7.

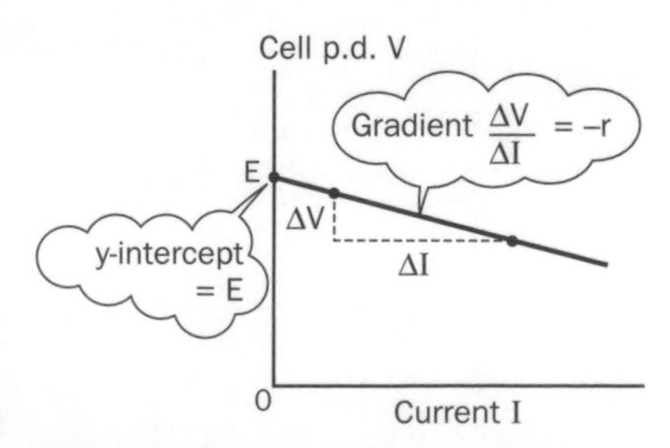

Fig. 14.7 Cell p.d. against current.

Worked example 14.2

A cell of unknown e.m.f., E, and internal resistance, r, was connected in series with a resistance box, R, and an ammeter, as shown in Fig. 14.8.

When the resistance of R was 8.0 Ω, the current was 0.50 A. When the resistance of R was changed to 4.0 Ω, the current was 0.90 A. Calculate the e.m.f. and the internal resistance of the cell.

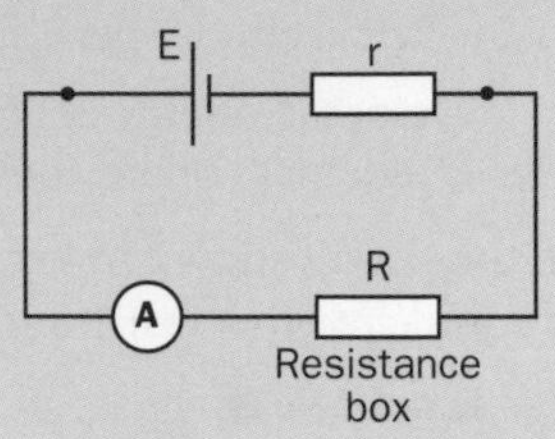

Fig. 14.8

Solution

Use $E = IR + Ir$

When $R = 8.0\ \Omega$, $I = 0.50$ A $\therefore E = (0.5 \times 8.0) + 0.5r = 4.0 + 0.5r$

When $R = 4.0\ \Omega$, $I = 0.90$ A $\therefore E = (0.90 \times 4.0) + 0.9r = 3.6 + 0.9r$

The two equations above are a pair of simultaneous equations that can be solved to find E and r as follows:

$E = 3.6 + 0.9r = 4.0 + 0.5r$

$\therefore 0.9r - 0.5r = 4.0 - 3.6$

$\therefore 0.4r = 0.4$, so $r = \dfrac{0.4}{0.4} = 1.0\ \Omega$ and $E = 4.0 + (0.5 \times 1.0) = 4.5$ V

Questions 14.1

1. A cell of e.m.f. 2.0 V and internal resistance 0.5 Ω is connected in series with an ammeter and a torch bulb. The ammeter reads 0.25 A.
   (a) Sketch the circuit diagram, including the internal resistance of the cell.
   (b) Calculate the p.d. across the bulb and the power delivered to the bulb.
   (c) Calculate the power supplied by the cell.
   (d) Explain why more power is supplied by the cell than is delivered to the bulb.
2. A cell of unknown e.m.f., E, and internal resistance, r, is connected in series with a resistance box. A voltmeter is then connected across the terminals of the resistance box. When the resistance of the resistance box was 16.0 Ω, the voltmeter read 1.20 V. When the resistance of the resistance box was reduced to 8.0 Ω, the voltmeter reading fell to 1.00 V.
   (a) Sketch the circuit diagram and explain why the voltmeter reading decreased as a result of increasing the resistance of the resistance box.
   (b) Calculate the e.m.f. and the internal resistance of the cell.

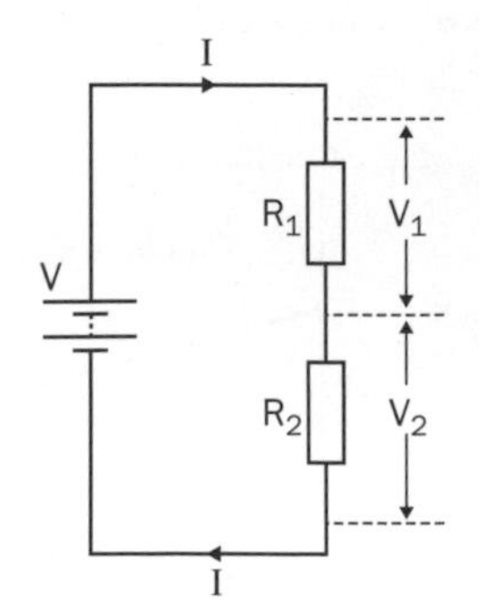

Fig. 14.9 The potential divider.

14.2 The potentiometer

The potential divider

A potential divider is used to supply a specified voltage from a fixed voltage supply. For example, a 5 V electronic circuit could be powered from a 9 V battery using a potential divider. Fig. 14.9 shows a potential divider consisting of two resistors connected to a fixed voltage supply. The potential difference across each resistor is a fixed fraction of the potential difference across the two resistors. Since the current, I

$$= \frac{V}{(R_1 + R_2)}$$

then the p.d., V_1, across resistor R_1

$$= IR_1 = \frac{R_1}{(R_1 + R_2)} V$$

and the p.d., V_2, across resistor R_2

$$= IR_2 = \frac{R_2}{(R_1 + R_2)} V$$

Worked example 14.3

A potential divider is to be used to supply a fixed potential difference of 5.0 V using a 10 kΩ resistor, a 9.0 V battery and a resistor R, as shown in Fig. 14.10.

(a) Calculate the current through the 10 kΩ resistor when the p.d. across it is 5.0 V.

(b) Calculate the potential difference across R when the p.d. across the 10 kΩ resistor is 5.0 V.

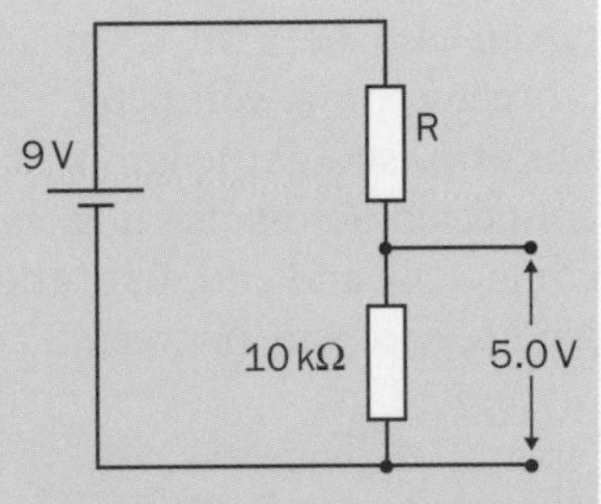

Fig. 14.10

(c) Calculate the resistance of R for a potential difference of 5.0 V across the 10 kΩ resistor.

Solution

(a) $I = \frac{V}{R} = \frac{5.0}{10 \times 10^3} = 5.0 \times 10^{-4}$ A

(b) P.d. across R = 9.0 − 5.0 = 4.0 V

(c) Use $R = \frac{V}{I}$ to give $R = \frac{4.0}{5.0 \times 10^{-4}} = 8000\ \Omega\ (= 8\ \text{k}\Omega)$

Using a potential divider in a sensor circuit

In Fig. 14.10, a different value for R would give a different voltage across each resistor. For example, a value of 10 kΩ, the same as the other resistor, would give a p.d. of 4.5 V across each resistor. The supply voltage is shared between the two resistors in proportion to their resistances. Altering the resistance ratio alters the proportions in which the supply voltage is shared between the two resistors.

If resistor, R, in Fig. 14.10 is replaced by a thermistor (a temperature-dependent resistor) or a light-dependent resistor (LDR), the p.d. across the other fixed resistor varies according to the temperature or the light intensity respectively. Fig. 14.11 shows the idea. The circuit in Fig. 14.11 supplies an 'output p.d.' across the fixed resistor that increases with temperature (or with light intensity if an LDR is used). The output p.d. can be used to operate a relay or to switch on a warning indicator, as explained in Unit 17.

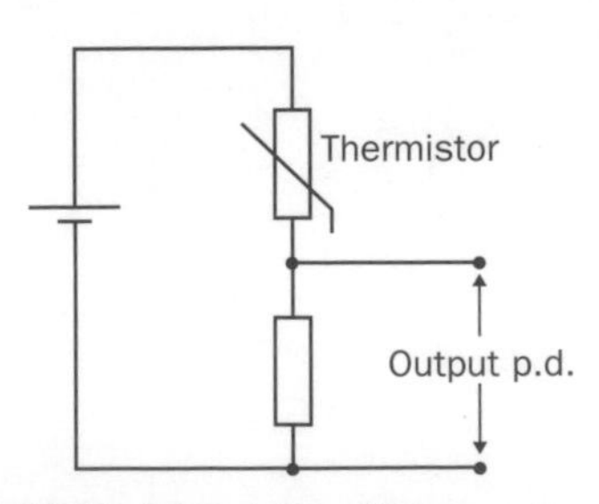

Fig. 14.11 Using a thermistor.

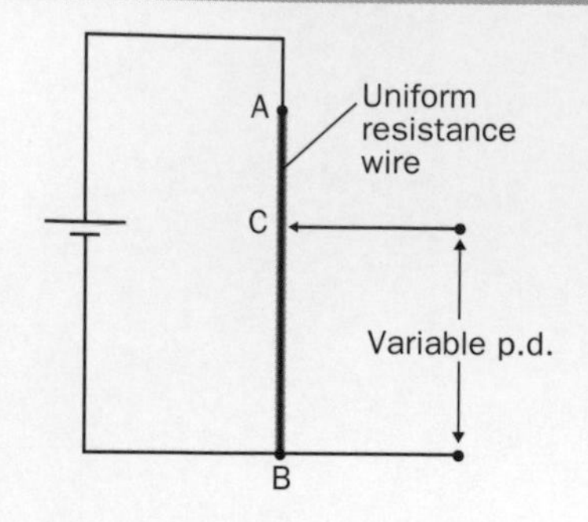

Fig. 14.12 A variable potential divider.

Supplying variable p.d.

This is achieved by connecting a battery across a fixed length of wire of uniform resistance. A sliding contact is moved along the wire, giving a variable p.d. between the sliding contact and either end of the wire, as in Fig. 14.12. The wire may be coiled around a circular track or wound around an insulated tube or replaced by a uniform, thin-layered track of carbon or other resistive material. In Fig. 14.12, the p.d. between B and C increases from zero up to the battery voltage as the sliding contact C is moved from B to A.

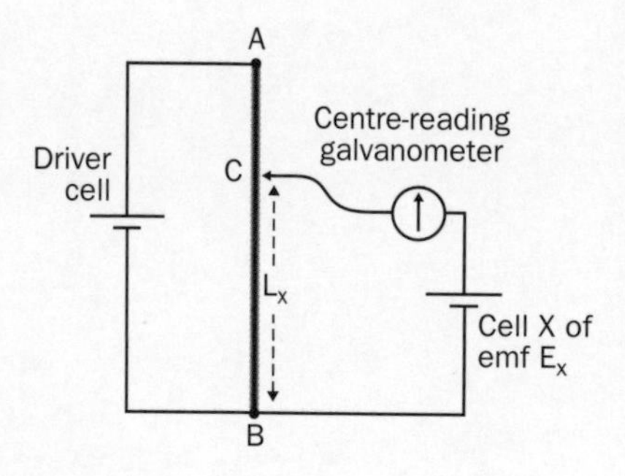

Fig. 14.13 The slide-wire potentiometer.

The principle of the potentiometer

A potentiometer is a potential divider used to compare or measure e.m.fs. It can also be used to measure internal resistance and to calibrate an ammeter or a voltmeter. The slide-wire potentiometer consists of a uniform, straight length of resistance wire with a 'driver' cell or battery connected across its ends, as in Fig. 14.13. The p.d. between the sliding contact, C, and end, B, of the wire, V_{BC}, is therefore in proportion to the length, L, of wire between B and C.

$$V_{BC} = kL$$

where k is a constant.

1. A test cell, X, and a centre-reading galvanometer in series are connected between B and C, as in Fig. 14.13. If the sliding contact, C, is moved down the wire from A to B, the meter deflection changes direction at some point between A and B. The position of C on the wire where the meter reading is exactly zero, the balance point, is located precisely. At this point, the potential difference between B and C opposes the e.m.f. of cell X exactly.

   $$E_x = kL_x$$

   where L_x is the length of wire between B and C needed to balance the e.m.f. of X.

   There is no lost voltage across the internal resistance of X when the meter reads zero because there is no current through X at the balance point.

2. The procedure is repeated using a standard cell, S, with an accurately known e.m.f., E_s, in place of X. The new balance point is located and the new balance length, L_s, is measured.

   $$E_s = kL_s$$

   provided the driver cell continues to supply the same p.d. across the wire. Hence

   $$\frac{E_x}{E_s} = \frac{kL_x}{kL_s} = \frac{L_x}{L_s} \text{ to give } E_x = \frac{L_x}{L_s} E_s$$

Thus E_x can be calculated if E_s is known.

Worked example 14.4

A potentiometer consisting of a metre length of uniform resistance wire with a driver cell connected across the wire was used to measure the e.m.f. of a cell, X. The cell was connected between one end of the wire and a sliding contact on the wire, as in Fig. 14.13. The balance point was located and the balance length measured at 480 mm. Cell X was then replaced by a standard cell of e.m.f. 1.50 V and the new balance point was located, giving a balance length of 655 mm. Calculate the e.m.f. of X.

Solution

Using $E_x = \frac{L_x}{L_s} E_s = \frac{0.480}{0.655} \times 1.5 = 1.1\ V$

Experiment to measure the e.m.f. of a cell

Use a standard cell of known e.m.f. to calibrate a slide-wire potentiometer. Replace the standard cell with the cell under test and measure the balance length for the test cell. Use the standard cell again to check the calibration measurement and confirm the p.d. across the wire has not changed. Use the measurements to calculate the e.m.f. of the test cell.

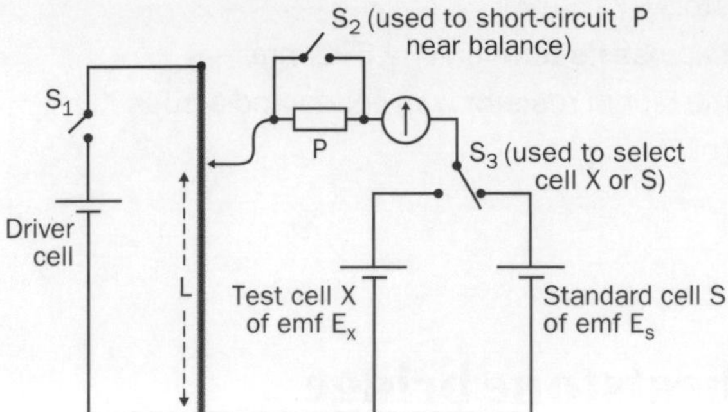

Fig. 14.14 Using a slide wire potentiometer.

Notes

1. Include a protective 1 kΩ resistor, P, in series with the centre-reading meter until the sliding contact is near the balance position. The balance point can then be located precisely by short-circuiting the protective resistor.
2. If the balance length for the standard cell differs when the calibration is repeated, use the average of the two measurements.
3. The internal resistance of cell X can be determined by measuring the balance length with a resistor of known resistance connected across X's terminals.

   The p.d. across X's terminals (cell p.d.), $V = k L_x'$, where L_x' is the balance length with R across the terminals of cell X.

   The current through resistor R at balance, $I = \frac{V}{R} = \frac{k L_x'}{R}$

   Since the e.m.f. of X, $E_x = k L_x$,
   then using $E = V + Ir$ gives

   $$r = \frac{(E_x - V)}{I} = \frac{(kL_x - k L_x')}{k L_x' / R} = \frac{(L_x - L_x')R}{L_x'}$$

Questions 14.2

1. For each potential divider shown in Fig. 14.15, calculate the cell current and the p.d. across resistor Y.

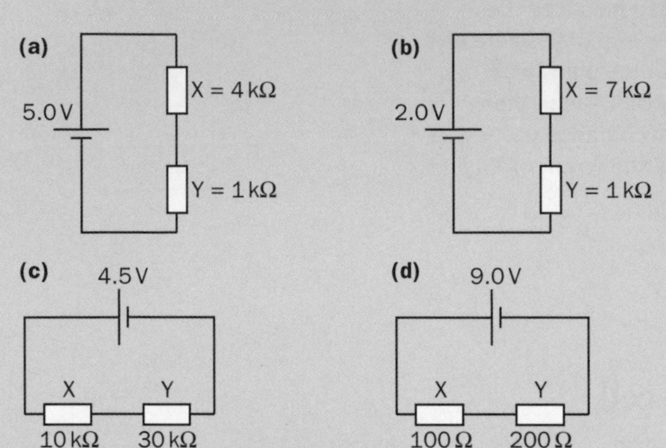

Fig. 14.15

2. A standard cell of e.m.f. 1.08 V was used to calibrate a slide-wire potentiometer. The standard cell was then replaced with a cell, X, of unknown e.m.f. and the balance length was then measured with and without a 5.0 Ω resistor connected across the terminals of the cell. The measurements are listed below.
   Balance length for the standard cell = 558 mm
   Balance length for the test cell without the 5 ohm resistor = 775 mm
   Balance length for the test cell with the 5 ohm resistor across its terminals = 692 mm
   (a) Explain why the balance length was smaller when the 5 ohm resistor was connected across X.
   (b) Calculate **(i)** the e.m.f., **(ii)** the internal resistance of X.

14.3 The Wheatstone bridge

Comparison of two resistances

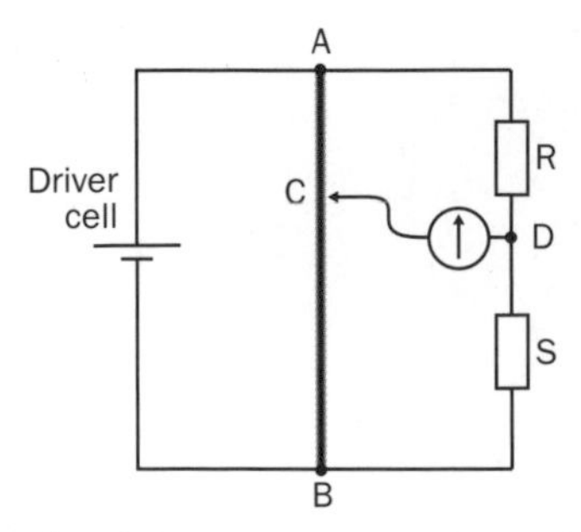

Fig. 14.16 Comparing two resistances.

Fig. 14.16 shows two resistors, R and S, in series with each other, connected in parallel with a fixed length of uniform resistance wire. A driver cell supplies a constant potential difference between points A and B at the ends of the wire and the resistors. A centre-reading meter is connected between a sliding contact, C, on the wire and fixed point, D, the junction of R and S. As the contact is moved along the wire, the meter reading changes, becoming zero where the p.d. between C and B is balanced by the p.d. between D and B.

At the balance point,

$$\frac{\text{the p.d. across R}}{\text{the p.d. across S}} = \frac{\text{the p.d. between A and C}}{\text{the p.d. between C and B}}$$

Since R and S form a potential divider, the ratio of the p.d. across R to the p.d. across S is equal to the resistance ratio R ÷ S. Also, the ratio of the p.d.

between A and C to the p.d. between C and B is the same as the length ratio AC ÷ CB since wire ACB is a potential divider.

$$\therefore \frac{R}{S} = \frac{L_{AC}}{L_{CB}}$$

If S is a standard resistance, then the value of R can be determined precisely by locating the balance point, measuring the lengths AC and CB, then using the above equation.

The theory of the balanced Wheatstone bridge

The circuit on page 194 is an example of a balanced Wheatstone bridge. Fig. 14.17 shows the same circuit, except the wire has been replaced with two resistors, P and Q.

The centre-reading meter reads zero when the p.d. between points C and D is zero. This occurs if

$$\frac{\text{p.d. across P}}{\text{p.d. across Q}} = \frac{\text{p.d. across R}}{\text{p.d. across S}}$$

The ratio of the p.d. across P to the p.d. across Q = P ÷ Q at balance, as the current through P is the same as the current through Q at balance.

Also, the ratio of the p.d. across R to the p.d. across S = R ÷ S at balance, as the current through R is the same as the current through S at balance. Therefore the resistance ratio P ÷ Q is equal to the resistance ratio R ÷ S.

$$\frac{P}{Q} = \frac{R}{S}$$

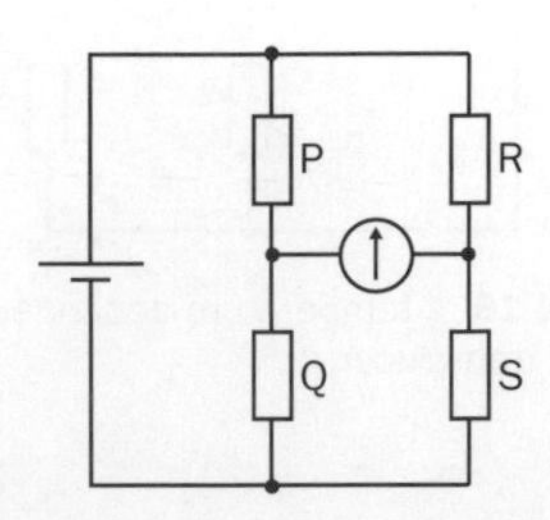

Fig. 14.17 The balanced Wheatstone bridge.

Worked example 14.5

An unknown resistance, R, was connected into a Wheatstone bridge circuit consisting of a metre length of uniform resistance wire and a standard 10.0 Ω resistor, S, as in Fig. 14.18. The balance point was located and the length of wire between the balance point and the end of the wire to which S was connected was measured at 658 mm.

(a) Calculate the resistance of R.

(b) The standard 10 ohm resistor was then replaced with a 5.0 Ω resistance. Determine the position of the new balance point.

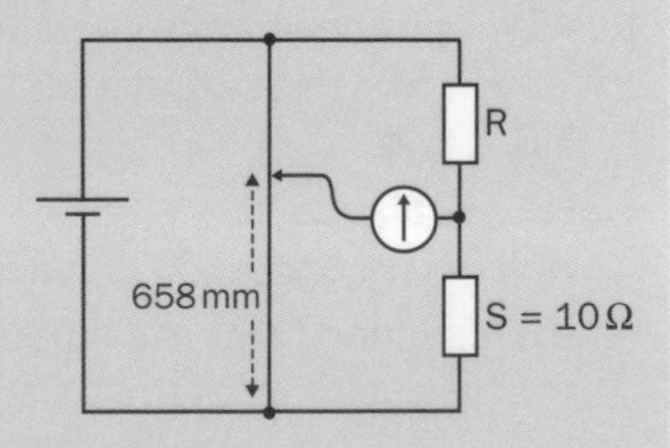

Fig. 14.18

Solution

(a) $\frac{R}{S} = \frac{L_1}{L_2}$ where $L_1 = (1000 - 658)$ mm = 0.342 m, and $L_2 = 0.658$ m

Hence $R = \frac{0.342}{0.658} \times 10.0 = 5.2\ \Omega$

(b) Let L = the balance length from the end at S; let L_w = the length of the wire (= 1.0 m).

$\therefore$ Using $\frac{R}{S} = \frac{(L_w - L)}{L}$ gives $\frac{(L_w - L)}{L} = \frac{R}{S} = \frac{5.2}{5} = 1.04$

Hence $L_w - L = 1.04\,L \therefore 2.04\,L = L_w$, so $L = \frac{L_w}{2.04} = \frac{1.0\text{ m}}{2.04} = 0.490$ m

Using the Wheatstone bridge

To measure an unknown resistance, R

As explained on p. 194, P and Q are replaced with a uniform length of resistance wire and S is a standard resistance. The balance point is located and the length of each section, L_p and L_Q, corresponding to P and Q, is measured. The value of S is chosen so the balance point is in the middle third of the wire. The unknown resistance is calculated using the equation

$$R = \frac{L_P}{L_Q} S$$

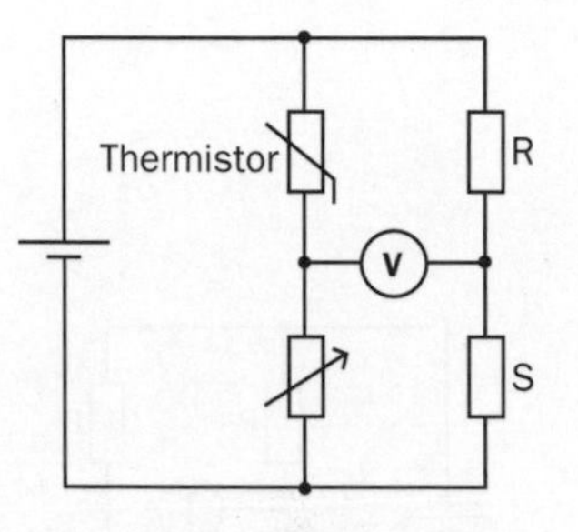

Fig. 14.19 A temperature-dependent output transducer.

To make an output transducer

A transducer produces an output voltage that changes in response to change of a physical variable. For example, Fig. 14.19 shows an output transducer based on a Wheatstone bridge, in which a thermistor is used in place of resistor P and a variable resistor in place of resistor Q. The centre-reading meter has been replaced by a voltmeter. The variable resistor is adjusted so the voltmeter reads zero. When the temperature of the thermistor alters, the voltmeter reading becomes either positive or negative, depending on whether the temperature of the thermistor rises or falls. The output voltage could be supplied to an electronic circuit to operate an alarm. Replacing the thermistor with a light-dependent resistor would give a light-dependent output transducer.

Questions 14.3

1. A Wheatstone bridge consists of a metre length of uniform resistance wire, a 50 Ω standard resistor, S, and an unknown resistor, X, as in Fig. 14.20.

(a) At balance, the length of wire from the balance point to the end at which X is connected is 0.452 m. Calculate the resistance of X.

(b) A second 50 Ω resistor is connected in parallel with S. Determine the new position of the balance point.

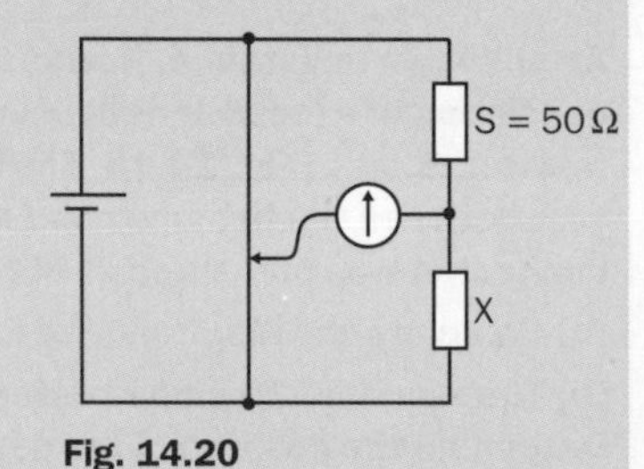

Fig. 14.20

2. Fig. 14.21 shows a light-dependent resistor, R, a variable resistor, VR, and two 10 kΩ resistors in a Wheatstone bridge. With the LDR exposed to daylight, the variable resistor is adjusted to make the output voltage zero.

(a) The light-dependent resistor is then covered. State and explain how this change affects the p.d. between X and Y.

(b) State and explain whether the resistance of the variable resistor should be increased or decreased to return the output voltage to zero with the LDR still in darkness.

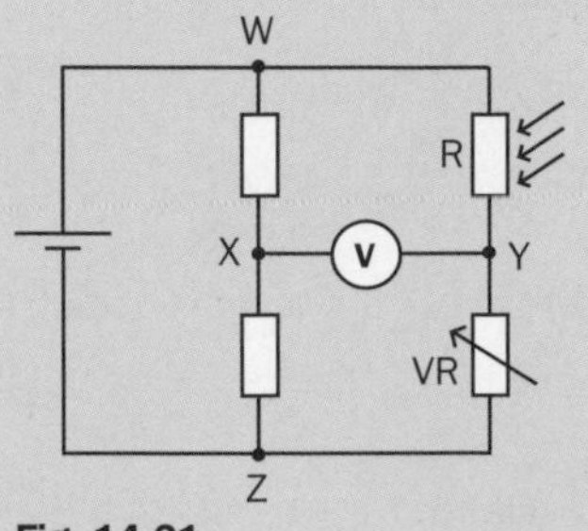

Fig. 14.21

14.4 Resistivity

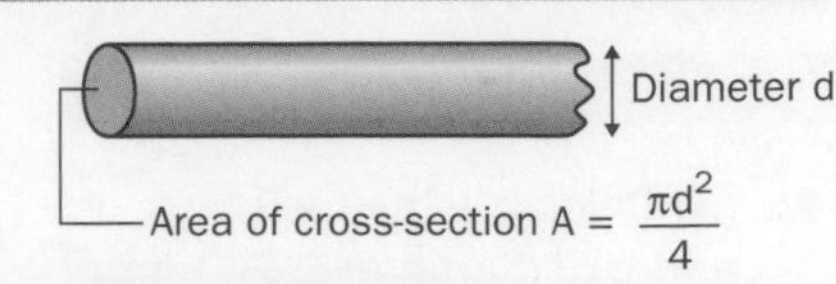

Fig. 14.22 Resistivity.

The resistance of a wire depends on the length and diameter of the wire as well as the material the wire is made from. Measurements show that the resistance R of a wire is

1. proportional to its length, L,

2. inversely proportional to its area of cross-section, A.

$$R = \text{constant} \times \frac{L}{A}$$

The constant in the above equation depends on the material of the wire and is referred to as the **resistivity** (symbol ρ, pronounced 'rho') of the material. Rearranging the above equation therefore gives the following equation for resistivity:

$$\rho = \frac{RA}{L}$$

Notes

1. The unit of resistivity is the ohm metre (Ω m).
2. The area of cross-section of a wire = $\pi d^2 \div 4$, where d is the diameter of the wire.
3. Resistivity values for different materials are given in Table 14.2.
4. The conductivity of a substance (symbol σ, pronounced ' sigma') is defined as $1 \div \rho$. The unit of conductivity is $\Omega^{-1}\,m^{-1}$.

Worked example 14.6

Calculate the resistance of a copper wire of length 2.50 m and uniform diameter 0.36 mm. The resisitivity of copper is 1.7×10^{-8} Ω m.

Solution

$$\text{Area of cross-section, } A = \frac{\pi d^2}{4} = \frac{\pi(0.36 \times 10^{-3})^2}{4} = 1.0 \times 10^{-7}\,m^2$$

$$\therefore R = \frac{\rho L}{A} = \frac{1.7 \times 10^{-8} \times 2.50}{1.0 \times 10^{-7}} = 0.42\,\Omega$$

Table 14.2 Resistivity values for different materials

Material	Resistivity/Ω m
copper	1.7×10^{-8}
constantan	5.0×10^{-7}
carbon	3.0×10^{-5}
silicon	2 300
PVC	$\sim 10^{14}$

Experiment to measure the resistivity of a wire

Use a Wheatstone bridge as in Fig. 14.16 to measure the resistance of different measured lengths of the wire under test. Use a micrometer to measure the diameter of the wire at different points to obtain an average value, d. Calculate the area of cross-section, A, using the equation $A = \pi d^2 \div 4$.

See www.palgrave.com/foundations/breithaupt for more information about this experiment.

Plot a graph of resistance, R, on the vertical axis against length, L, on the horizontal axis. The graph should be a straight line through the origin, in accordance with the equation

$$R = \frac{\rho L}{A}$$

Since the gradient of the line = $\rho \div A$, the resisitivity can be calculated from

$$\rho = \text{gradient} \times A$$

Questions 14.4

1. **(a)** Calculate the resistivity of a wire of diameter 0.24 mm and of length 5.50 m that has a resistance of 58 Ω.

   (b) Calculate the resistance of a 1.0 m wire of identical material that has a diameter of 0.35 mm.

2. A thin layer of carbon on a flat insulating surface is 1.0 mm wide and 10.0 mm long. The resistance from one end to the other is 15 Ω. Calculate the thickness of the layer. The resistivity of carbon is 3.0×10^{-5} Ω m.

14.5 Electrical measurements

Most electrical measurements involve using an ammeter to measure current, or a voltmeter to measure p.d., or an oscilloscope to display and measure an electrical waveform. A multimeter is used as either an ammeter a voltmeter according to its switch settings. All these instruments need to be checked periodically to ensure they give the correct measurement. An analogue ammeter or voltmeter is usually a moving coil meter in which a coil turns against a spring. If the spring gradually weakens, the readings gradually exceed the correct measurements. Digital multimeters and oscilloscopes contain electronic components that might deteriorate and give false readings. Periodic checks might result in an instrument needing to be recalibrated. This section considers how to check electrical instruments.

(a)

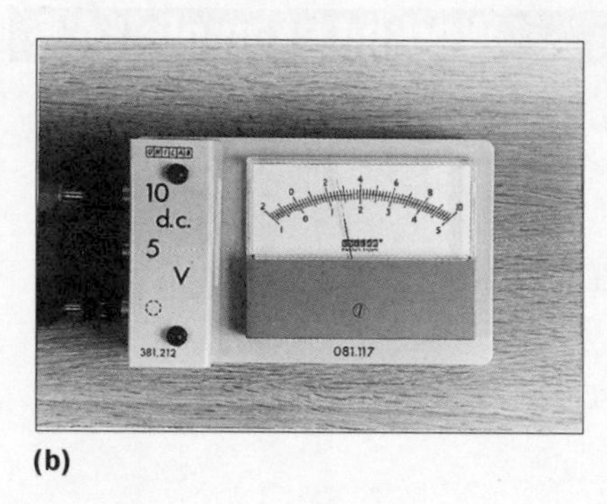

(b)

(c)

Fig. 14.23 Electrical instruments **(a)** ammeter, **(b)** voltmeter, **(c)** digital multimeter.

- The **ampere**, one of the base units of the SI system of units, is defined in terms of the force between two current-carrying conductors, as explained on p. 263. An ammeter can be checked by connecting it in series with a current balance which is designed to measure the force between two current-carrying conductors. However, accurate current balances are not readily available and are not easy to use.
- The **volt** is a derived unit, equal to the potential difference between two points in a circuit when 1 coulomb of charge delivers 1 joule of electrical energy on passing from one point to the other. Potential difference can be determined by measuring the electrical energy supplied to an electric heater by a measured amount of charge. The potential difference is the energy supplied per unit charge. Fig. 14.24 shows the idea.

  A potential difference determined in this way can then be used to calibrate a potentiometer and thus to measure accurately the e.m.f. of a

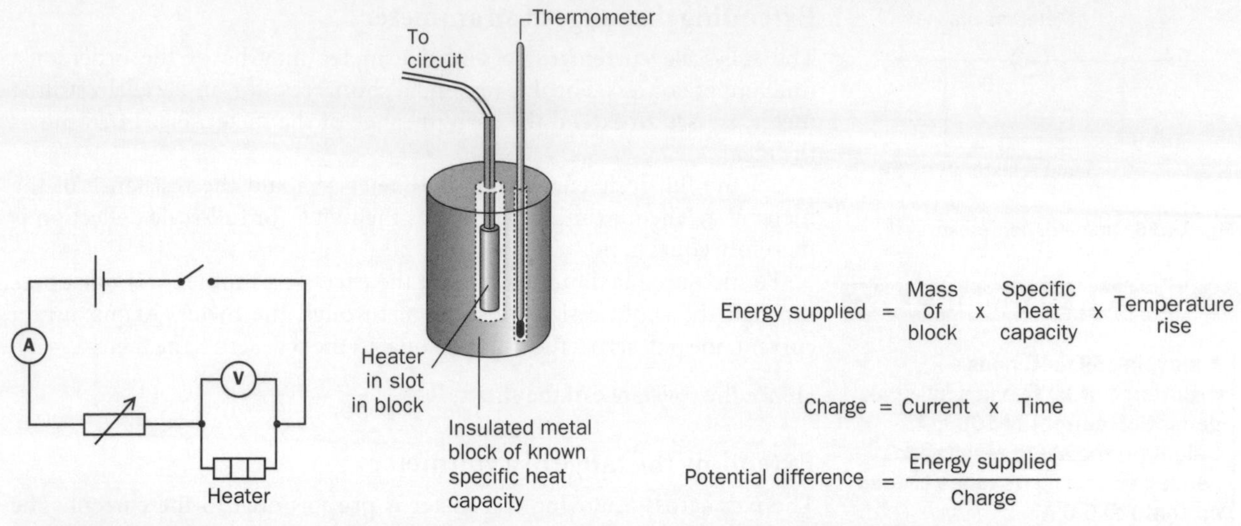

Fig. 14.24 Measuring potential difference.

standard cell. Although this work and other methods of determining potential difference are part of the activity of specialist standards laboratories, standard cells as accurate sources of e.m.f. are readily available as a result. A standard cell can be used to calibrate a potentiometer which can then be used to check the accuracy of a voltmeter, as in Fig. 14.25.

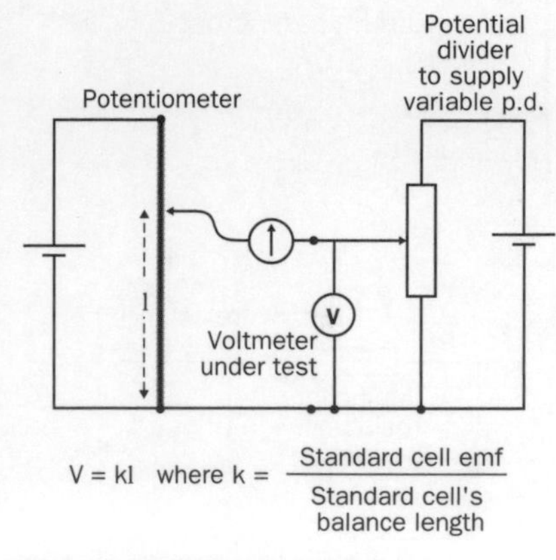

Fig. 14.25 Calibrating a voltmeter.

- The **ohm** is also a derived unit, equal to the resistance of a conductor carrying a current of 1 A when the p.d. across the conductor is exactly 1 V. The resistance of a conductor can be determined by measuring the p.d. across the conductor when a measured current is passed through it. The resistance of a standard resistance can therefore be determined using a calibrated voltmeter and a calibrated ammeter.

The electrical meters in a laboratory can therefore be checked if a standard cell and a standard resistance are available. As explained above, a standard cell may be used to calibrate a voltmeter. Standard resistances do not usually change and can therefore be used with a calibrated voltmeter to check the accuracy of an ammeter. In addition, a standard resistance can be used to measure other resistances using a Wheatstone bridge, as explained on p. 198.

Extending the range of a moving coil meter

A moving coil meter is a current-operated instrument with a linear scale. In other words, the deflection of the pointer from zero is in proportion to the current. If the current through a moving coil meter is increased, the deflection of the pointer increases. If the current exceeds the maximum reading on the scale, referred to as the 'full-scale' deflection, the coil of the instrument could be permanently damaged due to the heating effect of the current.

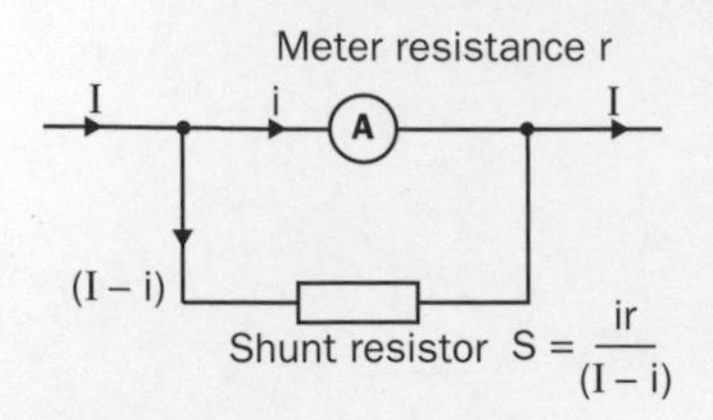

Fig. 14.26 Ammeter conversion.

Extending the range of an ammeter

The full-scale current for a sensitive meter may be of the order of a milliampere or less. For this reason, a 'shunt' resistor, in parallel with the meter, is used to extend the current range of the meter. Fig. 14.26 shows the idea.

Let the full-scale current for the meter = i and the resistance of the meter = r. The maximum p.d. across the meter for full-scale deflection is therefore equal to ir.

To measure maximum current I, the excess current (I − i) must pass through the shunt resistor instead of through the meter. At maximum current, the p.d. across the shunt is equal to the p.d. across the meter (= ir).

Hence the resistance of the shunt, $R = \frac{\text{p.d.}}{\text{current}} = \frac{ir}{(I - i)}$

Worked example 14.7

A moving coil meter has a resistance of 10 Ω and a full-scale deflection current of 100 mA. Calculate the shunt resistance needed to extend its range to read currents to 5.0 A.

Solution

The maximum current through shunt S = 5.0 − 0.1 = 4.9 A

The maximum p.d. across S = i R = 0.1 × 10 = 1.0 V

$\therefore S = \frac{1.0}{4.9} = 0.20\ \Omega$

Extending the range of a voltmeter

The p.d. across a moving coil meter is proportional to the current. The meter may be used to measure p.d. provided the current does not exceed the full-scale current. However, since the resistance of most moving coil meters is low, full-scale current is reached when the p.d. is no more than of the order of millivolts. To extend the range of a moving coil voltmeter, a 'multiplier' resistance is connected in series with the meter, as in Fig. 14.27.

Let the full-scale voltage across the meter = v and the resistance of the meter = r. Hence the maximum current through the meter, I = v ÷ r. This is also the maximum current through the multiplier, since the multiplier is in series with the meter.

To read voltages up to V, the excess voltage (V − v) must be dropped across the multiplier.

$$\therefore \text{the multiplier resistance, } R = \frac{(V - v)}{I} = \frac{(V - v)}{v}\, r$$

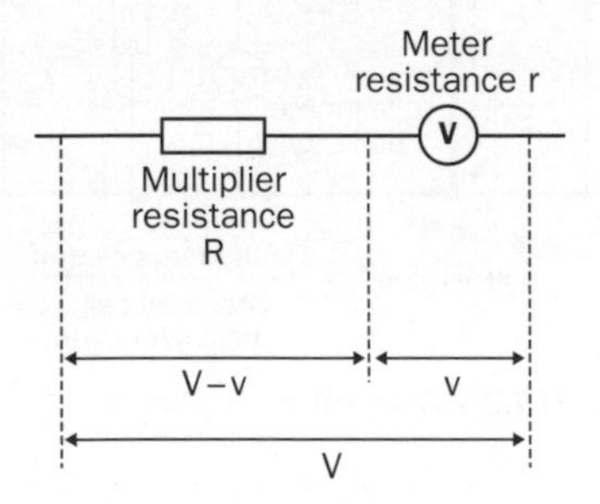

Fig. 14.27 Voltmeter conversion.

Worked example 14.8

A moving coil meter has a resistance of 1000 Ω and a full-scale deflection current of 0.1 mA. Calculate the resistance of the multiplier that must be connected in series with it to extend its range to read voltages up to 20 V.

Solution

Maximum current, I = 0.1 mA ∴ maximum p.d. across meter coil = 0.1 mA × 1000 Ω = 0.1 V

Hence excess voltage across multiplier = 20 − 0.1 = 19.9 V

$\therefore \text{multiplier resistance} = \frac{19.9\ \text{V}}{0.1\ \text{mA}} = 199\,000\ \Omega$

Multimeters

A multimeter can be used as an ammeter or a voltmeter, according to the position of a switch that connects a resistor in series or in parallel with the meter.

Digital meters

A digital multimeter is a digital voltmeter which can be used to measure voltage directly. If it is connected in series or in parallel with a resistor, it can be used to measure voltage or current, as explained below.

Digital voltmeters

Most digital voltmeters have a very high input resistance and therefore draw negligible current when connected across a component. To extend the range of a digital voltmeter, a suitable resistor is connected in series with the voltmeter, just as for a moving coil voltmeter. For example, suppose a digital voltmeter reads up to 2.0 V and has an input resistance of 1 MΩ. The maximum current is therefore 2 μA (= 2.0 V ÷ 1 MΩ). To read up to 20 V, the p.d. across the series resistor must be 18 V (= 20 − 2 V). Hence the resistance of the series resistor must be 9 MΩ (= 18.0 ÷ 2 μA).

Digital ammeters

A digital ammeter is a digital voltmeter with a resistor of low resistance connected across its input terminals, as in Fig. 14.28. For example, a digital voltmeter which reads up to 2.0 V and has an input resistance of 1 MΩ can be converted to read currents up to 2.0 A by connecting a 1.0 Ω resistor across its input terminals. The current through the actual meter is negligible, so the entire current of 2.0 A passes through the 1.0 Ω resistor to give a p.d. of 2.0 V across the meter's input terminals.

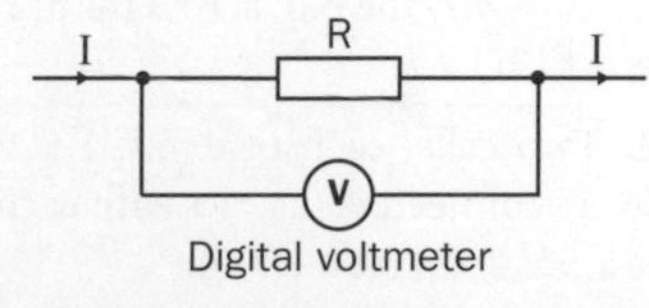

Fig. 14.28 A digital meter used as an ammeter.

Summary

- **The electromotive force (e.m.f.)** of a cell is the electrical energy per unit charge created in the cell.
- **The internal resistance** of a cell is the lost voltage per unit current in the cell.
- **The potential divider equation**

  For two resistors R_1 and R_2 in series with each other and a voltage supply V,

  the p.d. across $R_1 = \dfrac{R_1V}{(R_1 + R_2)}$

  and the p.d. across $R_2 = \dfrac{R_2V}{(R_1 + R_2)}$
- **Comparison of two e.m.fs using the slide-wire potentiometer**

  $$\frac{E_1}{E_2} = \frac{L_1}{L_2}$$
- **The Wheatstone bridge equation**

  $$\frac{P}{Q} = \frac{R}{S}$$
- **Resistivity**

  $$\rho = \frac{RA}{L}$$

Questions 14.5

1. A moving coil meter has a resistance of 100 Ω and a full-scale deflection current of 1.0 mA. How could the meter be adapted to read
   (a) a current up to **(i)** 100 mA, **(ii)** 10.0 A?
   (b) voltages up to **(i)** 10.0 V, **(ii)** 30.0 V?
2. A digital meter has an input resistance of 5.0 MΩ and a full-scale reading of 200 mV. How could the meter be adapted to read
   (a) a voltage up to **(i)** 2.0 V, **(ii)** 100 V?
   (b) a current up to **(i)** 200 mA, **(ii)** 10 A?

◆ **Meter conversion**

To extend the range of a meter of resistance r and f.s.d. current i to

1. maximum current I, shunt resistance $S = \frac{ir}{(I - i)}$
2. maximum voltage V, multiplier resistance $R = \frac{(V - v)}{v} r$, where $v = ir$

Revision questions

14.1. A cell of e.m.f. 2.0 V and internal resistance 0.8 Ω is connected to a 4.2 Ω resistor.
(a) (i) Sketch the circuit diagram, **(ii)** calculate the current and the p.d. across the 4.2 Ω resistor.
(b) Explain why the p.d. across the 4.2 Ω resistor is less than the cell e.m.f.

14.2. Two cells, each of e.m.f. 1.5 V and internal resistance 0.5 Ω are connected in series with each other, a variable resistor and a 2.5 V 0.25 A torch bulb.
(a) Sketch the circuit diagram.
(b) The variable resistor is adjusted so the bulb is at its normal brightness. Calculate **(i)** the power delivered to the bulb, **(ii)** the electrical energy per second produced by the two cells.
(c) Account for the difference between your answers to (b)(i) and (ii).

14.3. A cell of e.m.f. E, and internal resistance, r, was connected in series with an ammeter and a resistance box. When the resistance of the resistance box was 15.0 Ω, the current was 100 mA. When the resistance of the box was changed to 5.0 Ω, the current was 200 mA.
(a) Calculate the e.m.f. and the internal resistance of the cell.
(b) Calculate the current from the cell when the resistance of the box was changed to 3.0 Ω.

14.4. The potential divider in Fig. 14.29a consists of a resistor, R, of resistance 10 kΩ and a thermistor in series connected to a 12.0 V battery. Fig. 14.29b shows how the resistance of the thermistor varies with its temperature.
(a) Calculate the voltage across R at 20° C.
(b) Determine the temperature at which the voltage across R is 5.0 V.

14.5. Two 1 k Ω resistors are connected in series with each other and a cell of e.m.f. 2.0 V and negligible internal resistance. A voltmeter is connected across one of the resistors.
(a) Sketch the circuit diagram.
(b) Calculate the voltmeter reading if the voltmeter resistance is **(i)** infinite, **(ii)** 1 k Ω.

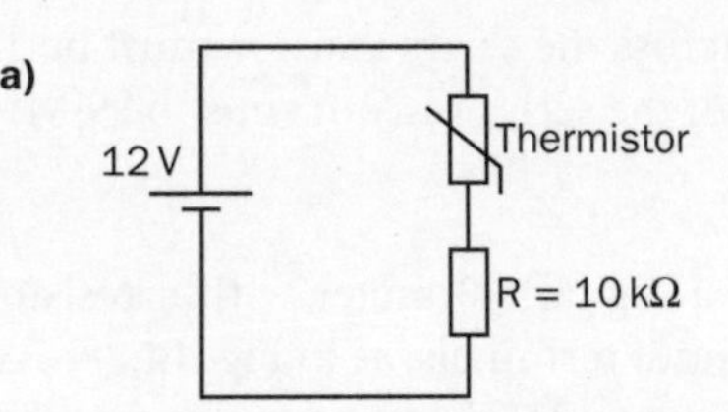

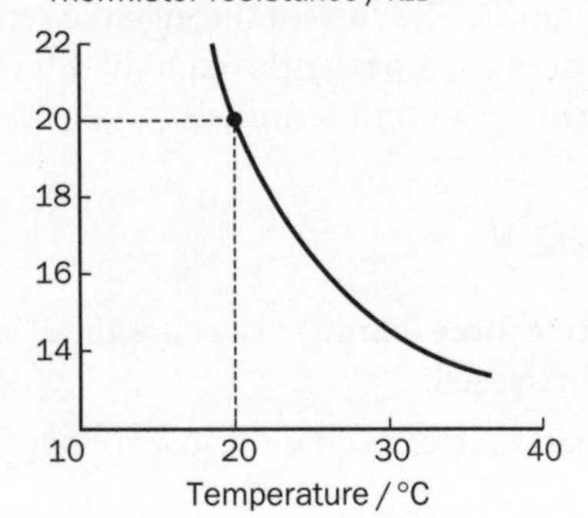

Fig. 14.29

14.6. In a slide-wire potentiometer experiment to measure the e.m.f. of a cell X, the wire was calibrated using a standard cell of e.m.f. 1.50 V. The balance length readings obtained were as follows:
1. Balance length for cell X = 651 mm
2. Balance length for the standard cell = 885 mm
(a) Calculate the e.m.f. of X.
(b) A resistor of resistance 2.0 Ω was connected across the terminals of the cell X when it was connected in the circuit, reducing the balance length to 542 mm. Calculate **(i)** the p.d. across the terminals of cell X at balance with the 2.0 Ω resistor connected across its terminals, **(ii)** the internal resistance of cell X.

14.7. In the Wheatstone bridge circuit in Fig. 14.30, calculate
(a) the resistance of R to balance the bridge when P = 5.0 Ω, Q = 12.0 Ω and S = 50.0 Ω,
(b) the resistance of Q to balance the bridge when P = 5.0 Ω, R = 20.0 Ω and S = 50.0 Ω,

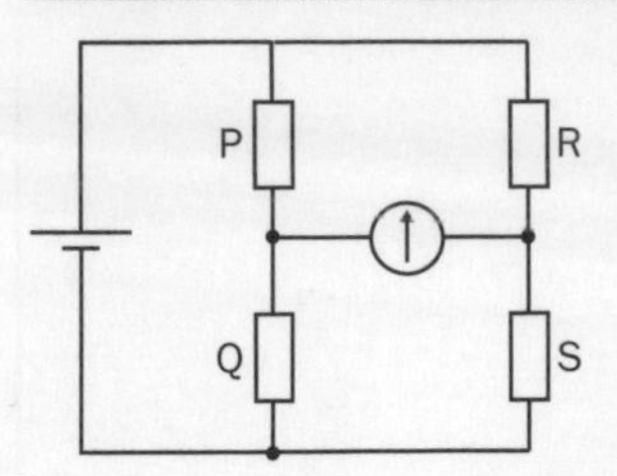

Fig. 14.30

(c) the resistance of P to balance the bridge when R = 20.0 Ω, Q = 12.0 Ω and S = 50.0 Ω.

14.8. A Wheatstone bridge consisting of a metre length of uniform resistance wire, a test resistor, R, and an resistance box, S, as in Fig. 14.31, was used to measure the resistance of the test resistor.

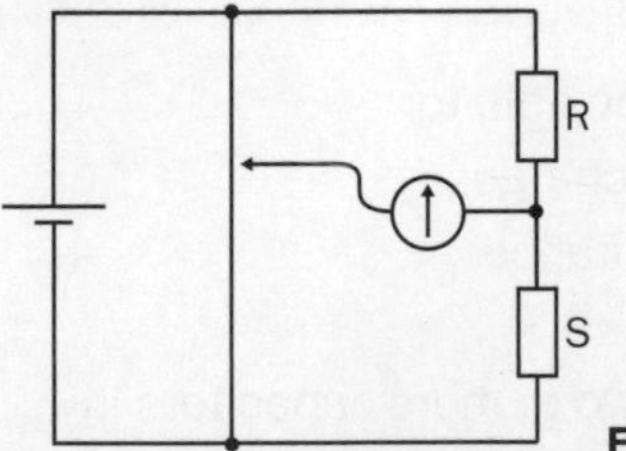

Fig. 14.31

(a) When S = 12.0 Ω, the balance point was 438 mm from the end of the wire where S was connected.

(i) Calculate the resistance of R. **(ii)** The balance point was located to a precision of 2 mm. Calculate the probable error in the resistance of R.

(b) Determine the new balance point if a 6.0 Ω resistor was connected in parallel with S.

14.9. (a) Calculate the resistance of a wire of length 1.50 m and diameter 0.36 mm which is made from material of resistivity 4.8×10^{-7} Ω m.

(b) A 10.0 m length of wire made of the same material as in (a) has a resistance of 105 Ω. Calculate the diameter of this wire.

(c) Calculate the length of a copper wire of diameter 2.0 mm and resistance 0.040 Ω. The resistivity of copper is 1.7×10^{-8} Ω m.

14.10. A moving coil meter has a resistance of 500 Ω and a full-scale deflection current of 0.10 mA. Explain how the meter can be adapted to read

(a) a maximum current of **(i)** 100 mA, **(ii)** 5.0 A,

(b) a maximum p.d. of **(i)** 1.0 V, **(ii)** 15.0 V.

Capacitors

Contents

Objectives

After working through this unit, you should be able to:

- ▶ describe how a capacitor stores electric charge
- ▶ define capacitance and the unit of capacitance
- ▶ measure the capacitance of a capacitor
- ▶ calculate the combined capacitance of two or more capacitors in series or in parallel
- ▶ calculate the energy stored in a charged capacitor
- ▶ calculate the change of p.d. and energy stored when two capacitors are connected together
- ▶ describe the factors that determine the capacitance of a parallel plate capacitor
- ▶ describe and calculate the change of p.d. across a capacitor charging or discharging through a fixed resistor

15.1 Storing charge

If you have ever received an electric shock from the body of a car, you will know that an insulated metal object is capable of storing charge. Storage of charge can be achieved by direct contact with a charged insulator or by connecting one terminal of a battery to the object and then connecting the other terminal to an earthed metal pipe. Electrons transfer to or from the insulated metal object, depending on whether it is negatively or positively charged.

A **capacitor** is a device designed to store charge. The simplest type of capacitor consists of two parallel metal plates opposite each other. When a battery is connected to the plates, as in Fig. 15.1, electrons from the negative terminal of the battery transfer onto the plate connected to that terminal. At the same time, electrons transfer from the other plate to the positive terminal of the battery, leaving this plate with a positive charge. The flow of electrons decreases as the plates charge up. When the plate p.d. is the same as the battery p.d., the flow of electrons ceases.

Fig. 15.1 Storing charge on a parallel plate capacitor **(a)** circuit, **(b)** symbol for capacitor.

The two plates store equal and opposite amounts of charge. If the charge on the positive plate is +Q, the charge on the negative plate will be −Q. The capacitor is said to store charge Q. The symbol for a capacitor is shown in Fig. 15.1b.

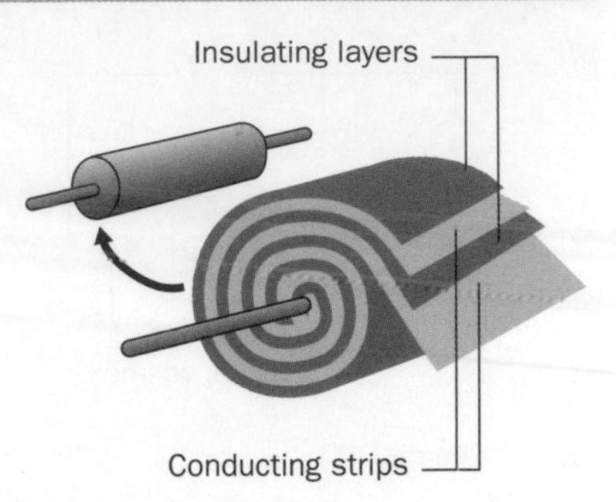

Fig. 15.2 Practical capacitor.

Capacitors in practice

A practical capacitor consists of two flexible conducting strips separated by an insulating layer. The strips are like the metal plates of the parallel plate capacitor but they are rolled up to make the capacitor much smaller in physical size. The insulating layer is a **dielectric** substance, which allows the capacitor to store much more charge for a given voltage. The molecules of the dielectric substance become polarised when a voltage is applied between the two conducting strips. This means that the electrons of each molecule are attracted slightly towards the positive plate. This causes the dielectric surface in contact with the positive plate to become negative and the other surface positive. The overall result is that the surface charge on the dielectric surface effectively cancels out some of the charge on the conducting strip, allowing more charge to be stored on the conducting strip. Fig. 15.3 shows the action of a dielectric substance.

A dielectric substance is an insulator because the electrons remain attached to its molecules, even if displaced slightly, when a voltage is applied between the two conducting strips. However, if the applied voltage exceeds a certain value that depends on the nature of the substance and the thickness of the insulating layer, the insulation breaks down. The maximum voltage rating of a capacitor should never be exceeded for this reason. In an electrolytic capacitor, the dielectric substance is created by electrolysis as a very thin layer at the interface between a conducting strip and the insulating material when the capacitor is first charged up. An electrolytic capacitor **must** be connected into the circuit according to the polarity indicated at its terminals, otherwise it might explode.

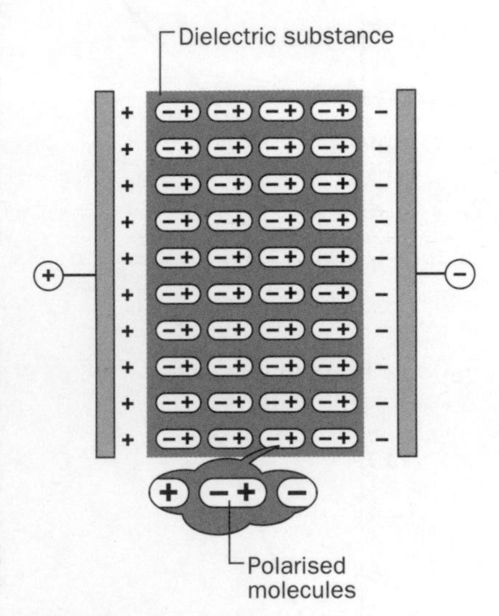

Fig. 15.3 The action of a dielectric substance.

Experiment to investigate charging a capacitor

1. Through a fixed resistance

Fig. 15.4 shows how to charge a capacitor through a fixed resistor. The current decreases to zero as the capacitor charges up. When the capacitor voltage is the same as the battery voltage, no more charge can be stored on the capacitor and the rate of flow of charge is then zero. Using a larger resistance reduces the rate of charging so the charging process takes longer.

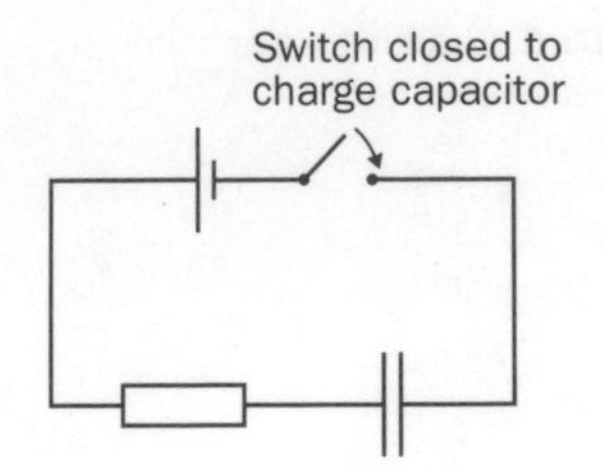

Fig. 15.4 Charging a capacitor through a fixed resistor.

2. At constant current

Fig. 15.5 shows how a capacitor can be charged at a constant rate. The variable resistor must be continually reduced to keep the charging current constant. When the variable resistance has been reduced to zero, the current falls to zero as the voltage across the capacitor is then the same as the battery voltage. A digital voltmeter is used to measure the capacitor voltage at fixed intervals as the capacitor charges up at constant current. The ammeter is used to measure the current. The results from such an experiment are

(a)

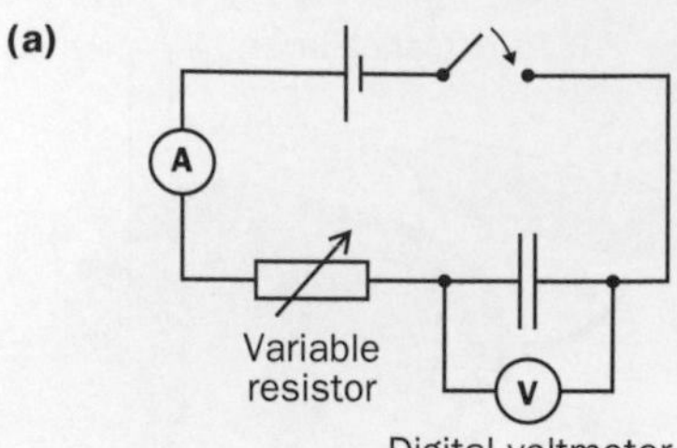

(b)

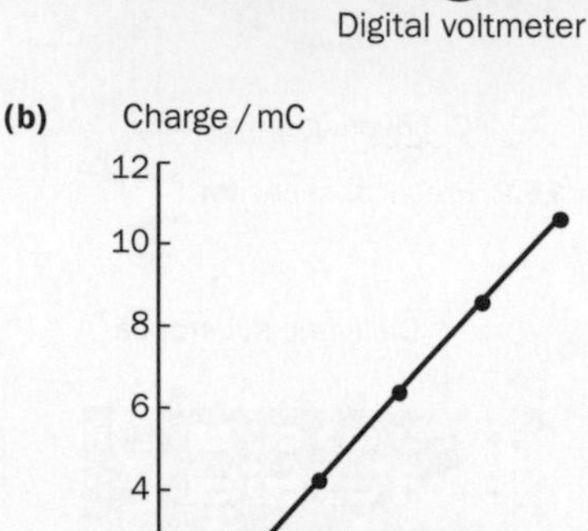

(c)

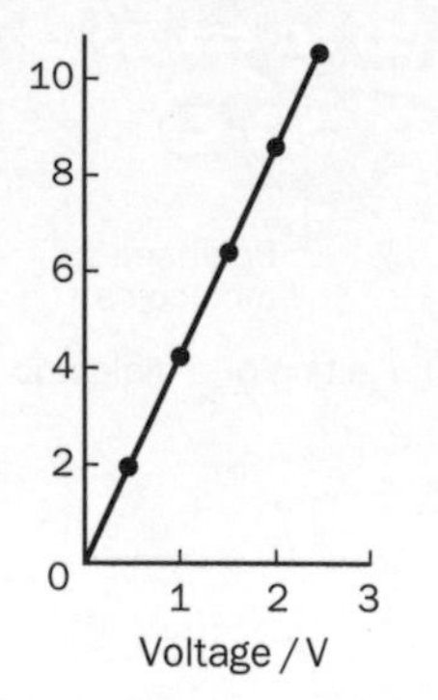

Fig. 15.5 Charging a capacitor at constant current.

Table 15.1 Charging a capacitor at constant current

Time/s	0.00	5.00	10.00	15.00	20.00	25.00
Capacitor voltage/V	0.00	0.45	0.91	1.37	1.81	2.27
Charge/mC	0.00	2.15	4.30	6.45	8.60	10.75

Current = 0.43 mA

shown in Table 15.1. The charge stored has been calculated using the equation 'charge = current × time'.

The results may be plotted as a graph of y = charge stored, Q, on the vertical axis against x = applied voltage, V, on the horizontal axis. The measurements in Table 15.1 are plotted in Fig. 15.5b. The results show that the charge stored is proportional to the voltage. In other words, the charge stored per unit voltage is constant.

The capacitance, C, of a capacitor is defined as the charge stored per unit voltage by the capacitor.

The unit of capacitance is the farad (F) which is equal to 1 coulomb per volt. In practice, capacitance values are usually expressed in microfarads (μF = 10^{-6}F) or picofarads (pF = 10^{-12}F). Show for yourself that the capacitance in Fig. 15.5 is 4700 μF.

$$\textbf{Capacitance } C = \frac{Q}{V}$$

where Q = charge stored, V = voltage.

> **Mathematical note**
>
> Rearranging $C = \frac{Q}{V}$ gives $Q = CV$ or $V = \frac{Q}{C}$

Worked example 15.1

A 4700 μF capacitor is connected in series with a switch, a variable resistor, an ammeter and a 3.0 V battery. When the switch is closed, the variable resistor is continually reduced to keep the current constant at 0.050 mA. Calculate (a) the charge stored by the capacitor when fully charged, (b) the time taken to reach full charge at a current of 0.050 mA.

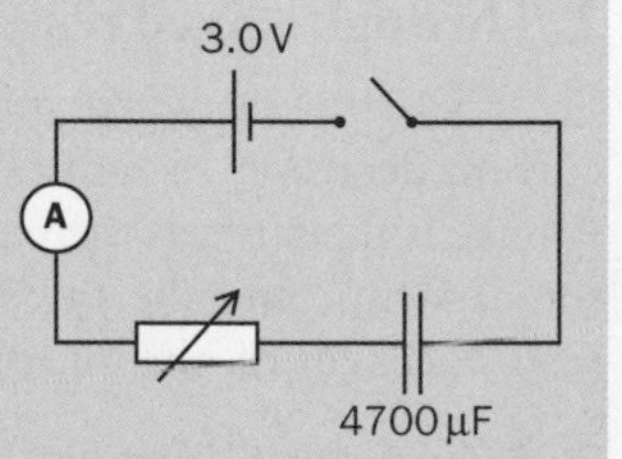

Fig. 15.6

Solution

(a) Rearrange $C = \frac{Q}{V}$ to give $Q = CV = 4700 \times 10^{-6} \times 3.0 = 1.4 \times 10^{-2}$ C

(b) Rearrange $Q = It$ to give $t = \frac{Q}{I} = \frac{1.4 \times 10^{-2}}{0.050 \times 10^{-3}} = 280$ s.

Questions 15.1

1. Copy and complete the table below by calculating the missing value in each column.

Capacitance / μF	10.00	0.22	?	1.00×10^{-3}
Voltage / V	6.00	?	4.50	?
Charge / μC	?	0.33	9.90×10^{4}	5.00×10^{-2}

2. A 2200 μF is charged at a constant current of 2.5 mA using a 9 V battery and a variable resistor. Calculate **(a)** the charge stored by this capacitor at 9.0 V, **(b)** the time taken to become fully charged at a constant current of 0.25 mA.

15.2 Capacitor combinations

Capacitors in series

Fig. 15.7 shows two capacitors, C_1 and C_2, connected in series to a battery. Each capacitor stores the same amount of charge, Q. The total charge stored is the same as the charge stored on each capacitor. This is because the charge on the two capacitor plates connected together is due to electrons from one of these two plates being attracted onto the other plate. Just as two resistors in series carry the same current, two capacitors in series carry the same charge.

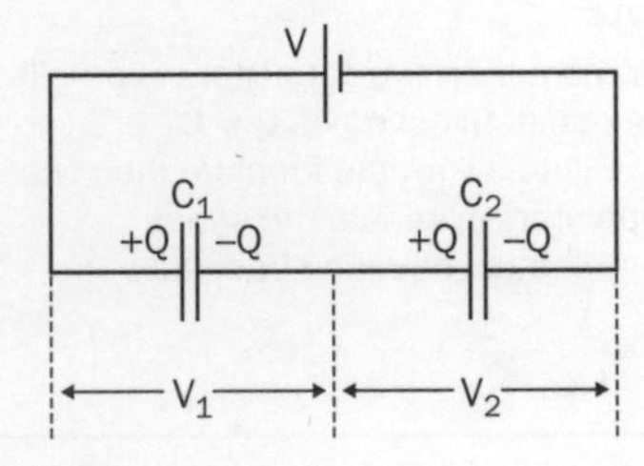

Fig. 15.7 Capacitors in series.

The p.d. across capacitor C_1, $V_1 = \frac{Q}{C_1}$

The p.d. across capacitor C_2, $V_2 = \frac{Q}{C_2}$

The battery voltage, $V = V_1 + V_2$;

hence $V = \frac{Q}{C_1} + \frac{Q}{C_2}$

Because the combined capacitance, $C = \frac{Q}{V}$

$$\frac{1}{C} = \frac{V}{Q} = \left(\frac{Q}{C_1} + \frac{Q}{C_2}\right) \div Q = \frac{1}{C_1} + \frac{1}{C_2}$$

The combined capacitance, C, of two capacitors, C_1, and C_2, in series is given by the equation

$$\mathbf{\frac{1}{C} = \frac{1}{C_1} + \frac{1}{C_2}}$$

Worked example 15.2

A 2 μF capacitor is connected in series with a 6 μF capacitor and a 6.0 V battery. Calculate
(a) the combined capacitance of these two capacitors in series,
(b) (i) the charge stored by each capacitor,
(ii) the voltage across each capacitor.

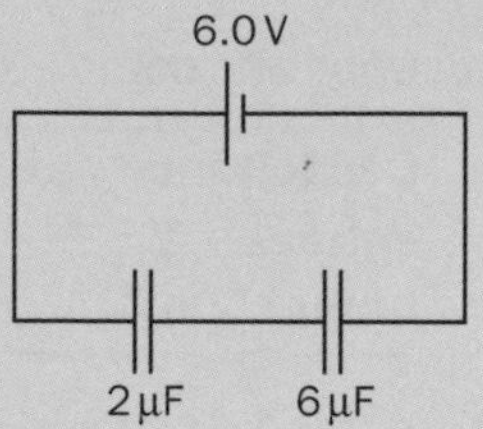

6.0 V

2 μF 6 μF

Fig. 15.8

Solution

(a) $\frac{1}{C} = \frac{1}{C_1} + \frac{1}{C_2} = \frac{1}{2} + \frac{1}{6} = \frac{6+2}{2 \times 6} = \frac{8}{12} \quad \therefore \quad C = \frac{12}{8} = 1.5\ \mu F$

(b) (i) $Q = CV = 1.5\ \mu F \times 6.0\ V = 9.0\ \mu C$,

(ii) $2\ \mu F;\ V = \frac{Q}{C} = \frac{9.0\ \mu C}{2.0\ \mu F} = 4.5\ V$

$6\ \mu F;\ V = \frac{Q}{C} = \frac{9.0\ \mu C}{6.0\ \mu F} = 1.5\ V$

Note that the voltages add up to the battery voltage.

Capacitors in parallel

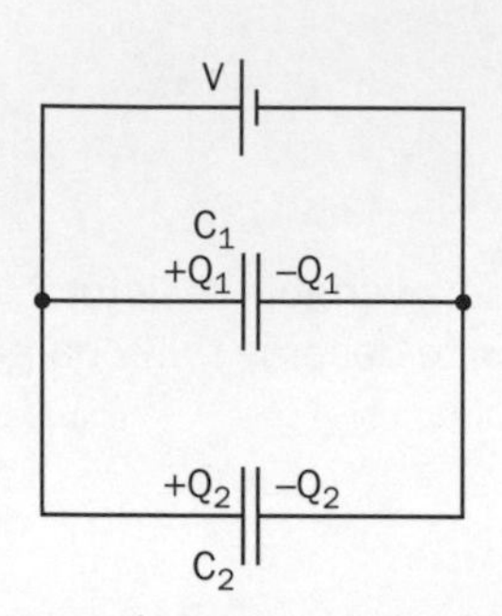

Fig. 15.9 Capacitors in parallel.

> **Note**
> For more than two capacitors in parallel, the equation becomes $C = C_1 + C_2 + C_3 +$, etc. Likewise, for more than two capacitors in series, the series combination equation becomes
> $\frac{1}{C} = \frac{1}{C_1} + \frac{1}{C_2} + \frac{1}{C_3}$ +, etc.

Fig. 15.9 shows two capacitors, C_1 and C_2, connected in parallel to a battery. Each capacitor has the same voltage, V, across its terminals because they are in parallel with each other.

Charge on capacitor C_1, $Q_1 = C_1V$
Charge on capacitor C_2, $Q_2 = C_2V$
Total charge stored, $Q = Q_1 + Q_2 = C_1V + C_2V$

Hence the combined capacitance, $C = \frac{Q}{V} = \frac{C_1V + C_2V}{V} = C_1 + C_2$

The combined capacitance, C, of two capacitors, C_1 and C_2, in parallel is given by the equation

$$\mathbf{C = C_1 + C_2}$$

Worked example 15.3

A 4 μF capacitor and a 6 μF capacitor in parallel are connected to a 3.0 V battery. Calculate (a) the combined capacitance of these two capacitors in parallel, (b) (i) the total charge stored, (ii) the charge stored on each capacitor.

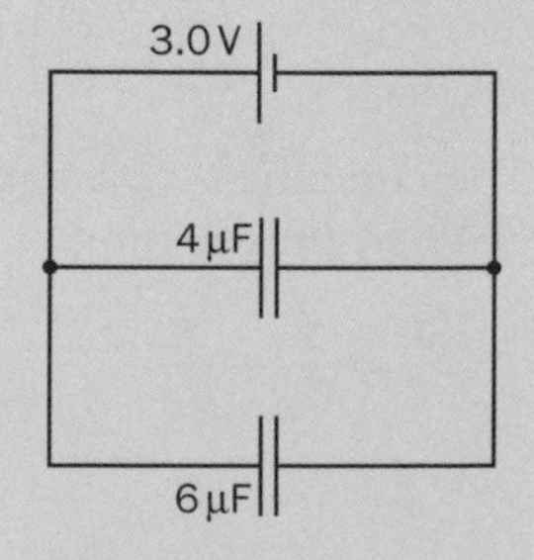

Fig. 15.10

Solution

(a) $C = C_1 + C_2 = 4 + 6 = 10\ \mu F$

(b) (i) $Q = CV = 10\ \mu F \times 3.0\ V = 30\ \mu C$

(ii) $4\ \mu F;\ Q = CV = 4\ \mu F \times 3.0\ V = 12\ \mu C$
$6\ \mu F;\ Q = CV = 6\ \mu F \times 3.0\ V = 18\ \mu C$

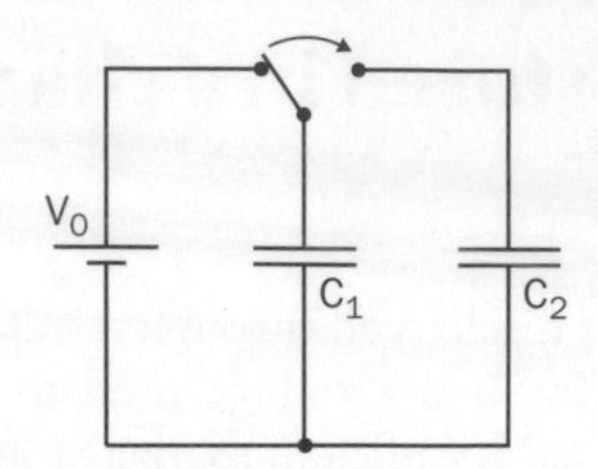

Fig. 15.11 Sharing charge.

Sharing charge

When an uncharged capacitor is connected in parallel with an isolated charged capacitor, the total charge is shared between the two capacitors. As a result, the voltage across the second capacitor falls.

In Fig. 15.11, a capacitor, C_1, is charged from a battery of voltage, V_o, using a two-way switch, S. The switch is then changed over to disconnect the battery and to connect a second capacitor, C_2, in parallel with C_1. Let V_f represent the final voltage across the two capacitors.

The initial charge stored on C_1, $Q = C_1V_o$
The final charge stored on C_1, $Q_1 = C_1V_f$
and the final charge stored on C_2, $Q_2 = C_2 V_f$
Since the total final charge stored = the inital charge stored,
$Q = Q_1 + Q_2$
Hence $C_1V_o = C_1V_f + C_2V_f = (C_1 + C_2)V_f$

Rearranging this equation gives

$$V_f = \frac{C_1}{(C_1 + C_2)} V_o = \frac{\text{initial charge } C_1V_o}{\text{combined capacitance } (C_1 + C_2)}$$

Worked example 15.4

A 3.0 V battery is briefly connected across a 10 μF capacitor to charge the capacitor. An uncharged 22 μF capacitor is then connected across the charged capacitor. Calculate (a) the initial charge stored on the 10 μF capacitor, (b)(i) the final p.d. across the two capacitors, (ii) the charge stored on each capacitor.

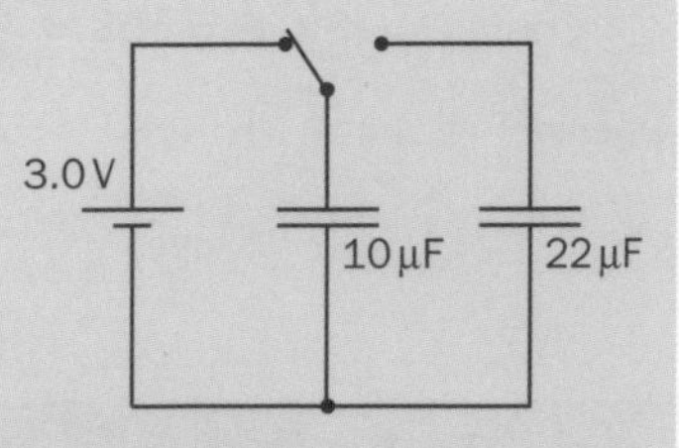

Fig. 15.12

Solution

(a) $Q = CV = 10\ \mu F \times 3.0\ V = 30\ \mu C$

(b) (i) Combined capacitance = $(C_1 + C_2) = 10 + 22 = 32\ \mu F$

$$\therefore \text{final p.d.} = \frac{\text{initial charge}}{\text{combined capacitance}} = \frac{30\ \mu C}{32\ \mu F} = 0.94\ V$$

(ii) 10 μF; $Q = CV = 10\ \mu F \times 0.94\ V = 9.4\ \mu C$
22 μF; $Q = CV = 22\ \mu F \times 0.94\ V = 20.6\ \mu C$

Questions 15.2

1. **(a)** Calculate the combined capacitance of a 3 μF capacitor and a 6 μF capacitor **(i)** in series, **(ii)** in parallel.
(b) For each circuit shown in Fig. 15.13, calculate **(i)** the combined capacitance, **(ii)** the total charge stored, **(iii)** the charge stored and the voltage for each capacitor.

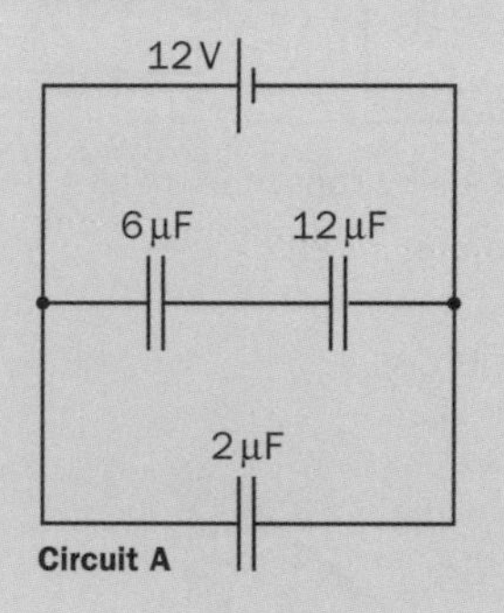

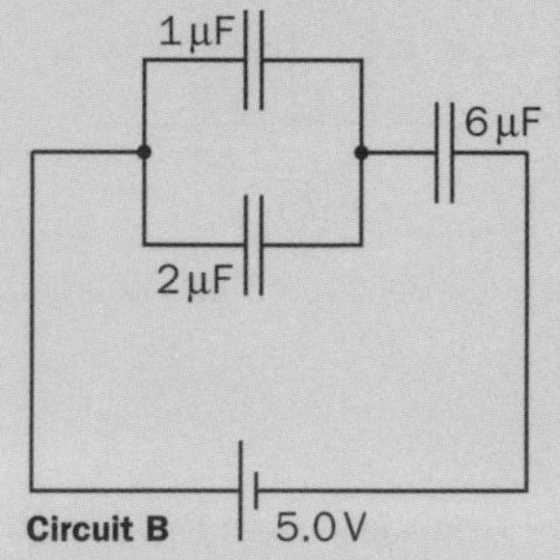

Fig. 15.13

2. A 4.7 μF capacitor was charged using a 6.0 V battery then disconnected from the battery. An uncharged 10 μF capacitor was then connected across the terminals of the 4.7 μF capacitor. Calculate **(a)** the total charge stored, **(b)** the final p.d. across each capacitor, **(c)** the final charge stored on each capacitor.

15.3 Energy stored in a charged capacitor

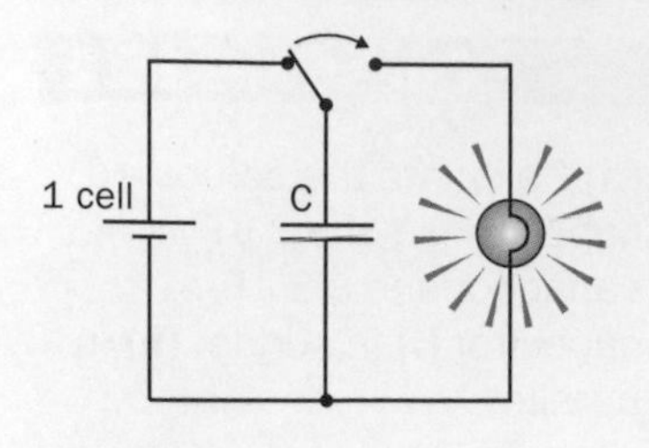

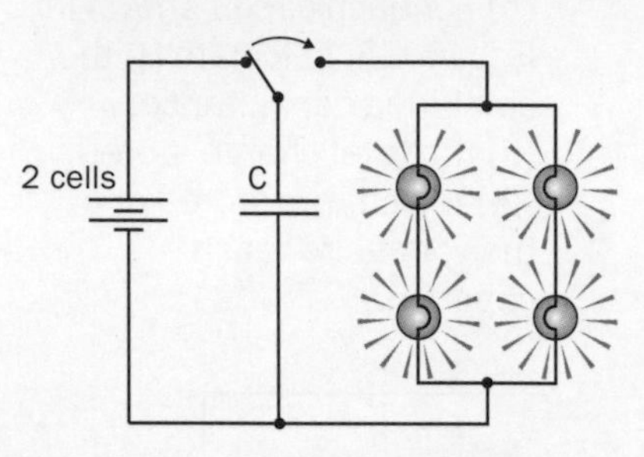

Fig. 15.14 Using capacitor energy.

If a charged capacitor is discharged through a suitable light bulb, the light bulb will light up briefly. If the voltage used to charge the capacitor is doubled, the energy stored is sufficient to make four light bulbs, connected as in Fig. 5.14, light up briefly. This is because four times as much energy is stored in a charged capacitor as a result of doubling the charging voltage. In other words, the energy stored in a charged capacitor is proportional to the square of the voltage.

Consider a brief interval of time, δt, during the process of charging a capacitor, C, to final voltage, V. During this brief interval, suppose the voltage across the capacitor rises from υ to $\upsilon + \delta\upsilon$ and the charge increases from q ($= C\upsilon$) to $q + \delta q$ ($= C(\upsilon + \delta\upsilon)$). During this time, extra charge, δq, is forced by the battery onto the plates against an average potential difference of $\upsilon + \frac{1}{2}\delta\upsilon$. Hence the work done on this extra charge,

$$\delta W = \delta q\ (\upsilon + \tfrac{1}{2}\delta\upsilon)$$

since potential difference is the work done per unit charge.

$$\delta W = (\upsilon\delta q) + (\delta q\ \tfrac{1}{2}\delta\upsilon) = \upsilon\delta q$$

as the second term is negligible.

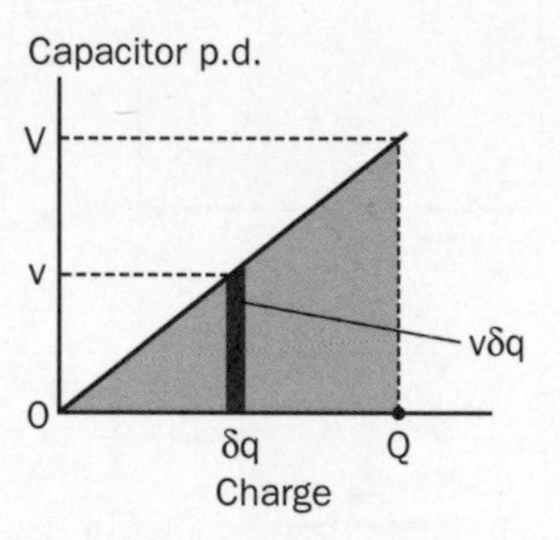

Fig. 15.15 Energy stored in a charged capacitor.

Fig. 15.15 shows how the p.d. across the capacitor increases as the charge on its plates rises. The work done in the interval δt ($= \upsilon\delta q$) is represented by the area of a narrow strip of width δq and height υ on this graph. It therefore follows that the total work done to charge the capacitor is represented by the total area under the line, which is half the area of the rectangle of width Q and height V. Hence the total work done, $W = \frac{1}{2}QV$. This is therefore the energy stored in the capacitor. Since the final charge $Q = CV$, then the energy stored $= \frac{1}{2}QV = \frac{1}{2}CV^2$.

Energy stored, $\mathbf{E = \frac{1}{2}QV = \frac{1}{2}CV^2}$

where Q = charge stored, V=voltage and C=capacitance.

See www.palgrave.com/foundations/breithaupt for more about energy stored in a capacitor.

Worked example 15.5

A 10 000 μF capacitor is charged using a 6.0 V battery. (a) Calculate the energy stored in this capacitor at 6.0 V. (b) Estimate for how long this capacitor could light a 6.0 V, 0.36 W torch bulb.

Solution

(a) $E = \frac{1}{2}CV^2 = 0.5 \times 10\,000 \times 10^{-6} \times 6.0^2 = 0.18\ \text{J}$

(b) Rearrange power $= \dfrac{\text{energy}}{\text{time}}$ to give time $= \dfrac{\text{energy}}{\text{power}} = \dfrac{0.18\ \text{J}}{0.36\ \text{W}} = 0.5\ \text{s}$

Questions 15.3

1. **(a)** Calculate the energy stored in **(i)** a 5.0 μF capacitor charged to a p.d. of 12.0 V, **(ii)** a 100 μF capacitor containing 1.8 mC of charge, **(iii)** a capacitor containing 30 mC of charge at a voltage of 6.0 V.

   (b) **(i)** Calculate the voltage at which a 20 μF capacitor would store 250 μJ of energy. **(ii)** Estimate the time for which a 50 000 μF capacitor charged to 6.0 V could light a 6.0 V, 0.2 W torch bulb.

2. Fig. 15.16 shows a two-way switch used to charge a 100 μF capacitor from a 6.0 V battery. The switch is then used to reconnect the charged capacitor to an uncharged 50 μF capacitor.

   (a) Calculate **(i)** the initial charge and energy stored in the 100 μF capacitor, **(ii)** the final p.d. across the two capacitors, **(iii)** the final charge and energy stored on each capacitor.

   (b) Account for the loss of stored energy as a result of reconnecting the switch.

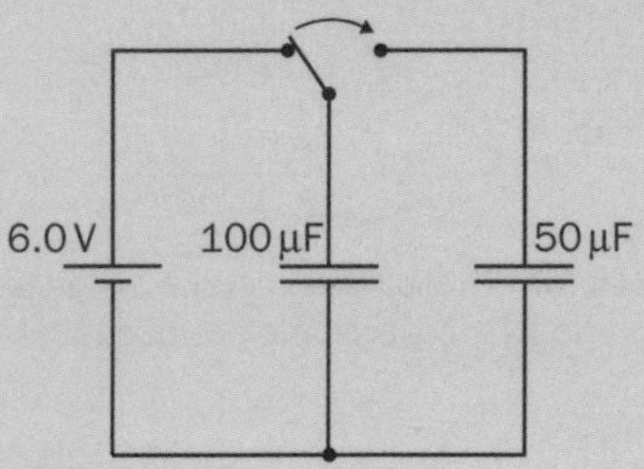

Fig. 15.16

15.4 Capacitor design factors

The capacitance of a capacitor depends on the surface area of the conducting strips, the spacing between the strips and the dielectric substance between the conducting strips. The larger the surface area, the greater the charge the capacitor can store for the same p.d. The closer the plates are together, the more charge the capacitor can store for the same p.d.

Experiment to investigate the factors that determine the capacitance of a parallel plate capacitor

Fig. 15.17 shows how these factors can be investigated using a parallel plate capacitor. The vibrating reed switch repeatedly charges and discharges the capacitor at a known frequency, f. Each time the switch connects the capacitor to the supply voltage V, the capacitor gains charge $Q = CV$. The

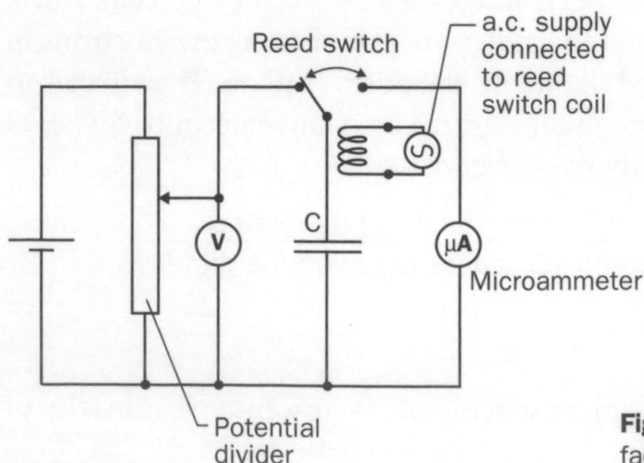

Fig. 15.17 Investigating capacitor factors.

switch then discharges the capacitor through the microammeter. Since this process occurs repeatedly, the microammeter registers a constant current $I = Q \div t = Qf$, where $t = 1 \div f$. Hence $I = CVf$. Rearranging this equation therefore gives

$$C = \frac{I}{Vf}$$

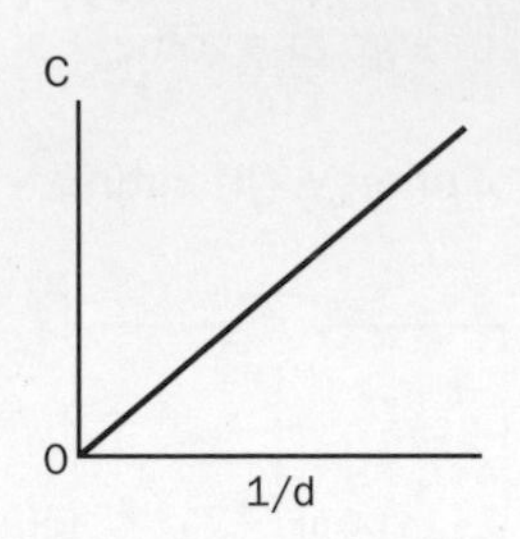

Fig. 15.18 Capacitance versus 1/ plate separation for a parallel plate capacitor.

1. **To investigate the effect of the plate separation, d, on the capacitance:** the current, I, is measured at a known p.d., V, with the reed switch operating at a measured frequency, f. Insulating spacers of known thickness may be used to keep the plates at a measured spacing. The capacitance, C, can therefore be calculated from the measurements of I, V and f for different spacings, d.

   Fig. 15.18 shows the results plotted on a graph of capacitance C on the vertical axis against $1 \div d$ on the horizontal axis. The results define a straight line through the origin, proving that the capacitance is proportional to $1 \div d$.

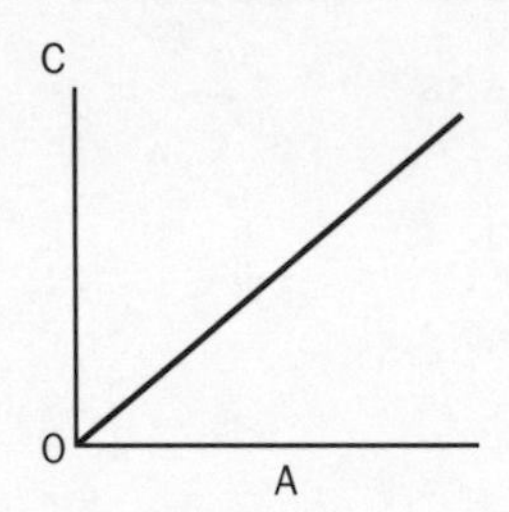

Fig. 15.19 Capacitance versus overlap area for a parallel plate capacitor.

2. **To investigate the effect of surface area, A, on the capacitance:** the overlap area, A, is changed, keeping the p.d. and the plate spacing the same. The current is measured for different measured overlap areas. Fig. 15.19 shows the results plotted on a graph of capacitance, C, on the vertical axis against A on the horizontal axis. The results ought to give a straight line through the origin, proving that the capacitance is proportional to the overlap area, A.

   Combining the two results gives the following equation for C,

$$C = \text{constant}\frac{A}{d}$$

   The constant of proportionality in this equation is called the **absolute permittivity of free space**. This constant is a measure of the charge per unit area, Q/A, on the plates needed to produce unit potential gradient, V/d, between the plates. The symbol for the absolute permittivity of free space is ε_0 (pronounced 'epsilon nought'). The value of ε_0 is 8.85×10^{-12} F m^{-1}.

3. **To investigate the effect of different materials between the plates on the capacitance:** the capacitance is measured without then with different materials between the plates. Any material that increases the capacitance is a dielectric substance. As explained on p. 205, the molecules of a dielectric substance between the plates are polarised as the electrons in each molecule are attracted slightly to the positive plate. The effect is to increase the capacitance of the capacitor. The relative permittivity, ε_r of a dielectric substance can therefore be written as

$$\varepsilon_r = \frac{\text{capacitance, C, with dielectric filling the plates}}{\text{capacitance, } C_0\text{, with no substance between the plates}}$$

$$\textbf{hence } \mathbf{C = \varepsilon_r C_0 = \frac{A\varepsilon_0\varepsilon_r}{d}}$$

where A = surface area, d = plate spacing, ε_0 = absolute permittivity of free space, and ε_r = the relative permittivity of the dielectric.

Worked example 15.6

Calculate the capacitance of a pair of conducting strips of metal foil of length 1.20 m and width 0.05 m, separated by a sheet of polythene of thickness 0.20 mm. The relative permittivity of polythene = 2.3. ($\epsilon_o = 8.85 \times 10^{-12}$ F m^{-1})

Solution

Using $C = \frac{A\varepsilon_o\varepsilon_r}{d}$ gives $C = \frac{1.2 \times 0.05 \times 8.85 \times 10^{-12} \times 2.3}{0.20 \times 10^{-3}} = 6.1 \times 10^{-9}$ F

Questions 15.4

$\varepsilon_o = 8.85 \times 10^{-12}$ F m^{-1}

1. A capacitor consists of two strips of metal foil of length 1.5 m and width 0.040 m separated by an insulating layer of thickness 0.01 mm and of relative permittivity 2.5.
   (a) Calculate the capacitance of this capacitor.
   (b) (i) The insulator breaks down and becomes a conductor if the p.d. across it exceeds 700 kV per millimetre thickness. Calculate the maximum p.d. allowable across the insulator if it is not to break down.
   (ii) Calculate the energy stored in the capacitor at maximum p.d.
2. A car is capable of storing charge in dry weather if it is insulated from the ground.
   (a) By considering a car as one plate of a parallel plate capacitor and the ground as the other plate, estimate the capacitance of a car which has a clearance of 0.10 m above the ground and an effective surface area of 6.0 m^2.
   (b) Calculate the charge and energy stored by the car for a p.d. of 1000 V.

15.5 Capacitor discharge

The power supply of a computer may be fitted with a large capacitor in parallel with its output terminals. If the power supply is interrupted, the capacitor continues to supply electricity to the computer until the power supply is restored. Without the capacitor, the loss of power to the computer would cause loss of data. Provided the interruption is not too long, the capacitor can supply enough power to enable the computer to operate normally.

Experiment to investigate the discharge of a capacitor through a resistor

An oscilloscope or a digital voltmeter may be used to measure the voltage across a capacitor as it discharges through a resistor. In Fig. 15.20a, the voltage across the capacitor is measured at regular intervals after the switch is opened. The results may be plotted as a graph of voltage on the vertical axis against time on the horizontal axis, as in Fig. 15.20b.

The voltage decreases with time at a decreasing rate. The shape of the curve in Fig. 15.20b is called an **exponential decay** curve. The voltage

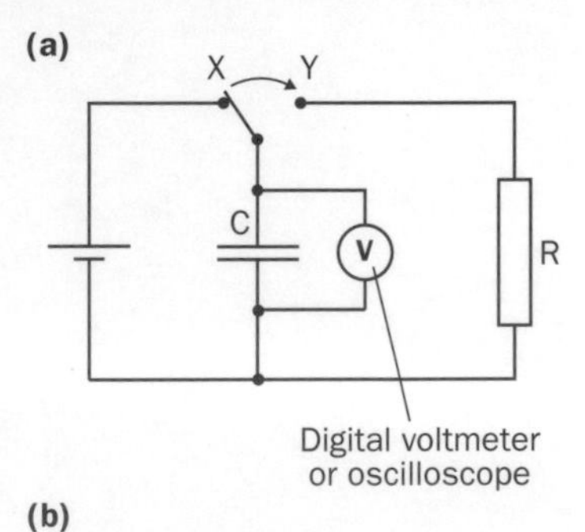

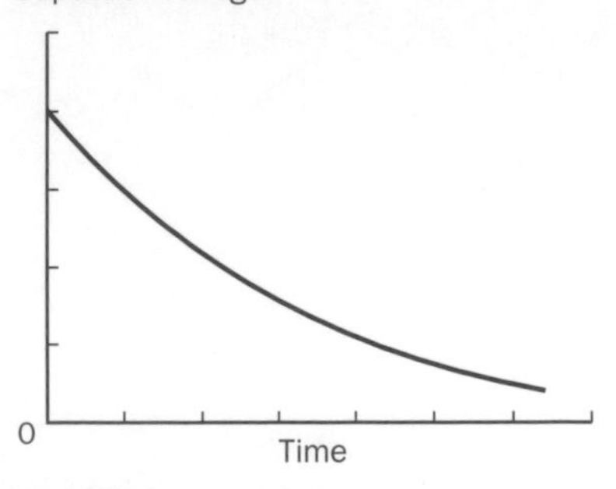

Fig. 15.20 Investigating capacitor discharge **(a)** circuit diagram, **(b)** exponential decay.

Table 15.2 Calculated results

Time from start / s	0.00	5.00	10.00	15.00	20.00	25.00	30.00	35.00
Voltage / v	10.00	9.00	8.10	7.29	6.56	5.90	5.31	4.78

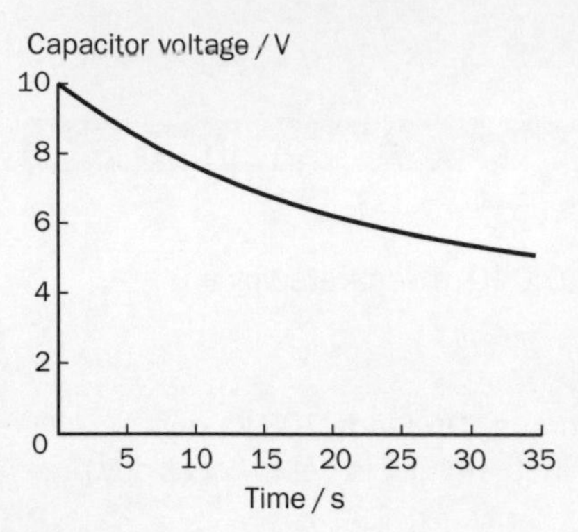

Fig. 15.21 Calculated voltage versus time.

continues to decrease but in theory never becomes zero. Suppose the time taken for the voltage to drop to 90% of its initial value is t_1, then the voltage drops to 90% of 90% of the initial voltage in time $2t_1$ and to 90% of 90% of 90% of the initial voltage in time $3t_1$, and so on. The voltage therefore never becomes zero, although in practice it eventually becomes negligibly small.

For example, if the voltage drops from 10.0 V initially to 9.0 V in 5.0 s, then it falls to 8.1 V (= 90% of 0.9 V) in a further 5 seconds and to 7.3 V (= 90% of 0.81 V) 5.0 s later. Table 15.2 shows how the voltage in this numerical example continues to decrease. The calculated values are plotted in Fig. 15.21 to give a curve with the same shape as the experimental curve in Fig. 15.20b.

Theory of capacitor discharge

Consider the capacitor at time, t, after the discharge has started. Let V represent the voltage at this instant.

1. The charge on the capacitor, $Q = CV$, where C is its capacitance.

2. The discharge current, I, through the resistor, R, is given by the equation

$$I = \frac{V}{R}$$

Combining the two equations gives

$$I = \frac{Q}{CR}$$

In a small time interval, Δt, the charge flow $\Delta Q = I\Delta t$, hence the current

$$I = \frac{\Delta Q}{\Delta t} = \frac{dQ}{dt}$$

where $\frac{d}{dt}$ means 'rate of change'. See Section 10.6 if necessary.

$$\therefore \frac{dQ}{dt} = \frac{-Q}{CR}$$

where the minus sign indicates loss of charge.

The solution of this equation is

$$Q = Q_0 e^{-t/CR}$$

where the function

$$e^x = 1 + x + \frac{x^2}{2} + \frac{x^3}{3 \times 2} + \frac{x^4}{4 \times 3 \times 2} +, \text{etc.}$$

This function is called the **exponential function**. Its shape is shown in Fig. 15.22. Its main property is that its rate of change

$$\frac{d}{dx}(e^x)$$

is proportional to the function itself. This is because the rate of change of any of the terms in the above formula equals the adjacent left-hand term. For example

$$\frac{d}{dx}\left(\frac{x^2}{2}\right) = x$$

Therefore, the exponential function appears in any physical process in which the rate of change of a quantity is proportional to the quantity itself.

Mathematical note on differentiation

The operation of determining an expression for the rate of change of a function is called **differentiation.** The reason why the rate of change of any term in the exponential function is equal to the adjacent left-hand term is because of the general rule that

$$\frac{d}{dx}(x^n) = nx^{n-1}.$$

To explain this general rule, consider the function $y = x^n$.

A small change of x to $x + \delta x$ causes a small change of y to $y + \delta y$, where $y + \delta y = (x + \delta x)^n$

Expansion of $(x + \delta x)^n$ gives $x^n + nx^{n-1}\delta x$ + terms in δx^2, δx^3, etc.

Hence $\delta y = (x + \delta x)^n - x^n = nx^{n-1}\delta x$ + terms in δx^2, δx^3, etc.

$$\therefore \frac{\delta y}{\delta x} = \frac{nx^{n-1}\delta x + \text{terms in } \delta x^2, \delta x^3, \text{etc.}}{\delta x} = nx^{n-1} + \text{terms in } \delta x, \delta x^2, \text{etc.}$$

In the limit $\delta x \rightarrow 0$, all the terms in δx, δx^2, etc. become zero and $\frac{\delta y}{\delta x} \rightarrow \frac{dy}{dx}$

$$\therefore \frac{dy}{dx} = nx^{n-1}$$

Notes on the exponential function

1. The inverse of the exponential function is called the natural logarithm function, denoted by either 'ln' or '$\log_e$'. In other words, if $y = e^x$ then $x = \ln y$.
2. A scientific calculator will have an exponential function button labelled 'e' or accessed by pressing 'inverse ln'.
3. The quantity, CR, is called the **time constant** of the circuit. It is equal to the time taken for the capacitor to discharge to $0.37\ Q_0$. This is because at time $t = CR$, $Q = Q_0\, e^{-1} = 0.37\ Q_0$.
4. Rearranging the equation

$$\frac{dQ}{dt} = -\frac{Q}{CR} \quad \text{gives} \quad \frac{dQ}{Q} = -\frac{1}{CR}\, dt$$

See www.palgrave.com/foundations/breithaupt to model capacitor discharge using a spreadsheet.

This rearranged equation means that the fractional decrease of charge

$$\frac{dQ}{Q}$$

in each interval of time, dt, is a constant, determined only by the product CR. For example, if C = 100 μF, R = 0.5 MΩ

(giving $\frac{1}{CR} = 0.02$) and dt = 5.0 s,

then $\frac{dQ}{Q} = -\frac{1}{CR}dt = -0.02 \times 5 = -0.1$.

The charge therefore decreases by 10% every 5 s. Work out for yourself that if the initial charge is 1000 μC, the charge falls to 900 μC after 5 s, to 810 μC after 10 s, to 729 μC after 15 s, etc. as in Table 15.2.

Worked example 15.7

A 2.0 μF capacitor is charged to a voltage of 5.0 V using the circuit in Fig. 15.22. The capacitor is then discharged through a 10 MΩ resistor by reconnecting the switch. Calculate (a) the initial charge stored by the capacitor, (b) the charge stored 30 s after the switch was reconnected, (c) the voltage after 30 s.

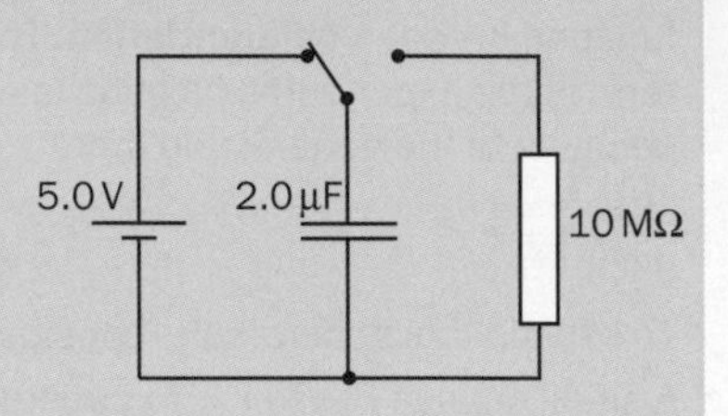

Fig. 15.22

Solution

(a) Initial charge, $Q_0 = CV_0$, where V_0 is the initial voltage.
Hence $Q_0 = 2.0\ \mu F \times 5.0\ V = 10\ \mu C$

(b) t = 30 s, CR = 2.0μF × 10 MΩ = 20 s $\therefore \frac{t}{CR} = \frac{30}{20} = 1.5$

hence $Q = Q_0\, e^{-t/CR} = 10\, e^{-1.5} = 2.2\ \mu C$

(c) $V = \frac{Q}{C} = \frac{2.2\ \mu C}{2.0\ \mu F} = 1.1\ V$

Questions 15.5

1. A 5.0 μF capacitor, charged using a 12.0 V battery, is discharged through a 0.5 MΩ resistor. Calculate **(a)** the initial charge and energy stored in the capacitor, **(b)** the charge and energy stored 5.0 s after the discharge started.
2. A 2200 μF capacitor is charged from a 6.0 V battery then discharged through a 100 kΩ resistor. Calculate **(a)** the initial charge stored in the capacitor, **(b)** the time constant of the circuit, **(c)** the charge stored and the capacitor voltage after a time equal to the time constant, **(d)** the current through the resistor at this instant.

Summary

- **Capacitance** is defined as charge stored per unit voltage.
- **The unit of capacitance** is the farad (F), equal to 1 coulomb per volt.
- **The time constant** of a capacitor discharge circuit = CR.
- **Equations**

Capacitance equation, $C = \frac{Q}{V}$

Combined capacitance
$C = C_1 + C_2$ for two capacitors in parallel
$\frac{1}{C} = \frac{1}{C_1} + \frac{1}{C_2}$ for two capacitors in series

Energy stored, $E = \frac{1}{2}CV^2$

Capacitance of parallel plates, $C = \frac{A\varepsilon_o\varepsilon_r}{d}$

Capacitor discharge equation, $\frac{dQ}{dt} = -\frac{Q}{CR}$

Solution of the capacitor discharge equation, $Q = Q_0\, e^{-t/CR}$

Revision questions

15.1. (a) A capacitor is connected in series with a light-emitting diode, a switch and a 1.5 V cell. When the switch is closed, the LED lights briefly. Explain **(i)** why the LED lights up initially, **(ii)** why the LED stops emitting light.

(b) A 5000 μF capacitor in series with a variable resistor, an ammeter and a 3.0 V battery is charged at a constant current of 0.15 mA. Calculate **(i)** the charge stored by the capacitor when it is fully charged, **(ii)** the time taken to charge fully with a constant charging current of 0.15 mA.

15.2. (a) Calculate the combined capacitance of a 3 μF capacitor and a 6 μF capacitor **(i)** in parallel with each other, **(ii)** in series with each other.

(b) Calculate **(i)** the total capacitance, **(ii)** the total charge stored, **(iii)** the charge and energy stored in each capacitor for each circuit in Fig. 15.23.

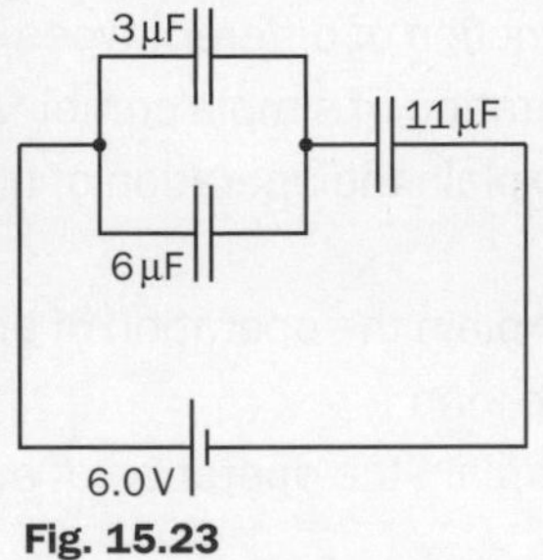

Fig. 15.23

15.3. (a) A parallel plate capacitor consists of two rectangular metal plates of dimensions 0.30 m × 0.25 m, separated by an insulator of thickness 1.5 mm and of relative permittivity 3.5. Calculate the capacitance of the capacitor.

(b) Calculate the charge and energy stored in the parallel plate capacitor in (a) when a voltage of 50 V is applied across its plates.

15.4. (a) In Fig. 15.24, a 4.7 μF capacitor is charged to a p.d. of 9.0 V then discharged through a 10 MΩ resistor. Calculate **(i)** the initial charge and energy stored in the capacitor, **(ii)** the time constant of the discharge circuit, **(iii)** the charge and energy stored in the capacitor 100 s after the discharge started.

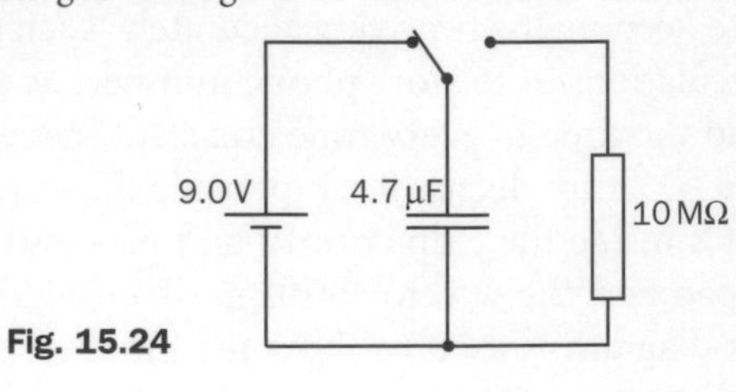

Fig. 15.24

(b) In Fig. 15.24, the capacitor is allowed to discharge fully and the 10 MΩ resistor is replaced by a 2.2μF capacitor. The 4.7μF capacitor is then charged to a p.d. of 9.0 V then isolated from the battery and reconnected to the 2.2μF capacitor. Calculate **(i)** the final p.d. across each capacitor, **(ii)** the energy stored in each capacitor at this p.d., **(iii)** the loss of stored energy as a result of connecting the capacitors together.

15.5. A security alarm circuit is powered by a 6.0 V voltage supply which has a 50 000 μF capacitor connected across its output terminals. The alarm circuit draws a current of 0.5 mA from the voltage supply when the alarm is not activated.

(a) Calculate the input resistance of the alarm circuit when it is not activated.

(b) If the voltage supply is interrupted, the capacitor continues to supply current to the alarm circuit, which can operate in its non-activated state provided the voltage supply does not fall below 5.0 V.

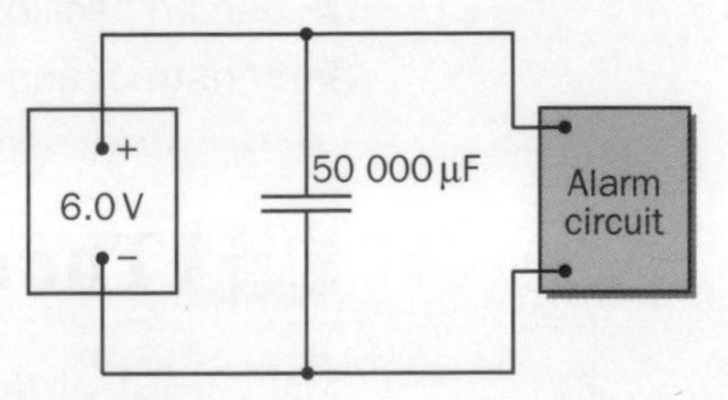

Fig. 15.25

(i) Calculate the time constant of the circuit.

(ii) Calculate the voltage across the capacitor 100 s after the 6.0 V voltage supply has been disconnected.

(iii) Calculate how long the alarm circuit in its non-activated state can operate after the voltage supply has been disconnected.

Electronics

Contents

Objectives

After working through this unit, you should be able to:

- represent an electronic system in block diagram form
- explain the difference between a digital system and an analogue system
- describe the function of different types of logic gates
- explain the operation of simple combinations of logic gates
- describe and explain the operation of different input and output transducers
- describe and explain the operation of an inverting op-amp and a non-inverting op-amp
- describe and explain the operation of a monostable and an astable multivibrator
- use an oscilloscope to display and measure waveforms
- construct and test a simple electronic system

16.1 The systems approach

Make a list of the functions that can be carried out using a calculator watch. In addition to keeping the time very accurately, such a watch can be used to perform calculations or to store phone numbers as well as keeping track of the date and the time in every time zone. All these functions and more are carried out by a tiny electronic chip inside the watch. Thousands of electronic switches inside the chip control and process the input voltages from the key pad and the selector buttons. To understand how the chip works, a block diagram is used to show the function of each part of the chip and how these parts relate to each other. This method of representing electronic circuits is referred to as a 'systems approach'. The function of each block must be known, rather than the circuits the block represents. Fig. 16.1 shows part of such a block diagram in which voltage pulses at exactly 1 pulse per second are supplied to a counter linked to a separate display chip.

(a)

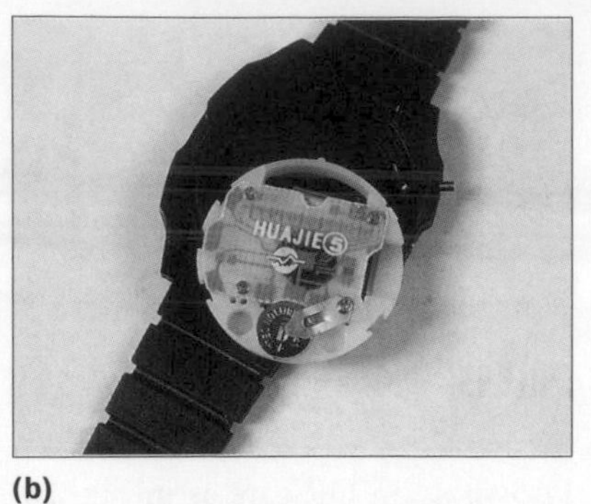

(b)

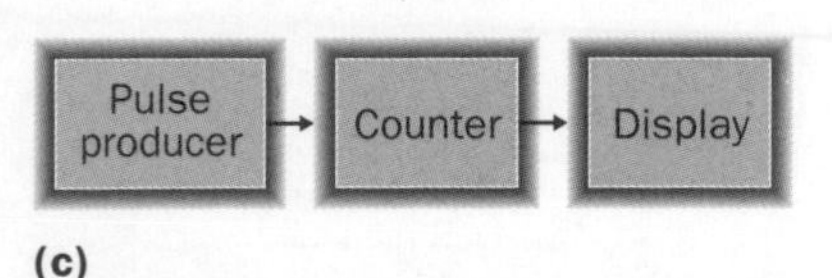

(c)

Fig. 16.1 An electronic system **(a)** a digital watch, **(b)** inside a digital watch, **(c)** a block diagram.

Electronic circuits are either **digital** or **analogue**. These terms are used widely in connection with telecommunications, particularly in connection with mobile phones and television broadcasting.

1. In a **digital circuit**, the voltage at any point in the circuit can have just two values, referred to as 'high' or 'low' or as 1 or 0. Usually, these two voltages are close to the voltage limits set by the power supply. For example, a digital electronic system connected to a 5.0 V power supply might operate at 4.5 V for its high voltage (1) and 1.0 V for its a low (0). No point in the system would be at a voltage between 1.0 and 4.5 V.
2. In an **analogue circuit**, the voltage at any point can vary over a continuous range within the limits set by the power supply. For example, the voltage from a microphone to an amplifier increases with loudness and the output voltage from the amplifier is proportional to the input voltage. The amplifier is an analogue circuit because the voltages in it can be at any level between the limits of the power supply, depending on the loudness of sound at the microphone.

Questions 16.1

1. Find out if each of the following electronic devices are digital or analogue.
   (a) A fax machine, **(b)** a microcomputer, **(c)** an audio amplifier, **(d)** a music cassette player.
2. Make a block diagram showing a microcomputer, keyboard, printer, mouse and visual display unit, indicating clearly the direction of the signals in each link.

16.2 Electronic logic

Everyone makes decisions every day. Each decision involves making a choice, usually determined by conditions relevant to the decision. A logical decision is one in which the choice made is the same whenever the same conditions arise. For example, suppose you are about to go out and you are thinking about whether or not to wear a coat. Your decision might depend on your answer to two questions:

Question A Is it raining?
Question B Is it cold outdoors?

A logical answer to these questions would be YES (i.e. take a coat) if A = YES **OR** B = YES.

The decision could be made for you by an electronic circuit consisting of a temperature sensor and a rain sensor connected to a logic circuit

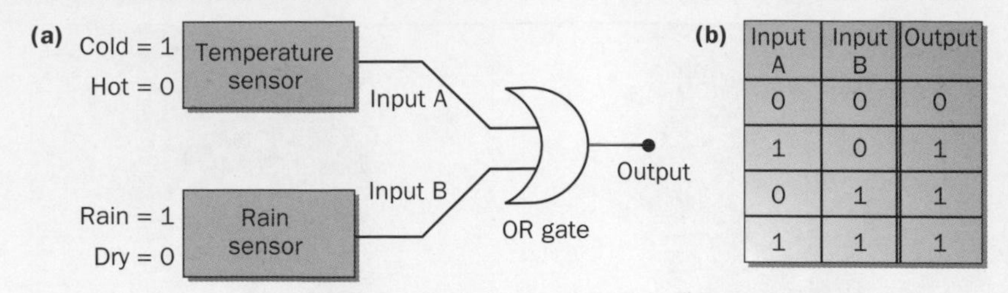

Input A	Input B	Output
0	0	0
1	0	1
0	1	1
1	1	1

Fig. 16.2 (a) using an OR gate, **(b)** the truth table of an OR gate.

Type of gate	Symbol	Truth table Input A	B	Output
OR		0 1 0 1	0 0 1 1	0 1 1 1
AND		0 1 0 1	0 0 1 1	0 0 0 1
NOR		0 1 0 1	0 0 1 1	1 0 0 0
NAND		0 1 0 1	0 0 1 1	1 1 1 0
NOT		0 1		1 0

Fig. 16.3 Logic gates.

known as an OR gate as in Fig. 16.2. A logic circuit is a digital circuit in which the output voltage is determined by the condition of the input voltage. All voltages are relative to the negative terminal of the voltage supply which is usually earthed. Each input or output voltage has two possible 'states', which are 'high' (i.e. 1) and 'low' (i.e. 0). The OR gate in Fig. 16.2 is designed to produce an output state of 1 if any of its input states is 1. The **truth table** in Fig. 16.2b shows how the output state depends on the input states for this type of logic gate.

The circuit symbols and truth tables for some common logic gates are shown in Fig. 16.3. In Fig. 16.2a, what would happen if an AND gate was used instead of an OR gate? You could be very cold on a dry day in winter or very wet on a wet day in summer!

Inside a logic gate

Logic gates are made in chip form. Fig. 16.4 shows a chip which contains four NAND gates, each with two inputs and an output. The chip contains tiny electronic switches called transistors as well as resistors and diodes. All these components are made on a small piece of pure silicon in a process that involves masking the silicon surface at different stages and 'doping' the exposed surface with different types of atoms. It is not necessary to understand the internal circuitry of a chip or the manufacturing process to understand its functions. Table 16.1 shows the main properties of TTL and CMOS chips which are the two main families of chips in common use at present.

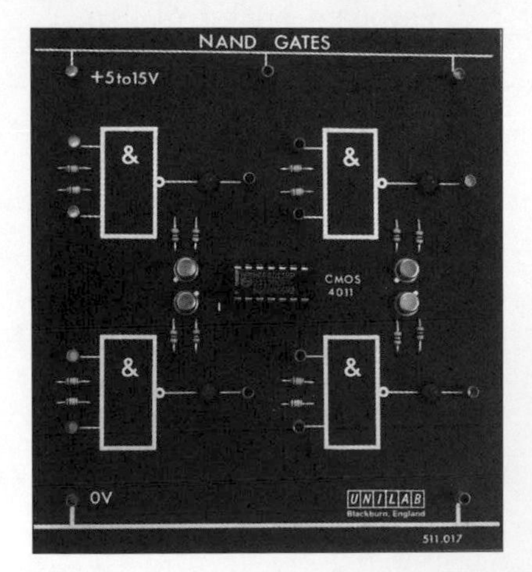

Fig. 16.4 A quad 2-input NAND chip.

Table 16.1 Comparison of TTL and CMOS chips

	Supply voltage V_s	Input voltage	Output voltage	Input resistance	Power consumption	Fan-out
TTL	5 V	Low = 0.8 V High = 2.0 V	Low = 0.4 V High = 2.4 V	$10^6\ \Omega$	milliwatts	10
CMOS	3–15 V	Low = 0.3 V High = 3.5 V for 5 V supply	Low = 0 High = 5.0 V for 5 V supply	$10^{12}\ \Omega$	microwatts	50

Notes

1. The fan-out is the number of inputs of other gates that can be 'driven' by the output of the gate.
2. Unconnected TTL inputs are always high (i.e. logic state = 1) whereas unconnected CMOS inputs can be high or low. Unused CMOS inputs should always be connected either to $+V_s$ or $-V_s$ according to the circuit.
3. The output voltage of a logic gate can be indicated using an LED in series with a suitable resistor, as in Fig. 16.5. The LED lights when a '1' is applied to the input terminal of the indicator.

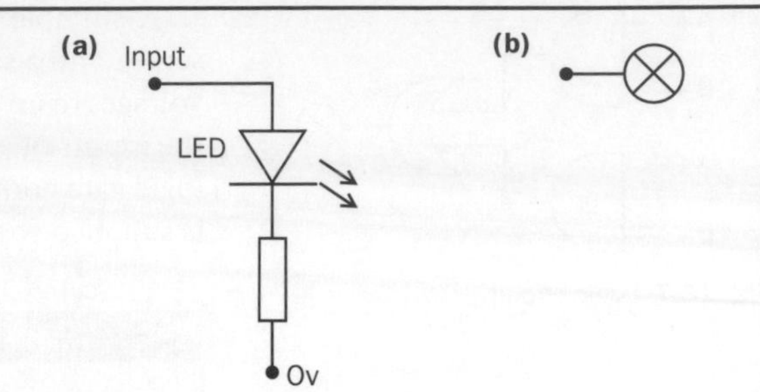

Fig. 16.5 An LED indicator **(a)** circuit, **(b)** symbol.

Logic gate combinations

In general, a logic system consists of input sensors connected to a logic circuit which drives an output device. Different types of input sensors include switch sensors, pressure sensors, light sensors and temperature sensors. Output devices include warning indicators driven directly by logic gates, and relay-operated electric motors, heaters and sound alarms. The logic circuit is a combination of logic gates linked together to activate the output device when the input sensors are in specified states.

For example, new cars are now fitted with a 'seat belt' warning light that indicates if either of the front seat belts is unfastened when the handbrake is on and either seat is occupied. Sensors are attached to each front seat and to the seat belts. These sensors are connected to the logic circuit which switches the warning indicator on or off according to the input conditions. Fig. 16.6 shows a block diagram and the truth table for this system. The warning light is on if the driver's seat is occupied AND the driver's seat belt is unfastened OR the front passenger's seat is occupied AND the passenger seat belt is unfastened.

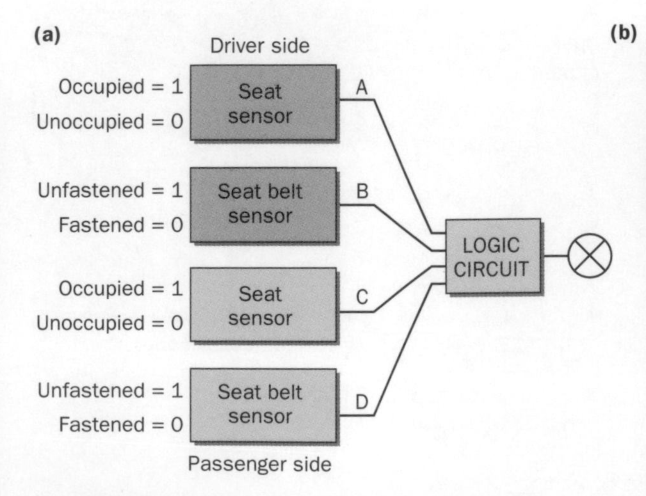

(b)

A	B	C	D	Output
0	0	0	0	0
0	0	0	1	0
0	0	1	0	0
0	0	1	1	1
0	1	0	0	0
0	1	0	1	0
0	1	1	0	0
0	1	1	1	1
1	0	0	0	0
1	0	0	1	0
1	0	1	0	0
1	0	1	1	1
1	1	0	0	1
1	1	0	1	1
1	1	1	0	1
1	1	1	1	1

Fig. 16.6 A seat-belt warning indicator **(a)** block diagram, **(b)** truth table.

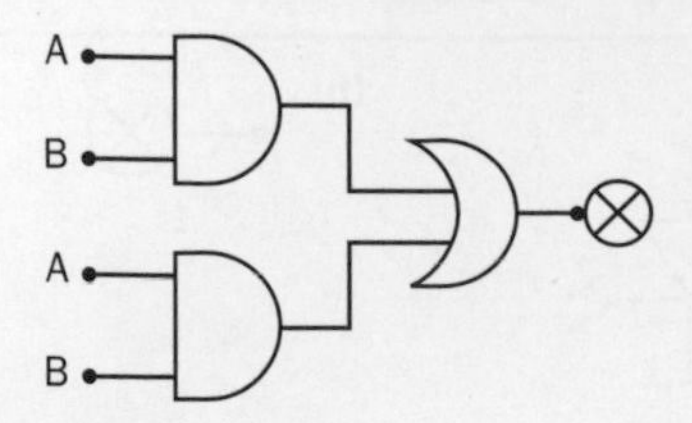

Fig. 16.7 Logic circuit.

A suitable logic circuit for this system is shown in Fig. 16.7. A seat sensor and a seat-belt sensor fitted to the each front seat supplies the input voltages to an AND gate. The outputs of the two AND gates are supplied to the inputs of an OR gate which is connected to the warning indicator. This logic gate combination activates the warning indicator if either AND gate is supplied with a 1 at each input.

Worked example 16.1

A home security alarm system consists of a control panel fitted with a key switch, K, to turn the system on or off, a main door sensor, D, and a window sensor, W. The alarm is activated if the system is on AND the main door is open OR the window sensor is open. Fig. 16.8 shows the block diagram and an incomplete truth table for the system.

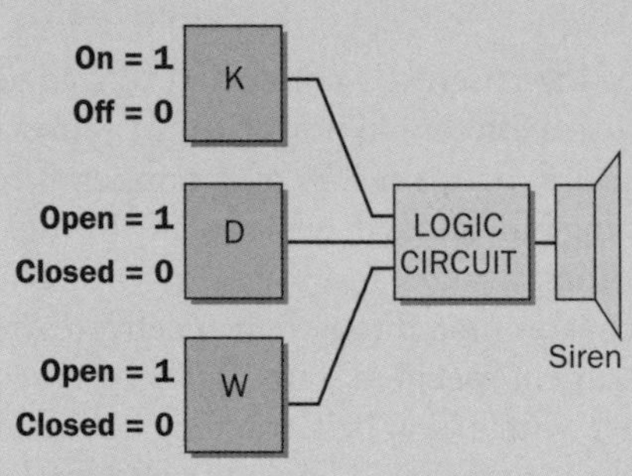

Fig. 16.8

(a) Complete the truth table for this system.

(b) Sketch a suitable combination of logic gates for this system.

Solution

(a)

Door sensor D	**Window sensor W**	**Key sensor K**	**Alarm siren**
open = 1 closed = 0	open = 1 closed = 0	on = 1 off = 0	on = 1 off = 0
0	0	0	**0**
1	0	0	**0**
0	1	0	**0**
1	1	0	**0**
0	0	1	**0**
1	0	1	**1**
0	1	1	**1**
1	1	1	**1**

(b) See Fig. 16.9.

K
D
W
Output

Fig. 16.9

Questions 16.2

1. Copy and complete the truth table for each of the combinations of logic gates illustrated in Fig. 16.10.

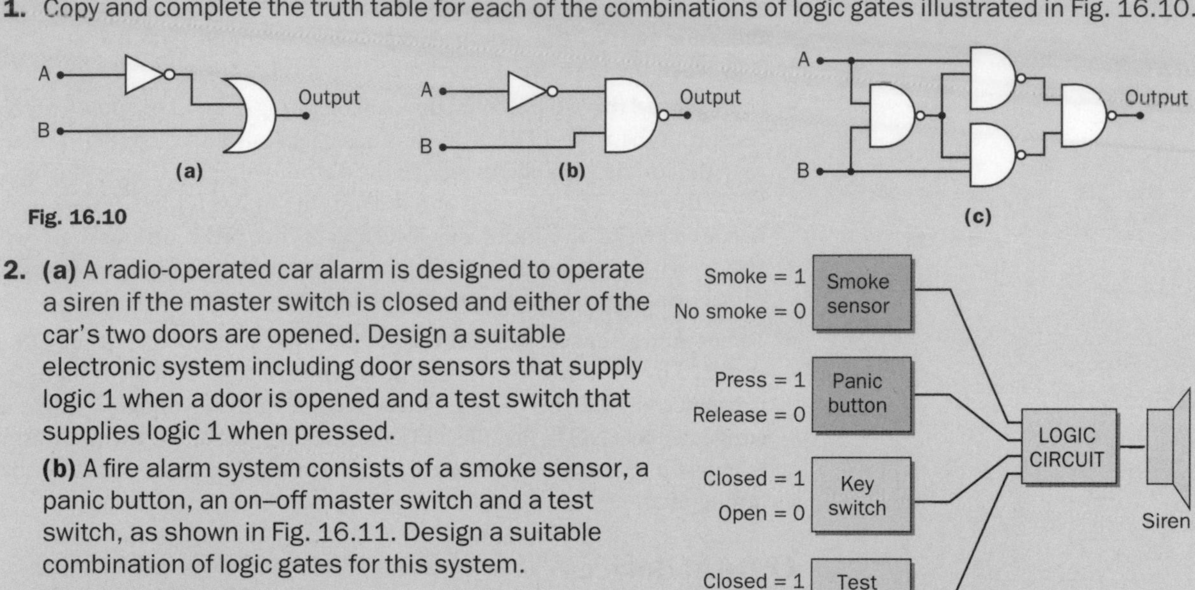

Fig. 16.10

2. **(a)** A radio-operated car alarm is designed to operate a siren if the master switch is closed and either of the car's two doors are opened. Design a suitable electronic system including door sensors that supply logic 1 when a door is opened and a test switch that supplies logic 1 when pressed.

(b) A fire alarm system consists of a smoke sensor, a panic button, an on–off master switch and a test switch, as shown in Fig. 16.11. Design a suitable combination of logic gates for this system.

Fig. 16.11

16.3 Electronics at work

Automated doors, card-operated cash machines, pressure-activated air bags and many more systems use logic gates which are supplied with input voltages from sensors and which control devices such as electric motors and valves. All these systems consist of one or more input systems connected to a logic circuit which operates an output device.

Input sensors

Fig. 16.12 shows the construction of some different input sensors.

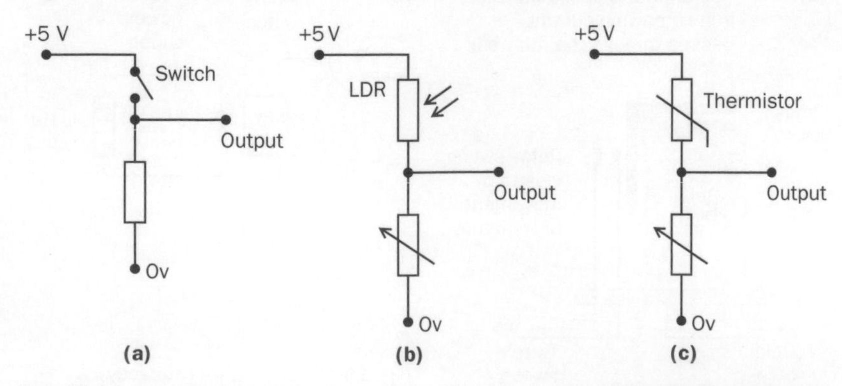

Fig. 16.12 Input sensors **(a)** switch sensor, **(b)** light sensor, **(c)** temperature sensor.

- **Switch sensor:** when the switch is closed, the output voltage is high. When the switch is open, no current passes through the resistor so the output voltage is low.
- **Pressure sensor:** the switch is a pressure-operated switch that closes only when the pressure is above a certain level.
- **Light sensor:** this is a potential divider consisting of an LDR and a variable resistor. When the LDR is in darkness, its resistance is much greater than that of the variable resistor so the output voltage is low. In daylight, the LDR resistance decreases and the output voltage is high. If the sensor is connected to the input of a NOT gate, the NOT output state will either be 0 or 1, according to whether or not the light intensity is above or below a value determined by the setting of the variable resistor.
- **Temperature sensor:** this is the same as a light sensor except the LDR is replaced by a thermistor. When the thermistor temperature is high, the resistance of the thermistor is low so the output voltage is high. If connected to a NOT gate, the NOT output state will be 0 if the thermistor is above a specified temperature and 1 if the thermistor is below that temperature.

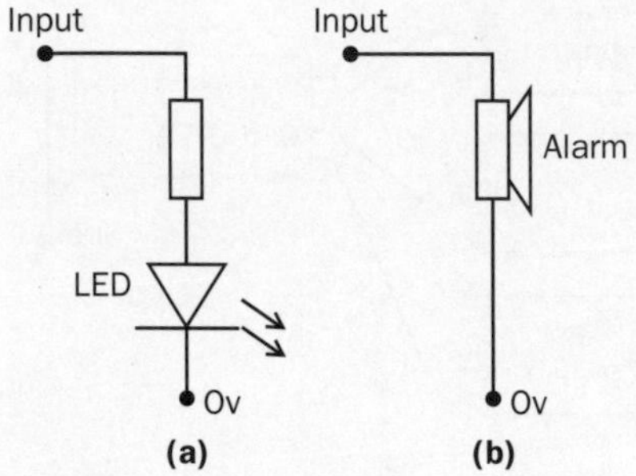

Input Input Alarm LED Ov Ov (a) (b)

Fig. 16.13 Warning indicators **(a)** an LED indicator, **(b)** an audio indicator.

Output devices

- A **warning indicator** can be visual (e.g. an LED indicator) or audible (e.g. a sound bleeper). Provided the current at the input terminal of a warning indicator is low enough to be supplied directly from the output terminal of a logic gate, no current amplifier is needed.
- A **relay** is a current-operated switch that can be used to switch on or off a high-current output device such as a mains heater. A relay is necessary if the output device needs more current at its input terminal than a logic gate could supply. A relay consists of an electromagnet that automatically closes a switch when a current passes through the relay coil. If the necessary coil current is sufficiently low, the relay can be operated directly by the output voltage of a logic gate. If not, the relay itself must be switched on or off by a transistor.
- A **transistor** is a semiconductor device that can function as an electronic

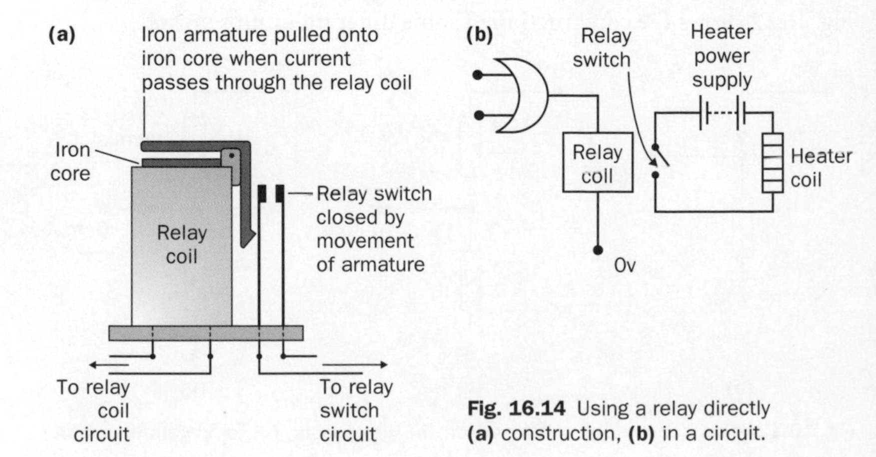

Fig. 16.14 Using a relay directly **(a)** construction, **(b)** in a circuit.

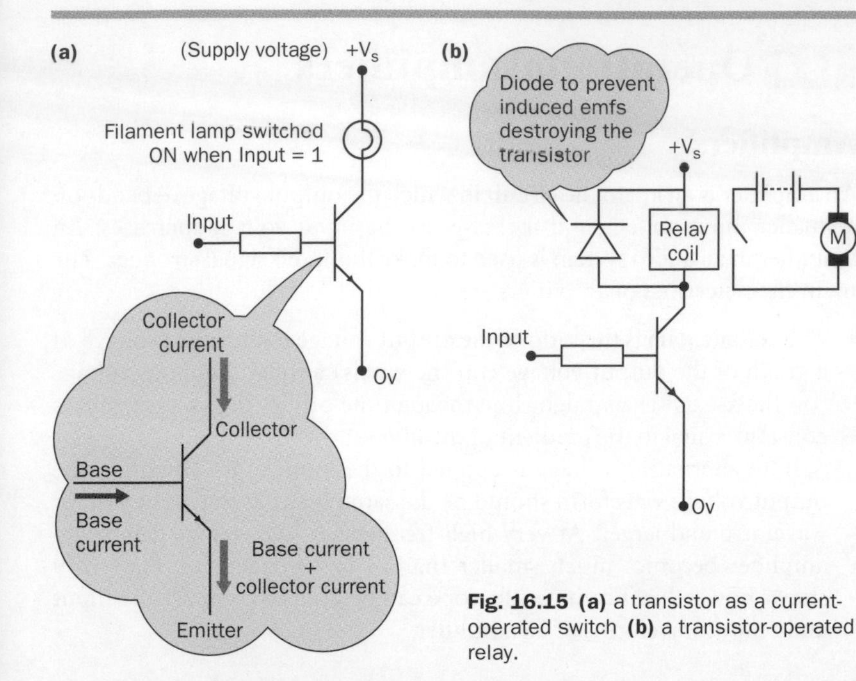

Fig. 16.15 **(a)** a transistor as a current-operated switch **(b)** a transistor-operated relay.

switch. Essentially, a transistor is a current amplifier in which a small current controls a much larger current. It can therefore be used as a current-operated switch, as shown in Fig. 16.15. A small current which enters the base and leaves at the emitter of the transistor, controls a much larger current which enters the collector and leaves also at the emitter. The transistor may therefore be used to switch on or off a device in series with the collector. If the collector current is insufficient to operate the device, a relay coil in series with the collector can be used to switch on or off the device operated from its own power supply.

Experiment to construct a light-operated electric motor

Construct and test the circuit in Fig. 16.16 which is designed to switch an electric motor on when the LDR is illuminated. The variable resistor is used to set the light level at which the motor is switched on.

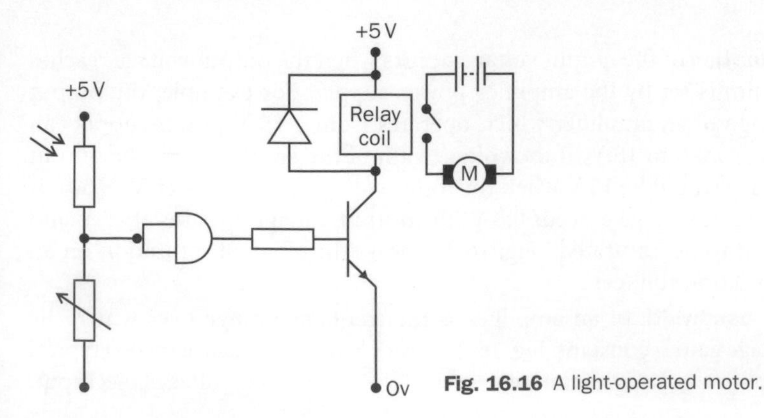

Fig. 16.16 A light-operated motor.

Questions 16.3

1. Fig. 16.17 shows a temperature-controlled fan which is fitted with a test switch, S.

   (a) Explain why the fan is switched on if the test switch is closed.

   (b) (i) Why does the fan switch on if the thermistor is above a certain temperature?

   (ii) Describe and explain the effect of reducing the resistance of the variable resistor.

2. Design an electronic system that would switch a mains electric heater on if the temperature in a room falls below a certain value at night. Your system should include a light sensor, a temperature sensor, a test switch, a suitable logic circuit, as well as a heater operated via a relay and a transistor.

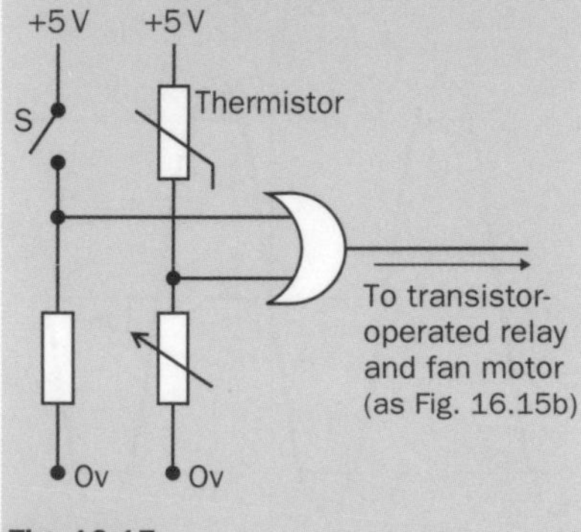

Fig. 16.17

16.4 Operational amplifiers

Amplifiers

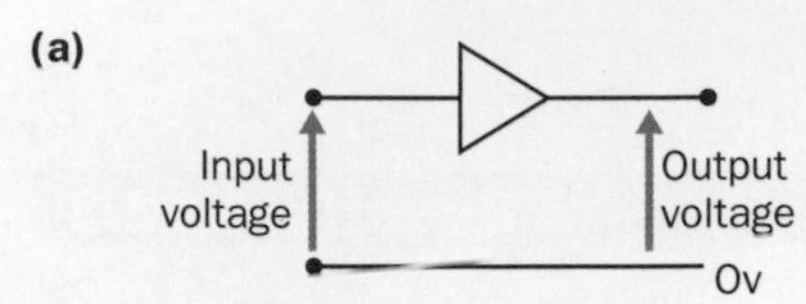

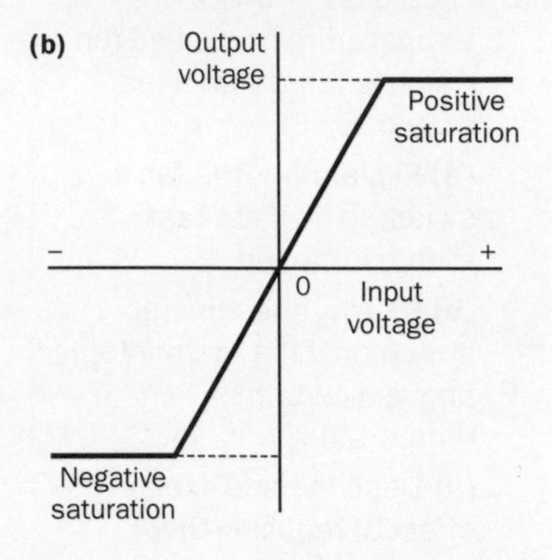

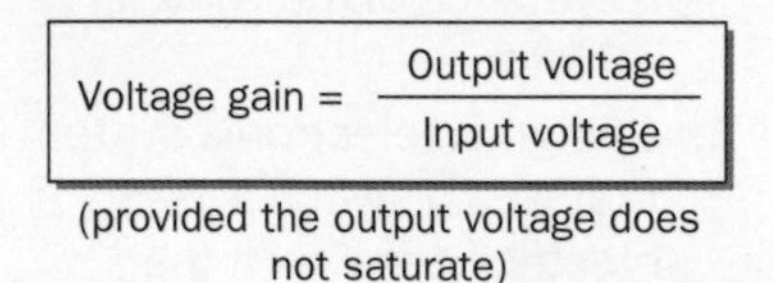

Fig. 16.18 Voltage gain **(a)** an amplifier, **(b)** output voltage versus input voltage.

An amplifier is an analogue circuit in which the output voltage depends on a smaller input voltage and increases as the input voltage increases. An amplifier in an audio system is used to make the audio signal stronger. The main characteristics of amplifiers are:

- The **voltage gain** is the ratio of the output voltage to the input voltage. If a graph of the output voltage (on the y-axis) against the input voltage (on the x-axis) is a straight line through the origin, the voltage gain is constant, equal to the gradient of the line.

  If an alternating voltage is applied to the input of an amplifier, the output voltage waveform should be the same shape as the input voltage waveform and larger. At very high frequencies, the voltage gain of an amplifier becomes much smaller than at low frequencies. Fig. 16.19 shows how a dual beam oscilloscope can be used to compare the input and output waveforms of an amplifier.

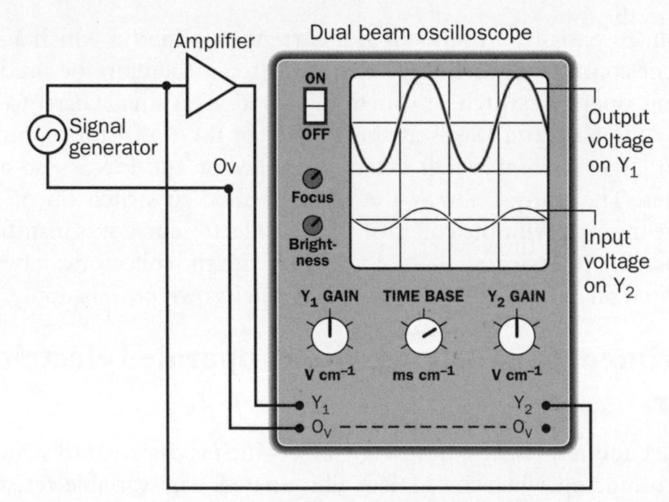

Fig. 16.19 Amplifying an alternating voltage.

- **Saturation** of the output voltage occurs when the output voltage reaches the limits set by the amplifier power supply. For example, thc output voltage of an amplifier which operates from a 15 V power supply can vary from 0 to 15 V. If the voltage gain of the amplier is 50, the output voltage would be 15 V when the input voltage is 0.3 V (= 15 V ÷ 50). If the input voltage exceeds 0.3 V, the output voltage remains at 15 V and is said to be 'saturated'. Fig. 16.20 shows the effect of saturation on an alternating voltage.
- The **bandwidth** of an amplifier is the frequency range over which the voltage gain is constant. Fig. 16.21 shows how the voltage gain of a typical amplifier varies with the frequency of the alternating voltage at its input.

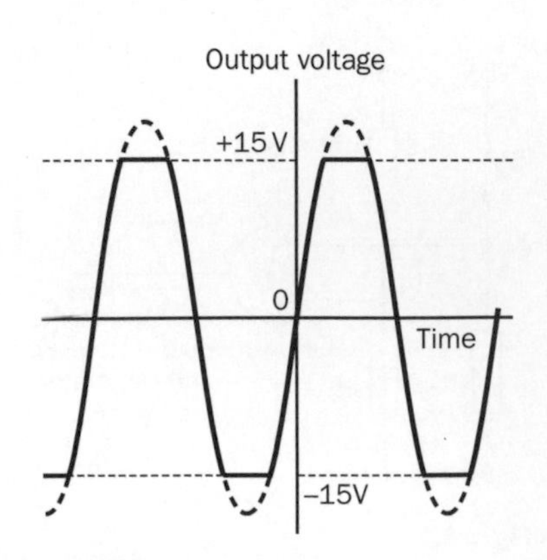

Fig. 16.20 Saturation.

In this example, an audio signal which contains a continuous range of frequencies up to about 18 kHz would be distorted because its frequency components above 10 kHz would not be amplified as much as the components below 10 kHz.

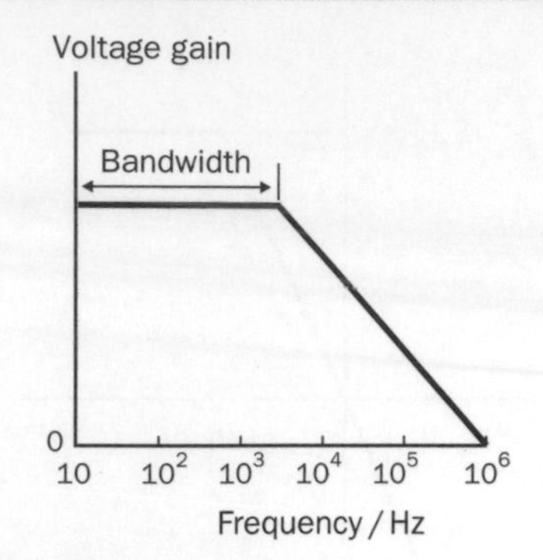

Fig. 16.21 Bandwidth.

- The **signal to noise ratio** of an amplifier determines how weak an input signal can be and still be detected at the output of the amplifier. Most amplifiers generate internal 'noise' caused by random motion of electrons in the resistors in the circuit. If the noise is too high, the input signal cannot be detected at the output. For example, a hiss due to internal noise is all that can be heard from an untuned radio.
- **Feedback** in an amplifier is where a fraction of the output voltage is fed back to the input terminals of the amplifier. The link between the output terminals and the input terminals of an amplifier is called a **feedback loop**. If no feedback exists, the amplifier is said to be on **open-loop**.

  1. Positive feedback is where the fraction of the output voltage fed back to the input terminals reinforces the input voltage, making the output larger. For example, this occurs when a microphone is placed too near the speaker of a sound system, creating an unpleasant screeching noise.

  2. Negative feedback is where the fraction of the output voltage fed back to the input terminals reduces the input voltage, making the output smaller. For example, a negative feedback loop would reduce the gain of an amplifier and make saturation less likely.

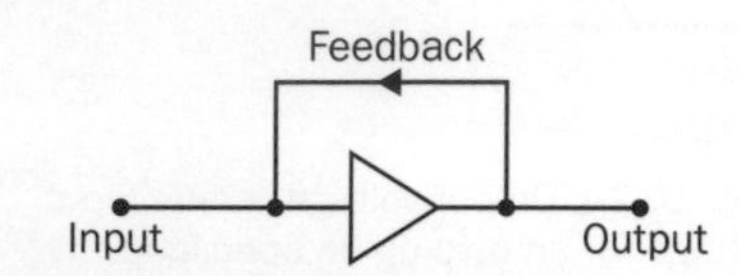

Fig. 16.22 Feedback.

Operational amplifiers

An operational amplifier is designed to amplify voltages or to add them together. By letting the voltage in an analogue circuit represent a mathematical variable, certain mathematical operations such as multiplication by a constant (i.e. amplification at constant voltage gain) or summation (i.e. adding two voltages together) can be carried out using operational amplifiers. For example, suppose an amplifier is required with a voltage gain of 10. This is achieved as explained later by connecting appropriate resistors to an operational amplifier. Choosing different resistors would give a different voltage gain. Unlike an audio amplifier, an operational amplifier can be adapted to perform exactly as needed.

An operational amplifier is a high-gain voltage amplifier with two input terminals and an output terminal. It is connected to a power supply with a positive terminal, $+V_S$, a negative terminal, $-V_S$, and an earthed terminal. The most common type of operational amplifier, the 741 chip, needs a ± 15 V power supply with an earthed terminal.

The output voltage, V_{OUT}, of an operational amplifier, the voltage between its output terminal and the zero voltage power supply terminal, can vary from $+V_S$ to $-V_S$, which are its supply voltage limits. The output voltage is proportional to the voltage difference between its two input terminals, which are labelled P and Q in Fig. 16.23.

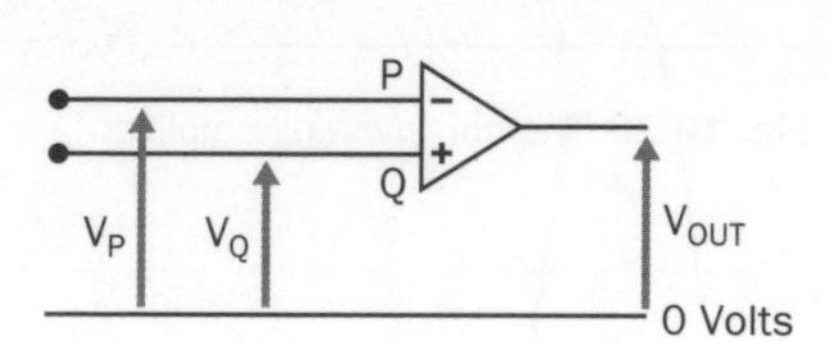

Fig. 16.23 An operational amplifier on 'open-loop'.

$$V_{OUT} = A\,(V_Q - V_P)$$

where A is the open-loop voltage gain.

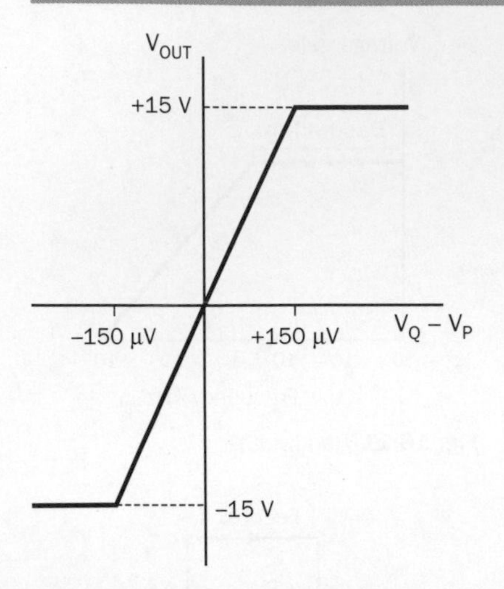

Fig. 16.24 Output voltage versus input voltage for an op-amp on open-loop.

- **Input Q** is called the **non-inverting input** because if P is earthed, the output voltage is the same polarity as the voltage applied to Q.
- **Input P** is called the **inverting input** because if Q is earthed, the output voltage is the opposite polarity to the voltage applied at P.
- The **input resistance** (i.e. the resistance between P and Q) for a 741 op-amp is of the order of 1 MΩ. Very little current therefore enters the op-amp at its input terminals.
- The **open-loop gain, A**, is very high, typically 100 000, varying slightly from one chip to another. Thus an op-amp on open-loop saturates if the voltage between P and Q exceeds 150 μV (= 15 V ÷ 100 000). The output voltage is therefore at positive saturation (i.e. $+V_S$) when V_Q exceeds V_P by 150 μV, or at negative saturation (i.e. $-V_S$) when V_P exceeds V_Q by 150 μV. Fig. 16.24 shows how the output voltage depends on the input voltage.
- The **comparator:** an op-amp on open-loop can therefore be used to compare two voltages by applying one voltage to input P and one voltage to input Q. If $V_{OUT} = +V_S$, then V_Q exceeds V_P; If $V_{OUT} = -V_S$, then V_P exceeds V_Q. Such comparators are used in computer circuits to open electronic gates and direct data along desired routes.
- The **voltage follower** is an op-amp in which the output terminal is directly connected to the inverting terminal, P. The input voltage is applied to the non-inverting terminal, Q. Provided the output does not saturate, P and Q are virtually at the same voltage. Hence the voltage at the output is the same as the voltage at the input. In other words, the output voltage follows the input voltage. Because the input resistance between P and Q is very high, the voltage follower draws very little current from the circuit its input terminal is connected to. It can therefore be used with a moving coil voltmeter across its output to measure voltage without drawing current.

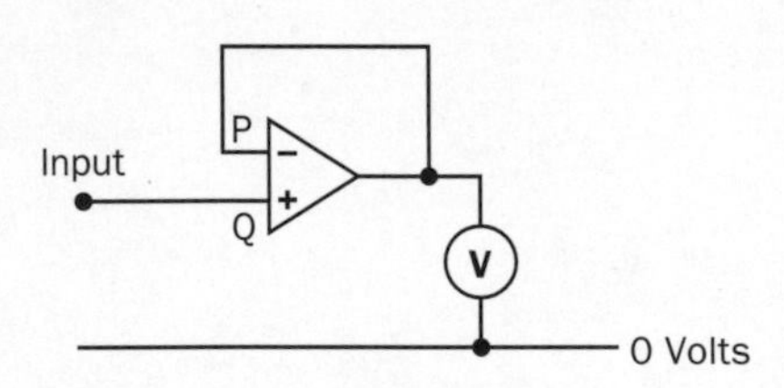

Fig. 16.25 The voltage follower.

The non-inverting amplifier

A potential divider between the output terminal and the 0 V line (i.e. earth) is used to supply a fraction of the output voltage to the inverting input, P. The input voltage, V_{IN}, is applied to the non-inverting input. Fig. 16.26 shows the arrangement.

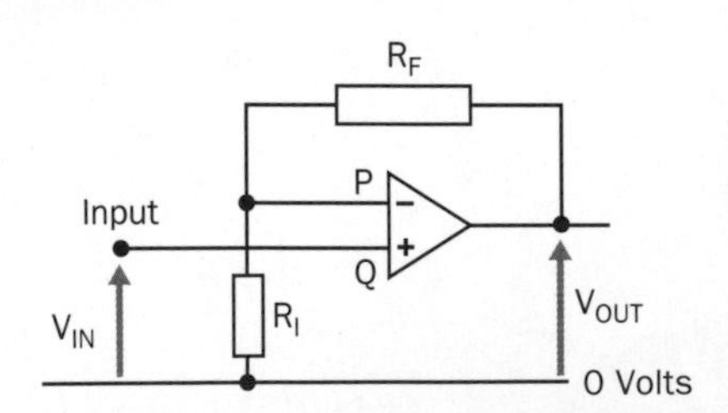

Fig. 16.26 The non-inverting amplifier.

Provided the output voltage does not saturate, V_Q and V_P are virtually at the same voltage, since $V_Q - V_P$ does not exceed 150 μV.

$$\text{Since } V_P = \frac{R_1}{R_F + R_1} V_{OUT}$$

(as R_1 and R_F form the potential divider) and

$$V_Q = V_{IN},$$

then

$$V_P = V_Q \quad \text{gives} \quad \frac{R_1}{R_F + R_1} V_{OUT} = V_{IN},$$

$$\therefore \textbf{the voltage gain } \frac{V_{OUT}}{V_{IN}} = \frac{R_1 + R_F}{R_1}$$

Note

The voltage gain is therefore determined only by the resistances, R_1 and R_F, and is independent of the open-loop gain of the chip.

Worked example 16.2

A non-inverting amplifier is required with a voltage gain of 20. Calculate (a) the feedback resistance if the resistance of R_1 is 0.5 MΩ, (b) the maximum input voltage if the output voltage is not to exceed 15 V.

Solution

(a) Voltage gain $= \frac{R_1 + R_F}{R_1} = 20$, and $R_1 = 0.5$ MΩ

$\therefore \frac{0.5 + R_F}{0.5} = 20$, so $0.5 + R_F = (20 \times 0.5) = 10$

Hence $R_F = 10 - 0.5 = 9.5$ MΩ

(b) For $V_{OUT} = 15$ V, $V_{IN} = \frac{V_{OUT}}{20} = 0.75$ V

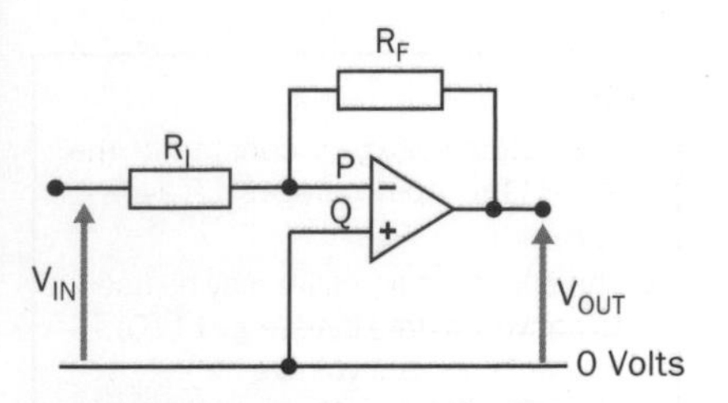

Fig. 16.27 The inverting amplifier.

The inverting amplifier

A potential divider between the output terminal and the input terminal is used to supply a fraction of the output voltage to the inverting input, P. The non-inverting input is connected to the 0 V line. Fig. 16.27 shows the arrangement.

Resistors R_1 and R_F form a potential divider between the output terminal to V_{OUT} and the input terminal at V_{IN}.

Hence the voltage between P and the input terminal

$$V_P - V_{IN} = \frac{R_1}{R_F + R_1}(V_{OUT} - V_{IN})$$

Provided the output voltage does not saturate, V_Q and V_P are virtually at the same voltage, since $V_Q - V_P$ does not exceed 150 μV. Because Q is earthed (i.e. $V_Q = 0$) then V_P is virtually at earth potential too (i.e. $V_P \approx 0$ V).

$$\text{Hence} \quad -V_{IN} = \frac{R_1}{R_F + R_1}(V_{OUT} - V_{IN}) \quad \text{as } V_P = 0$$

Rearranging this equation gives

$$-V_{IN}(R_F + R_1) = R_1(V_{OUT} - V_{IN})$$

$$\therefore -V_{IN}R_F - V_{IN}R_1 = R_1V_{OUT} - R_1V_{IN}$$

$$\text{hence } -V_{IN}R_F = R_1V_{OUT}$$

$$\therefore \textbf{the voltage gain } \frac{V_{OUT}}{V_{IN}} = -\frac{R_F}{R_1}$$

Notes

1. The voltage gain depends only on the resistances R_F and R_1, and is independent of the open-loop gain of the chip.
2. The output voltage has the opposite polarity to the input voltage.
3. Negligible current enters the op-amp at its input terminals P and Q because its input resistance is very high. Consequently the current through R_1 $= \frac{(V_P - V_{IN})}{R_1}$ is equal to the current through R_F $= \frac{(V_{OUT} - V_P)}{R_F}$. Since $V_P = 0$, then $\frac{V_{OUT}}{R_F} = -\frac{V_{IN}}{R_1}$ which gives the voltage gain formula.

The summing amplifier

Each voltage to be added is applied to an input terminal connected to the inverting input, P, via a resistor. The non-inverting input is earthed, as in Fig. 16.28.

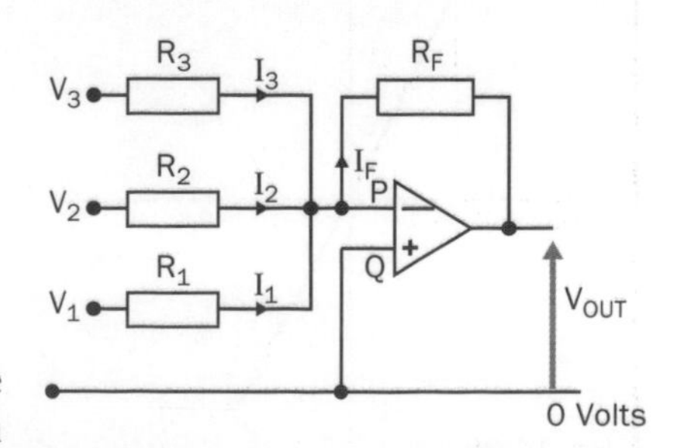

Fig. 16.28 The summing amplifier.

Since P is at virtual earth, assuming the output voltage is not saturated, the current through resistor R_1 can be written as

$$I_1 = \frac{V_P - V_1}{R_1} = -\frac{V_1}{R_1}$$

the current through resistor R_2 can be written as

$$I_2 = \frac{V_P - V_2}{R_2} = -\frac{V_2}{R_2}$$

the current through resistor R_3 can be written as

$$I_3 = \frac{V_P - V_3}{R_3} = -\frac{V_3}{R_3}, \text{ etc.}$$

Hence the total current through the feedback resistor R_F can be written as

$$I_F = I_1 + I_2 + I_3 +, \text{ etc.}$$

$$\therefore I_F = -\frac{V_1}{R_1} - \frac{V_2}{R_2} - \frac{V_3}{R_3}, \text{ etc.}$$

Since the output voltage, $V_{OUT} = I_F R_F$

then

$$\mathbf{V_{OUT}} = -\frac{\mathbf{R_F V_1}}{\mathbf{R_1}} - \frac{\mathbf{R_F V_2}}{\mathbf{R_2}} - \frac{\mathbf{R_F V_3}}{\mathbf{R_3}}, \text{ etc.}$$

Notes

1. The output voltage = constant × the sum of the input voltages if $R_1 = R_2 = R_3$, etc.
2. The summing amplifier may be used to convert a data byte (e.g. 1110) into an analogue voltage by choosing the resistances such that $V_{OUT} = -V_1 - 2V_2 - 4V_3 - 8V_4$, etc. and applying the least significant bit to V_1, the next bit to V_2, etc. Thus the data byte, 1110, would give an output voltage of −14 V if each bit is at 1 V.

Worked example 16.3

Fig. 16.29 shows an inverting operational amplifier and a graph of its output voltage against its input voltage. (a) Calculate (i) its voltage gain from the graph, (ii) the feedback resistance necessary to produce this voltage gain. (b) An alternating voltage is applied between the input terminal of this operational amplifier and its zero volt line. Sketch graphs to show how the output voltage varies with time if the peak-to-peak value of the input voltage is (i) 2.0 V, (ii) 4.0 V.

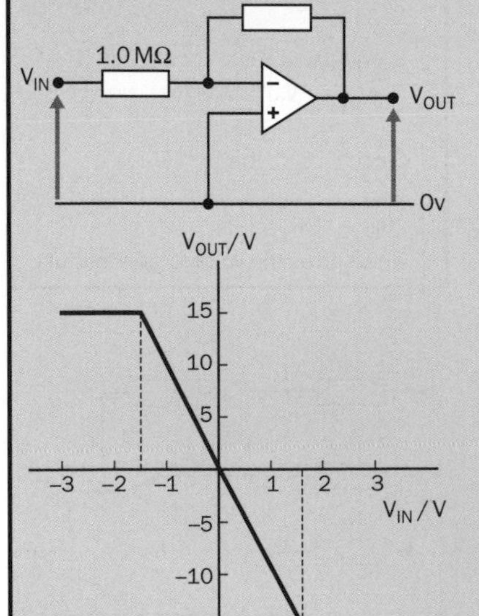

Fig. 16.29

Solution

(a) (i) Voltage gain = gradient of graph $= -\frac{15\text{ V}}{1.5\text{ V}} = -10$

(ii) Voltage gain $= -\frac{R_F}{R_1} = -10 \quad \therefore R_F = 10\,R_1 = 10\text{ M}\Omega$

(b) See Fig. 16.30.

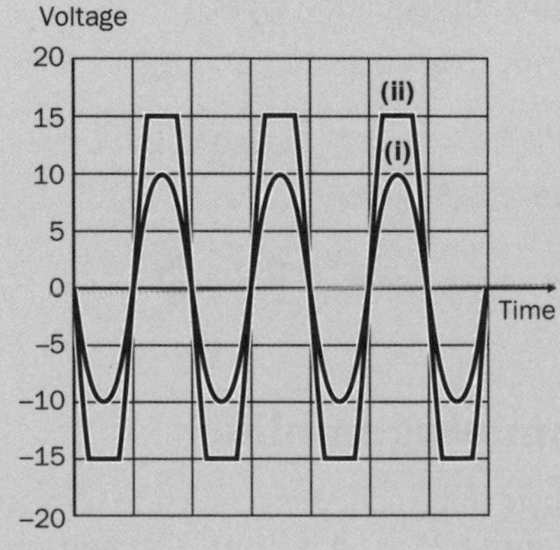

Fig. 16.30

Experiment to investigate the inverting amplifier

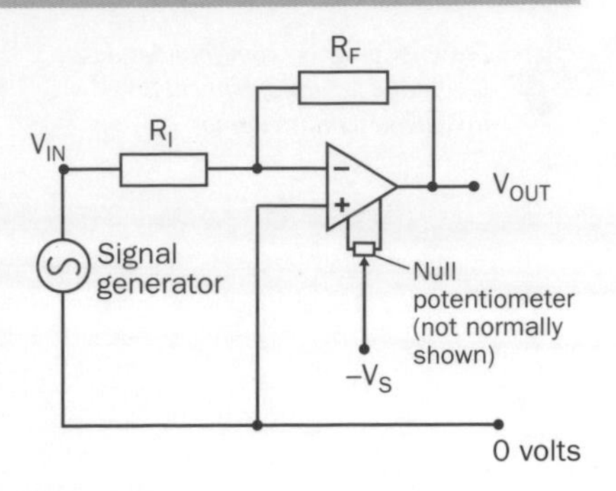

Fig. 16.31 Investigating the inverting amplifier.

Adjust the null potentiometer of the op-amp so that the output voltage is zero when the input voltage is zero. This is necessary to correct internal bias voltages within the op-amp. Use a signal generator to apply an alternating voltage to the input of an inverting amplifier. Display the output and input waveforms using a double beam oscilloscope. Note that the output wave should be amplified and inverted compared with the input wave.

1. Measure the amplitude of the input voltage and of the output voltage and calculate the voltage gain. Compare this result with the theoretical voltage gain $- R_f \div R_1$.
2. Increase the amplitude of the input voltage until the output wave just saturates. Observe the effect of increasing further the amplitude of the input wave.
3. Measure the voltage gain for different frequencies and plot a graph of voltage gain against frequency.

Questions 16.4

1. For each circuit in Fig. 16.32, calculate **(a)** the voltage gain, **(b)** the maximum input voltage if the output voltage is not to exceed 15 V.

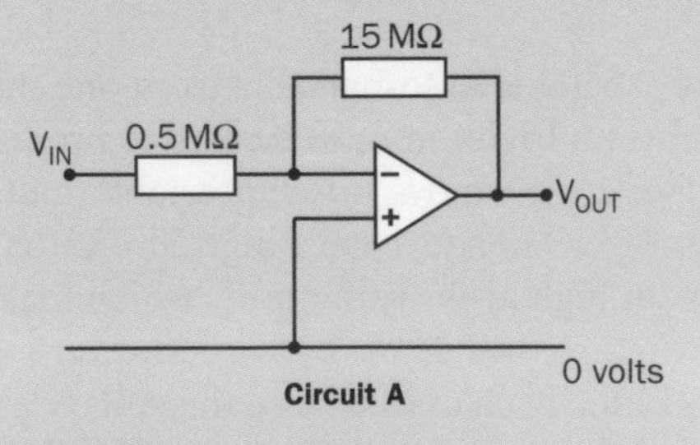

Fig. 16.32

2. Fig. 16.33 shows an operational amplifier on open-loop connected to a wheatstone bridge circuit which includes a thermistor and a variable resistor. The variable resistor is increased until the output voltage of the op-amp switches from +15 V to −15 V.

   (a) Explain why the output voltage switches from +15 V to −15 V when the variable resistor reaches a certain resistance.

   (b) If the thermistor is then cooled, the output voltage of the op-amp switches from −15V to +15 V. Explain why this happens at a certain temperature as the thermistor's temperature is lowered.

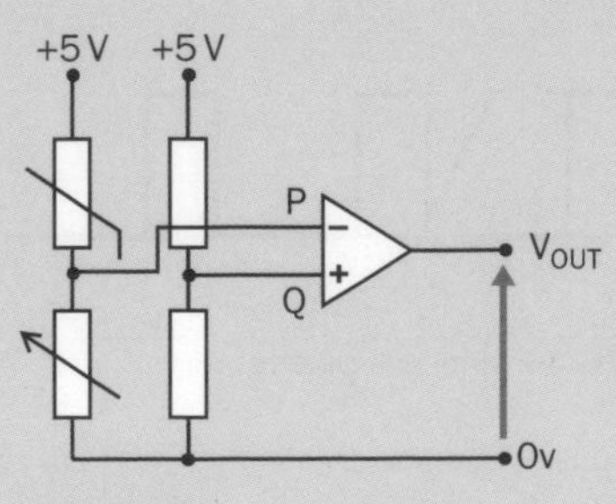

Fig. 16.33

See www.palgrave.com/foundations/breithaupt for an experiment to investigate a multivibrator.

16.5 Multivibrators

A multivibrator is a circuit in which the output voltage switches between two states, either automatically or as a result of receiving an input pulse.

Monostable multivibrator

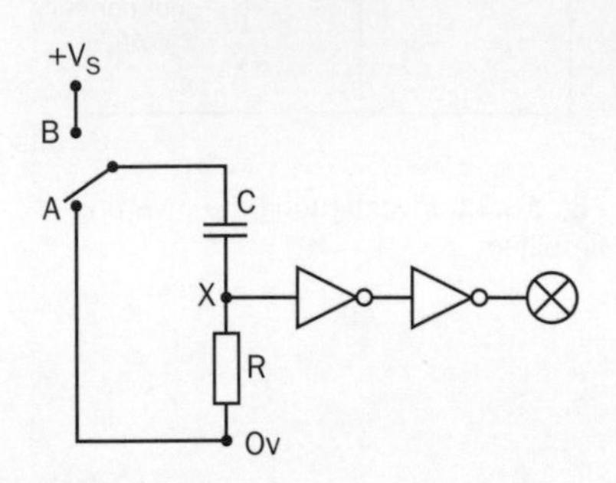

Fig. 16.34 A monostable multivibrator.

A monostable multivibrator has one stable output state. It can be forced into its unstable state by an input pulse but returns to its stable state after a certain time. A monostable multivibrator can therefore be used in a logic circuit to provide a time delay. For example, a home security alarm system is usually fitted with a delay circuit so the system can be switched off when someone with an alarm key enters by the front door.

Fig. 16.34 shows how a monostable multivibrator works. With the switch at A, the voltage at X is zero, so the output indicator is off. When the switch is moved from A to B, the voltage at X is raised to $+V_S$ without delay so the output indicator is switched on. However, as the capacitor charges up, the voltage at X decreases and at some point can no longer supply a high enough voltage for logic 1 to the first NOT gate. The input to the second NOT gate therefore becomes '1' at this point and the indicator at its output is therefore switched off.

The time delay depends on how fast the capacitor charges up. This depends on the time constant, CR, of the capacitor, C, in series with the resistor R.

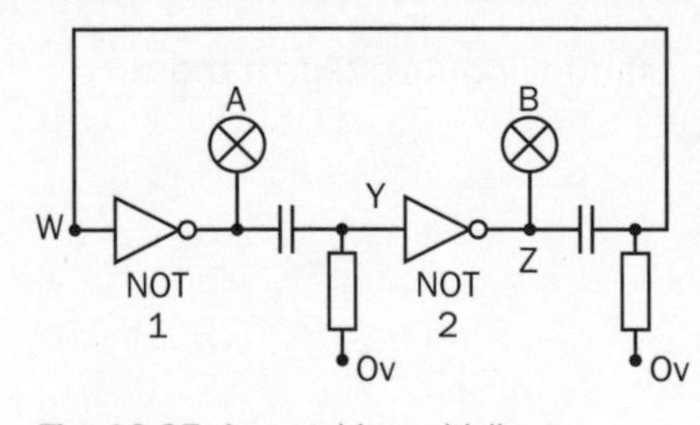

Fig. 16.35 An astable multivibrator.

Astable multivibrator

An astable multivibrator repeatedly switches automatically from one state to the other and back. Thus an astable multivibrator can be used to produce a continuous stream of pulses. Fig. 16.35 shows how an astable multivibrator works. When the output of NOT gate 1 goes high, indicator A switches on and the voltage at Y goes high at the same time, which makes the voltage at Z and W low, switching indicator B off. However, the capacitor at the output of NOT 1 gradually charges up and the voltage at Y therefore falls. When this voltage falls low enough, the output of NOT 2 at Z and at W goes high, which switches indicator B on and indicator A off. The capacitor at the output of NOT 2 now charges up gradually and the voltage at Z falls until it is low enough to make the output of NOT 1 high again. In this way, indicator A repeatedly switches on and off and indicator B repeatedly switches off and on.

Whenever A is on, B is off and vice versa. The 'on' time for each indicator is determined by the time constant, CR, of the capacitor and resistor it is connected to. Choosing different values of C and R at each NOT output would give pulses with unequal times for each state as shown in Fig. 16.36.

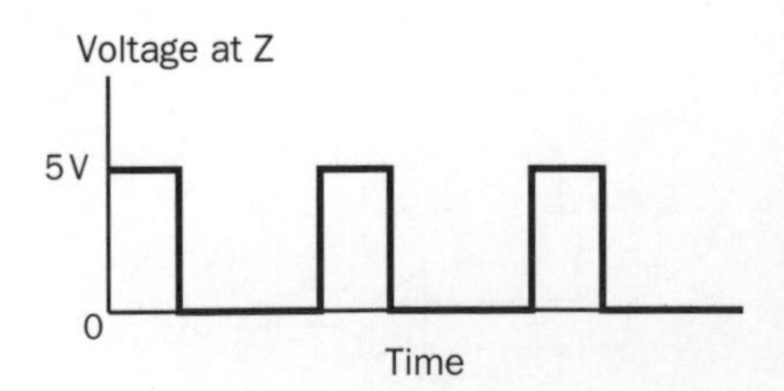

Fig. 16.36 Unequal pulses.

Summary

◆ **Digital circuits**

Logic levels:

Logic level 1 (high) is when the voltage is at or above a certain value
Logic level 0 (low) is when the voltage is at 0 volts

Logic gates:

NOT gate; Output is NOT equal to the input
AND gate; output = 1 if **all** inputs = 1
NAND gate; output = 0 if **all** inputs = 1
OR gate; output = 1 if **any** input = 1
NOR gate; output = 0 if **any** input = 1

Input sensors:

switch (i.e. position), light, temperature, pressure

Output devices:

Indicator LEDs and low-power sound alarms can be switched on or off directly by a logic gate. Transistor and/or relay used to switch high-current devices on or off.

◆ **Analogue circuits**

Amplifiers:

1. Voltage gain = output voltage ÷ input voltage
2. Saturation is when the output voltage reaches its limits set by the power supply.
3. Bandwidth is the frequency range over which the voltage gain is constant.
4. Signal-to-noise ratio is the ratio of the signal amplitude to the noise.
5. Feedback is where a fraction of the output voltage is fed back to the input.

Operational amplifiers:

1. A **comparator** is an open-loop op-amp used to compare two voltages.
2. A **voltage follower** $V_{OUT} = V_{IN}$; negligible current is drawn at the input.
3. The **inverting amplifier** Voltage gain $= \dfrac{-R_F}{R_1}$
4. The **non-inverting amplifier** Voltage gain $= \dfrac{(R_F + R_1)}{R_1}$
5. The **summing amplifier**

Output voltage, $V_{OUT} = -\dfrac{R_F V_1}{R_1} - \dfrac{R_F V_2}{R_2} - \dfrac{R_F V_3}{R_3}$, etc.

◆ **Multivibrators**

1. **Monostable:** one stable state which the output reverts to after a delay.
2. **Astable:** the output voltage repeatedly switches between two levels.

Questions 16.5

1. In the monostable circuit shown in Fig. 16.34, what would be the effect on the operation of the circuit of **(a)** increasing the capacitance of C, **(b)** decreasing the resistance of R.
2. **(a)** An astable multivibrator includes two 1 μF capacitors, each in series with a 1000 Ω resistor.
   (i) Calculate the time constant of a 1 μF capacitor in series with a 1000 Ω resistor. **(ii)** Estimate the frequency of the pulses produced by this astable multivibrator.
   (b) What would be the effect on the operation of the astable multivibrator in (a) of replacing one capacitor with a 4.7 μF capacitor, and the resistor connected to the other capacitor with a 10 kΩ resistor?

See www.palgrave.com/foundations/breithaupt for electronic memory.

Revision questions

16.1. Copy and complete the truth table for each combination of logic gates in Fig. 16.37.

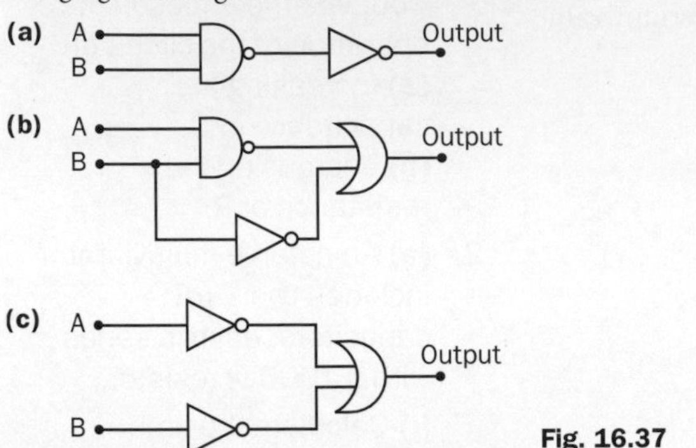

Fig. 16.37

16.2. Fig. 16.38 shows a temperature-operated electric motor used to open a vent in a greenhouse if the temperature in the greenhouse rises too much.

(a) (i) What is the purpose of the switch marked S?

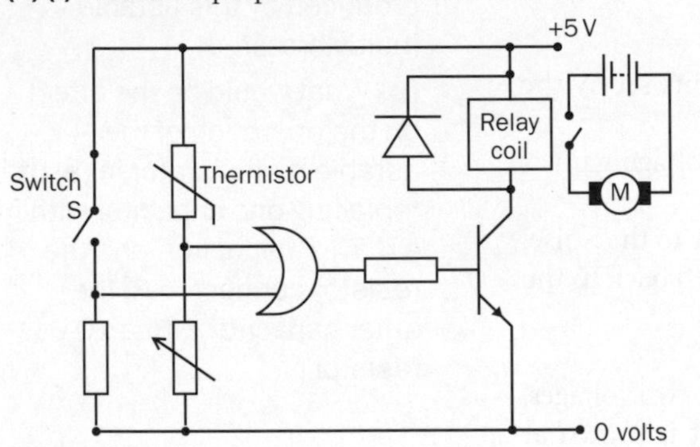

Fig. 16.38

(ii) The variable resistor can be adjusted to enable the system to switch a fan on at different temperatures. What adjustment is needed to enable the fan to switch on at a lower temperature?

(iii) What type of logic gate is needed in this system?

(b) The electric motor in Fig. 16.38 is operated via a transistor and a relay. Explain why the motor is switched on when the output of the logic circuit goes high.

16.3. (a) Design a logic system that switches an alarm on after a delay when either of the two main doors in a house are open and the alarm system is turned on. Include a test switch in your system.

(b) What would be the effect on the delay time of increasing the capacitance of the capacitor in the delay circuit? Explain your answer.

16.4. (a) An alternating voltage of peak-to-peak voltage 2.0 V is applied to the inverting input of an op-amp on open-loop, operating from a +15 V, 0, −15 V power supply, as shown in Fig. 16.39.

(i) The non-inverting input terminal, Q, is connected to earth. Sketch the output waveform produced by this circuit.

(ii) The non-inverting input, Q, is then connected to a potential divider and a constant voltage of 0.5 V is applied to Q. Fig. 16.40 shows the output waveform produced by this arrangement. Explain the shape of this waveform.

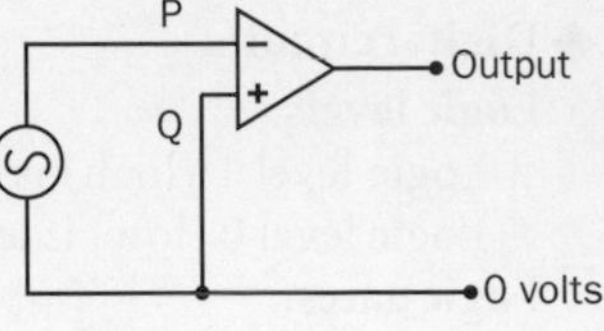

Fig. 16.39

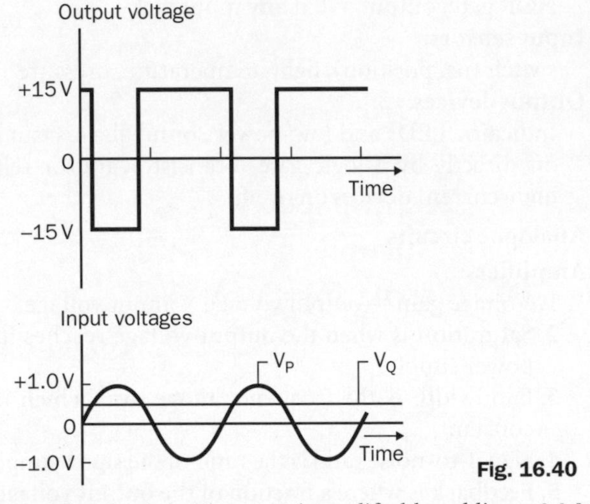

Fig. 16.40

(b) The circuit in Fig. 16.39 is modified by adding a 2.0 MΩ feedback resistor and a 0.4 MΩ resistor between the inverting input, P, and earth. The a.c. supply unit is disconnected from P and reconnected to Q.

(i) What type of operational amplifier circuit is this?

(ii) Calculate the voltage gain of this circuit.

(iii) Calculate the maximum input voltage for no saturation at the output.

16.5. (a) Fig. 16.41 shows an operational amplifier used to add two voltages, V_1 and V_2.

(i) Input 2 is connected to earth and a voltage of +2.0 V is applied to input 1. Calculate the output voltage.

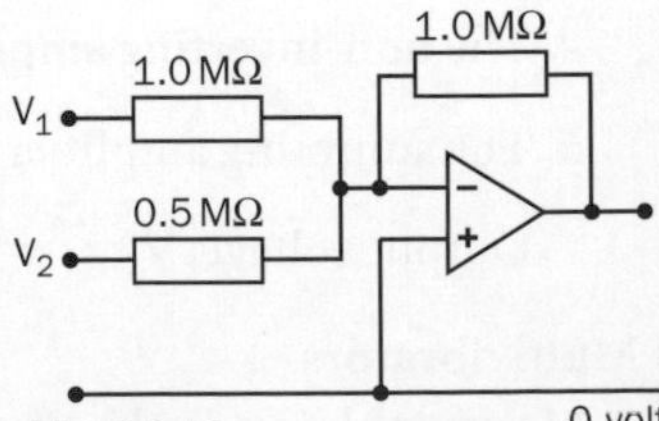

Fig. 16.41

(ii) Input 1 is now connected to earth and a voltage of 1.5 V is applied to input 2. Calculate the output voltage now.

(iii) Calculate the output voltage if a voltage of 2.0 V is applied to input 1 and a voltage of 1.5 V is applied to input 2.

(b) Design a summing amplifier which can convert three input voltages, either 0 or 1 V, representing a 3-bit binary number, into a voltage between 0 and 7 V.

PART 5

Fields

Part 5 concentrates on key concepts of electric and magnetic fields which underpin a wide range of appliances and devices used in the electrical industry to generate and supply electrical power. Before commencing this part, you ought to have studied Part 4, Electricity. The practical work in this part is closely related to the development of the principles in each unit and is intended to help you develop your understanding of the key concepts of electric and magnetic fields. Mathematical skills in algebra and graph work from earlier parts are used extensively in Part 5.

Electric Fields

Objectives

After working through this unit, you should be able to:

- ▶ use Coulomb's law to calculate the force between two point charges
- ▶ explain what is meant by a line of force and an equipotential surface
- ▶ sketch the pattern of lines of force and equipotentials in a uniform field, a radial field and near two point charges
- ▶ define electric field strength and electric potential
- ▶ calculate the electric field strength of a uniform field created by two oppositely charged parallel plates and relate it to the charge on the plates
- ▶ calculate the electric field strength and the electric potential at a given distance from a point charge

Contents

17.1 Electrostatic forces

Two charged objects exert a force on each other due to their electric charge. If the two objects carry the same type of charge, they exert equal and opposite forces of repulsion on each other. If the two objects carry opposite types of charge, they exert equal and opposite forces of attraction on each other.

Like charges repel; unlike charges attract.

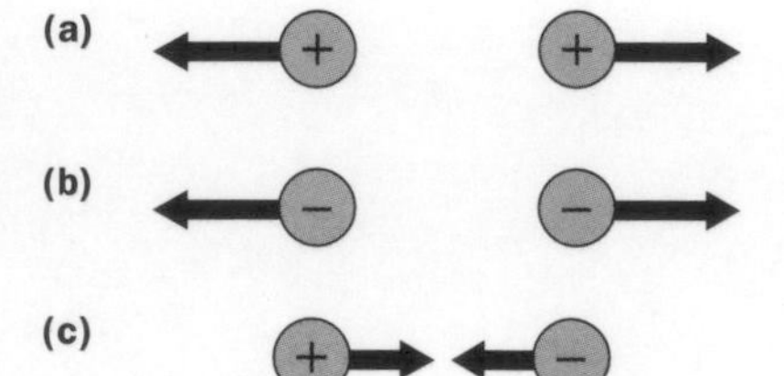

Fig. 17.1 Like charges repel; unlike charges attract **(a)** two positive charges repel, **(b)** two negative charges repel, **(c)** a positive and a negative charge attract.

Coulomb's investigations

The force between two point charges was investigated by Charles Coulomb in 1784. He devised a very sensitive torsion balance to measure the force between two charged pith balls, one on the end of a vertical rod and the other on the end of a needle suspended horizontally on a fine thread. The two balls were given the same type of charge, causing them to repel each other. By turning the torsion head at the top of the wire, the distance

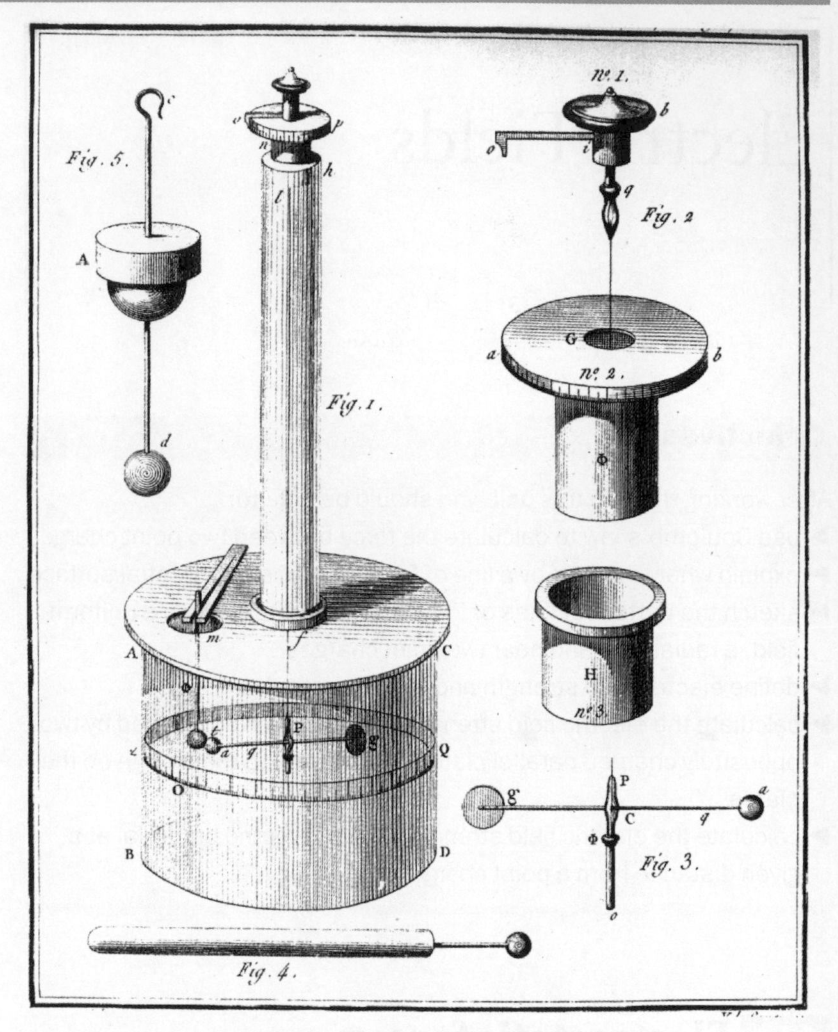

Fig. 17.2 Original drawing of Coulomb's torsion balance.

between the two balls could be set at any required value. The amount of turning needed to achieve that distance gave the force.

Some of Coulomb's measurements are in Table 17.1. Both variables were measured by Coulomb in degrees, as in the Table, and converted to the correct units subsequently.

Table 17.1 Coulomb's measurements

Distance, r	36	18	8.5
Force, F	36	144	567

The measurements in Table 17.1 show that as the distance decreases, the force increases. In fact, halving the distance from 36 to 18 causes the force to become 4 times larger and again from 18 to 8.5 (near enough 9). The measurements show that the force is inversely proportional to the square of the distance apart. In other words, doubling the distance causes the force to becomes 4 times weaker, trebling the distance causes the force to become 9 times weaker and so on. This relationship is known as **Coulomb's Law** and is usually expressed in the form of the equation below.

$$F = \frac{kQ_1Q_2}{r^2}$$

where F is the force between two point charges, Q_1 and Q_2, at distance, r, apart.

As explained later, the constant of proportionality

$$k = \frac{1}{4\pi\varepsilon_0}$$

where ε_0 is the absolute permittivity of free space. This quantity was introduced on p. 212 in the section of the parallel plate capacitor and we shall see later in this unit why it appears in Coulomb's law in the form of the constant $^1/_{4\pi\varepsilon}$. Hence Coulomb's Law is

$$F = \frac{1}{4\pi\varepsilon_0}\,\frac{Q_1Q_2}{r^2}$$

Note

F >0, due to Q_1 and Q_2 carrying the same type of charge; corresponds to a repulsive force.

F <0, due to Q_1 and Q_2 carrying the opposite type of charge; corresponds to an attractive force.

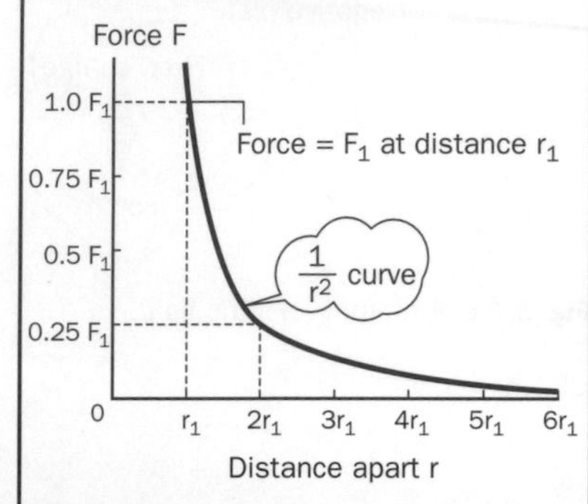

Fig. 17.3 The inverse square law of force.

Questions 17.1

1. Calculate the force between an electron and a proton at a distance apart of 0.10 nm. The charge on a proton = – the charge on an electron = 1.6×10^{-19} C.

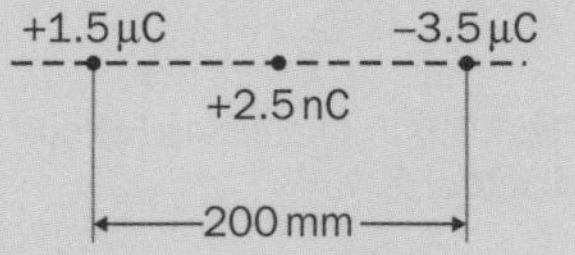

2. A +2.5 nC charge is placed midway between a +1.5 μC point charge and a −3.5 μC point charge which are at a separation of 200 mm. Calculate the magnitude and direction of the resultant force on the 2.5 nC charge.

+1.5 μC —— +2.5 nC —— −3.5 μC
200 mm

Fig. 17.4

Worked example 17.1

$\varepsilon_0 = 8.85 \times 10^{-12}$ F m^{-1}

Calculate the force between a +1.5 μC point charge and a −2.0 μC point charge which are at a distance of 0.050 m apart.

Solution

$$F = \frac{1}{4\pi\varepsilon_0} \times \frac{Q_1Q_2}{r^2}$$

$$= \frac{+1.5 \times 10^{-6} \times -2.0 \times 10^{-6}}{4\pi \times 8.85 \times 10^{-12} \times (0.050)^2}$$

$$= -10.8\ \text{N}$$

17.2 Electric field patterns

Lines of force

Two charged bodies exert equal and opposite forces on each other without being in direct contact. This 'action at a distance' can be considered the result of a charged body creating an **electric field** around itself. Any other charged body placed in this field experiences a force due to the field.

A **line of force** in an electric field is a line along which a small positive 'test' charge would move in the field if free to do so. The 'test' charge must be small enough not to change the electric field in which it is placed.

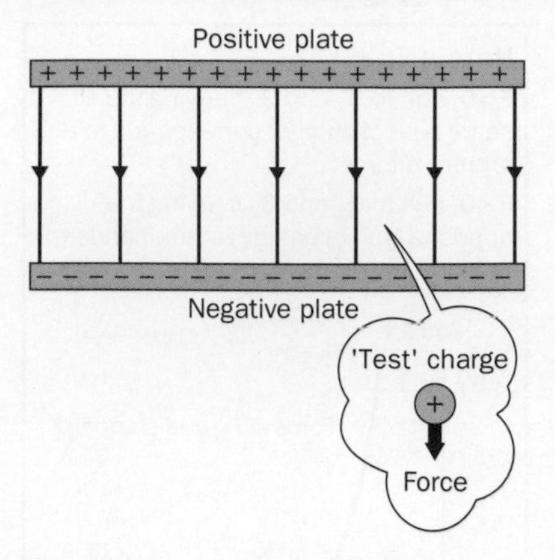

Fig. 17.5 A uniform electric field.

The patterns of the lines of force of two different electric fields are shown in Fig. 17.5 and 17.6.

1. A **uniform electric field** is created between two oppositely charged parallel plates. The lines of force always point from the positive plate to the negative plate because that is the direction of the force on a small positive 'test' charge anywhere between the plates.
2. A **radial electric field** exists near a point charge.
   - The lines of force created by a positive point charge, +Q, radiate from the charge outwards. This is because a positive test charge placed in this field would be forced away from the charge +Q.
   - The lines of force created by a negative point charge, −Q, point inwards. This is because a positive test charge placed in this field would be forced towards the charge −Q.

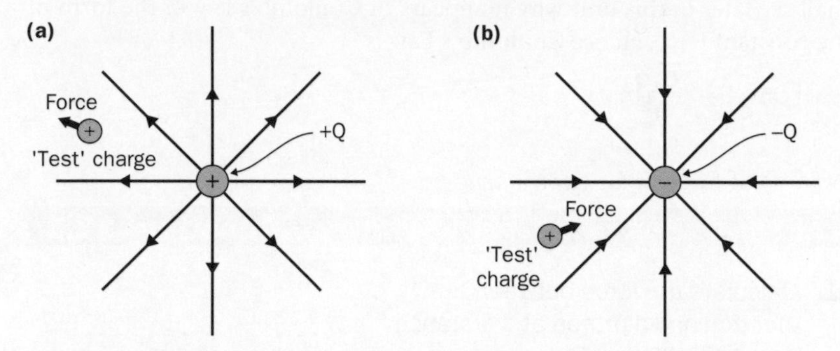

Fig. 17.6 Radial fields (a) due to a positive charge, +Q, (b) due to a negative charge, −Q.

Observing electric field patterns

The pattern of the lines of force of an electric field can be seen using the arrangement in Fig. 17.7. An electric field across the surface of a viscous, clear oil is set up between two metal conductors connected to

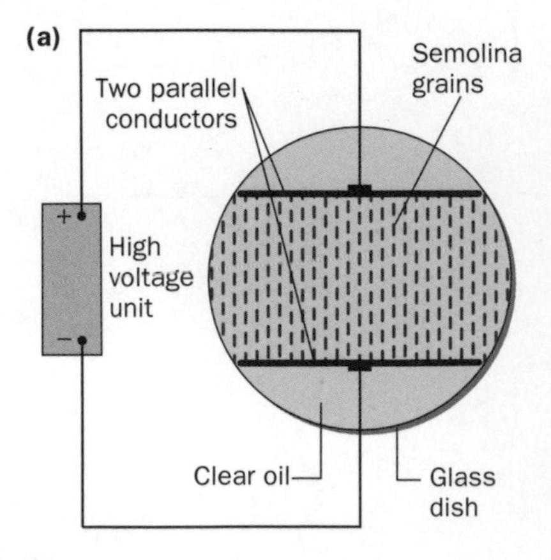

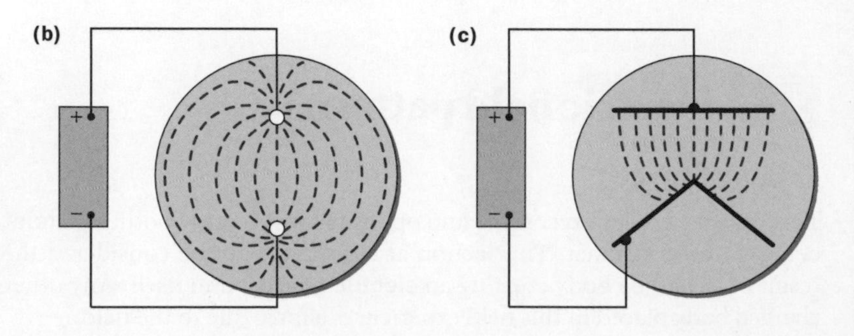

Fig. 17.7 Electric field patterns (a) top view, (b) point objects, (c) a V-shaped conductor near a straight conductor.

opposite poles of a high voltage supply unit. When grains of semolina are sprinkled on the oil surface, the grains align along the lines of force of the field.

1. **Between two oppositely charged parallel conductors**, the lines of force are parallel to each other and perpendicular to the conductors. The field is uniform, the same as between two oppositely charged parallel plates.
2. **Between two oppositely charged point objects**, the lines of force are concentrated at each point charge. A positive test charge released from an off-centre position would follow a curved path to the negative point charge.
3. **Between a V-shaped conductor and an oppositely charged straight conductor**, the lines of force spread out from the V onto the straight conductor. The lines are strongest where the V-shaped conductor is most curved. This is because the concentration of charge on the V-shaped conductor is greatest where the conductor is most curved.

Equipotentials

No work is needed to move a 'test' charge a small distance **at right angles** to a line of force in an electric field, even though a force acts on the test charge. This is because the direction of movement of the test charge is perpendicular to the direction of the force, so the distance moved in the direction of the force is zero. Since work done is defined as force × distance moved in the direction of the force, the work done is therefore zero. This is just the same as an object moving horizontally on a friction-free surface; its potential energy remains constant because there is no movement in the direction of the force of gravity.

The **electric potential** at a point in an electric field is the work that must be done per unit charge to move a small positive 'test' charge to that point in the field from infinity. The unit of electric potential is the volt, equal to 1 joule per coulomb. Effectively, the electric potential at a point in an electric field is the potential difference between that point and infinity. If W represents the work done to move a small 'test' charge, +q, from infinity to a point in an electric field, the potential, V, at that point is $\frac{W}{q}$.

To move a point charge, +q, from infinity to a point in an electric field where the potential is V, the work, W, that must be done is therefore qV.

$$W = qV$$

An **equipotential surface** is a surface on which the potential energy of a test charge is the same at any point on the surface. Fig. 17.8 shows the lines of force and the equipotentials of a radial electric field created by point positive charge, Q. The equipotentials are circles because the potential energy of a test charge does not change along one of these circles. The lines of force are perpendicular to the equipotentials. The equipotentials are just like contours on a map, because contours are lines of constant gravitational potential energy. The contours of a cone-shaped hill are circles like Fig. 17.8.

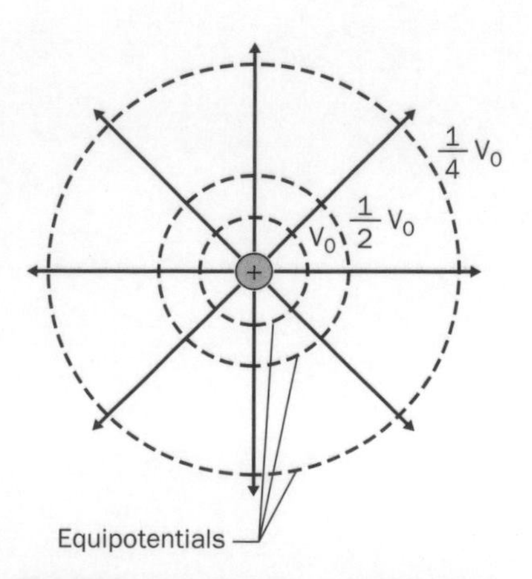

Fig. 17.8 Equipotentials in a radial field.

(a)

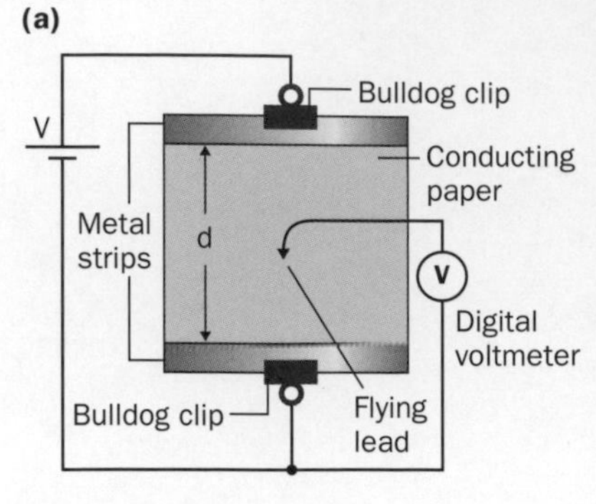

(b)

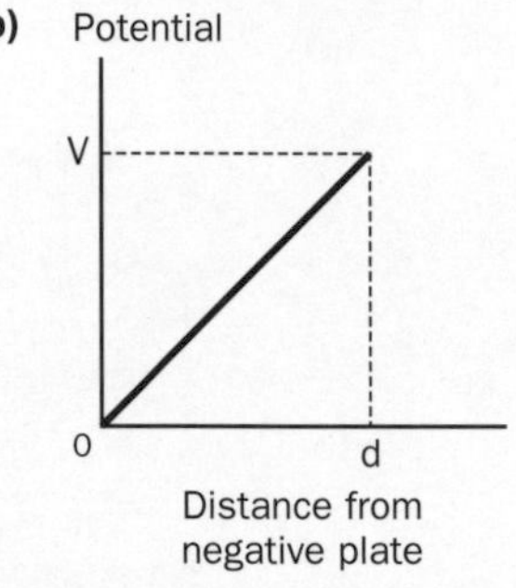

Fig. 17.9 Investigating equipotentials **(a)** practical arrangement, **(b)** potential versus distance.

> **Note**
>
> This formula gives the force on a test charge, q, at any position between two oppositely charged parallel plates. You will meet it again in Unit 21 Electrons and Photons.

Investigating the equipotentials in a uniform electric field

An electric field is set up along the surface of a sheet of conducting paper between two metal strips connected to a battery, as shown in Fig. 17.9a. The negative strip is earthed. The potential at different points on a sheet of conducting paper is measured relative to the negative strip, using a digital voltmeter. Points of equal potential can be traced out by moving the 'flying lead' across the surface so as to keep the voltmeter reading constant. The equipotentials between the metal strips are straight lines, parallel to the strips. The lines of force are therefore straight lines from the positive strip to the negative strip.

- The potential changes along a line of force by equal amounts in equal distances. Fig. 17.9b shows that the potential changes uniformly from the negative strip to the positive strip.
- Suppose a positive test charge, +q, is moved from the negative to the positive plate through a potential difference, V. The force exerted on +q due to the field is towards the negative plate. To overcome this force, an equal and opposite force, F, must be applied to +q to move it.

The work done by this force, W = Fd, where d is the distance between the two metal strips.

Since W = qV from the definition of V, it follows that Fd = qV.

$$\therefore \textbf{Force, } F = \frac{qV}{d}$$

Worked example 17.2

$g = 9.8\ \text{m s}^{-2}$

Two horizontal metal plates are fixed 4.0 mm apart, one above the other. A high voltage supply unit is connected to the two plates to make the top plate negative relative to the lower plate. A charged oil droplet of mass 8.8×10^{-15} kg, carrying a charge of $+4.8 \times 10^{-19}$ C, is observed falling vertically when the plate p.d. is 500 V.

Charged oil droplet

Fig. 17.10

(a) Calculate (i) the change of potential over a distance of 1.0 mm along a vertical line in the field, (ii) the change of electrical potential energy of the droplet as a result of falling through a distance of 1.0 mm, (iii) the change of gravitational potential energy of the droplet as a result of falling through a distance of 1.0 mm.

(b) Calculate the potential difference necessary to hold this droplet stationary between the plates.

Solution

(a) (i) The change of potential over 4.0 mm from one plate to the other is 500 V. For a distance of 1.0 mm, the change of potential is therefore $500 \div 4.0 = 125$ V.

Worked example 17.2 continued

(ii) The electrical potential energy increases because the droplet is positive and it is falling towards the positive plate. Over 1 mm, the increase of electrical potential energy = W = qV = $4.8 \times 10^{-19} \times 125 = 6.0 \times 10^{-17}$ J.

(iii) The gravitational potential energy falls. Over 1 mm, the decrease of gravitational potential energy = mgh = $8.8 \times 10^{-15} \times 9.8 \times 1.0 \times 10^{-3} = 8.6 \times 10^{-17}$ J.

(b) For the droplet to be at rest, the electric field force $\frac{qV}{d}$ must be equal and opposite to its weight, mg.

Rearranging $\frac{qV}{d} = mg$ gives $V = \frac{mgd}{q}$

$$= \frac{8.8 \times 10^{-15} \times 9.8 \times 4.0 \times 10^{-3}}{4.8 \times 10^{-19}}$$

= 720 V.

Questions 17.2

1. Fig. 17.11 shows some of the equipotentials of an electric field created by a V-shaped conductor and an oppositely charged straight conductor.

(a) Calculate the change of potential energy when a small object carrying a +2.5 nC charge is moved **(i)** from X to Y, **(ii)** from X to Z, **(ii)** from Z to Y.

(b) If the distance from X to Y is 5.0 mm, calculate the average force on the +2.5 nC when it is between X and Y.

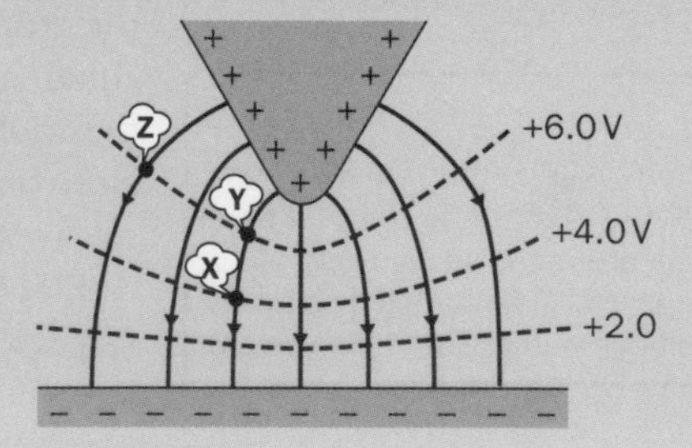

Fig. 17.11

2. A uniform electric field is created between two horizontal parallel plates 12 mm apart by applying a potential of +600 V to the upper plate and earthing the lower plate, as in Fig. 17.12. A charged oil droplet of mass 2.5×10^{-15} kg is held at rest in the field.

(a) (i) What type of charge does this droplet carry?

(ii) Calculate the charge on the droplet.

(b) Sketch a graph to show how the potential between the plates changes with distance from the earthed plate to the positive plate.

(c) Calculate the electrical potential energy of the droplet if its position is **(i)** 2.0 mm above the earthed plate, **(ii)** 6.0 mm above the earthed plate, **(iii)** 10 mm above the earthed plate.

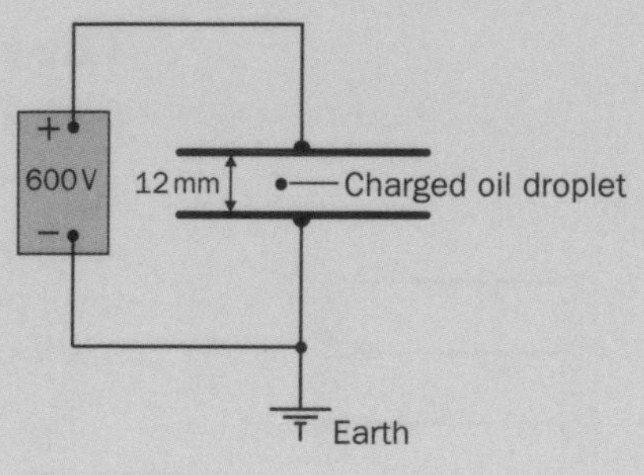

Fig. 17.12

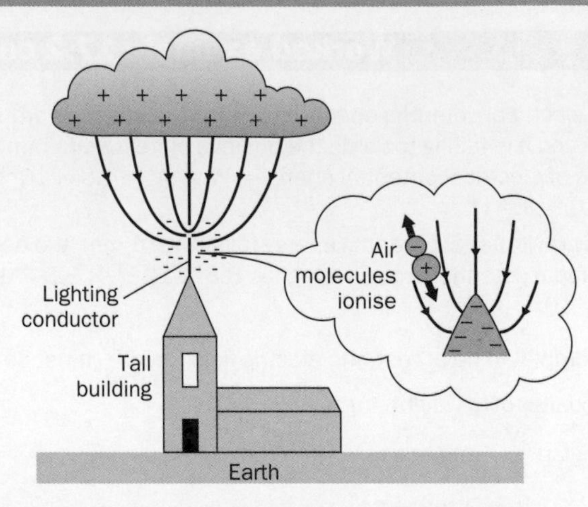

Fig. 17.13 A lightning conductor works by ionising air molecules.

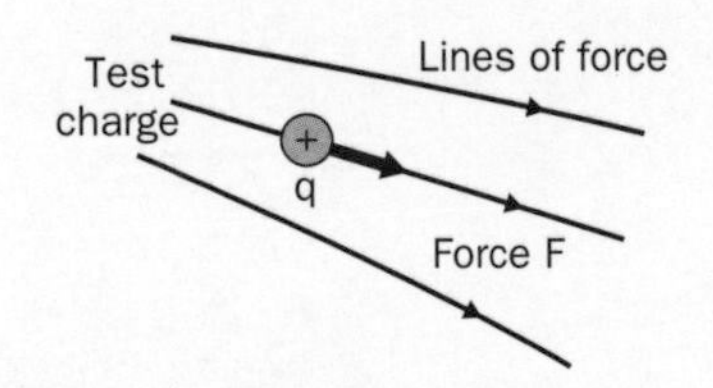

Fig. 17.14 Electric field strength.

17.3 Electric field strength

Notes

1. E is a vector quantity in the direction of the force on a positive 'test' charge.
2. The unit of E is the newton per coulomb (NC^{-1}) which is the same as the volt per metre ($V\ m^{-1}$). See below.

A lightning conductor discharges an overhead thundercloud because the electric field at the tip of the conductor is strong enough to ionise air molecules. The air becomes able to conduct and allows the charge on the cloud to discharge to the ground gradually. The lines of force concentrate at the tip because the thundercloud's charge attracts charge to the tip from the ground. An air molecule near the tip is acted upon by extremely strong equal and opposite forces due to the electric field. The electrons of the molecule are attracted towards the positive charge and one or more electrons become detached from the molecule which is then attracted towards the negative charge.

Electric field strength, E, at a point in an electric field is defined as the force per unit charge on a positive 'test' charge at that point.

$$E = \frac{F}{q}$$

The electric field between two oppositely charged parallel plates

This is a uniform field in which the lines of force are straight, from the positive to the negative plate. As explained on p. 242, the force, F, on a small positive 'test' charge, q, is related to the plate p.d., V, and the plate spacing, d, in accordance with the equation

$$F = \frac{qV}{d}$$

The force on q is the same at all points in the field. Since the electric field strength E is defined as

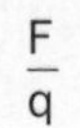

$$\frac{F}{q}$$

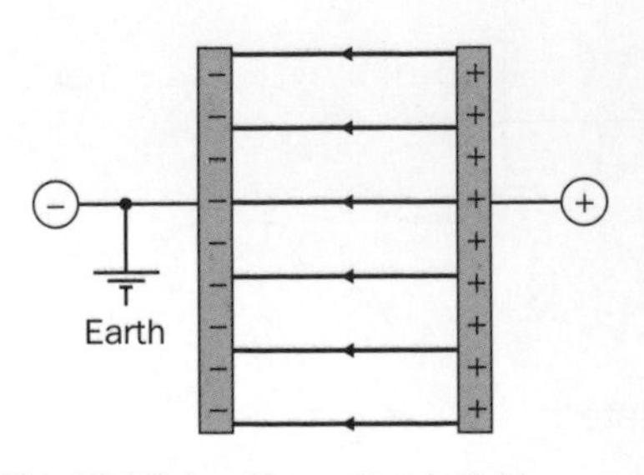

Fig. 17.15 A uniform electric field.

then

$$\mathbf{E} = \frac{\mathbf{V}}{\mathbf{d}}$$

Note
This equation shows that the unit of E can be given as the volt per metre.

For example, the electric field strength between two plates 50 mm apart at a p.d. of 100 V is 2000 V m^{-1}. Fig. 17.16 shows how the potential changes from the negative plate (which is earthed in this instance) to the positive plate.

The **potential gradient**, defined as the change of potential per unit distance, is constant, equal to

$$\frac{V}{d}$$

This is equal in magnitude and opposite in direction to the electric field strength as the potential rises from the − plate to the + plate, whereas the force on a positive test charge is towards the negative plate.

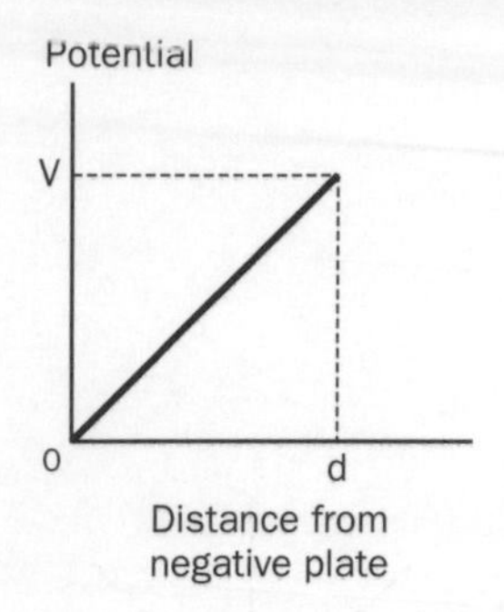

Fig. 17.16 A uniform potential gradient.

The absolute permittivity of free space, ε_o

In Fig. 17.15, the charge, Q, on each plate is related to the plate p.d. in accordance with the equation

$$\mathbf{Q} = \frac{\mathbf{A\varepsilon_o V}}{\mathbf{d}}$$

where ε_o is the absolute permittivity of free space, and A is the plate area. This equation originated on p. 212 as a result of finding that the charge, Q, on the plates of a parallel plate capacitor is proportional to the plate area, A, and the plate p.d., V, and inversely proportional to the plate spacing, d. Provided there is nothing between the plates, the absolute permittivity of free space is introduced as the constant of proportionality.

Since the electric field strength, E, can be written as

$$E = \frac{V}{d},$$

the equation for the charge, Q, can be written in terms of the electric field strength E as $Q = A\varepsilon_o E$. Hence the surface charge density (i.e. the charge per unit area)

$$\frac{Q}{A} = \varepsilon_o E$$

Charge per unit area, $\mathbf{\frac{Q}{A} = \varepsilon_o E}$

The permittivity of free space is therefore a measure of the charge per unit area on a parallel plate capacitor needed to create an electric field of strength 1 volt per metre. Since $\varepsilon_o = 8.85 \times 10^{-12}$ F m^{-1}, then a surface charge density of 8.85×10^{-12} C m^{-2} would give an electric field of strength 1 V m^{-1}.

Worked example 17.3

Two parallel plates 0.10 m × 0.10 m are at a fixed separation of 0.004 m, as shown in Fig. 17.17, at a p.d. of 250 V. Calculate (a) the electric field strength between the plates, (b) the surface charge density, (c) the charge on each plate, (d) the number of electrons per m² on the surface.

$\varepsilon_o = 8.85 \times 10^{-12}$ F m^{-1}; the charge of the electron, e = 1.6×10^{-19} C

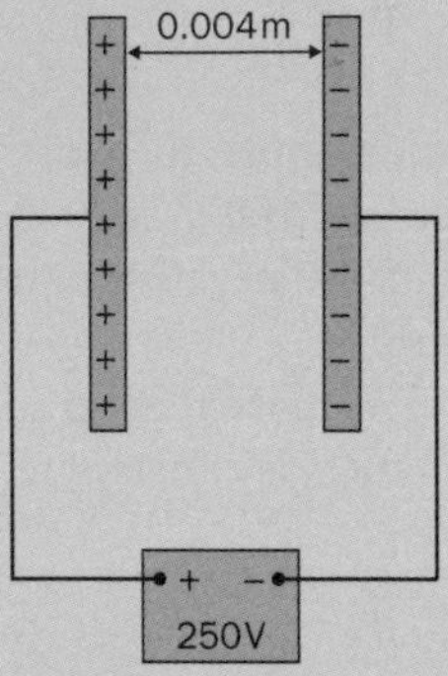

Fig. 17.17

Solution

(a) $E = \frac{V}{d} = \frac{250}{0.004} = 6.25 \times 10^4\ \text{V m}^{-1}$

(b) $\frac{Q}{A} = \varepsilon_o E = 8.85 \times 10^{-12} \times 6.25 \times 10^4 = 5.5 \times 10^{-7}\ \text{C m}^{-2}$

(c) $A = 0.1 \times 0.1 = 0.010\ \text{m}^2 \therefore Q = 5.5 \times 10^{-7} \times 0.010 = 5.5 \times 10^{-9}$ C

(d) For $A = 1.0\ \text{m}^2$, $Q = 5.5 \times 10^{-7}$ C

$\therefore$ the number of electrons per m$^2 = \frac{5.5 \times 10^{-7}}{1.6 \times 10^{-19}} = 3.5 \times 10^{12}$

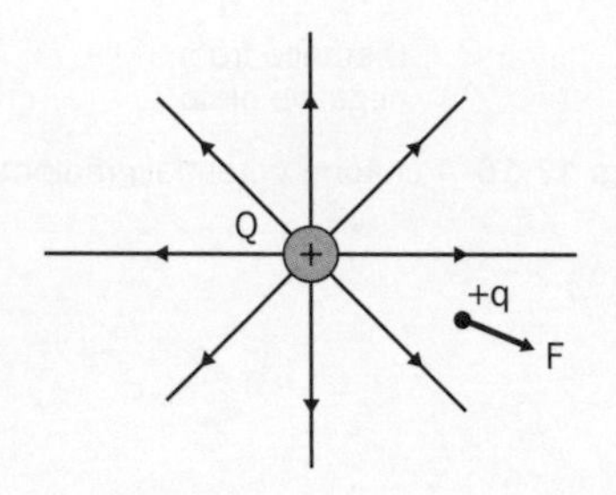

Fig. 17.18 The electric field near a point charge.

Electric field strength created by a point charge

The lines of force of the electric field created by a positive point charge, +Q, radiate away from the charge. This is because a positive test charge, +q, placed near Q experiences a force of repulsion directly away from +Q. This force depends on the distance, r, between the two charges, in accordance with Coulomb's law.

$$F = \frac{1}{4\pi\varepsilon_o} \frac{Qq}{r^2}$$

Because electric field strength, E, is defined as the force per unit charge on a positive test charge, then the electric field strength, E, at distance, r, from a point charge, Q, is given by the equation

$$\mathbf{E = \frac{F}{q} = \frac{Q}{4\pi\varepsilon_o r^2}}$$

This equation shows that the electric field strength due to a point charge is proportional to the inverse of the square of the distance to the charge, as shown by Fig. 17.19. For example, the electric field strength at distance, d, is four times stronger than at distance, 2d.

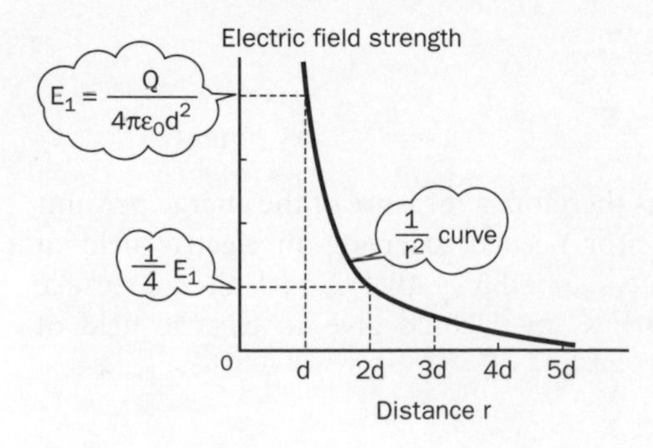

Fig. 17.19 E versus r for a point charge.

Gauss's law

Gauss was a famous 19th century mathematician who recognised that there is a common link between the electric field strength near a point charge and between oppositely charged parallel plates. For two oppositely charged parallel plates, the charge on the plates, $Q = A\varepsilon_o E$. Rearranging this formula gives $E = Q/A\varepsilon_o$. Gauss recognised that the field is uniform because the lines of force are parallel. At any position between the plates, the area that the lines of force pass through is unchanged, equal to the area of the plates. However, the lines of force due to a point charge spread out and the field becomes weaker with increased distance. At a certain distance, r, from a point charge, Q, the lines of force pass through a sphere of radius, r. Gauss realised that the formula for the electric field strength, E, at distance, r, from a point charge, Q

$$E = \frac{Q}{4\pi\varepsilon_o r^2}$$

may be written in the form

$$E = \frac{Q}{A\varepsilon_o}$$

because the surface area of a sphere of radius r is $4\pi r^2$. Gauss was able to develop his mathematical ideas much further but this limited glimpse of his ideas is sufficient to explain the underlying reason why the constant

$$\frac{1}{4\pi\varepsilon_o}$$

is introduced in Coulomb's law.

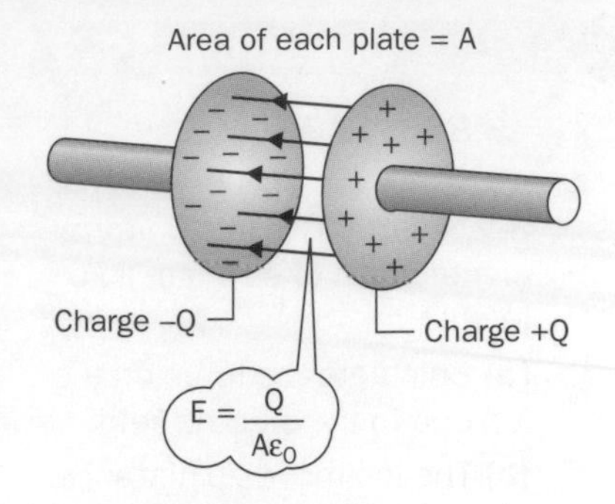

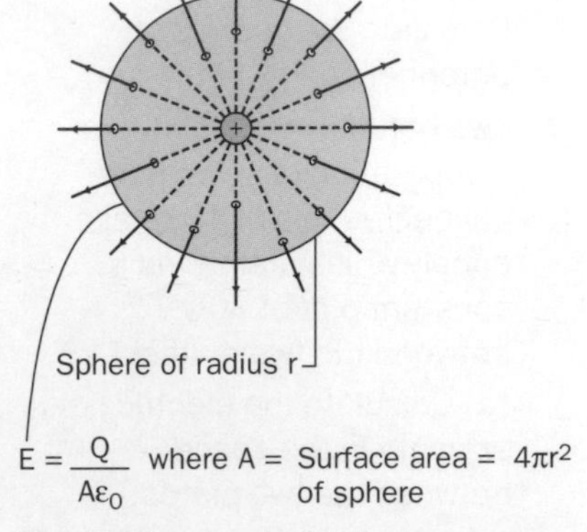

Fig. 17.20 Gauss's field ideas.

Worked example 17.4

Calculate the electric field strength at a distance of (a) 1.0 mm, (b) 10 mm from a small object carrying a + 4.5 pC (= 4.5×10^{-12} C) charge. $\varepsilon_o = 8.85 \times 10^{-12}$ F m^{-1}

Solution

(a) $E = \frac{Q}{4\pi\varepsilon_o r^2} = \frac{4.5 \times 10^{-12}}{4\pi \times 8.85 \times 10^{-12} \times (1.0 \times 10^{-3})^2} = 4.1 \times 10^4$ V m^{-1}

(b) At 10 mm, the electric field strength is 100 times weaker than at 1.0 mm because the distance is ten times greater. Hence E = 410 V m^{-1}.

17.4 Electric potential

When a charged object in an electric field is moved, its potential energy changes unless the object is moved perpendicular to the lines of force of the field. No change of potential energy occurs if the object is moved in a direction perpendicular to the lines of force.

The electric potential at a point in an electric field is defined as the work done to move a positive 'test' charge to that point from infinity.

Questions 17.3

$\varepsilon_0 = 8.85 \times 10^{-12}$ F m^{-1}

1. A positive ion of charge of 3.2×10^{-19} C is in a uniform electric field of strength 120 kV m^{-1}.

   (a) Calculate the force on the ion due to the electric field.

   (b) The ion moves a distance of 5.0 μm in the direction of the lines of force of the field. Calculate the change of potential energy of the ion.

2. Two parallel metal plates at a separation of 10 mm are connected to a high voltage supply which maintains a constant p.d. of 300 V between the two plates.

   (a) Calculate the electric field strength in the space between the two plates.

   (b) Calculate the charge per unit area on each metal plate.

   (c) Calculate the number of electrons per mm^2 of area on each plate.

Consider a point charge, q, in an electric field at a point where the electric potential is V.

Potential energy of q = qV

This is because qV is the work done on q to move it to a point at potential V from infinity, where its potential energy is zero. For example, if a +1 nC charge is moved from infinity to a point in a field where V = 1200 V, the potential energy of the charge at this point would be 1200 nJ (= 1 nC × 1200 V).

At distance, r, from a point charge, Q,

the electric potential, $V = \dfrac{Q}{4\pi\varepsilon_0 r}$

Fig. 17.21 shows how the potential changes with distance. Note that the potential does not decrease as fast as the electric field strength does, since V is proportional to 1/r, whereas E is proportional to 1/r^2.

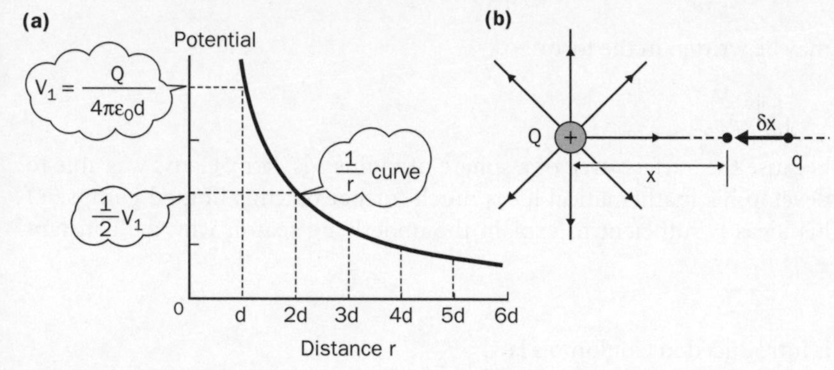

Fig. 17.21 Electric potential near a point charge **(a)** potential versus distance, **(b)** moving a test charge.

The formula for potential can be proved by considering a small positive 'test' charge, q, moved from infinity to distance, r, from charge, Q.

The work done to move q a small distance, δx, towards Q,

$$\delta W = F\delta x = \frac{qQ}{4\pi\varepsilon_0 x^2}\,\delta x$$

since the force between q and Q at separation x is $\dfrac{qQ}{4\pi\varepsilon_0 x^2}$ in accordance with Coulomb's law

$$\therefore \text{ the change of potential, } \delta V = \frac{\delta W}{q} = \frac{Q}{4\pi\varepsilon_0 x^2}\,\delta x$$

$$\text{Since } \frac{1}{x} - \frac{1}{(x+\delta x)} = \frac{\delta x}{x^2}, \text{ provided } \delta x \ll x,$$

$$\text{then } \delta V = \frac{Q}{4\pi\varepsilon_0\, x} - \frac{Q}{4\pi\varepsilon_0\,(x+\delta x)} = \text{potential at } x - \text{the potential at } x + \delta x.$$

By adding the change of potential for every step, δx, from infinity to r, the total change of potential from infinity is therefore

$$\frac{Q}{4\pi\varepsilon_0 r}$$

The charged sphere

The charge on a conducting sphere is spread evenly over its surface. The lines of force therefore radiate from the surface, giving the same pattern as if all the charge were concentrated at the centre. At and beyond the surface, the electric field strength and the electric potential at distance, r, from the centre are therefore given by the same formulae as for a point charge.

For $r \geq R_S$, the sphere radius

$$E = \frac{Q}{4\pi\varepsilon_0 r^2}$$

$$V = \frac{Q}{4\pi\varepsilon_0 r}$$

Inside the sphere, the electric field strength is zero as the lines of force radiate outwards from the surface. Since there is no force on a test charge inside the sphere, the electric potential inside the sphere is therefore constant, equal to the surface potential

$$\frac{Q}{4\pi\varepsilon_0 R_S}$$

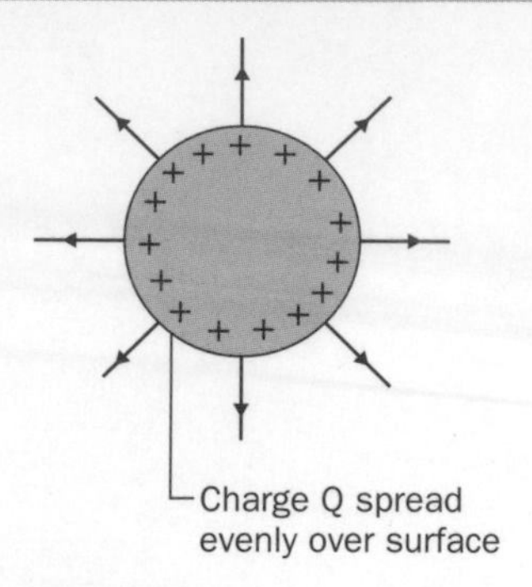

Fig. 17.22 E and V near a charged sphere.

Questions 17.4

$\varepsilon_0 = 8.85 \times 10^{-12}\ \mathrm{F\,m^{-1}}$

1. Calculate the electric field strength and the potential
   (a) at a distance of 5.0 mm from a point object carrying a charge of +8.5 pC,
   (b) midway between two parallel plates 40 mm apart, one plate being earthed and the other plate at a potential of +200 V.
2. A conducting sphere of radius 0.20 m is charged to a positive potential of 100 000 V. Calculate **(a)** the charge on the sphere, **(b)** the electric field strength at the surface, **(c)** the electric field strength and the potential at a distance of 2.0 m from the sphere.

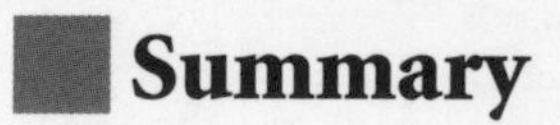

Summary

◆ **Coulomb's law**

$$F = \frac{Q_1 Q_2}{4\pi\varepsilon_0 r^2}$$

◆ **Electric field patterns**

1. A **line of force** is a line along which a positive 'test' charge would move if free to do so.
2. An **equipotential surface** is a surface on which the potential is constant.
3. A **uniform field** is a field in which the lines of force are parallel and the force on a 'test' charge is independent of its position.
4. A **radial field** is where the lines of force are straight lines towards or away from the point charge.

◆ **Electric potential**, **V** is the work done per unit charge to move a positive 'test' charge from infinity. The unit of V is the volt, equal to 1 joule per coulomb.

◆ **Electric field strength, E** is the force per unit charge on a positive test charge. The unit of E is the newton per coulomb or the volt per metre.

◆ **In a uniform field between two oppositely charged parallel plates**

1. The force on a charge q between the plates $= \dfrac{qV}{d}$
2. The electric field strength between the plates $= \dfrac{V}{d}$
3. The charge on each plate, $Q = \dfrac{A\varepsilon_0 V}{d}$

4. The potential gradient is equal and opposite to the electric field strength.

◆ **In a radial field**

1. Due to a point charge, Q,

$$E = \frac{Q}{4\pi\varepsilon_0 r^2}, \quad V = \frac{Q}{4\pi\varepsilon_0 r}$$

2. At or beyond the surface of a charged sphere, Q,

$$E = \frac{Q}{4\pi\varepsilon_0 r^2}, \quad V = \frac{Q}{4\pi\varepsilon_0 r}$$

◆ **Inside a charged sphere**

$$E = 0, V = \frac{Q}{4\pi\varepsilon_0 R_S}$$

Revision questions

$\varepsilon_0 = 8.85 \times 10^{-12}\ \mathrm{F\,m^{-1}}$

17.1. (a) Sketch the pattern of lines of force between two equal and opposite point charges.
(b) A +2.5 pC point charge is placed at a distance of 20 mm from a –7.2 pC point charge. Calculate the force between these two charges and state if it is an attractive or a repulsive force.
(c) Calculate **(i)** the electric field strength, **(ii)** the potential at the midpoint between these two point charges.

17.2. A uniform electric field was established between two horizontal plates which were spaced 5.0 mm apart using a voltage supply unit and a potential divider as shown in Fig. 17.23. A charged oil droplet of mass 5.6×10^{-15} kg was held stationary between the plates when the plate p.d. was adjusted to 430 V.

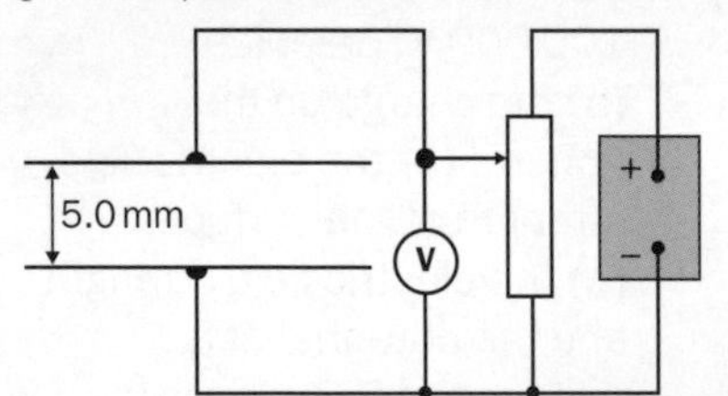

Fig. 17.23

(a) Calculate **(i)** the electric field strength between the plates at this potential difference, **(ii)** the charge carried by this droplet.
(b) The droplet coalesced with an uncharged droplet and the potential difference needed to be adjusted to 620 V. Calculate the mass of the uncharged droplet before it coalesced with the charged droplet.

17.3 (a) Calculate the force of attraction between a proton and an electron at a separation of 1.0×10^{-10}m.
(b) Calculate the work that must be done to separate a proton and an electron at a separation of 1.0×10^{-10} m to infinity. The charge of the electron, $e = -1.6 \times 10^{-19}$ C.

17.4. Two identical +1.6 nC point charges are placed 40 mm apart.
(a) Explain why the resultant electric field strength at the midpoint is zero.
(b) (i) Show that the total potential at the midpoint is 1440 V. **(ii)** Sketch a graph to show how the potential changes from infinity to the midpoint along a line that is equidistant from the two point charges.
(c) Calculate the work that must be done to move a particle carrying a charge of 1.5×10^{-17} C to the midpoint from infinity.

17.5. (a) A metal sphere is charged to a potential of 10 000 V. Calculate the electric field strength at the surface of the sphere if its radius is **(i)** 100 m, **(ii)** 1.0 mm.
(b) Use your calculation in (a) to explain why a charged conductor discharges into the surrounding air if there are sharp points on its surface.
(c) An uncharged metal sphere of radius 50 mm is briefly brought into contact with an isolated charged sphere of radius 100 mm, which is at a potential of 5000 V. Calculate **(i)** the initial charge on the 100 mm sphere, **(ii)** the charge on each sphere after they have made contact, **(iii)** the final potential of each sphere.

Magnetic Fields

Objectives

After working through this unit, you should be able to:

- use a plotting compass to plot magnetic lines of force
- sketch the magnetic field patterns for different- shaped magnets and current-carrying conductors
- describe the operation of an electromagnet and its use in a relay coil
- describe the motor effect and use it to explain the operation of an electric motor, a moving coil meter and a loudspeaker
- define magnetic flux density and calculate the force on a current-carrying wire in a magnetic field.
- describe and calculate the effect of a uniform magnetic field on a beam of charged particles
- explain the Hall effect and its use in measuring magnetic flux density
- describe the factors that determine the magnetic flux density at the centre of a solenoid and near a long straight wire
- explain why two parallel current-carrying wires interact and describe how the ampere is defined
- describe the magnetic properties of iron and steel and relate these properties to uses

Contents

18.1 Magnetic field patterns

Magnetic materials

A bar magnet is capable of attracting and holding iron filings or a steel paper clip. You can test a nail to find out if it is magnetic by seeing if it will attract iron filings or a steel paper clip. If it is magnetic, iron filings will stick to either end of the nail. The ends of a bar magnet are called **magnetic poles** because the magnetism seems to be concentrated at the ends.

The most common magnetic materials are iron and steel. An iron bar can be easily magnetised by 'stroking' it with a bar magnet, as shown in Fig. 18.32 on p. 267. Steel is less easy to magnetise but it retains its magnetism better than iron. Another way to magnetise an iron nail is to wrap an insulated wire around it and pass a direct current through the wire. The current produces its own magnetic field that magnetises the iron nail.

Magnetic lines of force

A magnetic compass is an essential piece of equipment for a hill-walker or a pilot on a boat as the compass needle always points north. The line along which a compass needle points is an example of a **magnetic line of force**. Fig. 18.1a shows how to plot the lines of force of the Earth's magnetic field. On a global scale, the lines of force of the Earth's magnetic field are concentrated at the magnetic poles, as shown in Fig. 18.1b.

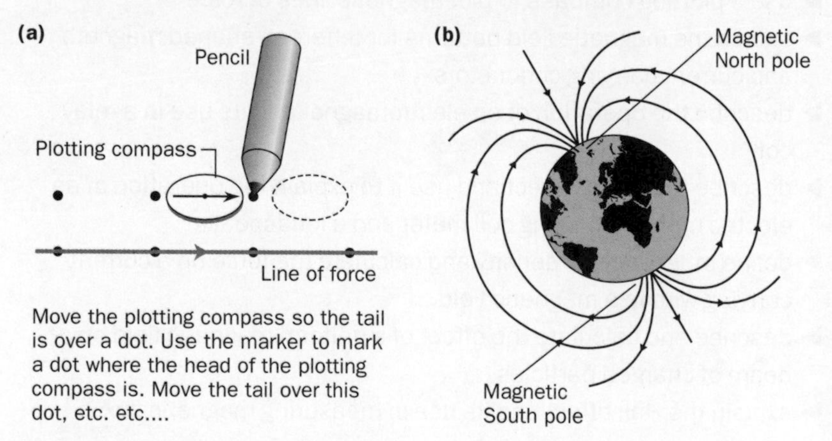

Fig. 18.1 The Earth's magnetic field **(a)** plotting lines of force, **(b)** the Earth's magnetic poles.

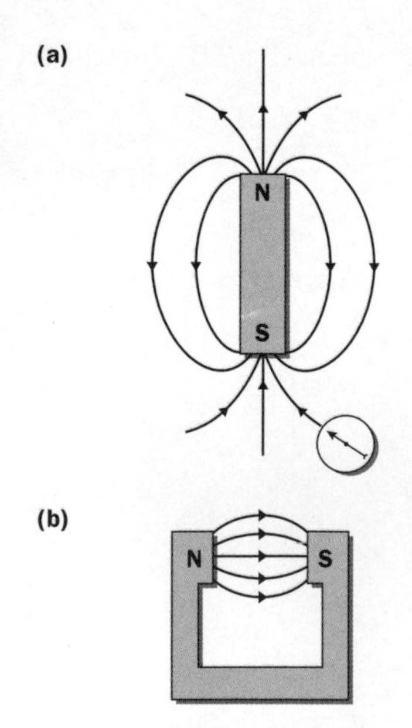

Fig. 18.2 Magnetic field patterns in **(a)** a bar magnet, **(b)** a U-shaped magnet.

A bar magnet suspended horizontally on a thread is a magnetic compass since it aligns itself along a line from north to south. The end that points north is called the north-seeking (N) pole and the other end the south-seeking (S) pole. If a N-pole of a second bar magnet is held near a N- pole of a suspended bar magnet, the two poles repel each other. The same happens if a S -pole is held near a S-pole. If, however, a N-pole is held near a S-pole, the two poles attract each other.

Like poles repel; unlike poles attract

The lines of force of a bar magnet form a characteristic pattern of loops from its N-pole to its S-pole, as shown in Fig. 18.2. The head of a plotting compass (i.e. its N-pole) is attracted to the S-pole of the bar magnet so the lines of force always point towards the S-pole of a bar magnet. The pattern of lines of force of a U-shaped magnet is also shown in Fig. 18.2. Note that the lines of force between the poles of the U-shaped magnet are parallel to each other, directed from the N-pole to the S-pole.

Investigating the magnetic field patterns between two magnets

Use a plotting compass to plot the lines of force between two bar magnets in each arrangement shown in Fig. 18.3. In (a), there is a **neutral point** between the two N-poles. This is where the compass needle is attracted equally by the two N-poles so the resultant magnetic field is zero at this position.

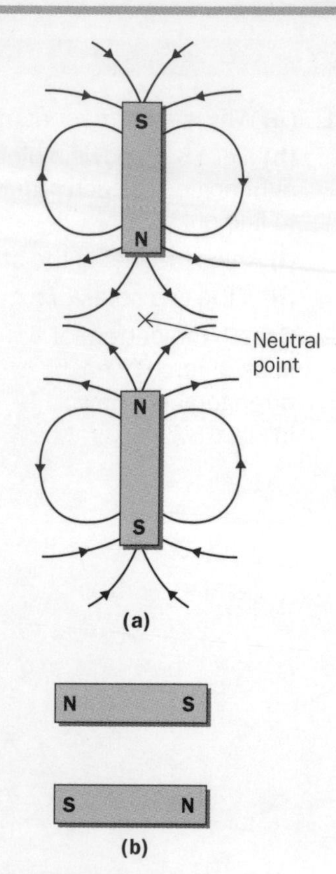

Fig. 18.3 More magnetic fields.

Electromagnetism

A magnetic field is created when a current is passed along a wire. This magnetic field disappears when the current is switched off.

1. A **long straight wire** carrying a constant current produces lines of force which are concentric circles around the wire in a plane perpendicular to the wire. Note that the lines of force around a current-carrying wire are continuous loops and do not originate or terminate like the lines of force of a permanent magnet.

   The direction of the lines of force depends on the direction of the current, as shown by Fig. 18.4. The 'corkscrew' rule is a convenient way of remembering how the direction of a line of force is related to the current direction.

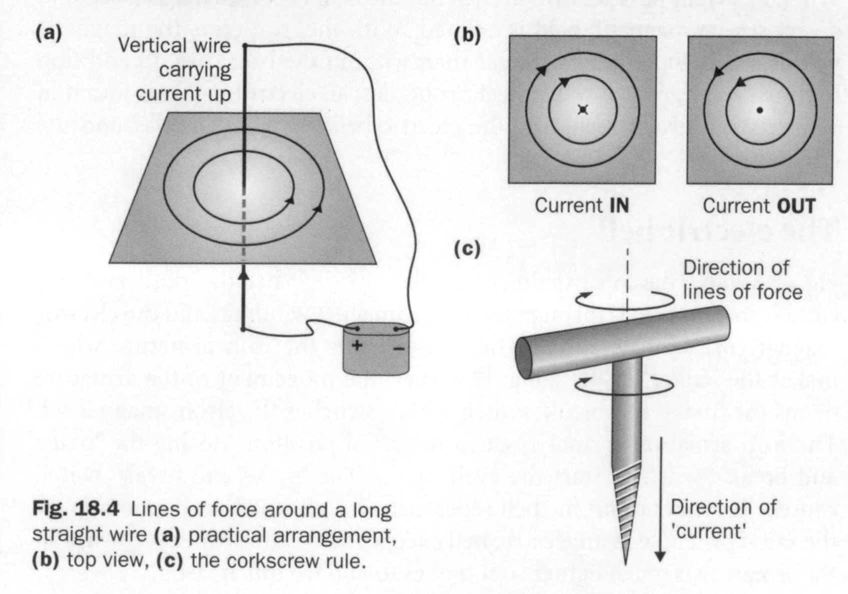

Fig. 18.4 Lines of force around a long straight wire **(a)** practical arrangement, **(b)** top view, **(c)** the corkscrew rule.

2. A **solenoid** carrying a constant current produces lines of force that emerge from one end of the solenoid and loop around to the other end. For an empty solenoid, the lines of force pass through the solenoid, parallel to its axis, as in Fig. 18.5. The end of the solenoid where the lines of force emerge is like the N-pole of a bar magnet and the other end is like the S-pole. The 'solenoid' rule is a convenient way of remembering how the direction of the lines of force at each end relates to the current direction.

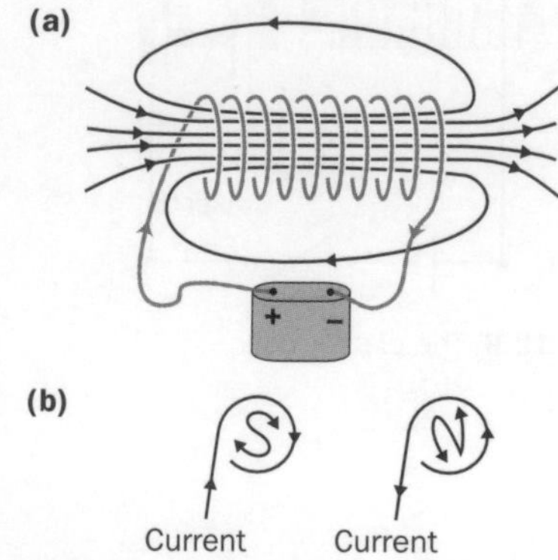

Fig. 18.5 **(a)** lines of force through and around an empty solenoid, **(b)** the solenoid rule.

Questions 18.1

1. **(a)** Why is a permanent magnet made from steel rather than iron?

   (b) Fig. 18.6 shows a plotting compass and two bar magnets, X and Y, at right angles to each other. Describe how you would use this arrangement to find out

   (i) which magnet is the stronger of the two,

   (ii) if the two poles of a magnet are the same strength.

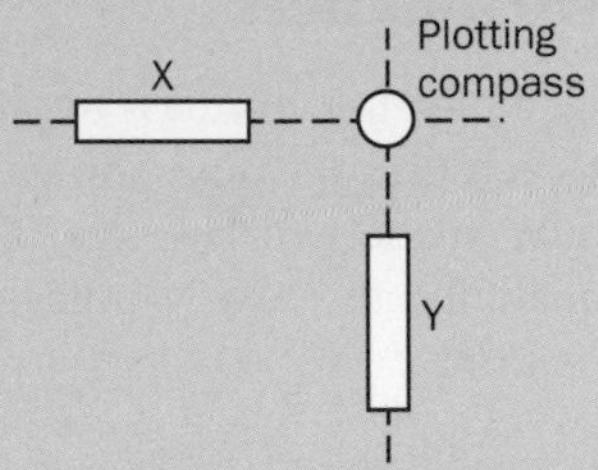

2. Sketch the pattern of lines of force of each arrangement shown in Fig. 18.7.

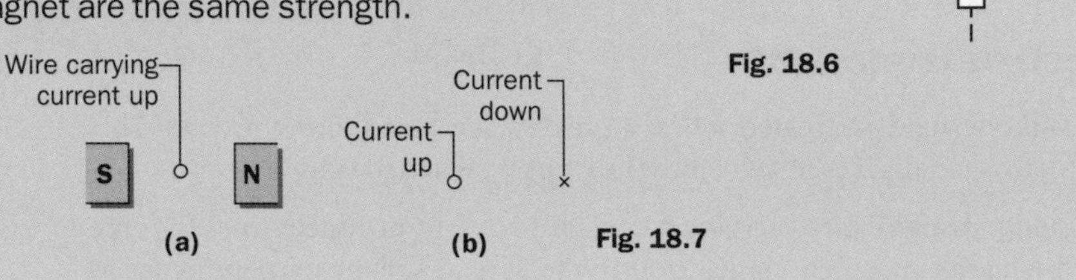

Fig. 18.7

Fig. 18.6

18.2 Electromagnets

An electromagnet is a coil of insulated wire wound around an iron core. When a current passes through the coil, the iron becomes magnetised and a very strong magnetic field is created. With the iron core, the magnetic field is about 2000 times stronger than without the iron core. In addition to its use in scrapyards to lift steel car bodies, an electromagnet is found in a variety of devices including the electric bell, the electric relay and the circuit breaker.

The electric bell

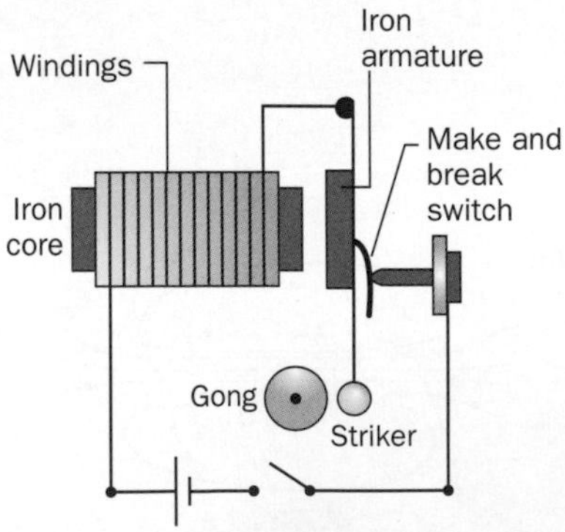

Fig. 18.8 shows the construction of an electric bell. When the 'push' switch is closed, current passes through the electromagnet windings and the electromagnet core is magnetised. The core attracts the iron armature which makes the striker hit the gong. However, the movement of the armature opens the 'make and break' switch which switches the electromagnet off. The iron armature springs back to its initial position, closing the 'make and break' switch to start the cycle again. The 'make and break' switch causes the striker to hit the bell repeatedly. An electric buzzer operates on the same principle as an electric bell except there is no striker or gong and the armature is much lighter so it moves to and fro much faster.

Fig. 18.8 The electric bell.

The electric relay

A relay is an electromagnet which opens and closes a switch, enabling a much larger current to be turned on or off by switching on or off a much smaller current through the electromagnet windings. Fig. 16.14 shows the construction of a relay. When current is passed through the electromagnet windings, the electromagnet core becomes magnetised and attracts the iron armature. The movement of the armature opens or closes the relay switch which is in a different circuit. The relay may be 'normally open' (NO) or 'normally closed' (NC), where 'normal' refers to the switch state when there is no current through the windings of the electromagnet.

The circuit breaker

- An **excess current circuit breaker** is a 'trip' switch opened by an electromagnet in the same circuit when the current through the windings exceed a certain value. Unlike a 'make and break' switch, a 'trip' switch is designed to stay open after it has been opened by the electromagnet. The trip switch is reset manually after the cause of excessive current has been removed.
- A **residual current circuit breaker** is a trip switch opened by an electromagnet which has two coils wound in opposite directions around its iron core. It is designed to disconnect an appliance from the mains supply if the current in the neutral wire and the live wire differ. This could happen if the insulation of the live wire deteriorates and current passes to earth instead of to the neutral wire. A residual current circuit breaker is designed to trip if the live and neutral currents differ by more than 30 mA.

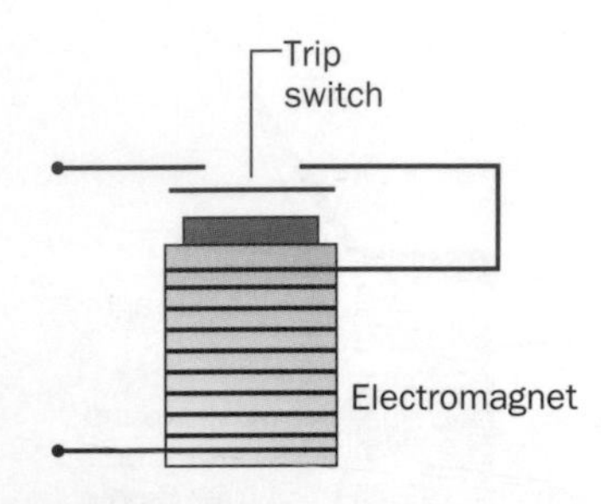

Fig. 18.9 A circuit breaker.

Questions 18.2

1. In an electric bell, give a reason why **(a)** the electromagnet core is iron, **(b)** the electromagnet windings are copper, **(c)** the make and break switch contacts are platinum.
2. **(a)** Explain how an excess current circuit breaker works.
   (b) State one advantage a circuit breaker has over a fuse.

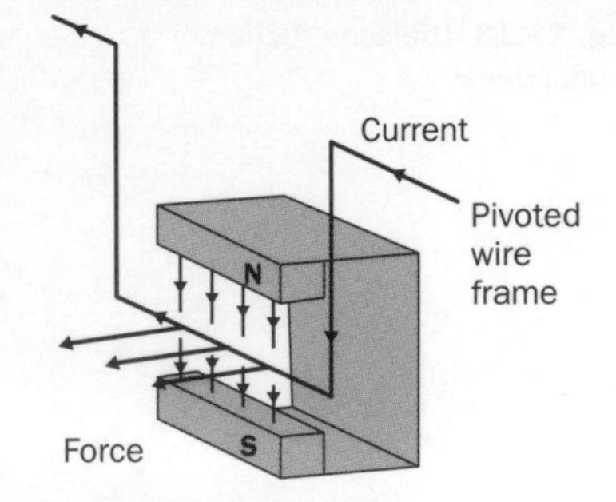

Fig. 18.10 The motor effect.

18.3 The motor effect

When a current is passed along a wire, a magnetic field is created around the wire. If the wire is between the poles of a U-shaped magnet, equal and opposite forces act between the wire and the magnet provided the wire is not parallel to the lines of force of the magnetic field. This effect, called the **motor effect**, can be readily demonstrated using the arrangement shown in Fig. 18.10. The force is greatest when the wire is at right angles to the lines of force of the magnet's field.

The direction of the force is at right angles to both the direction of the current and the direction of the lines of force of the magnetic field. The **left-hand rule** in Fig. 18.11 is a convenient way of working out the force direction from the other two directions. Note that if the direction of the current is reversed, the force direction is reversed provided the magnetic field is unchanged.

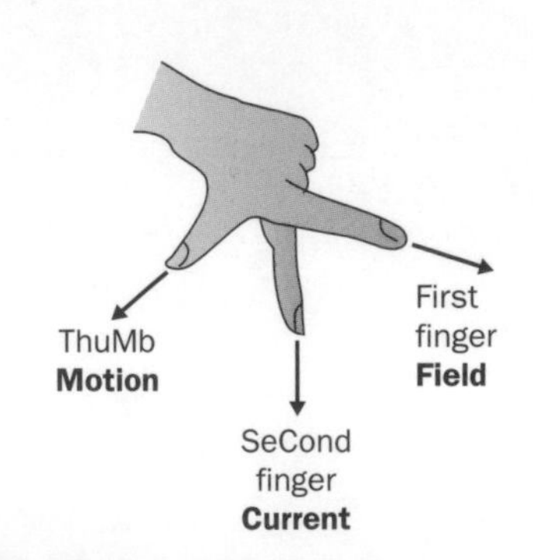

Fig. 18.11 The left hand rule.

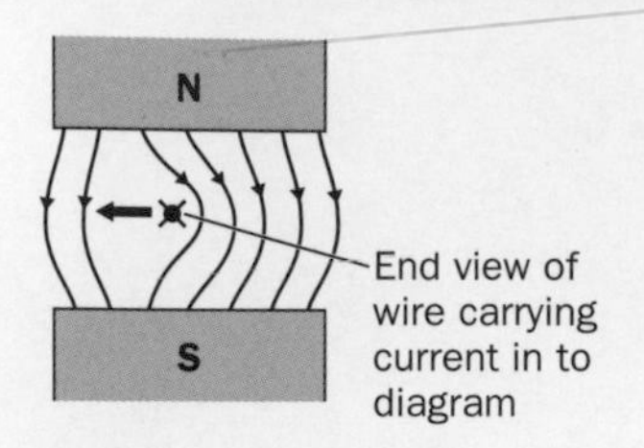

Fig. 18.12 Lines of force around a current-carrying wire between the poles of a U-shaped magnet.

Fig. 18.12 shows the magnetic field around a current-carrying wire between the poles of a U-shaped magnet. The magnetic field due to the wire and the magnetic field of the U-shaped magnet reinforce each other on one side of the wire and cancel each other on the other side. The force on the wire is towards the neutral point, which is where the two fields cancel.

The loudspeaker

When an alternating current is passed through the coil of a loudspeaker, the coil vibrates because it is in a magnetic field. In the first half of one full cycle of the alternating current, the current through the coil is in a certain direction so the wire is forced in a particular direction according to the left hand rule. When the current reverses its direction in the second half of the cycle, the force on the coil reverses its direction. Hence the coil experiences an alternating force which makes it vibrate, forcing the diaphragm to vibrate and create sound waves in the surrounding air. The coil consists of insulated wires wound on a plastic tube. The magnet's cross-section is shown in Fig. 18.13. Its poles are a central disc and an outer ring. The coil fits between the magnetic poles, oscillating freely when an alternating current is passed through the coil windings.

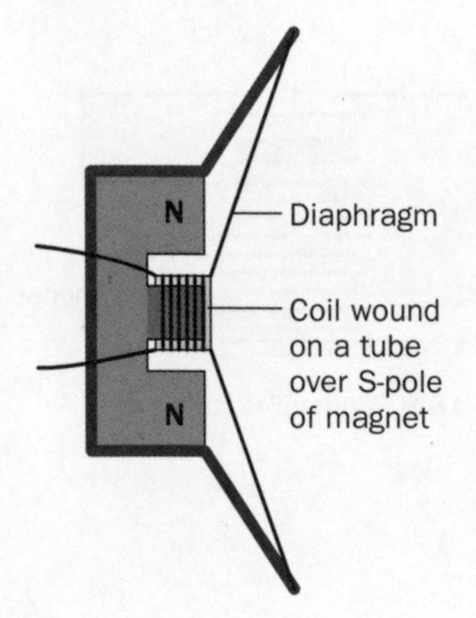

Fig. 18.13 The construction of a loudspeaker.

The direct current electric motor

Fig. 18.14 shows the construction of an electric motor. The armature coil spins between the poles of a U-shaped magnet when a direct current is passed through the windings of the armature coil. This occurs because current enters and leaves the coil via a **split-ring commutator** so each side of the coil carries a current at right angles to the lines of force of a uniform magnetic field. Because the two sides carry current in opposite directions, the force on one side is in the opposite direction to the force on the other side. The coil is therefore forced to turn.

When the coil reaches a position with its plane perpendicular to the

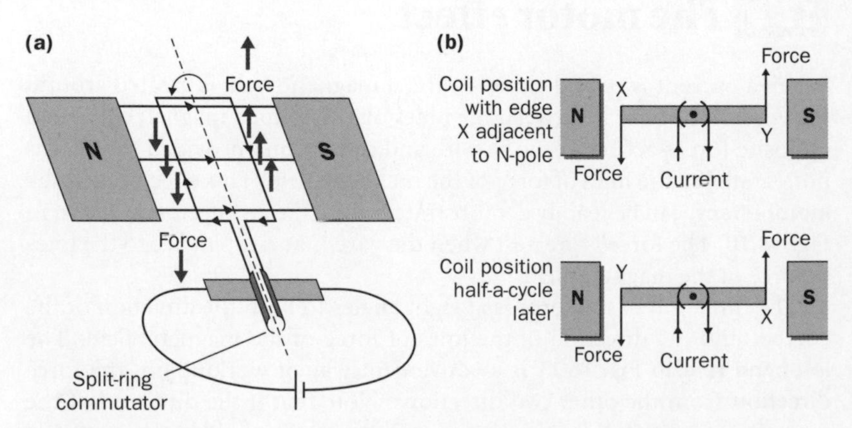

Fig. 18.14 An electric motor **(a)** construction and operation, **(b)** the action of the split-ring commutator.

lines of force, its spinning motion takes it past this position and reverses the connection of the split-ring commutator to the contact brushes. The current now passes around the coil in the reverse direction, but each side is now adjacent to the other pole to which it was adjacent previously. Consequently, the forces on the two sides continue to turn the coil in the same direction of rotation.

A practical d.c. electric motor has several identical coils wound on the same armature. The coils are evenly spaced, each coil connected to opposite sections of a multiple split-ring commutator. This design ensures that the armature spins evenly when a steady current passes through the coils.

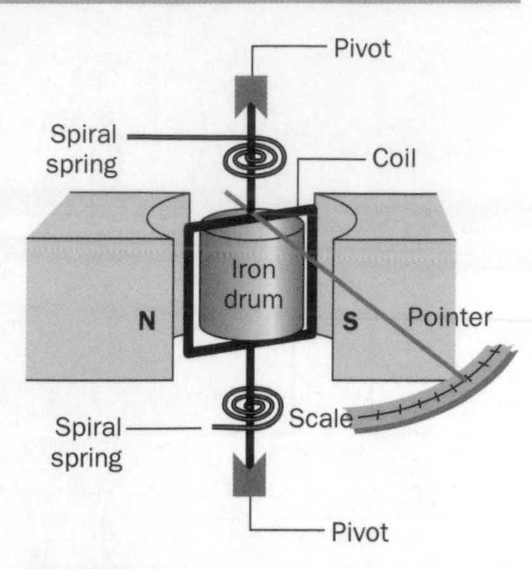

Fig. 18.15 The moving coil meter.

The moving coil meter

The construction of a moving coil meter is shown in Fig. 18.15. Current enters and leaves the coil via two spiral springs. The coil turns when current is passed through it because equal and opposite forces due to the magnetic field act on the sides of the coil. As the coil turns, the spiral springs tighten until the coil cannot turn anymore. The current is measured from the position on the scale of the pointer attached to the coil. When the current is switched off, the spiral springs return the coil to its initial position.

The scale is **linear** if the coil turns by a constant amount for equal increases of current. This occurs if the magnetic field is radial. This design feature is achieved by fitting an iron drum at the centre between curved pole pieces. The coil turns in the gap between the pole pieces and the iron drum. The radial field ensures the coil plane is always perpendicular to the lines of force, ensuring equal increases of current gives equal degrees of turning.

Deflection, $\theta = kI$

where I is the current and k is the deflection per unit current.

Questions 18.3

1. Fig. 18.16 is an end-view of the rotating coil of a d.c. electric motor.
   (a) In which direction is the coil rotating?
   (b) Explain why the contact brushes are connected to the coil via a split-ring commutator.
2. Two moving coil meters, X and Y, are identical except X has a weaker magnet than Y. The two meters are connected in series and a current is passed through them.
   (a) State and explain how the deflection of meter X compares with that of Y.
   (b) If the magnets of X and Y had been the same strength but X had a weaker spiral spring than Y, how would the deflection of X have compared with that of Y?

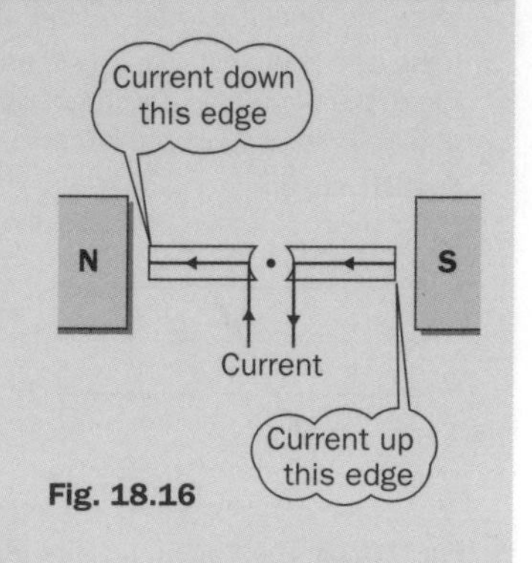

Fig. 18.16

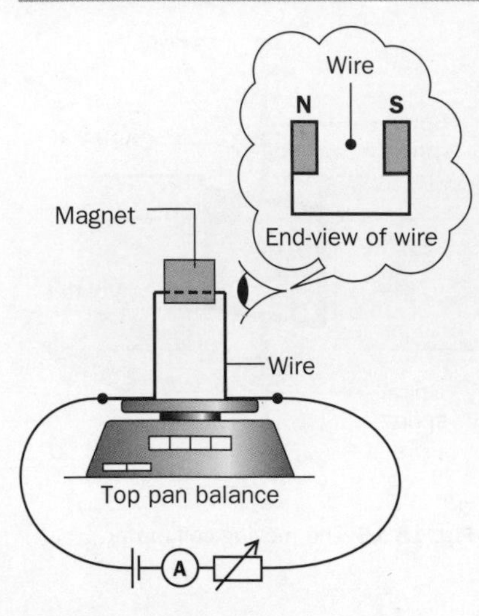

Fig. 18.17 Measuring the force on a current-carrying conductor.

18.4 Magnetic flux density

The force on a current-carrying conductor in a magnetic field is used to define the strength of the magnetic field. The stronger the magnetic field, the greater the force on a wire of fixed length carrying a constant current. The force can be measured directly by using a rigid wire fixed to a top pan balance, connected to the rest of the circuit via flexible leads, as in Fig. 18.17. The wire is positioned at right angles to the lines of force of the magnetic field. The force is measured from the change of the reading of the top-pan balance when the current is switched on. The measurements show that the force F is proportional to

1. the current, I
2. the length, L, of the wire in the field.

These results enable the strength of the magnetic field, the **magnetic flux density, B**, to be defined as the force per unit current per unit length on a wire placed perpendicular to the lines of force of the magnetic field.

Magnetic flux density, $B = \frac{F}{IL}$

The unit of B is the tesla (T), equal to 1 newton per ampere per metre. Rearranging the above equation therefore enables the force to be calculated if the current, length and magnetic flux density are known.

$F = BIL$

Notes

1. The direction of the force is given by the left-hand rule. See Fig. 18.11.
2. If the wire is parallel to the lines of force, no force is exerted on the wire.
3. If the wire is at angle θ to the lines of force, the perpendicular component of B (= B sin θ) gives the force.

$F = BIL \sin\theta$

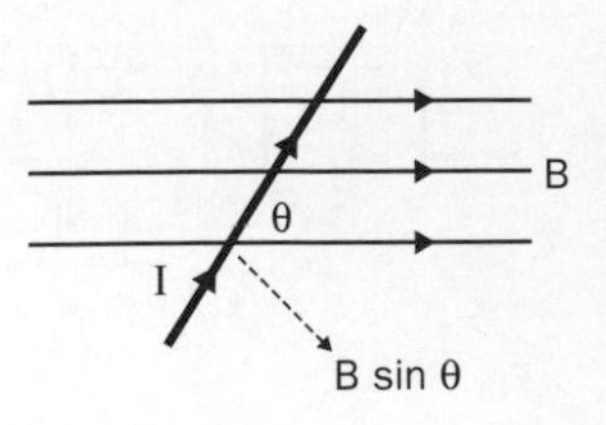

Fig. 18.18 The force on a wire at angle θ to the lines of force of a uniform magnetic field.

Worked example 18.1

A wire of length 0.040 m is placed perpendicular to a uniform magnetic field of magnetic flux density 0.30 T. Calculate (a) the force on the wire when the current is 5.0 A, (b) the current in the wire when the force is 0.050 N.

Solution

(a) $F = BIL = 0.30 \times 5.0 \times 0.040 = 0.060$ N

(b) Rearrange $F = BIL$ to give $I = \frac{F}{BL} = \frac{0.050}{0.30 \times 0.040} = 4.2$ A

The force on a moving charge in a uniform magnetic field

A charged particle moving across the lines of force of a uniform magnetic field is deflected by the field. A moving charged particle is effectively an electric current, since current is charge flow per second. The moving charged particle therefore experiences a force due to the magnetic field if its direction is not parallel to the lines of force of the magnetic field.

Consider a particle carrying charge Q, moving at constant speed, υ, in a direction perpendicular to the lines of force of a uniform magnetic field. In time t, it travels a length, $L = \upsilon t$ and is equivalent to a current $I = Q/t$ since current is charge flow per second. Hence the force due to the magnetic field, F, can be written as

$$F = BIL = B\frac{Q}{t}\upsilon t = BQ\upsilon$$

More generally, if the direction of motion of the particle is at angle θ to the lines of force,

$$\mathbf{F = BQ\upsilon \sin \theta}$$

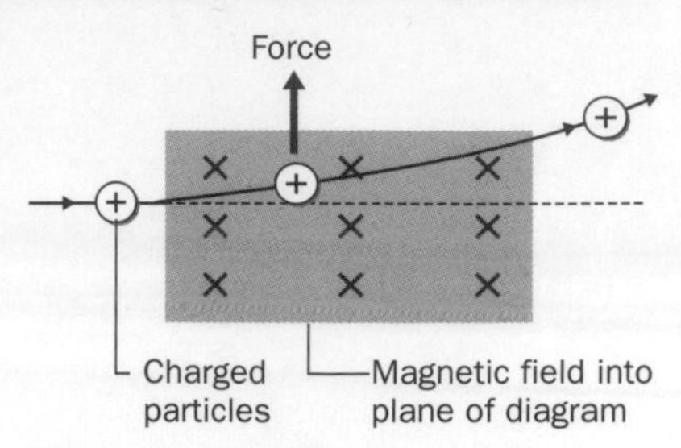

Fig. 18.19 The force on a moving charge in a magnetic field.

The Hall effect

When a current passes through a conductor or a semiconductor in a magnetic field, the charge-carrying particles are forced by the magnetic field to one face of the conductor. Consequently, a potential difference is created between this face and the opposite face. This effect is called the **Hall effect** after its discoverer. The p.d. between the two faces is called the **Hall voltage**.

See www.palgrave.com/foundations/breithaupt for an experiment to investigate the effect of a magnetic field on an electron beam.

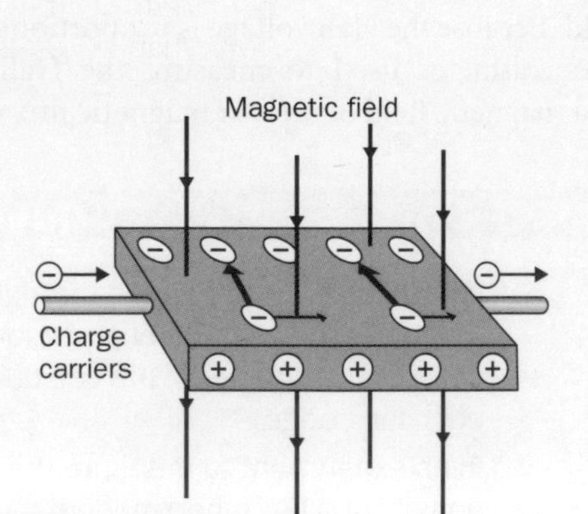

Fig. 18.20 Explanation of the Hall effect.

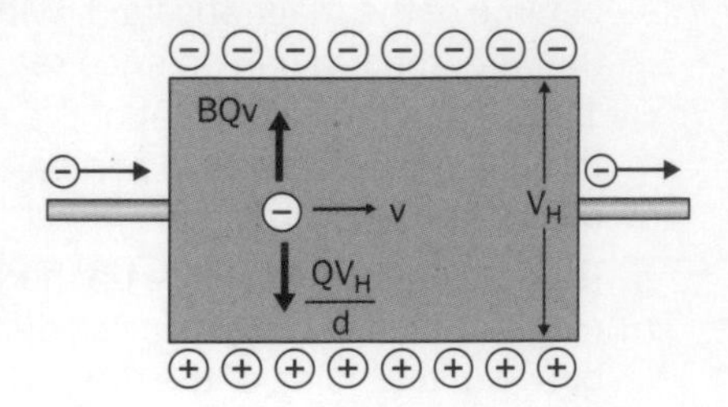

Fig. 18.21 The Hall voltage.

The magnetic field forces the charge carriers to one edge, causing an electric field to build up from one edge to the other edge which acts on the charge carriers against the magnetic force. Charge carriers therefore pass through the conductor without deflection because the magnetic force is balanced by the electric field force. For a carrier of charge Q, the electric field force $= QV_H/d$, where V_H is the Hall voltage and d is the distance between the two faces (see Fig. 18.21).

Since the magnetic field force $= BQ\upsilon$ where υ is the speed of the charge carriers through the conductor, then because the magnetic field force is equal and opposite to the electric field force

$$\frac{QV_H}{d} = BQ\upsilon$$

Hence $\mathbf{V_H = B\upsilon d}$

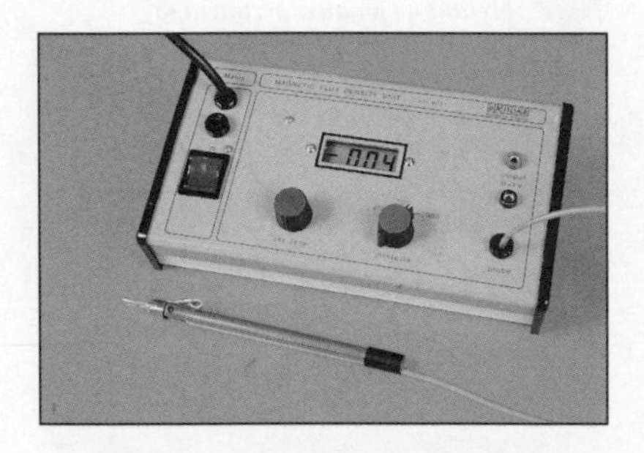

Fig. 18.22 A calibrated Hall probe.

The Hall voltage is therefore proportional to the magnetic flux density, B. The effect is much greater in a semiconductor than in a metal because the charge carriers move through a semiconductor much faster than through a metal for the same current. A Hall probe designed to measure magnetic flux density contains a small sample of a semiconductor through which a constant current is passed. The Hall voltage across opposite faces is therefore a measure of the magnetic flux density when the sample is placed in a magnetic field. Because the Hall voltage is proportional to the magnetic flux density, the voltmeter used to measure the Hall voltage can be calibrated using a magnetic field of known magnetic flux density.

Questions 18.4

1. **(a) (i)** Calculate the magnitude and direction of the force acting on a 5.2 cm vertical length of wire carrying a current of 3.2 A downwards in a uniform magnetic field of flux density 80 mT, acting horizontally from East to West.

   (ii) Calculate the current in a horizontal wire of length 40 mm aligned at right angles to a uniform horizontal magnetic field of flux density 250 mT with lines of force from west to east if the force on the wire is 15 mN vertically upwards. State the direction of the current.

   (b) A 30 mm × 40 mm rectangular coil consists of 80 turns of insulated wire wrapped on a plastic frame. The coil is placed in a uniform magnetic field of flux density 65 mT with its plane parallel to the lines of force, as in Fig. 18.23. A current of 7.2 A is passed through the coil. Calculate **(i)** the force on each long edge of the coil, **(ii)** the couple on the coil due to the force on each long edge.

   Uniform magnetic field — 7.2 A — 7.2 A

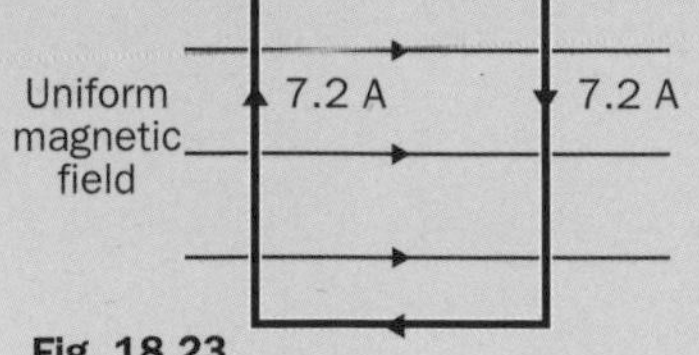

   Fig. 18.23

2. In an experiment to measure the magnetic flux density of a U-shaped magnet midway between its poles, the magnet was placed on a top pan balance. A stiff wire was held rigidly between the poles of the magnet at right angles to the lines of force of the magnetic field. With no current in the wire, the top pan balance reading was 105.38 g. When a current of 6.5 A was passed through the wire, the top pan balance reading decreased to 104.92 g.

   (a) Calculate **(i)** the force on the wire due to the magnet, **(ii)** the magnetic flux density midway between the poles of the magnet, if the length of the wire between the poles was 35 mm.

   (b) The wire was then turned horizontally through 30° with the current unchanged. Calculate the new reading of the top pan balance.

18.5 Magnetic field formulae

Investigating the magnetic field in a solenoid

When a solenoid carries a constant current, the lines of force of the magnetic field inside the solenoid are parallel to the axis of the solenoid. Fig. 18.24a shows how a Hall probe may be used to investigate the factors that determine the magnetic flux density in a 'slinky' solenoid.

1. The probe can be used to measure the magnetic flux density across the solenoid and along its axis. The results show that the magnetic flux density is unchanged across the solenoid and along its axis inside it, away from the ends. Fig. 18.24b shows how the magnetic flux density varies with position along the axis.
2. The variation of magnetic flux density with current, I, can be investigated by keeping the probe at a fixed position and measuring the magnetic flux density for different currents. The results may be plotted as a graph of magnetic flux density against current as in Fig. 18.24c. The results show that the magnetic flux density, B, is proportional to the current, I.
3. The magnetic flux density depends on the number of turns per unit length of the solenoid. For constant current, the magnetic flux density at a fixed position is measured with the coil stretched by different amounts. For each measurement, the number of turns, N, in a measured length, L, is counted. The results may be plotted as a graph of magnetic flux density, B, against the number of turns per unit length, N/L. The results show that the magnetic flux density is proportional to the number of turns per unit length.

The magnetic flux density, B, at and near the centre of a solenoid is proportional to

- the current, I,
- the number of turns per unit length, N/L

$$\therefore B = \text{constant} \times \frac{N}{L} \times I$$

The constant in the equation is referred to as the **permeability of free space**, μ_0. As explained on p. 263, the value of μ_o is $4\pi \times 10^{-7}$ tesla metres per ampere, as a result of how the ampere is defined.

$$\mathbf{B = \mu_0 nI}$$

where n = the number of turns per unit length, N/L.

> **Note**
> The magnetic flux density at the end of a solenoid is exactly half the value at or near the centre (i.e. $=0.5\ \mu_0 nI$).

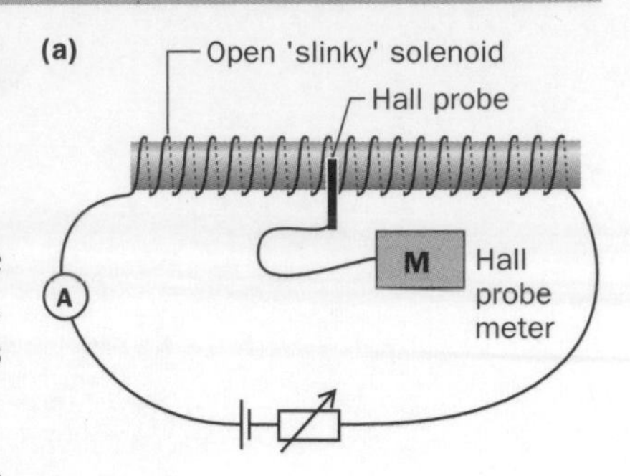

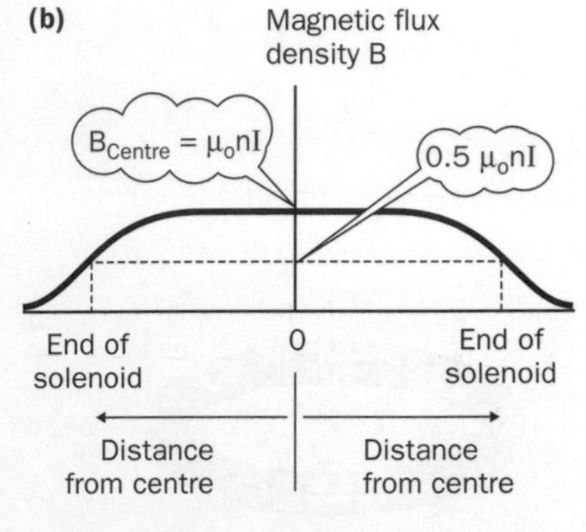

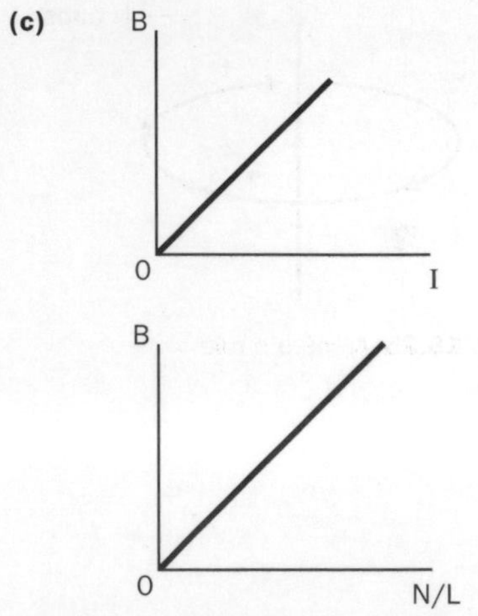

Fig. 18.24 Investigating a solenoid **(a)** using a Hall probe, **(b)** magnetic flux density at different positions, **(c)** factors affecting B.

Worked example 18.2

$\mu_0 = 4\pi \times 10^{-7}$ T m A^{-1}

A constant current of 4.0 A is passed through a solenoid of length 0.60 m and 1200 turns. Calculate (a) the magnetic flux density at its centre, (b) the current needed to produce a magnetic flux density of 250 mT at its centre.

Solution

(a) $B = \mu_0 nI = 4\pi \times 10^{-7} \times \frac{1200}{0.60} \times 4.0 = 0.010\,\text{T}$

(b) Rearrange $B = \mu_0 nI$ to give $I = \frac{B}{\mu_0 n} = \frac{0.25}{4\pi \times 10^{-7} \times (1200 \div 0.60)} = 100\,\text{A}$

Ampère's rule

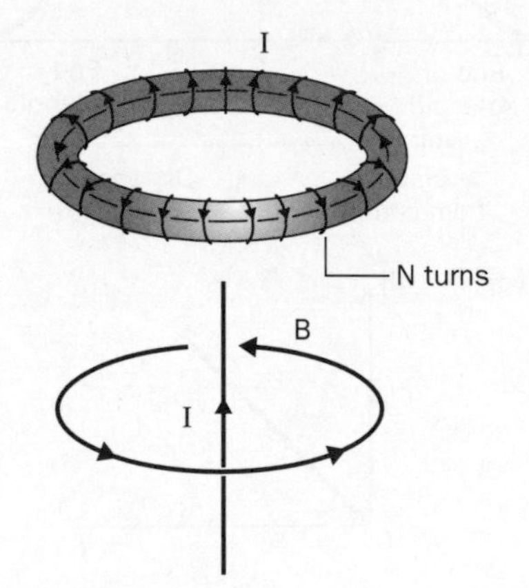

Fig. 18.25 Ampère's rule.

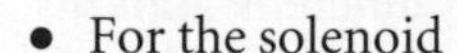

The lines of force of the magnetic field created by a current-carrying wire are circles centred on the wire in a plane perpendicular to the wire. See p. 253. A circular magnetic field can also be produced by joining the ends of a long solenoid together with the axis of the solenoid as a circle. Ampere recognised that these two fields could be described by a common formula because they were the same shape.

- For the solenoid

$$B = \frac{\mu_0 NI}{L} \text{ hence } BL = \mu_0 NI$$

- For a long straight wire, the length of the line of force once round a circle is equal to $2\pi r$, where r is the distance from the line of force to the centre of the wire. Also, only one wire emerges through the 'line of force' loop so $N = 1$. Hence $B \times 2\pi r = \mu_0 I$.

∴ at distance, r, from the centre of a long straight wire

$$B = \frac{\mu_0 I}{2\pi r}$$

The force between two parallel current-carrying wires

Two current-carrying wires interact with each other because each wire produces a magnetic field which acts on the other wire. Fig. 18.26 shows the magnetic field patterns around two wires carrying current in the same direction (in (a)) and in opposite directions (in (b)).

(a) When the two currents are in the same direction, the two wires attract. The magnetic field of either wire at the position of the other wire creates a force towards the first wire.

(b) When the two currents are in opposite directions, the two wires repel. The magnetic field of either wire at the position of the other wire creates a force away from the first wire. Let I_1 and I_2 represent the currents in the two wires. Hence the magnetic flux density, B_1, at distance, r, from the centre of the wire carrying current I_1 is given by

$$B_1 = \frac{\mu_0 I_1}{2\pi r}$$

The force, F, on length, L, of the wire carrying current I_2 due to magnetic flux density B_1, is given by $F = B_1 I_2 L$. Therefore the force per unit length, F/L, on this wire

$$= B_1 I_2 = \frac{\mu_0 I_1 I_2}{2\pi r}$$

The same expression is obtained for the force per unit length on the other wire.

$$\therefore \textbf{The force per unit length on either wire} = \frac{\mu_0 I_1 I_2}{2\pi r}$$

The definition of the ampere

One ampere is defined as that current which causes a force per unit length of 2.0×10^{-7} N m^{-1} in two infinitely long parallel wires of negligible thickness 1.0 m apart connected in series in a vacuum. Substituting $I_1 = I_2 =$ 1.0 A and $r = 1.0$ m into the formula for the force per unit length gives

$$\frac{\mu_0}{2\pi} = 2.0 \times 10^{-7}\ \text{N m}^{-1}\ \text{m A}^{-2}$$

Hence

$$\mu_0 = 4\pi \times 10^{-7}\ \text{T m A}^{-1}$$

The value of μ_0 therefore follows from the definition of the ampere.

Worked example 18.3

$\mu_0 = 4\pi \times 10^{-7}$ T m A^{-1}
The magnetic flux density due to a current-carrying wire at a distance of 40 mm from the wire is 84 μT. Calculate (a) the current in this wire, (b) the force per unit length on a parallel wire carrying a current of 5.0 A at a perpendicular distance of 20 mm from the first wire.

Solution

(a) Rearrange $B = \frac{\mu_0 I}{2\pi r}$ to give $I = \frac{2\pi r B}{\mu_0} = \frac{2\pi \times 40 \times 10^{-3} \times 8.4 \times 10^{-5}}{4\pi \times 10^{-7}}$

$= 16.8$ A

(b) At 20 mm from the first wire, $B = \frac{\mu_0 I}{2\pi r} = \frac{2.0 \times 10^{-7} \times 16.8}{0.020}$

$= 1.68 \times 10^{-4}$ T

$\therefore \frac{F}{L} = BI_2 = 1.68 \times 10^{-4} \times 5.0 = 8.4 \times 10^{-4}$ N m^{-1}

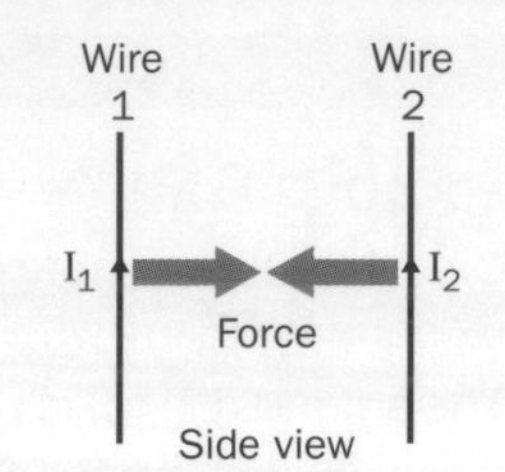

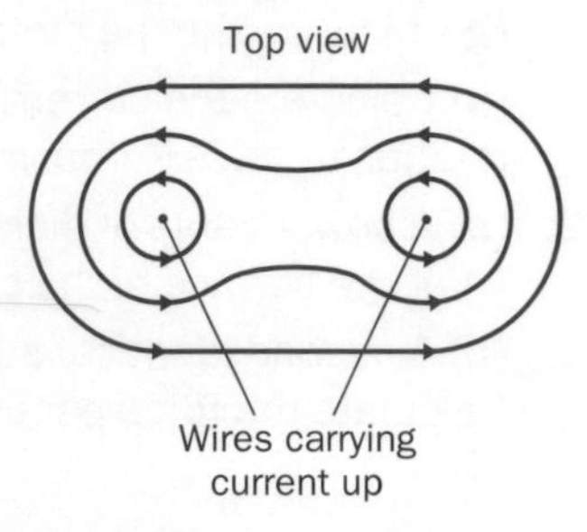

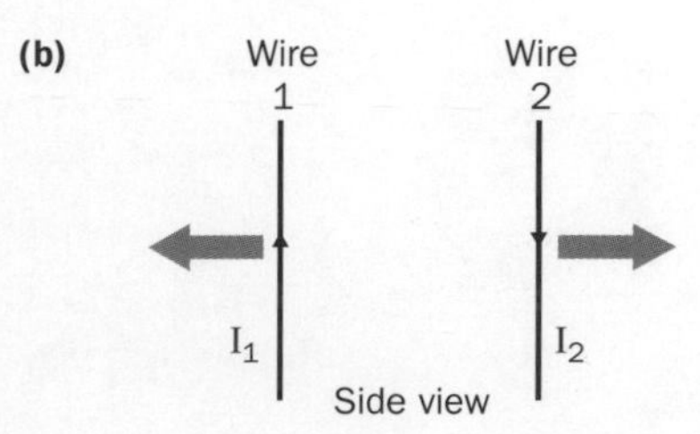

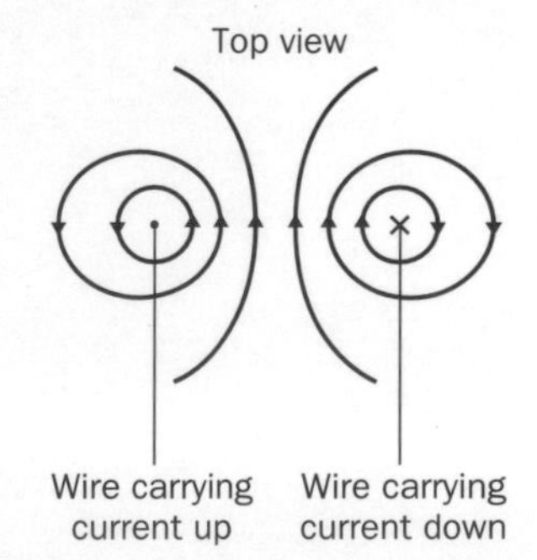

Fig. 18.26 Interaction between parallel wires **(a)** like currents attract, **(b)** unlike currents repel.

Questions 18.5

$\mu_0 = 4\pi \times 10^{-7}$ T m A^{-1}

1. **(a)** Calculate the number of turns needed for a solenoid of length 800 mm to produce a magnetic field of flux density 25 mT when the current is 8.0 A.

(b) The solenoid in (a) is placed with its internal magnetic field parallel to the lines of force of a uniform magnetic field of flux density 50 mT.

(i) Calculate the magnetic flux density in the solenoid when the current is 8.0 A,

(ii) Calculate the current needed to produce zero field in the solenoid and indicate its direction in relation to the external magnetic field.

2. **(a)** A power cable of diameter 25 mm carries a current of 1000 A. Calculate the magnetic flux density **(i)** at the surface of the cable, **(ii)** at a distance of 10.0 m from the cable.

(b) A second identical power cable is parallel to the first cable at a perpendicular distance of 10.0 m. Calculate the force per unit length between the two cables when they are each carrying a current of 1000 A.

18.6 Magnetic materials

Ferromagnetism

Materials such as iron, steel, cobalt and nickel can be permanently magnetised. Such materials are described as **ferromagnetic** because they retain their magnetism. The cause of ferromagnetism is the tiny magnetic fields created by electrons orbiting the nucleus of each atom. Each electron in its orbit is equivalent to an electric current and therefore creates a magnetic field. In ferromagnetic atoms, the electrons in each atom produce a resultant magnetic field; in all other types of atoms, the electrons produce zero resultant magnetic field. Ferromagnetic atoms are therefore like tiny magnets. When a bar of iron is placed in a direct current solenoid, the atomic magnets line up along the solenoid field, all pointing in the same direction. The result is a very strong magnetic field. When the solenoid current is switched off, the atomic magnets remain partly lined up.

Investigating ferromagnetism

Fig. 18.27 shows a bar of ferromagnetic material in a solenoid which is in series with a resistor, R, and an alternating current supply unit. The X-plates of an oscilloscope are connected across the resistor so the alternating voltage across the resistor makes the oscilloscope spot oscillate horizontally. Since the current is proportional to the voltage for a resistor, the horizontal position of the spot on the screen represents current in the circuit.

The end of the bar is positioned adjacent to the screen so the magnetic field of the bar deflects the electron beam either vertically up or vertically down.The magnetic field alternates at the same frequency as the alternating current through the solenoid. The vertical position of the spot represents the magnetisation of the bar. The oscilloscope tube must be unscreened if the magnetic field of the bar is to affect the beam.

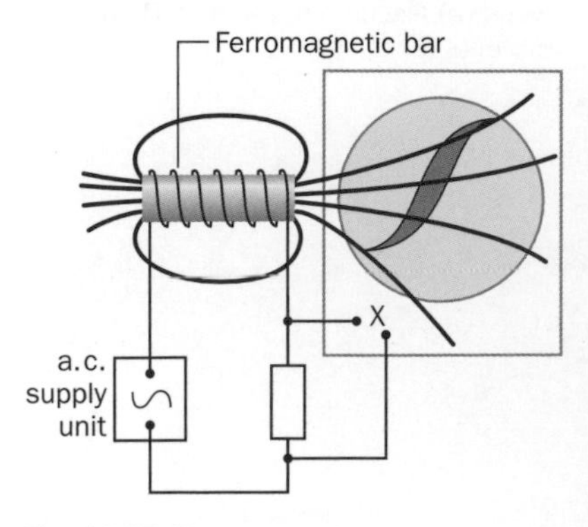

Fig. 18.27 Ferromagnetism.

The beam traces a loop on the screen as shown in Fig. 18.28. This loop is called a **hysteresis** loop. The magnetisation lags behind the current, changing direction after the current has changed direction, not at the same time. All ferromagnetic materials display the following features:

- **Saturation** occurs when the magnetisation reaches a maximum value which cannot be increased for that material, no matter how large the current. The atomic magnets are all aligned when the material is magnetically saturated.
- **Remanence** occurs where the current is zero and the material remains magnetised. The atomic magnets remain partly aligned.
- The **loop area** is a measure of the energy needed to magnetise and demagnetise the same in each cycle. Materials such as steel have a large loop area and are hard to magnetise and demagnetise. They are therefore described as magnetically **hard**. Materials such as iron have a much smaller loop area because they are much easier to magnetise. Such materials are described as magnetically **soft**.

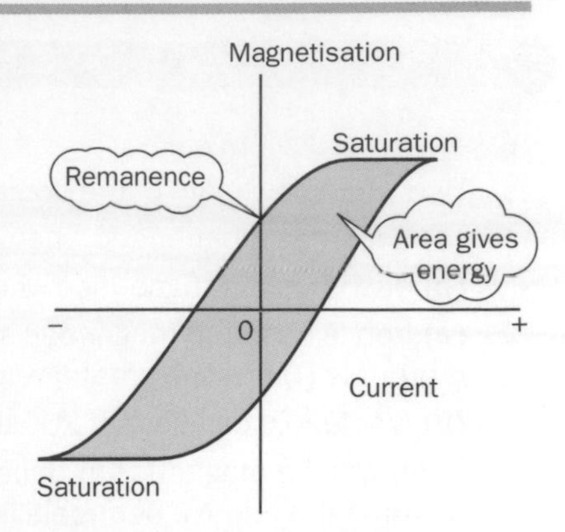

Fig. 18.28 A hysteresis loop.

The presence of a ferromagnet in a solenoid increases the magnetic flux density considerably. The greatest effect is in a toroidal (i.e. circular) solenoid which is completely filled with the material. This arrangement can be used to defined the **relative permeability**, μ_r, of the material as

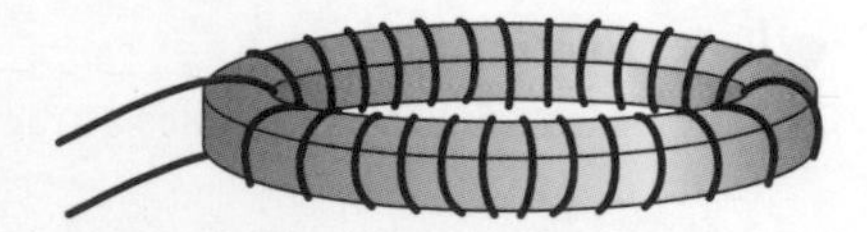

Fig. 18.29 A toroidal solenoid.

$$\frac{\text{the magnetic flux density with the material}}{\text{the magnetic flux density without the material}}$$

provided the material does not become magnetically saturated. Hence the magnetic flux density, B, in a toroidal solenoid with n turns per unit length wrapped around a ferromagnetic core

$$\mathbf{B = \mu_r \mu_0 n I}$$

For example, suppose the magnetic flux density in a solenoid is 0.10 mT in the absence of ferromagnetic material and 0.10 T with the unsaturated material present. Hence the relative permeability of the material is 1000 (= 0.10T ÷ 0.10 mT). However, if the material saturates at 0.15 T, then doubling the solenoid current would not double the field with the material present.

The domain theory of ferromagnetism

The atoms of a ferromagnetic material affect each other because they interact magnetically. When the material is unmagnetised, the atoms align in small regions called **domains**. The direction of magnetisation of the domains differs from one domain to the next in a random way. The result is that the overall magnetisation of the whole sample is zero. When the material is in an external magnetic field, the domains align and merge with each other to produce an overall magnetisation. When the external field is removed, domains are recreated. However, the material retains some magnetism because the domains remain partly aligned.

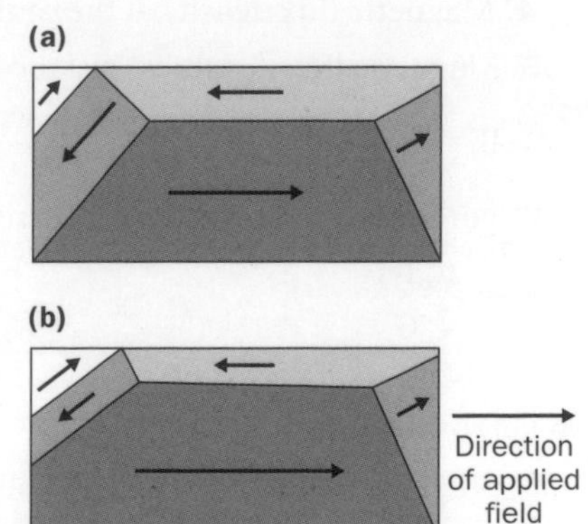

Fig. 18.30 Domains **(a)** unmagnetised, **(b)** partly magnetised.

Questions 18.6

$\mu_o = 4\pi \times 10^{-7}$ T m A^{-1}

1. Use your knowledge of the domain theory to explain why a ferromagnet
   (a) becomes magnetised when it is in a direct current solenoid,
   (b) becomes demagnetised when it is slowly withdrawn from an alternating current solenoid.
2. **(a)** Iron is easier to magnetise and demagnetise than steel. Which of these two ferromagnetic materials **(i)** has the greater remanence, **(ii)** has a larger hysteresis loop?
   (b) A solenoid consists of 200 turns of insulated wire wound around an iron ring of diameter 40 mm. Estimate the magnetic flux density in the ring when a current of 50 mA is passed through the wire. Assume the relative permeability of iron is 2000.

Summary

◆ Rules to remember

Like poles repel, unlike poles attract.

Lines of force:

- of a permanent magnet are from the north pole to the south pole
- around a long, straight, current-carrying wire are circles; see the corkscrew rule, Fig. 18.4
- in a direct current solenoid are parallel to the axis, emerging from one end and looping around to the other end. See the solenoid rule Fig. 18.5.

The left hand rule:

First finger = Field; seCond finger = Current; thuMb = Motion.

◆ Devices

Electromagnets are used in the electric bell, the relay and the circuit breaker.

The motor effect is used in the moving coil meter, the d.c. electric motor and the loudspeaker.

◆ Magnetic field formulae

1. Force on a current-carrying conductor, $F = BIL \sin\theta$
2. Force on a moving charge, $F = BQv \sin\theta$
3. Hall voltage, $V_H = Bvd$
4. Magnetic flux density at the centre of a solenoid, $B = \mu_o nI$, where n = number of turns per unit length
5. Magnetic flux density at distance r from the centre of a long straight current-carrying wire
   $$B = \frac{\mu_o I}{2\pi r}$$
6. Force between two parallel current carrying wires
   $$= \frac{\mu_o I_1 I_2}{2\pi r}$$
   where r = distance apart.

◆ Ferromagnetism

- Saturation occurs when the magnetisation of a ferromagnetic material is at maximum.
- Remanence is where a ferromagnet remains magnetised after the magnetising current has been switched off.
- The area of a hysteresis loop is a measure of the energy needed to magnetise and demagnetise a ferromagnetic material.

Revision questions

18.1. (a) Copy and complete the magnetic field patterns in Fig. 18.31.

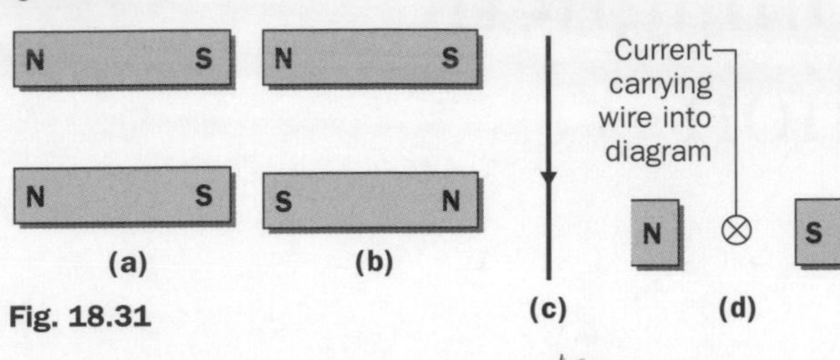

Fig. 18.31

(b) An iron nail can be magnetised by stroking it with the end of a bar magnet, as in Fig. 18.32. Use your knowledge of the domain theory of ferromagnetism to explain why **(i)** the iron nail becomes magnetised, **(ii)** the polarity of the iron nail is as shown in Fig. 18.32.

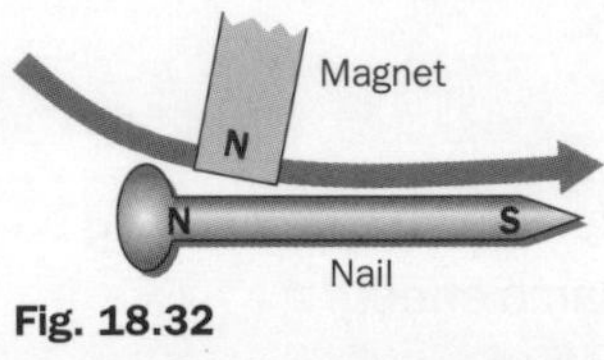

Fig. 18.32

18.2. (a) With the aid of a diagram, describe the construction and operation of a direct current electric motor.

(b) A direct current electric motor contains a rectangular armature coil 40 mm long by 30 mm wide, consisting of 200 turns of insulated wire. The coil spins in a magnetic field of flux density 120 mT about an axis parallel and midway between its long edges. Calculate the force on each long edge of the coil when the coil current is 5.0 A and the coil plane is **(i)** parallel to the lines of force of the field, **(ii)** perpendicular to the lines of force of the field.

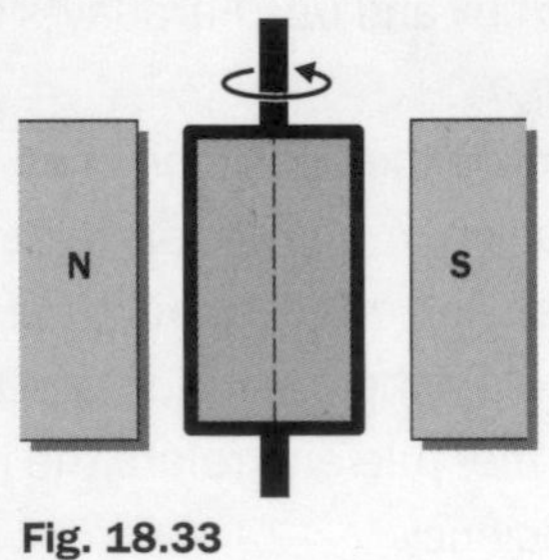

Fig. 18.33

18.3. A narrow beam of electrons at a speed of 2.8×10^7 m s^{-1} is directed into a uniform electric field between two oppositely charged parallel plates 50 mm apart. When the top plate is at a positive potential of 3500 V relative to the lower plate, the beam curves towards the top plate, as shown in Fig. 18.34. The deflection of the beam is cancelled out as a result of a uniform magnetic field applied perpendicular to the diagram.

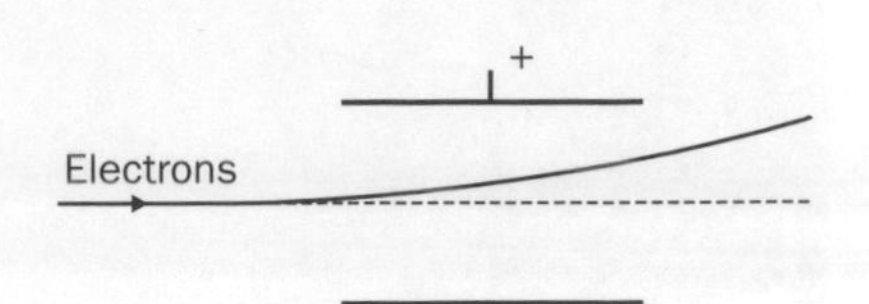

Fig. 18.34

(a) (i) State the direction of the magnetic field.

(ii) By considering the forces on each electron, show that the magnetic flux density

$$B = \frac{V}{\upsilon d}$$

where υ is the speed of the electrons in the beam, d is the spacing between the plates and V is the plate p.d.

(iii) Calculate the magnetic flux density of this magnetic field.

(b) The magnetic flux density is doubled. Calculate the potential difference between the plates necessary to make the beam pass through undeflected.

18.4. (a) A current of 6.5 A was passed through a 500 turn open solenoid of length 250 mm. Calculate the magnetic flux density **(i)** at the centre of this solenoid, **(ii)** at the end of this solenoid.

(b) The solenoid was aligned north-south with its axis parallel to the lines of force of the Earth's magnetic field, as shown in Fig. 18.35, at a location where the Earth's magnetic flux density was 60 μT. Calculate the current that needs to be passed through this solenoid to create zero magnetic flux density in its interior.

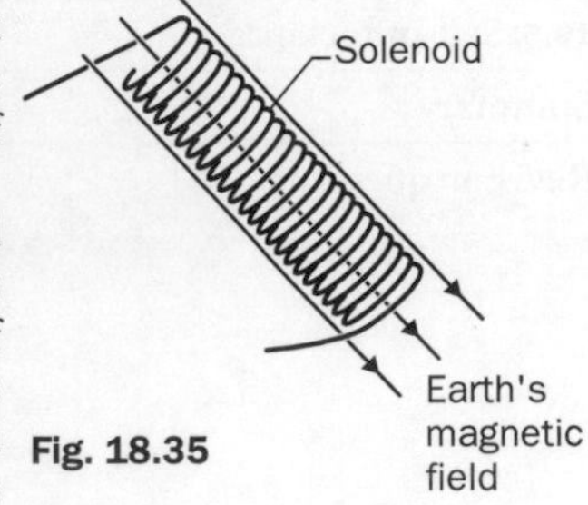

Fig. 18.35

18.5. A rectangular 50-turn coil of dimensions 60 mm × 40 mm is placed with its long edges parallel to a long, straight, vertical wire at a distance of 40 mm between the wire and the nearest edge of the coil. A current of 8.5 A is passed vertically up the wire. When the coil current is 2.0 A as shown in Fig. 18.36, calculate

(a) the force on each long edge of the coil,

(b) the magnitude and direction of the resultant force on the coil due to the wire.

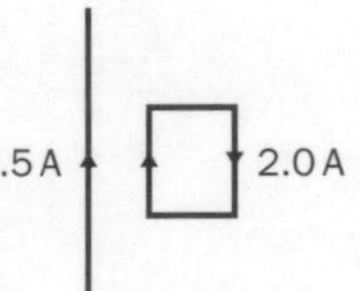

Fig. 18.36

Electromagnetic Induction

Contents

Objectives

After working through this unit, you should be able to:

- describe the dynamo effect
- explain the cause of an induced voltage in a conductor cutting the lines of force of a magnetic field
- state Lenz's law and relate the direction of an induced current to the change which causes it
- define magnetic flux and use Faraday's law to relate change of flux to induced voltage
- describe and explain the operation of an alternating current generator
- explain why a back e.m.f. is generated in an electric motor
- explain the cause and effect of eddy currents
- describe and explain the operation of a transformer
- use the transformer rule and relate the input power and the output power to the efficiency
- describe the main causes of inefficiency in a transformer
- explain why electrical power is transmitted using alternating currents at high voltages
- describe and explain self-inductance

19.1 The dynamo effect

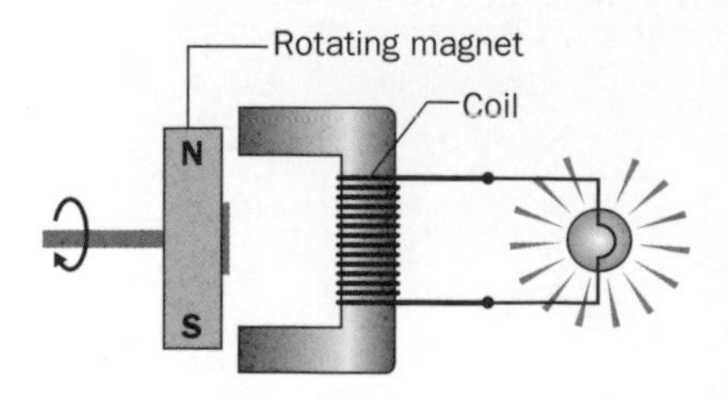

Fig. 19.1 Inside a dynamo.

A cyclist at night has to pedal harder to keep supplying electricity to the cycle lamps to light the way ahead and to be seen from behind, if the electricity for the lamps is supplied from a dynamo. The electricity from a dynamo is free, unlike that from a battery which needs to be replaced when it runs down. Some of the work done by the cyclist is used to turn the dynamo in order to generate electricity. Fig. 19.1 shows the internal construction of a simple dynamo. The magnet is forced to spin past the coil, causing the lines of force of the magnetic field to sweep past the coil windings. The effect is to induce an e.m.f. which forces current round the circuit through the lamp. The faster the cyclist pedals, the brighter the lamps are as more work is done each second by the cyclist.

Investigating the dynamo effect

Connect a length of insulated wire to a microammeter and observe the deflection of the meter when a section of the wire is passed between the poles of a U-shaped magnet. No deflection is produced if the wire is stationary in the field or if the wire is moved along the lines of force. The wire must cut across the lines of force of the magnetic field to produce a deflection. It does not matter if the magnet is moving and the wire is stationary or if the wire is moving and the magnet is stationary. As long as the wire cuts the lines of force of the magnetic field, an e.m.f. is induced in the wire.

Note the difference when the wire is moved in the opposite direction or when it is moved at a different speed. You should find that the direction of the induced current depends on the direction the wire is moved through the field. Also, the faster the wire is moved through the field, the greater the deflection.

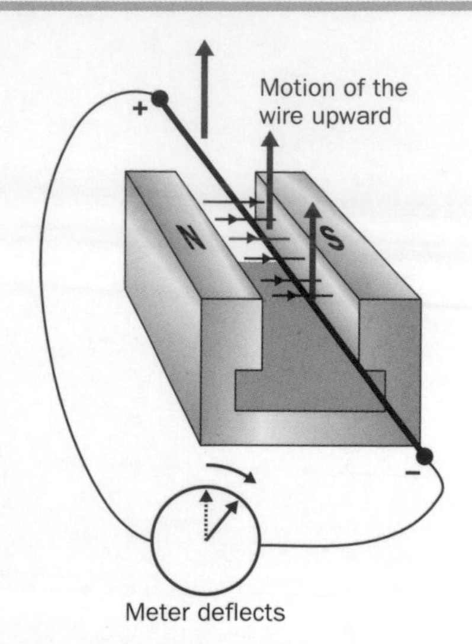

Fig. 19.2 Testing the dynamo effect.

To explain why an e.m.f. is induced in a wire when it cuts the lines of force of a magnetic field, consider Fig. 19.3 in which a straight wire moves at constant speed through a magnetic field at right angles to the lines of force of the field. The conduction electrons move at the speed of the wire inside the wire across the field. Just like electrons in a beam, the magnetic field forces the electrons in the wire down the wire. Consequently, the top of the wire is at positive potential and the bottom is at negative potential. If the wire is part of a complete circuit, the electrons are forced around the circuit just the same as if the wire had been replaced by a battery with its positive terminal at the top and its negative terminal at the bottom. Note that an e.m.f. is induced in the wire regardless of whether or not the wire is part of a complete circuit. The induced e.m.f. produces a current only if the wire is part of a complete circuit.

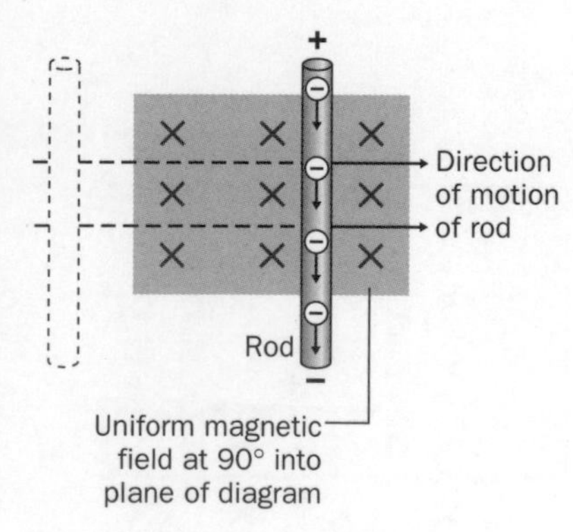

Fig. 19.3 Explaining the dynamo effect.

Lenz's law

The direction of the induced current in a circuit is always such as to oppose the change that causes it. This is known as Lenz's law and it can be demonstrated using the arrangement in Fig. 19.4. When the north pole of a bar magnet is pushed into the coil, the meter connected to the coil

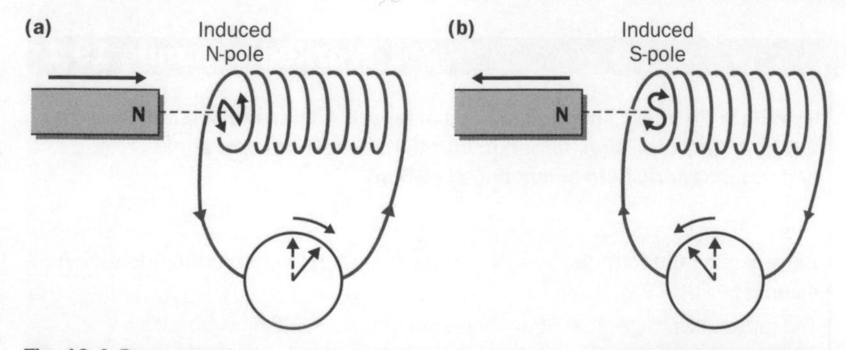

Fig. 19.4 Demonstrating Lenz's law **(a)** inserting an N-pole, **(b)** withdrawing an N-pole.

deflects. The current direction in the coil is such as to generate a magnetic field that opposes the incoming north pole of the bar magnet. A north pole is therefore induced at the end of the coil where the bar magnet is inserted. When the north pole of the bar magnet is withdrawn, a south pole is induced at this end of the coil, which makes the meter deflect in the opposite direction.

Lenz's law follows from the conservation of energy. Work must be done to generate an induced current, hence a force must be applied to do the necessary work. The induced current in the conductor in the magnetic field creates a reaction force due to the motor effect, which acts against the applied force. The force created by the induced current must be in the opposite direction to the applied force otherwise it would increase the speed of the conductor, which would generate more current, which would increase the speed further, etc. thus creating electrical energy and kinetic energy from nothing. Clearly, the principle of conservation of energy requires that the reaction force must be in the opposite direction to the applied force so that work done by the applied force is used to overcome the reaction force and create electrical energy.

The e.m.f. induced in a rod moving across a uniform magnetic field

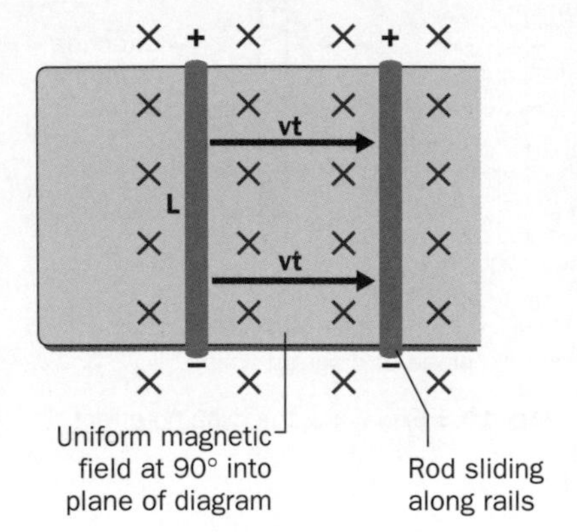

Fig. 19.5 The e.m.f. induced in a moving rod.

Fig. 19.5 shows a conducting rod of length, L, moving at speed, v, in a direction at right angles to the lines of force of a uniform magnetic field. If the rod is part of a complete circuit, the induced voltage, V, drives a current, I, around the circuit. A reaction force, F, acts on the rod due to this current.

- The electrical energy generated in time t = IVt
- The work done, W, to overcome the reaction force = Fvt, since the rod moves distance vt in time t.

Since F = BIL, where B is the magnetic flux density, then the work done, W = BILvt

According to the principle of conservation of energy

$IVt = BILvt$

$\therefore$ the induced voltage, **V = BLv**

Worked example 19.1

Calculate the e.m.f. induced in a rod of length 40 mm moving at a speed of 2.8 m s^{-1} in a direction perpendicular to its length at right angles to a uniform magnetic field of flux density 85 mT.

Solution

Before using the formula, V = BLv, make a sketch to ensure the question is clear. See Fig. 19.3.

The induced voltage, $V = BLv = 0.085 \times 0.040 \times 2.8 = 9.5 \times 10^{-3}$ V

Questions 19.1

1. Fig. 19.6 shows several examples of how an e.m.f. can be generated by moving a magnet relative to a wire or a coil of wire. In each case, state the polarity of the induced e.m.f. and the direction of the induced current.

(a)
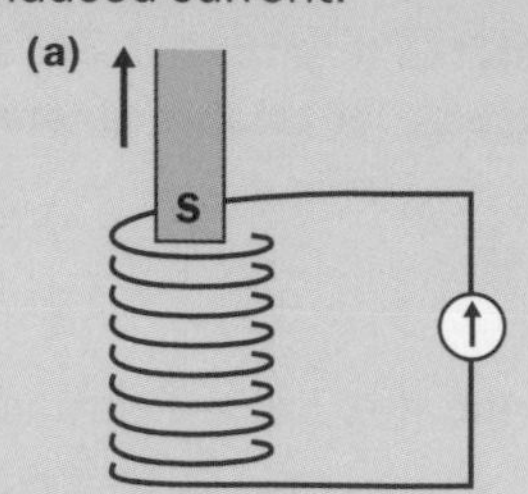

(b)
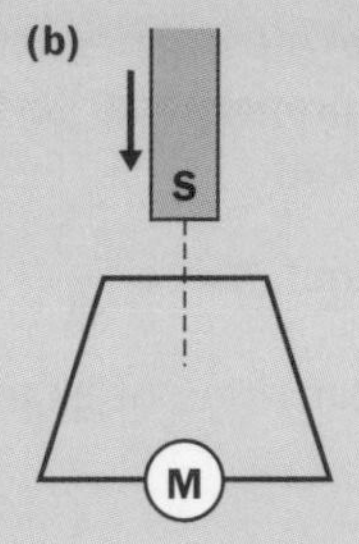

(c)
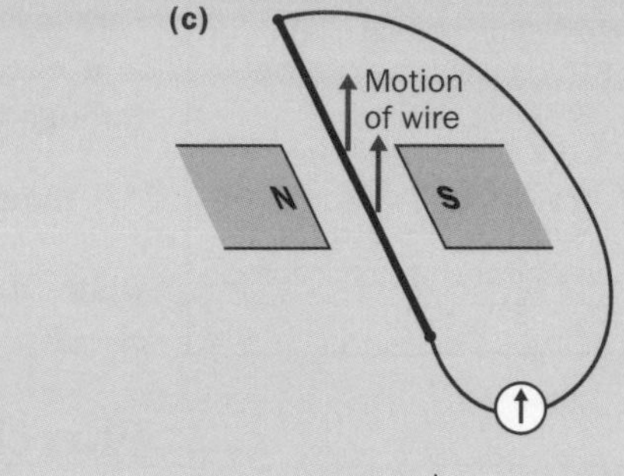

Fig. 19.6

2. An aircraft with a wing span of 22.0 m is in level flight, heading due north at a constant speed of 180 m s^{-1}. Calculate the e.m.f. induced between its wing tips due to the Earth's magnetic field, which is 60μT in a direction due North at 70° to the horizontal at the aircraft's location.

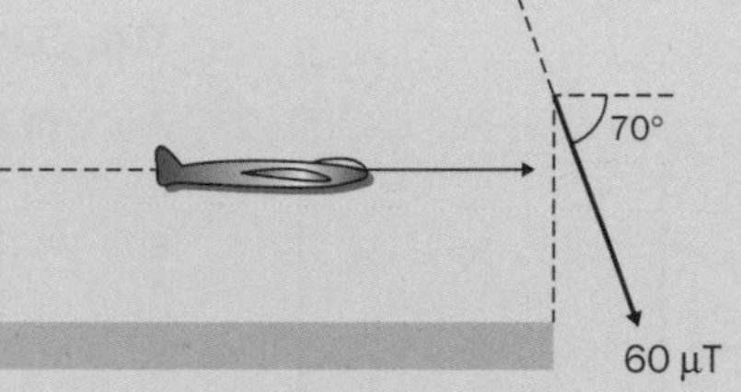

Fig. 19.7

19.2 Magnetic flux

The magnetic flux density of a U-shaped magnet can be increased by making the poles tapered. The magnet's field is concentrated between the poles as a result of using narrower poles. The lines of force are sometimes referred to as lines of 'flux', hence the term 'flux density' is a measure of how concentrated the lines are. Although, technically, magnetic flux density, B, is defined in terms of the force on a current-carrying wire, the term 'magnetic flux density' is used in preference to 'magnetic field strength' because it is easier to visualise lines of force. Conceptually, lines of force are no more than a useful way of picturing a magnetic field and should not be seen as the mechanism by which magnetic fields act on objects.

Consider again a moving rod of length, L, cutting through the lines of force of a uniform magnetic field, as in Fig. 19.8.

As explained on p. 270, the induced e.m.f is equal to BLv. The rod sweeps out an area equal to its length, L, × its speed, v, each second because it moves distance v each second. By defining magnetic flux, ϕ (pronounced 'phi') as the magnetic flux density, B × the area, A, swept out, the rod can be thought of as sweeping out magnetic flux as it cuts through the lines of force.

Magnetic flux, $\phi = BA$

through an area of cross-section, A, at right angles to a uniform magnetic field of flux density, B. The unit of magnetic flux is the weber (Wb), equal to 1 T m^2.

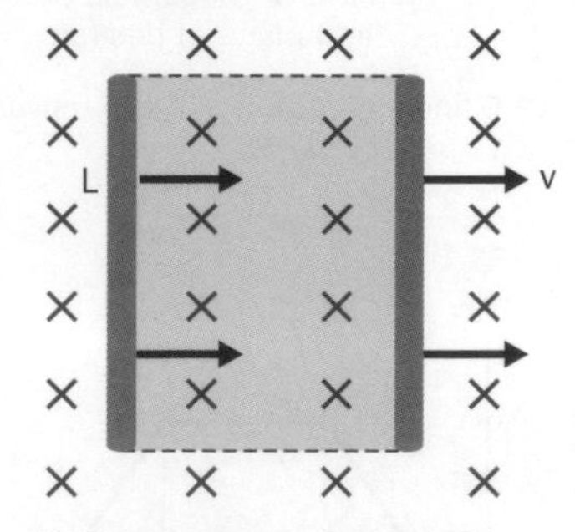

Fig. 19.8 Sweeping out magnetic flux.

Note

The flux linkage (or total magnetic flux) through a coil of area, A and N turns, at right angles to a uniform magnetic field of flux density, B, is defined as BAN. The symbol Φ (capital phi) is used for flux linkage. Hence $\Phi = BAN = N\phi$.

From the above definition of magnetic flux, it follows that the induced e.m.f. (= $BL\upsilon$) is equal in magnitude to the magnetic flux swept out per second. This relationship is known as Faraday's law of electromagnetic induction after Michael Faraday, who discovered how to generate electricity and established the theory of electromagnetic induction.

Notes

1. The constant of proportionality in Faraday's law is unity as a result of the definition of $\Phi = BAN$.
2. Lenz's law is incorporated in the equation by including the minus sign. This means that the induced e.m.f. acts against the change that causes it.

Faraday's law of electromagnetic induction states that the induced e.m.f. in a circuit is proportional to the rate of change of total magnetic flux through the circuit.

$$\textbf{Induced e.m.f, } V = -\frac{d\Phi}{dt}$$

where d/dt means rate of change and Φ = the flux linkage through the circuit.

Flux changes through a coil moved across a uniform magnetic field

An e.m.f. is induced in a conductor whenever it cuts across the lines of force of a magnetic field. In Fig. 19.9, a rectangular 1-turn coil of length, L and width, d, is moved at speed, υ, into then out of a uniform magnetic field of flux density, B.

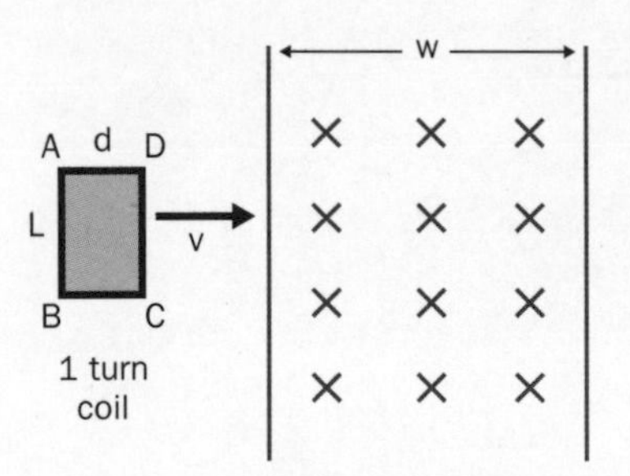

Fig. 19.9 Induced e.m.f. in a coil moved across a magnetic field.

1. An e.m.f equal to $BL\upsilon$ is induced in the leading edge, CD, as it passes through the field. In time, δt, as CD enters the field, the flux through the coil changes by an amount, $\delta\phi = BL\upsilon\delta t$, as CD moves a distance, $\upsilon\delta t$ and sweeps out area, $L\upsilon\delta t$ in time, δt. The flux change per second, $\delta\phi/\delta t$, is therefore equal to the induced e.m.f ($BL\upsilon$).
2. A reverse e.m.f equal to $-BL\upsilon$ is induced in the trailing edge, AB, as it passes through the field. Therefore, a net e.m.f. is created only when one of the edges, AB or CD, is moving in the field. When both edges are in the field, the net e.m.f. is zero. A non-zero e.m.f. is produced when the coil area is partly in or out of the field but not when it is totally inside the field. To generate a non-zero e.m.f., the magnetic 'flux' through the coil must change. Fig. 19.10 shows how the flux changes as the coil is moved into, across and out of the magnetic field.

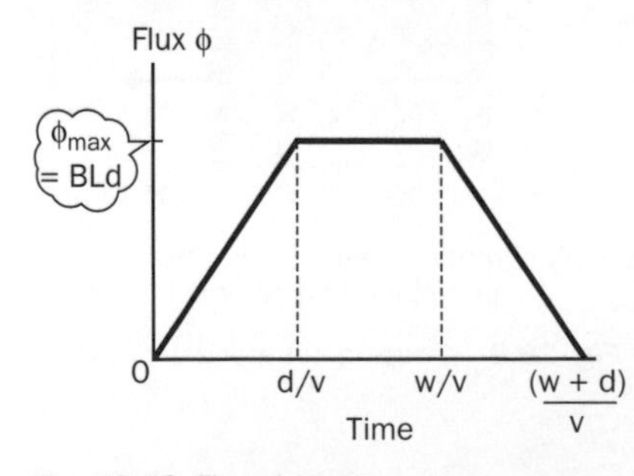

Fig. 19.10 Flux change.

Investigating flux changes

1. Connect a flat, circular coil of known diameter to a data recorder linked to a microcomputer or an oscilloscope. Measure the induced e.m.f when the coil is removed from between the poles of a U-shaped magnet. Display the measurements as a graph to show how the induced e.m.f. varies with time. This can usually be done automatically using the data recorder options menu. A typical graph is shown in Fig. 19.11.

The area under the curve represents the total change of magnetic flux through the coil.

The idea of area under a curve representing a physical quantity was introduced through speed v. time graphs on p 135. Just as the area under a graph of speed (= rate of change of distance) against time represents

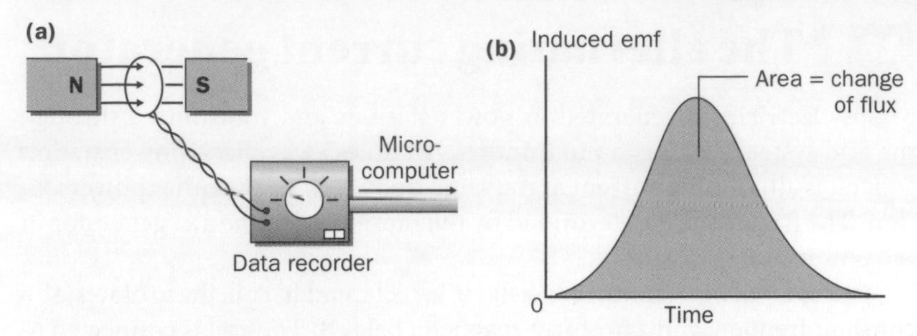

See www.palgrave.com/foundations/breithaupt for further investigations on flux changes using a data recorder.

Fig. 19.11 Investigating flux changes **(a)** using a data recorder, **(b)** induced e.m.f. versus time.

distance, so too the area under a graph of induced e.m.f. (= rate of change of magnetic flux) against time represents flux.

In formal mathematical terms,

$$\text{speed, } v = \frac{ds}{dt}$$

Hence s = area under the speed v. time graph. The induced e.m.f., V, can be written as

$$V = (-)\frac{d\Phi}{dt}$$

Hence Φ = area under induced e.m.f. v. time graph.

2. Measure the area under the curve in volt seconds. This is equal to the total initial flux (=BAN) through the coil, since the total final flux through it is zero. Hence calculate the magnetic flux density through the coil in its initial position.

Worked example 19.2

A circular coil of 150 turns and of diameter 20 mm connected to a data recorder, is placed with its face perpendicular to the lines of a uniform magnetic field. The coil is rapidly removed from the field. The data recorder is used to record and display a graph of the induced voltage with time as shown in Fig. 19.12. Calculate (a) the total flux change, (b) the initial magnetic flux density through the coil.

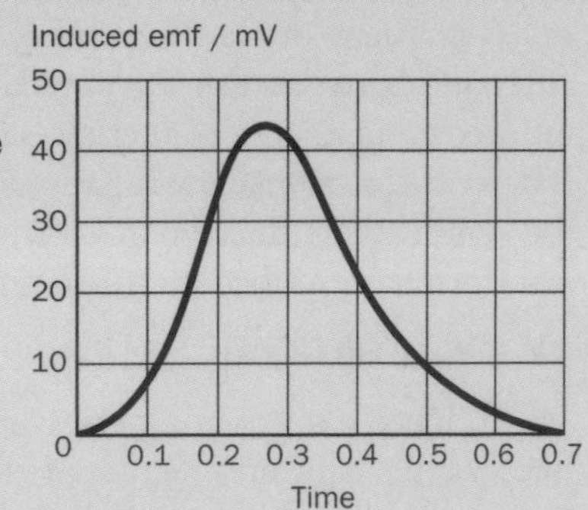

Fig. 19.12

Solution

(a) Each block of the grid corresponds to 10 mV for 100 ms. Hence each block represents a flux change of 1000 mV ms which is equal to 1.0×10^{-3} Wb. There are 12 blocks under the curve (counting part blocks as 1 if over half a block and zero if under half). Hence the total magnetic flux change is 1.2×10^{-2} Wb ($= 12 \times 1.0 \times 10^{-3}$ Wb)

(b) Total flux change = BAN

$$\therefore B = \frac{\text{total flux change}}{AN} = \frac{1.2 \times 10^{-2}}{\pi (10 \times 10^{-3})^2 \times 150} = 0.25\text{ T}$$

Questions 19.2

1. A horizontal rod of length 750 mm aligned east-west is dropped from rest at a location where the Earth's magnetic field is 60 μT due North at an angle of 70° to the horizontal. Calculate **(a)** the speed of the rod 0.5 s after being released, **(b)** the e.m.f. induced across the ends of the rod 0.5 s after being released.
2. A 50 turn square coil, 40 mm × 40 mm, is placed with its plane perpendicular to the lines of force of a uniform magnetic field of flux density 90 mT.

   (a) Calculate the total magnetic flux through the coil.

   (b) The coil is moved out of the magnetic field at a steady speed of 0.16 m s^{-1}, as shown in Fig. 19.13. Calculate **(i)** the time during which the flux decreases, **(ii)** the e.m.f. induced as the coil leaves the field.

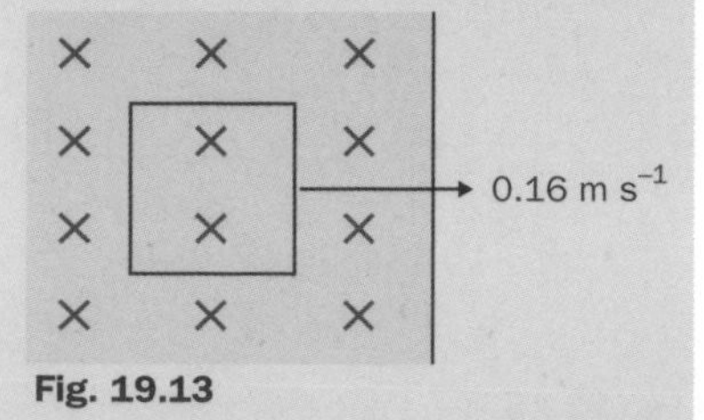

Fig. 19.13

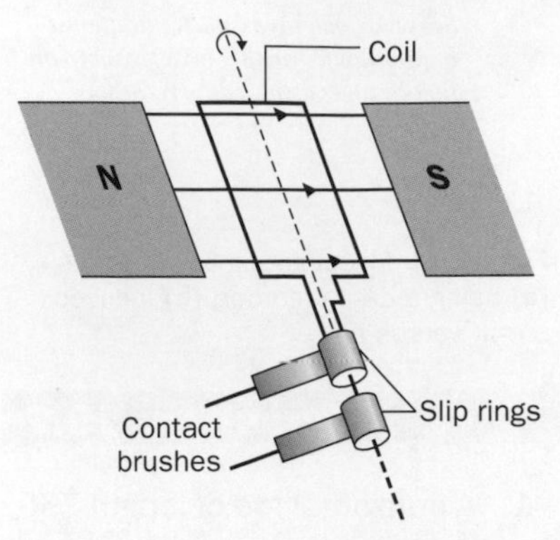

Fig. 19.14 The simple a.c. generator.

19.3 The alternating current generator

Mains electricity is generated in power stations and transmitted through the grid system to users up to hundreds of miles away. Each power station generates alternating current at the same frequency as any other power station. The frequency is determined by the rate at which the a.c. generator in the power station rotates.

A simple a.c. generator consists of a rectangular coil that rotates at a constant frequency in a uniform magnetic field, B. The coil is connected to an external circuit via two 'brushes', each in contact with its own slip ring. The magnetic flux through the coil changes as the coil position changes.

- The maximum flux linkage through the coil = BAN, where A is the coil area and N is the number of turns.
- When the angle between the coil plane and its position for maximum flux is θ, the magnetic flux linkage, Φ, through the coil is BAN cos θ, since the component of the flux density perpendicular to the coil plane is B cos θ.
- Fig. 19.15 shows how the flux linkage changes with position. The flux linkage alternates between +BAN and –BAN every half-cycle. The minus sign for the flux linkage corresponds to the lines of force passing through the coil in the opposite direction, which is what occurs when the coil turns through half a cycle.
- The induced e.m.f

  $$V = -\frac{d\Phi}{dt}$$

  = – gradient of the flux linkage v. time graph. Since the gradient alternates between a positive maximum and a negative maximum, the induced e.m.f. therefore alternates in a similar way. This type of variation is described as **sinusoidal** because its shape is a sine wave. Fig. 19.15 also shows how the induced e.m.f. varies with time. As explained below, the waveform is described mathematically by the equation

  $$V = V_{MAX} \sin (2\pi ft)$$

- The induced e.m.f is at a peak whenever the coil is parallel to the field lines, as this is where the rate of change of flux is greatest. The sides of the coil cut directly through the lines of force in this position, each wire producing a contribution to the induced e.m.f equal to BLv. The

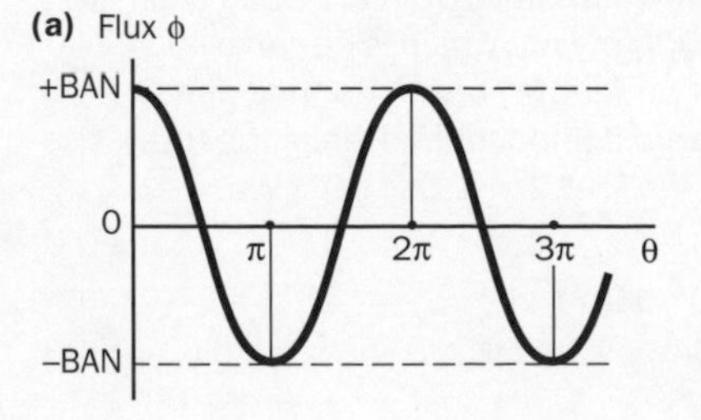

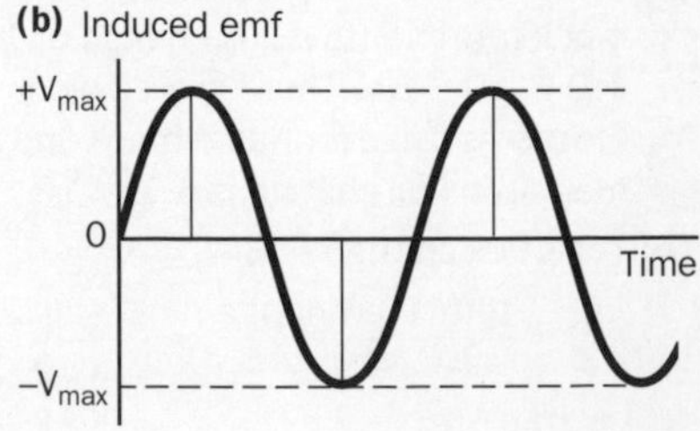

Fig. 19.15 e.m.f and flux changes **(a)** alternating flux, **(b)** alternating voltage.

maximum induced e.m.f., V_{MAX}, is therefore $2 \times (BL\upsilon) \times N$, taking account of the two sides and the N turns. The speed of each side, υ = circumference (πd) $\times$ the number of turns per second (f) = πdf, where f is the frequency of rotation and d is the width of the coil.

Hence $V_{MAX} = 2 \times (BL\upsilon) \times N = 2 \times (BL\pi fd) \times N = 2\pi f\ BAN$

where A = the coil area = Ld.

$$V_{MAX} = 2\pi f\ BAN$$

- At time, t, after passing through a flux peak, the coil has turned through a fraction f × t of one cycle. Since the angle it turns through for a full cycle is 2π radians, the angle it turns through in time t, in radians, is given by

  $\theta = 2\pi ft$

  Each side of the coil at this instant is moving in a direction at this angle to the field lines. The component of velocity of each coil side perpendicular to the field lines is therefore $\upsilon \sin \theta$. Hence the induced voltage at this instant

  $V = V_{MAX} \sin \theta = V_{MAX} \sin (2\pi ft)$

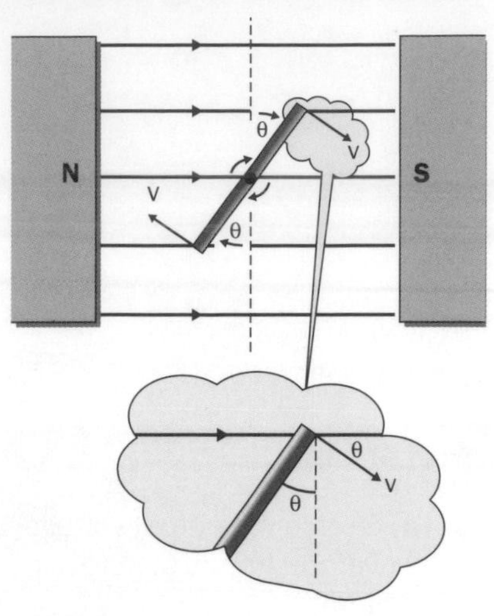

Fig. 19.16 Sinusoidal changes.

Worked example 19.3

A 200 turn rectangular coil of length 50 mm and width 30 mm spins at a constant frequency of 50 Hz in a uniform magnetic field of flux density 150 mT.

(a) Calculate the peak induced e.m.f. at this frequency of rotation.

(b) Sketch a graph to show how the induced e.m.f. varies with time.

(c) Calculate (i) the angle in radians which the coil turns through in 4 ms after a voltage zero, (ii) the induced e.m.f. at this instant.

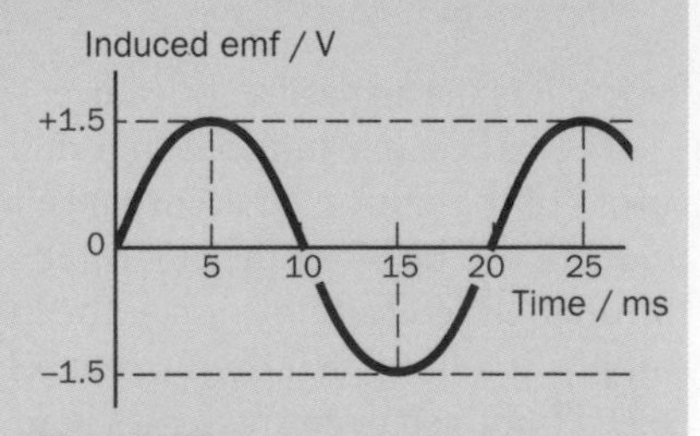

Fig. 19.17

Solution

(a) $V_{MAX} = 2\pi f\ BAN = 2\pi \times 50 \times 0.150 \times (0.050 \times 0.030) \times 200 = 14.1$ V

(b) See Fig. 19.17.

(c) (i) In 4 ms, $\theta = 2\pi ft = 2\pi \times 50 \times 4.0 \times 10^{-3} = 1.26$ radians

(ii) $V = V_{MAX} \sin (2\pi ft) = 14.1 \sin 1.26 = 13.4$ V

The power-station alternator

This consists of a direct current electromagnet, the rotor, that spins at a constant frequency between three pairs of 'stator' coils. The two coils in each pair are connected in series so that an alternating voltage is induced between the terminals of each pair of coils. The three pairs of coils are aligned at 120° to each other, so that each pair produces an alternating voltage one third of a cycle out of phase with the voltage produced by the

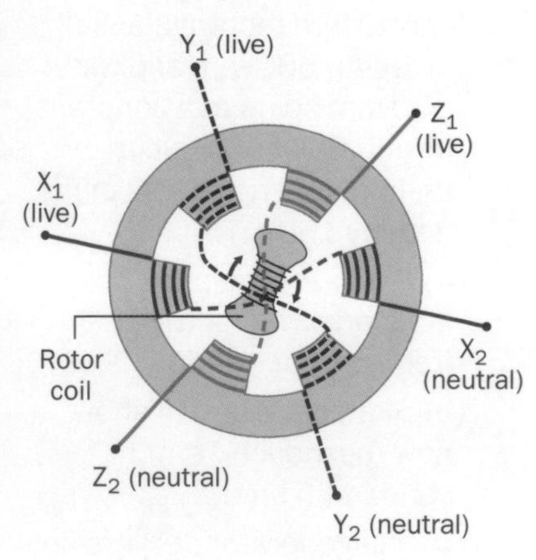

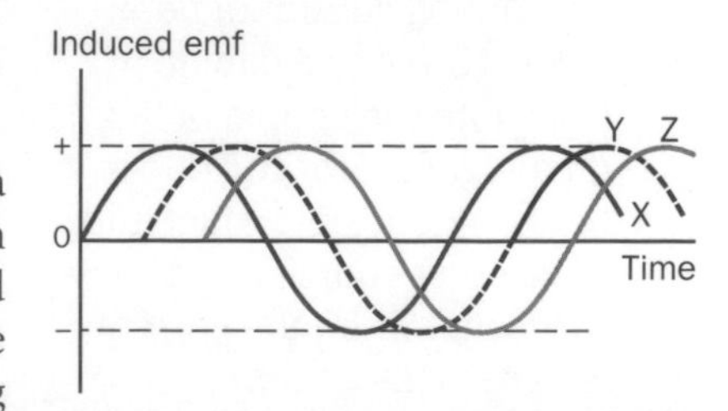

Fig. 19.18 The power-station alternator.

other coils. The rotor is supplied with direct current from a d.c. generator driven on the same shaft.

One wire from each pair of coils, the neutral wire, is earthed at the power station. The potential of the other wire, the live wire, alternates from positive to negative and back every cycle. Mains electricity is supplied from a power station along four cables, one for each separate live wire and the fourth cable as the neutral wire for each phase. Each phase is distributed at the local substation to a different group of users. Sometimes one phase switches off due to a fault but the other phases remain on. This is why sometimes a power cut affects some users served by a local substation but not others.

Questions 19.3

1. **(a)** Sketch the output voltage waveform from a simple a.c. generator.
   (b) Indicate on your sketch at which point in a cycle the coil plane is **(i)** perpendicular to, **(ii)** parallel to the lines of force of the magnetic field.
   (c) On the same axes, sketch the output voltage waveform from the same a.c. generator if its speed of rotation is halved.
2. A 1500 turn rectangular coil of length 500 mm and width 300 mm spins at a constant frequency of 50 Hz in a uniform magnetic field of flux density 150 mT.
   (a) Calculate the peak induced e.m.f. at this frequency of rotation.
   (b) Sketch a graph to show how the induced e.m.f. varies with time.
   (c) The generator delivers an alternating current of peak value 10.0 A to a circuit. Calculate the couple acting on the generator coil due to the magnetic field which the drive shaft couple must overcome when the current is at its peak value.

Back e.m.f

A simple electric motor consists of a coil rotating in a magnetic field as a result of passing a current through the coil. The rotating motion of the coil in the field generates an e.m.f. that acts against the voltage supply to the motor. This induced e.m.f. is referred to as a back e.m.f. as it opposes the voltage supply.

The voltage supply, V_S, must overcome the back e.m.f. V_B, to force current through the armature windings. Since the net e.m.f. is $V_S - V_B$, the current

$$I = \frac{V_S - V_B}{R}$$

where R is the armature resistance.

The back e.m.f induced in a d.c. electric motor is proportional to the speed of the motor, in accordance with Faraday's law of electromagnetic induction. The faster the coil rotates, the greater the rate of change of flux through the coil and the larger the induced e.m.f. acting against the voltage supply. It follows that the current is low when the motor is running at high speed but the current increases if an increased load on the motor makes it slow down. When this happens, the back e.m.f. becomes smaller and so the net e.m.f. and the current become larger.

19.4 The transformer

An e.m.f. is induced in a circuit if the magnetic flux through the circuit changes. So far we have considered changes of magnetic flux due to relative motion between the circuit and the source of the magnetic field. Another way of changing the magnetic flux through a circuit is to use an electric current to create the magnetic flux and change the magnetic flux by altering the current. The circuit that creates the magnetic flux is the **primary** circuit and the circuit in which the e.m.f. is induced is the **secondary** circuit. Transformers and induction coils are just two of a large number of devices that are based on this idea.

Changing current in the primary circuit
⇒ changing magnetic flux through the secondary circuit
⇒ induced e.m.f. in the secondary circuit

The operation of a transformer

A transformer consists of two coils of insulated wire wound around the same iron core, as shown in Fig. 19.19. When an alternating current is passed through the primary coil, an alternating magnetic field is created in the core through both coils. This alternating magnetic field induces an alternating e.m.f. in the secondary coil.

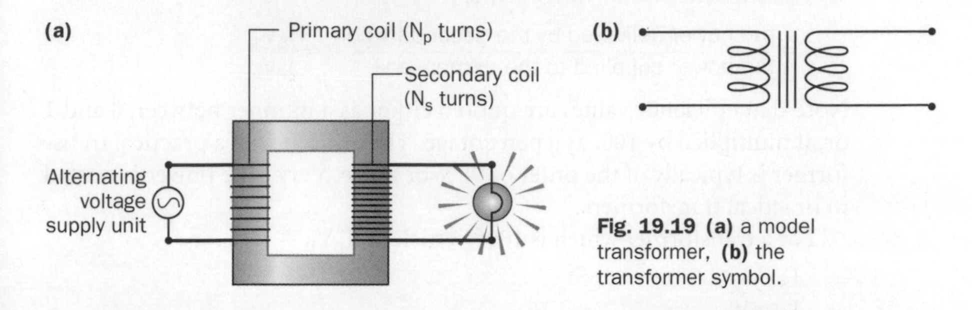

Fig. 19.19 (a) a model transformer, **(b)** the transformer symbol.

The peak voltage across the secondary coil, V_S, depends on the turns ratio of the two coils and on the peak voltage across the primary coil, V_P, according to the transformer rule

$$\frac{V_S}{V_P} = \frac{\text{number of turns in the secondary coil, } N_S}{\text{number of turns in the primary coil, } N_P}$$

1. A **step-up transformer** has more turns in the secondary coil than in the primary coil (ie. $N_S > N_P$) so $V_S > V_P$. The secondary voltage is greater than the primary voltage in a step-up transformer.
2. A **step-down transformer** has fewer turns in the secondary coil than in the primary coil (ie. $N_S < N_P$) so $V_S < V_P$. The secondary voltage is less than the primary voltage in a step-down transformer.

To understand the transformer rule, let ϕ represent the flux per turn in the core at some instant.

- The total flux linkage through the secondary coil is therefore $N_S\phi$. Using Faraday's law

  $$V = -\frac{d\Phi}{dt}$$

  therefore gives

  $$V_S = -N_S \frac{d\phi}{dt}$$

- The total flux linkage through the primary coil is $N_P\phi$, so the induced e.m.f. across the primary which the supply voltage must overcome (i.e. the primary voltage, V_P) is

  $$-N_P \frac{d\phi}{dt}$$

  The transformer rule therefore follows, since $d\phi/dt$ is the same for both coils provided no flux is lost from the core.

Transformer efficiency

The secondary coil of a transformer is used to deliver power to a circuit which it is connected to. When 'on load', the power delivered by the secondary coil = $I_S V_S$, where I_S is the current through the secondary coil.

The power supplied to the primary coil = $I_P V_P$, where I_P is the current through the primary coil.

The efficiency of the transformer

$$= \frac{\text{the power delivered by the secondary coil}}{\text{the power supplied to the primary coil}} = \frac{I_S V_S}{I_P V_P}$$

Note that efficiency values are quoted either as a number between 0 and 1 or, if multiplied by 100, as a percentage. The efficiency of a practical transformer is typically of the order of 98% or more. Very little power is wasted in practical transformers.

For a transformer which is 100% efficient, $I_P V_P = I_S V_S$, so

$$\frac{I_S}{I_P} = \frac{V_P}{V_S}$$

which means that if the voltage is stepped up, the current is stepped down and if the voltage is stepped down, the current is stepped up.

The main reasons for loss of efficiency are listed below.

1. **Resistance heating** occurs in the windings of the coils; copper wire is used to make the resistance of the windings as low as possible.
2. **Eddy currents** are induced in the core by the changing magnetic flux in the core. These currents are called 'eddy currents' because they swirl around in the core like eddy currents in water and they cause the core to heat up if they are excessive. In a practical transformer, the core is made of laminated strips of iron, separated by insulating layers which prevent eddy currents from travelling across the laminations.
3. **Hysteresis losses** occur because energy is used to magnetise and demagnetise the core each cycle. The core is made of iron because iron is easier than steel to magnetise and demagnetise.

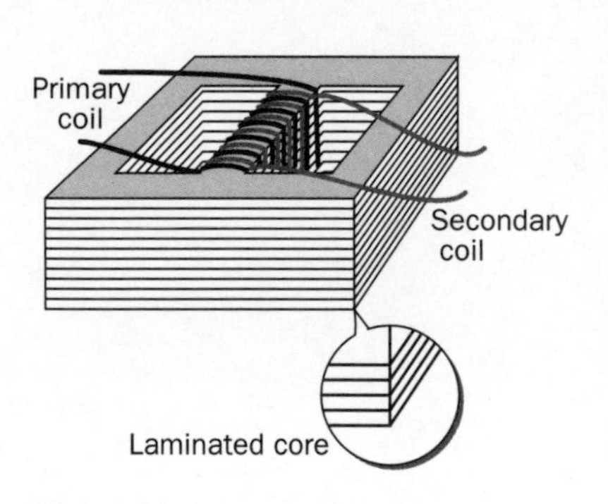

Fig. 19.20 A practical transformer.

Fig. 19.20 shows the construction of a practical transformer. The core is designed to ensure all the flux from the primary coil passes through the secondary coil.

The grid system

Electricity is supplied to your home from a power station through the grid system which is a national network of cables. The voltage at which the

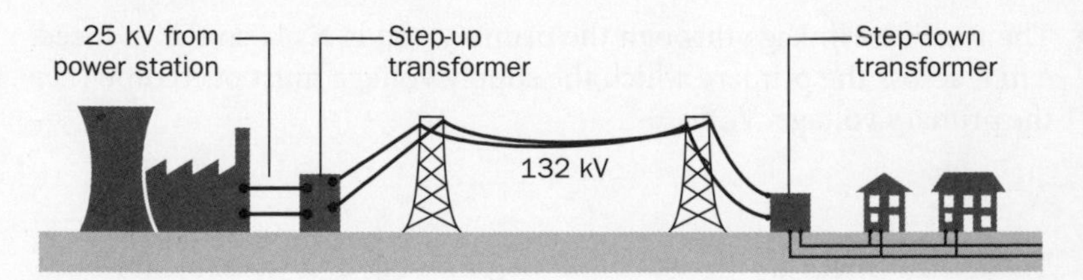

Fig. 19.21 The grid system.

Worked example 19.4

A transformer is to be used to step down an alternating voltage of 230 V to provide electrical power to a 12 V, 36 W light bulb connected to the secondary coil. Calculate (a) the number of turns necessary on the secondary coil if the primary coil has 2400 turns, (b) the maximum current in each circuit.

Solution

(a) Rearrange $\frac{V_S}{V_P} = \frac{N_S}{N_P}$ to give $N_S = \frac{N_P V_S}{V_P} = \frac{2400 \times 12}{230} = 125$ turns

(b) Power delivered by the secondary coil $I_S V_S = 36$ W, hence $I_S = \frac{36\text{ W}}{12\text{ V}} = 3.0$ A

Assuming 100% efficiency, $I_P V_P = I_S V_S$

Rearranging this equation gives $I_P = \frac{I_S V_S}{V_P} = \frac{3.0 \times 12}{230} = 0.16$ A

grid system operates, typically 132 kV, is much higher than power-station voltages, typically 25kV, which are much higher than the voltage of 230 V supplied to consumers. Transformers are used to step up the voltage from a power station onto the grid system and to step down the grid voltage at local substations to the mains voltage supplied to consumers.

The reason why voltages are stepped up between a power station and the grid system is that the current needed to transfer a certain amount of power through the grid system is reduced if the voltage is increased. The lower the current through the grid cables, the smaller the fraction of power wasted due to resistance heating of the grid cables. Alternating voltages are used rather than direct voltages because alternating voltages can be stepped up or stepped down using transformers, thus enabling efficient power transfer through the grid system.

Questions 19.4

1. A transformer consists of a 60 turn primary coil and a 1200 turn secondary coil wound on the same iron core. A 230 V, 100 W light bulb is to be connected to the secondary coil. Calculate
   (a) the primary voltage that needs to be applied to make the light bulb light normally,
   (b) the current through **(i)** the light bulb when it lights normally, **(ii)** the primary coil, assuming 100% efficiency.
2. An alternating current generator is used to supply electrical power in a building in the event of a mains power failure. The generator's output voltage of 110 V is stepped up to 230 V using a transformer. The generator is capable of supplying 20 kW of electrical power. Calculate
   (a) the turns ratio of a suitable transformer to connect the generator to the mains cable,
   (b) the maximum current the generator is able to supply,
   (c) the maximum current that could be drawn from the mains cables when a mains power failure occurs and the generator is in use.

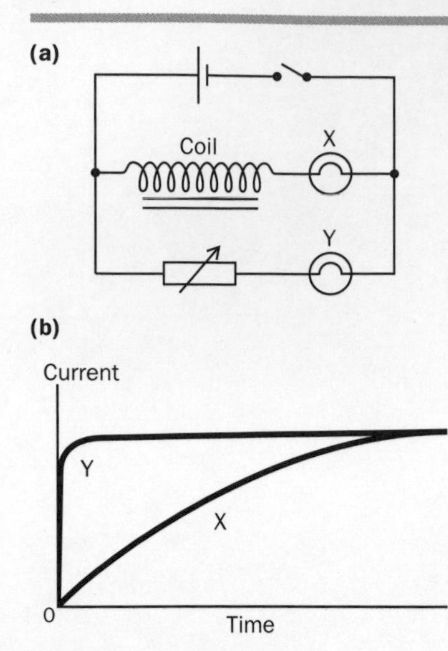

Fig. 19.22 Self-induction.

19.5 Self-inductance

When a current through a coil changes, the change of the magnetic field produced by the coil induces an e.m.f. in the coil itself which acts against the change that causes it. Changes of the magnetic field produced by the wires in any part of the coil induce an e.m.f. in the other parts of the coil. This effect is called **self-induction**. In a direct current circuit, self-induction slows current growth and prolongs current decay. In an alternating current circuit, self-induction limits the current through the coil as the coil reacts against the voltage supply.

Self-induction in a direct-current circuit

Fig. 19.22a shows two light bulbs in parallel, one in series with a coil and the other in series with a variable resistor. The variable resistor is adjusted so that the light bulbs become equally bright after the battery switch is closed. However, the light bulb in series with the coil does not light up as rapidly as the other one. This is because of the effect of self-induction in the coil. As the current increases through the coil, a self-induced e.m.f. is generated in the coil, which acts against the battery and slows the growth of current through the coil and the light bulb in series with it. Fig. 19.22b shows how the current changes with time in each bulb. If the iron core of the coil is removed, both bulbs light up equally rapidly because the magnetic field in the coil is much weaker without the iron core.

Because the magnetic flux density, B, in the coil is proportional to the current, I, through the coil, as explained on p. 261, the flux linkage, Φ, through the coil is proportional to the current, I, in the coil. This relationship is used to define the **self-inductance**, L, of the coil as the flux linkage per unit current.

Flux linkage, $\Phi = L I$

Using Faraday's law

$$V = -\frac{d\Phi}{dt}$$

therefore gives $-L\,dI/dt$ for the self-induced e.m.f. in a coil.

$$\textbf{Self-induced e.m.f., } V = -L\frac{dI}{dt}$$

The unit of self-inductance is the henry (H), equal to 1 ohm second (= 1 volt second per ampere).

Self-inductance and current growth

Consider a coil of self-inductance, L, connected to a battery and a switch. Suppose the total resistance of the circuit is R, as in Fig. 19.23. After the switch is closed, the current increases from zero to reach a maximum value, I_0. At any instant after the switch is closed, the battery p.d. is used to

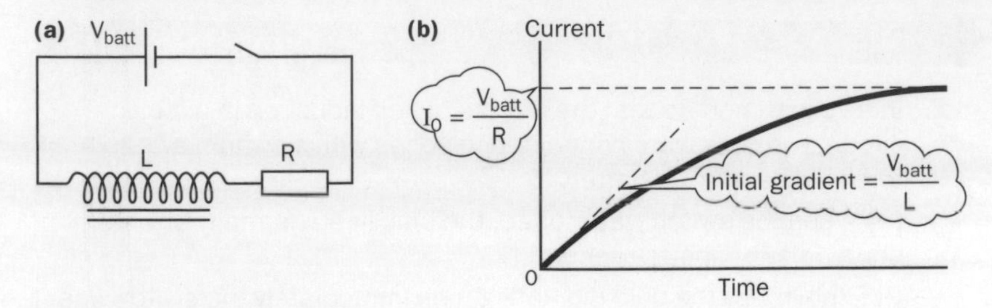

Fig. 19.23 Current growth.

overcome the self-induced e.m.f. to force current through the resistance of the circuit.

$$\therefore \text{ the battery p.d., } V_{BATT} = IR + L\frac{dI}{dt}$$

1. **Initially**, the current is zero so there is no p.d. due to resistance. Hence

$$L\frac{dI}{dt} = V_{BATT}$$

initially so the initial rate of growth of current

$$\left(\frac{dI}{dt}\right)_0 = \frac{V_{BATT}}{L}$$

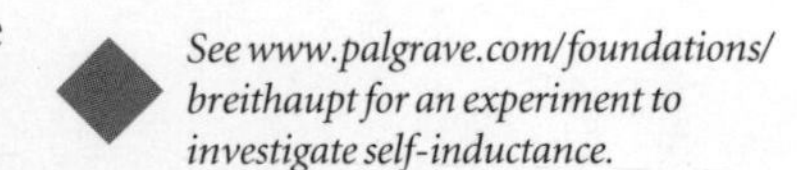

See www.palgrave.com/foundations/breithaupt for an experiment to investigate self-inductance.

2. **Eventually** the current reaches a maximum value and so dI/dt becomes zero. Hence the maximum current

$$I_0 = \frac{V_{BATT}}{R}$$

Energy stored in a self-inductance

When a current, I, passes through a coil of self-inductance, L, energy is 'stored' in the coil due to its self-inductance. This energy is stored in the magnetic field of the coil and is given by the equation

Energy stored, $E = \frac{1}{2} LI^2$

This formula can be understood by considering a small increase of the current from i to i + δi. The energy supplied for this increase, $\delta E = iV\delta t$.

However, $V\delta t = L\delta i$ for small changes since the magnitude of $V = Ldi/dt$.

$$\therefore \delta E = iL\delta i$$

Since $(i + \delta i)^2 = i^2 + 2i\,\delta i + \delta i^2$, then neglecting δi^2, as it is much smaller than the other two terms, gives

$$\delta E = \tfrac{1}{2}L(i + \delta i)^2 - \tfrac{1}{2}Li^2$$

It therefore follows that the energy stored increases during this interval from $\frac{1}{2}Li^2$ to $\frac{1}{2}L(i + \delta i)^2$, so the energy stored for current, I, must be $\frac{1}{2}LI^2$.

Questions 19.5

1. In an experiment to test the effects of self-induction in a d.c. circuit, two coils were connected in series with a switch, a 0.25 A torch bulb and a 3 V battery.

   (a) When the switch was closed, the bulb lit up normally after a short delay of about a second.

   (I) Explain why the bulb did not light up immediately the switch was closed.

   (ii) Estimate the self-inductance of the two coils.

   (b) When the connections to one of the coils was reversed and the switch was closed, the torch bulb lit up immediately. Explain why there was no delay in this circuit.

2. A coil of resistance 2.0 Ω and self-inductance 25 H is connected in series with a switch and a 12 V battery which has an internal resistance of 1.0 Ω. Calculate

   (a) the initial rate of growth of current in the circuit when the switch is closed,

   (b) (i) the final current in the circuit, **(ii)** the final energy stored in the circuit.

Summary

◆ **Magnetic flux** = BA through an area, A, at right angles to the lines of force of a magnetic field, B.

◆ **Magnetic flux linkage** through a coil of N turns = BAN.

◆ **Laws of electromagnetic induction**

1. **Lenz's law** states the induced current is always in a direction so as to oppose the change that causes it.
2. **Faraday's law** states that the induced e.m.f. is proportional to the rate of change of magnetic flux linkage.

◆ **Equations**

1. **Faraday's Law** $V = -\frac{d\Phi}{dt}$
2. **E.m.f. induced in a moving rod** $V = BLv$
3. **Voltage from an a.c. generator** $V = V_{MAX} \sin(2\pi ft)$ where $V_{MAX} = 2\pi fBAN$
4. **Transformer equation** $\frac{V_S}{V_P} = \frac{N_S}{N_P}$
5. **Efficiency of a transformer** $= \frac{I_S V_S}{I_P V_P} \times 100\%$
6. **Self-inductance** $L = \frac{\Phi}{I} = \frac{V}{(dI \div dt)}$

   Energy stored $= \frac{1}{2}LI^2$

Revision questions

19.1. The north pole of a bar magnet is pushed into one end of an open solenoid connected to a sensitive meter, as shown in Fig. 19.24.

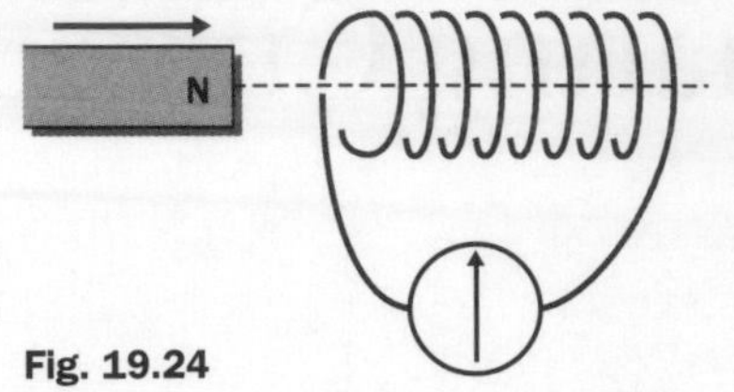

Fig. 19.24

(a) (i) What is the induced polarity at this end of the solenoid?
(ii) In which direction does the current pass through the meter?
(b) How does the deflection of the meter differ if **(i)** the north pole is inserted more slowly, **(ii)** a south pole is inserted more rapidly?

19.2. (a) An electromagnetic brake fitted to a heavy vehicle consists of a rotating magnet coupled to the drive shaft which spins near one end of a coil. The brake is applied by short-circuiting the coil terminals.
(i) Explain why the action of short-circuiting the coil terminals has a braking effect on the rotating magnet.
(ii) The braking effect reduces the kinetic energy of the vehicle. Describe the energy transformations that occur as a result of this loss of kinetic energy.
(b) (i) Explain why the current through an electric motor increases if the motor is slowed down by increasing the load on it.
(ii) Explain why the current through the primary coil of a transformer increases if a suitable load is connected across the secondary coil.

19.3. (a) An aircraft of wingspan 35.0 m, flying horizontally at a speed of 550 m s^{-1}, passes over a location at a height where the vertical component of the Earth's magnetic flux density is 80 μT. Calculate the induced e.m.f. between the wing tips of the aircraft.
(b) A 120-turn circular coil of diameter 25 mm is positioned mid-way between the poles of a U-shaped magnet with its plane parallel to the pole faces of the magnet. The terminals of the coil are connected to a data recorder which is used to record the induced e.m.f. in the coil when it is withdrawn from the magnet. Fig. 19.26 shows how the induced e.m.f. varies with time.
(i) Estimate the change of flux linkage through the coil from the area under the curve in Fig. 19.25.
(ii) Hence determine the magnetic flux density between the poles of the magnet.

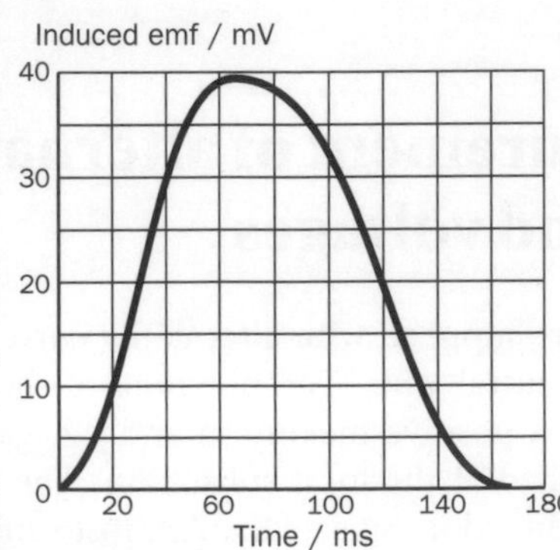

Fig. 19.25

19.4. (a) A primary coil and a secondary coil are wound on the same core. The primary coil is connected in series with a battery and a switch. The terminals of the secondary coil are connected to two rods, as in Fig. 19.26. When the switch is closed, a spark is generated between the ends of the two rods.

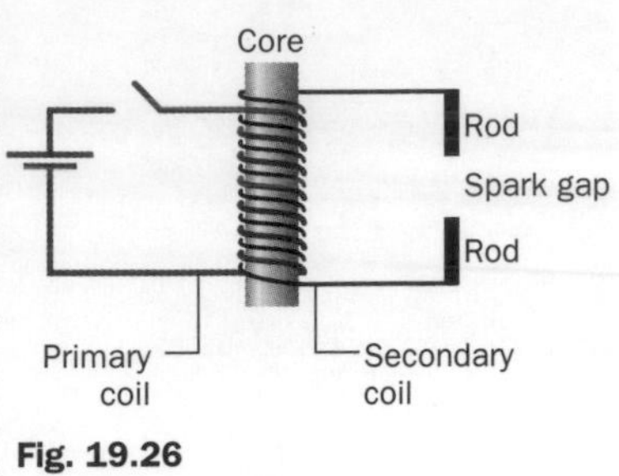

Fig. 19.26

(i) Explain the cause of this spark.
(ii) Explain why a spark is also produced when the switch is opened.
(iii) Why are no sparks produced when the switch is left closed?
(b) A transformer is to be used to step down the 230 V main voltage to 9 V, to provide electrical power to a 36 W light bulb.
(i) Calculate the number of turns needed in the secondary coil if the primary coil has 1200 turns.
(ii) Calculate the current through the secondary coil when the light bulb lights normally.
(iii) Assuming the transformer is 100% efficient, calculate the current in the primary coil when the light bulb lights normally.

19.5. An inductive coil consists of 80 turns of insulated wire wound tightly around a ferromagnetic ring of mean diameter 35 mm and of cross-sectional area 6.0×10^{-5} m^2.

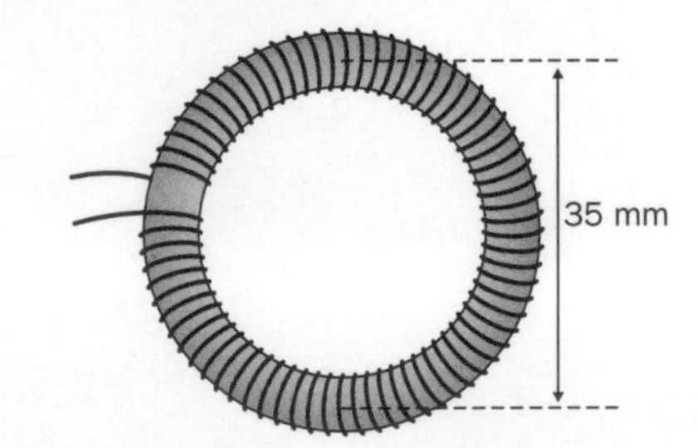

Fig. 19.27

(a) (i) The relative permeability, μ_r, of the ring is 2000. Use the formula, $B = \mu_r\mu_o nI$ (where n is the number of turns per unit length) to calculate the magnetic flux density in the ring when the current is 0.06 A. ($\mu_o = 4\pi \times 10^{-7}$ T m A^{-1})
(ii) Hence calculate the flux linkage through the coil when the current is 0.06 A, and the self-inductance of the coil.
(b) The ring is connected in series with a 1.5 V battery, a switch, an ammeter and a torch bulb.
(i) Calculate the initial rate of growth of current through the bulb when the switch is closed.
(ii) The ammeter reads 0.06 A when the current stops increasing after the switch is closed. Calculate the energy stored in the self-inductance.

Alternating Current

Contents

Objectives

After working through this unit, you should be able to:

- ▶ define and measure the peak value and the time period of an alternating voltage waveform
- ▶ calculate the frequency of an alternating current or voltage from its time period
- ▶ describe a sinusoidal waveform graphically and mathematically
- ▶ define the root mean square value of an alternating current or voltage and relate it to the peak value
- ▶ calculate the heating effect of an alternating current
- ▶ describe rectifier circuits to convert alternating current to direct current
- ▶ describe the effect of a capacitor or a coil in an a.c. circuit
- ▶ calculate and use reactance values for coils and capacitors
- ▶ explain how a reactive component differs from a resistive component
- ▶ describe applications of reactive components in a.c. circuits
- ▶ calculate currents and voltages for a.c. circuits containing reactive and resistive components in series

20.1 Measurement of alternating currents and voltages

Fig. 20.1 An alternating waveform.

Mains appliances are supplied with alternating current via two wires, the live wire and the neutral wire. The potential of the live wire alternates repeatedly between a positive maximum and a negative maximum. The neutral wire is earthed at the local substation. The current through the appliance reverses direction repeatedly. A transformer in an appliance is used to step the mains voltage down if necessary. Fig. 20.1 shows how the voltage from a step-down transformer connected to the mains supply varies. This type of waveform is described as 'sinusoidal', as explained on p. 274.

- The **peak value** of an alternating current or voltage is the maximum value. This is represented in Fig. 20.1 by the height of the waveform from the centre.
- One **complete cycle** of an alternating current or voltage waveform is the interval from a current or voltage peak in a certain direction to the next peak in that direction.
- The **time period** of an alternating current or voltage is the time taken for one complete cycle.
- The **frequency** of an alternating current or voltage is the number of complete cycles per second. The unit of frequency is the hertz (Hz), equal to 1 cycle per second. The frequency, f, of an alternating waveform can be calculated from its time period, T, using the equation

$$f = \frac{1}{T}$$

Using an oscilloscope to measure an alternating voltage

An oscilloscope is designed to measure voltages. An electron beam from an electron gun at the back of the tube hits the screen and creates a spot of light at the point of impact. The beam is deflected by two pairs of parallel plates, one pair (the X-plates) designed to deflect the beam horizontally and the other pair (the Y-plates) designed to deflect it vertically. In normal operation, a 'time base' voltage is applied to the X-plates which makes the spot move across the screen slowly from left to right, then returns it much more rapidly to start the next 'sweep'. At the same time, the waveform to be displayed on the screen is applied to the Y-plates, so the waveform is traced by the spot on the screen as it sweeps across from left to right.

The main controls for measurement purposes on an oscilloscope are:

- the **Y-gain**, which is the voltage per unit deflection in the vertical direction. For example, if the Y-gain control knob is set at 0.5 V cm^{-1}

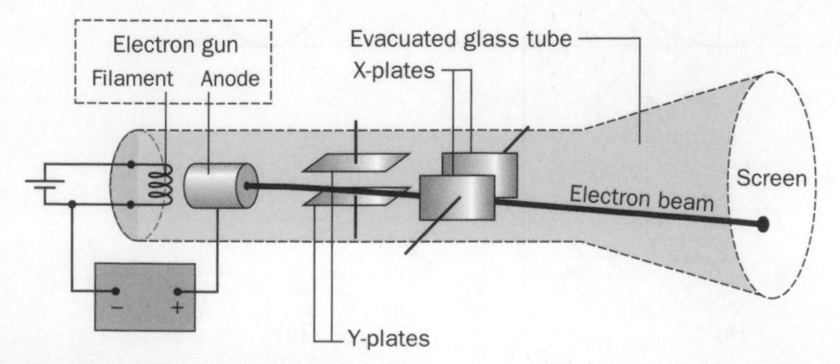

Fig. 20.2 Inside an oscilloscope tube.

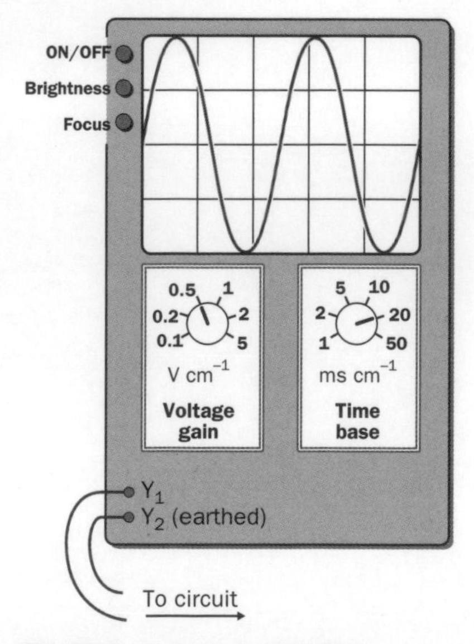

Fig. 20.3 Using an oscilloscope.

and the waveform is 40 mm from the bottom (negative) peak to the top (positive) peak, the voltage change from the negative to the positive peak is 2.0 V (= 0.5 V cm^{-1} × 4.0 cm). The peak voltage is therefore 1.0 V.

- the **time-base**, which is the time taken by the spot to move unit distance horizontally. For example, if the time-base control knob is set at 20 ms cm^{-1} and one cycle of the waveform measures a horizontal distance of 25 mm, the time period of the waveform must be 50 ms (= 20 ms cm^{-1} × 2.5 cm), giving a frequency of 20 Hz (= 1/0.020 s).

The equation for a sinusoidal waveform

The voltage from an a.c. generator is a sinusoidal waveform like Fig. 20.1. As explained on p. 274, the voltage, V, varies with time in accordance with the equation

$$\mathbf{V = V_0 \sin (2\pi ft)}$$

where V_0 is the peak voltage, f is the frequency of rotation of the generator and t is the time after the voltage is zero as it reverses from − to + polarity. The voltage at a given time can be either read off a waveform graph like Fig. 20.1 or it can be calculated using the above equation.

Another method of finding the voltage at a given time is to represent the peak value as a rotating vector or **phasor**, as in Fig. 20.4a. When the phasor is at angle θ to the + x-axis, the coordinates of its tip are $x = V_0 \cos\theta$ and $y = V_0 \sin\theta$. By considering the phasor rotating anticlockwise at constant frequency, f, at time, t, after it passes through the + x-axis, the phasor has turned through a fraction, ft, of a cycle and therefore the angle it makes in radians to the x-axis is 2πft (i.e. θ = 2πft). Hence its y-coordinate, $V_0 \sin\theta$, represents the voltage at time t.

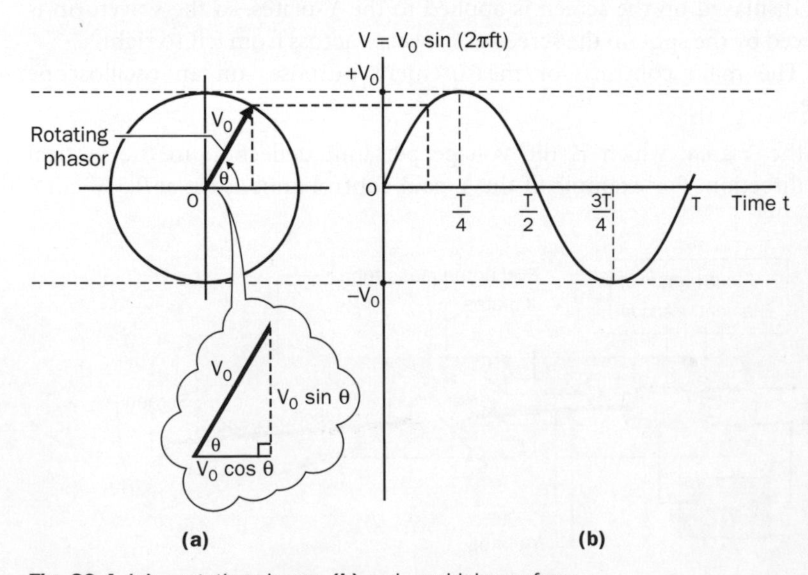

Fig. 20.4 **(a)** a rotating phasor, **(b)** a sinusoidal waveform.

Worked example 20.1

A sinusoidal alternating voltage with a peak value of 12 V has a frequency of 50 Hz. Calculate (a) its time period, (b) the voltage (i) 4 ms, (ii) 7 ms, (iii) 18 ms after the voltage was zero as it changed from negative to positive.

Solution

(a) $T = \frac{1}{f} = \frac{1}{50} = 0.020\ \text{s} = 20\ \text{ms}$

(b) Fig. 20.5 shows the position of the phasor for each case. The length of the phasor represents 12 V so the y-coordinate at each position gives the voltage at that position. The values read off from the phasor diagram should be the same as the calculated values as follows:

(i) $V = V_0 \sin(2\pi ft) = 12 \sin(2\pi \times 50 \times 0.004\ \text{radians}) = 11.4\ \text{V}$.

(ii) $V = V_0 \sin(2\pi ft) = 12 \sin(2\pi \times 50 \times 0.007\ \text{radians}) = 9.7\ \text{V}$.

(iii) $V = V_0 \sin(2\pi ft) = 12 \sin(2\pi \times 50 \times 0.018\ \text{radians}) = -7.1\ \text{V}$.

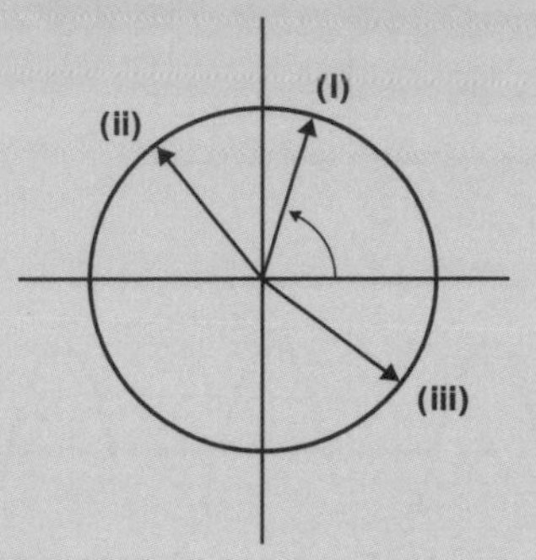

Fig. 20.5

Questions 20.1

1. Fig. 20.6 shows an oscilloscope screen displaying an alternating waveform. Calculate the peak voltage and the frequency of this waveform if the control knob settings are **(a)** a Y-gain of $0.2\ \text{V cm}^{-1}$ and a time base of $10\ \mu\text{s cm}^{-1}$,
   (b) a Y-gain of $5.0\ \text{V cm}^{-1}$ and a time base of $5\ \text{ms cm}^{-1}$.
2. An alternating voltage has a peak voltage of 4.0 V and a frequency of 200 Hz.
   (a) Construct a phasor diagram and use it to determine the voltage **(i)** 1.0 ms, **(ii)** 3.0 ms after the voltage was zero and changing from − to +.
   (b) Use the equation $V = V_0 \sin(2\pi ft)$ to calculate the voltage **(i)** 1.0 ms, **(ii)** 3.0 ms after the voltage was zero and changing from − to +.

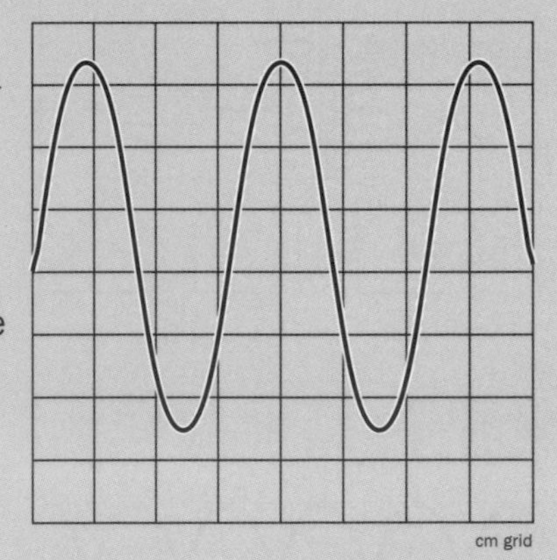

Fig. 20.6

20.2 Rectifier circuits

Electronic organs, transistor radios and portable computers are just three of many devices that can be powered from either a battery or a mains-operated, low voltage supply unit, which consists of a transformer and a rectifier circuit. The transformer steps the mains voltage down to the desired voltage and the rectifier circuit converts the alternating voltage from the transformer to a direct voltage.

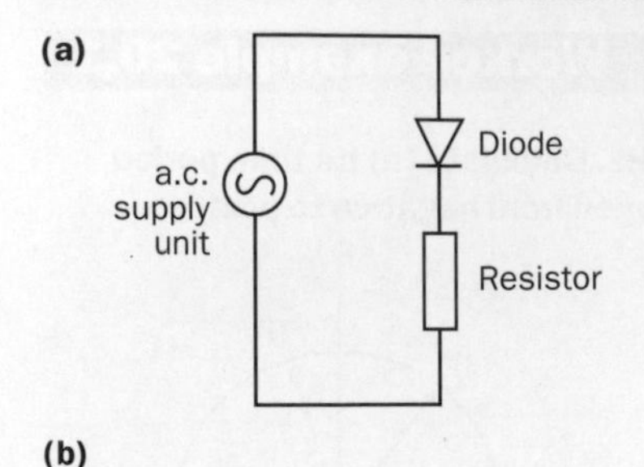

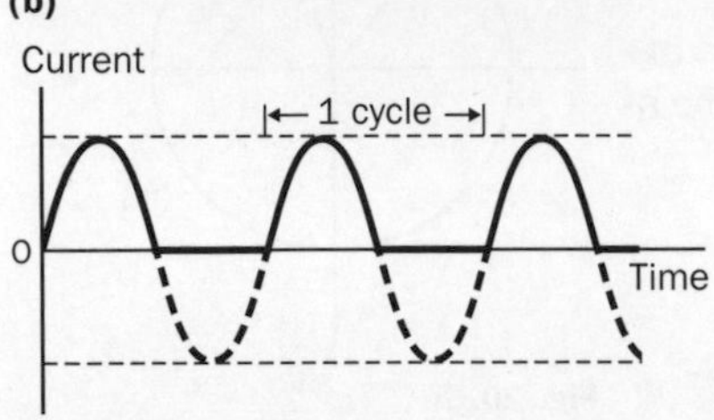

Fig. 20.7 Half-wave rectification **(a)** using a single diode, **(b)** a half-wave direct current.

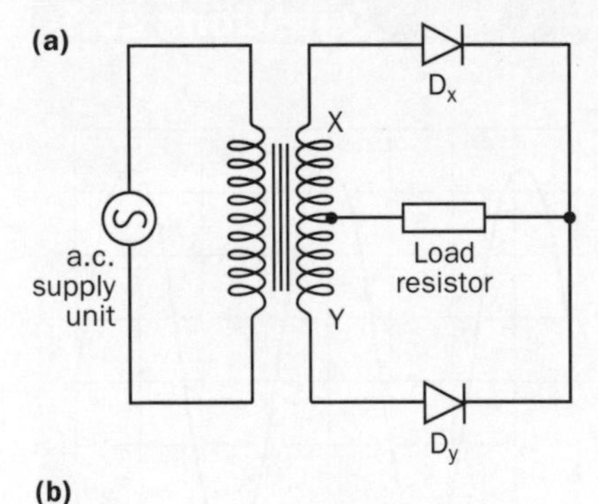

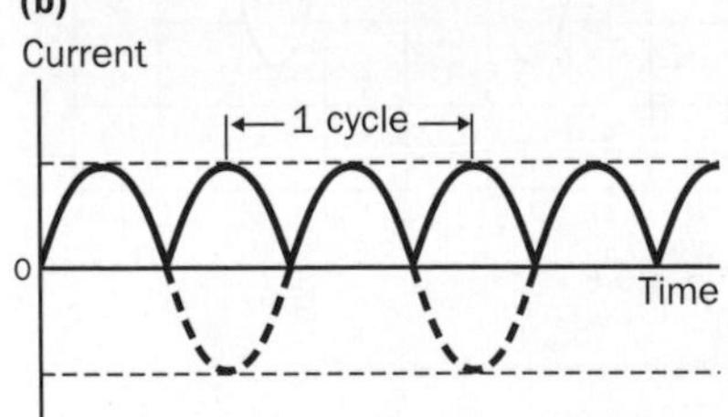

Fig. 20.8 Full-wave rectification **(a)** using two diodes, **(b)** a full-wave direct current.

Investigating rectifier circuits

1. A **single diode in series with a resistor** and an a.c. supply unit allow current through every other half-cycle when the supply polarity is such that the diode is forward-biased. This is referred to as **half-wave rectification**. The current waveform is shown in Fig. 20.7b. Since the voltage across the resistor is proportional to the current, the voltage across the resistor also has a 'half-wave' waveform like Fig. 20.7. This can be displayed on an oscilloscope screen by connecting the Y-input of the oscilloscope across the resistor.

2. **Two diodes and a transformer** connected as in Fig. 20.8 give **full-wave rectification**, in which the voltage varies from zero to a peak value and back to zero every half-cycle. This can be achieved by tapping off the output voltage from the centre of the transformer's secondary coil and through a diode connected at each end of the secondary coil. In Fig. 20.8, when the alternating supply connected to the primary coil makes end X of the secondary coil positive (and the other end negative), the diode at this end, D_X, conducts so current passes through the load resistor back to the central terminal of the secondary coil. When the supply reverses polarity, the other end of the secondary coil, end Y, is positive so the diode at this end, D_Y, now conducts and current passes through the load resistor in the same direction as before, back to the central terminal of the transformer's secondary coil.

   The current waveform and the resistor voltage waveform are shown in Fig. 20.8b. This waveform is described as a 'full-wave' rectified waveform because the current is non-zero and in the same direction in both halves of each cycle.

3. The **bridge rectifier** consists of four diodes as in Fig. 20.9, connected directly to the a.c. supply to be rectified. On one half-cycle of each cycle, diodes P and R conduct, so current passes through the load resistor frrom left to right. On the other half-cycle, diodes Q and S conduct and current passes through the load resistor left to right again. Full-wave rectification is once again produced as the current through the resistor and the resistor voltage do not change direction.

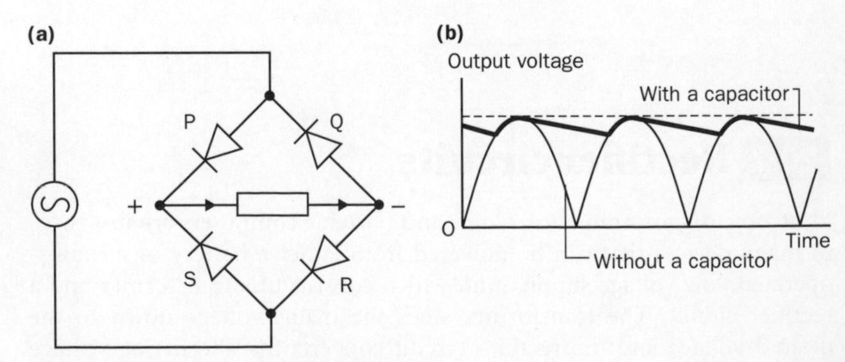

Fig. 20.9 **(a)** using a bridge rectifier, **(b)** capacitor smoothing.

Capacitor smoothing

A large capacitor connected across the output terminals of a full-wave rectifier will charge up in the first half-cycle then release its charge very slowly after the supply voltage peaks, thus maintaining the output voltage from the full-wave rectifier. Fig. 20.9 shows the idea. The capacitor must be large, so the time constant RC (see p. 217) of the circuit is much longer than the time period of the alternating supply voltage.

20.3 Alternating current and power

Imagine an electric heater operating at a very low frequency. The heating element would glow at peak current and fade at zero current every half-cycle, regardless of the direction in which the current passed through. The power supplied to the heater would peak every half-cycle when the current was at its peak in either direction. The heating effect is due to the resistance of the heating element.

The power supplied to a resistor, R, carrying current, $I = IV = I^2R$, since the resistor p.d., $V = IR$.

For a sinsoidal current

$I = I_0 \sin 2\pi ft$

1. the resistor voltage

$V = IR = I_0 R \sin 2\pi ft$

2. the power supplied

$P = IV = I_0^2 R \sin^2 2\pi ft$

The power supplied varies from zero to a peak value of I_0^2R and back to zero every half-cycle. Fig. 20.12 shows how the power supplied varies with time over a few cycles. This power curve is symmetrical about $\frac{1}{2} \times$ the peak power $(= \frac{1}{2} I_0^2R)$.

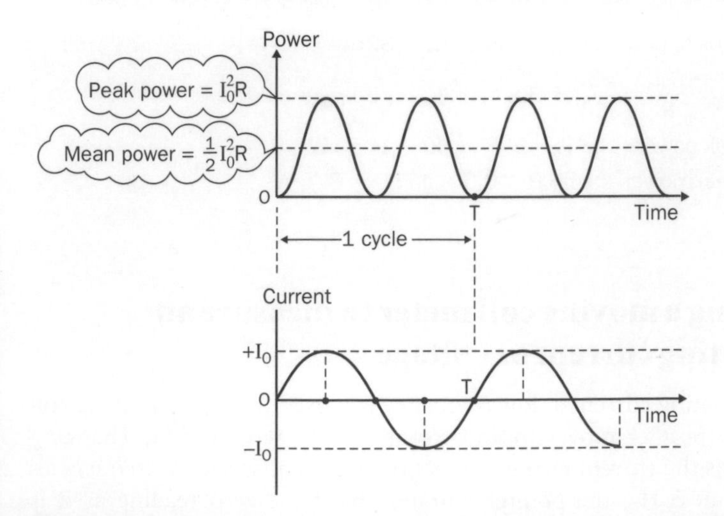

Fig. 20.12 Power versus time for a sinusoidal current through a resistor.

Questions 20.2

1. Fig. 20.10 shows a bridge rectifier comprising four diodes, A, B, C and D, connected to an a.c. supply unit.

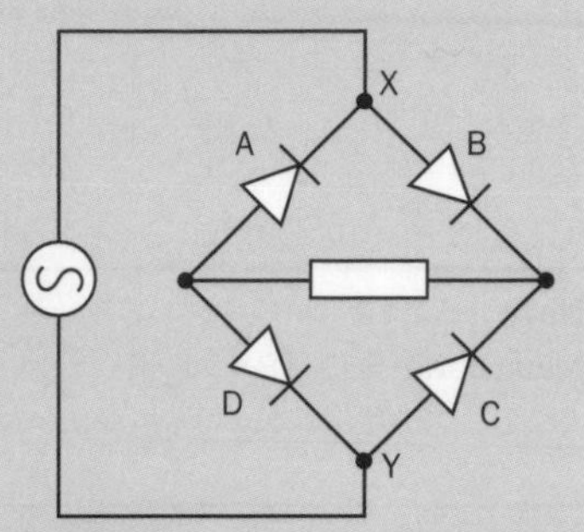

Fig. 20.10

(a) (i) In which direction does current pass through the load resistor?

(ii) When terminal X of the a.c. supply unit is negative, which diodes conduct?

(b) Sketch the current waveform if diode A fails and becomes non-conducting.

2. Fig. 20.11 shows two diodes in a circuit in opposite directions connected to an a.c. supply unit. Two unequal resistors are also connected in the circuit as shown.

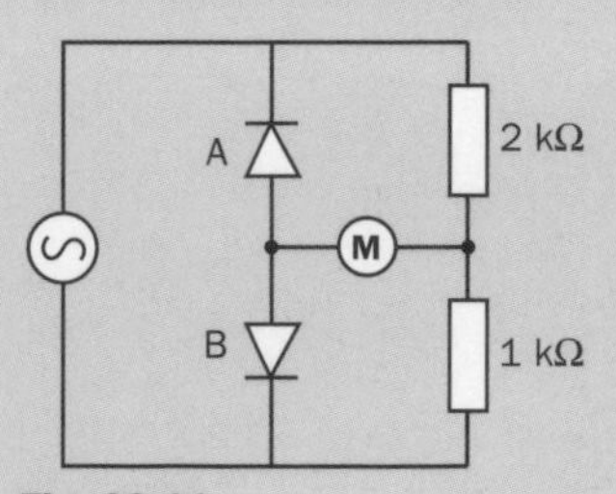

Fig. 20.11

(a) In which direction does current pass through the meter?

(b) With the aid of a waveform sketch, explain how the supply current varies with time over several cycles.

The root mean square (r.m.s.) value of an alternating current (or voltage) is the value of the direct current (or voltage) that gives the same heating effect in a given resistor as the alternating current (or voltage).

See www.palgrave.com/foundations/breithaupt for more about mains electricity.

For a sinusoidal alternating current of peak value I_0, its r.m.s. value, I_{rms}, is therefore related to the peak current by equating the d.c. power (I_{rms}^2R) to the mean a.c. power ($\frac{1}{2}I_0^2R$).

i.e. $I_{rms}^2R = \frac{1}{2}I_0^2R$

Cancelling R therefore gives

$$I_{rms}^2 = \frac{1}{2}I_0^2$$

Taking the square root of both sides,

$$I_{rms} = \frac{1}{\sqrt{2}}I_0$$

Note

Mean power $= I_{rms}^2R = \frac{1}{2}I_0^2R$
$= I_{rms}V_{rms} = \frac{1}{2}I_0V_0$

Similarly, the r.m.s. voltage V_{rms} is related to the peak voltage of a sinusoidal alternating voltage waveform by the equation

$$V_{rms} = \frac{1}{\sqrt{2}}V_0$$

Worked example 20.2

A sinusoidal alternating voltage which has an r.m.s. value of 230 V is connected to a 60 Ω heating element designed to operate at an r.m.s. voltage of 230 V. Calculate

(a) (i) the peak voltage, (ii) the voltage change from a positive peak to a negative peak.

(b) (i) the r.m.s. current, (ii) the peak current.

(c) (i) the peak power, (ii) the mean power.

Solution

(a) (i) Rearrange $V_{rms} = \frac{1}{\sqrt{2}}V_0$ to give $V_0 = \sqrt{2}\,V_{rms} = \sqrt{2} \times 230 = 325$ V

(ii) Peak to peak voltage $= 2V_0 = 2 \times 325 = 650$ V

(b) (i) $I_{rms} = \frac{V_{rms}}{R} = \frac{230}{60} = 3.8$ A, (ii) $I_0 = \sqrt{2}\,I_{rms} = \sqrt{2} \times 3.8 = 5.4$ A

(c) (i) Peak power $= I_0^2R = 5.4^2 \times 60 = 1750$ W

(ii) Mean power $= \frac{1}{2}I_0^2R = 0.5 \times 1750 = 875$ W

Adapting a moving coil meter to measure an alternating current or voltage

If an alternating current at low frequency is passed through a moving coil meter, the pointer moves to and fro about the zero reading, changing direction as the current changes direction every half-cycle. At frequencies above about 5 Hz, the pointer vibrates about the zero reading as it is unable to keep up with the repeated changes of direction.

The pointer can be made to deflect in one direction by connecting a diode in series with the meter or by using a bridge rectifier, as in Fig. 20.13. In both cases, the pointer gives a non-zero reading which is less than the peak current. Since the reading is proportional to the peak current and hence proportional to the r.m.s. current, the meter scale is usually calibrated so the pointer position gives the r.m.s. value.

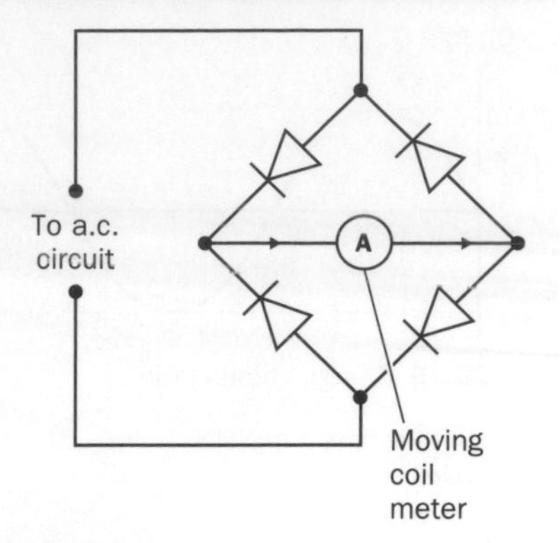

Fig. 20.13 Adapting a moving coil meter.

Questions 20.3

1. An oscilloscope is used to display and measure an alternating voltage waveform. The waveform measures 72 mm vertically from the negative peak to the positive peak on the screen, when the Y-gain is at 0.2 V cm^{-1}.
   (a) Calculate **(i)** the peak to peak voltage V_{pp}, **(ii)** the peak voltage V_o, **(iii)** the r.m.s. voltage.
   (b) This alternating voltage is the result of a sinusoidal current of r.m.s. value 120 mA passing through a resistor. Calculate **(i)** the resistance of this resistor, **(ii)** the mean power dissipated in this resistor.
2. A mains heater carries the label '230 V r.m.s, 1000 W'. Calculate
   (a) the peak voltage across the heating element,
   (b) (i) the r.m.s. current, **(ii)** the peak current, **(iii)** the resistance of the heating element.

20.4 Capacitors in a.c. circuits

Reactive and resistive components

Coils and capacitors are described as reactive components because they react against alternating current passing through without dissipating power. In comparison, a resistive component opposes an alternating current due to its resistance, which causes it to dissipate power. A resistive component is one which dissipates power when a current passes through it.

As explained in Section 20.3, in a simple a.c. circuit consisting of a resistor connected to an a.c. power supply unit, the current and the potential difference across the resistor alternate in phase, which means that they rise and fall and change direction together. The result is that power is dissipated in the resistor every half-cycle. As explained later in this section, in a simple circuit containing either a capacitor or a coil connected to an a.c. supply unit, the current and p.d. are out of phase by a quarter of one cycle. As a result, power is supplied to the component then returned to the power supply unit every half-cycle so the mean power supplied is zero. The term reactance is used for a capacitor or a coil because these components react against the power supply unit without taking power from it.

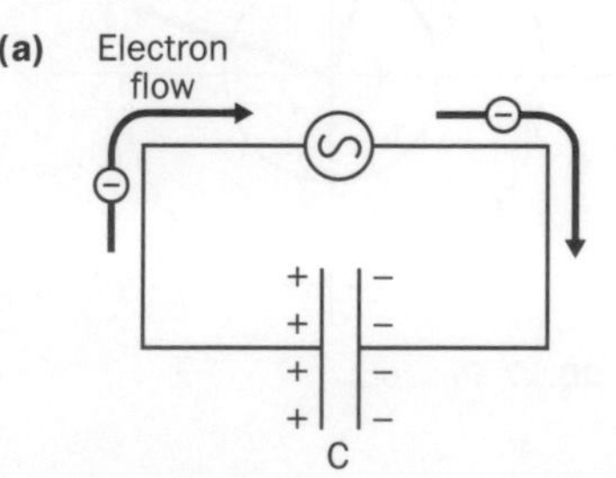

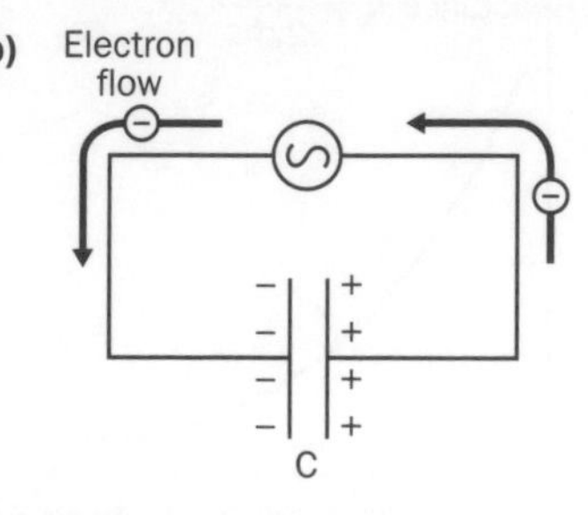

Fig. 20.14 A capacitor in an a.c. circuit **(a)** at a given instant and **(b)** half a cycle later.

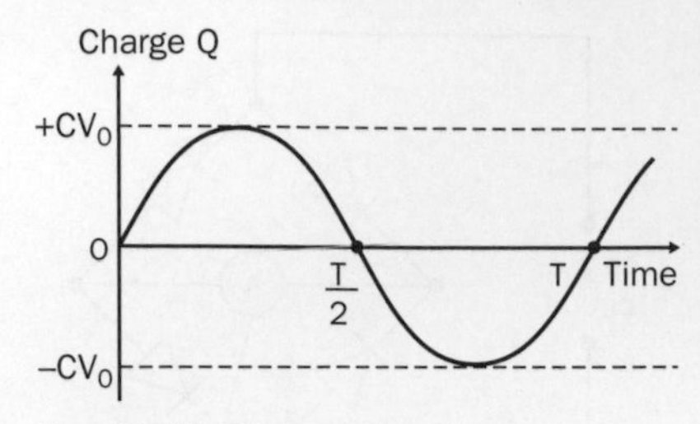

Fig. 20.15 Charge versus time.

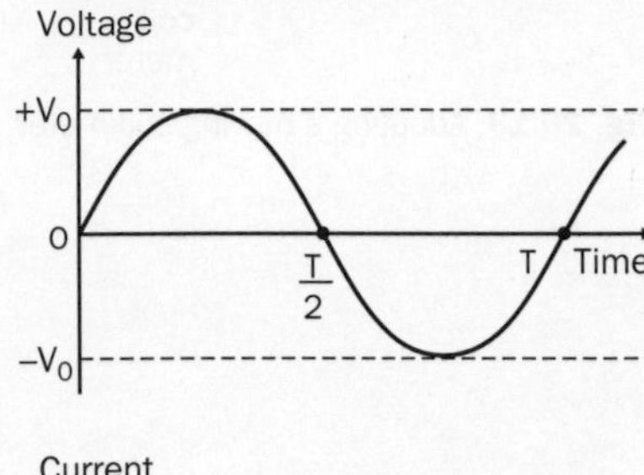

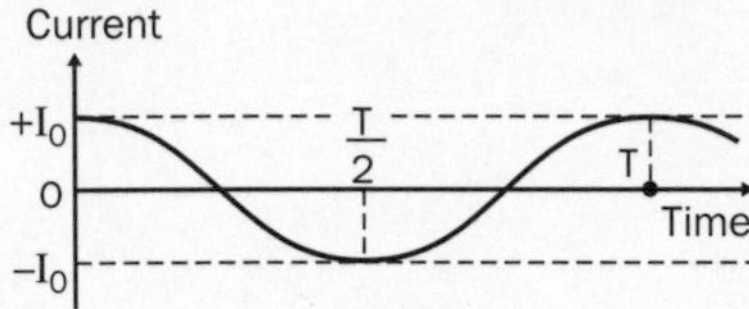

Fig. 20.16 Current versus time.

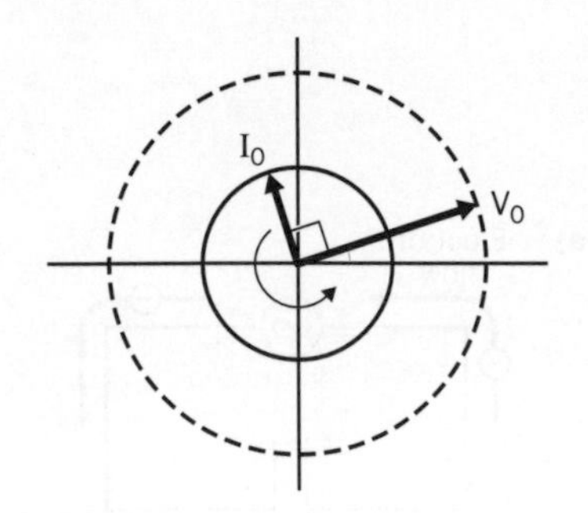

Fig. 20.17 Phasors.

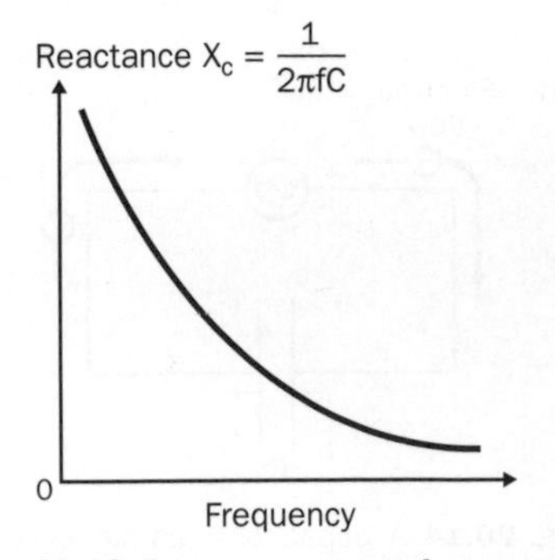

Fig. 20.18 Reactance versus frequency for a capacitor.

The reactance of a capacitor

Consider an a.c. supply unit connected to a single capacitor. As the potential difference between the capacitor plates alternates, electrons flow onto one plate then back to the supply unit again, then onto the other plate and back to the supply unit again.

- The flow of electrons to and from the supply is an alternating current, therefore the capacitor allows an alternating current in the circuit, whereas it would block a direct current if the supply unit was a d.c. unit.
- The faster the frequency of the a.c. supply, the greater the flow of charge per second. The charge flows to and from the capacitor at a faster rate if the frequency is increased. Therefore, the current is greater the higher the frequency, even though the peak supply voltage is constant. Conversely, the lower the frequency, the smaller the current.

For any capacitor, the charge stored, $Q = CV$, where V is the p.d. across its terminals and C is its capacitance. In an alternating current circuit, the charge and the potential difference are therefore in phase with each other. In mathematical terms, if $V = V_0 \sin(2\pi ft)$, then $Q = CV_0 \sin(2\pi ft)$. Figs 20.15 and 20.16 show how the voltage and charge vary over a few cycles.

Since the current, I = the rate of flow of charge, $dQ \div dt$, which is the gradient of the charge v. time graph, the current at any point is given by the gradient of the charge v. time graph. Fig. 20.16 shows how the current (i.e. the gradient of the charge v. time graph) varies with time when the charge varies sinusoidally as in Fig. 20.15. The current is one quarter of a cycle ahead of the potential difference and charge. Essentially, the rate of flow of charge on the capacitor is at its peak at the instant the capacitor is uncharged and about to charge up.

As explained on p. 293, if $Q = CV_0 \sin(2\pi ft)$

$$\text{then } I = \frac{dQ}{dt} = \frac{d}{dt}(CV_0 \sin(2\pi ft)) = 2\pi fCV_0 \cos(2\pi ft)$$

1. The peak current $I_0 = 2\pi fCV_0$ since the peak value of $\cos(2\pi ft) = 1$.
2. The current is ahead of the p.d. by one quarter cycle (= 90°). The phasor diagram for the current and voltage for a capacitor in an a.c. circuit is shown in Fig. 20.17.
3. The **reactance, X_C**, of a capacitor is defined as $\frac{1}{2\pi fC}$

$$\text{Since } I_0 = 2\pi fCV_0, \text{ then } \frac{1}{2\pi fC} = \frac{V_0}{I_0}$$

The unit of reactance is therefore the ohm, the same as the unit of resistance.

$$\mathbf{X_C = \frac{V_0}{I_0} = \frac{1}{2\pi fC}}$$

Note that the reactance decreases as the frequency increases. Fig. 20.18 shows how the reactance decreases with increasing frequency.

4. Fig. 20.19 shows how the power (= current × potential difference) varies with time. Every other quarter cycle, power is returned to the supply unit as electrons are forced into the negative terminal. In these quarter cycles, the current and the potential are in opposite 'directions' so IV is negative and power returns to the supply unit. In between these quarter cycles, the current and the potential difference are in the same direction, so IV is positive, so power is delivered by the supply unit. As a result, no power is delivered by the supply over a complete cycle.

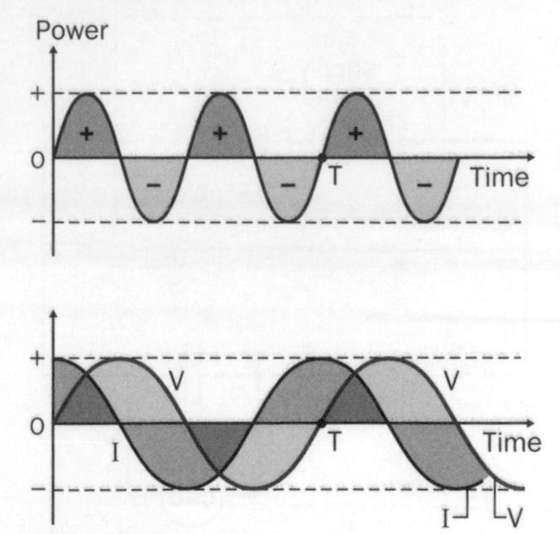

Fig. 20.19 The power curve for a capacitor.

Differentiation of a sine function

Proof of $\frac{d}{dt}[\sin(2\pi ft)] = 2\pi f\cos(2\pi ft)$

In Fig. 20.20, OP is a rotating phasor of length, r, at angle, $\theta = 2\pi ft$ to the + x-axis at time, t, after passing through the x-axis in an anticlockwise direction. A short time, δt, later, at time $t + \delta t$, after passing through the + x-axis, its angle to the x-axis has increased by $\delta\theta = 2\pi f\delta t$ and its tip moves through a distance $\delta s = r\delta\theta$.

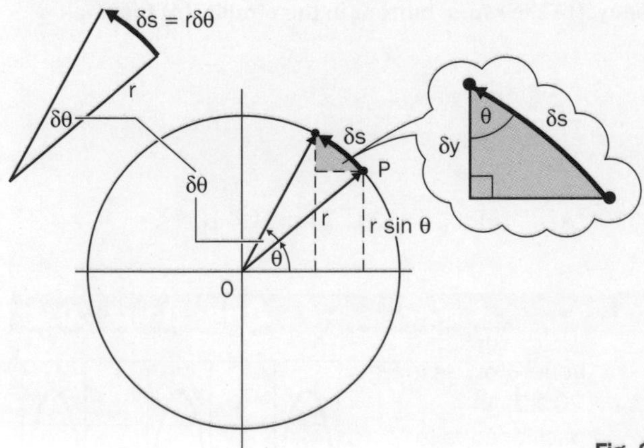

Fig. 20.20 Small changes.

The y-coordinate of the tip of the vector OP is given by $y = r\sin\theta$, where r = length OP. In the time interval from t to $t + \delta t$, the y-coordinate increases by δy. However, since δy and δs almost form a right-angled triangle, then $\delta y = \delta s\cos\theta$, provided δs is small.

Since $\delta s = r\,\delta\theta$ and $\delta\theta = 2\pi f\,\delta t$, then $\delta y = \delta s\cos\theta = 2\pi f\,r\,\delta t\cos\theta$.

Dividing δt into both sides of $\delta y = 2\pi f\,r\,\delta t\cos\theta$ therefore gives

$$\frac{\delta y}{\delta t} = \frac{2\pi f\,r\,\delta t}{\delta t}\cos\theta$$

Hence in the limit $\delta t \to 0$,

$$\frac{dy}{dt} = 2\pi f\,r\cos\theta$$

$$\therefore \frac{d}{dt}[r\sin(2\pi ft)] = 2\pi fr\cos(2\pi ft)$$

Cancelling r on both sides therefore gives the required equation.

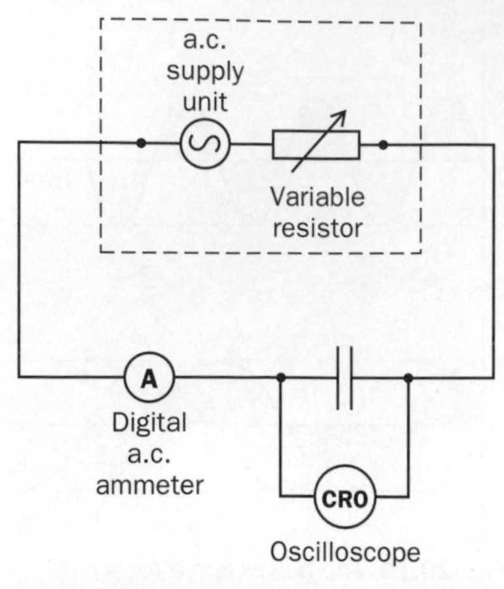

Fig. 20.21 Measuring reactance.

Measuring the reactance of a capacitor

1. Connect the capacitor in series with a digital ammeter and an a.c. supply unit. Connect an oscilloscope across the capacitor terminals, as shown in Fig. 20.21, to display the voltage waveform. Keeping the frequency constant, measure the r.m.s. current using the ammeter for different peak to peak voltages. For each current, measure the peak voltage and calculate the r.m.s. voltage. Use the oscilloscope to measure the time period of the waveform and hence calculate the frequency.
2. Record the measurements and use them to plot a graph of the capacitor voltage against the capacitor current. The gradient of this graph is the reactance of the capacitor. Hence determine the capacitor reactance and use its value to calculate the capacitance from the equation

$$X_C = \frac{1}{2\pi f C}$$

Worked example 20.3

A 2.2 μF capacitor is connected to a 50 Hz a.c. power supply unit which has an r.m.s. output p.d. of 6.0 V. Calculate (a) the reactance of this capacitor at this frequency, (b) the r.m.s. current in the circuit, (c) the peak current in the circuit.

Solution

(a) $X_C = \frac{1}{2\pi f C} = \frac{1}{2\pi \times 50 \times 2.2 \times 10^{-6}} = 1.45 \times 10^3\ \Omega$

(b) Since $X_C = \frac{V_0}{I_0} = \frac{V_{rms}}{I_{rms}}$, $I_{rms} = \frac{V_{rms}}{X_C} = \frac{6.0}{1.45 \times 10^3} = 4.1 \times 10^{-3}\ \text{A}$ **(c)** $I_0 = \sqrt{2}\, I_{rms} = 5.9 \times 10^{-3}\ \text{A}$

Questions 20.4

1. In an experiment to measure the capacitance of a capacitor, the capacitor current was measured for different p.d.s applied across the capacitor at a constant frequency of 2000 Hz. The results are in the Table below.

r.m.s. voltage/V	0.0	2.0	4.0	6.0	8.0	10.0
r.m.s. current/mA	0.0	12.1	24.0	35.5	48.5	60.0

(a) Plot a graph of the r.m.s. voltage against the r.m.s. current.

(b) Use your graph to calculate the reactance of the capacitor.

(c) Hence calculate the capacitance of this capacitor.

2. A capacitor is connected in series with an a.c. ammeter and a variable frequency signal generator, as in Fig. 20.22. An oscilloscope is connected across the capacitor terminals. The oscilloscope trace is also shown in Fig. 20.22.

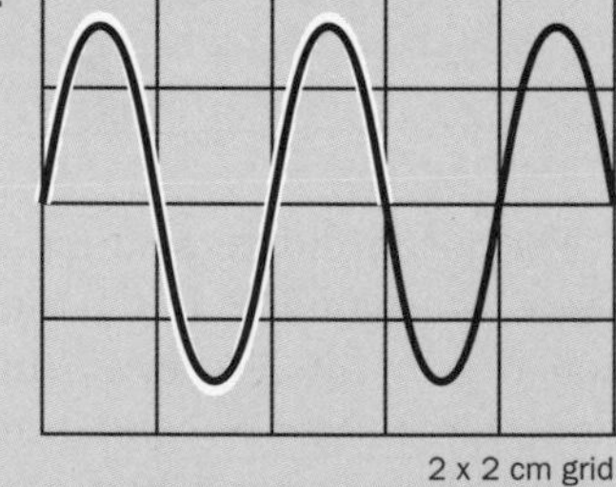

Fig. 20.22

(a) The voltage gain of the oscilloscope was 2.0 V cm^{-1}. Calculate **(i)** the peak voltage across the capacitor, **(ii)** the r.m.s. voltage across the capacitor.

(b) The time base of the oscillscope was 0.5 ms cm^{-1}. Calculate **(i)** the time period, **(ii)** the frequency of the alternating voltage.

(c) The r.m.s. current was 0.48 mA, measured using the ammeter. Calculate **(i)** the reactance, **(ii)** the capacitance of the capacitor.

20.5 Coils in a.c. circuits

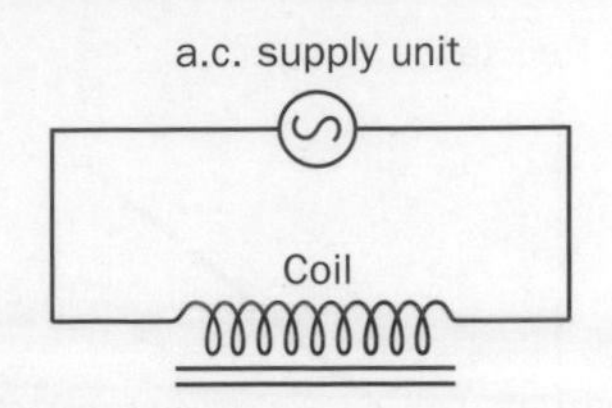

Fig. 20.23 An inductive coil in an a.c. circuit.

Consider an a.c. supply unit connected to a single coil of negligible resistance. As the current through the coil alternates, the changing magnetic field due to the alternating current induces an e.m.f. in the coil which acts against the alternating voltage applied to the coil.

The faster the frequency of the a.c. supply, the greater the rate of change of current and hence the greater the induced e.m.f. The voltage supply must therefore be made greater to keep the current the same if the frequency is raised. Therefore, the smaller the current, the higher the frequency if the supply voltage is unaltered. Conversely, the lower the frequency, the larger the current.

In mathematical terms, if $I = I_0 \sin(2\pi ft)$, and the self-inductance of the coil is L, then the induced e.m.f. V, can be written as

$$V = L\frac{dI}{dt} = L\frac{d}{dt}[I_0 \sin(2\pi ft)]$$

Because

$$\frac{d}{dt}[\sin(2\pi ft)] = 2\pi f \cos(2\pi ft)$$

as explained on p. 293, then

$$V = 2\pi fLI_0 \cos(2\pi ft)$$

As the induced e.m.f. acts against the supply voltage, the supply voltage must therefore be equal to the induced e.m.f. to force current around the circuit.

$$V = 2\pi fLI_0 \cos(2\pi ft)$$

1. The peak voltage, $V_0 = 2\pi fLI_0$, since the peak value of $\cos(2\pi ft) = 1$.
2. The p.d. is ahead of the current by one quarter cycle ($= 90^o$) as shown in Fig. 20.24. The phasor diagram for the current and voltage for a coil in an a.c. circuit is also shown in Fig. 20.24.
3. The **reactance, X_L**, of a coil of inductance, L, is defined as $2\pi fL$. Since $V_0 = 2\pi fLI_0$, then

$$2\pi fL = \frac{V_0}{I_0}$$

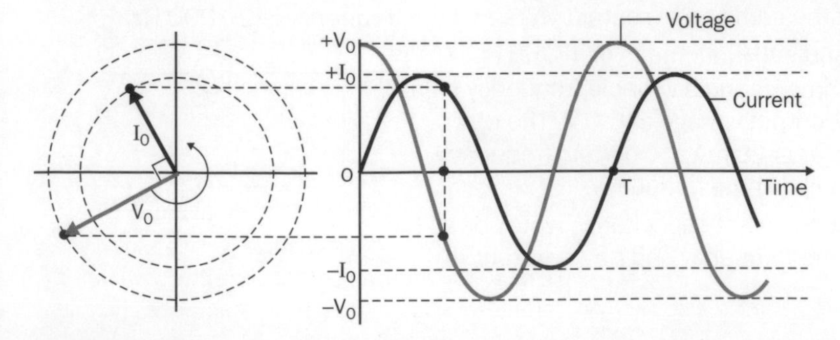

Fig. 20.24 The current and p.d. waveforms for a coil in an a.c. .circuit.

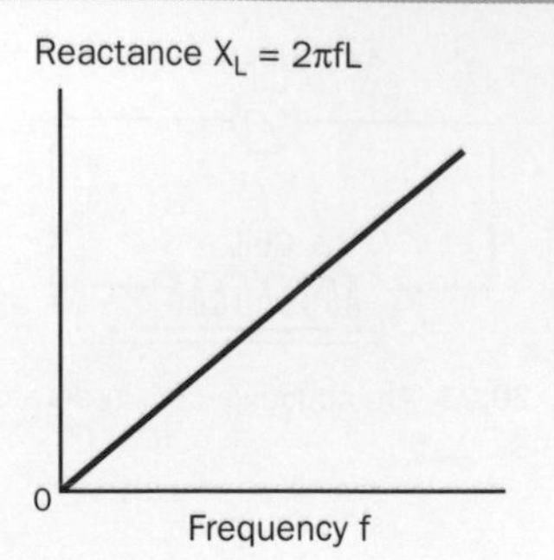

Fig. 20.25 Reactance v frequency for an inductor.

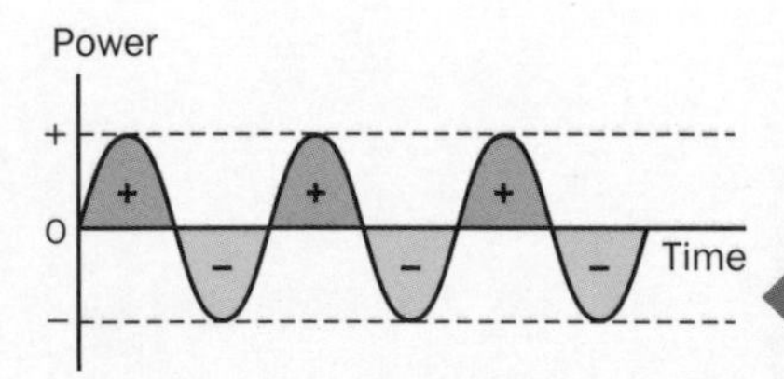

Fig. 20.26 The power curve for an inductor.

See www.palgrave.com/foundations/breithaupt for more information about measuring reactance of a coil.

The unit of reactance is therefore the ohm, the same as the unit of resistance.

$$X_L = \frac{V_0}{I_0} = 2\pi fL$$

Note that the reactance increases as the frequency increases. Fig. 20.25 shows how the reactance increases with increasing frequency.

4. Fig. 20.26 shows how the power (= current × potential difference) varies with time. Every other quarter cycle, the current and the potential difference are in the same direction, so IV is positive, so power is delivered by the supply unit. In between these quarter cycles, the current and the potential are in opposite 'directions' so IV is negative and power returns to the supply unit. As a result, no power is delivered from the supply unit over a complete cycle.

Worked example 20.4

A coil of self inductance 28 mH and negligible resistance is connected across the terminals of a 2000 Hz a.c. supply unit which has an r.m.s. output voltage of 1.5 V. Calculate (a) the reactance of the coil, (b) the r.m.s. current.

Solution

(a) $X_L = 2\pi fL = 2\pi \times 2000 \times 28 \times 10^{-3} = 352\ \Omega$

(b) $X_L = \frac{V_0}{I_0} = \frac{V_{rms}}{I_{rms}} \therefore I_{rms} = \frac{V_{rms}}{X_L} = \frac{1.5}{352} = 4.26 \times 10^{-3}$ A

Questions 20.5

1. A coil of negligible resistance is connected in series with an ammeter and a 1500 Hz a.c. supply unit with an r.m.s. output voltage of 12.0 V. The r.m.s. current in the circuit is 110 mA.
   (a) Calculate **(i)** the reactance of the coil, **(ii)** the self-inductance of the coil.
   (b) Calculate the r.m.s. current for the same r.m.s. output voltage if the frequency is 20 000 Hz.
2. **(a)** A coil of negligible resistance and self-inductance of 45 mH is connected in series with an a.c. ammeter and a variable frequency signal generator which supplies an r.m.s. output voltage of 6.0 V. The r.m.s. current measured using the ammeter is 28 mA at a certain frequency. Calculate **(i)** the reactance of the coil, **(ii)** the frequency.
   (b) If the coil is replaced by a capacitor, what capacitance would pass the same current as the coil for the same frequency and r.m.s. output voltage?

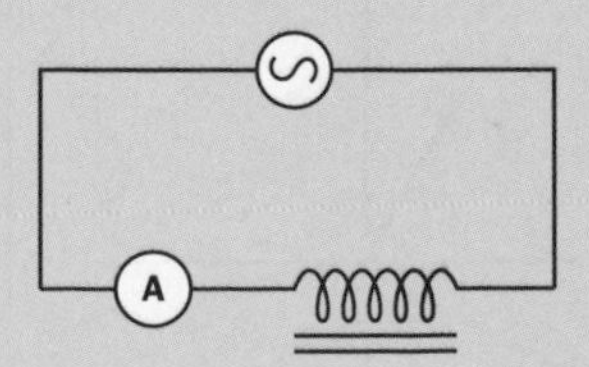

Fig. 20.27

20.6 Resonant circuits

Coils and capacitors have opposite effects in a.c. circuits.

- The reactance of a capacitor decreases with increasing frequency, whereas the reactance of a coil increases as the frequency increases.
- The current through a capacitor is 90° ahead of the voltage, whereas the current through a coil is 90° behind the voltage.

The series LCR circuit

Consider a circuit in which a coil of self-inductance, L, is in series with a capacitor of capacitance, C, and an a.c. supply unit. Let R represent the total resistance of the circuit, including resistors and internal resistance of the a.c. supply unit.

Suppose the current is represented by the equation $I = I_O \sin(2\pi ft)$.

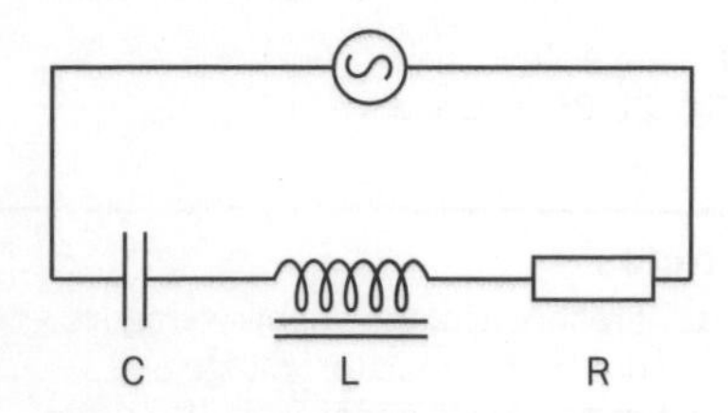

Fig. 20.28 A series LCR circuit.

1. The peak resistor voltage, $V_R = I_O R$ and the resistor voltage is in phase with the current.
2. The peak capacitor voltage

   $$V_C = I_0 X_C = I_0 \frac{1}{2\pi fC}$$

   and the capacitor voltage is behind the current one-quarter of a cycle.
3. The peak inductor voltage, $V_L = I_0 X_L = I_0 2\pi fL$ and the inductor voltage is ahead of the current by one-quarter of a cycle.

Fig. 20.29 shows the phasor diagram for these voltages in relation to the current phasor I_0. The capacitor voltage is 180° out of phase with the inductor voltage and therefore V_C and V_L are in opposite directions. Their combined voltage is therefore the difference between V_C and V_L. The peak supply voltage, V_0, is therefore the resultant of V_R and $(V_C - V_L)$ which are at 90° to each other. Using Pythagoras' theorem therefore gives

$$V_0^2 = V_R^2 + (V_C - V_L)^2$$

Since $V_R = I_0 R$, $V_C = I_0 X_C$ and $V_L = I_0 X_L$, then

$$V_0^2 = I_0^2 R^2 + (I_0 X_C - I_0 X_L)^2$$

Hence the circuit impedance, **Z, defined as** $\frac{V_0}{I_0}$ is given by

$$Z^2 = \frac{V_0^2}{I_0^2} = \frac{I_0^2 R^2 + (I_0 X_C - I_0 X_L)^2}{I_0^2} = R^2 + (X_C - X_L)^2$$

$$\therefore \mathbf{Z = [R^2 + (X_C - X_L)^2]^{1/2}}$$

Also, the angle between the current phasor and the supply voltage phasor ϕ is given by

$$\tan\phi = \frac{(V_C - V_L)}{V_R} = \frac{(X_C - X_L)}{R}$$

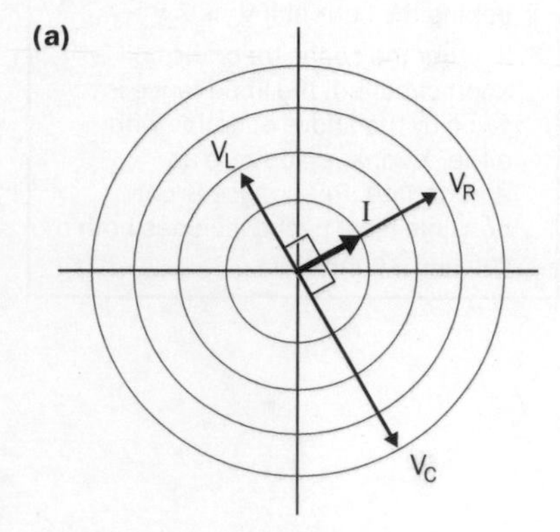

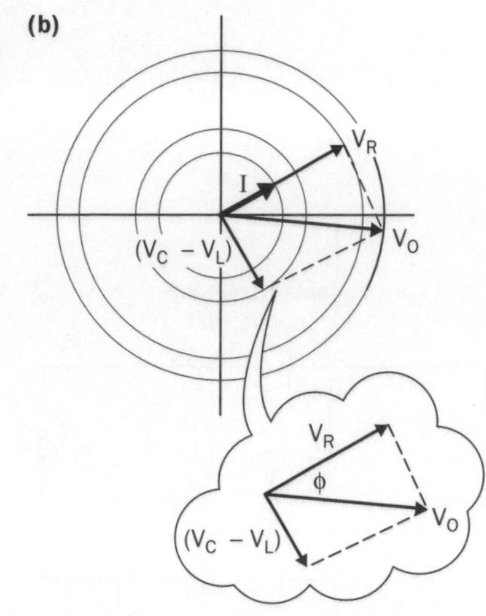

Fig. 20.29 The phasor diagram for a series LCR circuit **(a)** V_R, V_L, V_C phasors, **(b)** simplified phasor diagram.

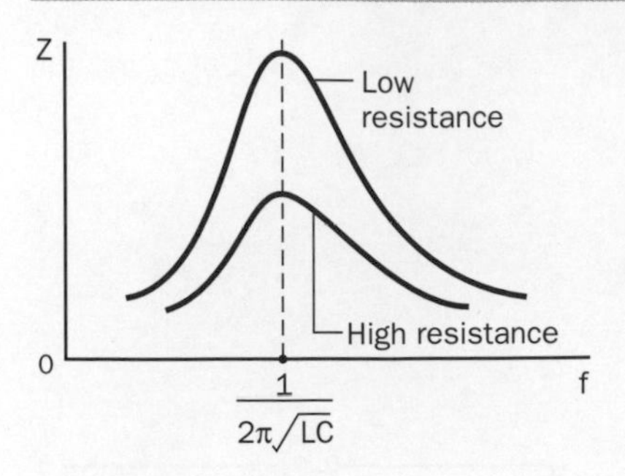

Fig. 20.30 Resonance.

The circuit impedance is a measure of the opposition of the whole circuit to the a.c. supply unit. Fig. 20.30 shows how the impedance of a series LCR circuit varies with frequency. The impedance is a minimum when $X_C = X_L$, corresponding to the capacitor voltage being exactly equal and opposite to the inductor voltage. The circuit is said to be in **resonance** at this frequency. The resonant frequency, f_O, is given by the equation

$$2\pi f_O L = \frac{1}{2\pi f_O C}$$

Rearranging this equation gives the following equation for the resonant frequency

$$\mathbf{f_O = \frac{1}{2\pi\sqrt{LC}}}$$

Notes

1. At resonance, the supply voltage is equal to the resistor voltage and therefore is in phase with the current.
2. Off resonance, the supply voltage is ahead of the current if $V_C < V_L$ and behind the current if $V_C > V_L$.
3. If either the capacitor or the coil is short-circuited, the impedance is given by the above equation with either X_C or X_L made zero as appropriate. Resonance is only possible if the circuit includes both a coil and a capacitor.

Worked example 20.5

A 2.2 μF capacitor was connected in series with a coil, an a.c. ammeter, and a variable frequency signal generator. The frequency was increased, keeping the output voltage from the signal generator at an r.m.s. voltage of 5.0 V. The r.m.s. current reached a maximum of 40 mA when the frequency was 1550 Hz. Calculate

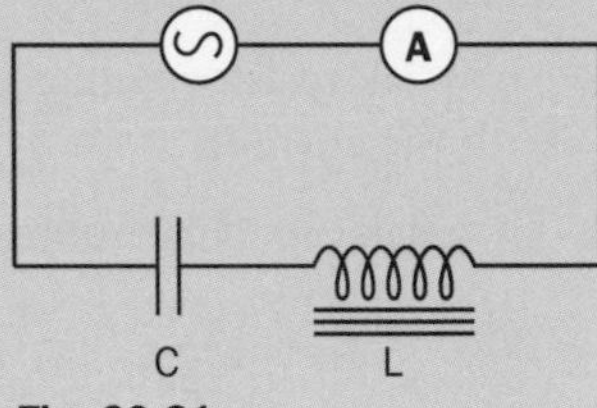

Fig. 20.31

(a) the impedance of the circuit at resonance, (b) the capacitor reactance at this frequency, (c) the self-inductance of the coil.

Solution

(a) $Z = \frac{V_O}{I_O} = \frac{V_{rms}}{I_{rms}} = \frac{5.0\text{ V}}{0.040\text{ A}} = 125\ \Omega$

(b) $X_C = \frac{1}{2\pi fC} = \frac{1}{2\pi \times 1550 \times 2.2 \times 10^{-6}} = 47\ \Omega$

(c) $X_L = X_C$ at resonance, so $2\pi fL = X_C$ $\therefore L = \frac{X_C}{2\pi f} = \frac{47}{2\pi \times 1550}$

$= 4.8 \times 10^{-3}$ H

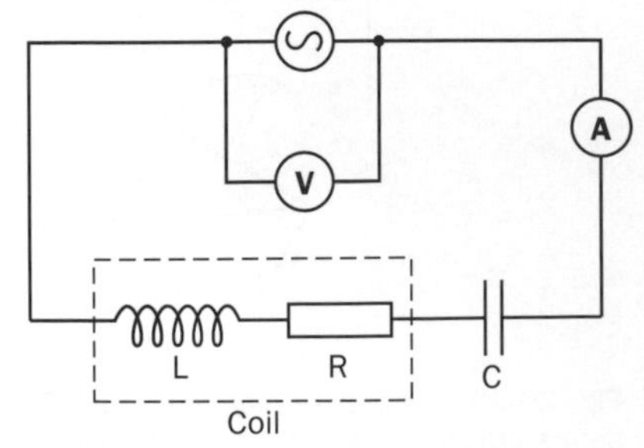

Fig. 20.32 Investigating resonance.

Investigating resonance of a series LCR circuit

1. Connect a coil and a capacitor of known capacitance in series with a digital ammeter and a variable frequency supply unit. Use a digital voltmeter to measure the r.m.s. supply voltage. Measure the frequency from the frequency dial of the supply unit. For different measured frequencies, measure the supply voltage and the current.
2. Use the results to plot a graph of the impedance, Z, against the frequency. Hence determine the resonant frequency and calculate the self-inductance of the coil.

Questions 20.6

1. A series LCR circuit consists of a 10 μF capacitor and a coil of self-inductance 0.48 H and resistance 55 Ω in series with a 50 Hz a.c. supply unit with an r.m.s. output voltage of 6.0 V.

(a) Calculate **(i)** the reactance of the capacitor, **(ii)** the reactance of the coil, **(iii)** the circuit impedance, **(iv)** the r.m.s. current.

(b) Calculate the r.m.s. voltage across **(i)** the resistance, **(ii)** the self-inductance, **(iii)** the capacitance.

(c) **(i)** Sketch the phasor diagram for this circuit. **(ii)** Calculate the angle between the current and the supply voltage phasors.

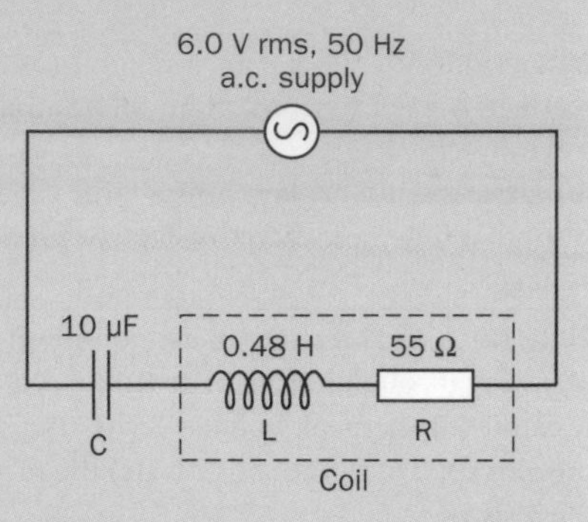

Fig. 20.33

2. A circuit consists of a coil of self-inductance 0.15 mH in series with a capacitor of capacitance 0.47 μF, an a.c. ammeter and a variable frequency signal generator.

(a) Calculate **(i)** the resonant frequency of this circuit, **(ii)** the reactance of the capacitor at resonance.

(b) At resonance, the r.m.s. current is 350 mA when the r.m.s. supply voltage from the signal generator is 5.0 V.

(i) Calculate the resistance of the circuit.

(ii) Calculate the voltage across the capacitor at resonance.

(iii) Sketch the phasor diagram for this circuit at resonance.

Summary

◆ **Frequency** $f = \dfrac{1}{\text{time period T}}$

◆ **R.m.s. value** $= \dfrac{\text{peak value}}{\sqrt{2}}$ for sinusoidal a.c.

◆ **Power** $= I_{rms} V_{rms} = \dfrac{1}{2} I_0 V_0$

◆ **Components in a.c. circuits**

Type of circuit	V_O/I_O	Phase	Power
Resistance only	R	0	$\frac{1}{2} I_0^2 R$
Capacitance only	$X_C = \dfrac{1}{2\pi fC}$	I ahead of V by 90°	zero
Inductance only	$X_L = 2\pi fL$	I behind V by 90°	zero
Series LCR	$Z = [R^2 + (X_C - X_L)^2]^{\frac{1}{2}}$	$\tan\phi = \dfrac{(X_C - X_L)}{R}$	$\frac{1}{2} I_0^2 R$

◆ **Resonance**

1. Current and supply voltage in phase as $X_L = X_C$

2. Resonant frequency, $f_O = \dfrac{1}{2\pi\sqrt{(LC)}}$

Revision questions

20.1. **(a)** With the aid of a diagram, describe how a bridge rectifier is used to convert an alternating voltage into a full-wave direct voltage.
(b) Describe and explain how a large capacitor may be used to smooth a full-wave direct voltage to produce a constant direct voltage.

20.2. **(a)** A 47 Ω resistor in an a.c. circuit has an r.m.s. current of 0.75 A through it. Calculate **(i)** the peak current through the resistor, **(ii)** the peak voltage across the resistor, **(iii)** the peak power supplied to the resistor, **(iv)** the mean power supplied to the resistor.
(b) An a.c. supply unit has a sinusoidal output voltage of peak value 12.0 V and of frequency 200 Hz. Sketch a phasor diagram to show the voltage phasor when **(i)** the voltage is zero and changing from negative to positive, **(ii)** 2.0 ms after its position in (i), **(iii)** 6.0 ms after its position in (i). State the voltage in (ii) and (iii).

20.3. In an experiment to measure the capacitance of a capacitor, an a.c. ammeter was connected in series with the capacitor and an a.c. supply unit. An oscilloscope was connected across the capacitor. Fig. 20.34 shows the circuit and the oscilloscope screen.

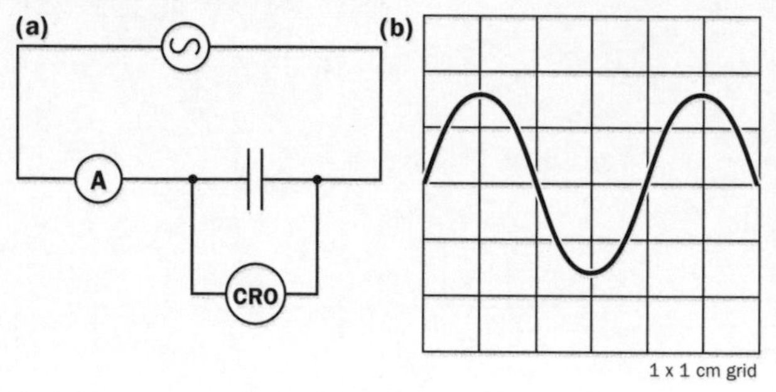

Fig. 20.34 **(a)** circuit diagram, **(b)** waveform.

(a) The Y-gain of the oscilloscope was set at 2.0 V cm^{-1} and the time base was set at 5 ms cm^{-1}.
Calculate **(i)** the frequency of the alternating current, **(ii)** the peak voltage across the capacitor, **(iii)** the r.m.s. voltage across the capacitor.
(b) The r.m.s. current was 2.8 mA, measured using the a.c. ammeter. Calculate **(i)** the reactance of the capacitor, **(ii)** the capacitance.

20.4. **(a)** A coil of negligible resistance was connected in series with an a.c. ammeter and a 50 Hz. a.c. supply unit. The r.m.s. current was 0.35 A, measured using the a.c. ammeter, when the output voltage of the a.c. supply unit was adjusted to an r.m.s. value of 6.0 V. Calculate **(i)** the reactance of the coil, **(ii)** the self-inductance of the coil.
(b) Calculate the capacitance of a capacitor that would give the same ammeter reading at this frequency for the same output voltage.
(c) If the frequency of the a.c. supply unit was changed to 100 Hz, what difference would it make to the current for the same output voltage **(i)** in (a) above, **(ii)** in (b) above.

20.5. A 2.2 μF capacitor and a coil of unknown resistance R and self-inductance L were connected in series with an a.c. ammeter and a variable frequency signal generator. The r.m.s. output voltage of the supply unit was maintained at 6.0 V and the frequency was adjusted until the r.m.s current, measured using the ammeter, was at a maximum of 65 mA. This occurred at a frequency of 750 Hz.
(a) Explain why the r.m.s. current is at a maximum at 750 Hz.
(b) Calculate **(i)** the reactance of the capacitor at this frequency, **(ii)** the self-inductance of the coil, **(iii)** the resistance of the coil.
(c) Calculate the r.m.s. voltage at this frequency across **(i)** the resistance, **(ii)** the capacitance, **(iii)** the inductance.
(d) Sketch the phasor diagram and explain why the current is in phase with the supply voltage at this frequency.

Atomic and Nuclear Physics

In Part 6, the properties of electrons in and out of the atom are studied in depth, linking up to the nature of light and why optical spectra can be used to identify elements. The principle of operation of the X-ray tube is developed before moving on to consider the properties and nature of radioactive substances. The section moves on to consider how energy is released from the nucleus and how the structure of matter at its most fundamental level has been discovered. The section includes essential practical work and makes use of mathematical skills developed earlier. Before commencing this part, it is useful to have studied Unit 4, Electromagnetic Waves; Unit 5, Matter and Molecules; Unit 12, Energy and Power; Unit 17, Electric Fields and Unit 18, Magnetic Fields.

Electrons and Photons

Objectives

After working through this unit, you should be able to:

- ▶ describe and explain how a beam of electrons is produced
- ▶ calculate the speed of the electrons in a beam from the accelerating voltage
- ▶ measure the specific charge of the electron, e/m
- ▶ measure the charge of the electron, e
- ▶ describe and explain the photoelectric effect
- ▶ carry out calculations using the photon energy equation, $E = hf$
- ▶ describe and explain ionisation and excitation of atoms
- ▶ use energy level diagrams to explain line spectra
- ▶ describe the principle of operation of an X-ray tube and the intensity distribution of its output
- ▶ describe how X-rays are used in medicine

Contents

21.1 Electron beams

Cathode rays

An atom is the smallest particle of an element that is characteristic of the element. The atomic theory of matter was first proposed by Democritus in Ancient Greece. The theory was developed on a scientific basis by John Dalton and fellow chemists in the 19th century to describe how elements form compounds as a result of atoms joining together to form molecules. The molecular theory of matter successfully explains the general features of solids, liquids and gases. Dalton put forward the idea that atoms are indestructible and showed that the hydrogen atom is lighter than any other type of atom. The atomic theory was a very successful theory because it explained many observed reactions and correctly predicted many new reactions.

Imagine the surprise when it was discovered that atoms contain much

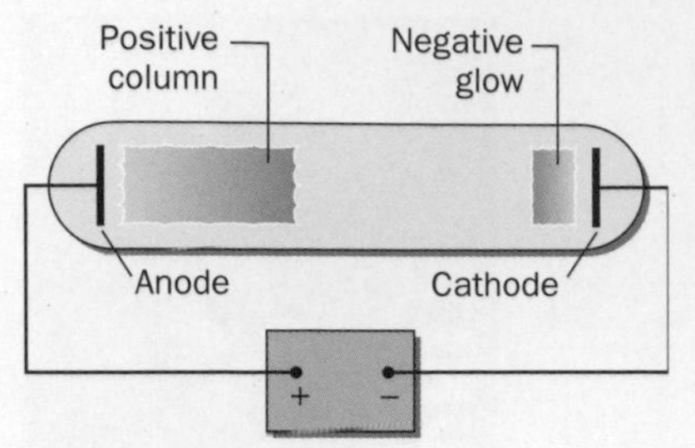

Fig. 21.1 A discharge tube.

smaller particles which carry a negative charge. These particles became known as **electrons** because they carry electric charge. They were discovered as a result of investigations into electrical conduction through gases at low pressures. Fig. 21.1 shows a discharge tube used for such an investigation.

With a very high voltage between the two electrodes, the gas is evacuated from the tube using a vacuum pump. At a sufficiently low pressure, the gas remaining inside the tube glows with light. This occurs because electrons within each atom are pulled out by the strong electrical field created by the high voltage. The atoms become positively charged ions as a result of losing electrons and are attracted to the **cathode**, the negative electrode in the tube. The electrons are attracted to the **anode**, the positive electrode. The gas glows with light because positive ions moving to the cathode recombine with electrons moving to the anode, releasing light when they combine.

When these investigations were first carried out, the cause of the glow was a mystery. It was realised that the glowing gas contained charged particles moving towards each end because a magnet held near the tube distorted the glow. This is the same effect as the force on a current-carrying wire in a magnetic field which is used to explain the operation of an electric motor. The investigations on gas discharges became possible as a result of the invention of high-voltage generators and better vacuum pumps. It was thought that radiation from the cathode, referred to as **cathode rays**, caused ionisation in the gas.

Scientists used different gases in redesigned tubes to find out more about the nature of cathode rays and the charged particles produced in the discharge tube. By using an electrode with a small hole in it, some of the charged particles attracted to the electrode passed through the hole to form a beam. Such beams could then be deflected by a magnetic field, or by an electric field created by two oppositely charged plates, as in Fig. 21.2.

The end of the tube forms a screen coated internally with fluorescent paint. The impact of the beam on this screen creates a spot of light at the point of impact. The effect of a magnetic or electric field on the beam can therefore be measured. The results showed the following:

1. The deflection of the positive ions by an electric or a magnetic field depends on the gas. For example, positive ions from chlorine gas are

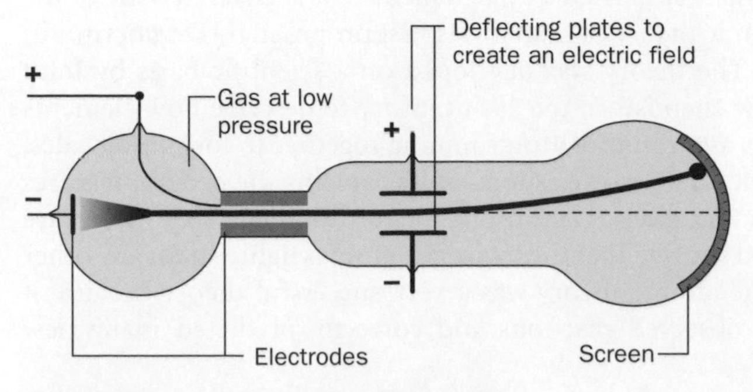

Fig. 21.2 Testing a beam of charged particles.

deflected less easily than positive ions from oxygen gas. It was realised that the heavier the mass of a molecule of a gas, the smaller the deflection. The positive ions were therefore identified as gas molecules that had become positively charged in the discharge tube.

2. The deflection of the cathode rays does not depend on the type of gas in the tube. The exact nature of the cathode rays was disputed and some scientists thought that these rays were waves like X-rays which ionised the gas, whereas others thought they were definitely composed of negatively charged particles. The dispute was settled in 1897 by J.J. Thomson, who proved conclusively that cathode rays are composed of identical, negatively charged particles, referred to subsequently as **electrons**. Thomson carried out a series of detailed experiments using electric and magnetic fields to deflect a beam of cathode rays and discovered that the electron's charge to mass ratio, its specific charge, e/m, is 1800 times greater than the hydrogen ion's specific charge, the largest known specific charge at that time. This was known to be 9.6×10^7 C kg^{-1} from electrolysis experiments (see Section 13.2) in which it was found that 9.6×10^7 C of charge was needed to release 1 kg of hydrogen gas. Thomson's measurements on cathode rays gave a specific charge of 1.8×10^{11} C kg^{-1}.

Fig. 21.3 Joseph John Thomson (1856–1940).

The reason why the specific charge of the electron is much higher than that of the hydrogen ion could be due either to the electron having a much greater charge or a much smaller mass than the hydrogen ion. Thomson had no firm evidence on whether a much greater charge or a much smaller mass was the reason for the much greater specific charge. Nevertheless, he proved conclusively that cathode rays are sub-atomic, negative particles. In 1915 in the United States, Robert Millikan measured the charge of the electron (see p. 308) and showed that electrons and hydrogen ions carry equal and opposite charges, thus proving conclusively that the electron is much lighter than the atom.

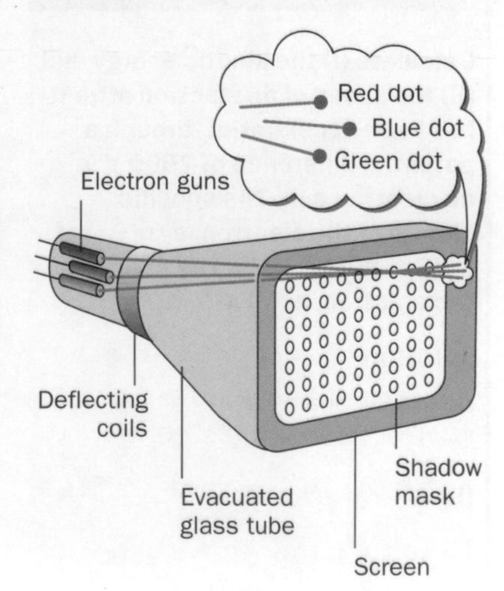

Fig. 21.4 Inside a TV tube.

Thermionic emission

Every television and microcomputer VDU uses one or more beams of electrons to build up a picture on its screen. An electron beam in a television is controlled by a pair of electromagnets which are supplied with a varying electric current from a control circuit. The beam creates a spot of light where it hits the screen. The picture is built up as a result of the beam scanning the screen. The beam itself is produced by an effect known as **thermionic emission**. This was discovered by Thomas Edison in 1903, who demonstrated that electrons could be attracted onto a positive metal plate from a hot wire filament in a vacuum. This discovery led to the invention of the electron gun which produces a narrow beam of electrons in every TV tube.

Fig. 21.5 shows how an electron gun in an evacuated tube works. The filament glows as a result of an electric current passing through it. At this temperature, conduction electrons in the filament possess sufficient kinetic energy to leave the filament. These thermionically emitted electrons are

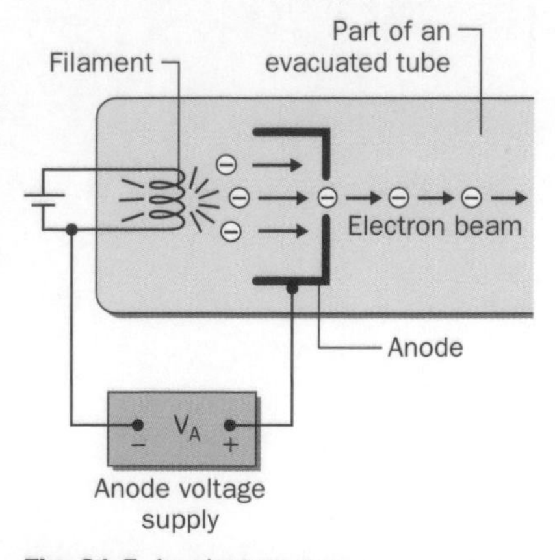

Fig. 21.5 An electron gun.

Notes

1. **One electron volt (eV)** is defined as the work done on an electron when it is moved through a potential difference of 1.0 V. Since e = 1.6×10^{-19}C, then 1 electron volt is equal to 1.6×10^{-19} J.
2. The kinetic energy formula $\frac{1}{2}mv^2$, is only valid at speeds much less than the speed of light ($= 3 \times 10^8$ m s^{-1}). See p. 354.

attracted to the metal anode because it is positive, relative to the filament. The electrons gain more kinetic energy as they are attracted to the anode. Some of these electrons pass through a small hole in the anode to form a beam of electrons. The presence of gas molecules would prevent the electrons reaching the anode so the apparatus must be contained in a vacuum tube.

By definition, the potential difference between the filament and the anode is the work done per unit charge moved between the filament and the anode. The work done on each electron attracted from the filament to the anode is therefore eV_A, where V_A is the potential difference between the anode and the filament (i.e. the anode voltage) and e is the charge of the electron. As the work done on each electron increases its kinetic energy very significantly, its kinetic energy at the anode is therefore eV_A. The speed, v, of each electron in the beam is therefore given by the following equation

$$\frac{1}{2}mv^2 = eV_A$$

Worked example 21.1

Calculate (i) the kinetic energy and (ii) the speed of an electron after it has been accelerated through a potential difference of 2500 V in an electron gun. The specific charge of the electron, e/m, = 1.76×10^{11} C kg^{-1}. The charge of the electron, e = 1.6×10^{-19} C.

Solution

(i) Kinetic energy = $eV_A = 1.6 \times 10^{-19} \times 2500 = 4.0 \times 10^{-16}$ J

(ii) $\frac{1}{2}mv^2 = eV_A$, hence $v^2 = \frac{2eV_A}{m}$

$= 2 \times 1.76 \times 10^{11} \times 2500$

$= 8.8 \times 10^{14}$ m^2 s^2

$\therefore v = (8.8 \times 10^{14})^{1/2}$

$= 3.0 \times 10^7$ m s^{-1}

Measurement of the specific charge of the electron, e/m

Fig. 21.6 shows a vacuum tube in which a narrow beam of electrons from an electron gun is directed between two oppositely charged, parallel plates. The path of the beam between the plates is made visible by a vertical fluorescent screen in line with the beam.

- With a constant potential difference, V_P, between the parallel plates, the electron beam is attracted towards the positive plate.

The force due to the plate voltage $= \frac{(eV_P)}{d}$

where d is the separation between the plates. See p. 242 for a proof of this equation.

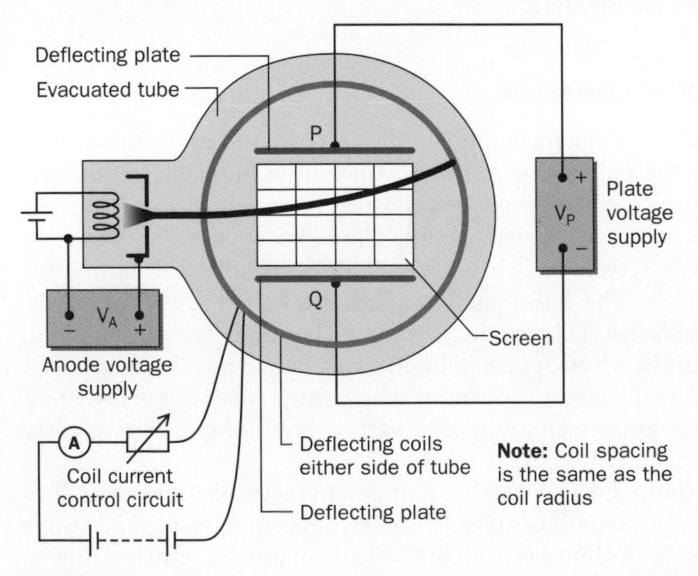

Fig. 21.6 Measuring e/m.

- A uniform magnetic field applied perpendicular to the plane of the screen deflects the beam down or up depending on the magnetic field direction. Two current-carrying coils in series are used to apply the magnetic field. The strength of the magnetic field and hence the deflection of the beam depends on the current in the coils.

  The force due to the magnetic field = $Be\upsilon$

  where υ is the speed of the electrons and B is the magnetic flux density. See p. 259 for a proof of this equation.

The deflection due to the plate voltage can be cancelled out by the magnetic field. To do this, the beam is first deflected by a constant plate voltage and then the coil current is gradually increased so that the deflection becomes less and less until the beam passes straight through. This occurs when the force due to the plates is equal and opposite to the force due to the magnetic field. Hence the beam passes through undeflected when the magnetic flux density is such that

$$Be\upsilon = \frac{(eV_P)}{d}$$

Thus the speed of the electrons can be calculated from the equation

$$\upsilon = \frac{V_P}{Bd}$$

The electron gun equation, $\frac{1}{2}m\upsilon^2 = eV_A$, can then be used to calculate the specific charge of the electron.

Note

The magnetic flux density can be calculated from the coil current using the equation $B = 0.9 \times 10^{-6} NI \div R$, where N is the number of turns of each coil, R is the coil radius and I is the coil current. The unit of B is the tesla (T).

Worked example 21.2

In an experiment to measure the specific charge of the electron, e/m, a beam of electrons was produced from an electron gun operating at an anode voltage of 3500 V. The beam was directed between two plates at a potential difference of 2800 V and a separation of 50 mm. This deflection was cancelled by a uniform magnetic field of magnetic flux density 1.6 mT. Calculate (a) the speed of the electrons in the beam, (b) the specific charge of the electron, e/m.

Solution

(a) $\upsilon = \frac{V_P}{Bd} = \frac{2800}{1.6 \times 10^{-3} \times 50 \times 10^{-3}}$

$= 3.5 \times 10^7 \text{ m s}^{-1}$

(b) Rearranging $\frac{1}{2}m\upsilon^2 = eV_A$ gives

$e/m = \frac{\upsilon^2}{2V_A} = \frac{(3.5 \times 10^7)^2}{2 \times 3500}$

$= 1.75 \times 10^{11} \text{ C kg}^{-1}$

Questions 21.1

The specific charge of the electron, $e/m = 1.76 \times 10^{11} \text{C kg}^{-1}$.

1. Calculate the speed of electrons accelerated from rest in a vacuum through a potential difference of **(a)** 100 V, **(b)** 4000 V.
2. A beam of electrons from an electron gun operating at an anode voltage of 3200 V is directed between two parallel plates as in Fig. 21.5. A potential difference of 4200 V is applied between the plates which are spaced 40 mm apart. The deflection of the beam is cancelled when a uniform magnetic field is applied to the beam. Calculate **(a)** the speed of the electrons in the beam, **(b)** the magnetic flux density of this magnetic field.

Fig. 21.7 Robert Andrews Millikan (1868–1953).

21.2 The charge of the electron

An electric current in a wire is a flow of charge carried by electrons forced along the wire by the source of e.m.f. A wire carrying a current of 1 ampere has 6.25 million million million electrons passing along its length every second. The fact that each electron carries a charge of 1.6×10^{-19}C was discovered by Robert Millikan, as a result of measuring the charge carried by tiny oil droplets. He also deduced that the charge of the electron is the least possible amount of charge and that the charge of any charged body is always a multiple of the charge of the electron. The method Millikan used over 80 years ago to determine the charge of the electron is described in this section. The result of this measurement was of profound importance, as all matter is now known to be composed of quarks in combination with each other (see p. 356) and electrons.

Balancing an oil droplet

Millikan observed oil droplets from an oil spray as they fell between two parallel, horizontal metal plates, one above the other, in a beam of light. When a potential difference was applied between the two plates, Millikan discovered that the many of the droplets were affected by the electric field created between the plates. Some droplets fell faster, whereas some reversed their motion and rose.

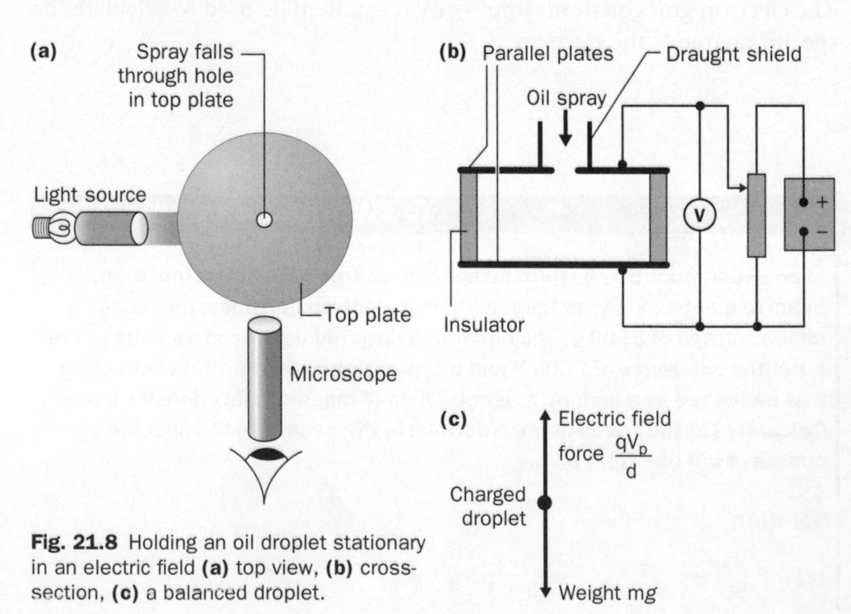

Fig. 21.8 Holding an oil droplet stationary in an electric field **(a)** top view, **(b)** cross-section, **(c)** a balanced droplet.

By adjusting the plate p.d., Millikan was able to hold individual droplets, one at a time, in the electric field. This happens if the droplet carries an opposite charge to the top plate which causes it to be attracted by the top plate with a force equal and opposite to its weight.

As explained on p. 242, the electrostatic force on a charged droplet

$= QV_P/d$, where d is the separation of the plates and Q is the charge of the droplet.

Hence $QV_P/d = mg$, where m is the mass of the droplet.

If the droplet mass is known, the droplet charge can therefore be calculated from the equation,

$$Q = \frac{mgd}{V_P}$$

Millikan discovered that the charge on an oil droplet is always a whole number $\times 1.6 \times 10^{-19}$C, and that the least possible charge is 1.6×10^{-19}C. He concluded that the charge on the electron is 1.6×10^{-19}C and that this is the **quantum** of electric charge. Any charged body has either gained or lost a number of electrons and therefore its charge is a whole number $\times$ e, the charge of the electron.

Worked example 21.3

A droplet of mass 6.2×10^{-15} kg was held stationary between two parallel plates spaced 5.0 mm apart when the plate p.d. was 950 V. Calculate (a) the weight of the droplet, (b) the charge of the droplet.

Solution

(a) Weight $= mg = 6.2 \times 10^{-15} \times 9.8 = 6.1 \times 10^{-14}$ N

(b) Rearranging $\frac{QV_P}{d} = mg$ gives $Q = \frac{mgd}{V_P} = \frac{6.1\times10^{-14} \times 5.0\times10^{-3}}{950}$

$= 3.2\times10^{-19}$ C

Measuring the mass of an oil droplet

With the plate p.d. switched off, each droplet falls at a constant speed, being acted upon by a drag force due to the air. When the plate p.d. is first switched off, the droplet accelerates due to gravity. As the droplet moves faster and faster, the drag force increases until it becomes equal and opposite to the force of gravity on the droplet. The resultant force is then zero and the droplet moves at its terminal speed. The build-up to terminal speed is almost instant so the terminal speed can be measured by timing the droplet as it falls through a measured distance. The terminal speed, v, can therefore be calculated from

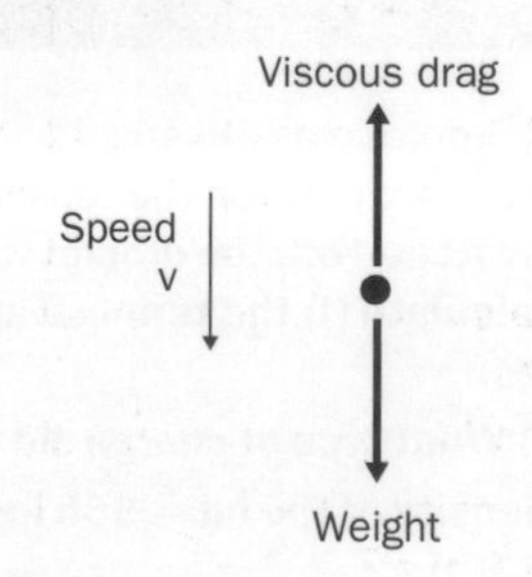

Fig. 21.9 Terminal speed.

$$v = \frac{s}{t}$$

where s is the measured distance and t is the time taken.

Assuming the droplet is a sphere of radius r:

1. Its volume $= \frac{4}{3}\pi r^3$, hence its mass, $m = \frac{4}{3}\pi r^3 \rho$, where ρ is the density of the oil.
2. The viscous drag, $F = 6\pi\eta r v$, where η is the viscosity of air, in accordance with Stokes' law.

At the terminal speed υ, the droplet weight, mg = the viscous drag, F. Hence

$$\frac{4}{3}\pi r^3 \rho g = 6\pi\eta r \upsilon$$

assuming buoyancy is negligible.

The droplet radius can therefore be calculated by rearranging this equation to give

$$r^2 = \frac{9\eta\upsilon}{2\rho g}$$

The droplet mass can then be calculated using the equation

$$m = \frac{4}{3}\pi r^3 \rho$$

To determine the charge of a charged oil droplet

1. The plate p.d. is adjusted and measured when the droplet is stationary.
2. The plate p.d. is then switched off and the droplet is timed as it falls through a measured distance.
3. The droplet's terminal speed is calculated. $\left(\text{using } \upsilon = \frac{s}{t}\right)$
4. The droplet radius is then calculated. $\left(\text{using the equation } r^2 = \frac{9\eta\upsilon}{2\rho g}\right)$
5. The droplet mass is then calculated. (using the equation $m = \frac{4}{3}\pi r^3 \rho$)
6. The droplet charge is then calculated. $\left(\text{using } Q = \frac{mgd}{V_P}\right)$

Worked example 21.4

An oil droplet was observed between two horizontal plates which were at a separation of 4.2 mm. The droplet was held stationary when the top plate was at a positive potential of 235 V relative to the lower plate. When the plate p.d. was switched off, the droplet fell at a constant speed, taking 11.3 s to fall through a distance of 1.0 mm.
(a) Calculate (i) the terminal speed of the droplet, (ii) the droplet radius, (iii) the droplet mass, (iv) the droplet charge.
(b) (i) What type of charge did the droplet carry? (ii) How many electrons were responsible for the droplet's charge?
The density of the oil = 950 kg m^{-3}; the viscosity of air = 1.8×10^{-5} N s m^{-1}; the acceleration due to gravity, $g = 9.8$ m s^{-2}.

Solution

(a) (i) $\upsilon = \frac{s}{t} = \frac{1.0 \times 10^{-3}}{11.3} = 8.8 \times 10^{-5}$ m s^{-1}

(ii) $r^2 = \frac{9\eta\upsilon}{2\rho g} = \frac{9 \times 1.8 \times 10^{-5} \times 8.8 \times 10^{-5}}{2 \times 950 \times 9.8}$

$= 7.7 \times 10^{-13}$ m^2

$\therefore r = 8.8 \times 10^{-7}$ m

(iii) $m = \frac{4}{3}\pi r^3 \rho = \frac{4}{3}\pi (8.8 \times 10^{-7})^3 \times 950$

$= 2.7 \times 10^{-15}$ kg

(iv) $Q = \frac{mgd}{V_P} = \frac{2.7 \times 10^{-15} \times 9.8 \times 4.2 \times 10^{-3}}{235}$

$= 4.7 \times 10^{-19}$ C

(b) (i) negative, **(ii)** $3 \left(= \frac{4.7 \times 10^{-19}}{1.6 \times 10^{-19}}\right)$

Questions 21.2

$g = 9.8\ \text{m s}^{-2}$

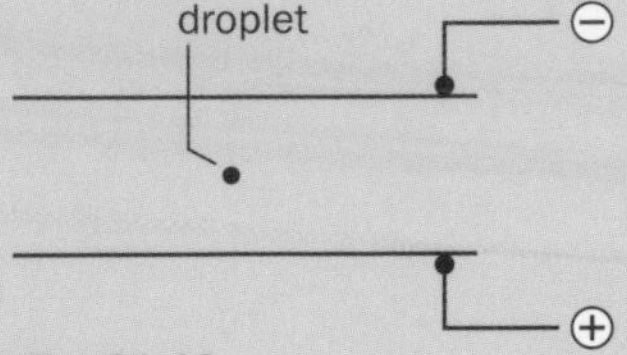

1. A charged oil droplet of mass 3.8×10^{-15} kg was observed between two oppositely charged parallel plates, one above the other, at a separation of 50 mm. With the upper plate negative, the p.d. between the two plates was adjusted to make the droplet stationary. The p.d. needed to do this was 595 V.

   (a) State the sign of the charge on the droplet.

   (b) Calculate the charge carried by this droplet.

Charged oil droplet

⊖

⊕

Fig. 21.10

2. In an experiment to measure the mass and charge of an oil droplet, the following readings were taken:

   Plate p.d. to hold the droplet stationary = 375 V

   Plate separation = 40 mm

   Time taken by the droplet to fall a distance of 1.0 mm = 16.5 s for zero plate p.d.

   (The density of the oil = 960 kg m^{-3}, the viscosity of air = 1.8×10^{-5} N s m^{-1})

   (a) Calculate **(i)** the terminal speed of the droplet when it fell in zero electric field, **(ii)** the droplet's radius, **(iii)** the droplet's mass, **(iv)** the droplet's charge.

   (b) Without calculation, **(i)** explain why the droplet fell at steady speed when the plate p.d. was zero, **(ii)** describe and explain the droplet's motion if the plate p.d. was suddenly reversed when the droplet was stationary.

21.3 Photoelectricity

Electrons are emitted from a metal plate as a result of illuminating the metal plate with ultraviolet light. Fig. 21.11 shows how this can be demonstrated. The metal plate is attached to a gold leaf electroscope and charged negatively. When the plate is illuminated with ultraviolet light, the gold leaf gradually falls. If a glass sheet is inserted between the metal plate and the ultraviolet light source, the leaf stops falling because glass does not transmit ultraviolet light. Removing the glass sheet causes the leaf to continue its fall. The leaf falls because electrons are emitted from the metal plate when the plate is illuminated with ultraviolet light.

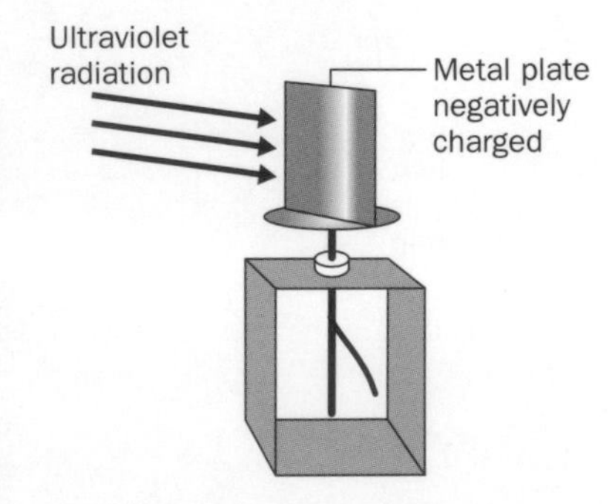

Fig. 21.11 Photoelectricity.

This effect, called **photoelectric emission**, was discovered over a century ago. It led to a completely new theory of light and a drastic rethink about the nature of matter and energy. The key discovery was that photoelectric emission from a metal surface does not occur unless the frequency of the incident radiation exceeds a minimum value, referred to as the **threshold frequency**, that depends on the metal. For example, in Fig. 21.11, photoelectric emission does not occur with visible light even if the incident light is extremely intense, yet ultraviolet light at very low intensity is capable of producing photoelectric emission. Investigations on photoelectricity carried out at the end of the 19th century produced the following results:

1. Photoelectric emission only occurs if the frequency of the incident radiation exceeds a threshold value that depends on the metal.

2. Increasing the intensity of the incident radiation increases the number of photoelectrons per second emitted by the metal plate, provided the frequency exceeds the threshold frequency. If the frequency is less than the threshold frequency, no photoelectric emission occurs no matter how intense the incident radiation is.
3. Emission of electrons occurs instantly when the incident radiation is directed at the metal surface.

The results could **not** be explained using the wave theory of light. According to wave theory, every wavefront arriving at the metal surface ought to share its energy with all the conduction electrons on or near the surface. The electrons would therefore gain enough energy to leave the metal plate, regardless of the frequency of the incident radiation. Photoelectric emission ought to occur at all frequencies according to the wave theory of light. The results demonstrated the existence of a threshold frequency and this could not be explained by the wave theory of light.

The photon theory of light

Einstein explained the photoelectric effect by putting forward the theory that electromagnetic radiation consists of packets of waves called **photons**, each carrying energy, E, in proportion to the frequency, f, of the radiation in accordance with the equation

Energy of a photon, $E = hf$

where h, known as the Planck constant, has a numerical value of 6.63×10^{-34} Js.

The idea that atoms emit and absorb radiation in discrete amounts or 'quanta' was first advanced by Planck to explain black body radiation (see p. 51). Einstein realised that the radiation itself is quantised and that when an atom emits electromagnetic radiation, it releases a short 'burst' of electromagnetic waves that leave the atom in one direction rather than spreading out in all directions.

According to Einstein, photoelectric emission from a metal can only occur if an electron at the surface of the metal gains sufficient energy from a single photon to leave the metal. Since each photon incident on the metal surface is absorbed by a single electron, a photoelectron (i.e. an electron that has escaped from the surface) must have gained energy equal to hf as a result of absorbing a photon. Because the electron needs to use some of this energy to leave the metal, its maximum kinetic energy on leaving the metal is equal to $hf - \phi$, where the **work function, ϕ**, of a metal is defined as the work that must be done by an electron to leave the metal completely.

Maximum kinetic energy of an emitted electron $= hf - \phi$

According to Einstein's theory, electrons cannot escape from the metal surface if the photon energy, hf, is less than the work function of the metal, ϕ.

$$\therefore hf > \phi$$

for photoelectric emission to occur.

The threshold frequency, f_o, the minimum frequency of the incident radiation for photoelectric emission to occur, is therefore given by the condition $hf_o = \phi$.

$$\therefore f_o = \frac{\phi}{h}$$

Photoelectric emission can be stopped by making the metal surface sufficiently positive, relative to the metal terminal where the photoelectrons are collected. When the metal surface is at a positive potential, V, relative to the collecting terminal, each electron must do extra work, equal to eV, to move from the metal plate to the collecting terminal. Photoelectric emission is therefore stopped if the potential difference is such that $eV_S = hf - \phi$, where V_S is referred to as the **stopping potential**.

$$eV_S = hf - \phi$$

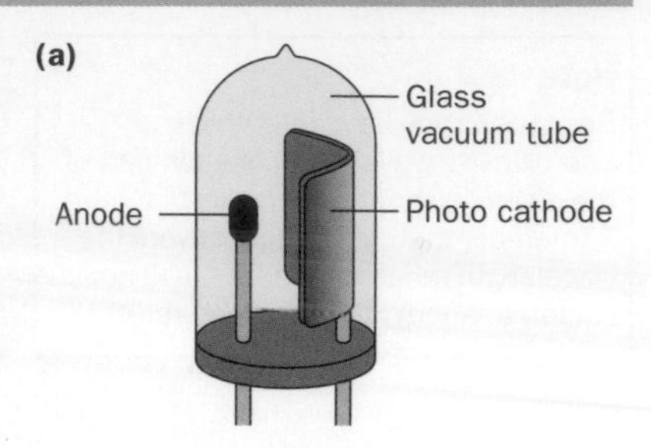

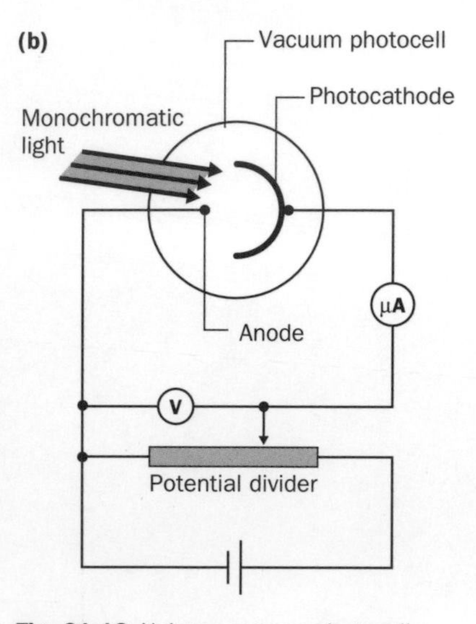

Fig. 21.12 Using a vacuum photocell **(a)** construction, **(b)** measuring stopping potential.

Investigating photoelectric emission using a vacuum photocell

A vacuum photocell contains a clean metal plate, the photocathode, capable of emitting electrons when illuminated with visible radiation. A smaller metal terminal near the plate, the anode, is necessary to collect the emitted photoelectrons.

Fig. 21.12 shows the photocell in series with a microammeter which registers the current due to the photoelectrons. A potential divider and a voltmeter are used to apply a variable positive potential to the photocathode, relative to the anode. When the photocathode is illuminated with light of known frequency, the potential divider is adjusted to increase the potential of the photocathode until the current is reduced to zero. The stopping potential is then measured from the voltmeter. The experiment is repeated using different light frequencies to illuminate the photocathode. Note that if the wavelength, λ, of the light is known, the frequency, f, can be calculated using the equation $f\lambda = c$, where c is the speed of light.

The results may be plotted as a graph of stopping potential, V_S, on the vertical axis against frequency on the horizontal axis, as shown in Fig. 21.13. The results define a straight line intercepting the +x-axis.

According to Einstein's explanation of photoelectricity,

$$eV_S = hf - \phi$$

$$\therefore V_S = \frac{hf}{e} - \frac{\phi}{e}$$

In comparison with the general equation for a straight line, $y = mx + c$,

1. the gradient of the line, $m = \frac{h}{e}$

2. the y-intercept, $c = -\frac{\phi}{e}$

3. the x-intercept (i.e. when $V_S = 0$) is therefore the threshold frequency

$$f_o = \frac{\phi}{h}$$

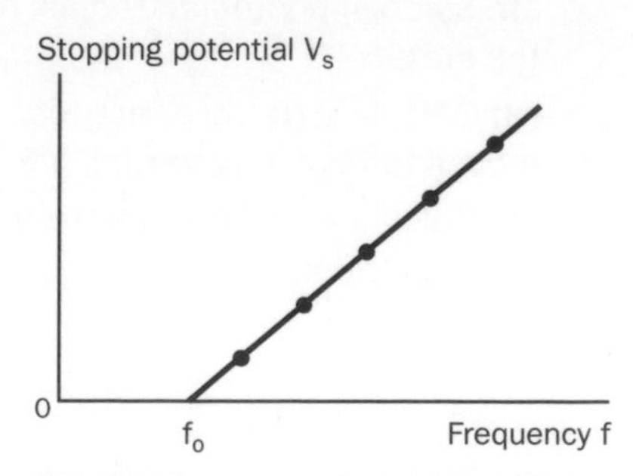

Fig. 21.13 Stopping potential against frequency.

Note

See p. 38 if you have forgotten the approximate wavelength of each part of the electromagnetic spectrum. Remember that visible light covers the wavelength range from about 400 nm (violet) to about 700 nm (deep red).

Thus the results from the experiment confirm Einstein's theory and can be used to measure the Planck constant, h, and the work function, ϕ, of the photocathode.

Worked example 21.5

The photocathode of a vacuum photocell in series with a microammeter is illuminated with light of wavelength 430 nm. The microammeter reading is reduced to zero when the potential of the photocathode is increased to 1.0 V relative to the anode. Calculate (a) the energy of a photon of wavelength 430 nm, (b) the work function of the photocathode, (c) the maximum wavelength at which photoelectric emission can occur from the photocathode when it is at zero potential.

$h = 6.6 \times 10^{-34}$ Js, $e = 1.6 \times 10^{-19}$C, $c = 3.0 \times 10^{8}$ m s^{-1}

Solution

(a) $E = hf = \frac{hc}{\lambda} = \frac{6.6 \times 10^{-34} \times 3.0 \times 10^{8}}{430 \times 10^{-9}} = 4.6 \times 10^{-19}$ J

(b) Rearrange $eV_S = hf - \phi$ to give

$\phi = hf - eV_S = 4.6 \times 10^{-19} - (1.6 \times 10^{-19} \times 1.0) = 3.0 \times 10^{-19}$ J

(c) The threshold frequency $f_o = \frac{\phi}{h} = \frac{3.0 \times 10^{-19}}{6.6 \times 10^{-34}} = 4.5 \times 10^{14}$ Hz

Maximum wavelength $\lambda_o = \frac{c}{f_o} = 6.7 \times 10^{-7}$ m

Questions 21.3

$h = 6.6 \times 10^{-34}$ Js, $e = 1.6 \times 10^{-19}$C, $c = 3.0 \times 10^{8}$ m s^{-1}

1. **(a)** Calculate the energy of a photon of wavelength **(i)** 600 nm, **(ii)** 100 nm.

   (b) A certain metal surface has a work function of 0.64 eV. When this metal surface is at zero potential, calculate the maximum kinetic energy of a photoelectron when the surface is illuminated with photons of wavelength **(i)** 600 nm, **(ii)** 100 nm.

2. A metal surface is illuminated with light of wavelength 550 nm causing it to emit photoelectrons. The emission of photoelectrons is reduced to zero as a result of applying a positive potential of 0.85V to the surface.

   (a) Calculate **(i)** the energy of a photon of wavelength 550 nm, **(ii)** the work done by an electron that moves through a potential difference of 0.58 V, **(iii)** the work function of the metal.

   (b) Calculate the maximum wavelength capable of producing photoelectrons from this surface at zero potential.

21.4 Electrons in the atom

The electrons in an atom are arranged in shells, as explained on p. 74, each shell capable of holding a certain number of electrons at fixed energies. To escape from an atom, an electron must be given energy to enable it to overcome the attraction of the positively charged nucleus of the atom. If an electron is removed from an uncharged atom, the atom becomes a positively charged ion.

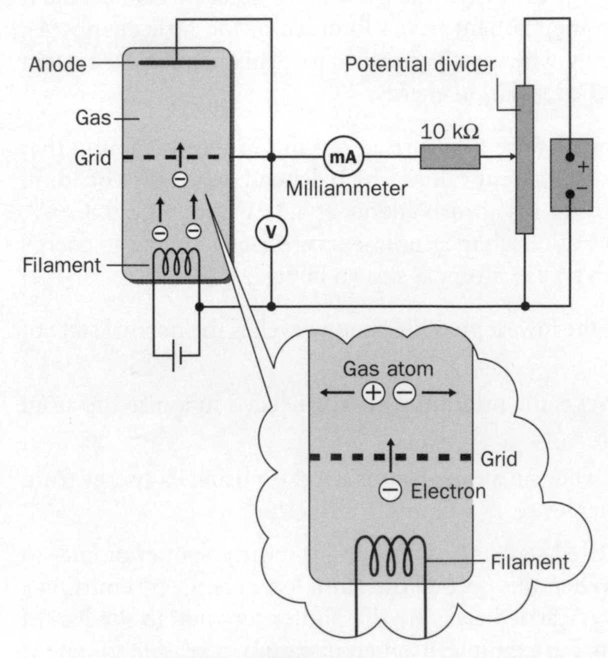

Fig. 21.14 Ionisation by collision.

- **Ionisation**, the process of creating ions, occurs when radioactive particles travel through matter. A single radioactive particle passing through air ionises millions of gas atoms. A more controlled way to ionise gas atoms is to direct a beam of electrons into the gas. If the electrons in the beam have sufficient kinetic energy, they will knock electrons out of the gas atoms.

Fig. 21.14 shows a tube used to demonstrate ionisation by collision. Electrons are emitted from a hot-wire filament by the process of thermionic emission. As the anode voltage is increased from zero, little current is registered on the milliammeter until the voltage reaches a certain value. At this voltage, some of the electrons from the filament arrive at the grid and pass through the grid wires with just sufficient kinetic energy to ionise the gas atoms. The beam electrons knock electrons out of the gas atoms at and above a certain anode voltage. The current increases because the conductivity of the gas increases due to ions being created. Fig. 21.15 shows the results for a tube filled with xenon, an inert gas, which ionises at an anode voltage of 12.1 V. The **ionisation energy of xenon** (i.e. work done to ionise a xenon atom) is therefore 12.1 eV.

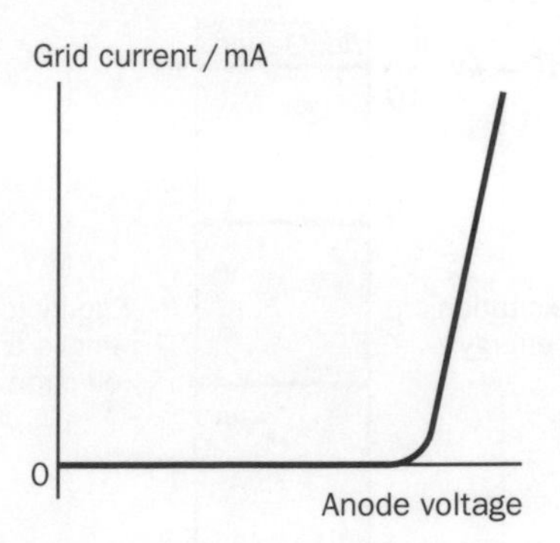

Fig. 21.15 Ionisation results for a tube filled with xenon.

Note

1 eV = 1.6×10^{-19} J. See p. 306 if necessary.

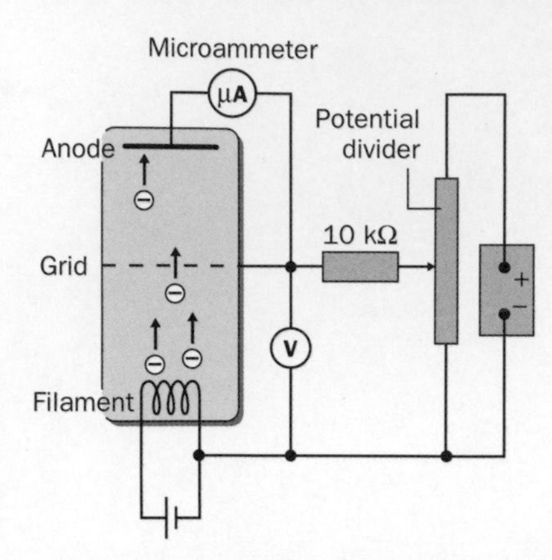

Fig. 21.16 Excitation by collision.

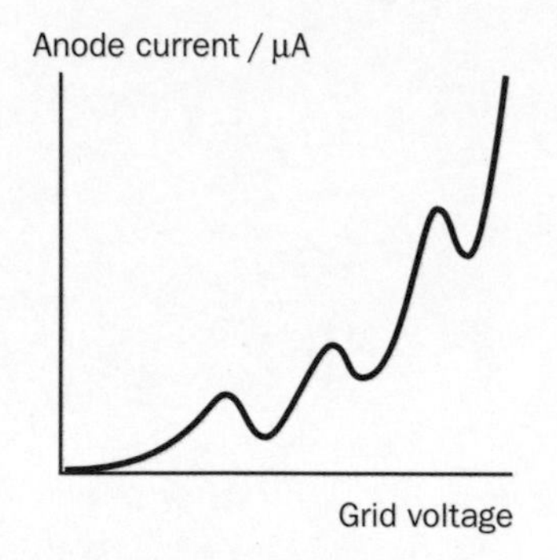

Fig. 21.17 Excitation results.

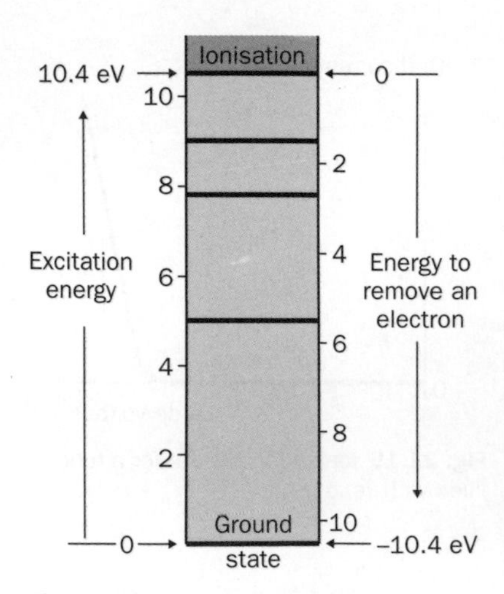

Fig. 21.18 The energy levels of a mercury atom.

- **Excitation** of an atom is possible without ionisation. In other words, atoms can absorb energy without losing electrons. Fig. 24.16 shows the tube in Fig. 21.14 in a modified circuit in which a microammeter is used to measure the number of electrons per second reaching the anode.

  As the anode voltage is increased from zero, more and more electrons from the filament shoot through the grid to reach the anode, causing the anode current to rise. However, at certain voltages, the anode current drops sharply because the electrons from the filament have just enough kinetic energy to 'excite' the gas atoms. No excitation occurs if the electrons from the filament have too much or too little energy. An atom absorbs energy when excitation occurs. This happens at certain energies less than the ionisation energy.

When these experiments were first carried out in 1914, it was found that atoms absorb energy at different values which depend on the type of atom. For example, mercury atoms absorb energy at 4.9 eV, 7.6 eV and 8.9 eV, and they ionise at 10.4 eV. Such measurements are used to make an energy level diagram for each type of atom, as shown in Fig. 21.18.

1. The ground state, the lowest possible energy level, is the normal state of the atom.
2. The ionisation level is the minimum energy needed to ionise the atom in its ground state.
3. Excitation occurs when an atom absorbs energy, raising its energy from one level to a higher level.
4. An atom in an excited state is unstable and it returns sooner or later to its ground state. When this occurs, the atom loses energy by emitting a photon. The energy carried away by the photon is equal to the loss of energy of the atom. For example, if a mercury atom in an excited state at 4.9 eV returns to its ground state directly, it releases a photon of energy 4.9 eV.

When excitation experiments were first carried out on mercury, it was discovered that ultraviolet radiation of wavelength 250 nm was emitted from the mercury atoms. Prove for yourself that the energy of a photon of wavelength 250 nm is 4.9 eV. Other wavelengths are emitted by mercury atoms, corresponding to photons of different energies released when the atoms move down the energy level ladder in other ways. The energy level picture of each type of atom can therefore be worked out from the energies of the photons released by that type of atom.

Electron shells and energy levels

The electrons in an atom move about the nucleus, prevented from leaving the atom by the attractive force of the nucleus. Each electron has a fixed amount of energy which depends on the shell it occupies.

- When an atom absorbs energy, it moves to a higher energy level because an electron in one of its shells has moved to a shell of higher energy. This can occur as a result of collisions between atoms or as a result of an incoming radioactive particle or electron colliding with an electron in the atom.

- When an atom releases energy, an electron in the atom moves to a shell of lower energy and releases a photon as it does so. Because the electrons in an atom occupy shells of well-defined energies, the photons released by an atom when it loses energy are of well-defined energies.

  Photon energy, hf = $E_1 - E_2$

  where E_1 = initial energy level of electron and E_2 = final energy level.

The emission spectrum of light from a discharge tube or a vapour lamp consists of discrete coloured lines, each line corresponding to a definite wavelength. The pattern of spectral lines from the atoms of a given element are characteristic of that element and can be used to identify the element. See p. 53 for details of how the wavelengths of light in a line spectrum can be measured.

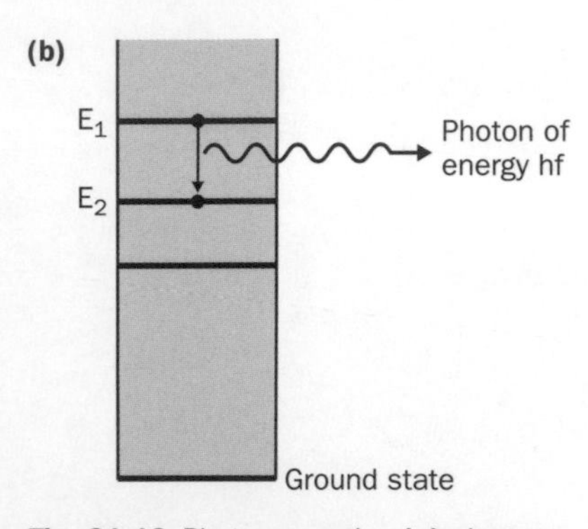

Fig. 21.19 Photon energies **(a)** electron shells, **(b)** energy levels.

An atomic model

The hydrogen atom is the simplest type of atom. It consists of a single electron and a proton. The energy levels of the hydrogen atom can be calculated correctly using **quantum mechanics** which is beyond the scope

Worked example 21.6

Fig. 21.20 shows some of the energy levels of a certain type of atom.

(a) (i) Calculate the energy and the wavelength of the photon released when an electron moves from a shell at an energy of 7.2 eV to a shell at 2.6 eV.
(ii) Identify to which part of the electromagnetic spectrum this photon belongs.

(b) Identify the electron transition responsible for photons of wavelength 476 nm emitted by this type of atom.

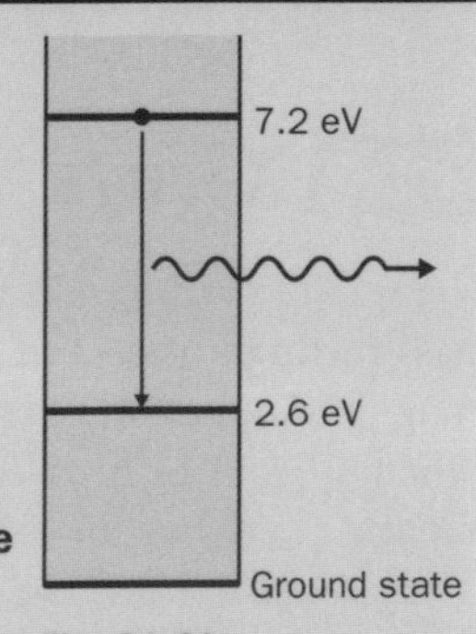

Fig. 21.20

Solution

(a) (i) Photon energy = $E_1 - E_2 = 7.2 - 2.6 = 4.6\ \text{eV} = 4.6 \times 1.6 \times 10^{-19}$
$= 7.4 \times 10^{-19}\ \text{J}$

Rearrange photon energy $E = hf = \dfrac{hc}{\lambda}$ to give

$$\lambda = \frac{hc}{E} = \frac{6.6 \times 10^{-34} \times 3.0 \times 10^{8}}{7.4 \times 10^{-19}} = 2.7 \times 10^{-7}\ \text{m}$$

(ii) Ultraviolet.

(b) Photon energy, $E = hf = \dfrac{hc}{\lambda} = \dfrac{6.6 \times 10^{-34} \times 3.0 \times 10^{8}}{476 \times 10^{-9}}$

$= 4.2 \times 10^{-19}\ \text{J} = 2.6\ \text{eV}$

The transition is therefore from the energy level at 2.6 eV to the ground state.

See www.palgrave.com/foundations/breithaupt for an experiment to investigate the spectrum of hydrogen.

of this book. In addition, the shape of each shell can be worked out. The energy of each shell is given by the formula

$$E = -\frac{13.6}{n^2}\ \text{eV}$$

where n is a positive integer referred to as the principal quantum number.

Questions 21.4

1. **(a)** Calculate the energy and wavelength of the photon released as a result of an electron moving from a shell at 7.2 eV to the ground state.
   (b) Calculate the wavelength of the photon released when a mercury atom moves from an energy level of 8.9 eV to an energy level of **(i)** 7.6 eV, **(ii)** 4.9 eV.
2. **(a)** Use the energy level formula $E = \frac{-13.6}{n^2}$ eV to calculate the energy levels in eV of the n = 2, 3, 4 and 5 shells of the hydrogen atom.
   (b) (i) Construct an energy level diagram for the hydrogen atom.
   (ii) Determine the wavelength emitted when an electron in a hydrogen atom moves from n = 5 to n = 4.
   (iii) Identify to which part of the electromagnetic spectrum this photon belongs.

21.5 X-rays

X-rays are used to take photographs of bones in the body because X-rays pass through tissue but not through bone. X-rays are photons of electromagnetic radiation of wavelength about 1 nm or less. They are produced when a beam of high-energy electrons is stopped by a metal target. The kinetic energy of the electrons in the beam is converted to X-ray energy.

Fig. 21.21 shows the construction of an X-ray tube. The electrons are emitted from a hot wire filament and attracted directly to the metal anode

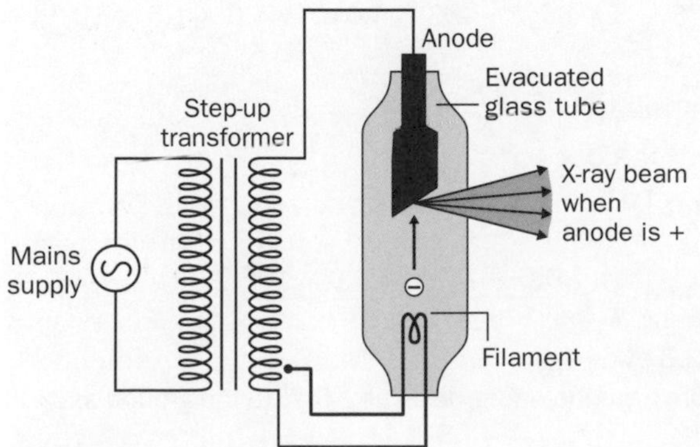

Fig. 21.21 An X-ray tube.

which is at a high positive potential relative to the filament. X-rays are produced at the point of impact on the anode surface. The glass tube is evacuated and the anode surface is at 45° to the beam direction so that the X-rays emerge through the side of the tube. A step-up transformer is used to apply a large alternating voltage between the anode and the filament. Electrons are attracted to the anode every other half-cycle of the alternating supply when the anode is positive relative to the filament.

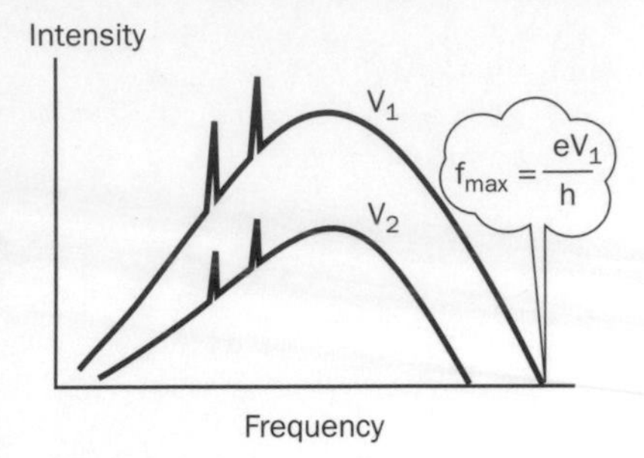

Fig. 21.22 Intensity distribution.

- The **intensity distribution** of the X-ray beam changes continuously with frequency up to a maximum, as shown by Fig. 21.22. This is because the maximum energy of an X-ray photon is equal to the kinetic energy of an electron in the beam, corresponding to all the energy of a single electron being used to produce one photon.

For a beam of electrons produced by an electron gun operating at an anode potential V_A, the kinetic energy of each electron in the beam is eV_A. Hence the maximum frequency, f_{MAX}, of the X-rays produced when the beam is stopped is given by equating the photon energy hf_{MAX} to the kinetic energy of a single electron.

$$hf_{MAX} = eV_A$$

$$\therefore f_{max} = \frac{eV_A}{h}$$

The X-rays produced as a result of stopping a beam of electrons have a spread of frequencies up to f_{MAX}, Consequently, the spread of wavelengths is continuous down to a minimum value

$$\lambda_{MIN} = \frac{c}{f_{max}} = \frac{hc}{eV_A}$$

Fig. 21.22 shows the intensity distribution for two different tube voltages. The higher the voltage, the higher the maximim frequency and the greater the intensity of the X-ray beam.

- **Intensity spikes** are produced at certain frequencies that are characteristic of the metal target. Each spike occurs because beam electrons knock electrons out of the inner shells of the target atoms. These inner shells are at very large negative energies because they are deep within each atom. The vacancies created in these shells are filled by electrons from the outer shells or by conduction electrons. The photons released are therefore high energy photons in the X-ray part of the electromagnetic spectrum. The spikes are characteristic of the target atoms because they correspond to the possible electron transitions between the energy levels of the target atoms.
- The **power** used by an X-ray tube can be calculated by multiplying the beam current by the anode potential. Only a small proportion of the beam energy is converted into X-ray energy. Typically, a tube operating at an anode potential of 20 000 V and a beam current of 40 mA would use 800 W of electrical power (= beam current × anode potential), perhaps producing about 20 W of X-ray energy per second. The rest of the power delivered to the tube would be converted to internal energy at

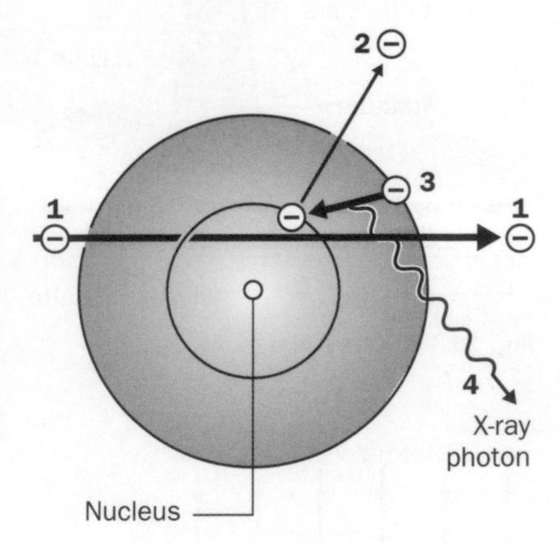

1 Incoming electron knocks an atomic electron out

2 Vacancy created in shell

3 Outer electron fills vacancy and emits X-ray photon **4**

Fig. 21.23 Characteristic X-rays.

the anode which therefore becomes very hot. For this reason, the anode is usually a tungsten plate (tungsten has a very high melting point) set in a copper block (copper is a good heat conductor).

Worked example 21.7

An X-ray tube operates at an anode potential of 40 kV and a beam current of 5 mA. Calculate (a) the minimum wavelength of the X-rays produced, (b) (i) the number of electrons striking the anode per second, (ii) the X-ray energy produced per second if the tube efficiency is 1%.

Solution

(a) $\lambda_{MIN} = \frac{hc}{eV_A} = \frac{6.6 \times 10^{-34} \times 3.0 \times 10^{8}}{1.6 \times 10^{-19} \times 40\,000} = 3.0 \times 10^{-11}$ m

(b) (i) Beam current, I = no. of electrons per second n × e

hence $n = \frac{I}{e} = 3.1 \times 10^{16}$

(ii) Electrical power used by the tube $= I V_A = 5 \times 10^{-3} \times 40\,000 = 200$ W

∴ X-ray energy per second = 1% × 200 W = 2 W

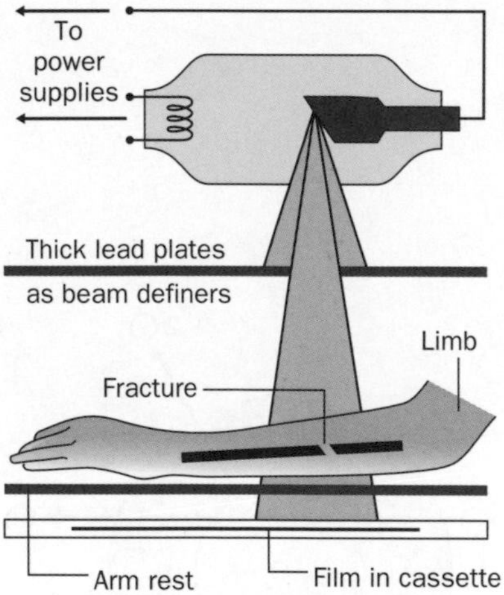

Fig. 21.24 X-rays in use.

X-rays in medicine

Fig. 21.24 shows an X-ray tube used to take an X-ray picture of a broken limb. A shadow of the limb is formed on the photographic film because bone absorbs X-rays whereas flesh does not. To produce a clear picture, the beam must originate from as small an area of the target as possible. However, if this target area is too small, the beam might melt the target at the point of impact. As X-rays are harmful due to their ionising effect, the tube is surrounded by lead shielding to prevent X-rays reaching people in the vicinity of the tube.

- **Beam definers** are used to limit the area of the patient exposed to the beam. The lead plates absorb X-rays which would otherwise reach other parts of the patient.
- An **aluminium metal plate** is placed in the path of the beam between the patient and the tube. This absorbs low-energy X-rays that would otherwise be absorbed by the soft tissue of the patient.
- A **metal grid** between the patient and the photographic film prevents X-rays scattered by the patient from reaching the film. Only those X-rays that pass through the patient without scattering travel through the narrow tubular holes in the grid block. This device therefore improves the contrast of the image.
- A **contrast medium** is used to obtain X-ray photographs of organs or blood vessels. For example, before taking an X-ray photograph of the stomach, the patient is given a barium drink which lines the stomach. Barium is used because it absorbs X-rays and eventually passes out of the patient without any harmful effects.

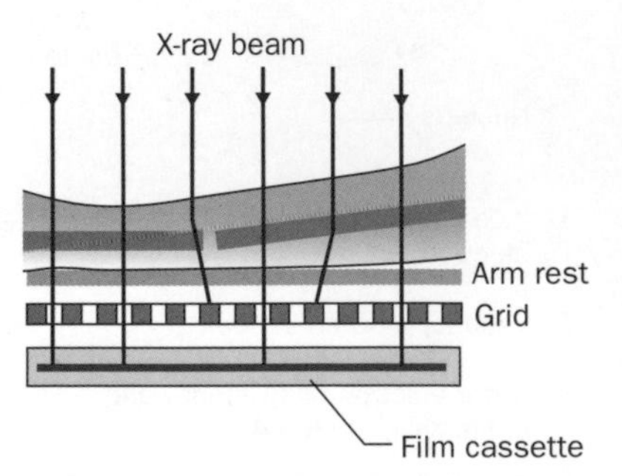

Fig. 21.25 Improving contrast.

- A **film badge** must be worn by all personnel in an X-ray unit. The film is sealed in a light-proof container covered in different parts by different materials. When the film is developed, the exposure of the wearer to different types of ionising radiation can be measured. Strict limits are imposed by law on the maximum permitted exposure of radiation workers to ionising radiation. This is because ionising radiation is harmful, either killing living cells or causing mutation.

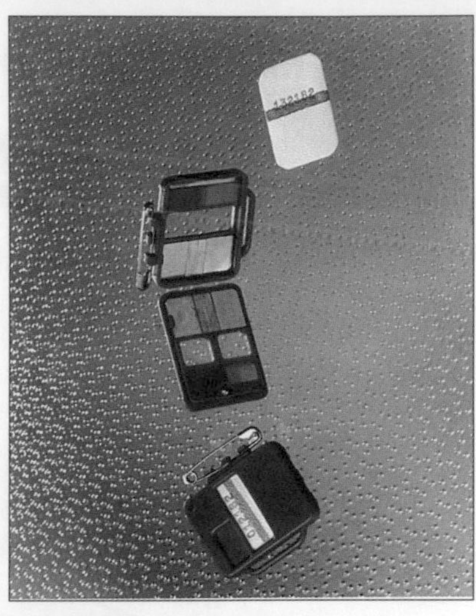

Fig. 21.26 A film badge. The film is shown out of its case and sealed inside. This allows monitoring of the exposure of the wearer to radiation.

Questions 21.5

1. An X-ray tube operates at a potential of 25 kV and a beam current of 30 mA with an efficiency of 2%. Calculate **(a)** the minimum wavelength of the X-ray photons from this tube, **(b)** the electrical energy used by the tube each second, **(c)** the heat produced per second at the target.
2. Describe how an X-ray picture is made of a broken limb, explaining the steps taken to ensure good contrast between the bones and the surrounding soft tissues.

Summary

The electron gun equation

$\frac{1}{2}mv^2 = eV_A$

Electrons in fields

1. Force, $F = \frac{eV_P}{d}$ for an electron between two oppositely charged parallel plates.
2. Force, $F = Bev$ for an electron moving at speed, v, perpendicular to a magnetic field, B.

Millikan's experiment

1. $\frac{QV_P}{d} = mg$ for a charged droplet at rest.
2. $F = 6\pi\eta rv$ gives the viscous force on a droplet of radius, r, moving at speed, v.
3. $Q = ne$, where n is a whole number.

Photons

1. Energy of a photon, $E = hf = \frac{hc}{\lambda}$
2. Maximum k.e. of a photoelectron $= hf - \phi$, where ϕ is the work function.
3. $eV_S = hf - \phi$ gives the stopping potential V_S.

Energy level equation

$hf = E_1 - E_2$

X-ray intensity distribution

$f_{MAX} = \frac{eV_A}{h}$

Revision questions

$h = 6.6 \times 10^{-34}$ Js, $e = 1.6 \times 10^{-19}$C, $c = 3.0 \times 10^{8}$ m s^{-1}

21.1. (a) A beam of electrons is produced by an electron gun operating at an anode potential of 4200 V. Calculate **(i)** the kinetic energy in joules, **(ii)** the speed of the electrons in the beam.

(b) The beam in (a) is directed between two oppositely charged parallel plates 50 mm apart, along a line midway between the plates. When the top plate is at a positive potential of 5000 V relative to the lower plate, the beam curves as shown in Fig. 21.27.

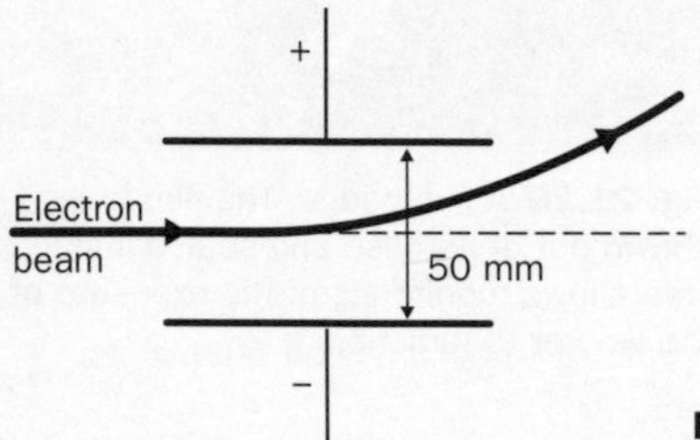

Fig. 21.27

(i) Calculate the force on an electron in the beam when it is between the plates.

(ii) The force due to the electric field is cancelled out if a suitable magnetic field is applied at 90° to the plane of the diagram. State whether this magnetic field should be directed into or out of the plane of the diagram and calculate the magnetic flux density necessary for the beam to be undeflected.

21.2. In an experiment to measure the charge on an oil droplet, an oil droplet was held stationary between two parallel plates 4.0 mm apart at a potential difference of 590 V. When the plate p.d. was switched off, the droplet fell at steady speed, moving through a distance of 1.2 mm in 14.6 s.

(a) (i) The top plate was positive relative to the lower plate. What type of charge was carried by the droplet?

(ii) Why did the droplet fall at constant speed when the plate p.d. was switched off?

(b) Calculate **(i)** the terminal speed of the droplet, **(ii)** the radius of the droplet, **(iii)** the droplet's mass, (iv) the charge on the droplet. The viscosity of air = 1.8×10^{-5} N s m^{-1}, the density of the oil = 960 kg m^{-3}

21.3. (a) Calculate the frequency and the energy of a photon of wavelength **(i)** 500 nm, **(ii)** 50 nm.

(b) In a photoelectricity experiment, light of wavelength 500 nm is directed at the surface of the photocathode in a vacuum photocell. The photoelectric current is reduced to zero as a result of applying a positive potential of 0.36 V to the photocathode, relative to the anode. Calculate **(i)** the work done by an electron moved through a potential difference of 0.36 V, **(ii)** the work function of the photocathode surface in electron volts, **(iii)** the maximum wavelength of incident radiation capable of causing photoelectric emission from this surface when it is at zero potential.

21.4. (a) The energy levels of the hydrogen atom are given by the equation

$$E = \frac{-13.6}{n^2}\text{ eV}.$$

Calculate the wavelength of the photon emitted by an electron in a hydrogen atom when it transfers from **(i)** n = 4 to n = 3, **(ii)** n = 2 to n = 1.

(b) Fig. 21.28 shows the energy levels of a certain type of atom known to emit photons of wavelengths 565 nm and 430 nm. Calculate the energy of a photon of each wavelength and identify the transition responsible for each photon.

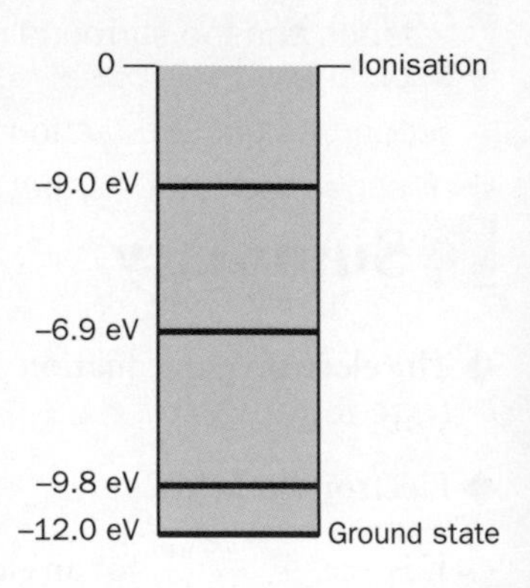

Fig. 21.28

21.5. (a) (i) Sketch curves showing how the intensity of the X-rays in an X-ray beam varies with frequency for two different tube voltages.

(ii) Explain why the intensity is very high at certain wavelengths that depend on the target element.

(b) Calculate the minimum wavelength of the X-rays produced from an X-ray tube operating at **(i)** 25 kV, **(ii)** 100 kV.

Radioactivity

Objectives

After working through this unit, you should be able to:

- ▶ describe the structure of the atom in terms of electrons, protons and neutrons
- ▶ explain what the term 'isotope' means
- ▶ explain what a radioactive substance is, and describe the main properties of the radiation from radioactive substances
- ▶ describe the changes that occur when an unstable nucleus emits radiation
- ▶ carry out experiments to measure background radiation and to investigate the absorption of radiation from radioactive substances
- ▶ describe and explain the operation of a Geiger–Müller tube and a cloud chamber
- ▶ outline Rutherford's alpha-scattering experiment and how the results led Rutherford to deduce the nuclear model of the atom
- ▶ explain what is meant by radioactive decay and the term 'half-life'
- ▶ carry out half-life calculations

Contents

22.1 Inside the atom

An atom is the smallest particle characteristic of an element. Until near the end of the 19th century, the atom was thought to be indivisible and indestructible. J.J. Thomson's experiments on the electron proved that atoms contain these much lighter particles. In this unit, we will look at how the discovery of radioactivity during the same period led to the conclusion that some atoms are unstable and that every atom consists of electrons surrounding a nucleus composed of protons and neutrons. Before considering radioactivity and the properties of radioactive substances, here are some reminders about the structure of the atom.

Electrons, protons and neutrons form the building blocks of every atom. Electrons are negatively charged particles of mass about 1/2000 th of

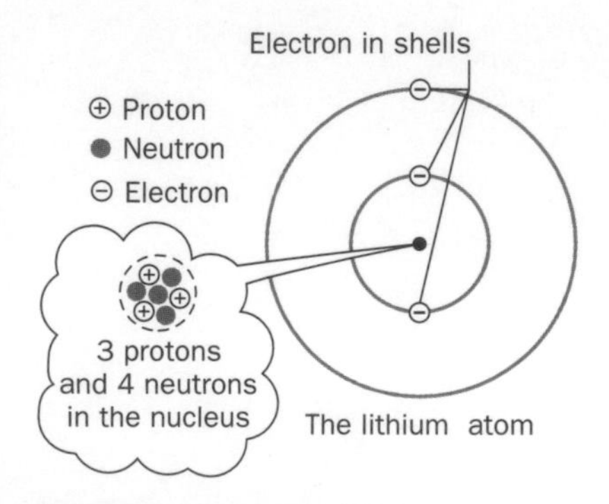

Fig. 22.1 Inside the atom.

the mass of a hydrogen atom. The protons and neutrons in an atom are concentrated in the nucleus which has a diameter about 10 millionths of the diameter of the atom. Almost all the mass of an atom is located in its nucleus. Some further key points about the structure of the atom are listed below.

- The electrons occupy shells at fixed energy which surround the nucleus of the atom.
- The proton carries a positive charge, equal and opposite to the charge of the electron, and has a mass almost equal to that of a hydrogen atom.
- The neutron is uncharged and has a mass about the same as that of the proton.
- Every atom of a given element contains the same number of protons.
- The number of neutrons in the nucleus of an atom is not the same for all the atoms of an element.

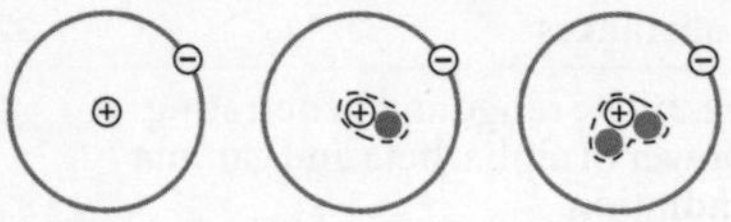

An isotope of an element consists of atoms of that element that have the same number of neutrons.

Normal hydrogen $^{1}_{1}H$ Deuterium $^{2}_{1}H$ Tritium $^{3}_{1}H$

Fig. 22.2 The three isotopes of hydrogen.

For example, there are three isotopes of hydrogen. Most hydrogen atoms contain a nucleus consisting of a proton only. Deuterium is a rare isotope of hydrogen in which the nucleus of every atom consists of a proton and a neutron. Tritium, the third isotope of hydrogen, is even rarer than deuterium and consists of atoms which each have a nucleus consisting of a proton and two neutrons.

Numbers and symbols

The proton number, Z, of an atom is the number of protons in the nucleus of the atom.

Each atom of a given element contains Z protons. The proton number of an atom is also referred to as the atomic number of the element as this is the order number of the element in the Periodic Table (see p. 74).

The mass number, A, of an atom is the number of protons and neutrons in its nucleus.

The lightest known atom is the hydrogen atom, which has a single proton as its nucleus. The neutron has about the same mass as the proton. Electrons are much lighter than protons or neutrons. Therefore, the mass of an atom relative to the mass of a proton is approximately equal to its mass number, A.

An isotope is represented by the symbol $^{A}_{Z}X$

where X is the symbol for the chemical element. An isotope is sometimes referred to by the name of the element followed by the mass number. For example, uranium 235 (or more simply U 235) is the isotope of uranium which has a mass number of 235.

- There are Z protons in the nucleus of an atom of this isotope.
- The nucleus also contains (A − Z) neutrons.
- An uncharged atom has Z electrons in shells surrounding the nucleus.

The unit of atomic mass, 1 u, is defined as $\frac{1}{12}$th of the mass of an atom of $^{12}_{6}C$ (i.e. carbon 12).

Since there are 12 neutrons and protons in the nucleus of carbon 12, the mass of a proton or a neutron is approximately equal to 1 u.

The Avogadro constant, N_A, is defined as the number of atoms in exactly 12 g of carbon 12.

The accepted value of N_A is 6.02×10^{23}.

- A carbon-12 atom has a mass of $\frac{12}{N_A}$ grams or $\frac{0.012}{N_A}$ kilograms. Since this is equal to 12 u by definition, then it follows that

$$1 \text{ u} = \frac{1}{N_A} \text{ grams} = \frac{0.001}{N_A} \text{ kilograms} = 1.67 \times 10^{-27} \text{ kg}.$$

- The mass of 1 atom of mass number A is approximately equal to A in atomic mass units. For example, the mass of a uranium-235 atom is approximately equal to 235 u.

A mole is the quantity of matter in N_A particles of a substance.

The molar mass of a substance is the mass of 1 mole of the substance. The unit symbol for the mole is mol.

Questions 22.1

$N_A = 6.02 \times 10^{23}$ mol^{-1}

1. Calculate the number of protons and neutrons in one atom of each of the following isotopes:
   (a) $^{235}_{92}U$, **(b)** $^{14}_{6}C$, **(c)** $^{37}_{17}Cl$.
2. Calculate the number of atoms in
   (a) 1.0 kg of uranium 235,
   (b) 1 g of helium 4.

22.2 Radiation from radioactive substances

Becquerel's discovery

Radioactivity was discovered by Henri Becquerel in 1896, when he was investigating materials that glow in X-ray beams and ultraviolet light. He wanted to find out if uranium salts glow in strong sunlight so he prepared a sample and placed it in a drawer ready for a sunny day. The sample was placed on a key on some wrapped photographic plates which he was using to test the penetrating power of X-rays. When he developed the plates, he saw an image of the key and realised this was due to radiation emitted by the uranium salts. The uranium salts were referred to as 'radioactive'. As he was more interested in X-rays than this curious radiation from uranium salts, he passed the task of investigating radioactivity onto Marie Curie, one of his research students. She discovered other natural substances which also emit radiation, including two new elements, polonium and radium. Radium is over a million times more radioactive than uranium.

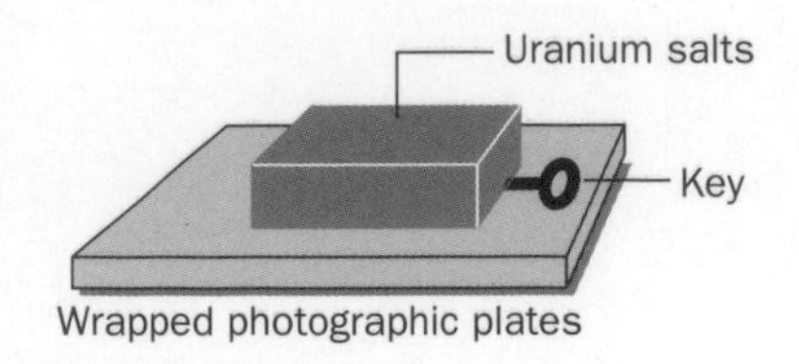

Fig. 22.3 Becquerel's discovery.

The nature of the radiation from radioactive substances

Ernest Rutherford showed that there were two types of radiation from radioactive substances. One type, which he called **alpha** (α) radiation, is easily absorbed and another, more penetrating component, he called **beta**

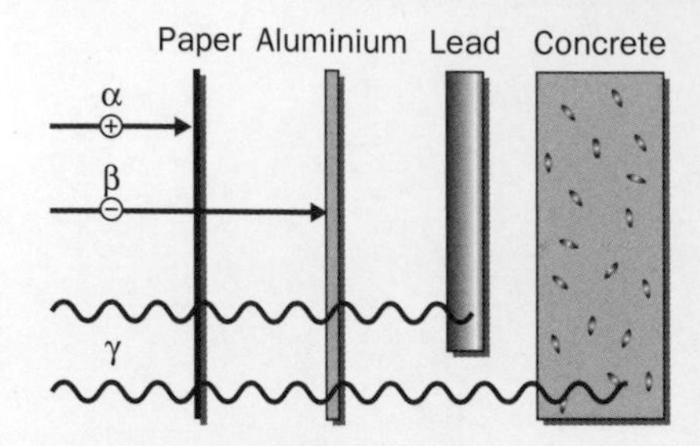

Fig. 22.4 Absorption of alpha, beta and gamma radiation.

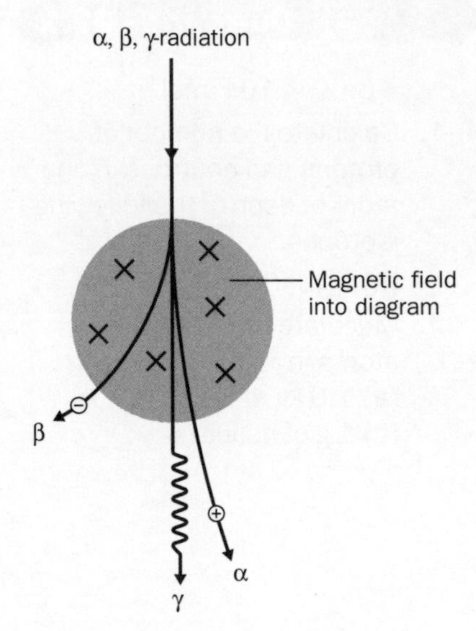

Fig. 22.5 The separation of α, β and γ radiation

(**β**) radiation. A third type, called **gamma** (**γ**) radiation, which is even more penetrating than beta radiation, was discovered later. Fig. 22.4 shows the effect of different materials on each type of radiation.

Further investigations on the nature and properties of radioactive substances were carried out as outlined below.

Deflection by a magnetic field

A magnetic field deflects moving charged particles. Investigations showed that alpha radiation consists of positively charged particles since a beam of alpha radiation is deflected as shown in Fig. 22.5. This type of investigation was also used to show that beta radiation consists of negative particles which are more easily deflected than alpha particles. Gamma radiation is unaffected by a magnetic field hence gamma radiation does not consist of charged particles.

The nature of each type of radiation

A radioactive substance is an isotope composed of atoms with unstable nuclei. Alpha, beta or gamma radiation is emitted when an unstable nucleus becomes stable.

- **Alpha particles** were shown to be identical to the nuclei of helium atoms, each composed of two protons and two neutrons. An alpha particle therefore has a mass of 4u and a charge of +2e. This was proved by showing that the spectrum of light from a discharge tube containing alpha particles is the same as that from a helium discharge tube.

  A nucleus that emits an alpha particle loses two protons and two neutrons so its proton number decreases by 2 and its mass number decreases by 4. This change is represented below:

  $${}^{A}_{Z}X \rightarrow {}^{4}_{2}\alpha + {}^{A-4}_{Z-2}Y$$

- **Beta particles** are fast-moving electrons, emitted by unstable nuclei which have too many neutrons. The identity of beta particles as electrons was made by measuring the path of the particles in a uniform magnetic field. A beta particle is created and emitted from a nucleus as a result of a neutron in the nucleus changing into a proton. The proton number of the nucleus increases by 1 and the mass number of the nucleus is unchanged. This change is represented below:

  $${}^{A}_{Z}X \rightarrow {}^{0}_{-1}\beta + {}_{Z+1}^{\ \ A}Y$$

- **Gamma radiation** is electromagnetic radiation of wavelength 10^{-12} m or less. An unstable nucleus of a gamma-emitting isotope emits a high energy photon when it becomes stable. The mass number and the proton number of an unstable nucleus are unchanged when a gamma photon is emitted by the nucleus.

Ionisation

An ion is a charged atom, formed as a result of the atom gaining or losing one or more electrons. All three types of radiation emitted by radioactive substances are capable of creating ions. Fig. 22.6 shows how the ionising effect of each type of radiation can be measured.

The radiation ionises air molecules in the ionisation chamber so the

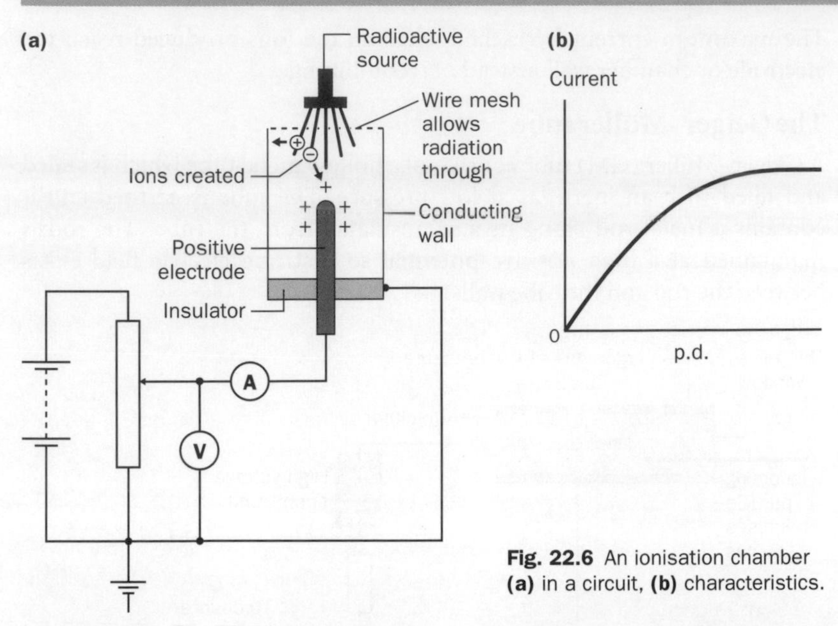

Fig. 22.6 An ionisation chamber **(a)** in a circuit, **(b)** characteristics.

positive ions created are attracted to the negative surface of the chamber and the negative ions to the positive electrode. The positive ions gain electrons and become neutral at the negative surface. The negative ions lose electrons and become neutral at the positive electrode. As a result, an ionisation current passes around the circuit and this is measured using a microammeter. The current is proportional to the number of ions created each second in the chamber.

- **Alpha radiation** passing through matter is capable of producing much more ionisation than beta or gamma radiation. This is because an alpha particle is much more likely to knock electrons out of the shells of an atom than are gamma radiation or a beta particle. An alpha particle on passing through matter therefore loses its kinetic energy more rapidly than a beta particle or gamma radiation.
- **Beta radiation** is more strongly ionising than gamma radiation but less strongly ionising than alpha radiation. A beta particle is a fast-moving electron created in and emitted from the nucleus. To ionise an atom, a beta particle must knock an electron out of one of the shells of an atom. A beta particle is much lighter than an alpha particle so it has less effect on an atom with which it collides than does an alpha particle.
- **Gamma radiation** is much less ionising than alpha or beta radiation. Gamma radiation is uncharged and therefore has less effect on the electrons in an atom than either alpha or beta radiation.

Detectors of ionising radiation

The ionisation chamber

The operation of an ionisation chamber is explained above. The ionisation current increases and reaches a maximum value as the potential difference between the chamber wall and the central electrode is increased.

The maximum current is reached when all the ions produced reach the electrode or chamber wall instead of recombining.

The Geiger–Müller tube

A Geiger–Müller (GM) tube consists of a hollow metal tube which is sealed and filled with an inert gas at low pressure. The tube is earthed and it contains a metal rod along its axis, insulated from the tube. The rod is maintained at a high positive potential so a strong electric field exists between the rod and the tube walls.

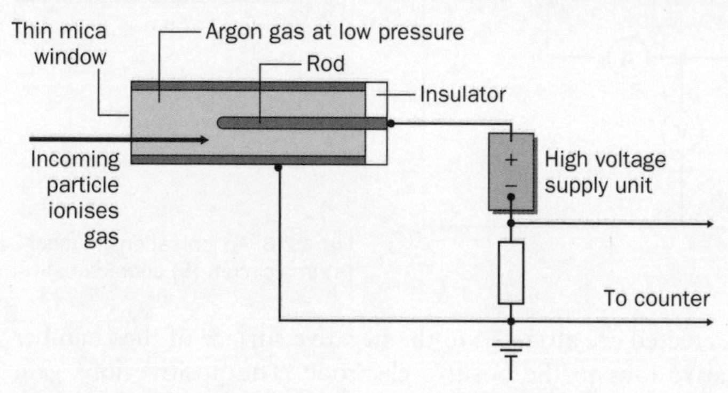

Fig. 22.7 The Geiger–Müller tube.

Note

As explained later on p. 336, radioactive emission is a random process and the number of counts in a given time from a source at constant distance will fluctuate. For this reason, the count rate is more accurate, the longer the time taken. Also, to ensure timing errors do not occur, the number of counts in a given time is repeated several times to give an average number. The count rate is then calculated from the average number of counts divided by the time taken.

When an alpha particle, a beta particle or a gamma photon enters the tube, ion pairs are created and the ions are attracted to the appropriate oppositely charged electrode. The ions gain kinetic energy due to this attraction and create further ion pairs as a result of collisions with uncharged gas molecules. These ions then produce further ionisation and an 'avalanche' of ions is therefore triggered from a single radiation particle. When the ions reach the electrodes, they are discharged and a pulse of charge passes round the circuit through the resistor in series with the GM tube. The voltage pulse created across the resistor is then registered as a single count by the electronic counter. The gas in the tube returns to its non-conducting state less than a millisecond after the particle enters the tube. The GM tube and the counter can therefore be used to measure the number of ionising particles entering the tube in a certain time. The **count rate**, the number of counts per unit time, can then be calculated by dividing the number of counts by the time taken.

The count rate depends on the tube voltage, as shown by Fig. 22.8. Below a certain voltage, the threshold voltage, no counts are detected because the tube voltage is not high enough to trigger further ionisation after an ionising particle enters the tube. As the voltage is increased above the threshold voltage, the count rate increases sharply to a plateau. The tube voltage is usually set so the tube is operating at the 'plateau' away from the steep part of this curve.

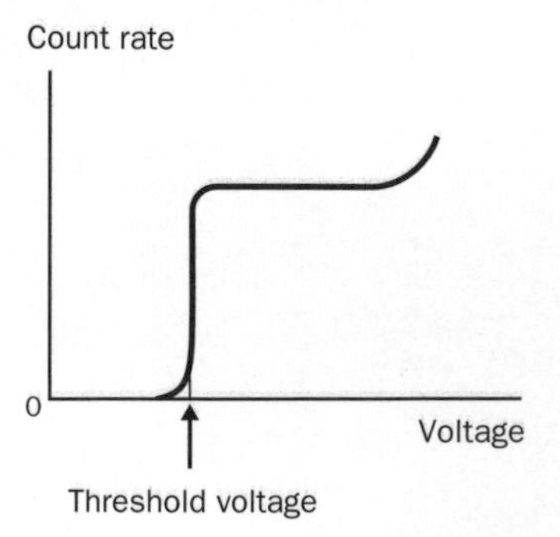

Fig. 22.8 The characteristics of a GM tube.

The cloud chamber

The tracks of alpha particles can easily be seen in a cloud chamber because the air space in the cloud chamber contains a supersaturated vapour.

When an ionising particle passes through this vapour, tiny droplets form from the vapour along the track of the particle because the ions formed along the track initiate condensation.

- **Alpha tracks** are clear, straight and of the same length for a given isotope.
- **Beta tracks** are irregular and poorly defined. This is because beta particles are less ionising and more easily deflected by air molecules than alpha particles.
- **Gamma particles** do not produce tracks as they produce insufficient ionisation.

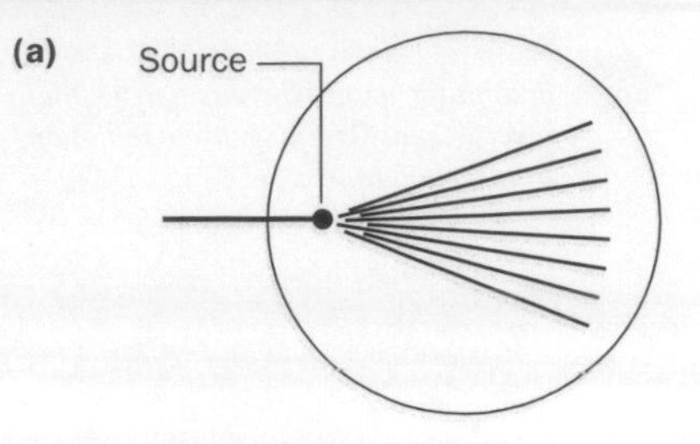

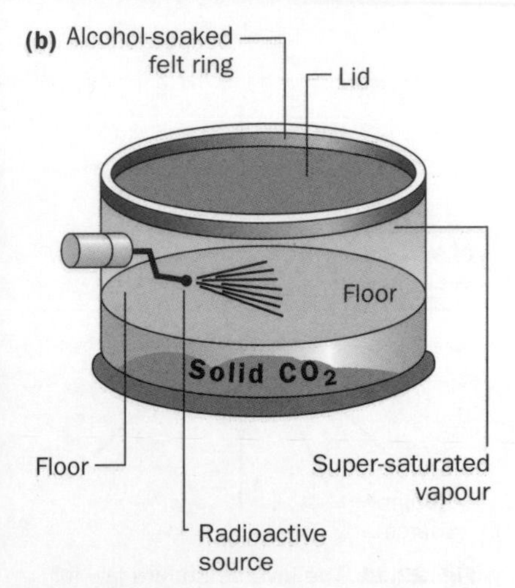

Fig. 22.9 The cloud chamber **(a)** α tracks, **(b)** construction.

Questions 22.2

1. Copy and complete the following equations
   (a) $^{238}_{92}U \rightarrow \; ^{4}_{[\,]}\alpha + \; ^{[\,]}_{90}Po$
   (b) $^{60}_{[\,]}Co \rightarrow \; ^{[\,]}_{-1}\beta + \; ^{[\,]}_{28}Ni$
2. **(a)** State one similarity and two differences between alpha and beta radiation.
   (b) Describe the change that occurs in an unstable nucleus as a result of emitting **(i)** an alpha particle, **(ii)** a beta particle.

22.3 The range and penetrating power of alpha, beta and gamma radiation

Fig. 22.10 shows how to investigate the range in air or the penetrating power of alpha or beta or gamma radiation. A GM tube is used to detect the radiation from a radioactive isotope in a small container. A different radioactive source is used to investigate each type of radiation. The background count rate (i.e. the count rate with no source present) must first be measured. This is subtracted from each count rate measured with the source present to give the corrected count rate.

1. **To investigate the range**, the count rate is measured with the tube at different distances from the source. Because the particles spread out as they move away from the source, the count rate decreases with increasing distance from the source.

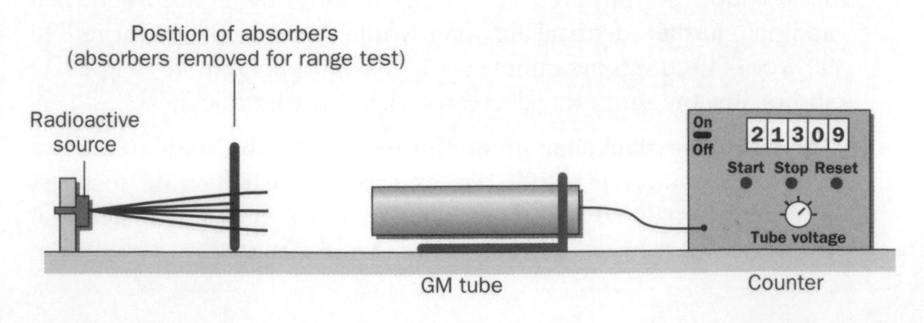

Fig. 22.10 Investigating range and penetrating power.

See www.palgrave.com/foundations/breithaupt for an experiment to test the inverse square law for gamma radiation from a point source.

- **Alpha particles** penetrate only a certain distance in air so the count rate suddenly falls to the background level when the tube is moved beyond the range of the alpha particles.
- **Beta particles** have a range in air up to no more than about 0.5 m, depending on their initial energy. The count rate decreases as the tube is moved away from the source because the particles are absorbed by air as well as spreading out.
- **Gamma radiation** is scarcely absorbed by air and therefore the corrected count rate, C, decreases with distance, r, from the source in accordance with the inverse square law

$$C = \frac{k}{r^2} \quad \text{where } k \text{ is a constant.}$$

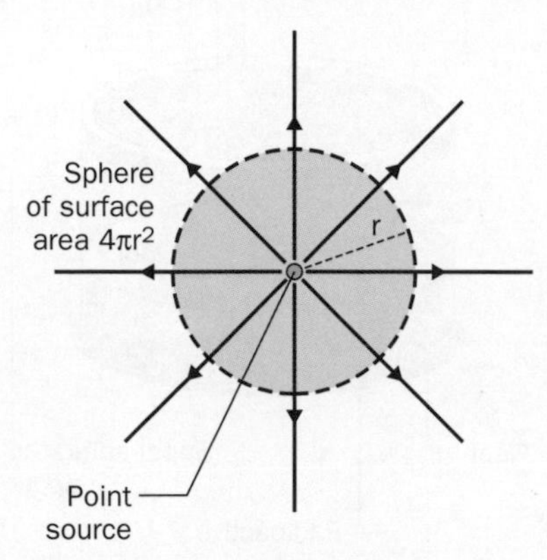

Fig. 22.11 The inverse square law for gamma radiation.

This equation assumes that no gamma radiation is absorbed by air. At distance, r, from the source, all the radiation passes through the surface of a sphere of radius r and surface area equal to $4\pi r^2$. If the source emits N gamma photons per second, the number of gamma photons entering the tube is therefore

$$\frac{NA}{4\pi r^2}$$

where A is the area of the tube facing the source. Hence the corrected count rate is given by the above equation with $k = NA \div 4\pi$.

2. To investigate the penetrating power, the distance between the source and the tube is kept constant. The count rate is measured with different thicknesses of each material to be tested.

- **Alpha radiation** is absorbed by paper and metal foil.
- **Beta radiation** penetrates paper and metal foil. The count rate decreases with increasing thickness of the absorber. A metal plate more than a few millimetres thick will absorb the beta radiation completely.
- **Gamma radiation** is even more penetrating than beta radiation. A lead plate of several centimetres thickness or a thick concrete block will stop gamma radiation. The count rate decreases **exponentially** with increasing thickness. This means that every extra millimetre thickness reduces the intensity of radiation penetrating the absorber by the same percentage. For example, if the count rate with no absorber is 1600 counts per minute and this is reduced to 90% of 1600 counts per minute (= 1440 per minute) by an absorber, then adding a **further** identical absorber would decrease the count rate to 90% of 1440 counts per minute (= 1296 counts per minute). Fig. 22.12 shows how the count rate decreases with absorber thickness.
- **The half-value thickness** of an absorber is the thickness needed to reduce the count rate to 50%. For example, if the half-value thickness of a certain material is 2.0 mm, then an absorber of this material of thickness 4.0 mm would reduce the count rate to 25% (= 50% of 50%).

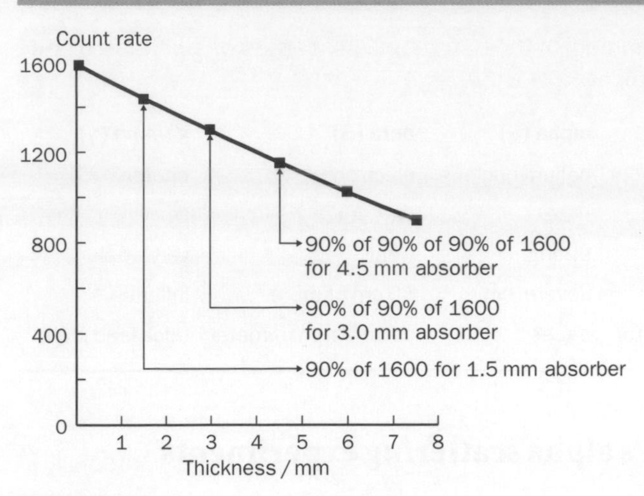

Fig. 22.12 Count rate versus thickness.

- **For beta radiation**, the count rate also decreases with increasing thickness in a similar way to gamma radiation, except that beta radiation is totally absorbed beyond a certain thickness whereas gamma radiation continues to decrease exponentially.

Worked example 22.1

A Geiger–Müller tube is used to measure the count rate due to a certain radioactive source which emits gamma radiation. Without the source present, the count rate due to background radioactivity is 28 counts per minute.

(a) When the GM tube is 100 mm from the source, the count rate is 950 per minute. Estimate the count rate if the distance between the tube and the source is increased to 200 mm.

(b) With the tube at 100 mm from the source, an aluminium plate is placed between the tube and the source, reducing the count rate to 655 counts per minute. Estimate the count rate if another identical absorber is placed between the source and the tube.

Solution

(a) Corrected count rate at 100 mm = 950 – 28 = 922 counts per minute.

According to the inverse square law, if the distance is doubled, the count rate is reduced to one quarter.

∴ Corrected count rate at 200 mm = 0.25×922 = 231 counts per minute.

The observed count rate at 200 mm would therefore be about 260 counts per minute (approx equal to 231 + 28 counts per minute).

(b) Corrected count rate with single absorber = 655 – 28 = 627 counts per minute.

This is 68% (= $627 \div 922 \times 100\%$) of the corrected count rate without the absorber.

∴ the corrected count rate with two identical absorbers = 68% of 627 = 426 counts per minute.

The observed count rate would therefore be 454 counts per minute (= 426 + 28 counts per minute).

Table 22.1 Summary of the nature and properties of alpha, beta and gamma radiation

Property	alpha (α)	beta (β)	gamma (γ)
Nature	helium nucleus	electron	photon
Charge	+2e	–e	0
Ionisation	strong	weak	very weak
Range in air	several cm	50 cm or more	infinite
Absorption by matter	paper	several mm of metal	thick lead plate

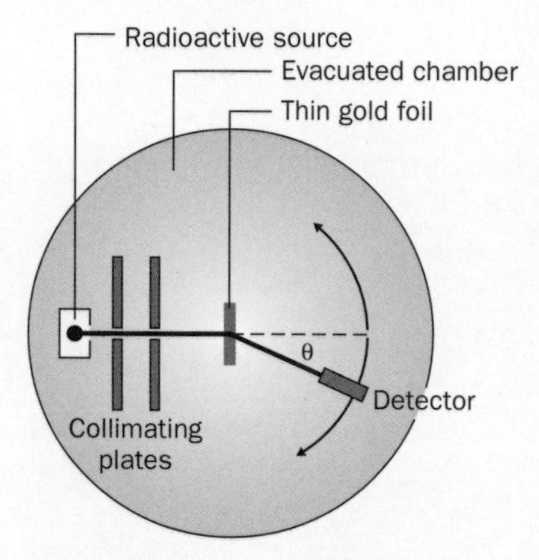

Fig. 22.13 Rutherford's alpha scattering experiment.

Rutherford's alpha scattering experiments

The nuclear model of the atom was proved by Rutherford. He devised an experiment in which a narrow beam of alpha particles was directed at a thin gold foil in a vacuum. Rutherford measured the number of alpha particles per second scattered in different directions by the foil.

He made the following observations:

1. Most alpha particles passed through the foil with little or no deflection,
2. A very small proportion of the alpha particles rebounded from the foil.

These results are akin to finding that bullets aimed at cardboard occasionally bounce back from the cardboard. If that happened, you would conclude that there must be some small but heavy objects in the cardboard which a bullet might hit.

Rutherford deduced from his measurements that:

1. the mass of every atom is concentrated in a comparatively small nucleus. This explained why most alpha particles passed through the foil with little or no scattering.
2. the nucleus of the atom is positively charged. This explains why an alpha particle can rebound from an atom. Such an alpha particle must have collided directly with the nucleus. Since the nucleus and the alpha particle are both positively charged, the alpha particle is repelled by the nucleus.

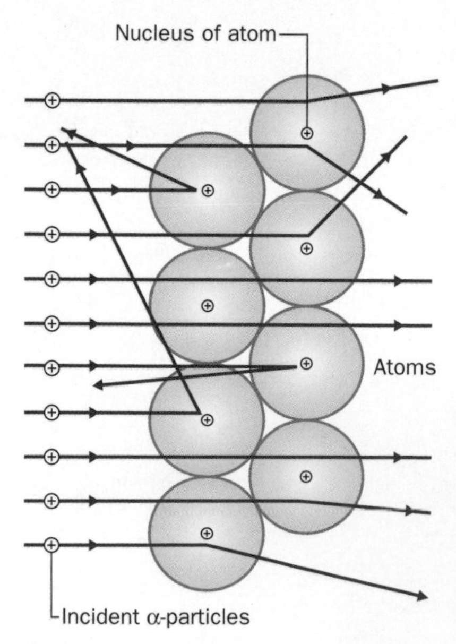

Fig. 22.14 The paths of scattered alpha particles.

Rutherford used the laws of mechanics and Coulomb's law of force to show that the number of alpha particles per second scattered by the foil through an angle, θ, from the original direction should be proportional to

$$\frac{1}{\sin^4 \theta/2}$$

His experimental results fitted this formula, thus proving conclusively the nuclear model of the atom.

Questions 22.3

1. In an experiment to determine the type of radiation from a radioactive source, the following measurements were made with different absorbers between a GM tube and the source.

   count rate with no absorber and no source present = 26 counts per minute;

   count rate with source present and no absorber = 628 counts per minute;

   count rate with paper between the tube and the source = 620 counts per minute;

   count rate with a 4.0 mm aluminium plate between the tube and source = 30 counts per minute.

   (a)(i) What type of radiation was emitted from the source? Give a reason for your answer.

   (ii) What further test could be carried out to confirm your answer to (i)?

   (b) In the above experiment, the count rate with a 0.5 mm metal plate between the same source and the tube at the same distance was 482 counts per minute.

   Calculate **(i)** the corrected count rate due to the source, with and without the 0.5 mm metal plate present, **(ii)** the percentage of the incident radiation transmitted by this metal plate, **(iii)** the expected count rate if an identical metal plate had also been placed in the path of the beam.

2. A GM tube was used to measure the count rate due to gamma radiation from a point source. The following readings were made:

   (i) Number of counts in 5 minutes without the source present = 143, 124, 136.

   (ii) Number of counts in 3 minutes with the source at a distance of 0.40 m from the tube = 725, 746, 738.

   (a) Calculate the corrected count rate due to the source.

   (b) Estimate the count rate if the distance between the tube and the source was reduced to **(i)** 0.20 m, **(ii)** 0.15 m.

22.4 Radioactive decay

Randomness

In a cloud chamber, the tracks created by alpha particles from a small radioactive source can easily be seen. The tracks occur at random with no indication of the direction of the next one. This is because the decay of an unstable nucleus is unpredictable. This is also why the number of counts by a Geiger counter in a given time, from a source at constant distance, fluctuates randomly.

The number of unstable nuclei in a sample of a radioactive isotope decreases gradually with time. This is because each decay of an unstable isotope means that there is one less unstable nucleus waiting to decay. Because the decay process is unpredictable, the number of unstable nuclei that decay each second (i.e. the rate of decay) is proportional to the number of unstable nuclei present. The rate of decay therefore slows down because the number of unstable nuclei decreases with time. Fig. 22.15 shows how the count rate from a radioactive isotope decreases with time.

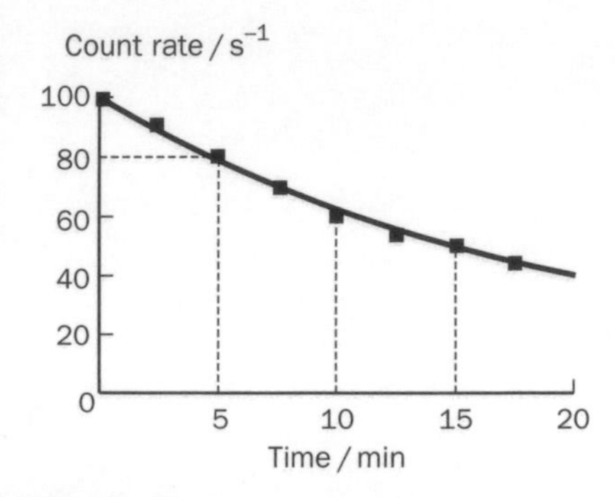

Fig. 22.15 An exponential decay curve.

The curve is called an **exponential decay** curve because the rate of decay is proportional to the number of unstable nuclei remaining. For example, suppose 10% of the unstable nuclei of a certain radioactive isotope decay

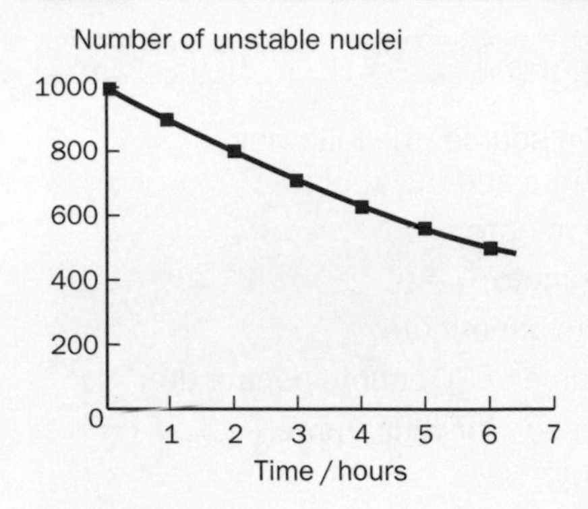

Fig. 22.16 Number of nuclei versus time.

Table 22.2 Numbers of unstable nuclei

Time at the start of each interval/hours	0	1	2	3	4	5	6
Number of unstable nuclei present initially	1000	900	810	729	629	567	510
Number of unstable nuclei that decay	100	90	81	73	63	57	51
Number of unstable nuclei remaining	900	810	729	656	566	510	459

every hour. For every 1000 unstable nuclei of this isotope, there would be 900 unstable nuclei (= 90% of 1000) present 1 hour later and 810 unstable nuclei (= 90% of 90% of 1000) present a further hour later. Table 22.2 shows how the number of unstable nuclei continues to decrease in this example. Fig. 22.16 shows the number of unstable nuclei plotted on the y-axis with time plotted on the x-axis. The shape of the curve is the same as Fig. 22.15. Prove for yourself that the number of unstable nuclei halves every 5 hours approximately.

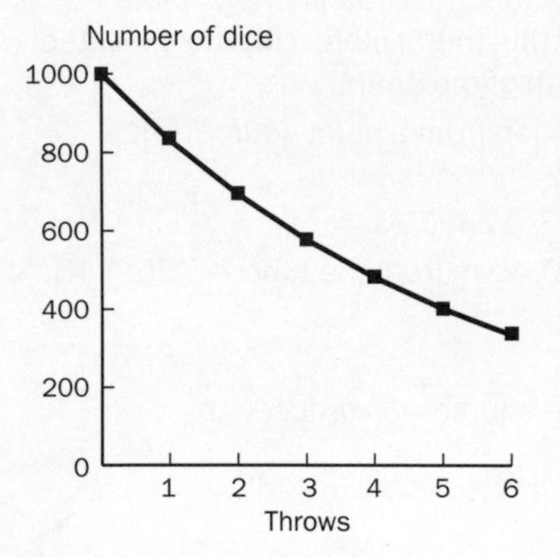

Fig. 22.17 A game of chance.

A model of radioactive decay

1. Suppose 1000 dice are thrown and all the dice with a six face-up are removed. Since the probability of a six is the same as any other number, the number of sixes should be 167, equal to 1/6 th of the total thrown. The number of dice remaining would be 833 (= 1000 − 167)
2. If the throw is repeated again and again with the dice not showing a six after each throw, the number of dice decreases by 1/6 th each time so the number remaining decreases exponentially, as shown in Table 22.3 and Fig. 22.17.

See www.palgrave.com/foundations/breithaupt to model radioactive decay using a spreadsheet.

Table 22.3 Numbers of sixes thrown

Throw number	1	2	3	4	5	6	7
Initial number of dice	1000	833	694	578	482	402	335
Number of sixes	167	139	116	96	80	67	56
Number of dice remaining	833	694	578	482	402	335	279

Activity

The activity of a sample of a radioactive isotope is the number of particles or photons emitted per unit time from the sample.

The unit of activity is the becquerel (Bq), equal to 1 disintegration per second.

Since each unstable nucleus emits either an alpha particle, a beta particle or a gamma photon when its nucleus becomes more stable, the activity is also the number of unstable nuclei that decay (i.e. become more stable) per unit time.

Because the number of unstable nuclei decreases with time, the activity also decreases with time. As explained earlier, the decrease is exponential because the number of nuclei that decay per unit time depends only on the number of nuclei present at that time. The time taken for the activity of a radioactive isotope to halve is defined as the **half-life**.

The half life of a radioactive isotope is the time taken for the activity of the isotope to decrease to 50% of its initial activity.

This is the same as the time taken for the number of unstable nuclei to decrease to 50% of the initial number.

For example, the half life of strontium 90 is 28 years. Thus a sample of strontium 90 with an initial activity of 80 kBq would have an activity of 40 kBq after 28 years and an activity of 20 kBq after a further 28 years. Fig. 22.18 shows how the activity of such a sample would decrease. The number of nuclei of strontium 90 and also the mass of strontium 90 would decrease at a similar rate because the activity is proportional to the number of strontium 90 nuclei present.

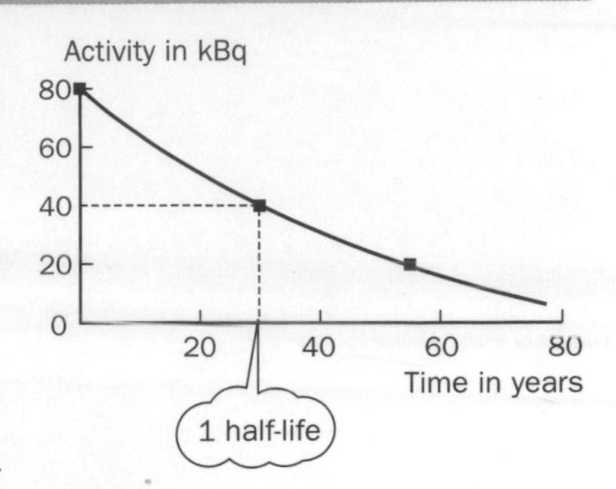

Fig. 22.18 Activity versus time for strontium 90.

See www.palgrave.com/foundations/breithaupt for applications of radioactivity.

Measurement of the half -life of protoactinium 234

Protoactinium 234 is a beta emitter produced by radioactive decay from thorium 234, which has a much longer half-life. A sealed bottle containing an organic solvent is used to separate protoactinium from thorium to enable the activity of the protoactinium to be measured at intervals. The bottle is first shaken to mix the solvent layer and the aqueous layer, then placed at rest to allow the solvent layer to form above the aqueous layer. The end of a GM tube is held adjacent to the solvent layer. The activity of the protoactinium isotope in the solvent layer is determined at intervals by measuring the count rate at 30 second intervals. The half life can be determined from a graph of count rate v. time as in Fig. 22.18.

Questions 22.4

1. A solvent contains a fixed quantity of a radioactive isotope with a half life of 68 seconds. The isotope decays to form a stable product. The activity of the isotope when the solvent layer was formed was measured using a GM tube, which gave a corrected count rate of 420 counts per minute.
   (a) Calculate the corrected count rate **(i)** two half-lives (i.e. 136 s), **(ii)** three half-lives (i.e. 204 s) after the solvent layer formed.
   (b) Estimate how long it takes for the activity to decrease so the corrected count rate falls below 25 counts per minute.
2. Strontium 90 is a beta-emitting isotope with a half-life of 28 years. A certain sample of strontium 90 has an activity of 160 kBq.
   (a) Calculate the activity **(i)** 28 years later, **(ii)** 56 years later.
   (b) Estimate the activity of the sample 100 years later.

22.5 The mathematics of radioactive decay

Consider a pure sample of a radioactive isotope that decays to form a stable isotope. Let N_0 represent the number of unstable nuclei initially present and suppose there are N unstable nuclei present at time, t, later. Since radioactive decay is a random process, the number, ΔN, of unstable nuclei that decay in a certain time interval, Δt, is proportional to

1. the number of unstable nuclei present, N;
2. the time interval, Δt.

Hence $\Delta N = -\lambda N \Delta t$, where λ is called the **decay constant** of the isotope. The minus sign means that the change of N is a decrease.

Since $\Delta N/N = -\lambda \Delta t$, and $\Delta N/N$ is the probability of decay of a single nucleus, it follows that λ is the probability of decay per unit time. Rearranging this equation gives $\Delta N/\Delta t = -\lambda N$.

This is usually written in the form

$$\frac{dN}{dt} = -\lambda N$$

where dN/dt, the rate of change of the number of nuclei, is the activity of the isotope.

The mathematical solution to this equation is written

$$N = N_0 \exp(-\lambda t)$$

where exp is referred to as the **exponential function**. The function is usually written as $e^{-\lambda t}$. The special feature about this function is that its rate of change is proportional to the function itself.

Mathematical notes

1. For any function of the form $y = x^n$, $\frac{dy}{dx} = nx^{n-1}$.
   See section 15.5, p. 217 if necessary.
2. The exponential function $e^x = 1 + x + \frac{x^2}{2} + \frac{x^3}{3 \times 2} + \frac{x^4}{4 \times 3 \times 2}$, etc.
3. Hence $\frac{d}{dx}(e^x) = 0 + 1 + x + \frac{x^2}{2} + \frac{x^3}{3 \times 2}$, etc. $= e^x$
   Thus the rate of change (with respect to x) of e^x is e^x.
4. If $y = e^x$, then $x = \log_e y$, since $\log_e$ is the inverse function of the exponential function.
   Note that most calculator keys denote $\log_e$ by the symbol ln.
5. To prove that $N = N_0 e^{-\lambda t}$ is the solution of $\frac{dN}{dt} = -\lambda N$

   $\frac{d}{dt}N = N_0 \frac{d}{dt}(e^{-\lambda t}) = \lambda N_0 \frac{d}{dx}(e^{-x})$ where $x = \lambda t$

   Since $\frac{d}{dx}(e^{-x}) = -e^{-x}$, then $\frac{dN}{dt} = -\lambda N_0 e^{-\lambda t} = -\lambda N$.

Half-life and the decay constant

The half-life, $T_{1/2}$, is the time taken for the number of nuclei of a radioactive isotope to decrease by 50%

Consider a radioactive isotope in which there are initially N_0 unstable nuclei. The number of unstable nuclei at time, t, later, $N = N_0 e^{-\lambda t}$.

At $t = T_{1/2}$, $N = \frac{1}{2} N_0$, hence $\frac{1}{2} N_0 = N_0 e^{-\lambda T_{1/2}}$.
Cancelling N_0 from both sides gives $\frac{1}{2} = e^{-\lambda T_{1/2}}$.

$\therefore e^{\lambda T_{1/2}} = 2$
$\boldsymbol{\lambda T_{1/2} = \log_e 2}$

Activity and the decay constant

The activity of a radioactive isotope, A = the magnitude of dN/dt, the number of disintegrations per second of the nuclei of the isotope.

Since

$$\frac{dN}{dt} = -\lambda N,$$

then the activity, A, can be written as

$$\mathbf{A = \lambda N}$$

Also, since $N = N_0 e^{-\lambda t}$, then the activity, $A = \lambda N_0 e^{-\lambda t} = A_0 e^{-\lambda t}$ where A_0, the initial activity, $= \lambda N_0$.

Note
The key for e on some calculators is inv ln.

Worked example 22.2

Strontium 90 is a radioactive isotope with a half-life of 28 years. Calculate (a) the decay constant of this isotope, (b) the activity of a 1.0 milligram sample of strontium 90, (c) the activity of a 1.0 milligram sample in 100 years time. The Avogadro constant, $N_A = 6.02 \times 10^{23}$ mol^{-1}.

Solution

(a) $\lambda T_{1/2} = \log_e 2$, hence $\lambda = \frac{\log_e 2}{T_{1/2}} = \frac{0.693}{28 \text{ yrs}} = 0.0248 \text{ yr}^{-1}$

$= \frac{0.0248}{365.25 \times 24 \times 3600 \text{ s}} = 7.8 \times 10^{-10} \text{ s}^{-1}$

(b) 90 g of strontium 90 contains 6.02×10^{23} atoms of the isotope.

Hence 1.0 mg contains $\frac{6.02 \times 10^{23}}{90 \times 1000}$

$= 6.7 \times 10^{18}$ atoms.

Hence the activity of 1.0 mg, $A_0 = \lambda N_0 = 7.8 \times 10^{-10} \times 6.7 \times 10^{18}$
$= 5.2 \times 10^9$ Bq

(c) Using $A = A_0 e^{-\lambda t}$ gives $A = 5.2 \times 10^9 e^{-(0.0248 \times 100)} = 4.4 \times 10^8$ Bq

Questions 22.5

The Avogadro constant $= 6.02 \times 10^{23}$ mol^{-1}

1. Polonium 210 is a radioactive isotope with a half-life of 140 days. Calculate
(a) the decay constant of this isotope,
(b) the number of atoms in 1.0 mg of pure polonium 210,
(c) the activity of 1.0 mg of this isotope,
(d) the activity of this isotope one year later.

2. Cobalt 60 is a radioactive isotope with a half-life of 1940 days. A sample of this isotope has an activity of 2.0 MBq. Calculate
(a) the decay constant,
(b) the number of atoms of the isotope,
(c) the time taken for the activity to decrease to 0.1 MBq.

Summary

◆ **Alpha (α) radiation**

- consists of helium 4 nuclei, each containing two protons and two neutrons. An α-particle is emitted from an unstable nucleus with too many protons and neutrons
- is highly ionising, absorbed by paper and has a well-defined range in air of a few centimetres.

◆ **Beta (β) radiation**

- consists of fast-moving electrons. A β-particle is emitted from an unstable nucleus with too many neutrons. A neutron in the nucleus changes into a proton and a β-particle is created and emitted at the instant of change
- is less ionising than α-radiation, absorbed by metal plates of thickness a few millimetres and has a range in air of more than 50 cm.

◆ **Gamma (γ) radiation**

- consists of high energy photons, emitted by an unstable nucleus with excess energy
- is less ionising than beta radiation, absorbed by lead plates of several centimetres thickness and has an unlimited range in air. The intensity from a point source follows the inverse square law.

◆ **Definitions**

- An isotope comprises atoms with the same number of protons and neutrons.
- The activity of a radioactive isotope is the number of disintegrations per second.
- The half-life $T_{½}$, of a radioactive isotope is the time for half the atoms of the isotope to disintegrate.

◆ **Equations**

Decay constant, $\lambda = \frac{\log_e 2}{T_{½}}$

Activity, A = magnitude of $\frac{dN}{dt} = \lambda N$

Radioactive decay equation, $\frac{dN}{dt} = -\lambda N$

Solution of the radioactive decay equation, $N = N_0 e^{-\lambda t}$

Revision questions

The Avogadro constant = 6.02×10^{23} mol^{-1}

22.1. Each of the following equations represents the decay of an unstable nucleus. Copy and complete each equation.

(a) $^{63}_{[\,]}\text{Ni} \rightarrow {}^{[\,]}_{29}\text{Cu} + {}^{0}_{[\,]}\beta$

(b) $^{[\,]}_{84}\text{Po} \rightarrow {}^{4}_{[\,]}\alpha + {}^{210}_{[\,]}\text{Pb}$

22.2. In an experiment to determine the type of radiation emitted by a radioactive source, the following measurements were made.,

Background count in 5 minutes = 128, 136, 138;

Counts in 3 minutes with the source at 10 cm from the GM tube

- with no absorber between the tube and the source = 565, 572, 552;
- with a 0.8 mm thick metal plate between the tube and the source as shown in Fig. 22.19 = 384, 368, 372.

(a) Calculate the average background count rate in counts per minute.

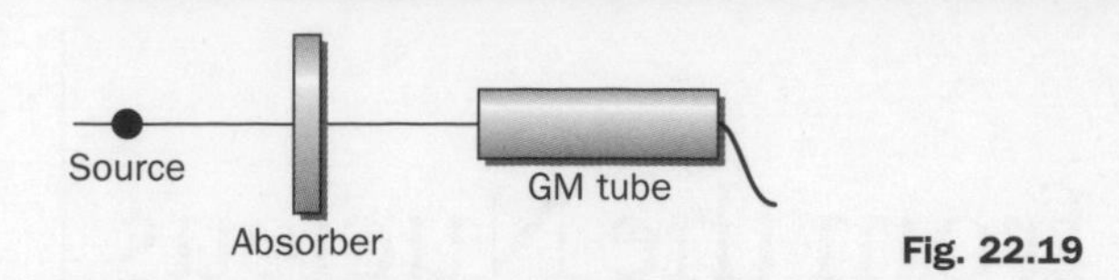

Fig. 22.19

(b) The source is known to emit one type of radiation only. What type of radiation does it emit? Give a reason for your answer.
(c) What percentage of the radiation incident on the plate is transmitted by the plate?
(d) Calculate the expected count rate if a further identical plate was placed in the path of the beam.

22.3. (a) A GM tube was used to measure the count rate due to gamma radiation from a point source inside a sealed container, as shown in Fig. 22.20. When the distance between the tube and the container was 90 mm, the corrected count rate was 1284 counts per minute. When this distance was increased to 190 mm, the corrected count rate was 330 counts per minute.

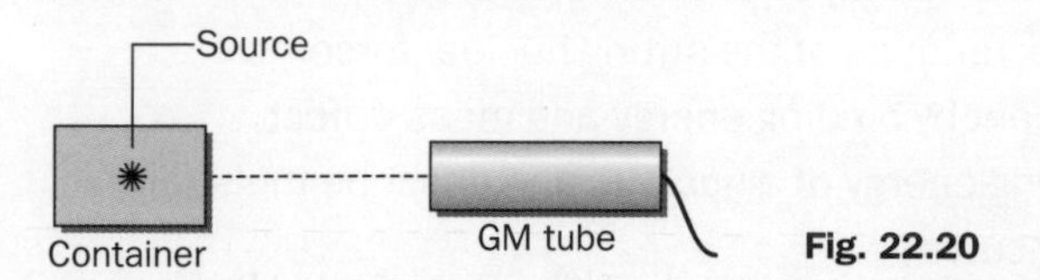

Fig. 22.20

(i) Use these results to show that the source was 10 mm inside the container.
(ii) Calculate the corrected count rate if the distance between the tube and the source had been increased to 240 mm.
(b) The source in (a) was replaced by a different source. Describe how you would identify the type of radiation emitted by the source, assuming it emits only one type.

22.4. Sodium 24 is a radioactive isotope with a half-life of 14.8 hours. Calculate
(a) the decay constant for this isotope,
(b) the number of atoms in a 1.0 mg sample of this isotope,
(c) the activity of a 1.0 mg pure sample of this isotope,
(d) the activity of such a sample exactly 24 hours later.

22.5. Plutonium 239 is a radioactive isotope with a half-life of 100 years.
(a) Calculate **(i)** the decay constant of this isotope, **(ii)** the number of atoms in 1.0 g of this isotope.
(b) A pure sample of this isotope has an activity of 1000 MBq. Calculate **(i)** its mass, **(ii)** its activity in 10 years time.

Energy from the Nucleus

Contents

Objectives

After working through this unit, you should be able to:

- ▶ describe the characteristics of the strong nuclear force
- ▶ explain what is meant by binding energy and mass defect
- ▶ calculate the binding energy of a nucleus and describe the shape of the binding energy curve
- ▶ use the binding energy curve to explain why energy is released as a result of fission and fusion
- ▶ describe the main features of a nuclear reactor
- ▶ discuss the advantages and disadvantages of nuclear power
- ▶ describe the use of high-energy accelerators to probe the nucleus
- ▶ describe the quark model of the nucleus

23.1 The nature of force

Fundamental forces

The force of gravity holds us on to the Earth and keeps the Earth on its orbit round the Sun. The force between charged objects, the electrostatic force, holds the atoms together in the human body and prevents us from disappearing into the Earth. Most of the forces acting on objects we can see are either due to gravity or due to the electrostatic force. For example, the tension in a cable supporting a lift is due to the electron bonds between the atoms of the cable. The force of gravity and the electrostatic force, more generally known as the electromagnetic force, are two of the fundamental forces of nature.

- The **force of gravity** acts between any two material objects. The force of gravity between any two objects is always attractive and varies with distance apart according to the inverse square law. See p. 398.
- The **electromagnetic force** acts between any two charged objects. It was shown in the 19th century that electrostatic forces and magnetic forces

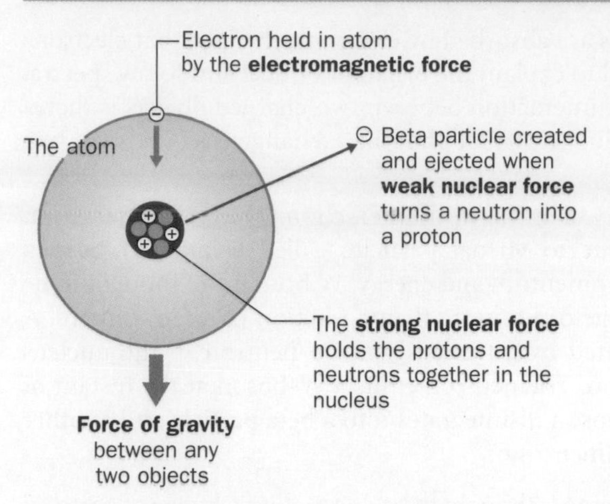

Fig. 23.1 Fundamental forces.

are essentially the same, differing only because the electrostatic force acts when charges are at rest and the magnetic force acts when they are in relative motion. The electromagnetic force varies with distance according to the inverse square law. See p. 239.

- The **strong nuclear force** is an attractive force acting between protons or neutrons over a short range of no more than a few femtometres. (1 femtometre = 1 fm = 10^{-15} m).
- The **weak nuclear force** is the force responsible for beta decay, which occurs when a neutron in a nucleus changes into a proton. The phenomenon of beta decay cannot be explained in terms of gravity or the electromagnetic force or the strong nuclear force. Recent research has led to the conclusion that the weak nuclear force and the electromagnetic force are different aspects of the same fundamental force, referred to as the **electroweak** force.

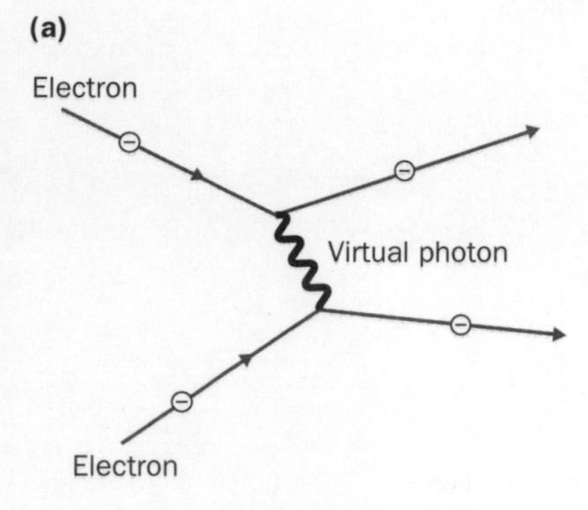

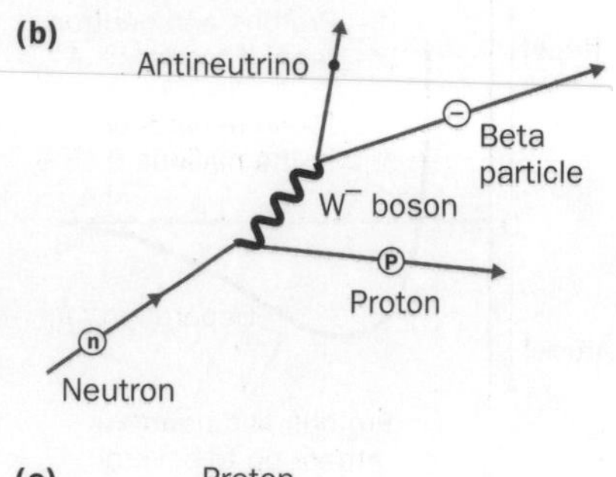

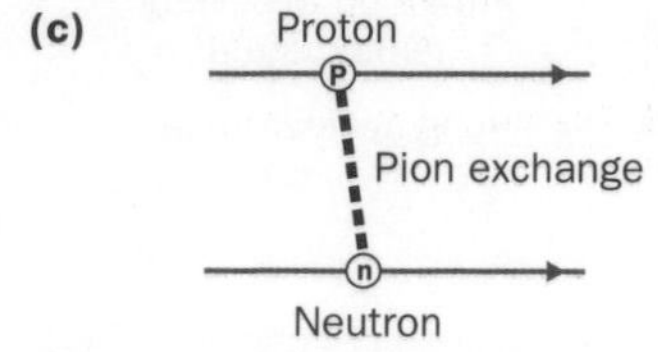

Fig. 23.2 Force carriers **(a)** coulomb repulsion, **(b)** the weak interaction, **(c)** the strong nuclear force.

Force carriers

When two objects interact, they exert equal and opposite forces on each other. Since force is defined as change of momentum per second, an interaction causes transfer of momentum if the objects are free to move. Each fundamental type of force is thought to act through exchange of quanta between the two objects. The quantum of each type of force may be thought of as the carrier of the force. The quanta may be described as 'virtual' because they cannot be detected without altering the interaction. Nevertheless, the idea of virtual quanta is supported by firm experimental evidence for all the fundamental forces except gravity (up to the point at which this text was written).

- The **electroweak force** is thought to be due to the exchange of:

  1. virtual photons in the case of electromagnetic interactions. Photons are massless, have an infinite range and carry energy and momentum. The idea that electrons absorb photons was established in the photoelectric effect where photons are absorbed by electrons; the idea that electrons emit photons is used to explain the origin of optical and X-ray spectra. The electromagnetic interaction between two charged objects is therefore thought to be due to virtual photons exchanged between the two charged objects.

2. W bosons in the case of the weak interaction. The weak interaction is thought to be due to virtual particles called **weak (W) bosons** which carry mass, momentum and energy. W bosons are thought to be unstable, with a range of no more than a fraction of a femtometre. A beta particle is emitted by a nucleus when a neutron in the nucleus changes into a proton. The neutron emits a W-boson at the instant of change and the W-boson disintegrates into a beta particle and another particle called an antineutrino.

- The **strong nuclear force** is thought to be due to the exchange of particles called **pions** between protons and neutrons in the nucleus. Pions were discovered as a result of exposing photographic plates to cosmic radiation at high altitude. Cosmic radiation consists of fast-moving particles, such as protons from the Sun. When such a high-energy particle hits a nucleus, it can knock pions as well as protons or neutrons out of the nucleus. Pions are unstable and therefore have a limited range.
- **Gravity** could be the result of the exchange of massless particles called **gravitons.** Experimental evidence for gravitational radiation was first obtained in 1975 from two neutron stars in orbit about each other. However, there is no firm evidence at present for the graviton or any other quantised form of gravity as the exchange particle of the gravitational force.

The strong nuclear force

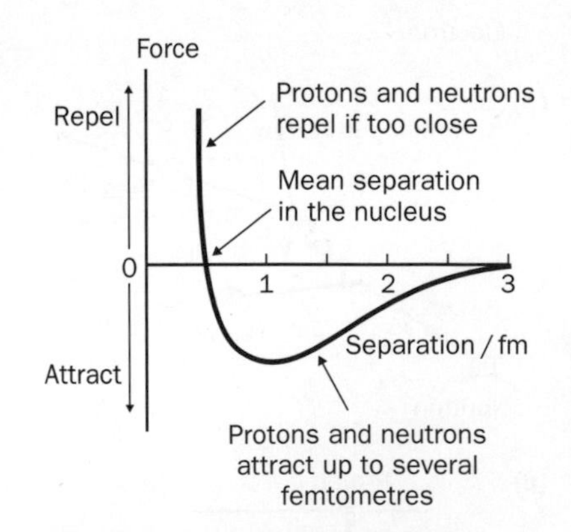

Fig. 23.3 The strong nuclear force.

The nucleus of an atom consists of protons and neutrons packed together at the centre of the atom. The diameter of the nucleus is of the order of a few femtometres, approximately 10 millionths of the diameter of the atom. The protons carry like charges, so they tend to repel each other due to the electrostatic force of repulsion between like charges. The fact that most nuclei are stable must mean that the protons and neutrons in a nucleus are held together by an attractive force which acts against the electrostatic force of repulsion. This attractive force is called the **strong nuclear force.**

- The **strength** of the strong nuclear force can be estimated from the fact that it must be sufficiently strong to overcome the electrostatic nuclear force between two protons at a distance of about 1 fm apart. Using Coulomb's Law (see p. 239) for the electrostatic force therefore gives

$$F = \frac{Q_1 Q_2}{4\pi\varepsilon_0 r^2} = \frac{1.6 \times 10^{-19} \times 1.6 \times 10^{-19}}{4\pi\varepsilon_0 \times (10^{-15})^2} = 230 \text{ N}$$

The strong nuclear force must be of the order of 100 N or more to overcome the electrostatic force of repulsion. The mass of a proton is of the order of 10^{-27} kg, so its weight (= mg) is about 10^{-26} N. The strong nuclear force is more than 10^{28} times the weight of a proton.

- The **range** of the strong nuclear force is no more than a few femtometres, unlike the electrostatic force which stretches to infinity. The largest nuclei, those of the heaviest elements, are of diameters of the order of no more than 10 fm. If the strong nuclear force extended much beyond a few femtometres, nuclei would be much larger than even the largest known nuclei.

Questions 23.1

1. State the fundamental force responsible for
   (a) the path of a satellite round the Earth,
   (b) release of energy in **(i)** alpha decay, **(ii)** beta decay,
   (c) preventing you from falling through the floor.
2. When alpha particles from a radioactive source pass through air, some of the particles collide directly with the nuclei of the atoms.
   (a) Calculate the electrostatic force between an alpha particle and a nitrogen $^{14}_{7}N$ nucleus at a separation of **(i)** 10 fm, **(ii)** 2 fm. ($\varepsilon_0 = 8.85 \times 10^{-12}$ F m^{-1}).
   (b) Use your knowledge of the strong nuclear force to explain why an alpha particle directed at a nitrogen nucleus is repelled by the nucleus unless it has sufficient kinetic energy.

23.2 Binding energy

Energy from reactions

Chemical energy is released when a fuel such as oil burns in air. The fuel molecules react with oxygen molecules to release energy. The atoms of each fuel molecule break away from each other and combine with oxygen atoms to form carbon monoxide, carbon dioxide and water vapour. Sufficient energy is released when the atoms from a fuel molecule combine with oxygen to enable more fuel molecules to break up. The energy is released when the bonds between the fuel atoms and oxygen atoms are formed. Such bonds involve electrons in the atoms becoming trapped, releasing energy in the process. Burning fuel releases about 50 MJ of heat per kilogram of fuel burned, which is equivalent to about 10^{-19} J per atomic mass unit. This is about the same as the work done to move a single electron through a potential difference of 1 volt.

The **electron volt (eV)** is the work done to move an electron through a potential difference of exactly 1 volt. Since the charge of an electron is 1.6×10^{-19} C and 1 volt is defined as work done per unit charge, then $1 \text{ eV} = 1.6 \times 10^{-19}$ J. Note that $1 \text{ MeV} = 1.6 \times 10^{-13}$ J.

Nuclear reactions involve energy changes of the order of millions of electron volts (MeV). Imagine pulling a nucleus apart, pulling off the protons and neutrons one at a time. Work needs to be done on a proton or a neutron to pull it out of a nucleus. This is necessary to overcome the strong nuclear force. Even though this force acts over an exceedingly short distance of no more than a few femtometres, its strength (of the order of 100 N or more) is so great that the work done is of the order of 10^{-13} J ($= 100 \text{ N} \times 1 \text{ fm}$). Thus the energy changes in nuclear reactions are of the order of a million times greater than the energy changes in chemical reactions.

The binding energy of a nucleus is defined as the work done to separate a nucleus into its constituent neutrons and protons.

Mass and energy

Whenever energy is gained by a body, the body's mass increases. When a body loses energy, its mass decreases. Transfer of energy without transfer of matter to or from a body causes its mass to change. The exact mechanism of how this process works is not yet known. However, the scale of the changes was determined by Einstein who proved that the total energy, E, of a body is related to its mass, m, in accordance with the now-famous equation

$$\mathbf{E = mc^2}$$

where c is the speed of light in free space ($= 3.0 \times 10^8 \text{ m s}^{-1}$).

For example, the mass of a sealed torch emitting light becomes smaller as it loses light energy. The scale of the change is insignificant where energy is released due to chemical changes. Prove for yourself that the mass loss when a 10 W torch is left on for 10 hours is about 4×10^{-12} kg ($= 10 \times 36\,000 \text{ J} \div c^2$).

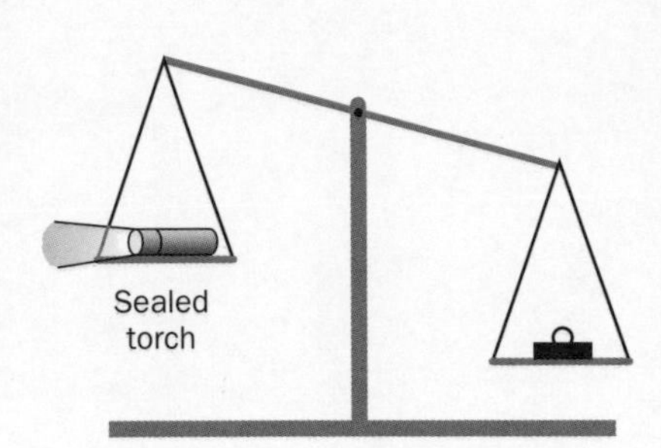

Fig. 23.4 Energy and mass – the mass of a sealed torch is reduced as it loses light energy.

However, in nuclear changes, the mass changes are significant as the energy changes are much greater than for chemical reactions. Prove for yourself that 10^{-13} J of energy released by a nucleus corresponds to a loss of mass of about 0.001 u ($= 10^{-30}$ kg approximately) which is about 0.1% of the mass of a proton.

When a nucleus is formed from separate neutrons and protons, energy is released equal to the binding energy of the nucleus. The mass of the nucleus is less than the mass of its constituent protons and neutrons because energy is released. This mass difference can be measured and is known as the mass defect of the nucleus.

Note

1 u is equivalent to 931.3 MeV.
This can be proved using $E = mc^2$, with $m = 1 \text{ u} = 1.660 \times 10^{-27}$ kg and $c = 2.998 \times 10^8 \text{ m s}^{-1}$, which gives $E = 1.492 \times 10^{-10}$ J or 931.3 MeV, since $1 \text{ MeV} = 1.602 \times 10^{-13}$J.

The mass defect of a nucleus is the difference between the mass of a nucleus and the mass of its constituent protons and neutrons.

A nucleus of an isotope ${}^{A}_{Z}X$ consists of Z protons and (A − Z) neutrons. Its mass defect, Δm, is therefore given by the equation

$$\mathbf{\Delta m = Zm_p + (A - Z)m_n - M}$$

where M is the mass of the nucleus, m_p is the mass of a proton ($= 1.00728$ u) and m_n is the mass of a neutron ($= 1.00867$ u).

The binding energy of a nucleus can therefore be calculated from the mass defect, Δm, using the equation $E = mc^2$ in the form

$$\textbf{Binding energy, BE} = \Delta mc^2$$

Worked example 23.1

The mass of a nucleus of ^4_2He is 4.00150 u. Calculate (a) its mass defect, (b) its binding energy. The mass of a proton = 1.00728 u and the mass of a neutron = 1.00867 u.

Solution

(a) $Z = 2, A = 4$

$\therefore \Delta m = Zm_p + (A - Z)m_n - M$

$= 2 \times 1.00728 + 2 \times 1.00867 - 4.00150 = 0.0304$ u

(b) $BE = \Delta mc^2 = 0.0304 \times 931 = 28.3$ MeV

The binding energy curve

The binding energy of a nucleus is the work needed to separate the nucleus into its constituent protons and neutrons. The protons and neutrons in a nucleus are collectively referred to as **nucleons**. The binding energy of a nucleus depends on the number of nucleons in a nucleus. The stability of different types of nuclei can be compared by calculating the binding energy per nucleon of each type of nucleus. This is the total binding energy of the nucleus divided by the number of nucleons in the nucleus. For example, if the binding energy per nucleon of nucleus, X, is 8.0 MeV per nucleon and the binding energy per nucleon of a nucleus, Y, is 8.5 MeV per nucleon, then Y is more stable than X because more work needs to be done to remove a nucleon from Y than from X.

$$\text{Binding energy per nucleon of a nucleus} = \frac{\text{binding energy of the nucleus}}{\text{total number of nucleons}}$$

The binding energy per nucleon of a nucleus ^A_ZX can be calculated as follows, if the mass of the nucleus is known.

Step 1 Calculate the mass defect of the nucleus using the equation

$$\Delta m = Zm_p + (A - Z)m_n - M$$

Step 2 Calculate the binding energy of the nucleus using the equation

$$BE = \Delta mc^2$$

Step 3 Calculate the binding energy per nucleon using the equation

$$\text{BE per nucleon} = \frac{BE}{A}$$

Fig. 23.5 shows how the binding energy per nucleon for every known type of nucleus varies with nucleon number. The graph shows that the binding energy per nucleon increases as A increases, up to about A = 60, then decreases gradually.

- The most stable nuclei occur at about A = 60.
- The helium nucleus (i.e. the alpha particle) is very stable compared to other light nuclei.

Note

If the atomic mass of X is given, the mass, M, of the **nucleus** is calculated by subtracting the mass of Z electrons from the atomic mass.

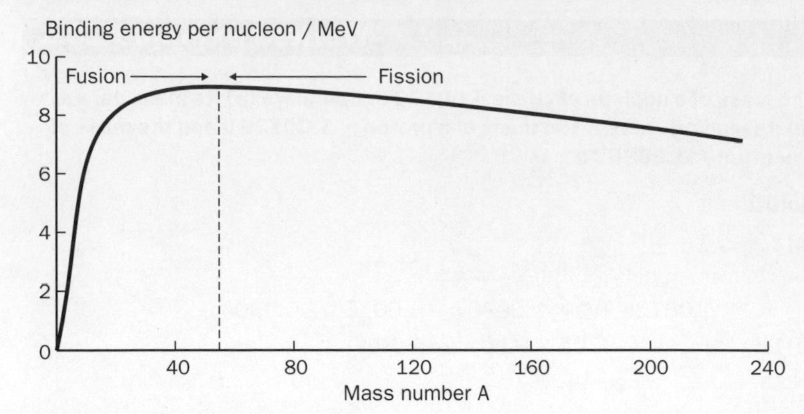

Fig. 23.5 The binding energy curve.

- Certain heavy nuclei, such as U 235, can become more stable by splitting into two approximately equal nuclei. This process is known as **nuclear fission**. When a heavy nucleus fissions, the binding energy per nucleon rises by about 1 MeV per nucleon. The energy released from a single fission event is therefore of the order of 200 MeV (= 1 MeV per nucleon × approximately 200 nucleons in a heavy nucleus).
- Light nuclei can be forced to fuse together if at very high temperatures. This process is known as **nuclear fusion**. For example, this occurs in the Sun's core where hydrogen is converted into helium. When nuclear fusion occurs, the binding energy per nucleon rises, so energy is released as a result.

Worked example 23.2

Calculate the binding energy per nucleon, in MeV, of (a) a $^{12}_{6}C$ nucleus, (b) a $^{56}_{26}Fe$ nucleus. (Atomic masses: Carbon 12 = 12.00000 u, Iron (Fe) 56 = 55.93493 u, electron mass = 0.00055 u, proton mass = 1.00728 u, neutron mass = 1.00867 u) 1 u = 931 MeV.

Solution

(a) Mass defect, $\Delta m = Zm_p + (A - Z)m_n - M$

$= 6 \times 1.00728 + 6 \times 1.00867 - (12.00000 - 6 \times 0.00055)$

$= 0.099$ u

$\therefore$ BE of C 12 nucleus $= 0.099 \text{ u} \times \dfrac{931 \text{ MeV}}{\text{u}} = 92.1$ MeV

$\therefore$ BE per nucleon $= \dfrac{BE}{A} = \dfrac{92.1}{12} = 7.7$ MeV per nucleon

(b) Mass defect, $\Delta m = Zm_p + (A - Z)m_n - M$

$= 26 \times 1.00728 + 30 \times 1.00867 - (55.93493 - 26 \times 0.00055)$

$= 0.529$ u

$\therefore$ BE of Fe 56 nucleus $= 0.529 \text{ u} \times \dfrac{931 \text{ MeV}}{\text{u}} = 492$ MeV

$\therefore$ BE per nucleon $= \dfrac{BE}{A} = \dfrac{492}{56} = 8.8$ MeV per nucleon

Q values

When a nuclear change occurs, binding energy changes cause the total mass of the products to differ from the total initial mass.

- If the total initial mass exceeds the total final mass, energy is released as a result of the change.
- If the total initial mass is less than the total final mass, energy is needed to make the change occur.

The **Q value** of a nuclear process is the energy released by the process ($Q > 0$) or needed for the process to occur ($Q < 0$).

Alpha decay

$${}^{A}_{Z}X \rightarrow {}^{A-4}_{Z-2}Y + {}^{4}_{2}\alpha + Q$$

$\therefore$ Q = Mass of nucleus of ${}^{A}_{Z}X$ − [mass of nucleus of ${}^{A-4}_{Z-1}X$ + mass of α particle]. Since an atom of X has Z electrons, an atom of Y has Z − 2 electrons and a helium atom is an alpha particle + 2 electrons,
Q = atomic mass of ${}^{A}_{Z}X$ − [atomic mass of ${}^{A-4}_{Z-2}Y$ + atomic mass of helium 4].

Beta decay

$${}^{A}_{Z}X \rightarrow {}_{Z+1}^{A}Y + {}_{-1}^{0}\beta$$

$\therefore$ Q = mass of nucleus of ${}^{A}_{Z}X$ − [mass of nucleus of ${}_{Z+1}^{A}Y$ + mass of β particle]. Since an atom of X has Z electrons, an atom of Y has Z+1 electrons and a beta particle is an electron,
Q = atomic mass of ${}^{A}_{Z}X$ − atomic mass of ${}_{Z+1}^{A}Y$.

Questions 23.2

Data: electron mass = 0.00055 u, proton mass = 1.00728 u, neutron mass = 1.00867 u, 1 u = 931 MeV.

1. **(a)** Calculate the binding energy per nucleon of **(i)** an ${}^{16}_{8}O$ nucleus, **(ii)** an ${}^{90}_{38}Sr$ nucleus.
   (b) Calculate the change of mass of **(i)** an atom when it releases a light photon of wavelength 600 nm, **(ii)** a nucleus when it releases a gamma photon of wavelength 100 fm.
   (Atomic masses: oxygen 16 = 15.99492, strontium 90 = 89.90730)
2. **(a)** The uranium nucleus ${}^{238}_{92}U$ decays by α-particle emission to form a nucleus of an isotope of thorium (Th).
   (i) Copy and complete the equation for this decay.
   (ii) Calculate the Q value of this decay.
   (Atomic masses: U 238 = 238.05076, Th 232 = 232.03821, Th 233 = 233.04143, Th 234 = 234.04357, He 4 = 4.00260)
   (b) The cobalt nucleus ${}^{60}_{27}Co$ decays by beta emission to form a nucleus of an isotope of nickel (Ni).
   (i) Copy and complete the equation for this decay.
   (ii) Calculate the Q value of this decay.
   (Atomic masses: Co 60 = 59.93381, Ni 59 = 58.93434, Ni 60 = 59.93078, Ni 61 = 60.93105)

23.3 Nuclear Power

Induced fission

Uranium and other heavy nuclei are unstable and usually decay by emitting alpha particles. However, these very heavy nuclei can be made to fission (i.e. split into two approximately equal fragments) as a result of bombardment by neutrons. This process is known as **induced fission.** Fast neutrons at kinetic energies of the order of 1 MeV are needed to fission most fissionable nuclei, although U 235 can also be fissioned by slow neutrons at kinetic energies of the order of 0.1 eV or less.

In a fission process, two or three neutrons are released, which may then go on to produce further fission. Energy is released in the fission process as kinetic energy of the fission fragments and the released neutrons, as well as in the form of gamma radiation. The fission fragments themselves are neutron-rich isotopes and therefore emit beta particles.

Fig. 23.6 shows how the **liquid drop model** of the nucleus is used to describe induced fission. The oscillating nucleus is similar to a vibrating liquid drop. At its most distorted, the strong nuclear force between the nucleons at the neck of the distorted nucleus is sufficient to prevent the two halves of the nucleus breaking away due to their electrostatic repulsion. If the nucleus is struck by a neutron at the point where it is most distorted, the added energy from the incident neutron is just enough to split the nucleus.

The Q value of a fission reaction can be estimated from the binding energy curve. Each nucleon has a binding energy of about 7.5 MeV per nucleon in the parent nucleus and about 8.5 MeV per nucleon in the fission fragments. Hence the binding energy per nucleon rises by about 1 MeV per nucleon, and about 1 MeV per nucleon is released. Thus the energy released by a single fission event is about 200 MeV (= 200 nucleons × 1 MeV per nucleon). Most of this energy is carried away by the fission fragments as recoil kinetic energy.

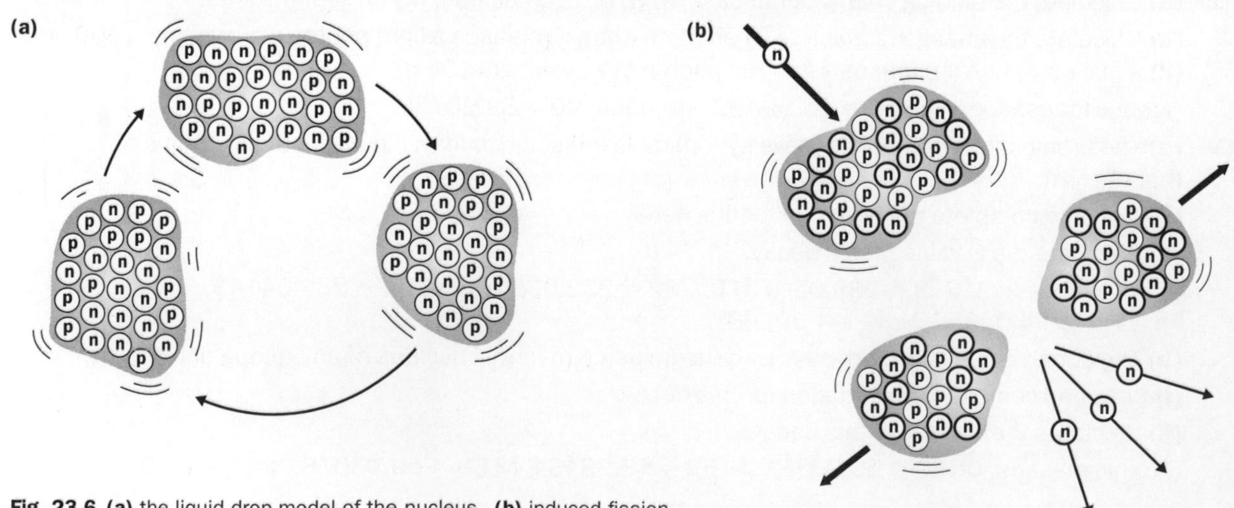

Fig. 23.6 **(a)** the liquid drop model of the nucleus, **(b)** induced fission.

Further fission

The neutrons released when a heavy nucleus is fissioned are capable of producing further fission in the same material, only if the material is either uranium 235 or plutonium 239. All other fissionable isotopes absorb fast neutrons without fission instead of undergoing further fission.

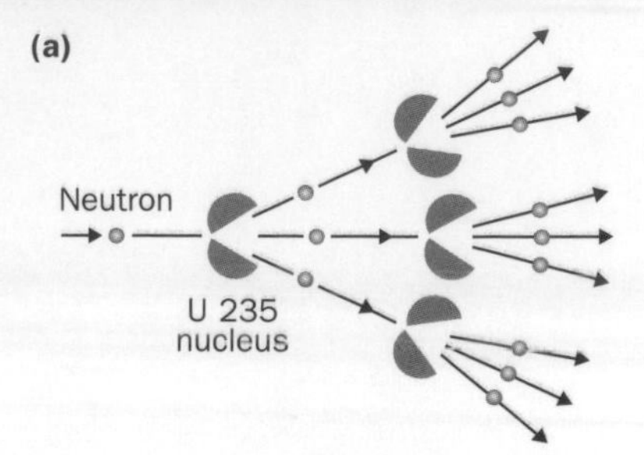

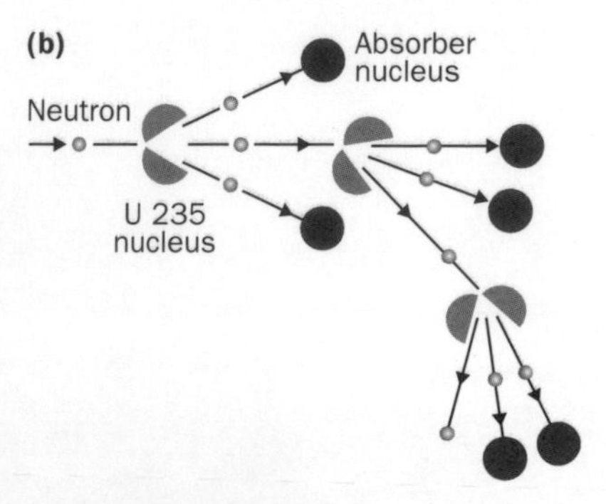

Fig. 23.7 Chain reactions **(a)** uncontrolled **(b)** controlled.

Uranium 235

Uranium 235 is one of the two isotopes in natural uranium. Natural uranium is more than 99% U 238 and less than 1% U 235.

- In pure U 235, when a nucleus fissions, the released neutrons will produce further fission if they do not escape from the material first. Further fission would result in more neutrons being released and the number of fission events would escalate in an uncontrollable **chain reaction**, releasing all the energy in a very short time and thus causing a massive explosion.
- Enriched uranium containing 2 or 3% of U 235 is used in a nuclear reactor. Without enrichment, the fission neutrons are more likely to be absorbed by U 238 than to produce further fission of U 235. Even so, the fission neutrons need to be slowed down to produce further fission of U 235. This is achieved in the **thermal nuclear reactor** by the presence of a moderator, as explained on p. 350. In addition, rods made of boron are used to control the chain reaction. The control rods absorb neutrons to ensure a steady rate of fission events. The control rods are inserted into the nuclear reactor core at just sufficient depth to ensure an average of exactly one neutron per fission goes on to produce further fission.

Plutonium 239

Plutonium 239 is an artificial isotope produced in a thermal nuclear reactor as a result of neutron absorption by U 238 to form U 239, which decays by beta emission to form the isotope neptunium 239. This change occurs rapidly as the half-life of uranium 239 is 24 minutes. Neptunium 239 decays by beta emission with a half-life of 56 hours, to form plutonium 239, which has a half-life of 100 years.

Fast neutrons cause plutonium 239 to fission, resulting in the release of further fast neutrons which then produce further fission if the total mass of plutonium 239 present is sufficiently large. The **fast breeder reactor** is designed to use plutonium 239 as its fuel. The plutonium core is surrounded by a blanket of U 238 which absorbs neutrons that escape from the core. The U 238 nuclei in the blanket are converted into plutonium 239 nuclei as a result of absorbing neutrons from the core. In this way, the reactor 'breeds' its own fuel from the surrounding uranium. No moderator is needed in a fast breeder reactor. The chain reaction is controlled by control rods in the same way as in the thermal nuclear reactor.

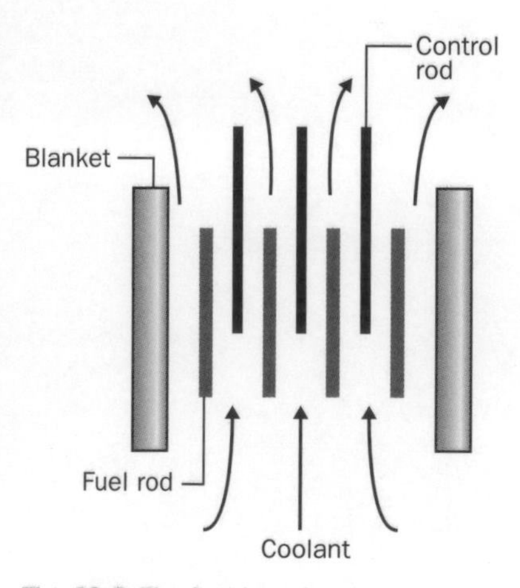

Fig. 23.8 The fast breeder reactor.

The thermal nuclear reactor

Enriched uranium, used as the fuel in a thermal nuclear reactor, contains about 2% U 235. As explained on p. 348, about 200 MeV of energy is released when a U 235 nucleus fissions. Thus the energy released by 1 mole (ie. 0.235 kg) of U 235 is 2×10^{13} J ($= 200 \times 1.6 \times 10^{-13}$J $\times 6 \times 10^{23}$). The total energy released as a result of 1 kg of U 235 fissioning is therefore about 8×10^{13} J ($= 2 \times 10^{13}$ J $\div$ 0.235 kg approx.). Since 50 kg of enriched uranium contains about 1 kg of U 235, the energy released as a result of fissioning all the U 235 nuclei in 1 kg of enriched uranium is about 1.6×10^{12}J ($= 8 \times 10^{13}$ J $\div$ 50 kg). In comparison, the energy released from burning 1 kg of oil is about 30 MJ. In other words, 1 kg of enriched uranium releases the same amount of energy as 50 000 kg of oil.

The fuel in a thermal nuclear reactor is contained in cylindrical metal tubes called fuel rods. The fuel rods are inserted into the moderator which may be water, as in the pressurised water reactor (PWR), or graphite, as in the advanced gas cooled reactor (AGR). The moderator is enclosed in a thick-walled steel vessel, the reactor core, inside a concrete chamber. Fig. 23.9 shows a cross-section of a thermal nuclear reactor.

- Each fission neutron enters the moderator at high speed and loses kinetic energy as a result of repeated collisions with the moderator atoms. Eventually, one fission neutron per fission re-enters a fuel rod at sufficiently slow speed to cause a further fission. The moderator atoms must be light enough to absorb kinetic energy readily from the fission neutrons. For this reason, water or graphite is used.

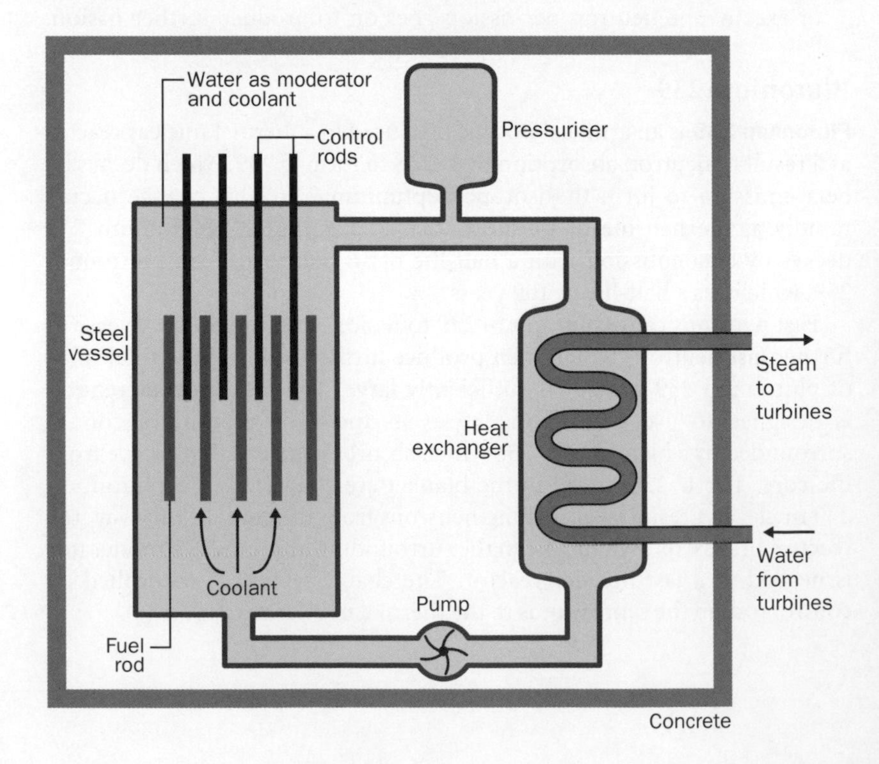

Fig. 23.9 The PWR thermal nuclear reactor.

- The moderator heats up as a result of gaining thermal energy from the fuel pins, which become very hot as the fission fragments in the fuel pins transfer their kinetic energy to the surrounding atoms. The moderator also gains thermal energy as a result of absorbing the kinetic energy of neutrons from the fuel pins and absorbing gamma radiation released from the fuel pins.
- A coolant is pumped through the steel vessel over the fuel rods, then through pipes to a heat exchanger, before re-entering the steel chamber. The coolant must flow easily and be non-corrosive. Water under pressure is used in the PWR as both the coolant and the moderator. In the AGR, the coolant is carbon dioxide gas, which is pumped through channels which contain the fuel pins and the control rods in the graphite moderator block. The hot coolant from the reactor core is pumped through a heat exchanger where it transfers its thermal energy before returning to the reactor core.
- Boron control rods are inserted into the moderator to ensure the neutron density remains constant. Boron nuclei easily absorb neutrons without fissioning. The depth of the rods in the moderator is controlled to ensure a steady rate of fissioning occurs.

Advantages and disadvantages of nuclear power

Advantages

1. The energy released from 1 kilogram of uranium fuel is about the same as from 50 000 kg of fossil fuel.
2. World oil and gas reserves are unlikely to last much beyond 2050 at the present rate of usage. Uranium reserves are sufficient to keep the present generation of thermal reactors running for 100 years or more. Extracting plutonium from the used fuel from thermal nuclear reactors could effectively extend world reserves of nuclear fuel from 100 years to many centuries.
3. Nuclear power stations do not produce so-called greenhouse gases such as carbon dioxide, which are thought to be responsible for atmospheric warming.

Disadvantages

1. The radiation from the core of a nuclear reactor is harmful. For this reason, the core of a reactor is enclosed in concrete which is intended to absorb the radiation. All operations in the core, such as adjusting the control rods or replacing spent fuel rods, must be carried out by remote control from a control room outside the concrete casing. Operating personnel must wear film badges to record exposure to radiation. If exposure exceeds the maximum permitted level, the operator is not allowed to continue.
2. The core is extremely hot and the coolant must remove the thermal energy of the core, otherwise meltdown could occur. Thermal energy could build up if the coolant pumps break down or the coolant leaks

Fig. 23.10 Chernobyl: an aerial view of a helicopter over the remains of the Chernobyl nuclear power station in the Ukraine during the early stages of making the reactor safe. The helicopter is taking radioactivity measurements.

from the core or heat exchanger. When the reactor is shut down by inserting the control rods fully into the core, thermal energy continues to be produced for some time afterwards. If the cooling system ceases to function, the temperature of the core would rise and meltdown would occur. The Chernobyl disaster in 1986 is thought to have been caused through failure of the cooling system.

3. The spent fuel from a nuclear reactor is highly radioactive. This is because spent fuel contains highly radioactive fission fragments, as well as uranium 238 and plutonium 239. The metal containers are also radioactive as a result of exposure to neutrons. Each nucleus of a metal atom that absorbs a neutron becomes unstable and subsequently emits a beta particle. Some of these radioactive processes have long half-lives, so the spent fuel rods remain hot because thermal energy continues to be generated as a result of radioactivity in the fuel rods.

   The spent fuel rods are moved by remote control as they are too radioactive and too hot to handle. After removal from the reactor core, the rods are stored in water-filled cooling ponds for some years after removal from the reactor. When sufficiently cool, the rods are cut open and the spent fuel is removed and reprocessed. The unused uranium 238 and plutonium 239, both highly radioactive isotopes, are separated and stored for possible use in fast breeder reactors. The remaining waste contains isotopes that stay radioactive for many years. This **high level waste** and the reprocessed isotopes need to be stored for many years underground in sealed containers, at sites which are secure and geologically stable.

4. Materials used in the reactor, such as remote handling equipment and the coolant, also become radioactive. This **medium level waste** is stored in sealed containers at the same sites as the high level waste. **Low level radioactive waste**, such as clothing, is disposed of and stored in sealed containers which are buried at restricted sites.

5. The first generation of thermal reactors were designed to last about 30 years. At the end of its designated life, a thermal reactor must be decommissioned and then monitored for many years, since its core continues to be radioactive and hot for some time. The process of dismantling a nuclear reactor is likely to be very costly, hazardous and a possible danger to the environment. At the present time, redundant nuclear reactors are being monitored, with dismantling not likely to commence for many years yet. The cost of decommissioning, monitoring and dismantling nuclear reactors is likely to prove a burden for future generations and is the main reason why the nuclear power programme in many countries has been wound down for the time being.

Questions 23.3

1. **(a)** Explain what is meant by **(i)** induced fission, **(ii)** a chain reaction of fission events.
   (b) (i) Why is it not possible for a chain reaction to occur if the mass of fissionable material is less than a critical amount?
   (ii) Describe how the rate of fission events in a thermal nuclear reactor is controlled to ensure energy is released at a steady rate.
   (c) (i) What is the function of a moderator in a thermal nuclear reactor?
   (ii) When a neutron collides elastically with a moderator atom, the percentage of the neutron's kinetic energy transferred to the moderator atom depends on the atomic mass of the moderator. For graphite, up to 25% of the kinetic energy of a neutron can be transferred in a single collision. How many collisions are necessary to reduce the kinetic energy of a neutron from 1 MeV to 0.01 eV?
2. **(a)** A kilogram of uranium releases as much thermal energy as 50 000 kg of oil. Calculate the mass of fuel used per day in a 1000 MW power station which is 25% efficient and which uses **(i)** oil, **(ii)** uranium. Assume 30 MJ of energy is released when 1 kg of oil is burned.
   (b) (i) Explain why spent fuel rods are highly radioactive.
   (ii) Why is spent fuel reprocessed?
   (iii) Why is a moderator not necessary in a fast breeder reactor?

23.4 Probing the nucleus

The discovery of the neutron

Rutherford established the nuclear model of the atom using alpha particles to probe atoms of thin metal foils. He used his measurements to show that the nucleus carries a charge of + Ze, where Z is the atomic number of the element (i.e. its order number in the Periodic Table of the Elements). He deduced that nuclei are composed of two types of particles, protons and neutrons. The proton is the nucleus of a hydrogen atom. Rutherford put forward the idea that nuclei also contained uncharged particles, neutrons, of about the same mass as protons. However, he had no direct experimental evidence for the existence of neutrons.

The existence of the neutron was proved about 20 years later by Chadwick, one of Rutherford's former students. He knew that alpha particles from polonium directed at beryllium foil caused radiation to be emitted from the foil. Some scientists thought the radiation was electromagnetic in nature, but Chadwick discovered that the radiation knocked protons out of a wax plate placed in the path of the beam. He made further measurements and proved from his measurements that the radiation consisted of neutral particles of approximately the same mass as the proton.

Antimatter

Chadwick's discovery provided direct evidence that the nucleus of the atom is composed of protons and neutrons. To explain the structure of any type of atom, no more than three types of particles are needed, namely the electron, the proton and the neutron. About the same time as Chadwick discovered the neutron, the first evidence for antimatter was obtained from cloud chamber studies by Carl Anderson in America. He photographed cloud chamber tracks produced by cosmic rays. These rays consist of high-energy particles from the Sun and other stars. When such a high energy particle crashes into the Earth's atmosphere, it creates a cascade of particles as a result of nuclear collisions. Anderson discovered how to photograph the cloud chamber at the instant each cascade reached the ground. By applying a strong magnetic field to the cloud chamber, charged particles passing through the chamber created curved tracks. Anderson found some tracks that were like beta particle tracks but curved in the opposite direction. He concluded that these were due to positively charged electrons known as positrons, predicted by Dirac some years earlier.

Note

The rest mass of an electron, in atomic mass units, is 0.00055 u. The energy produced when an electron and a positron annihilate each other is therefore at least 1.02 MeV (= 2 × 0.00055 × 931 MeV). The rest mass of a proton, equal to 1.00728 u, is equivalent to about 1000 MeV or 1 GeV.

Dirac put forward the theory that for every particle, there is an antimatter particle with the opposite charge which annihilates itself and the particle when they collide with each other. The discovery of the positron was the first experimental evidence that antimatter exists. Dirac's predictions about antimatter, all of which have been confirmed, included the following key ideas:

1. **Annihilation:** radiation is produced when a particle meets its anti-particle. Matter is converted into radiation energy in this process which is known as annihilation. The scale of conversion is in accordance with Einstein's equation $E = mc^2$.

2. **Pair production:** radiation of sufficiently high energy can produce particle and antiparticle pairs. In this process, known as pair production, radiation energy is converted into mass in accordance with $E = mc^2$. For example, a gamma photon can produce an electron and a positron if the energy of the gamma photon exceeds $2\ m_0c^2$, where m_0 is the rest mass of the electron.

Gamma photon energy, $hf \geq 2m_oc^2$

Acceleators

Charged particles can be accelerated to speeds approaching the speed of light by using electric or magnetic fields to accelerate them. No particle or antiparticle can reach the speed of light. When a particle of charge, q, is accelerated through a potential difference, V, in a vacuum, the particle gains kinetic energy equal to qV and its mass increases in accordance with $E = mc^2$. If the particle is repeatedly accelerated, its mass becomes ever larger and larger as its speed, υ, approaches c in accordance with Einstein's relativistic mass formula

$$m = m_o\left[1 - \left(\frac{\upsilon}{c}\right)^2\right]^{-\frac{1}{2}}$$

where m_o is the rest mass of the particle.

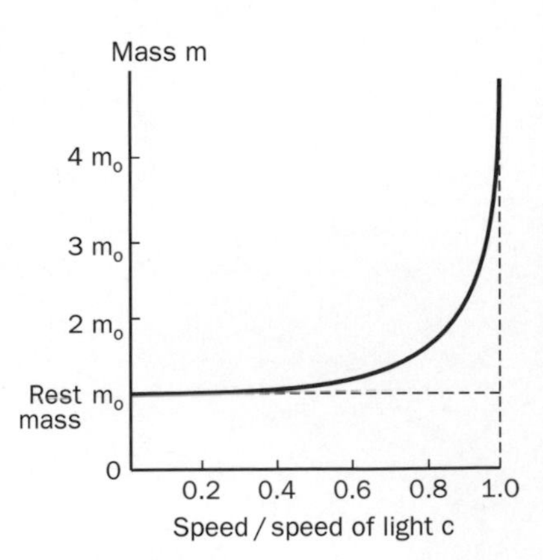

Fig. 23.11 Relationship between mass and speed.

More and more energy is needed to make a particle approach closer and closer to c. Fig. 23.11 shows how the mass of a particle rises as it approaches the speed of light. The speed of light is impossible to reach because the energy needed would be infinite.

For the past half century, fundamental research into the nature of matter has been conducted using larger and larger accelerators to hurl fast moving particles into each other and then study the debris from the collisions. From such experiments, scientists now know that all matter is composed of

- **quarks**, which make up neutrons and protons, and
- **leptons**, which include electrons and positrons.

Quarks

Cosmic ray studies and accelerator experiments up to about 1960 resulted in the discovery of a large number of very short-lived sub-atomic particles. The properties of these new particles were measured from the tracks the particles created in photographic emulsions, cloud chambers and bubble chambers. Bubble chambers contain liquid under high pressure which is released at the instant charged particles pass through it. This process creates tiny bubbles along the particle track, thus making the track visible.

For example, pions were discovered from tiny tracks created in the emulsion of photographic plates exposed to cosmic rays at high altitude. **Strange** particles were discovered in photographs of cloud chambers containing absorber plates. These particles were strange because they were created in short-lived impacts with nuclei but they decayed so slowly that they left tracks much longer than expected. In addition, they were created in pairs so they were assigned a **strangeness** number, S, on the basis that strangeness is conserved in an interaction involving the strong nuclear force.

The properties of all the new and existing particles were used to classify the particles in groups.

- Particles heavier than the proton (and the neutron) were called **baryons**.
- Particles lighter than the proton and heavier than the electron were called **mesons**.
- Particles lighter than the electron (including the electron) were called **leptons**.

Table 23.1 is a summary of the properties of baryons and mesons discovered by 1960.

Patterns emerged as a result of plotting each group on a grid of charge against strangeness. Fig. 23.12 shows the pattern for very short-lived baryons. This particular group forms a pattern which appears incomplete. Its completion required the existence of a negatively charged particle with a strangeness of –3 and a rest mass of 1.79 u. This particle was called the **omega minus** (Ω^-) and its discovery in 1963 led to the quark model to explain the patterns.

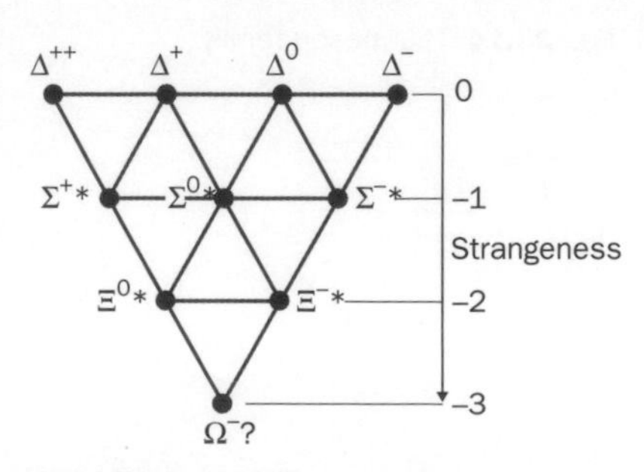

Fig. 23.12 Short-lived baryons.

The quark model was originally based on the existence of three types of quarks, referred to as the up quark, the down quark and the strange quark. The rules of the model are simple:

Table 23.1 Baryons and mesons

Name+symbol		**Charge/e**	**Mass/u**	**Strangeness**	**Lifetime/s**
Pion (π-meson)	π^+ π^-	±1	$^1/_7$	0	10^{-8}
	π^0	0	$^1/_7$	0	10^{-6}
K-meson	$K^\pm$	±1	$^1/_2$	±1	10^{-8}
	K^0	0	$^1/_2$	+1	10^{-8}
	$\bar{K}_o$	0	$^1/_2$	−1	10^{-8}
Proton	p	+1	1	0	infinite
Neutron	n	0	1	0	15 min
Sigma	$\Sigma^\pm$	±1	1.2	−1	10^{-10}
	Σ^0	0	1.2	−1	10^{-20}
Lambda	Λ^0	0	1.1	−1	10^{-10}
Xi	Ξ^0	0	1.3	−2	10^{-10}
	Ξ^-	−1	1.3	−2	10^{-10}
Delta	Δ^{++}	+2	1.33	0	10^{-23}
	Δ^+	+1	1.33	0	10^{-23}
	Δ^0	0	1.33	0	10^{-23}
	Δ^-	−1	1.33	0	10^{-23}
Sigma star	$\Sigma^{\pm *}$	±1	1.49	−1	10^{-23}
	Σ^{0*}	0	1.49	−1	10^{-23}
Xi-star	Ξ^{0*}	0	1.64	−2	10^{-23}
	Ξ^{-*}	−	1.64	−2	10^{-23}

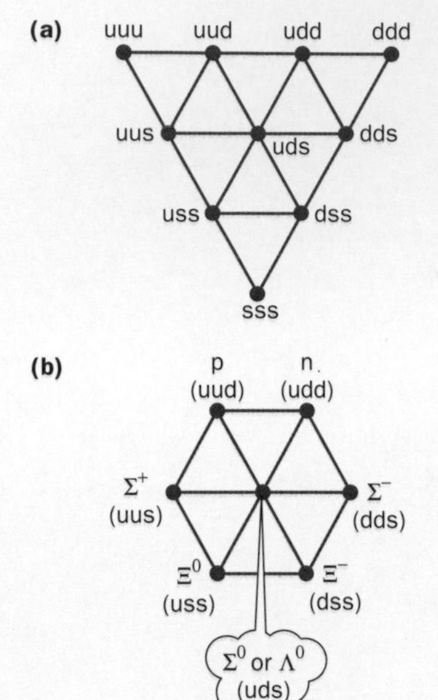

Fig. 23.13 The baryon family **(a)** short-life baryons, **(b)** long-life baryons.

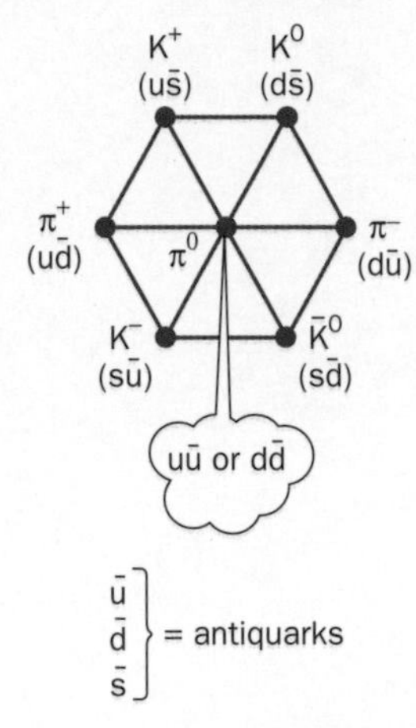

Fig. 23.14 The meson family.

1. The up quark has a charge of $+\frac{2}{3}$ e and strangeness 0. The down quark has a charge of $-\frac{1}{3}$ e and strangeness 0. The strange quark has a charge of $-\frac{1}{3}$ e and strangeness −1.
2. A baryon consists of three quarks and an antibaryon consists of three antiquarks. There are 10 possible combinations of three quarks. These are shown in Fig. 23.13. These 10 combinations explain the 10 particles of the short-lived bayon group. The quark composition of the longer lived baryons, including the neutron and the proton, is also shown in Fig. 23.13.
3. A meson consists of a quark and an antiquark. The quark composition of each meson in Table 23.1 is shown in Fig. 23.14.

Further experimental work and successful predictions have led to the discovery of three more members of the quark family. The six known quarks are thought to constitute the complete quark family. Whether there is a fundamental link between the quark family and the lepton family remains to be seen.

Questions 23.4

1. **(a)** An alpha particle is capable of knocking a neutron out of a beryllium $^9_4$Be nucleus. Write down the equation for this process.

   (b) The up quark has a charge of $+\frac{2}{3}$e and strangeness, 0, and the down quark has a charge of $-\frac{1}{3}$e and a strangeness, 0. The strange quark has a charge of $-\frac{1}{3}$e and a strangeness of −1.

   (i) A proton consists of two up quarks and a down quark. A neutron consists of two down quarks and an up quark. Show that the charge of a proton is +1e and the charge of a neutron is 0.

   (ii) What is the quark composition of a baryon that has a charge of +1e and a strangeness of −1?

2. **(a)** Fig. 23.15 represents a photograph of a pair production event in a bubble chamber. The particle and antiparticle curve around because a magnetic field was applied.

   (i) Why are there two tracks in opposite directions?

   (ii) Why do the tracks spiral?

   (b) When a proton annihilates an antiproton, two gamma photons are released in opposite directions. Calculate the energy of the gamma photons produced when a proton of kinetic energy 2 GeV collides with an antiproton of the same energy moving in the opposite direction. The rest mass of a proton = 1 GeV.

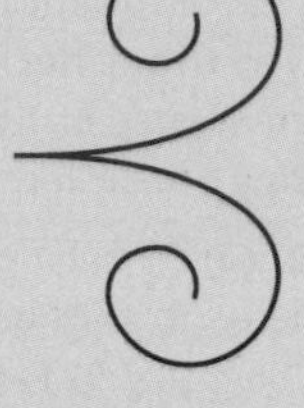

Fig. 23.15

Summary

◆ **Forces**

- The strong nuclear force: acts between protons and neutrons, attractive, ranges up to 2 or 3 fm.
- The electromagnetic force: acts between any two charged objects, attractive or repulsive, extends to infinity.
- The weak nuclear force: responsible for turning neutrons into protons, ranges no more than about 10^{-18}m.
- The force of gravity: acts between any two masses, attractive, extends to infinity.

◆ **Energy**

- The electron volt (1 eV) is the work done when an electron moves through 1 volt. 1 eV = 1.6×10^{-19}C.
- Binding energy is the work done to separate a nucleus into its constituent neutrons and protons.
- Mass defect is the difference between the mass of a nucleus and its constituent neutrons and protons.
- The Q value of a nuclear process is the energy released or needed.

◆ **Fission**

- U 235 is fissioned by slow neutrons; U 238 absorbs neutrons without fission.
- One fission neutron per fission goes on to produce further fission in a steady chain reaction.
- The moderator in a thermal reactor reduces the kinetic energy of the fission neutrons to enable them to produce further fission of U 235.

◆ **Quarks**

- A baryon consists of three quarks. An antibaryon consists of three antiquarks. A meson consists of a quark and an antiquark.
- The up quark has a charge of $+\frac{2}{3}$e and a strangeness of 0. The down quark has a charge of $-\frac{1}{3}$e and a strangeness of 0. The strange quark has a charge of $-\frac{1}{3}$e and a strangeness of −1.

◆ **Equation** relating energy and mass, $E = mc^2$.

Revision questions

$e = 1.6 \times 10^{-19}$, $\varepsilon_0 = 8.85 \times 10^{-12}$ F m^{-1}, 1 u ≡ 931 MeV, the mass of a proton = 1.00728 u, the mass of a neutron = 1.00867 u, the mass of an electron = 0.00055 u.

23.1. (a) The electrostatic potential energy of two charged particles at distance r apart is $\frac{Q_1Q_2}{4\pi\varepsilon_0 r}$. Calculate the closest distance that an alpha particle of kinetic energy 5 MeV can approach a beryllium $^{9}_{4}Be$ nucleus, assuming it is only affected by the electrostatic force.
(b) The strong nuclear force has a range of a few femtometres. Discuss whether or not a 5 MeV alpha particle is capable of causing a nuclear reaction when it hits a beryllium 9 nucleus.

23.2. (a) Calculate the binding energy per nucleon of **(i)** a $^{12}_{6}C$ nucleus, **(ii)** a $^{206}_{82}Pb$ nucleus. (Atomic masses: C 12 = 12.0000 u, Pb 206 = 205.97446 u)
(b)(i) Sketch a graph to show how the binding energy per nucleon varies with the mass number of the nucleus.
(ii) Use your graph to explain why energy is released when a heavy nucleus undergoes fission.

23.3. (a) Polonium 210 is an unstable alpha-emitting nucleus that contains 84 protons. It decays to form an isotope of lead (Pb).
(i) Write down the equation representing the decay of polonium 210.
(ii) Calculate the Q value of the reaction. The atomic mass of polonium 210 is 209.98287 u and the atomic mass of Pb 206 is 205.97446 u.
(b) Sodium 24 is an unstable beta-emitting nucleus that contains 11 protons. It decays to form an isotope of magnesium (Mg).
(i) Write down the equation that represents this decay.
(ii) Calculate the Q value of this decay. The atomic mass of sodium 24 is 23.99097 u and the atomic mass of magnesium 24 is 23.98505 u.

23.4. (a) (i) What is meant by a chain reaction in a nuclear reactor?
(ii) Describe how a chain reaction is controlled in a nuclear reactor.
(b) (i) What is the purpose of the moderator in a thermal nuclear reactor?
(ii) What physical properties are necessary for a moderator?
(c) (i) Why is spent fuel from a nuclear reactor dangerous?
(ii) Plutonium 239 has a half life of 100 years. Calculate the activity of 1 kg of this isotope. The Avogadro constant = 6.02×10^{23} mol^{-1}.

23.5. (a) In an accelerator called a synchrotron, charged particles are accelerated periodically as they travel on a circular path around the synchrotron.
(i) How can charged particles be forced to go around in a circle?
(ii) Charged particles radiate electromagnetic waves when they are forced around a circular path. The energy per second radiated by a particle increases with its speed. Explain why this process limits the kinetic energy of a charged particle in a synchrotron.
(b) An electron is accelerated until its kinetic energy is 20 GeV.
(i) Calculate its mass at this energy.
(ii) Explain why its speed is limited, no matter how much kinetic energy it acquires.
(c) Determine the charge and strangeness of **(i)** a uds baryon, **(ii)** a u $\bar{s}$ meson.
(d) Determine the quark composition of **(i)** a baryon of charge +2e and strangeness 0, **(ii)** a meson of charge 0 and strangeness +1.

Further Physics

Unit 24. Gases
Unit 25. Thermodynamics
Unit 26. Uniform Circular Motion
Unit 27. Gravitation
Unit 28. Simple Harmonic Motion

Part 7 provides opportunities to develop your grasp of physics principles to give you confidence in your ability to use the principles of the subject. The unit on gases enables you to develop your understanding of thermal physics from unit 6 by considering temperature in depth and its link to energy at a molecular theory. The unit on thermodynamics develops simple ideas on energy transfer from unit 12 into a sophisticated appreciation of the limits of efficiency of heat engines. The final units on circular motion, gravitation and simple harmonic motion build on the earlier units on mechanics, enabling key concepts and mathematical skills to be developed to a high level.

Gases

Objectives

After working through this unit, you should be able to:

- ▶ state and use Boyle's law and Charles' law.
- ▶ explain what is meant by an ideal gas
- ▶ state and use the ideal gas equation and the molar gas constant, R
- ▶ explain gas pressure using the kinetic theory of matter
- ▶ state and use the kinetic theory equation for gas pressure
- ▶ explain what is meant by the root mean square speed of the molecules of a gas.
- ▶ use the kinetic theory equation to prove Avogadro's hypothesis

Contents

24.1 The gas laws

Boyle's law

If a fixed amount of gas at constant temperature is compressed into a smaller volume, its pressure increases. The relationship between the pressure of a gas and its volume was first investigated in the 17th century by Robert Boyle. He discovered that the pressure multiplied by the volume of a gas is always the same, provided the temperature of the gas is unchanged. This relationship is known as **Boyle's law**.

pV = constant

where p is the pressure and V is the volume of a fixed amount of gas at constant temperature.

Fig. 24.1 shows how Boyle's law can be investigated, using a pump to compress some trapped air in a thick-walled glass tube. Each time the volume of the trapped air is altered, its pressure is measured using the pressure gauge and its volume is measured from the length of the trapped air column. Typical results are shown in Fig. 24.2. The product, pV, is unchanged provided the gas temperature stays the same. Any pair of

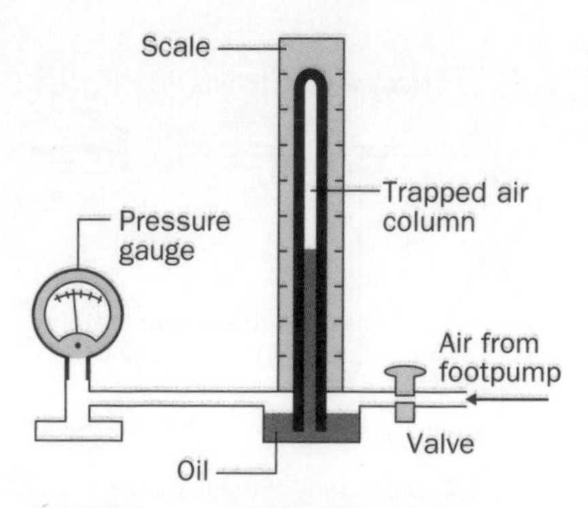

Fig. 24.1 Investigating Boyle's Law.

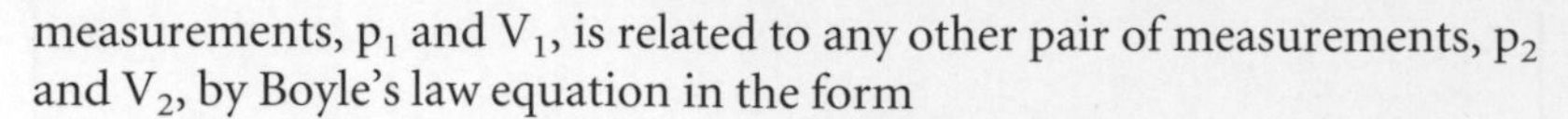

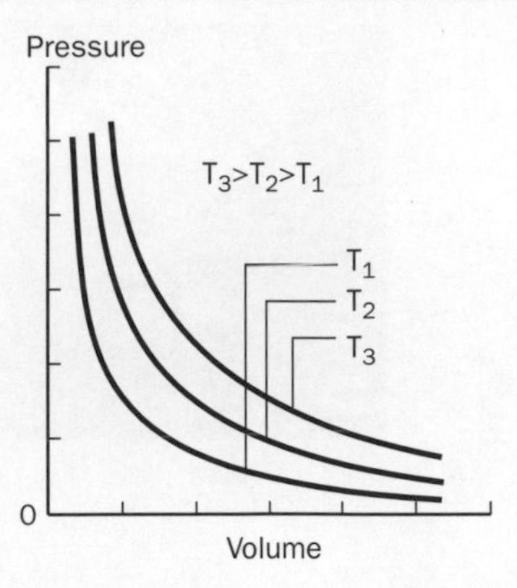

Fig. 24.2 Pressure versus volume.

measurements, p_1 and V_1, is related to any other pair of measurements, p_2 and V_2, by Boyle's law equation in the form

$$\mathbf{p_1V_1 = p_2V_2}$$

The measurements may be plotted on a graph of y = pressure against x = volume, as shown in Fig. 24.2. If the measurements are repeated for different fixed temperatures, each set of measurements gives a different curve according to the temperature at which those measurements were obtained. These curves are referred to as **isothermals** because each curve is a constant temperature curve.

The measurements may also be plotted on a graph of y = pressure against x = 1 ÷ volume, as shown in Fig. 24.3. Because the pressure × the volume (p × V) is constant, the points for each set of measurements define a straight line in accordance with the equation

$$p = \text{constant} \times \frac{1}{V}$$

if Boyle's law is obeyed. A gas that obeys Boyle's law is said to be an **ideal** gas.

> **Note**
> An isothermal change is a change at constant temperature.

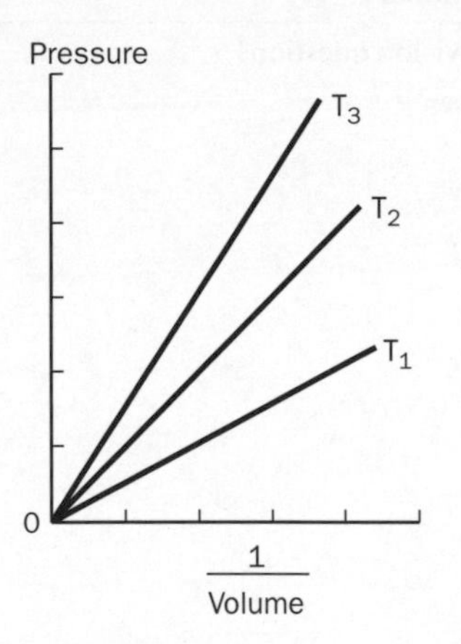

Fig. 24.3 Pressure versus 1÷ volume.

Worked example 24.1

A cylinder contained some trapped air which could be compressed using a piston. The volume of the trapped air could be reduced from a maximum volume of 0.0100 m³ when its pressure was 100 kPa, to 0.0020 m³. Assuming no change of temperature, calculate the pressure in the cylinder when the volume was 0.0020 m³.

Solution

$pV = \text{constant} \therefore p \times 0.0020 = 100 \times 10^3 \times 0.0100$

hence $p = \dfrac{100 \times 10^3 \times 0.0100}{0.0020} = 500 \times 10^3\ \text{Pa} = 500\ \text{kPa}$

Using Boyle's law to measure the volume of a powder

The air in a sealed flask of known volume, V_F, connected to a pressure gauge is compressed using a pump, as in Fig. 24.4.

1. With the flask empty, the pressure is measured before compression (p_0) and after compression (p_1).
Using Boyle's law gives $p_1V_F = p_0(V_F + V_P)$ where V_P is the volume of the pump. Rearranging this equation gives $p_0V_P = (p_1 - p_0)V_F$.

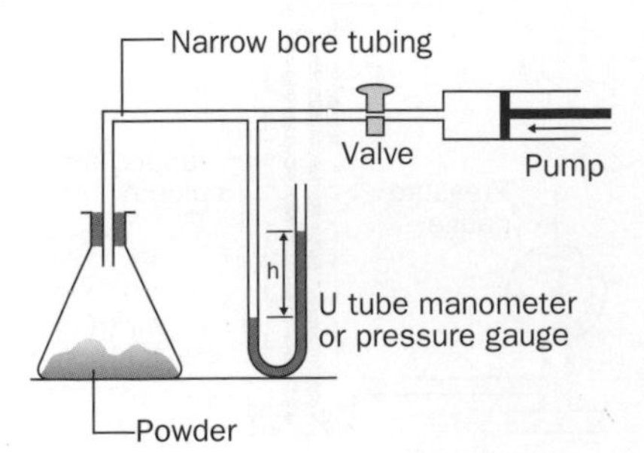

Fig. 24.4 Measuring the volume of a powder.

2. With the flask containing powder of unknown volume, υ, the pressure is measured before compression (p_0, which is the same as in 1) and after compression (p_2).
Using Boyle's law therefore gives $p_2(V_F - \upsilon) = p_0(V_F - \upsilon + V_P)$ since the volume of the air was $(V_F - \upsilon)$ before compression and $(V_F - \upsilon - V_P)$ after compression.
Rearranging this equation gives $p_0V_P = (p_2 - p_0)(V_F - \upsilon)$.
Combining the two rearranged equations therefore gives
$(p_2 - p_0)(V_F - \upsilon) = (p_1 - p_0)V_F$.
Rearranging this equation then gives the following expression from which the powder volume, υ, is calculated.

$$\upsilon = V_F - V_F\frac{(p_1 - p_0)}{(p_2 - p_0)} = \frac{(p_2 - p_1)}{(p_2 - p_0)}V_F$$

Questions 24.1a

1. A cycle pump of volume 4.5×10^{-5} m^3 (= 45 cm^3) was used to inflate a tyre of volume 1.50×10^{-3} m^3 (= 1500 cm^3), initially at a pressure of 110 kPa.
   (a) Calculate the pressure of the air in the tyre after one stroke.
   (b) Show that the pressure after each stroke was 3% higher than before the stroke.
   (c) Calculate the pressure after 50 strokes of the pump.
2. In an experiment to measure the volume of powder in a flask of volume 2.50×10^{-4} m^3 (= 250 cm^3), the pressure of the air in the flask was raised using a hand pump with, then without powder in the flask.
   (a) Without powder present, the pump raised the pressure in the flask from 100 kPa to 118 kPa. Calculate the volume of air in the pump initially.
   (b) With the flask approximately half full of powder, the test was repeated, raising the pressure in the flask from 100 to 140 kPa. Calculate the volume of the powder.

Charles' law

When a gas is heated, its volume increases if it is free to expand. Fig. 24.5 shows how this can be investigated. The gas under test is air trapped in the capillary tube between the sealed end and a thread of liquid. The length of the trapped air column is a measure of the volume of air and it increases if the trapped air is heated. This can be achieved using a water bath.

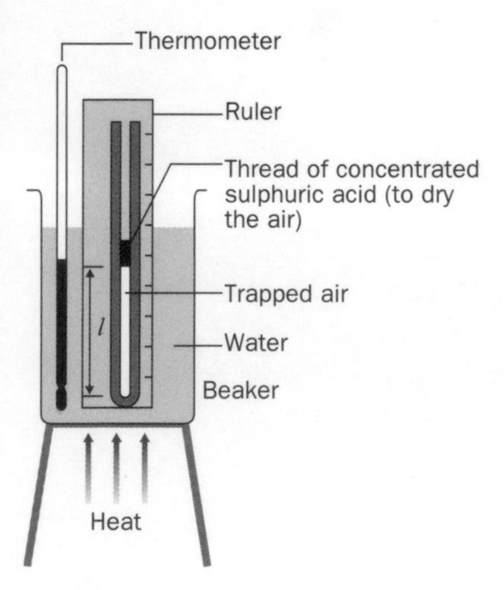

Fig. 24.5 Investigating the expansion of a gas.

The volume of the gas is measured at 0°C, the temperature of pure melting ice, and at 100°C, the temperature of steam at atmospheric pressure. These two measurements may be plotted on a graph to find the temperature at which the volume of the gas would become zero if it was cooled. Provided the gas under test is an ideal gas (i.e. obeys Boyle's law over the temperature range 0° to 100°C), this temperature is always −273°C, regardless of which gas is used or how much of the gas is present, and is referred to as the **absolute zero of temperature**. This is the lowest possible temperature.

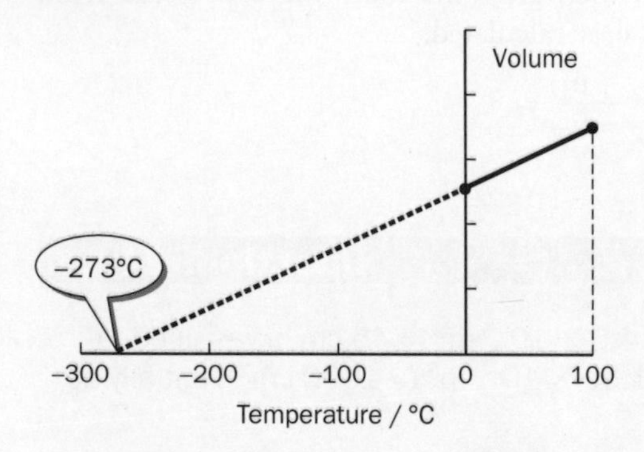

Fig. 24.6 Absolute zero.

The **absolute scale of temperature**, T, in kelvins, is related to the Celsius scale, t, in °C, by the equation

$$\mathbf{T = t + 273}$$

The volume, V, of an ideal gas is therefore proportional to the absolute temperature, T, of the gas, in accordance with the equation

$$V = \frac{V_0 T}{T_0}$$

where V_o is the volume of the gas at ice point, $T_0 = 273$ K. In the following form, this equation is known as Charles' law:

$$\frac{\mathbf{V}}{\mathbf{T}} = \mathbf{constant}$$

Investigating the variation of pressure with temperature for a gas at constant volume

The pressure of a fixed mass of gas at constant volume increases if the gas is heated. This can be investigated, using a U-tube mercury manometer to measure the pressure of air in the sealed flask. A water bath is used to change the temperature of the air in the flask. Each pressure measurement is made at constant temperature which is measured using a thermometer in the water bath. As explained on p. 107, the pressure of the air in the flask, p, is given by the equation

$$p = h\rho g + p_0$$

where h is the height difference between the mercury levels in the manometer, ρ is the density of mercury and p_0 is atmospheric pressure (measured separately using a barometer; see p. 108).

The results may be plotted on a graph of y = pressure against x = absolute temperature. For an ideal gas, the points define a straight line that passes through absolute zero at zero pressure. The relationship between the pressure, p, and the absolute temperature, T, of the air in the flask is therefore given by the following equation

$$\frac{p}{T} = \text{constant}$$

provided the volume and mass of gas stays fixed.

The combined gas law

The three separate gas laws can be combined into a single law which links together the pressure, p, volume V, and the absolute temperature, T, of any ideal gas, provided the mass of gas is constant.

$$\frac{pV}{T} = \text{constant}$$

For example, if the pressure, volume and temperature of a gas are p_1, V_1 and T_1 under one set of conditions and p_2, V_2 and T_2 under a second set of conditions, then

$$\frac{p_2 V_2}{T_2} = \frac{p_1 V_1}{T_1}$$

Worked example 24.2

In an electrolysis experiment, 25 cm³ (2.5×10^{-5} m³) of a gas is collected at a pressure of 103 kPa and a temperature of 20°C. Calculate the volume of this quantity of gas at 0°C and a pressure of 102 kPa.

Solution

$p_1 = 103$ kPa,
$V_1 = 25$ cm³,
$T_1 = 20 + 273 = 293$ K

$p_2 = 102$ kPa,
$V_2 = ?$,
$T_2 = 273$ K

Using $\frac{p_2 V_2}{T_2} = \frac{p_1 V_1}{T_1}$ gives

$$\frac{102 \times 10^3 \times V_2}{273} = \frac{103 \times 10^3 \times 2.5 \times 10^{-5}}{293}$$

Hence V_2

$$= \frac{103 \times 10^3 \times 2.5 \times 10^{-5} \times 273}{293 \times 102 \times 10^3}$$

$= 2.35 \times 10^{-5}$ m³ $= 23.5$ cm³
$= 23.5$ cm³

Questions 24.1b

1. Use the combined gas law to complete the missing data in each row of the Table below.

p_1 / kPa	V_1 / m³	T_1 / K	p_2 / kPa	V_2 / m³	T_2 / K
100	0.05	300	110	**(a)**	350
105	0.24	400	101	0.12	**(b)**
0.35	0.85	350	**(c)**	0.58	250
101	0.42	**(d)**	101	0.38	300
110	**(e)**	290	101	0.16	273

2. **(a)** In a chemistry experiment, 20 cm³ of gas released from a reaction is collected in a syringe at a pressure of 105 kPa and a temperature of 288 K. Calculate the volume of this amount of gas at 0°C and 101 kPa pressure.

   (b) A sealed can contains 30 cm³ of air at 20°C and 101 kPa pressure. Calculate the pressure in the can at 100°C if its volume increases to 31 cm³.

24.2 The ideal gas equation

Molar mass

- The **Avogadro constant, N_A** is defined as the number of atoms present in exactly 12 g of carbon 12, and is equal to 6.02×10^{23}.
- **One mole** of a substance is defined as N_A particles of the substance. The number of moles in a certain quantity of a substance is its **molarity**. The unit of molarity is the mol.
- The **molar mass, M,** of a substance is the mass of N_A particles of the substance. This is the same as the mass of 1 mole of the substance. The unit of molar mass is kg mol^{-1}.

The molar gas constant, R

Equal volumes of ideal gases at the same temperature and pressure contain equal numbers of moles.

This rule was established as a result of measuring the volume of gas released or used in a chemical reaction. Further measurements showed that one mole of any ideal gas at 0°C and a pressure of 101 kPa has a volume of 0.0224 m^{-3}, regardless of the gas. In other words, the molar volume of any gas at standard temperature and pressure (0°C and 101 kPa) is always 0.0224 m^3. The value of PV ÷ T for one mole is known as **the molar gas constant, R,** also referred to as the universal gas constant. Hence

$$R = \frac{101 \times 10^3 \text{ Pa} \times 0.0224 \text{ m}^3}{273} = 8.31 \text{ J mol}^{-1} \text{ K}^{-1}$$

The combined gas law may then be written in the form

$$\frac{pV_m}{T} = R$$

where V_m is the volume of 1 mole. Rearranging this equation gives the **ideal gas equation** below

$$\mathbf{pV_m} = \mathbf{RT} \text{ for 1 mole}$$
$$\text{or } \mathbf{pV} = \mathbf{nRT} \text{ for n moles}$$

Notes

1. The unit of R is the same as the unit of PV/T. Since 1 pascal = 1 N m^{-2}, then the unit of pV is the newton metre (Nm) or the joule (J). Hence the unit of R is J K^{-1} mol^{-1}.
2. The equation can also be written $pV = nMR_gT$ where M is the molar mass and $R_g = R \div M$ (which depends on the gas).

Questions 24.2

$N_A = 6.02 \times 10^{23}$ mol^{-1},
R = 8.31 J mol^{-1} K^{-1}

1. A sealed gas container fitted with a safety valve contains 200 cm^3 of gas at a pressure of 120 kPa and a temperature of 15°C.
   (a) Calculate the number of moles of the gas in the container.
   (b) The safety valve releases gas if the pressure rises above 150 kPa. Calculate the lowest temperature at which the valve would open.
   (c) If the container is heated above the lowest temperature at which the valve opens, the valve releases gas to keep the pressure at 150 kPa. If the container temperature is raised to 100°C, calculate **(i)** the number of moles that would be lost from the cylinder, **(ii)** the fraction of gas that would be lost.
2. The molar mass of nitrogen is 0.028 kg. Calculate the molar volume and density of nitrogen gas at **(i)** 0°C and 101 kPa pressure, **(ii)** 100°C and 101 kPa pressure.

Worked example 24.3

$N_A = 6.02 \times 10^{23}$ mol^{-1}, $R = 8.31$ J mol^{-1} K^{-1}

A sealed flask of volume 60 cm^3 contains a gas at a pressure of 10 kPa and a temperature of 27°C. Calculate (a) the number of moles of the gas in the bulb, (b) the number of molecules per cm^3 in the bulb.

Solution

(a) Rearrange $pV = nRT$ to give $n = \frac{pV}{RT} = \frac{10 \times 10^3 \times 60 \times 10^{-6}}{8.31 \times (273 + 27)}$

$= 2.4 \times 10^{-4}$ mol

(b) Number of molecules present $= n N_A = 2.4 \times 10^{-4} \times 6.02 \times 10^{23}$ $= 1.45 \times 10^{20}$

$\therefore$ number of molecules per cm^3 $= \frac{1.45 \times 10^{20}}{60} = 2.4 \times 10^{18}$

24.3 The kinetic theory of gases

Brownian motion

Smoke particles in a light beam observed using a microscope appear as quivering specks of light, moving about unpredictably and erratically. The motion of the particles is called **Brownian** motion after its discoverer, the botanist Robert Brown, who observed pollen grains in water moving in the manner described above. Fig. 24.7 shows an arrangement to observe the motion of smoke particles.

Brownian motion is caused by repeated and continual impacts on each smoke particle by fast-moving air molecules, too small to see but nevertheless able to make the smoke particles move about noticeably. The erratic motion of each smoke particle occurs because the molecules bombard the smoke particle unevenly and at random. Consequently, the overall force of the impacts on each smoke particle continually changes direction at random. Brownian motion provides direct evidence that a gas consists of very small, fast-moving molecules moving about at random.

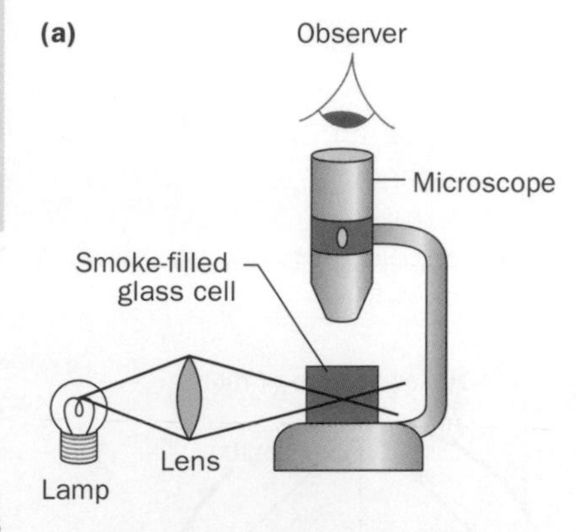

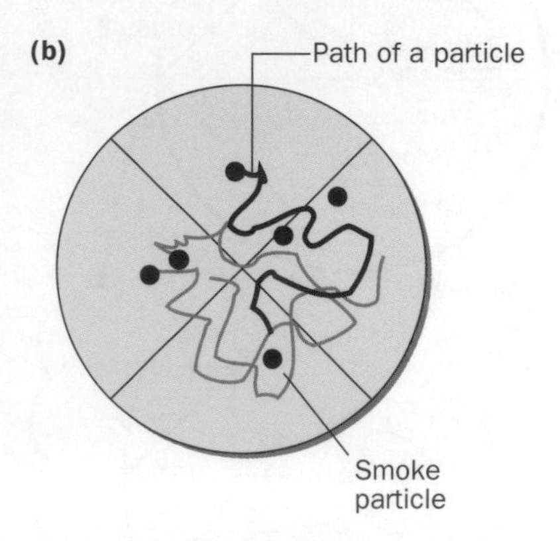

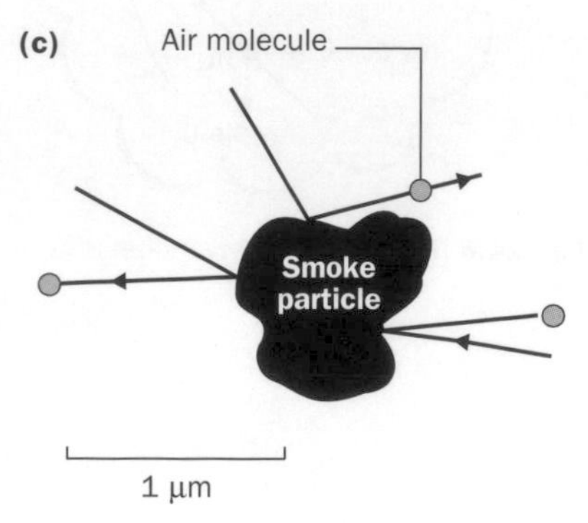

Fig. 24.7 Brownian motion **(a)** using a microscope, **(b)** observation, **(c)** explanation.

The cause of gas pressure

The pressure that a gas exerts on any surface exposed to the gas is due to the impacts of the gas molecules on the source. Each impact exerts a tiny force on the surface. The pressure is measureable because the number of impacts per second per unit area is very large, even though the force of each impact is very small.

- The pressure of a gas rises when its volume is reduced because more impacts take place each second. This is because the size of the gas container is reduced when the gas volume is reduced. Therefore, the molecules hit the surface of the container more frequently, thus raising the pressure of the gas.

- The pressure of the gas rises when its temperature is raised. This is because each impact of a molecule on the surface of the container is more forceful because the molecules move faster at higher temperature. In addition, the number of impacts per second increases because the molecules are moving faster and therefore take less time to move across the container. The pressure is therefore greater because there are more impacts per second and each impact is more forceful.

The kinetic theory model

The following assumptions form the basis of the kinetic theory model of a gas:

1. The gas consists of identical point molecules (i.e. the molecules themselves have negligible volume).
2. The molecules are in continual random motion.
3. Collisions between the molecules and the container surface or between molecules are elastic (i.e. there is no loss of kinetic energy in a collision).
4. The molecules do not attract each other.
5. When a molecule collides, the duration of the collision is much less than the time between successive collisions.
6. The mean kinetic energy of the molecules is proportional to the absolute temperature of the gas.

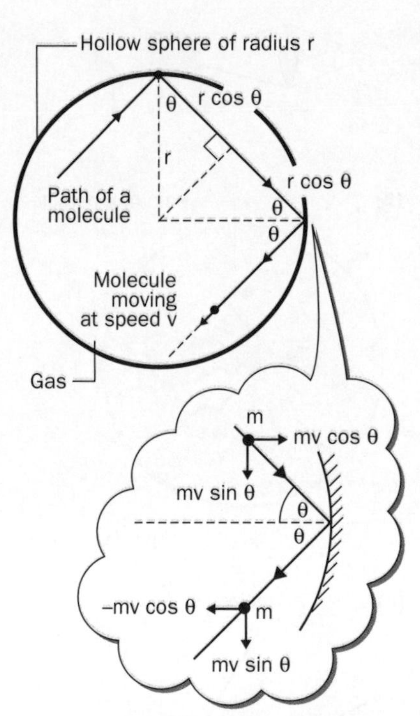

Fig. 24.8 The kinetic theory model of a gas.

Consider a single gas molecule of mass, m, moving at speed, υ, in a spherical container of internal radius, r.

- The molecule repeatedly collides with the surface at the same angle each time. For each collision, its momentum component perpendicular to the surface reverses from $m\upsilon \cos\theta$ to $-m\upsilon\cos\theta$. Its change of momentum = momentum after impact − momentum before impact = $2m\upsilon\cos\theta$.
- The time between successive impacts = $\dfrac{\text{distance}}{\text{speed } \upsilon} = \dfrac{2r\cos\theta}{\upsilon}$
- Impact force = $\dfrac{\text{change of momentum}}{\text{time taken}} = \dfrac{2m\upsilon\cos\theta}{\left(\dfrac{2r\cos\theta}{\upsilon}\right)} = \dfrac{m\upsilon^2}{r}$
- Area of internal surface of the sphere = $4\pi r^2$

$$\therefore \text{pressure, } p = \frac{\text{force}}{\text{area}} = \frac{\left(\dfrac{m\upsilon^2}{r}\right)}{4\pi r^2} = \frac{m\upsilon^2}{4\pi r^3}$$

Since the volume of the container $V = \dfrac{4\pi r^3}{3}$, then $p = \dfrac{m\upsilon^2}{3V}$

Now consider N identical molecules in the container:

- the pressure of molecule 1, p_1, can be written as $p_1 = \dfrac{m{\upsilon_1}^2}{3V}$

where υ_1 is the speed of the molecule

- the pressure of molecule 2, p_2, can be written as

$$p_2 = \frac{m\upsilon_2^2}{3V}$$

where υ_2 is the speed of the molecule

- the pressure of molecule 3, p_3, can be written as

$$p_3 = \frac{m\upsilon_3^2}{3V}$$

where υ_3 is the speed of the molecule . . . etc . . . etc

- the pressure of molecule N, p_N, can be written as

$$p_N = \frac{m\upsilon_N^2}{3V}$$

where υ_N is the speed of the molecule

- ∴ the total pressure, $p = p_1 + p_2 + p_3 + \ldots + p_N$

$$p = \frac{m\upsilon_1^2}{3V} + \frac{m\upsilon_2^2}{3V} + \frac{m\upsilon_3^2}{3V} + \ldots + \frac{m\upsilon_N^2}{3V}$$

$$\therefore p = \frac{Nm\upsilon_{RMS}^2}{3\,V} \quad \text{where } \upsilon_{RMS}^2 = \frac{\upsilon_1^2 + \upsilon_2^2 + \upsilon_3^2 + \ldots + \upsilon_N^2}{N}$$

This equation is known as the kinetic theory equation. Multiplying each side by V gives the equation in the form

$$\mathbf{pV = \frac{Nm\upsilon_{RMS}^2}{3}}$$

Notes

1. υ_{RMS} is referred to as the **root mean square** (rms) speed of the molecules of the gas.

$$\upsilon_{RMS} = \left(\frac{\upsilon_1^2 + \upsilon_2^2 + \upsilon_3^2 + \ldots + \upsilon_N^2}{N}\right)^{1/2}$$

2. The density, $\rho = \frac{\text{mass}}{\text{volume}} = \frac{Nm}{V}$

$$\therefore p = \frac{Nm\upsilon_{RMS}^2}{3V} = \tfrac{1}{3}\rho\upsilon_{RMS}^2$$

Worked example 24.4

$R = 8.31\ \text{J mol}^{-1}\ \text{K}^{-1}$

The molar mass of oxygen is 0.032 kg. Calculate (a) the density of oxygen at a temperature of 20°C and a pressure of 101 kPa, (b) the rms speed of oxygen molecules at this density and pressure.

Solution

(a) Use $pV_m = RT$ to calculate the volume of 1 mole at 20°C and 101 kPa pressure

Hence $V_m = \frac{RT}{p} = \frac{8.31 \times (273 + 20)}{101 \times 10^3} = 2.41 \times 10^{-2}\ \text{m}^3$

$\therefore$ density $\rho = \frac{\text{mass}}{\text{volume}} = \frac{\text{molar mass}}{\text{molar volume}} = \frac{0.032\ \text{kg}}{2.41 \times 10^{-2}\ \text{m}^3} = 1.33\ \text{kg m}^{-3}$

(b) Rearrange $p = \frac{1}{3}\rho\upsilon_{RMS}^2$ to give $\upsilon_{RMS}^2 = \frac{3p}{\rho} = \frac{3 \times 101 \times 10^3}{1.33}$

$= 2.28 \times 10^5\ \text{m}^2\ \text{s}^{-2}$

$\therefore \upsilon_{RMS} = 478\ \text{m s}^{-1}$

Kinetic energy and temperature

The total kinetic energy of all the molecules of gas

$$= \tfrac{1}{2}m\upsilon_1^2 + \tfrac{1}{2}m\upsilon_2^2 + \tfrac{1}{2}m\upsilon_3^2 + \ldots + \tfrac{1}{2}m\upsilon_N^2$$

$$= \tfrac{1}{2}Nm\frac{(\upsilon_1^2 + \upsilon_2^2 + \upsilon_3^2 + \ldots + \upsilon_N^2)}{N}$$

$$= \tfrac{1}{2}Nm\upsilon_{RMS}^2$$

By assuming the total kinetic energy of the molecules of a gas

$$\tfrac{1}{2}Nm\upsilon_{RMS}^2 = \tfrac{3}{2}nRT,$$

the kinetic theory equation $pV = \frac{1}{3}Nm\upsilon_{RMS}^2$ becomes $pV = nRT$. The ideal gas equation, $pV = nRT$, can therefore be derived by making certain assumptions (listed on p. 367) about the molecules of an ideal gas.

The mean kinetic energy of an ideal gas molecule

$$= \frac{\text{total kinetic energy}}{\text{number of molecules, N}}$$

$$= \frac{\frac{1}{2}Nm\upsilon_{RMS}^2}{N} = \tfrac{1}{2}m\upsilon_{RMS}^2$$

Also, since the total kinetic energy of all the molecules

$$= \tfrac{1}{2}Nm\upsilon_{RMS}^2 = \tfrac{3}{2}nRT$$

the mean kinetic energy of an ideal gas molecule

$$\tfrac{1}{2}m\upsilon_{RMS}^2 = \frac{3nRT}{2N} = \frac{3kT}{2}$$

where k, the Boltzmann constant

$$= \frac{R}{N_A} = \frac{nR}{N} = 1.38 \times 10^{-23}\ \text{J mol K}^{-1}.$$

Mean kinetic energy of a gas molecule = $\frac{3}{2}kT$

Worked example 24.5

$1\ eV = 1.6 \times 10^{-19}$ J, $k = 1.38 \times 10^{-23}\ J\ K^{-1}$

Calculate the mean kinetic energy of a gas molecule at 20°C in (a) J, (b) electron volts.

Solution

(a) Mean kinetic energy of a gas molecule $= \frac{3}{2}kT$
$= 1.5 \times 1.38 \times 10^{-23} \times (273 + 20)$
$= 6.1 \times 10^{-21}$ J,

(b) 6.1×10^{-21} J
$= \frac{6.1 \times 10^{-21}}{1.6 \times 10^{-19}}$ eV
$= 0.038$ eV

The distribution of molecular speeds in a gas

The molecules in an ideal gas possess a continuous range of speeds from zero upwards. The distribution of the number of molecules at different speeds is shown in Fig. 24.9. The peak of this distribution curve is the most probable speed of a molecule. The rms speed is not the same as the most probable speed. The distribution curve depends on temperature. At a higher temperature, more molecules move at higher speeds so the curve is broader and flatter.

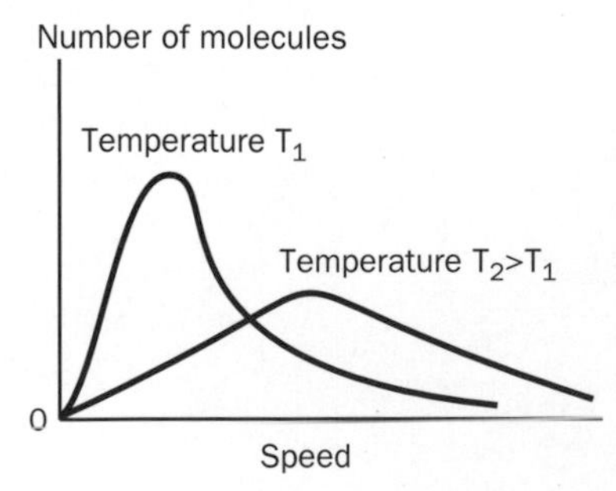

Fig. 24.9 Distribution of molecular speeds.

Avogadro's hypothesis

This hypothesis states that equal volumes of gas at the same temperature and pressure contain equal numbers of molecules.

Consider equal volumes, V of two gases X and Y at the same temperature, T, and pressure, p. According to the kinetic theory equation

$$pV = \frac{Nm{v_{RMS}}^2}{3} = NkT \text{ since } \tfrac{1}{2} m{v_{RMS}}^2 = \tfrac{3}{2} kT$$

$\therefore$ For gas X, $pV = N_X kT$

and for gas Y, $pV = N_Y kT$

$\therefore N_X kT = N_Y kT$

so $\mathbf{N_X = N_Y}$

in accordance with Avogadro's hypothesis.

Questions 24.3

$R = 8.31\ J\ K^{-1}$, $N_A = 6.02 \times 10^{23}\ mol^{-1}$, $k = 1.38 \times 10^{-23}\ J\ K^{-1}$

1. **(a)** Calculate **(i)** the density, **(ii)** the rms speed of molecules of nitrogen gas at 100° C and a pressure of 150 kPa.
   (b) (i) Calculate the mean kinetic energy of an ideal gas molecule at 100°C.
   (ii) Calculate the temperature at which the mean kinetic energy of an ideal gas molecule is double its mean kinetic energy at 100°C.
   The molar mass of nitrogen = 0.028 kg.
2. For hydrogen gas at **(a)** 0°C and **(b)** 100°C, calculate **(i)** the mean kinetic energy, **(ii)** the rms speed of molecules of the gas.
   The molar mass of hydrogen = 0.002 kg.

Summary

- **Absolute zero of temperature** is the lowest possible temperature, equal to −273°C.
- **Absolute temperature in Kelvins, T = temperature in °C + 273.**
- **The gas laws**

  Boyle's law pV = constant for constant mass and temperature

  Charles' law $\frac{V}{T}$ = constant for constant mass and pressure

  Pressure law $\frac{p}{T}$ = constant for constant mass and volume

  Combined gas law $\frac{pV}{T}$ = constant for constant mass

- **The ideal gas equation** $pV = nRT$
- **The kinetic theory equation** $pV = \frac{1}{3} Nm\upsilon_{RMS}^2$ or $p = \frac{1}{3} \rho\upsilon_{RMS}^2$
- **The root mean square speed** $\upsilon_{RMS} = \left(\frac{\upsilon_1^2 + \upsilon_2^2 + \ldots + \upsilon_N^2}{N}\right)^{1/2}$
- **Mean kinetic energy of a gas molecule** $= \frac{3}{2} kT$

Revision questions

$R = 8.31\ J\ mol^{-1}\ K^{-1}$, $k = 1.38 \times 10^{-23}\ J\ K^{-1}$,
$N_A = 6.02 \times 10^{23}\ mol^{-1}$

24.1. (a) In an electrolysis experiment, 40 cm^3 of oxygen gas is collected at 15°C and a pressure of 103 kPa. Calculate **(i)** the volume this gas would occupy at 0°C and 101 kPa pressure, **(ii)** the number of moles of this gas and its mass.
The molar mass of oxygen is 0.032 kg.
(b) An air bubble trapped in a brake pipe has a volume of 1.2 cm^3 at a pressure of 101 kPa and a temperature of 10°C.
(i) Calculate the volume of this air bubble at a pressure of 200 kPa and a temperature of 20°C.
(ii) Calculate the mass of this air bubble, assuming the molar mass of air is 0.029 kg.

24.2. (a) A tyre contains 1600 cm^3 of air at a pressure of 150 kPa and a temperature of 25°C. Calculate **(i)** the pressure of this amount of air at 0°C and a volume of 1500 cm^3, **(ii)** the mass of air in the tyre, assuming the molar mass of air is 0.029 kg.
(b) A capillary tube sealed at one end contained a column of air trapped by a thread of mercury of length 150 mm. When the tube was vertical, the length of the air column was 120 mm when the sealed end was below the open end. When the tube was inverted with the sealed end above the open end, the air column length at the same temperature was 180 mm. Calculate the atmospheric pressure in millimetres of mercury.

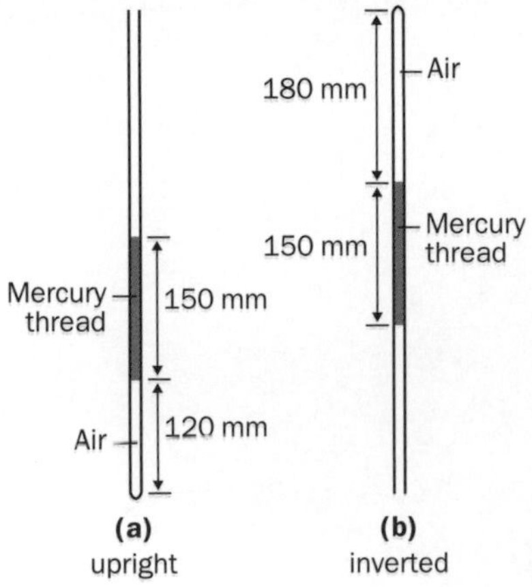

Mercury thread, 150 mm, Air, 120 mm — (a) upright
Air, 180 mm, Mercury thread, 150 mm — (b) inverted

Fig. 24.10

24.3. (a) Sketch graphs to show how the product of pressure and volume, pV, for 1 mole of an ideal gas varies with the temperature of the gas in **(i)** kelvins, **(ii)** °C.
(b) Show on each graph how pV varies with temperature for 0.5 moles.

24.4. (a) A gas cylinder contains 500 cm^3 of nitrogen at a pressure of 140 kPa and a temperature of 20°C. Calculate **(i)** the mass of gas in the cylinder, **(ii)** the number of molecules of gas in the cylinder, **(iii)** the r.m.s. speed of the molecules. The molar mass of nitrogen = 0.028 kg.
(b) The cylinder is fitted with a safety valve that operates at a pressure of 170 kPa. Calculate **(i)** the maximum temperature the gas can be heated to without the safety valve opening, **(ii)** the mass of gas lost if the gas temperature is raised to 100°C, **(iii)** the rms speed of the molecules at 100°C.

24.5. (a) Explain why the pressure of a gas falls if **(i)** its volume is increased at constant temperature, **(ii)** its temperature is reduced at constant volume.
(b) (i) Calculate the rms speed of hydrogen molecules at 20°C.
(ii) Calculate the rms speed of oxygen molecules at 20°C.
(iii) By comparing the two calculations in (i) and (ii), explain why the Earth's atmosphere retains oxygen but not hydrogen.
The molar mass of hydrogen = 0.002 kg, the molar mass of oxygen = 0.032 kg.

Thermodynamics

Objectives

After working through this unit, you should be able to:

- ▶ describe the use of different types of thermometers, including calibration on the Celsius and absolute scales.
- ▶ state and use the first law of thermodynamics
- ▶ explain what is meant by an isothermal change and an adiabatic change
- ▶ calculate the work done when the volume of a gas changes
- ▶ calculate the change of internal energy of an ideal gas when its temperature changes
- ▶ define the molar heat capacities of a gas at constant volume, C_V, and at constant pressure, C_P
- ▶ relate C_P and C_V to each other and to the atomicity of the molecules of a gas
- ▶ explain what is meant by enthalpy
- ▶ state and explain the second law of thermodynamics
- ▶ explain what a heat engine is and calculate its maximum possible efficiency
- ▶ define the thermodynamic scale of absolute temperature and relate it to the ideal gas scale

Contents

25.1 Temperature

As explained in Unit 6, a scale of temperature is defined by assigning numerical values to 'fixed points' which are standard degrees of hotness that are easily reproducible. Any thermometer works on the basis of a physical property, its **thermometric property**, that varies smoothly with change of temperature. Figs 25.2 and 25.3 show three different types of thermometers.

- The **liquid-in-glass thermometer** contains a liquid that expands when the thermometer bulb warms up. The liquid expands up the stem and

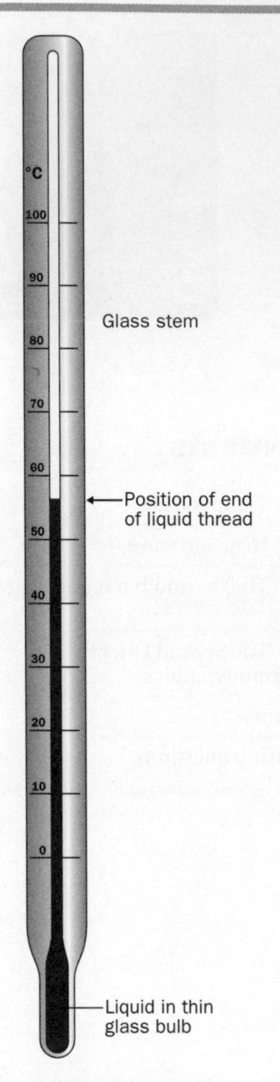

Fig. 25.1 The liquid-in-glass thermometer.

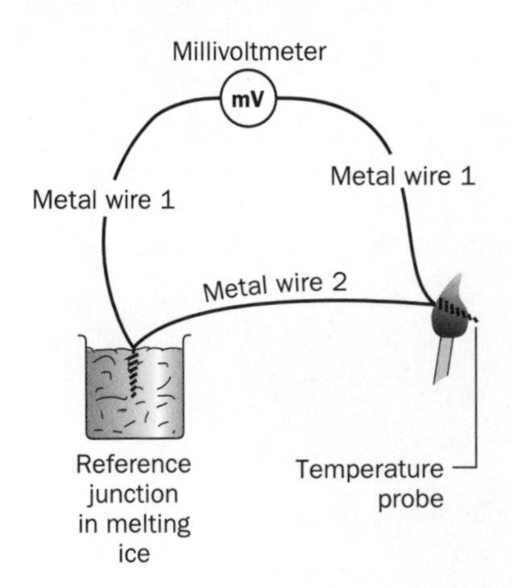

Fig. 25.2 The thermocouple thermometer.

the temperature is measured from the position of the end of the liquid thread in the stem against a calibrated scale. The thermometric property is the thermal expansion of the liquid.

- The **thermocouple thermometer** makes use of the fact that a potential difference exists between two dissimilar metals in contact. This p.d. depends on the temperature of the contact junction. In practice, two wires of identical materials are joined to opposite ends of a dissimilar wire, to give two junctions in series in opposition to each other, as in Fig. 25.2. A net p.d. is thus only produced when the two junctions are at different temperatures. One junction is maintained at ice point and the other is the 'temperature probe'. The temperature can then be measured from a suitably calibrated meter.
- The **constant volume gas thermometer** contains a bulb of dry air connected to a mercury manometer. The pressure of the air in the bulb is measured using the manometer. When the bulb temperature is changed, the open arm of the manometer is adjusted to keep the mercury level in the closed arm at the same position throughout. The height difference, h, between the mercury level in the closed arm and the open arm is measured and used to calculate the temperature, t, in °C from the formula

$$t = \frac{h - h_0}{h_{100} - h_0} \times 100°C$$

where h_0 is the height difference at ice point and h_{100} is the height difference at steam point.

Calibrating a thermometer

At temperatures away from the fixed points, different thermometers give different readings for the same temperature. The gas thermometer is

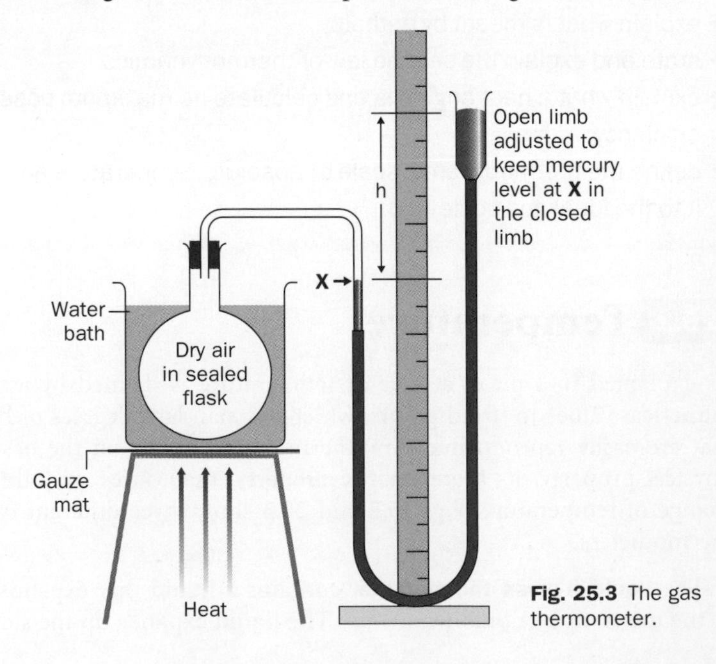

Fig. 25.3 The gas thermometer.

chosen as the standard against which other types of thermometers are calibrated between the fixed points. Temperatures according to the gas thermometer are referred to as **ideal gas temperatures**.

- The **Celsius scale** is achieved in practice by calibrating a gas thermometer at ice point and at steam point then using the equation on p. 374. The absolute zero of temperature on this scale is −273.15°C.

- The **absolute scale** is achieved in practice by calibrating a gas thermometer at the **triple point** of water which is the point at which water, water vapour and ice coexist in thermal equilibrium. Since this temperature is 0.01°C higher than ice point, the triple point temperature on the absolute scale is defined as 273.16 K. Temperatures on the absolute scale may therefore be measured from the readings on a gas thermometer in accordance with the equation

$$T = \frac{(pV)_{p \to 0}}{(pV_{Tr})_{p \to 0}} \times 273.16 \text{ K}$$

where $(pV)_{p \to 0}$ is the limit of pressure × volume of the gas at zero pressure at temperature T and $(pV_{Tr})_{p \to 0}$ is the same quantity at the triple point of water.

By defining the absolute scale of temperatures in this way, the link between the two scales is given by the following equation

T (in kelvins) = t (in °C) + 273.15

Thus the reading of any type of thermometer, which has been calibrated in °C against a gas thermometer, can be converted into kelvins by adding 273.15 to its reading.

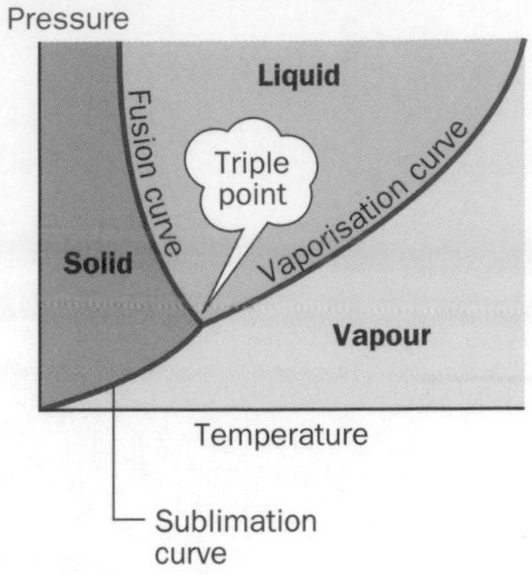

Fig. 25.4 The triple point of water.

Note

For any thermometer (other than a gas thermometer) which is calibrated at ice point and steam point, if the interval between the two fixed points is divided into 100 equal degrees, the scale is then a **centigrade** scale, not the Celsius scale. The unit °C should not be used for readings taken from a centigrade scale as °C means 'degree Celsius'.

Questions 25.1

1. **(a)** A constant volume gas thermometer was calibrated at ice point and steam point, then used to measure the temperature of a beaker of hot water. The readings for the height difference, h, between the mercury level in the closed limb and the open limb were −50 mm for ice point, +220 mm for steam point and +105 mm for the temperature of the water. Calculate the temperature of the hot water in °C.

(b) When the position of the mercury in a mercury-in-glass thermometer in some hot water was exactly midway between ice point and steam point, the height difference between the mercury levels of the constant volume gas thermometer in (a) was +80 mm. Calculate the temperature **(i)** on the centigrade scale of the mercury-in-glass thermometer, **(ii)** in °C.

2. The e.m.f., E, of a certain thermocouple varies with the temperature in °C in accordance with the equation $E = at - bt^2$, where $a = 40.5\ \mu V\ °C^{-1}$ and $b = 0.065\ \mu V\ °C^{-2}$.

(a) Calculate the e.m.f. at $t = 0, 20, 40, 60, 80$ and 100°C.

(b) (i) Plot a graph to show how the e.m.f. varies with temperature.

(ii) Use your graph to determine the temperature in °C when the e.m.f. is midway between its value at ice point and its value at steam point.

25.2 Heat and work

Work is done by a force when it moves its point of application in the direction of the force. Energy is the capacity of a body to do work. Energy can be transferred from one body to another by two methods:

1. Work is energy transferred by means of a force moving its point of application.
2. Heat is energy transferred by means other than a force. A temperature difference is said to exist between two bodies if heat transfer between the two bodies could occur.

If no heat transfer occurs between two bodies in thermal contact, the two bodies must be at the same temperature. This is known as the **zeroth law of thermodynamics**.

The **internal energy** of a body is the energy of the body (i.e. its capacity to do work) regardless of its position (i.e. its potential energy) or its motion (i.e. its kinetic energy). In effect, the internal energy of a body is the potential energy due to the bonds between its atoms and molecules and the kinetic energy due to the random motion of its atoms and molecules. When energy is transferred overall to a fixed body, its molecules gain either kinetic energy or potential energy or both. Increasing the kinetic energy of the molecules of a body raises its temperature. Increasing the potential energy of the molecules changes its state.

For example, suppose a sealed can of air is placed in a beaker of hot water. The air gains internal energy and the water loses internal energy due to heat transfer through the can. Thus the capacity of the air to do work has increased because its internal energy has increased. For example, if the can is pierced suddenly, air would be forced out due to the high pressure inside. The outflow of air could be used to do work. Two further examples of the use of heat to do work are shown in Figs 25.5 and 25.6. A device that uses heat to do work is called a **heat engine**.

- The four-stroke petrol engine uses the heat from the combustion of petrol and oxygen in air to do work. The sequence of strokes is shown in Fig. 25.5.
- The jet engine in Fig. 25.6 burns fuel in air to raise the temperature of air drawn into the engine by a compressor. The hot air is expelled at high speed from the exhaust, creating the thrust that drives the engine forward.

The First Law of Thermodynamics

The internal energy of a body can be changed, either by heat transfer in or out of the body, or by work done on or by the body. For example, suppose a body does 1000 J of work and its internal energy decreases by 800 J. From the principle of conservation of energy, it therefore follows that the body must have gained 200 J of heat. Using conservation of energy to equate the

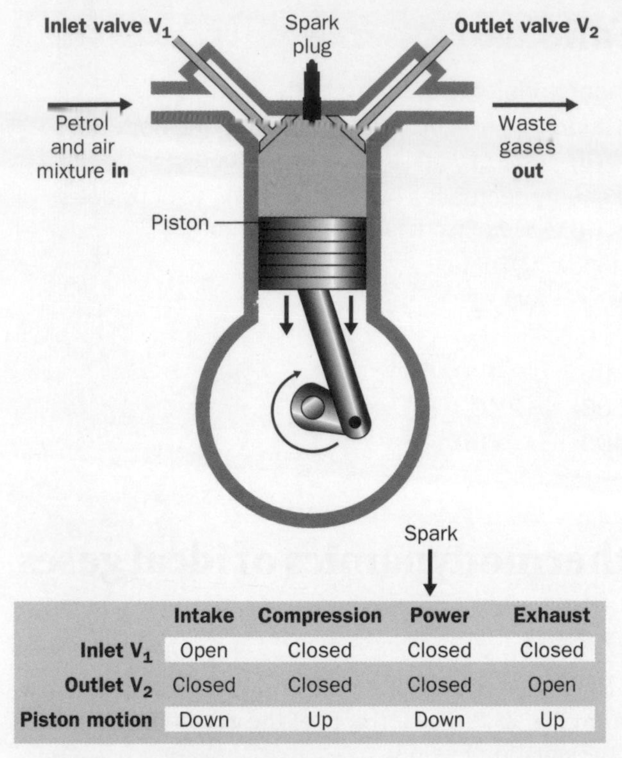

	Intake	Compression	Power	Exhaust
Inlet V_1	Open	Closed	Closed	Closed
Outlet V_2	Closed	Closed	Closed	Open
Piston motion	Down	Up	Down	Up

Fig. 25.5 The four-stroke petrol engine.

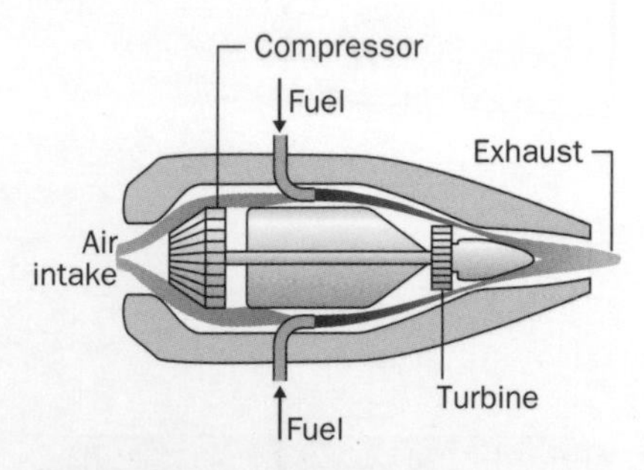

Fig. 25.6 The jet engine.

overall transfer of energy to the change of internal energy therefore gives rise to the following statement, known as the **First Law of Thermodynamics**.

For a body of internal energy, U, the heat entering the body, ΔQ = its change of internal energy, ΔU + the work done by the body, ΔW.

$$\Delta Q = \Delta U + \Delta W$$

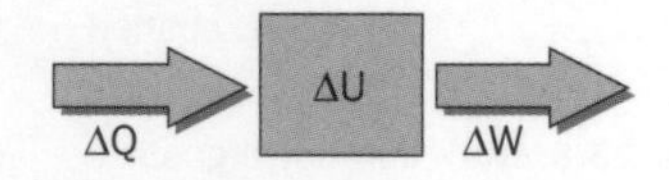

$\Delta Q = \Delta U + \Delta W$

Fig. 25.7 The First Law of Thermodynamics.

Notes

A sign convention is necessary to take account of the direction of energy transfer and change of internal energy.

1. A positive value for ΔU means an increase of internal energy and a negative value means a decrease.

2. A positive value for ΔQ means heat entering and a negative value means heat leaving.

3. A positive value for ΔW means work done by the gas (i.e. expansion) and a negative value means work done on the gas (i.e. compression).

4. An **adiabatic change** is defined as a change in which there is no heat transfer, i.e. $\Delta Q = 0$. For an adiabatic change $\Delta U = -\Delta W$.

Worked example 25.1

Calculate the change of internal energy of a gas as a result of 100 J of work done on the gas and 80 J of heat (a) entering the gas, (b) leaving the gas.

Solution

$\Delta Q = \Delta U + \Delta W$

(a) $\Delta Q = +80$ J, $\Delta W = -100$ J
$\therefore \Delta U = \Delta Q - \Delta W = 80 - (-100) = 180$ J increase.

(b) $\Delta Q = -80$ J, $\Delta W = -100$ J
$\therefore \Delta U = \Delta Q - \Delta W$
$= -80 - (-100)$
$= 20$ J increase.

Questions 25.2

1. Calculate the heat transfer to or from a body when
(a) its internal energy increases by 1000 J and **(i)** the body does 300 J of work, **(ii)** 300 J of work is done on the body,
(b) its internal energy decreases by 500 J and **(i)** the body does 200 J of work, **(ii)** 200 J of work is done on the body.

2. Complete the following Table

Δ**U** / J	Δ**Q** / J	Δ**W** / J
2 000	400	**(a)**
−2 000	**(b)**	2 400
(c)	−800	−2 400
−1 000	−400	**(d)**

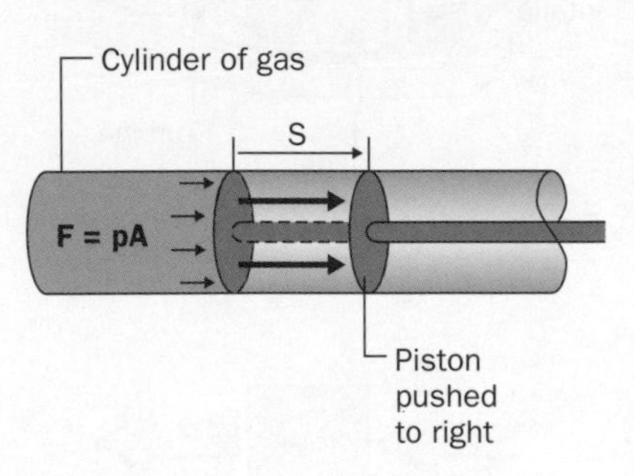

Fig. 25.8 Work done by a gas.

Note
For an ideal gas at constant pressure,
- before expansion, $pV = nRT$,
- after expansion $p(V+\Delta V) = nR(T+\Delta T)$.

Therefore subtracting the first equation from the second equation gives $p\Delta V = nR\Delta T$. Hence the work done by an ideal gas when it expands at constant pressure is given by $\Delta W = nR\Delta T$.

25.3 The thermodynamics of ideal gases

Work done by a gas

Consider a gas in a cylinder at pressure, p, held in by a piston of cross-sectional area, A, as in Fig. 25.8. The force, F, of the gas on the cylinder is therefore given by the equation $F = pA$.

Suppose the gas pushes the piston outwards by a distance. Δs. at constant pressure. The work done by the gas, $\Delta W = F\Delta s = pA\Delta s$.

Since the increase of volume of the gas $\Delta V = A\Delta s$, it therefore follows that $\Delta W = p\Delta V$.

Work done by a gas to expand at constant pressure

$$\Delta W = p\Delta V$$

Molar heat capacities

When heat is supplied to a gas, its temperature change depends on whether or not it is allowed to expand. The First Law of Thermodynamics for an ideal gas may be expressed in the form

$$\Delta Q = \Delta U + p\Delta V$$

- At **constant volume**, no work is done by or on a gas as $\Delta V = 0$, hence the heat supplied to an ideal gas at constant volume to raise its temperature is given by $\Delta Q = \Delta U$.

The molar heat capacity at constant volume of a gas, C_V, is defined as the heat supplied at constant volume to raise the temperature of 1 mole by 1 K.

The heat supplied, ΔQ, to raise the temperature of n moles by ΔT at constant volume is therefore given by

$$\Delta Q = nC_V\Delta T$$

Hence the change of internal energy of a gas when its temperature changes by ΔT is given by

$$\Delta U = nC_v\Delta T$$

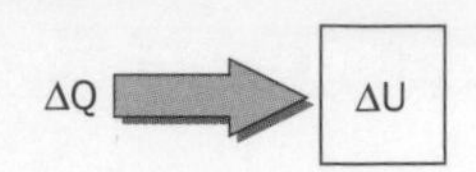

Fig. 25.9 Heating a gas at constant volume.

- At **constant pressure**, work is done by or on a gas as ΔV is not zero.

The molar heat capacity at constant pressure, C_p, of a gas is defined as the heat supplied at constant pressure to raise the temperature of 1 mole by 1 K.

The heat supplied, Q, to raise the temperature of n moles by ΔT at constant pressure is therefore given by

$$\Delta Q = nC_p\Delta T$$

Since $\Delta Q = \Delta U + p\Delta V$, and $p\Delta V = nR\Delta T$, then $nC_p\Delta T = nC_v\Delta T + nR\Delta T$. By considering $n = 1$ mole and $\Delta T = 1$ K, it follows that $C_p = C_v + R$.

$$C_p - C_v = R$$

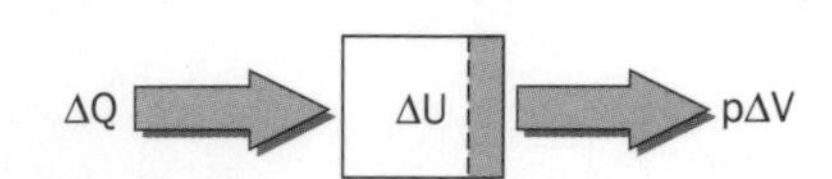

Fig. 25.10 Heating a gas at constant pressure.

Atomicity and degrees of freedom

A **monoatomic molecule** has three degrees of freedom, each degree corresponding to motion which is independent of the motion associated with any other degree of freedom. Thus a monoatomic molecule can move in any one of three perpendicular directions without altering its motion in the other two directions. Since the mean kinetic energy of a point molecule is $\frac{3}{2}$ kT, then the mean kinetic energy associated with each degree of freedom is $\frac{1}{2}$ kT.

The total kinetic energy of n moles of monoatomic molecules is $\frac{3}{2}$ nRT, since n moles contain $nR \div k$ molecules (as $k = R \div N_A$, then 1 mole contains $R \div k$ molecules).

Hence the internal energy, U, of n moles of an ideal gas of monoatomic molecules is $\frac{3}{2}$ nRT.

$$U = \frac{3}{2} nRT$$

If the temperature of an ideal gas changes by ΔT, the change of its internal energy, ΔU, is therefore equal to $\frac{3}{2} nR\Delta T$. Therefore, to raise the temperature of 1 mole by 1 K, the internal energy must increase by $\frac{3}{2}$ R. Since this is equal to the heat supplied at constant volume, it follows that

$$C_v = \frac{3}{2} R$$

Also, since $C_p = C_v + R$, then

$$C_p = \frac{5}{2} R$$

Therefore, the ratio of molar heat capacities, γ, can be written as

$$\gamma = \frac{C_p}{C_v} = \frac{5}{3} = 1.67$$

for a monoatomic gas.

A **diatomic molecule** has five degrees of freedom, three of which are due to its motion in any of three perpendicular directions. The other two are due to its rotational motion about two perpendicular directions at right angles to its axis. The mean kinetic energy of a diatomic molecule

(a)

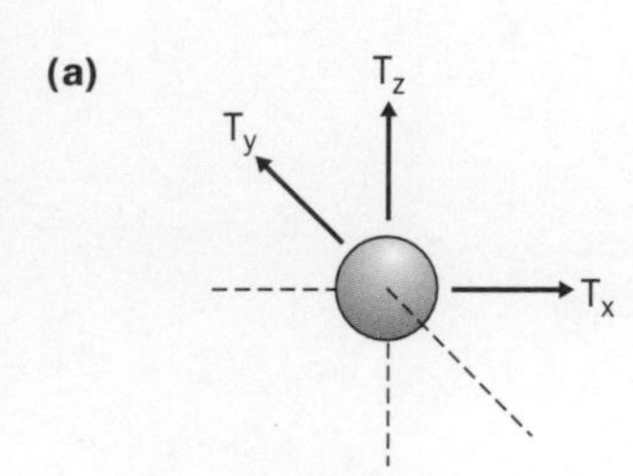

(b)

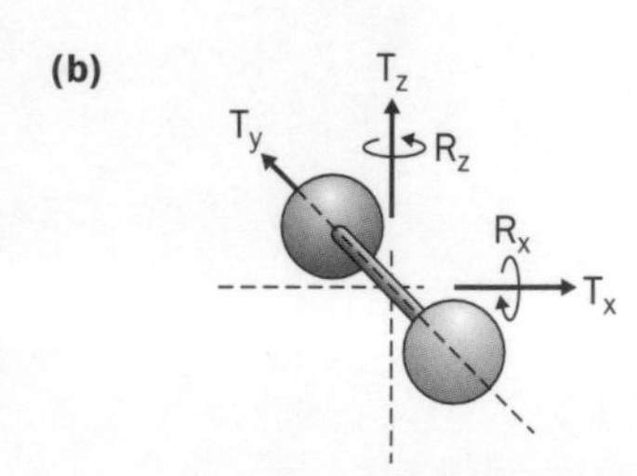

(c)

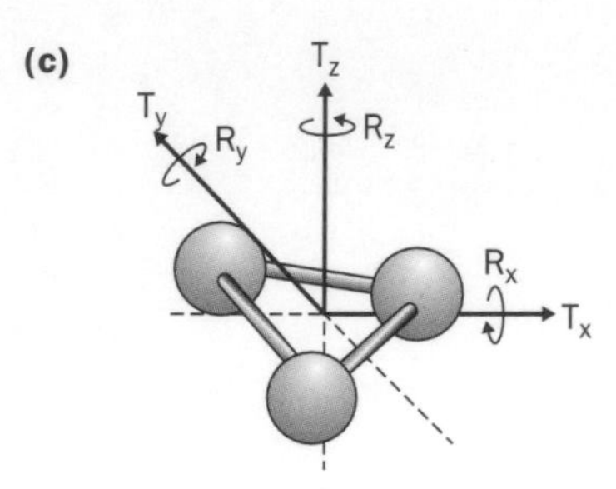

Fig. 25.11 Degrees of freedom **(a)** a monoatomic molecule, **(b)** a diatomic molecule, **(c)** a polyatomic molecule.

is therefore $\frac{5}{2}$kT. The total kinetic energy of 1 mole of diatomic molecules is therefore $\frac{5}{2}$RT.

Therefore, the total kinetic energy of n moles of diatomic molecules is $\frac{5}{2}$ nRT.

Hence the internal energy, U, of n moles of an ideal gas of diatomic molecules is $\frac{5}{2}$ nRT.

$$\mathbf{U = \frac{5}{2} nRT}$$

If the temperature of an ideal gas changes by ΔT, the change of its internal energy, ΔU, is therefore equal to $\frac{5}{2}$ nRΔT. Therefore, to raise the temperature of 1 mole by 1 K, the internal energy must increase by $\frac{5}{2}$ R. Since this is equal to the heat supplied at constant volume, it follows that

$$\mathbf{C_v = \frac{5}{2} R.}$$

Also, since $C_p = C_v + R$, then

$$\mathbf{C_p = \frac{7}{2} R}$$

Therefore, the ratio of molar heat capacities, γ, can be written as

$$\gamma = \frac{C_p}{C_v} = \frac{7}{5} = 1.4$$

for a diatomic gas.

A **polyatomic molecule** (i.e. a molecule with more than two atoms) has six degrees of freedom, five of them being the same as for a diatomic molecule. The sixth is due to rotation in the third possible perpendicular direction. The mean kinetic energy of a polyatomic molecule is therefore $\frac{6}{2}$kT.

Therefore, the total kinetic energy of n moles of polyatomic molecules is $\frac{6}{2}$ nRT.

Hence the internal energy, U, of n moles of an ideal gas of polyatomic molecules is $\frac{6}{2}$ nRT.

$$\mathbf{U = \frac{6}{2} nRT}$$

If the temperature of an ideal gas changes by ΔT, the change of its internal energy, ΔU, is therefore equal to $\frac{6}{2}$ nRΔT. Therefore, to raise the temperature of 1 mole by 1 K, the internal energy must increase by $\frac{6}{2}$ R. Since this is equal to the heat supplied at constant volume, it follows that

$$\mathbf{C_v = \frac{6}{2} R}$$

Also, since $C_p = C_v + R$, then

$$\mathbf{C_p = \frac{8}{2} R}$$

Therefore, the ratio of molar heat capacities, γ, can be written as

$$\gamma = \frac{C_p}{C_v} = \frac{8}{6} = 1.33$$

for a polyatomic gas.

Adiabatic changes of an ideal gas

An adiabatic change is defined as one in which no heat transfer occurs. In practice, this could be due to a change being very rapid or due to perfect insulation.

- If an ideal gas expands adiabatically, it does work and therefore its internal energy decreases. It therefore cools.
- If an ideal gas is compressed adiabatically, work is done on the gas and therefore its internal energy rises. Its temperature therefore increases. Fig. 25.12 shows how the pressure and volume of a fixed quantity of an ideal gas vary for both adiabatic changes and isothermal changes (i.e. constant temperature). The adiabatic curves are steeper than the isothermal curves because the temperature changes when an adiabatic expansion or compression occurs.
- The area under a pressure against volume curve gives the work done. This is because the work done in each small volume change, δV, is equal to $p\delta V$, which is represented by the strip of width, δV, under the curve. Therefore the total work done is represented by the total area under the curve.
- The pressure and volume of an ideal gas before and after an adiabatic change are related by the following equation:

  pV^{γ} = constant

  where γ is the ratio of molar heat capacities.

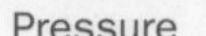

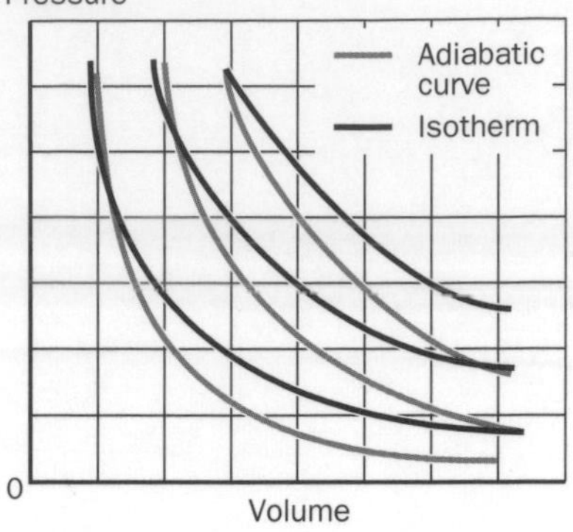

Fig. 25.12 Comparison of adiabatic and isothermal changes.

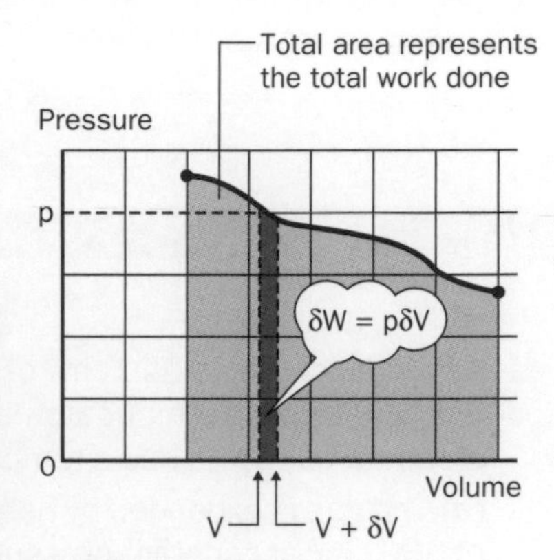

Fig. 25.13 Area under a p V curve.

Proof of pV^{γ} = constant

To prove this equation, consider the 1st Law of Thermodynamics in the form $\Delta Q = nC_v\Delta T + p\Delta V$.

For an adiabatic change, $\Delta Q = 0$, thus $nC_v\Delta T + p\Delta V = 0$.

$\therefore p\Delta V = -nC_v\Delta T$ **(1)**

From the ideal gas equation, $pV = nRT$, then $(p + \Delta p)(V + \Delta V) = nR(T + \Delta T)$

By subtraction, $V\Delta p + p\Delta V = nR\Delta T$

$\therefore V\Delta p = nR\Delta T - p\Delta V = nR\Delta T + nC_v\Delta T = nC_p\Delta T$

$V\Delta p = nC_p\Delta T$ **(2)**

Combining equations **1** and **2** gives $V\Delta p = \frac{C_p}{C_v} nC_v\Delta T = \gamma nC_v\Delta T = -\gamma p\Delta V$

Therefore $V\Delta p = -\gamma p\Delta V$

$\therefore \frac{\Delta p}{p} = \frac{-\gamma\Delta V}{V}$

Integrating both sides gives $\log p = -\gamma \log V +$ constant

(since the integral of $\frac{\Delta x}{x} = \log x$).

Hence $\log pV^{\gamma}$ = constant

$\therefore$ **pV^{γ} = constant**

Worked example 25.2

$\gamma = 1.4$ for air

A cylinder contains 100 cm^3 of air at a pressure of 100 kPa and a temperature of 20°C. The air is compressed adiabatically to 50 cm^3. Calculate (a) the pressure of the air after compression, (b) the temperature of the air after compression.

Solution

(a) $pV^{\gamma} = \text{constant}$, hence $p \times 50^{1.4} = 100 \times 100^{1.4}$

$\therefore p \times 239 = 100 \times 631$

$\therefore p = \dfrac{100 \times 631}{239} = 264\ \text{kPa}.$

(b) $pV = nRT \quad \therefore \dfrac{pV}{T} = \text{constant, so } \dfrac{264 \times 50}{T} = \dfrac{100 \times 100}{(273 + 20)}$

$T = \dfrac{264 \times 50 \times 293}{100 \times 100} = 387\ \text{K} = 114°\text{C}$

Questions 25.3

$R = 8.31\ \text{J mol}^{-1}\ \text{K}^{-1}$

1. A cylinder contains 240 cm^3 of a monoatomic gas ($\gamma = 1.67$ for a monoatomic gas) at a pressure of 80 kPa and a temperature of 17°C.

(a) The gas is expanded by heating to a volume of 360 cm^3 at constant pressure.

(i) Calculate the temperature of the gas at the end of this change.

(ii) Calculate the work done, change of internal energy and heat transfer that takes place.

(iii) Represent the change on a graph of pressure against volume.

(b) (i) If the gas in (a) had expanded to a volume of 360 cm^3 at constant temperature, what would its final pressure have been?

(ii) Represent this change on the same graph, labelling it 'constant temperature'.

(iii) Use the graph to estimate the work done by the gas as a result of this change. Hence determine the heat transfer for this expansion.

(c) (i) If the gas in (a) had expanded adiabatically to a volume of 360 cm^3, what would its final pressure and temperature have been?

(ii) Represent this change on your graph, labelling it 'adiabatic change'.

(iii) Calculate the change of internal energy in c (i) and hence the work done by the gas.

2. (a) A cylinder contains 1200 cm^3 of air at a pressure of 110 kPa and a temperature of 15°C. Calculate **(i)** the number of moles of air in the cylinder, **(ii)** the mass of this air, **(iii)** the internal energy of this air.

(b) The air in (a) is compressed adiabatically to a volume of 150 cm^3. Calculate **(i)** the pressure and the temperature of the air immediately after compression, **(ii)** the change of internal energy of the air, **(iii)** the work done on the air to compress it adiabatically.

The molar mass of air = 0.029 kg, $\gamma = 1.4$ for air.

Enthalpy

In a chemical reaction or a physical change of state, changes usually take place at constant pressure. The heat transfer that takes place is not equal to the change of internal energy of the substances involved because work is done as a result of volume changes.

The enthalpy, H, of a substance is defined by the equation $H = U + pV$.

For a change at constant pressure, the change of enthalpy

$$\Delta H = \Delta U + p\Delta V.$$

This change of enthalpy is therefore equal to the heat transfer, ΔQ, since $\Delta Q = \Delta U + p\Delta V$ from the 1st Law of Thermodynamics. The unit of enthalpy is the joule.

Notes

1. In **chemical reactions**, the enthalpy change is usually expressed in kilojoules per mole ($kJ\ mol^{-1}$). Where a reaction can take place by different routes, the enthalpy change is the same whichever route the reaction takes.

2. In a **physical change of state**, latent heat is released when a vapour condenses or a liquid freezes. Latent heat must be supplied to vaporise a liquid or to melt a solid. The enthalpy of the substance changes when it changes its physical state. The change of enthalpy is the latent heat involved in the change. Thus the fact that the specific latent heat of ice is 336 $J\ g^{-1}$ means that the enthalpy of fusion of water is 336 $J\ g^{-1}$ or 6.05 $kJ\ mol^{-1}$ (since the molar mass of water = 0.018 kg).

25.4 The Second Law of Thermodynamics

Energy changes

Energy tends to become less useful when it changes from any one form into other forms. For example, fuels contain energy in concentrated form. When a fuel burns, chemical energy is transformed into heat and light. Some, but not all, of this energy can be stored and used later. For example, the fuel could be used to drive an engine which raises a weight. The potential energy of the weight can be used to do work later. However, the amount of work that the weight could do (i.e. useful energy) is less than the energy released from the fuel. Although energy is conserved in any change, the amount of useful energy is less after the change than before because some energy is wasted (e.g. due to friction).

A falling ball is another example where energy is changing from one form into other forms, namely from potential energy to kinetic energy as it falls. If the ball is perfectly elastic, it returns to the height it was released from. Such reversible changes require perfect elasticity and zero friction. In practice, such conditions as zero friction are usually unattainable and therefore energy is wasted in most transformations. For example, the energy of a bouncing ball would eventually be converted to internal energy of the surroundings.

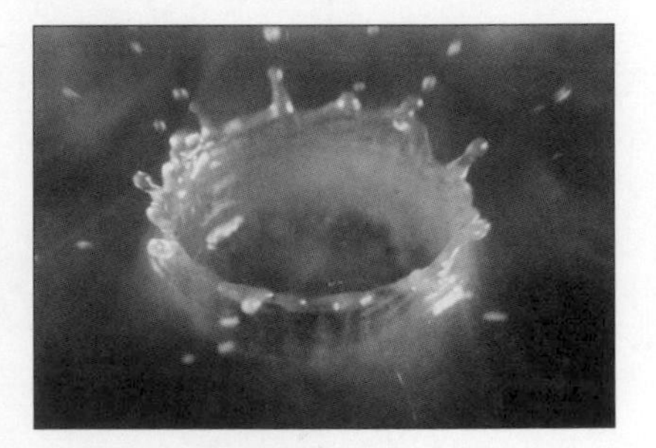

Fig. 25.14 Energy spreading out.

Engines at work

An engine is designed to make objects move. The fuel supplied to an engine enables it to do work. In most engines, fuel is burned to release chemical energy in the form of heat and light. The engine uses heat from the fuel to raise the temperature of a fluid to make the fluid do work. A high temperature 'reservoir' is essential to heat the fluid. A low temperature reservoir is also essential to 'draw' the fluid through the engine. Thus some of the heat transfer from the high temperature reservoir must be supplied by the hot fluid to the low temperature reservoir if the engine is to work. No heat engine can therefore be 100% efficient. It is therefore impossible for a

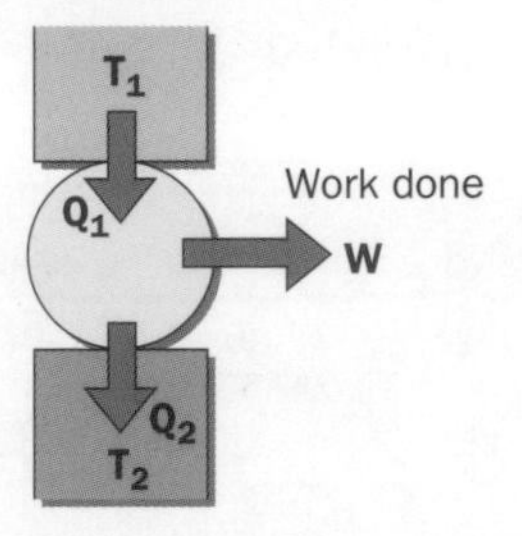

Fig. 25.15 A heat engine.

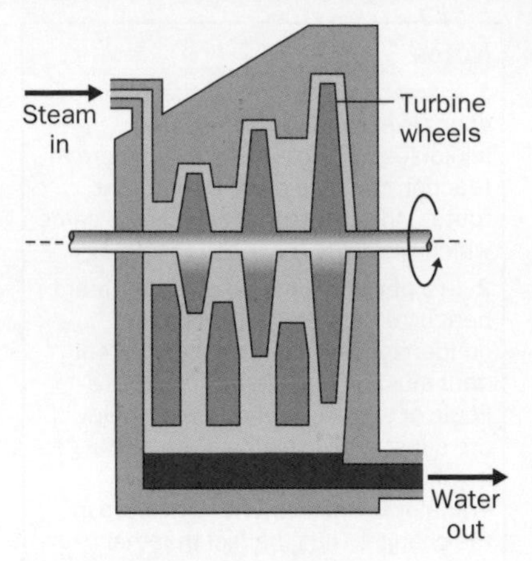

Fig. 25.16 A steam engine.

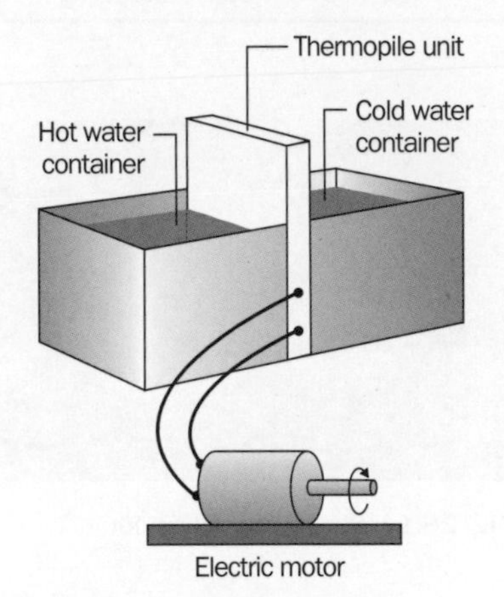

Fig. 25.17 A thermopile engine.

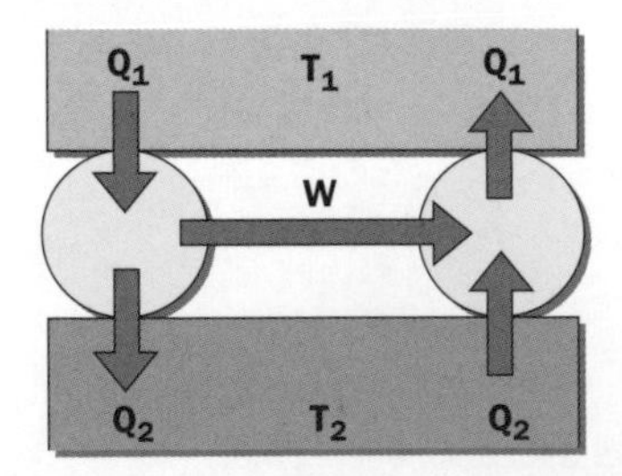

Fig. 25.18 Reversible engines.

heat engine to convert all the heat supplied to it into work. This statement is known as the **Second Law of Thermodynamics**.

The efficiency of a heat engine, η, can be written as

$$\eta = \frac{\text{work done by the engine, W}}{\text{heat supplied to the engine, } Q_1}$$

Since the heat lost to the low temperature reservoir, $Q_2 = Q_1 - W$, then $W = Q_1 - Q_2$.

$$\text{Hence } \eta = \frac{W}{Q_1} = 1 - \frac{Q_2}{Q_1}$$

Examples of heat engines at work

A steam turbine

Steam is raised as a result of using a fuel to heat water. Steam jets at high pressure are used to make the turbine wheels rotate. Steam at low pressure emerges from the turbine engine and this is condensed back to water which is then reused. A low temperature 'reservoir' is therefore essential to keep the steam moving through the engine. Without the low temperature reservoir provided by the cooling water, the steam would not condense and the inflowing steam would be unable to do work.

A thermopile engine

A thermopile generates an e.m.f. when there is a temperature difference between its two faces. Fig. 25.17 shows a thermopile between a reservoir of hot water and a reservoir of cold water. The thermopile may be used to operate an electric motor provided the cold water remains colder than the hot water. Heat flow to the cold water from the thermopile must therefore take place which means that not all the heat supplied to the thermopile from the hot water is converted to work.

The internal combustion engine and the jet engine are two further examples of heat engines. In both cases, the surroundings act as the low temperature reservoir. If an internal combustion engine or a jet engine was insulated from its surroundings, it would heat up and stop working sooner or later.

Reversible engines

The most efficient type of engine is a **reversible engine**, a theoretical engine which doesn't waste energy and can be operated in reverse. Suppose a reversible engine operating between a high temperature reservoir, T_1, and a low temperature reservoir, T_2, is used directly to drive a similar engine operating in reverse. Fig. 25.18 shows the idea. No net heat transfer occurs from either reservoir and so no energy is wasted.

If a more efficient engine were possible, it could be used to cause a net heat transfer from the cold reservoir to the hot reservoir without any overall work being done. Since such heat transfer is impossible without work being done, an engine more efficient than a reversible engine is not possible.

The efficiency of a heat engine

When a reversible engine does work, W, and heat transfer, Q_1, from the hot reservoir occurs, the heat transfer to the cold reservoir, $Q_2 = Q_1 - W$. The ratio Q_2:Q_1 depends on the ratio of the reservoir temperatures and can be shown to be equal to the ratio of absolute temperatures of the reservoirs, as below:

$$\frac{T_2}{T_1} = \frac{Q_2}{Q_1}$$

where T_1 is the temperature of the hot reservoir and T_2 is the temperature of the cold reservoir.

Therefore the efficiency of a reversible engine, η_R, can be written as

$$\eta_R = \frac{W}{Q_1} = 1 - \frac{Q_2}{Q_1} = 1 - \frac{T_2}{T_1} = \frac{T_1 - T_2}{T_1}$$

This is the maximum possible efficiency of any engine operating between temperatures T_1 and T_2.

Therefore the efficiency of an engine operating between temperatures T_1 and T_2

$$\eta \leq \frac{T_1 - T_2}{T_1}$$

Worked example 25.3

A heat engine operates between reservoirs at 750 K and 300 K, accepting heat from the hot reservoir at a rate of 2000 W and doing work at a rate of 600 W. Calculate (a) the heat wasted each second, (b) the efficiency of this engine, (c) The maximum possible effiency an engine could have, operating between 750 K and 300 K.

Solution

(a) $Q_2 = Q_1 - W = 2000 - 600 = 1400$ W

(b) Efficiency $= \frac{600}{2000} = 0.30$

(c) Maximum possible efficiency $= \frac{750 - 300}{750} = \frac{450}{750} = 0.60$

Questions 25.4

1. (a) A vehicle uses fuel at a rate of 15 kilometres per litre when its speed is 30 m s^{-1}.

(I) Calculate how long it takes to use a litre of fuel at this rate.

(ii) The fuel releases 20 MJ of heat per litre when it is burned. How much energy is released by the fuel per second when it is used at the above rate?

(b) The output power of the vehicle at 30 m s^{-1} is 6 kW. Calculate the overall efficiency of the vehicle engine and transmission system.

2. A heat engine with a power output of 40 W operates between reservoirs at temperatures of 400 K and 300 K. The cooling system removes heat from the engine at a rate of 160 W.

(a) Calculate the heat supplied per second by the high temperature reservoir.

(b) Calculate **(i)** its efficiency, **(ii)** the maximum efficiency of an engine operating between these two temperatures.

Summary

◆ Temperature scales

Absolute temperature, $T = \dfrac{pV}{(pV)_{Tr}} \times 273.16\ K$

Celsius temperature, $t = T - 273.15$

◆ The Laws of Thermodynamics

Zeroth law: If no heat transfer occurs between two bodies in thermal equilibrium, they are at the same temperature.

1st Law: $\Delta Q = \Delta U + \Delta W$

2nd Law: It is not possible for a heat engine to convert all the heat supplied to it into work.

◆ Thermodynamic changes

An isothermal change is a change at constant temperature.

An adiabatic change is a change in which no heat transfer occurs.

An enthalpy change is heat supplied or released at constant pressure.

◆ Thermodynamics of gases

1. Work done by a gas $= p\Delta V$

2. Molar heat capacities: $C_P - C_V = R$

where C_V is the heat supplied to raise the temperature of 1 mole by 1 K at constant volume, and C_P is the heat supplied to raise the temperature of 1 mole by 1 K at constant pressure.

3. For a monoatomic gas, $C_V = \frac{3}{2}RT$ and $C_P = \frac{5}{2}RT$,

$$\gamma = \frac{C_P}{C_V} = \frac{5}{3}$$

For a diatomic gas, $C_V = \frac{5}{2}RT$ and $C_P = \frac{7}{2}RT$,

$$\gamma = \frac{C_P}{C_V} = \frac{7}{5}$$

For a polyatomic gas, $C_V = \frac{6}{2}RT$ and $C_P = \frac{8}{2}RT$,

$$\gamma = \frac{C_P}{C_V} = \frac{8}{6}$$

4. For an adiabatic change of an ideal gas, $pV^{\gamma} = \text{constant}$

◆ Heat engines

$$\text{Efficiency} = \frac{W}{Q_1} = \frac{Q_1 - Q_2}{Q_1}$$

$$\text{Maximum possible efficiency} = \frac{T_1 - T_2}{T_1}$$

Revision questions

25.1. (a) The pressure of the gas in a constant volume gas thermometer, relative to atmospheric pressure was −6 kPa at ice point and +29 kPa at steam point. When the gas pressure was +20 kPa, what was the temperature of the gas thermometer in **(i)** °C, **(ii)** K?

(b) The e.m.f. of a thermocouple was amplified using a d.c. amplifier. The amplifier was adjusted so its output voltage was zero when the thermocouple was at ice point and 100 mV when the thermocouple was at steam point. The output voltage, V, was found to vary with the Celsius temperature, t, of the thermocouple in accordance with the equation $V = 1.25\ t\ (1 - 20 \times 10^{-4}\ t)$ mV. When the Celsius temperature was 50°C, what was **(i)** the output voltage, **(ii)** the centigrade reading of this thermocouple thermometer?

25.2. Complete the following table for the energy changes of an ideal gas.

ΔU / J	ΔQ / J	p /kPa	ΔV / m³
1000	**(a)**	100	$+2.0 \times 10^{-3}$
(b)	−600	150	-2.0×10^{-3}
−800	200	100	**(c)**
1200	400	**(d)**	-5.0×10^{-3}

25.3. A glass bulb contains 60 cm³ of inert gas consisting of monoatomic molecules at a pressure of 10 kPa and a temperature of 290 K.

(a) Calculate **(i)** the number of moles of gas in the bulb, **(ii)** the internal energy of this gas.

(b) The gas is heated to a temperature of 380 K. Calculate **(i)** its pressure at this temperature, **(ii)** its increase of internal energy.

25.4. (a) (i) Calculate the number of moles of an ideal gas in a volume of 100 cm³ ($= 1.00 \times 10^{-4}$ m³) at a pressure of 101 kPa and a temperature of 20°C.

(ii) Air consists of diatomic molecules of nitrogen and oxygen. Calculate the internal energy of the air in a cylinder of volume 100 cm³ at the above temperature and pressure.

(b)(i) The air in the cylinder in (a)(ii) is compressed adiabatically from its initial pressure of 101 kPa to a volume of 20 cm³. Calculate the pressure and temperature of the air in this cylinder after the compression.

(ii) Calculate the internal energy of the air in the cylinder at this new temperature.

(iii) Hence calculate the work done on the air when it was compressed adiabatically.

25.5. (a) An engine operating between 450 K and 300 K is supplied with heat at a rate of 2000 J s⁻¹ and delivers power at a rate of 600 W. Calculate **(i)** the heat supplied per second by the engine to the low temperature reservoir, **(ii)** the efficiency of the engine, **(iii)** the maximum possible efficiency of any heat engine working between these two temperatures.

(b) In a certain type of heat engine, air, initially at a pressure of 100 kPa, a volume of 0.0020 m³ and a temperature of 300 K, is heated at constant volume to a temperature of 1000 K. The air is then allowed to expand adiabatically to a pressure of 100 kPa. It is then compressed at constant pressure back to its initial volume. Fig. 25.19 shows the process.

(i) Show that the pressure is 333 kPa at the end of the first stage and the temperature at the end of the second stage is 706 K.

(ii) Calculate the maximum possible efficiency of any heat engine operating between 1000 K and 300 K.

(iii) Use the graph to estimate the work done by the engine in one cycle.

(iv) Calculate the heat gain by the gas.

(v) Estimate the efficiency of the engine.

(γ for air = 1.4)

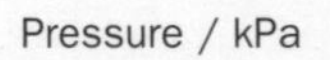

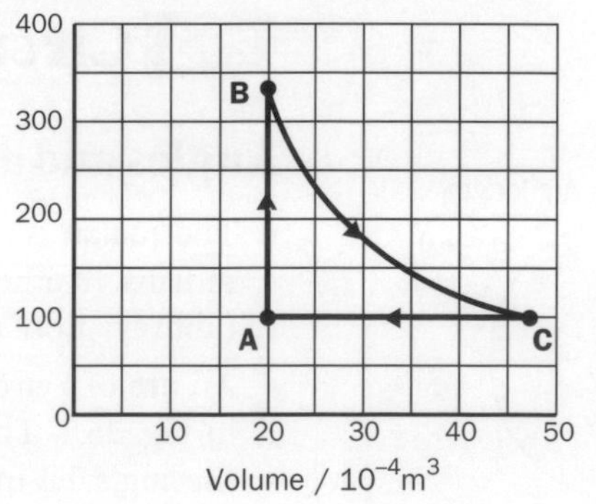

Fig. 25.19

Uniform Circular Motion

Contents

Objectives

After working through this unit, you should be able to:

- ▶ convert from degrees to radians
- ▶ calculate the frequency and the angular speed of an object in uniform circular motion from its time period
- ▶ calculate its speed from its time period and the radius of rotation
- ▶ explain why an object in uniform circular motion experiences a centripetal acceleration
- ▶ calculate the centripetal acceleration of an object in uniform circular motion from its speed and the radius of rotation
- ▶ analyse the motion of an object in uniform circular motion in terms of the forces acting on the object

26.1 Circular measures

Angles and arcs

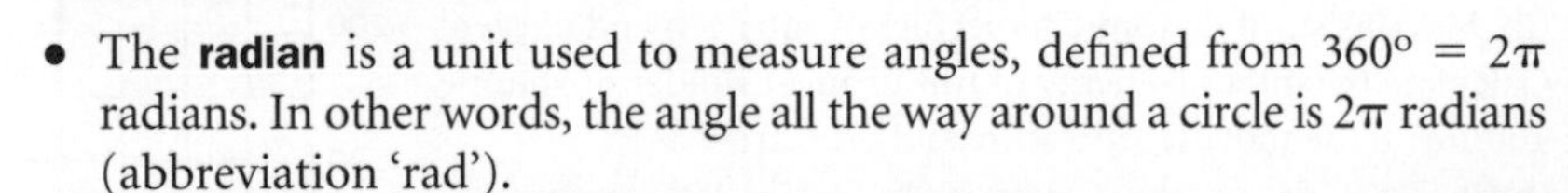

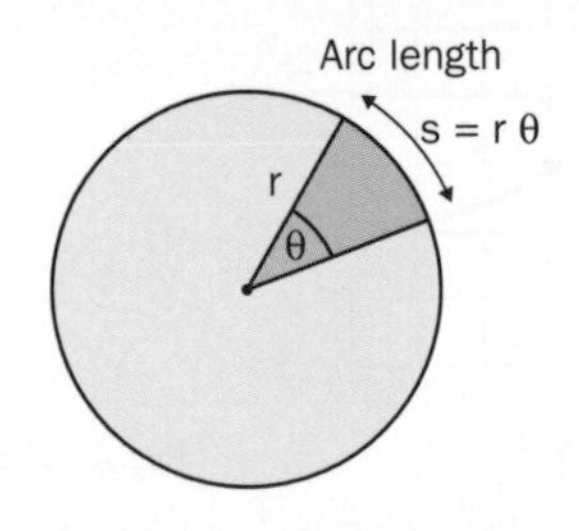

Fig. 26.1 Angles and arcs.

- The **radian** is a unit used to measure angles, defined from $360^\circ = 2\pi$ radians. In other words, the angle all the way around a circle is 2π radians (abbreviation 'rad').
- An **arc** of a circle is the curved edge of a segment of the circle, as shown in Fig. 26.1. The arc subtends an angle, θ, to the centre of the circle. If the angle θ is in radians, the length, s, of the arc is $r\theta$.

Arc length, $s = r\theta$

If the angle θ is small, the segment is almost the same as a triangle in which the length of the smallest side is equal to $r \sin \theta$. It therefore follows that for small angles

$\sin \theta = \theta$ in radians

This is known as the small angle approximation and it holds up to about 10°. This approximation will be used later in this unit. Table 26.1 shows

Table 26.1 Values for sin θ and θ in radians up to 20°

θ in degrees	0.0000	5.0000	10.0000	15.0000	20.0000
θ in radians	0.0000	0.0873	0.1745	0.2618	0.3491
sin θ	0.0000	0.0872	0.1736	0.2588	0.3420

some values of sin θ and θ in radians upto 20°. The difference between sin θ and θ in radians becomes too large to ignore over 10°.

Angular speed

Consider a point object, P, moving on a circular path of radius r.

- The **frequency** of rotation, f, is defined as the number of revolutions per second made by the object.

$$f = \frac{1}{T}$$

where T is the time for 1 rotation. The unit of frequency is the hertz (Hz), equal to 1 revolution per second.

- The **speed** of the object, υ, can be written as

$$\upsilon = \frac{\text{circumference}}{\text{time period}} = \frac{2\pi r}{T}$$

- The **angular speed,** ω, is defined as the angle swept out per second by the radial line OP. Since this line turns through 2π radians in time T,

$$\omega = \frac{2\pi}{T}$$

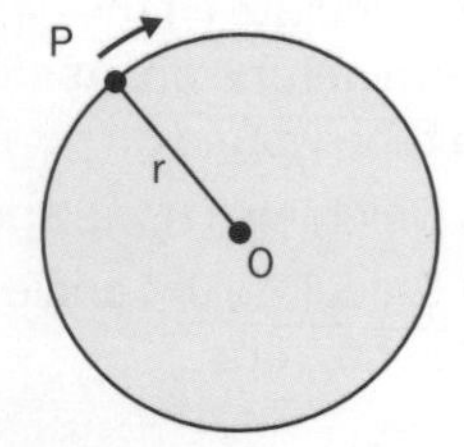

Fig. 26.2 Angular speed.

Notes
1. The unit of ω is the radian per second (rad s^{-1}).
2. The speed, υ, and the angular speed, ω, are related by the equation $\upsilon = \omega r$.

Worked example 26.1

An object on the equator goes round once every 24 hours. Assuming the Earth is a sphere of radius 6350 km, calculate (a) the speed, (b) the angular speed of a point on the equator, (c) the distance moved by such a point in exactly 1 hour.

Solution

(a) $\upsilon = \frac{2\pi r}{T} = \frac{2\pi \times 6.35 \times 10^6 \text{ m}}{24 \times 60 \times 60 \text{ s}} = 462 \text{ m s}^{-1}$

(b) $\omega = \frac{2\pi}{T} = \frac{2\pi \text{ rad}}{24 \times 60 \times 60 \text{ s}} = 7.27 \times 10^{-5} \text{ rad s}^{-1}$

(c) $s = \upsilon t = 462 \times 3600 = 1.66 \times 10^6 \text{ m}$

Questions 26.1

1. The drum of a spin drier has a diameter of 0.40 m and its maximum rate of rotation is 800 revolutions per minute. When it is turning at maximum speed, calculate
(a) the time it takes to turn through one revolution, **(b)** its angular speed, **(c)** the speed of an object on the drum surface.
2. An electric motor rotates at 3000 revolutions per minute. Calculate **(a)** the time taken for 1 rotation, **(b)** the angular speed of the motor.
3. A 2.0 m length of cassette tape is wound on to a spool at a steady rate of rotation of 2 turns per second. The spool diameter is 40 mm.
(a) Calculate **(i)** the circumference of the spool, **(ii)** the number of turns the spool makes to wrap this length of tape on. Assume the number of turns makes no difference to the diameter of the spool with the tape wound on.
(b) Calculate **(i)** the speed of the tape, **(ii)** the time it takes to wind onto the spool.
(c) Calculate the angle between each end of the tape and the centre of the spool **(i)** in radians, **(ii)** in degrees.

26.2 Centripetal acceleration

Speed and velocity

- **Speed** is rate of change of distance with time. Speed is a scalar quantity because it has no direction.
- **Velocity** is speed in a certain direction. Velocity is a vector because it has a direction.

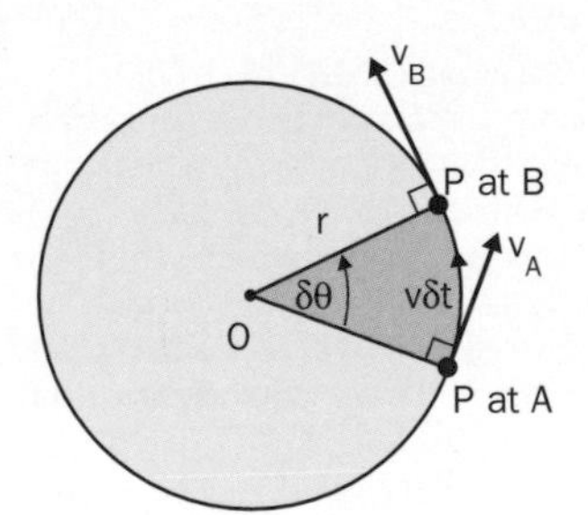

Fig. 26.3 Velocity directions.

For an object moving at constant speed on a circular path, its velocity continually changes as it changes direction. Since acceleration is defined as rate of change of velocity, it follows that an object in uniform circular motion experiences acceleration because its velocity changes direction.

The velocity of an object in uniform circular motion is along a tangent to the circle. As Fig. 26.3 shows, if the object moves around the circle through a certain angle, the velocity direction moves through the same angle.

Consider the object, P, at point A, at time t, and at point B, at time $t + \delta t$, where δt is a small interval of time.

- The radial line, OP, turns through angle $\delta\theta$ in time δt. Hence the angular speed, $\omega = \delta\theta \div \delta t$, which therefore gives $\delta\theta = \omega\delta t$.
- The arc length, $AB = \delta s = \upsilon\delta t$ (using distance = speed × time) $= \omega r\delta t$ since $\upsilon = \omega r$.
- Let the velocity at A be represented by vector υ_A and the velocity at B be represented by vector υ_B. If the two velocity vectors form a triangle as in Fig. 26.4, the shortest side of the triangle represents the change of velocity, $\delta\upsilon = \upsilon_B - \upsilon_A$ which is **towards** the centre of the circle. The length of the two other sides both represent the speed υ.
- The acceleration of the object is referred to as 'centripetal' (which means 'towards the centre'). This is because the acceleration direction is

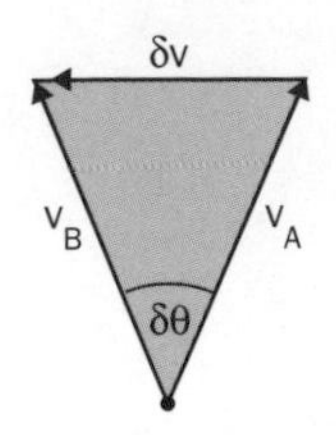

Fig. 26.4 Change of velocity.

towards the centre of the circle since acceleration is change of velocity per unit time.

- Because the velocity vector triangle is similar to triangle AOB (i.e. both have two equal sides and the same angle between the two equal sides), the ratio of the short side to either long side is the same for both triangles, i.e.

$$\frac{\delta v}{v} = \frac{\delta s}{r}$$

$$\text{hence } \delta v = v\frac{\delta s}{r} = v\frac{v\delta t}{r} = \frac{v^2\delta t}{r}$$

Therefore centripetal acceleration, a, can be written as

$$a = \frac{\text{change of velocity } \delta v}{\text{time taken } \delta t} = \frac{v^2}{r}$$

towards the centre of the circle. The equation may be written

$$a = -\frac{v^2}{r}$$

where the minus sign signifies 'towards the centre'.

Since $v = \omega r$, then centripetal acceleration,

$$a = -\frac{v^2}{r} = -\frac{\omega^2 r^2}{r} = -\omega^2 r$$

$$\mathbf{a = -\frac{v^2}{r} = -\omega^2 r}$$

Worked example 26.2

Calculate the centripetal acceleration of a car moving at a constant speed of 12 m s^{-1} around a roundabout of radius 50 m.

Solution

The centripetal acceleration, $a = \frac{v^2}{r} = \frac{12 \times 12}{50}$

$= 2.88$ m s^{-2} towards the centre.

Notes

1. The kinetic energy of an object in uniform circular motion is constant, even though the object is acted upon by the centripetal force. Because the force is towards the centre of the circle and the velocity is tangential, no work is done by the force because there is no movement towards or away from the centre of the circle.

2. If the centripetal force was suddenly removed from an object in uniform circular motion, the object would move off at a tangent to the circle. This is because the velocity at any point is tangential.

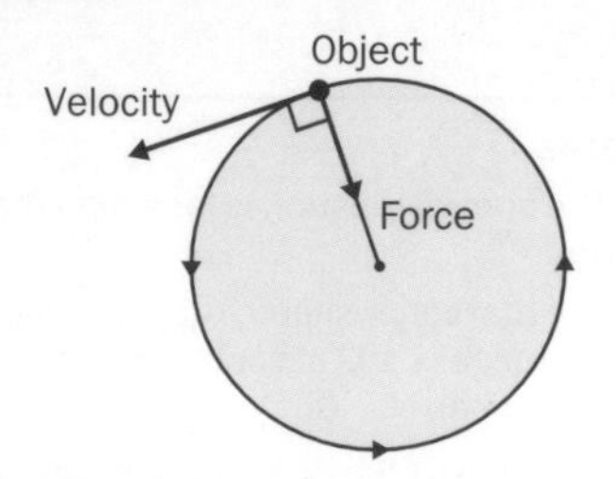

Fig. 26.5 Force and velocity.

Centripetal force

An object in uniform circular motion must be acted on by a resultant force towards the centre of the circle because its acceleration is towards the centre of the circle. The resultant force is referred to as the **centripetal** force because its direction is towards the centre. Using Newton's second law in the form F = ma therefore gives,

$$\textbf{Centripetal force, } \mathbf{F = ma = -\frac{mv^2}{r}}$$

Questions 26.2

1. A vehicle of mass 800 kg moves at a speed of 20 m s^{-1} on a circular path of radius 30 m around a bend. Calculate **(a)** the centripetal acceleration of the vehicle, **(b)** the centripetal force on the vehicle.
2. Calculate the maximum speed at which the vehicle in Q1 could travel safely on this path if the maximum frictional force for no slip between the tyres and the road is 0.5 × the vehicle's weight.
3. The centripetal force on a satellite in circular orbit above the Earth is provided by the force of gravity (i.e. its weight). This can be expressed by means of the equation

$$mg = \frac{mv^2}{r},$$

where m is the mass of the satellite, v is its speed and r is the radius of orbit.

(a) Prove from the above equation that $v = \sqrt{gr}$.

(b) Hence calculate the speed of a satellite in an orbit at a height of 100 km above the Earth's surface, assuming $g = 9.8$ m s^{-2} at this height. The Earth's mean radius is 6400 km.

26.3 Reactions and rides

Fairground rides are designed to take account of the forces acting on the moving parts and on passengers. The structures and materials used must be capable of withstanding the effects of circular motion.

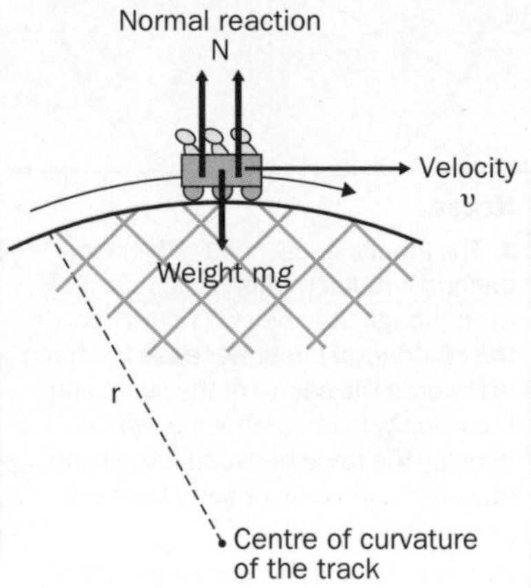

The roller coaster

Over the top

When a carriage travels over a convex section of a track of radius of curvature, r, the normal reaction of the track on the carriage is less than the weight of the carriage. At the highest point, the normal reaction, N, is directed vertically upwards, in the opposite direction to the weight, mg. The resultant force on the carriage is therefore mg − N towards the centre of curvature of the track.

Using F = ma therefore gives

$$mg - N = \frac{mv^2}{r} \text{ since } a = \frac{v^2}{r}$$

towards the centre. Hence the normal reaction

$$N = mg - \frac{mv^2}{r}$$

Normal reaction N, Velocity v, Weight mg, r, Centre of curvature of the track

Fig. 26.6 Over the top.

Notes

1. The normal reaction is less than the weight.

2. The maximum speed, v_{max}, at which the carriage is still in contact with the track is when N = 0.

Hence $\frac{mv_{max}^2}{r} = mg$, giving $v_{max} = \sqrt{gr}$

In a dip

When a carriage travels along a concave section of a track of radius of curvature, r, the normal reaction of the track on the carriage is more than the weight of the carriage. At the lowest point, the normal reaction, N, is directed vertically upwards, in the opposite direction to the weight, mg.

The resultant force on the carriage is therefore N − mg towards the centre of curvature of the track.

Using F = ma therefore gives

$$N - mg = \frac{mv^2}{r} \text{ since } a = \frac{v^2}{r}$$

towards the centre. Hence the normal reaction

$$N = mg + \frac{mv^2}{r}$$

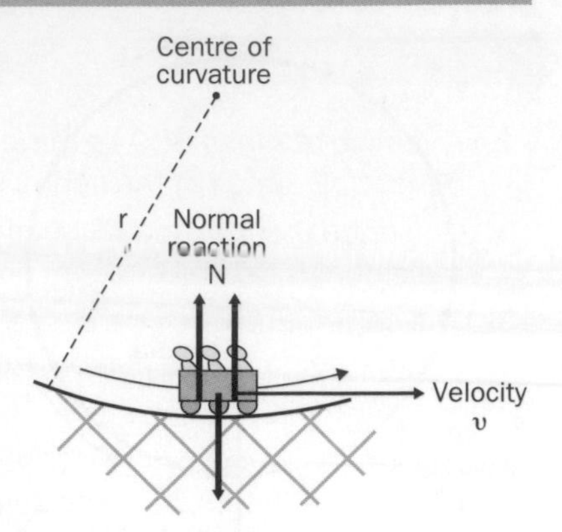

Fig. 26.7 In a dip.

> **Notes**
> 1. The normal reaction is more than the weight.
> 2. The extra '*g*-force' on the carriage is $mv^2 \div r$. A person in the carriage will experience an extra support force equal to $mv^2 \div r$ from the carriage, equivalent to increasing *g* by $v^2 \div r$.

The gravity wheel

The wheel starts rotating with its plane horizontal and the passengers standing around the inside of the rim. When its frequency of rotation is sufficiently high, the wheel axis is tilted gradually until it points in a fixed horizontal direction. The passengers now move on a vertical circle, held against the inside of the wheel's rim by the combined effect of weight and the normal reaction of the wheel.

Fig. 26.8 The gravity wheel.

At the top

The normal reaction, N_1, on each passenger is downwards, in the same direction as the passenger's weight, mg. The resultant force is therefore $N_1 + mg$ towards the centre of rotation. Using F = ma and $a = v^2 \div r$ therefore gives

$$N_1 + mg = \frac{mv^2}{r}$$

where v is the speed of the rim and r is its radius.

$$N_1 = \frac{mv^2}{r} - mg$$

> **Notes**
> 1. For a passenger at the top to remain in contact with the rim, $N_1 > 0$. Therefore the minimum speed, v_{min}, for this condition is given by
> $$\frac{mv_{min}^2}{r} - mg$$
> which gives $v_{min} = \sqrt{gr}$
> 2. If the speed becomes less than $\sqrt{gr}$ while the wheel is vertical, the passengers will fall off the inside of the rim before reaching the top unless strapped in.

Fig. 26.9 At the top.

At the bottom

The normal reaction, N_2, on each passenger is upwards, in the opposite direction to the passenger's weight, mg. The resultant force is therefore

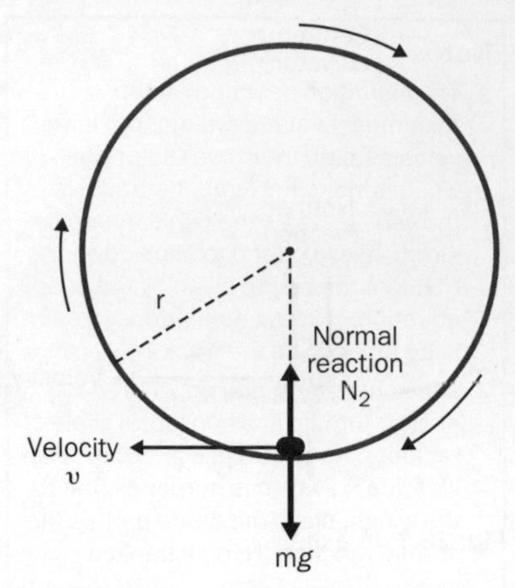

Fig. 26.10 At the bottom.

N_2 − mg towards the centre of rotation. Using F = ma and a = $\upsilon^2 \div r$ therefore gives

$$N_2 - mg = \frac{m\upsilon^2}{r}$$

where υ is the speed of the rim and r is its radius.

$$N_2 = \frac{m\upsilon^2}{r} + mg$$

> **Notes**
>
> **1.** The normal reaction on a passenger at the bottom exceeds the passenger's weight. This extra '*g*-force' is equal to $m\upsilon^2 \div r$, effectively increasing *g* by $\upsilon^2 \div r$.
>
> **2.** The maximum speed of rotation is governed by how much extra '*g*-force' a passenger can withstand without any adverse health effects.

The very long swing

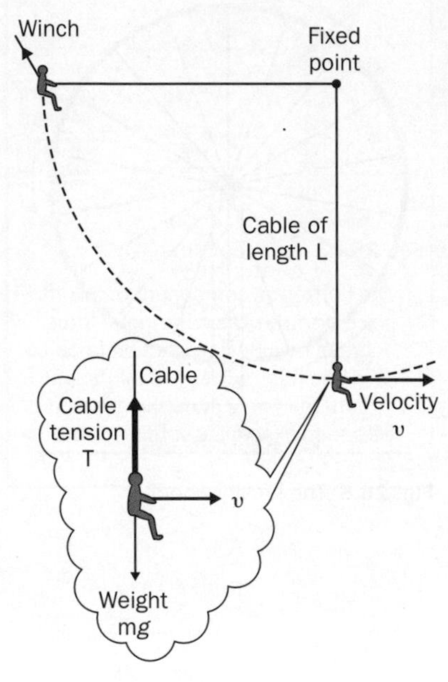

Fig. 26.11 A big swing.

In this ride, one or two passengers are suspended from the end of a long cable and then winched until the cable is almost horizontal. The cable is then released and the passengers swing from side to side with a gradually decreasing amplitude.

- The speed, υ, at the lowest point of a passenger of mass, m, may be calculated by equating the gain of kinetic energy to the loss of potential energy, i.e. $\frac{1}{2} m\upsilon^2 = mgh$, where h is the height drop.
- At the lowest point, the tension, T, in the cable acts directly upwards on the passengers, in the opposite direction to their weight. The resultant force is therefore T − Mg towards the centre of rotation, where M is the total mass suspended from the cable. Using F = ma and a = $\upsilon^2 \div L$ therefore gives

  $$T - Mg = \frac{M\upsilon^2}{L}$$

  where L, the cable length, is equal to the radius of rotation.

Hence the tension in the cable in this position is given by the equation

$$T = Mg + \frac{M\upsilon^2}{L}$$

This is the maximum tension in the cable.

Around the curves

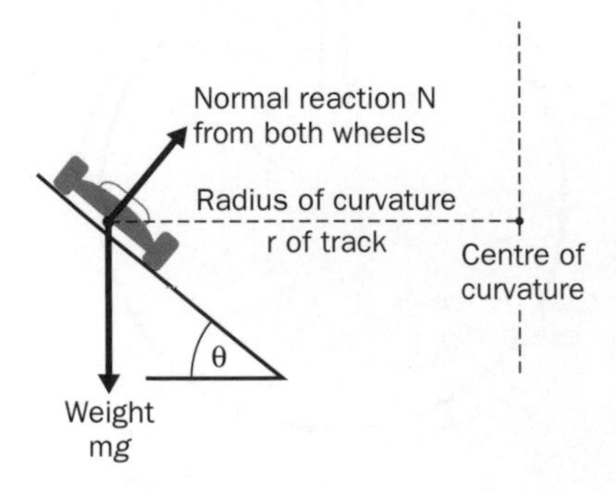

Fig. 26.12 Around a curve.

The track of a roller coaster is banked where it curves horizontally. This is so that the carriages can travel around the curves at high speed without flying off the curve. The same principle is used for rail tracks and motorways.

Fig. 26.12 shows the front view of a vehicle on a banked track of radius of curvature, r. The track is at angle θ to the horizontal. Assuming there is no sideways friction between the vehicle and the track, the horizontal component of the normal reaction, N, of the track on the vehicle, N sin θ, provides the centripetal force on the vehicle. The vertical component of N, N cos θ, is equal and opposite to the weight of the vehicle, mg.

$$N \sin\theta = \frac{mv^2}{r}$$

$$N \cos\theta = mg$$

$$\text{Hence } \tan\theta = \frac{\sin\theta}{\cos\theta} = \frac{mv^2}{mgr} = \frac{v^2}{gr}$$

Notes

1. The condition described here assumes no sideways friction. If the vehicle takes the curve faster, then it would tend to move up the track. If the vehicle takes the curve more slowly,it would tend to slide down the track. At the speed given by the above equation, there is no tendency to slide up or down the track.

2. The same equations apply to an aircraft turning in a horizontal circle of radius, r. As shown in Fig. 26.13, the lift force, L, which is perpendicular to the wings, plays the same part as the normal reaction, N, on a banked track.

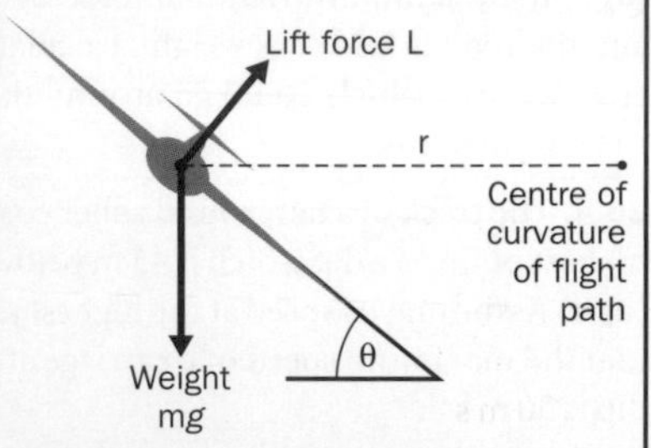

Fig. 26.13 An aircraft banking.

The horizontal component of the lift force provides the centripetal force. There is no sideways friction force as on a banked track so the pilot must ensure the angle of banking matches its speed exactly, according to the above equation.

Questions 26.3

1. A truck of total mass 2000 kg travelling at a speed of 15 m s^{-1}, goes over an arched bridge of radius of curvature 25 m.

(a) Calculate the force of the truck on the bridge when the truck is at the top of the arch.

(b) Calculate the maximum speed at which the truck could pass over the bridge without losing contact with the road.

2. A light aircraft pulls out of a descent, having reached a maximum speed of 95 m s^{-1} at the lowest point of the descent. The radius of curvature of its descent was 450 m. Calculate the centripetal acceleration of the aircraft at the lowest point of the descent and determine the maximum extra '*g*' force suffered by the pilot.

Summary

- Arc length, $s = r\theta$, where θ is in radians
  $360° = 2\pi$ radians
- Angular speed, $\omega = \frac{2\pi}{T} = 2\pi f$
- Centripetal acceleration, $a = \frac{-v^2}{r} = -\omega^2 r$
- Centripetal force, $F = ma = \frac{-mv^2}{r} = -m\omega^2 r$

Revision questions

26.1. A satellite moves around the Earth once every 2 hours in a circular orbit of radius 8000 km. Calculate **(a)** its angular speed, **(b)** its speed, **(c)** its centripetal acceleration.

26.2. The armature coil of an electric motor has a diameter of 0.08 m. When it rotates at a frequency of 50 Hz, calculate **(a)** its angular speed, **(b)** its speed, **(c)** its centripetal acceleration.

26.3. A vehicle of mass 1200 kg travels around a roundabout of radius 60 m.

(a) Calculate the centripetal acceleration of the vehicle when it travels at a speed of 15 m s^{-1} around the roundabout.

(b) If the maximum frictional force between the vehicle tyres and the road is 0.5 × its weight, calculate the maximum speed at which the vehicle could go around the roundabout without sliding outwards.

26.4. The track of a fairground roller coaster descends from the highest point to a dip which is 45 m below.

(a) (i) Assuming its speed at the highest point is negligible, show that the maximum speed of a carriage at the lowest point of the dip is 30 m s^{-1}.

(ii) What other assumption is it necessary to make in (a)(i)?

(b) At the dip, the track is concave in shape and has a radius of curvature of 25 m.

(i) Calculate the centripetal acceleration of a carriage passing through the dip at a speed of 30 m s^{-1} at the lowest point.

(ii) Calculate the ratio of the centripetal force to the weight when the carriage is at the lowest point of the dip.

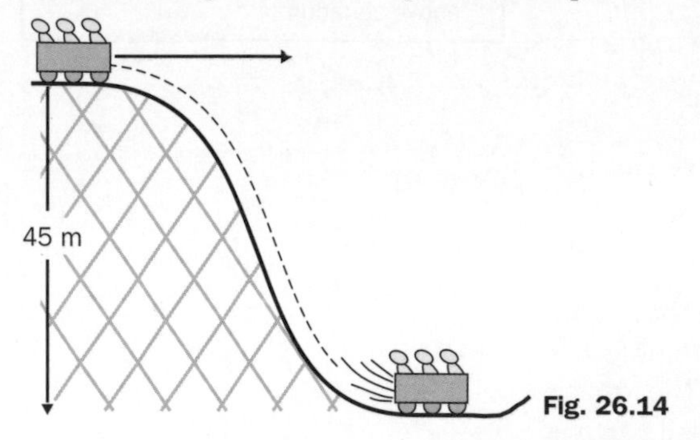

Fig. 26.14

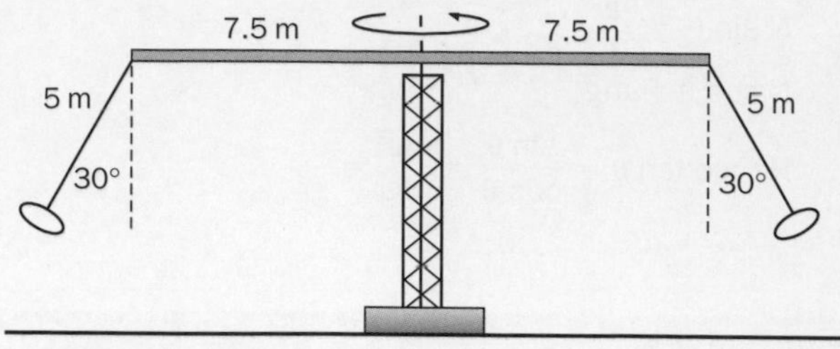

Fig. 26.15

26.5. A fairground roundabout consists of eight, two-seat 'aircraft', each suspended from a steel cable of length 5.0 m which is attached at its upper end to the end of a horizontal steel beam. Each beam extends 7.5 m from a vertical steel pillar which is along the axis of rotation of the roundabout. When the roundabout rotates at its maximum frequency of rotation, the steel cables are at an angle of 30° to the vertical.

(a) Calculate the radius of rotation of each 'aircraft' at this angle.

(b) Show that the speed of each aircraft, when the cable is at this angle, is 7.5 m s^{-1}.

(c) Calculate the time taken for one rotation at this speed.

26.6. (a) An aircraft travelling at a constant speed of 210 m s^{-1} travels horizontally on a circular path of radius 6400 m. Calculate the angle of banking of its wings when it turns in this way.

(b) A different aircraft performs a vertical loop of diameter 500 m.

(i) At its highest point, why is the direction of the lift force on its wings downwards?

(ii) If the lift force at its highest point in the loop is to equal its weight, show that its speed at the top of the loop must be 70 m s^{-1}.

Gravitation

Objectives

After working through this unit, you should be able to:

- ▶ explain the cause of an object's weight
- ▶ state and use Newton's law of gravitation to calculate the force between two masses
- ▶ define gravitational field strength and state its unit
- ▶ relate acceleration due to gravity, *g*, to gravitational field strength
- ▶ describe the variation of *g* with distance from the centre of a spherical planet
- ▶ calculate the potential energy of an object in a gravitational field
- ▶ explain what is meant by escape speed and calculate the escape speed from a planet
- ▶ explain the motion of a satellite in a circular orbit and relate its time period to the radius of its orbit
- ▶ carry out calculations relating the time period of a satellite to the radius of its orbit

Contents

27.1 Newton's theory of gravity

Release an object from rest above the ground and it falls down, pulled by the Earth's gravity. The Earth attracts the object, so unless the object is supported above the ground, it moves towards the Earth. The object attracts the Earth with an equal and opposite force but because the Earth is so much more massive, the Earth is not moved by the object.

Gravity acts between any two objects, causing a force of attraction which tries to pull the two objects together. The universal nature of the force of gravity, acting between any two objects, was first realised by Isaac Newton in the 17th century. He devised a mathematical theory of gravity which he used to explain a wide range of phenomena, including the motion of the planets and why comets reappear. He was able to use the theory to make many predictions, including the times of forthcoming tides, eclipses and reappearance of comets.

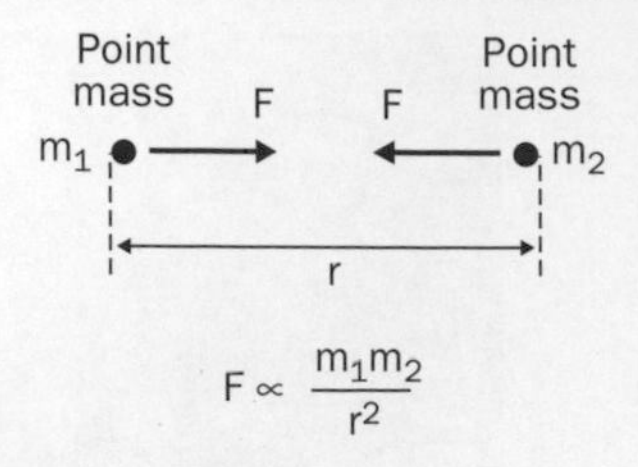

Fig. 27.1 Newton's law of gravitation.

Newton's theory of gravitation states that a force of attraction due to gravity exists between any two objects and that the force of gravity between two point objects is

1. proportional to the product of the masses of the objects,

2. inversely proportional to the square of their distance apart.

Thus for two point masses, m_1 and m_2, at distance, r, apart, the force of attraction, F, can be written as

$$F \propto \frac{m_1 m_2}{r^2}$$

This proportionality relationship is made into an equation by introducing a constant of proportionality, G, the Universal Constant of Gravitation. This equation, as stated below, is referred to as Newton's law of gravitation.

$$F = G\frac{m_1 m_2}{r^2}$$

where F = the force of gravitational attraction and r = the distance between point masses m_1 and m_2.

Notes

1. Newton extended his theory to objects which are not points. He showed that the equation gives the force of gravity between any two objects, provided r is the distance between their centres of gravity.
2. The equation is an example of an inverse square law of force in which the force is proportional to the inverse of the square of the distance. For example, if the distance is doubled, the force decreases by a factor of one quarter.
3. The unit of G is the newton metre^2 kilogram^{-2}, corresponding to $G = Fr^2 \div m_1 m_2$ from the equation. Note that mass must be in kilograms and distances in metres when using Newton's law of gravitation.
4. The value of G is 6.67×10^{-11} N m^2 kg^{-2}. This was first determined by Henry Cavendish in the 18th century. He made a torsion balance consisting of a horizontal rod which carried a small metal ball at either end, as shown in Fig. 27.2. The rod was suspended by a thin wire. Two large lead spheres were brought near either end of the rod so as to attract the metal balls and make the rod twist.

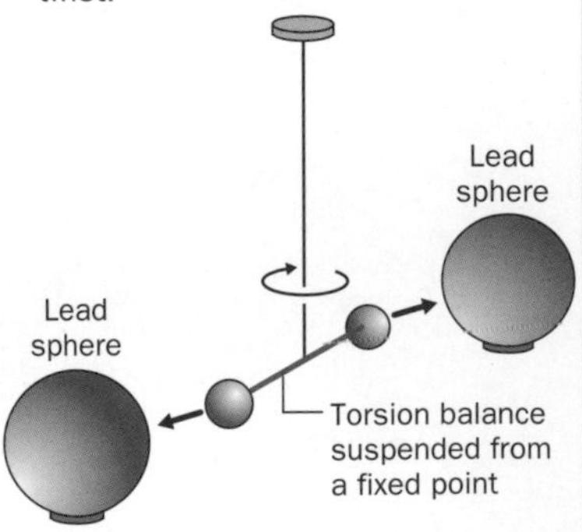

Fig. 27.2 Measurement of G.

Worked example 27.1

Calculate the force of attraction between a lead sphere of mass 2.00 kg and (a) an object of mass 6.00 kg when the centres of gravity are 0.100 m apart, (b) the Earth when the object is on the surface of the Earth. The Earth has a mass of 5.98×10^{24} kg and a radius of 6380 km. $G = 6.67 \times 10^{-11}$ N m^2 kg^{-2}

Solution

(a) $F = G\frac{m_1 m_2}{r^2} = \frac{6.67 \times 10^{-11} \times 2.00 \times 6.00}{(0.100)^2} = 8.00 \times 10^{-8}$ N

(b) $F = G\frac{m_1 m_2}{r^2} = \frac{6.67 \times 10^{-11} \times 2.00 \times 5.98 \times 10^{24}}{(6380 \times 10^3)^2} = 19.6$ N

The nature of gravity

The force of gravitational attraction exists between any two objects unless they are without mass. Each object is surrounded by a gravitational field due to its own mass. Any other object in this field experiences a force due to the field. For example, the gravitational field of the Earth causes an object at rest on the Earth's surface to be held on the Earth.

The Earth's gravitational field stretches to infinity. According to Newton's Law of Gravitation, no matter how far apart two objects are, the force of gravity between them is never zero. The force of gravitational attraction between any two objects becomes weaker and weaker as the two objects move further and further apart. The force tends to zero as the distance apart becomes larger and larger, but is nevertheless non-zero except at infinite separation. Thus it is not possible, in theory, to escape from the Earth's gravitational field or the gravitational field of any other object,

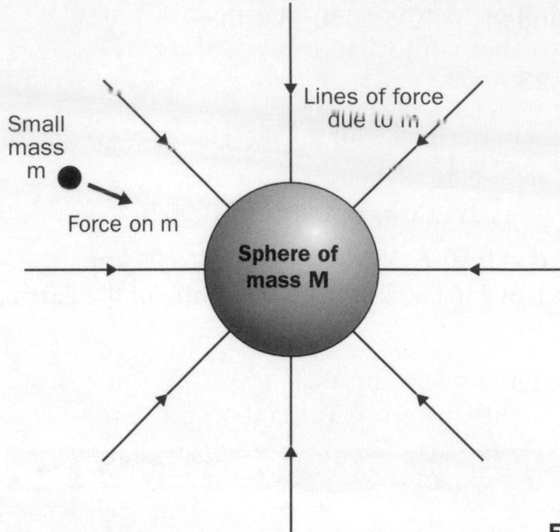

Fig. 27.3 In a gravitational field.

although in practice the force would eventually become negligibly small if the distance from the Earth became ever larger.

Zero gravity

The Earth's gravity stretches to infinity, but it isn't necessary to go to infinity to experience zero gravity. An object at any point along the line between the Earth and the Moon would experience a force due to the Earth and a force in the opposite direction due to the Moon. At one particular point along this line, as shown in Fig. 27.4, the two forces are equal and opposite.

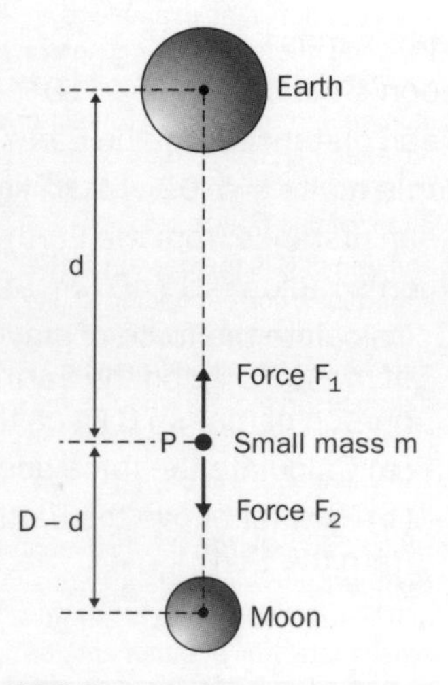

Fig. 27.4 Zero gravity.

Let D represent the distance between the centre of the Earth and the centre of the Moon. Let d represent the distance from the centre of the Earth to point P, the position where the force due to the Earth is equal and opposite to the force due to the Moon.

The force, F_1, due to the Earth on a small object of mass, m, at P, can be written as

$$F_1 = \frac{GM_1m}{d^2}$$

where M_1 = the mass of the Earth. The force, F_2, due to the Earth on a small object of mass, m, at P, can be written as

$$F_2 = \frac{GM_2m}{(D-d)^2}$$

where M_2 = the mass of the Moon. Since $F_1 = F_2$ at point P, then

$$\frac{GM_1m}{d^2} = \frac{GM_2m}{(D-d)^2}$$

Hence

$$\frac{(D-d)^2}{d^2} = \frac{M_2}{M_1}$$

Note

A spring balance may be used to measure the force of gravity on an object. If a spring balance supporting an object is released so the spring balance and the object fall freely, the reading on the spring balance becomes zero as soon as it is released. The object is unable to exert a force on the spring balance when they are both in free fall. In effect, this is the same as if the force of gravity due to the Earth vanished. An object in free fall above the Earth is therefore effectively in zero gravity.

Since $M_1 = 5.98 \times 10^{24}$ kg and $M_2 = 7.35 \times 10^{22}$ kg, then

$$\frac{M_2}{M_1} = \frac{7.35 \times 10^{22}}{5.98 \times 10^{24}} = 1.23 \times 10^{-2}$$

$$\therefore \frac{D-d}{d} = \sqrt{(1.23 \times 10^{-2})} = 0.11$$

$\therefore D-d = 0.11\,d$, which gives $1.11\,d = D$, thus $d = 0.90\,D$.

Since D = 384 000 km, then d = 0.90 × 384 000 km = 346 000 km.
Point P is therefore a distance of 346 000 km from the centre of the Earth.

Questions 27.1

$G = 6.67 \times 10^{-11}$ N m^2 kg^{-2}

Use the following data in the questions below:

Earth's mass = 5.98×10^{24} kg, Earth's radius = 6380 km.

Moon's mass = 7.35×10^{22} kg.

Mean distance from the Earth to the Moon = 384 000 km.

Sun's mass = 1.99×10^{30} kg.

Mean distance from the Earth to the Sun = 150 000 000 km.

Moon's radius = 1740 km, Sun's radius = 696 000 km.

1. Calculate the force of gravitational attraction between **(a)** the Earth and a person of mass 70 kg on the Earth, **(b)** the Earth and the Moon, **(c)** the Moon and a person of mass 70 kg on the Earth.
2. **(a)** Calculate the force due to the Sun on an object of mass 1.0 kg on the Earth.
   (b) How far above the Earth would an object need to be to experience an equal and opposite force from the Sun?

Fig. 27.5

Notes

1. The object must be small enough not to affect the distribution of mass that creates the field.
2. Gravitational field strength is measured in newtons per kilogram. For example, the gravitational field strength of the Earth on its surface is 9.8 N kg^{-1}.
3. Gravitational field strength is a vector quantity because it has magnitude and direction.

27.2 Gravitational field strength

A gravitational field is a region of space where an object experiences a force due to its mass. For example, the gravitational field of the Earth causes a force on any object near the Earth. The gravitational field of the Earth is due to its own mass and the field acts on any other mass.

The strength of a gravitational field, *g*, is defined as the force per unit mass on a small object placed in the field.

$$\textbf{Gravitational field strength } g = \frac{F}{m}$$

where F = force of gravity on a small object of mass m.

The weight of an object is the force of gravity on it.

For an object of mass m, its weight, W, is therefore given by

$$W = mg$$

where g is the gravitational field strength at the position of the object.

For example, a person of mass 70 kg would weigh approximately 700 N on the surface of the Earth since $g = 9.8\ \text{N kg}^{-1}$ at the Earth's surface. However, the same person on the Moon, where the gravitational field strength is just $1.6\ \text{N kg}^{-1}$, would weigh just 112 N ($= 1.6\ \text{N kg}^{-1} \times 70\ \text{kg}$).

- An **object in free fall** above the Earth accelerates due to the force of gravity on it. The acceleration of free fall is equal to g, the gravitational field strength. This is because acceleration = force ÷ mass = $mg \div m = g$.
- A **line of force** of a gravitational field is the line along which a small mass would move if it were free to move. Fig. 27.6 shows the lines of force surrounding a sphere. The lines point towards the centre of the sphere because any small mass in the field would be attracted towards the centre of the sphere. This is an example of a radial field because the lines of force point to the centre.

The force, F, on a small mass, m, at distance, r from the centre of a sphere of mass M is given by Newton's Law of Gravitation,

$$F = \frac{GMm}{r^2}$$

Since gravitational field strength, g, is defined as force per unit mass, then

$$g = \frac{F}{m} = \frac{GM}{r^2} \text{ at distance, r, from the centre of the sphere.}$$

$$\boldsymbol{g = \frac{GM}{r^2}} \text{ at distance, r, from the centre of a sphere of mass, M.}$$

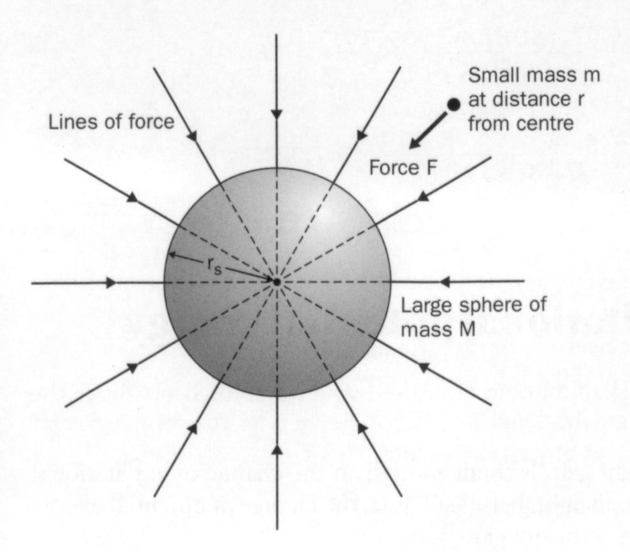

Fig. 27.6 Lines of force.

Notes

1. The formula for g is the same as if the sphere's mass, M, was concentrated at the centre of the sphere.
2. The gravitational field strength at the surface of the sphere, g_s, can be written as

$$g_s = \frac{GM}{r_s^2}$$

where r_s is the radius of the sphere.
3. The variation of g with distance, r, is according to the inverse square law. Fig. 27.7 shows how g decreases with increasing distance, r.

Since $GM = g_s r_s^2$, then $g = \frac{g_s r_s^2}{r^2}$

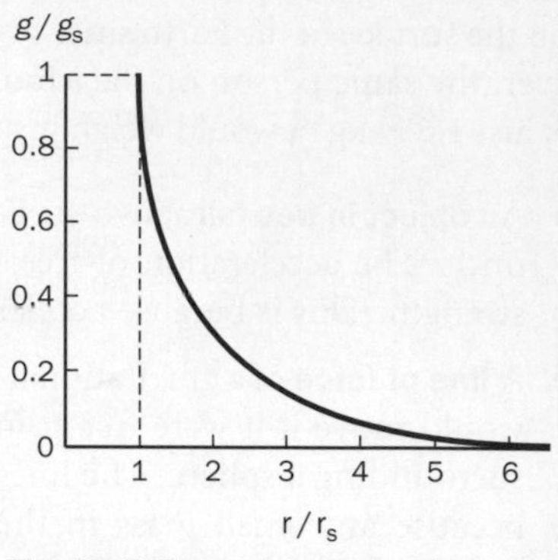

Fig. 27.7 The variation of g with distance from the centre of a spherical object.

Questions 27.2

Use the data in the question box on p. 400 for the following questions.

1. Calculate the gravitational field strength due to the Sun **(a)** at the surface of the Sun, **(b)** at the surface of the planet Mercury, a distance of 57.9 million kilometres from the Sun.
2. Jupiter has a mass of 1.90×10^{27} kg and moves around the Sun at a distance of 778 million kilometres. Calculate the distance from the Sun at which the gravitational field strength due to the Sun is equal and opposite to the gravitational field strength due to Jupiter.

Worked example 27.2

$G = 6.67 \times 10^{-11}$ N kg^{-2} m^2

(a) Show that the gravitational field strength of the Earth on its surface is 9.8 N kg^{-1}.

(b) Calculate the gravitational field strength of the Earth at a height of 1000 km above the Earth.

Earth's mass = 5.98×10^{24} kg, Earth's radius = 6380 km.

Solution

(a) $g_{surface} = \frac{GM}{r_E^2} = \frac{6.67 \times 10^{-11} \times 5.98 \times 10^{24}}{(6380 \times 10^3)^2} = 9.80 \text{ N kg}^{-1}$

(b) r = 6380 + 1000 km = 7380 km

$\therefore g = \frac{GM}{r^2} = \frac{6.67 \times 10^{-11} \times 5.98 \times 10^{24}}{(7380 \times 10^3)^2} = 7.32 \text{ N kg}^{-1}$

27.3 Gravitational potential energy

The potential energy of an object is raised when it is lifted up above the Earth's surface.

- Provided its height gain is small enough so the change of gravitational field strength over height, h, is negligible, thc change of potential energy = force of gravity × height gain = mgh.

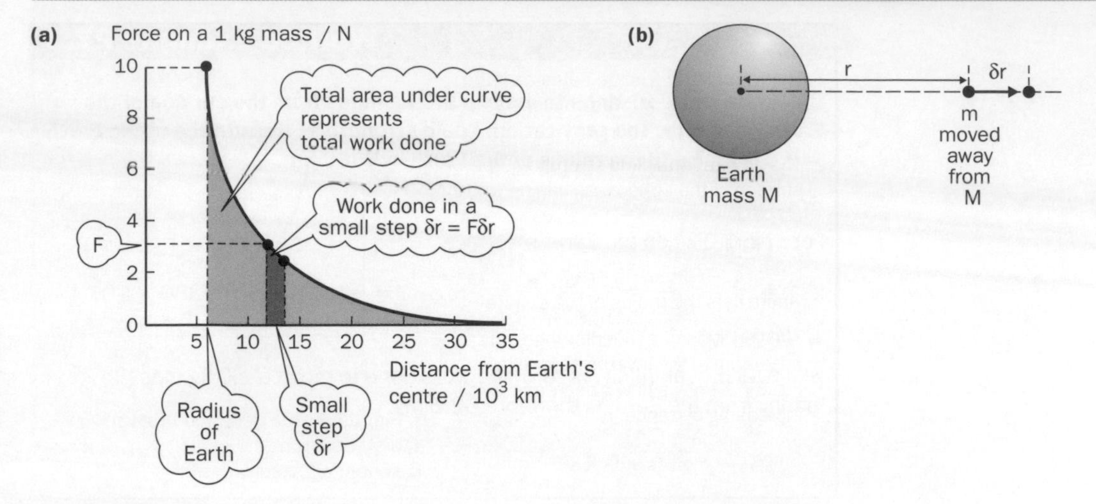

Fig. 27.8 Doing work **(a)** Graph of force versus distance near the Earth, **(b)** gaining potential energy.

- If the change of gravitational field strength is not negligible, the force of gravity on an object decreases with height. Fig. 27.8 shows how the force, F, varies with distance, r, from the centre of the Earth, in accordance with Newton's Law of Gravitation.

Consider an object of mass m that moves away from the centre of the Earth from distance r to distance r + δr. Provided δr is much less than r, the force of gravity on the object, F, can be written as

$$F = mg = \frac{GMm}{r^2}$$

where M is the mass of the Earth. The change of potential energy of the object, δ (p.e.) can be written as

$$\delta\,(\text{p.e.}) = F\delta r = \frac{GMm}{r^2}\,\delta r$$

However

$$\frac{1}{r} - \frac{1}{r + \delta r} = \frac{r + \delta r - r}{r\,(r + \delta r)} = \frac{\delta r}{r^2}$$

when δr is much less than r. Therefore the change of potential energy

$$= \frac{GMm}{r} - \frac{GMm}{r + \delta r}$$

Since the change of potential energy = the potential energy at r + δr − the potential energy at r, the potential energy at $r = -\frac{GMm}{r}$ and the potential energy at r + δr can be written as $-\frac{GMm}{(r + \delta r)}$

$$\textbf{Potential energy} = \frac{-GMm}{r}$$

where r = the distance between the centres of mass.

Notes

1. The potential energy varies as the inverse of the distance, r. As this distance increases, the potential energy becomes less negative and tends to zero at infinity.
2. The work done to move a mass, m, from distance, r, to infinity is equal to its change of potential energy from $-GMm/r^2$ at distance r to zero at infinity.
3. The area under the curve in Fig. 27.8a represents the work done to move mass, m, from distance, r, to infinity. This is because the work done in a small step, δr, is equal to Fδr, the area of a strip under the curve. So the total work done is represented by the area of all the strips.
4. Since the surface gravitational field strength, $g_S = GM \div r_S^2$, where r_s is the radius of the Earth, then $GM = g_S\, r_S^2$, hence the potential energy at r can be written as $\frac{-mg_S r_S^2}{r}$

Worked example 27.3

Calculate the work done to move a 1.0 kg mass from the surface of the Earth to infinity. The gravitational field strength at the surface of the Earth = 9.8 N kg^{-1} and the radius of the Earth = 6380 km.

Solution

For m = 1.0 kg on the Earth's surface,

its p.e. $= \frac{-GMm}{r_S} = -\frac{mg_S r_S^2}{r_S} = -mg_S r_S = -1.0 \times 9.8 \times 6380 \times 10^3$

$= -62.5$ MJ

Since its p.e. at infinity is zero, the work done to move a 1.0 kg mass to infinity from the Earth is therefore 62.5 MJ.

Escape speed

When a rocket is launched into space from the Earth, it is given sufficient kinetic energy to enable it to overcome the Earth's gravity. If the rocket does not gain sufficient kinetic energy, it falls back to the Earth's surface.

The escape speed, v_{esc}, of a projectile is the minimum speed it must be given to escape from the Earth to infinity.

For a projectile of mass, m, to escape, its initial speed, v, must be large enough so that its initial kinetic energy, $\frac{1}{2}mv^2$, exceeds GMm/r_s, the work done to escape from the surface to infinity. Hence

$$\frac{1}{2}mv_{esc}^2 = \frac{GMm}{r_S}$$

$$\therefore \text{ the escape speed, } v_{esc} = \sqrt{\left(\frac{2GM}{r_S}\right)} = \sqrt{(2g_S r_S)}$$

Black holes

A black hole is a massive object which has a gravitational field so strong that not even light can escape. Since no object can travel faster than light, it follows that nothing can escape from inside a black hole.

The **event horizon** of a black hole is the closest an object can approach the black hole without becoming trapped.

The escape speed from the surface of a sphere of mass, M, and radius, R, is given by the equation

$$v_{esc}^2 = \frac{2GM}{R}$$

Since the escape speed cannot exceed c, the speed of light in free space, the sphere would be a black hole if 2GM ÷ R exceeds c^2. In other words, for a sphere to be a black hole, its radius must be less than or equal to 2GM ÷ c^2.

Worked example 27.4

Show that a sphere of mass equal to the mass of the Earth ($= 5.98 \times 10^{24}$ kg) would need have a radius of 9 mm to be a black hole.
$c = 3.0 \times 10^8$ m s^{-1}, $G = 6.67 \times 10^{-11}$ N kg^{-2} m^2

Solution

$$R = \frac{2GM}{c^2} = \frac{2 \times 6.67 \times 10^{-11} \times 5.98 \times 10^{24}}{(3.0 \times 10^8)^2} = 8.9 \times 10^{-3}\text{m}$$

Questions 27.3

Use the data in the question box on p. 400 for the following questions.

1. Calculate the work needed to move a 1.0 kg mass from **(a)** the Moon's surface to infinity, **(b)** the Sun's gravitational field at a point on the Earth's orbit to infinity.
2. Calculate the radius to which the Sun would need to be compressed to become a black hole.
   $c = 3.0 \times 10^8$ m s^{-1}.

27.4 Satellite motion

Consider a satellite in a circular orbit around the Earth as in Fig. 27.9. The centripetal force needed to keep the satellite on its circular path is provided by the force of gravity on the satellite due to the Earth.

If the force of gravity suddenly vanished, the satellite would fly off at a tangent. A satellite in a circular orbit moves around the Earth at constant speed because the force on it is perpendicular to its direction of motion and therefore no work is done on the satellite.

The direction of motion of the satellite changes continuously as it moves around the Earth. At any point on its orbit, its velocity is tangential and its acceleration is towards the centre, in the same direction as the force of gravity on the satellite.

For a satellite of mass, m, moving at steady speed, υ, on a circular orbit of radius, r, about the Earth, its acceleration is given by the centripetal acceleration formula, $a = \upsilon^2 \div r$, from p. 391. The force on the satellite is given by Newton's law of gravitation, $F = GMm/r^2$, where M is the mass of the Earth. Using Newton's second law of motion,

$$F = ma$$

$$\therefore \frac{GMm}{r^2} = \frac{m\upsilon^2}{r}$$

Rearranging this equation gives

$$\upsilon^2 = \frac{GM}{r}$$

$$\text{Since } \upsilon = \frac{\text{circumference}}{\text{time for 1 complete orbit, T}} = \frac{2\pi r}{T}$$

$$\text{then } \frac{(2\pi r)^2}{T^2} = \frac{GM}{r}$$

$$\text{Hence } \mathbf{r^3 = \frac{GM}{4\pi^2} T^2}$$

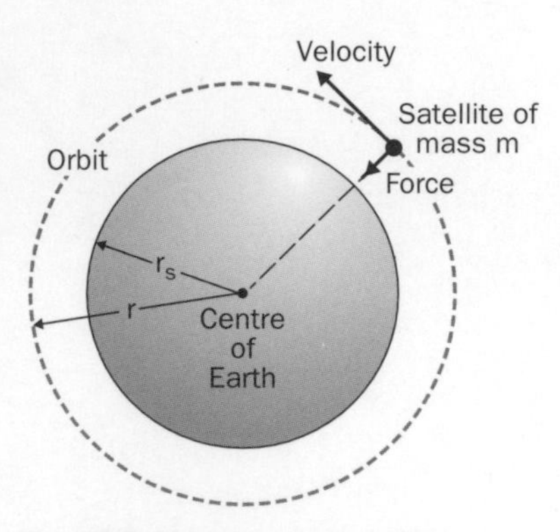

Fig. 27.9 The orbit of a satellite.

Notes

1. The equation opposite applies to any situation where a satellite is in orbit about a much more massive spherical body of mass, M.
2. Given values of r and T, the mass, M, of the central body can be calculated.
3. Since the surface gravitational field strength, $g_S = GM \div r_S^2$, where r_S is the radius of the central body, then the equation can be written as
   $$r^3 = \frac{g_S r_S^2}{4\pi^2} T^2$$

Worked example 27.5

$G = 6.67 \times 10^{-11}$ N m^2 kg^{-2}, radius of the Earth = 6380 km

The first artificial satellite was launched into an orbit 200 km above the Earth, moving around the Earth once every 88 minutes. Use this data to calculate the mass of the Earth.

Solution

Rearrange

$$r^3 = \frac{GM}{4\pi^2}T^2 \text{ to give } M = \frac{4\pi^2 r^3}{GT^2} = \frac{4\pi^2 (6580 \times 10^3)^3}{6.67 \times 10^{-11} \times (88 \times 60)^2} = 6.0 \times 10^{24} \text{ kg}$$

Kepler's Laws of Planetary Motion

Johannes Kepler was a 16th century astronomer who observed and measured the motion of the planets and established three laws describing their motion. The same laws also describe the motion of comets in the Solar System.

- **Keplers's 1st law** states that each planet moves around the Sun on an elliptical orbit.

  Fig. 27.10 shows an elliptical orbit. The Sun is at one of the two 'focal points' of the ellipse. The degree of ellipticity of the orbit of each planet differs from an almost circular orbit for the Earth to the orbit for Pluto, the outermost planet. Pluto's orbit is so elliptical that it moves closer to the Sun than its nearest neighbour, Neptune, for part of its orbit.

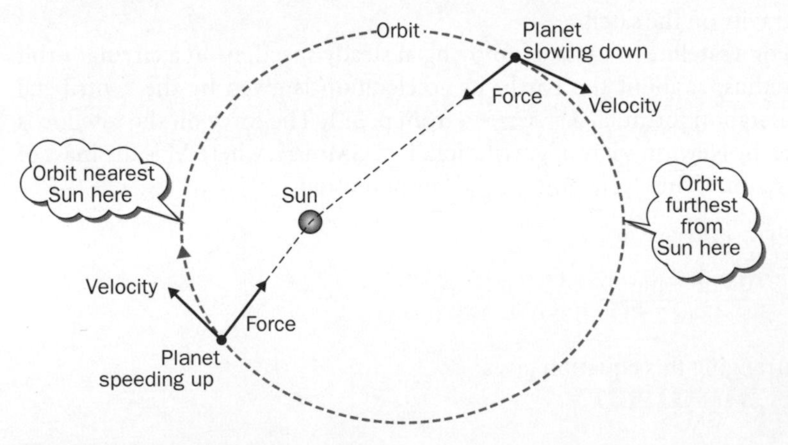

Fig. 27.10 An elliptical orbit.

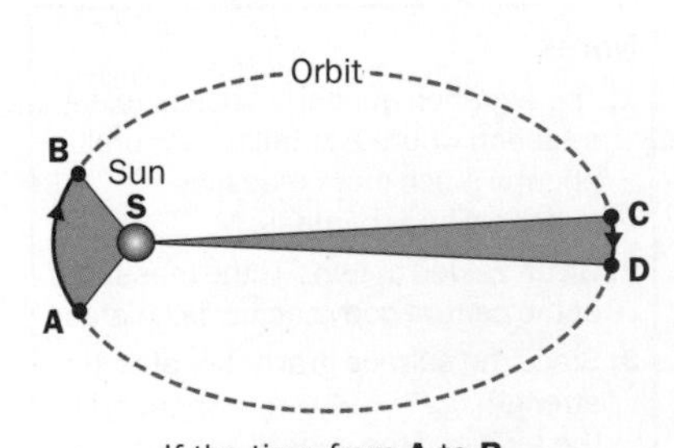

Fig. 27.11 Kepler's second law.

- **Kepler's 2nd law** states that the line from the Sun to a planet sweeps out equal areas in equal times as the planet moves around the Sun. Fig. 24.11 shows the idea. A comet moves faster as it approaches the Sun because the force of gravity on it increases. It slows down as it moves away from the Sun because the force of gravity decreases.

- **Kepler's 3rd law** states that the cube of the mean distance from a planet to the Sun is proportional to the square of the time it takes to move around the Sun. See Table 24.1. Newton devised the theory of gravitation to explain Kepler's 3rd law.

Note
One astronomical unit (A.U.) is defined as the mean distance from the Earth to the Sun and is equal to 1.496×10^8 km.

Table 27.1 Planetary data

	Mercury	Venus	Earth	Mars	Jupiter	Saturn	Uranus	Neptune	Pluto
Mean distance from Sun, r, in A.U.	0.39	0.72	1	1.52	5.2	9.54	19.2	30.1	39.4
Time taken for each complete orbit, T, in years	0.24	0.62	1	1.88	11.9	29.5	84	165	248
r^3/T^2 (in A.U.^{3} yr^{-2})	1	1	1	1	1	1	1	1	1

Geostationary satellites

A geostationary satellite is a satellite that takes exactly 24 hours to go around a circular orbit directly above the equator. Such a satellite is always directly above the same point of the equator as it moves around at the same rate as the Earth spins.

Geostationary satellites are used for communications and satellite TV broadcasting because transmitters and receivers on the Earth do not need to be realigned once they have been pointed towards the satellite.

The height, H, of a geostationary satellite can be calculated using the equation

$$r^3 = \frac{g_S r_S^2}{4\pi^2} T^2$$

from p. 405, where g_S is the surface gravitational field strength of the Earth and r_S is the radius of the Earth. Substituting $T = 24$ hours $= 24 \times 3600$ s, $g_S = 9.8$ N kg^{-1} and $r_S = 6380 \times 10^3$ m, gives

$$r^3 = \frac{g_S r_S^2}{4\pi^2} T^2 = \frac{9.8 \times (6380\times10^3)^2 \times (24\times3600)^2}{4\pi^2} = 7.54\times10^{22}\ \text{m}^3$$

Hence $r = 4.225 \times 10^7$ m $= 42\,250$ km,

$\therefore H = 42\,250 - 6380 = 35\,870$ km

Questions 27.4

Use the data in the question box on p. 400 for the following questions.

1. A satellite orbits the Earth in a circular orbit.
   (a) Calculate its speed and its time period if its height is 1000 km.
   (b) Calculate its speed and height if its time period is exactly 4 hours.
2. **(a)** Calculate the time period of a satellite in circular orbit around the Moon at a height of 100 km.
   (b) Calculate the height and time period of a satellite moving at a speed of 1.0 km s^{-1} in a circular orbit around the Moon.

Summary

◆ **Newton's law of gravitation** $F = \frac{G m_1 m_2}{r^2}$

◆ **Gravitational field strength** $g = \frac{F}{m}$

◆ **At distance, r, from the centre of a spherical mass, M**

1. $g = \frac{GM}{r^2}$

2. PE of a small mass, m $= \dfrac{-GMm}{r}$

3. Escape speed $= \sqrt{\dfrac{2GM}{r}} = \sqrt{(2gr)}$

◆ **For a satellite in a circular orbit round a spherical body of mass, M**

1. Its speed, $\upsilon = \sqrt{\left(\dfrac{GM}{r}\right)} = \dfrac{2\pi r}{T}$

2. $r^3 = \dfrac{GM\,T^2}{4\pi^2}$ where r = radius of orbit and T = time per orbit.

Revision questions

$G = 6.67 \times 10^{-11}$ N m^2 kg^{-2}, 1 A.U. = 150×10^6 km

Use the following data, where appropriate, in the questions below:

Earth's mass = 5.98×10^{24} kg, Earth's radius = 6380 km, g_S = 9.80 N kg^{-1}

Moon's mass = 7.35×10^{22} kg, mean distance from the Earth to the Moon = 384 000 km

Sun's mass = 1.99×10^{30} kg, mean distance from the Earth to the Sun = 150 000 000 km

Moon's radius = 1740 km, Sun's radius = 696 000 km

27.1. When a solar eclipse occurs, the Moon is directly between the Sun and the Earth, as in Fig. 27.12.

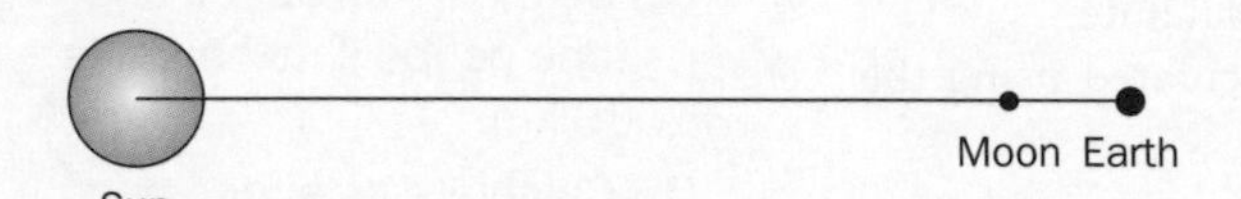

Fig. 27.12

(a) Calculate the force on the Moon due to **(i)** the Earth, **(ii)** the Sun.
(b) Hence calculate the overall force on the Moon when it is directly between the Sun and the Earth.

27.2. The mass of Jupiter is 318 × the mass of the Earth and its mean distance from the Sun is 5.20 A.U.
(a) Calculate the gravitational field strength of Jupiter at a distance of 1 A.U. from the Sun.
(b) When Jupiter is at opposition, it lies in the opposite direction to the Sun.
(i) Show that the gravitational field strength of Jupiter is equal and opposite to that of the Earth at a distance of 0.22 A.U. from the Earth.
(ii) Calculate the gravitational field strength of the Sun at this position.

27.3. (a) (i) The potential energy per unit mass of an object on the Earth is −62.5 MJ kg^{-1} due to the Earth's gravity. Calculate the escape speed from the Earth.
(ii) The potential energy per unit mass of an object on the surface of the Moon is −2.8 MJ kg^{-1}. Calculate the escape speed from the Moon.
(b) A spacecraft of mass 1000 kg travelled from the Moon's surface to the Earth. Assuming it was launched from the Moon with just sufficient kinetic energy to enable it to escape, calculate
(i) its kinetic energy just before reaching the Earth,
(ii) its speed just before reaching the Earth.

27.4. Mars has a mass of 0.108 × the mass of the Earth, and a radius of 3400 km.
(a) Calculate **(i)** the gravitational field strength at the surface of Mars, **(ii)** the escape speed from Mars.
(b) Mars has two moons, Phobos and Deimos.
(i) Phobos orbits Mars in a circular orbit once every 7.65 hours. Calculate the height of Phobos above the surface of Mars.
(ii) Deimos orbits Mars in a circular orbit at a height of 20 100 km above the surface of Mars. Calculate the time it takes to complete one orbit.

27.5. (a) Neptune has two moons, Triton and Nereid. Triton orbits Neptune at a distance of 354 000 km once every 5.88 days.
(i) Calculate the mass of Neptune.
(ii) Nereid orbits Neptune once every 359 days. Calculate the distance from Neptune to Nereid.
(b) A satellite orbits the Earth at a height of 1680 km in a circular orbit that takes it over the Earth's poles.
(i) Calculate its time period.
(ii) Calculate the distance along the equator between successive transits in the same direction across the equator.

Simple Harmonic Motion

Objectives

After working through this unit, you should be able to:

- ▶ describe examples of oscillating systems
- ▶ explain what is meant by the amplitude and the time period of an oscillating system
- ▶ describe without calculations how the acceleration, velocity and displacement of an oscillating system change with time
- ▶ explain what is meant by simple harmonic motion and relate it to circular motion
- ▶ define angular frequency and use it to relate the acceleration to the displacement for an object in simple harmonic motion
- ▶ use formulae for the time period of a mass oscillating on a spring and a simple pendulum
- ▶ distinguish between free and damped oscillations
- ▶ describe what is meant by forced oscillations and resonance

Contents

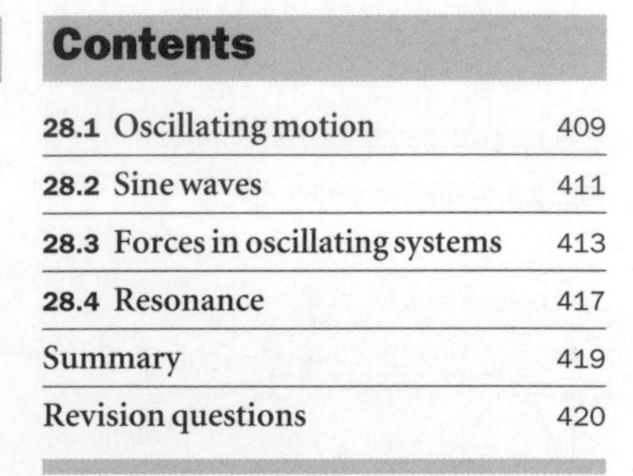

28.1 Oscillating motion

Oscillating systems

Any oscillating system moves to and fro repeatedly. Two simple systems are described below. By studying such simple systems, we can understand the motion of more complicated oscillating systems.

- If a pendulum is displaced from its equilibrium position and released, it oscillates to and fro, repeatedly passing back and forth through its equilibrium position. Eventually, the oscillations die away and it stops moving altogether. Each complete cycle of its oscillating motion takes it from one side of equilibrium to the other side and back again.
- A weight suspended on a spring will oscillate if it is pulled down from equilibrium then released. Each complete cycle of oscillation takes it from its lowest position to its highest position then back to its lowest

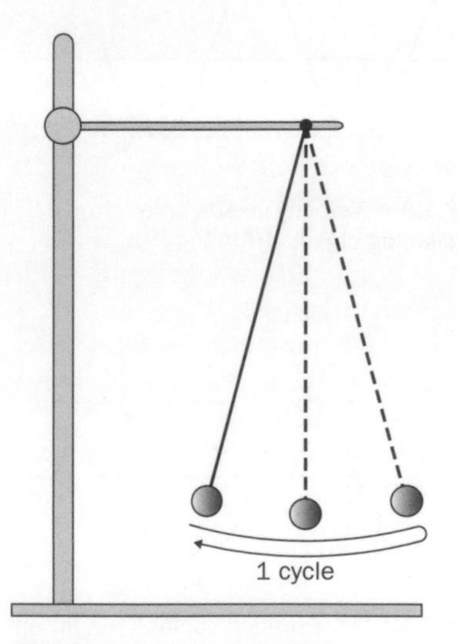

Fig. 28.1 A simple pendulum.

position. In practice, the oscillations die away gradually and it eventually stops moving.

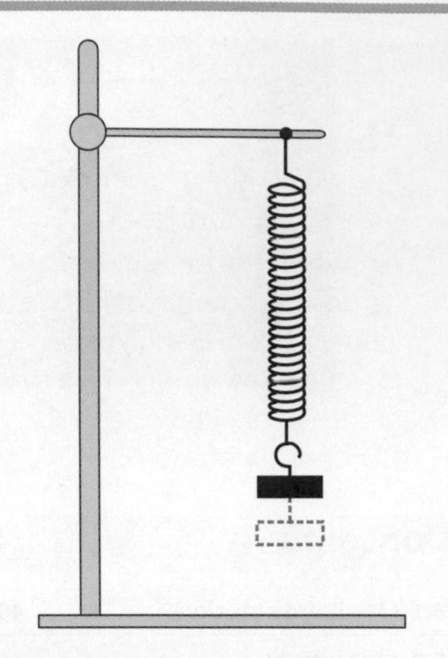

Fig. 28.2 An oscillating spring.

Measuring oscillations

Any oscillating object oscillates about a certain position which is therefore the centre of the oscillating motion. This position is usually the equilibrium position of the object, the point where the object would stay if released at rest at that point.

- The **displacement** of an oscillating object at any instant is the distance and direction of the object from its equilibrium position. Fig. 28.3 shows how the displacement of an oscillating object varies with time. The displacement is positive in one direction and negative in the opposite direction.
- The **amplitude of oscillations** is the maximum displacement from equilibrium. In the example of the simple pendulum and the oscillating spring, the amplitude gradually decreases due to air resistance. If there was no air resistance, the oscillations would continue indefinitely. Air resistance causes friction on the moving parts, gradually reducing the amplitude of the oscillations to zero.
- The **time period of oscillations** is the time for one complete cycle of oscillations. This is the least time taken by an oscillating object to move from a certain position and velocity back to the same position and velocity. For example, the time period of a simple pendulum is equal to the time taken by the pendulum to pass in a certain direction through equilibrium to the next pass through equilibrium in the same direction.

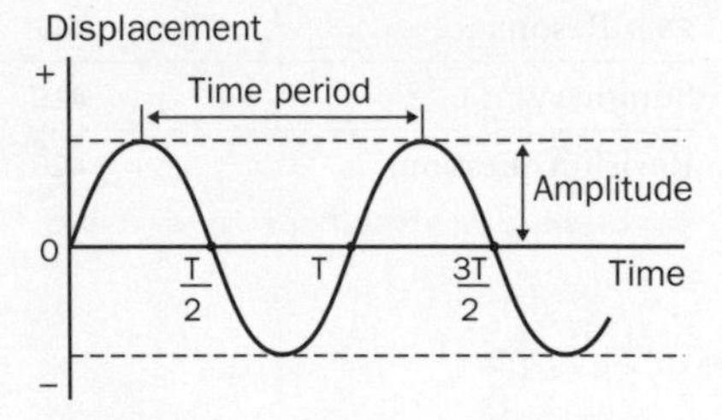

Fig. 28.3 Displacement versus time for an oscillating object.

Velocity and acceleration

When an oscillating pendulum is moving through its equilibrium position, its speed is at its greatest and its acceleration is zero. At maximum displacement from equilibrium, its speed is zero and the magnitude of its acceleration is at its greatest.

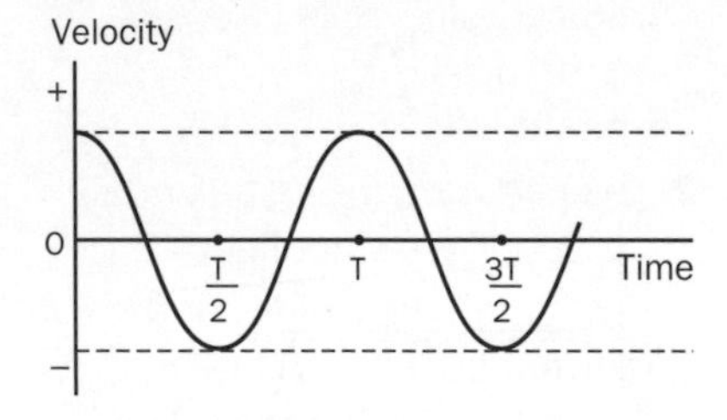

Fig. 28.4 Velocity versus time for an oscillating object.

Velocity v. time

Since velocity is defined as rate of change of displacement, the velocity of an oscillating object can be determined from the gradient of the displacement v. time graph for the object. Fig. 28.4 shows the velocity v. time graph corresponding to Fig. 28.3.

- At zero displacement, the gradient of Fig. 28.3 is a maximum, either in the positive direction or the negative direction, corresponding to the two possible directions of the object as it moves through equilibrium.
- At maximum displacement, the gradient of Fig. 28.3 is zero, corresponding to zero speed at maximum displacement in either direction. The object changes its direction at maximum displacement so the velocity changes from + to −, or − to +.

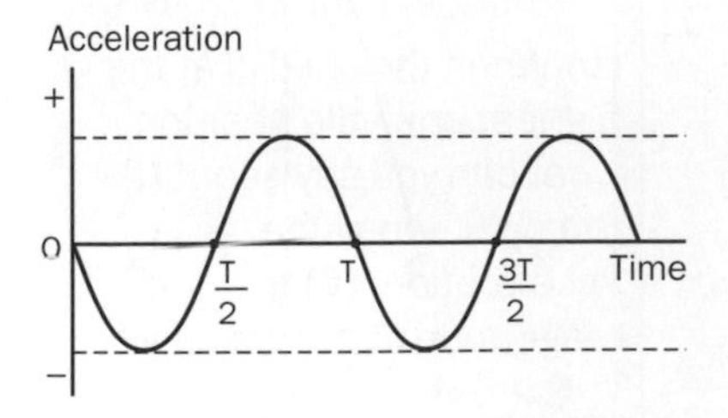

Fig. 28.5 Acceleration versus time for an oscillating object.

Acceleration v. time

Since acceleration is rate of change of velocity, the acceleration of an oscillating object can be determined from the gradient of the velocity v. time graph for the object. Fig. 28.5 shows the acceleration v. time graph corresponding to Fig. 28.4

- At zero velocity, the gradient of Fig. 28.4 is a maximum, either in the positive direction or the negative direction, corresponding to maximum displacement in either direction where the object turns around.
- At maximum velocity, the gradient of Fig. 28.4 is zero, corresponding to zero acceleration which occurs when the object passes through equilibrium.
- Compare Fig. 28.5 and Fig. 28.3. When the displacement is positive, the acceleration is negative. When the displacement is negative, the acceleration is positive. In other words, when the displacement is in one direction, the acceleration is in the opposite direction.

28.2 Sine waves

Making a sine wave

Consider point P, moving on a circular path, as shown in Fig. 28.7. Thus the coordinates of point P in Fig. 28.7 are $x = r\cos\theta$ and $y = r\sin\theta$, where r is the radius of the circle and θ is the angle between the line OP and the x-axis.

Fig. 28.7 also shows how its y-coordinate ($= r\sin\theta$) changes as θ increases from 0 to 360° as P moves around. The curve of Fig. 28.7 is called a sine curve because it is generated by the sine function. Its shape is described as 'sinusoidal'.

This type of curve describes the oscillating motion of a loaded spring and small oscillations of a simple pendulum. Oscillating motion described by a sinusoidal curve is called **simple harmonic motion.**

The link between circular motion and simple harmonic motion can be seen if the shadow of an object, P, moving on a circular path is observed adjacent to the shadow of an oscillating object, Q, as in Fig. 28.8. The shadows are projected on to a vertical screen perpendicular to the plane of

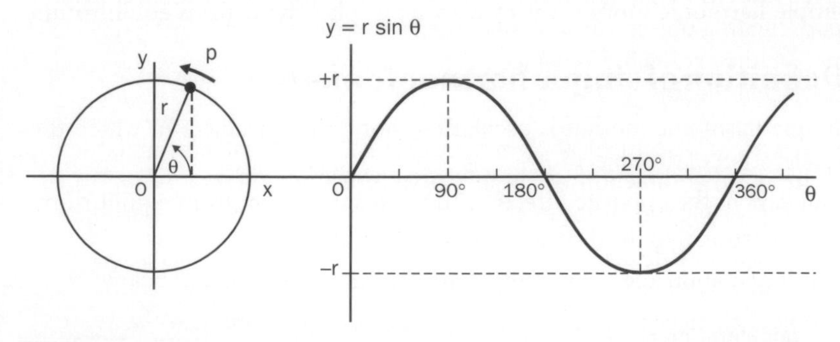

Fig. 28.7 Making a sine curve.

Questions 28.1

1. A pendulum oscillates in a line which lies in a vertical plane from north to south.

   (a) When the pendulum is at its northernmost position, what is the direction of **(i)** its displacement from the centre of oscillations, **(ii)** its acceleration?

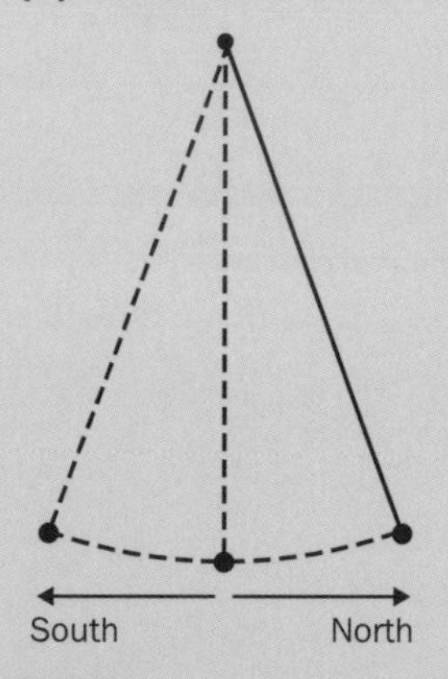

Fig. 28.6

   (b) What will be the direction of the displacement and the velocity of the pendulum a quarter of a cycle after it is at its northernmost position?

2. A child is bouncing up and down on a trampoline.

   (a) When the child is at the lowest possible position, what can you say about **(i)** the direction of the acceleration, **(ii)** the magnitude of the velocity?

   (b) When the child is at the highest possible position, what can you say about **(i)** the direction of the acceleration, **(ii)** the magnitude of the velocity?

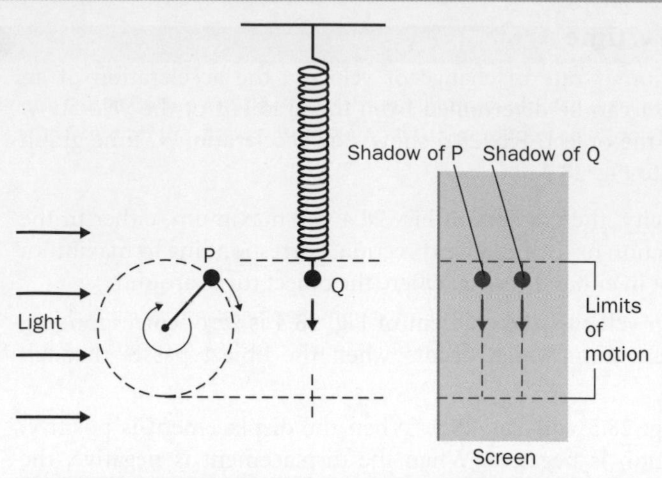

Fig. 28.8 Linking circular motion and simple harmonic motion.

the circle. The frequency of rotation of object P is adjusted to match the frequency of oscillation of object Q. The two shadows keep up with each other, demonstrating that the motion of the oscillating object is the same as the y-coordinate of the object in circular motion (i.e. sinusoidal).

To explore how the acceleration of an oscillating object is related to its displacement, consider the acceleration of object P in circular motion in Fig. 28.8.

The acceleration of P depends on the speed, υ, of the object and the circle radius, r, in accordance with the equation for centripetal acceleration

$$a_P = \frac{-\upsilon^2}{r}$$

where the minus sign indicates that the acceleration is towards the centre.

This may be written as $a_P = -\omega^2 r$, where the angular speed, ω, of object P is defined as

$$\omega = \frac{2\pi}{T} = \frac{\upsilon}{r}$$

and T is the time for 1 complete oscillation.

Since the vertical component of the acceleration $= a_P \sin\theta$ and $y = r\sin\theta$, then the vertical component of the acceleration $= -\omega^2 r \sin\theta = -\omega^2 y$. This last equation therefore describes how the acceleration of an object in simple harmonic motion varies with its displacement from equilibrium.

Notes

1. The acceleration is therefore always negative when the displacement is positive and is always positive when the displacement is negative.
2. The maximum speed, $\upsilon_{MAX} = \omega r$ where r is the amplitude of the oscillations. This follows from Fig. 28.8 because the oscillating object moves as fast as the object moving around the circle.
3. The displacement varies sinusoidally with time as shown in Fig. 28.7.
4. The time period, T, is independent of the amplitude for any object oscillating in simple harmonic motion.
5. The unit of ω is radian second^{-1} (rad s^{-1}) since 2π is the angle around a circle in radians and T is in seconds. In the context of simple harmonic motion, ω is called the **angular frequency**.

Definition of simple harmonic motion

Simple harmonic motion is oscillating motion of an object in which the acceleration of the object is

1. always in the opposite direction to the displacement from equilibrium,
2. proportional to the displacement from equilibrium.

The acceleration and displacement are linked by the equation below:

$$\textbf{acceleration} = -\omega^2 \times \textbf{displacement}, \quad \text{where } \omega = \frac{2\pi}{T}$$

Worked example 28.1

An object oscillates in simple harmonic motion with a time period of 5.0 seconds and an amplitude of 0.040 m. Calculate (a) its angular frequency,(b) its maximum speed, (c) its maximum acceleration.

Solution

(a) $\omega = \frac{2\pi}{T} = 1.26 \text{ rad s}^{-1}$

(b) $v_{MAX} = \omega r = 1.26 \times 0.040 = 0.050 \text{ m s}^{-1}$

(c) Maximum acceleration, a_{MAX}, is at maximum displacement which is 0.040 m.
Hence $a_{MAX} = -\omega^2 \times \text{maximum displacement} = -1.25^2 \times 0.040 = 6.3 \times 10^{-2} \text{ m s}^{-2}$

Questions 28.2

1. A mass on a spring oscillates along a vertical line, taking 12 s to complete 10 oscillations.
   (a) Calculate **(i)** its time period, **(ii)** its angular frequency.
   (b) Its height above the floor varies from a minimum of 1.00 m to a maximum of 1.40 m. Calculate **(i)** its amplitude, **(ii)** its maximum velocity, **(iii)** its acceleration when it is at its lowest position.
2. The motion of a vibrating blade is 'frozen' by illuminating it with a stroboscope (a flashing light). The least stroboscope frequency at which this occurs is 40 Hz.
   (a) Explain why the blade appears stationary at this frequency.
   (b) Calculate **(i)** the time period, **(ii)** the angular frequency of the vibrations.
   (c) The free end of the blade vibrates with an amplitude of 8.0 mm. Calculate the maximum speed of the free end of the blade.

28.3 Forces in oscillating systems

When an object is oscillating, the forces acting on it try to restore it to its equilibrium position. When the object is moving away from its equilibrium position, the forces on it slow it down and reverse its direction of motion. When the object is moving towards equilibrium, the forces on it speed it up and make it overshoot the equilibrium position.

If the object is oscillating with simple harmonic motion, its acceleration, a, depends on its displacement, s, in accordance with the equation

$$a = -\omega^2 s, \text{ where } \omega = \frac{2\pi}{T}$$

Applying Newton's second law of motion in the form 'force = mass × acceleration' gives the following equation for the restoring force, F, on an oscillating object:

$$F = -m\omega^2 s = -ks \text{ where } k = m\omega^2$$

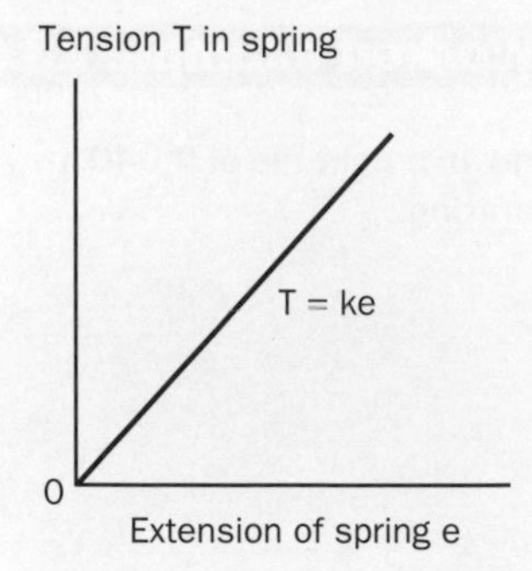

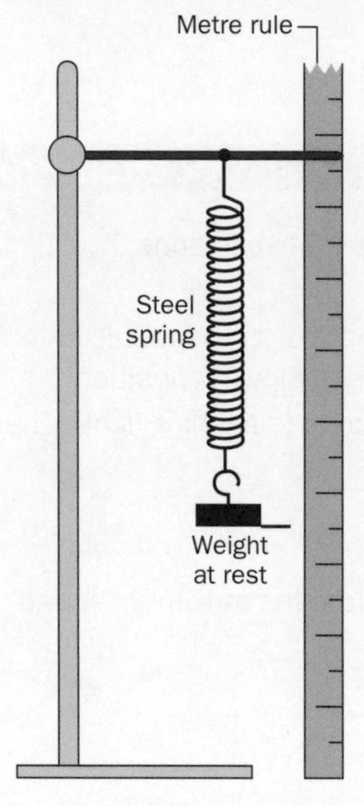

Fig. 28.9 Testing Hooke's Law.

It therefore follows that any oscillating system moves with simple harmonic motion if the restoring force, F, is

1. proportional to the displacement from the equilibrium position;
2. always directed towards the equilibrium position.

The oscillations of a loaded spring

Hooke's Law states that the tension, T, in a spring is proportional to its extension from its natural length. This may be expressed as $T = ke$, where e is the extension and k is a constant, referred to as the spring constant.

Consider a spring suspended vertically from a fixed point, supporting an object of mass, m, at its lower end, as in Fig. 28.9.

- When the object is at rest at its equilibrium position, the tension in the spring at equilibrium, T_0, is equal and opposite to the weight, mg, of the object.

  $$T_0 = mg$$

- When the object is oscillating, the tension in the spring changes as the length of the spring changes. When the object is at displacement, s, from equilibrium, the restoring force is provided by the change of tension from equilibrium, ΔT. Applying Hooke's Law to this change of tension gives $\Delta T = -ks$, where the minus sign signifies that the change of tension is in the opposite direction to the displacement. The acceleration, a, can be written as

  $$a = \frac{\text{restoring force}}{\text{mass}} = \frac{\Delta T}{m} = \frac{-k}{m}s$$

This is the same as the equation for simple harmonic motion, $a = -\omega^2 s$, where $\omega^2 = k \div m$. A loaded spring therefore oscillates with simple

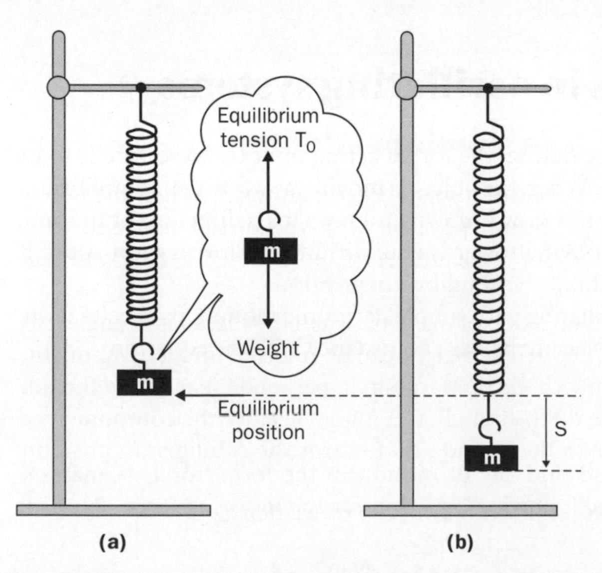

Fig. 28.10 The oscillations of a loaded spring (a) at rest, (b) oscillating.

harmonic motion. Furthermore, its time period, T, is related to the mass and the spring constant by the following equation:

$$T = \frac{2\pi}{\omega} = 2\pi\sqrt{\frac{m}{k}}$$

Worked example 28.2

$g = 9.8\ m\ s^{-2}$

A steel spring suspended from a fixed point supports a scale pan of mass 0.050 kg at equilibrium.

(a) The scale pan descends 40 mm to a new equilibrium position when a 1.0 N weight is placed on it. Calculate (i) the spring constant, (ii) the total mass of the scale pan and the 1.0 N weight.

(b) The scale pan with the 1.0 N weight on it is pulled a distance of 15 mm downwards from equilibrium then released. Calculate (i) the angular frequency of the oscillations, (ii) the time period of the oscillations, (iii) the maximum speed of the scale pan.

Solution

(a)(i) Rearranging F = ke gives $k = \frac{F}{e} = \frac{1.0}{0.040} = 25\ N\ m^{-1}$

(ii) Mass of 1.0 N weight $= \frac{weight}{g} = \frac{1.0}{9.8} = 0.10\ kg$

$\therefore$ Total mass = 0.15 kg

(b) (i) $\omega = \sqrt{\left(\frac{k}{m}\right)} = \left(\frac{25}{0.15}\right)^{\frac{1}{2}} = 13\ rad\ s^{-1}$

(ii) $T = \frac{2\pi}{\omega} = 0.49\ s$

(iii) The amplitude of the oscillations = 15 mm = 0.015 m

$\therefore v_{MAX} = \omega r = 13 \times 0.015 = 0.195\ m\ s^{-1}$

The oscillations of a simple pendulum

A simple pendulum consists of a small bob on a thread. When the bob is displaced from its equilibrium position with the thread taut and then released, it oscillates about its equilibrium position in a fixed vertical plane.

Consider the forces on a bob of mass, m, when the thread is at angle θ to the vertical, as in Fig. 28.11. The weight can be resolved into a component, $mg \cos \theta$, parallel to the thread and $mg \sin \theta$, perpendicular to the thread.

The restoring force on the bob, $F = -mg \sin \theta$, since the component of weight at right angles to the thread acts towards the equilibrium position of the bob.

Hence the acceleration of the bob, a, can be written as

$$a = \frac{\text{restoring force}}{\text{mass}} = \frac{-mg \sin \theta}{m} = -g \sin \theta$$

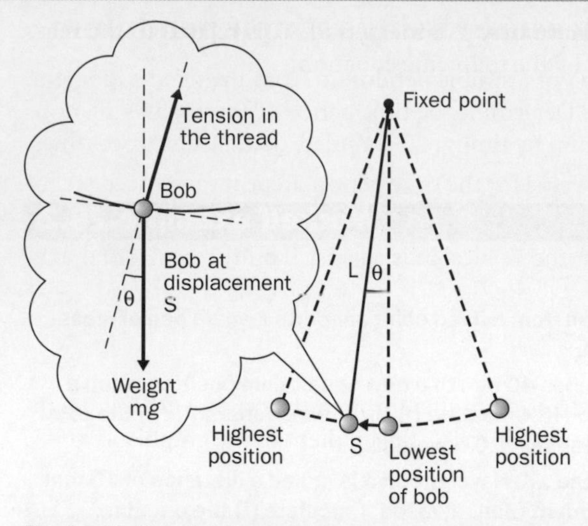

Fig. 28.11 The simple pendulum.

From Fig. 28.11,

$$\sin\theta = \frac{x}{L}$$

where l is the length of the thread and x is the horizontal displacement of the bob from equilbrium.

However, $x = s$, provided θ does not exceed about 10°.

$$\therefore \sin\theta = \frac{s}{L}$$

for small oscillations, hence

$$a = -g\sin\theta = -\frac{g}{L}s = -\omega^2 s, \text{ where } \omega^2 = \frac{g}{L}$$

A simple pendulum therefore oscillates with simple harmonic motion, provided θ does not exceed about 10°. Furthermore, its time period, T, is related to the length of the thread by the following equation:

$$\mathbf{T = \frac{2\pi}{\omega} = 2\pi\sqrt{\frac{L}{g}}}$$

Worked example 28.3

$g = 9.8\ \mathrm{m\,s^{-2}}$

Calculate the length of a pendulum that has a time period of exactly 1 second.

Solution

Rearrange $T = 2\pi\sqrt{\frac{L}{g}}$ to give $L = \frac{gT^2}{4\pi^2} = \frac{9.8 \times 1^2}{4\pi^2} = 0.25\ \mathrm{m}$

Experiment to measure *g* using a simple pendulum

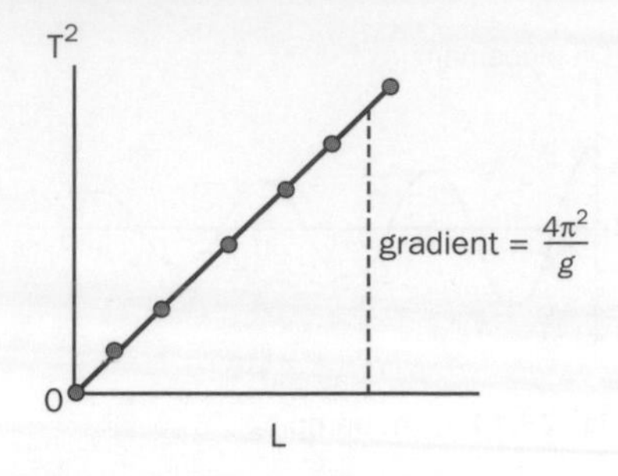

Fig. 28.12 Measurement of *g*.

1. Measure the length, L, of a simple pendulum from the fixed support to the centre of the bob. Determine the time period, T, of small oscillations of the simple pendulum by timing 20 complete oscillations three times.
2. Calculate an average value for the time period. Repeat the procedure for five further lengths.
3. Plot a graph of T^2 on the vertical axis against L on the horizontal axis. Since

$$T^2 = \frac{4\pi^2}{g} L$$

the graph should give a straight line through the origin with a gradient equal to $4\pi^2 \div g$. Hence $g = 4\pi^2 \div$ the gradient of the graph.

Questions 28.3

$g = 9.8 \text{ m s}^{-2}$

1. A steel spring suspended from a fixed point supports a 0.20 kg mass hung from its lower end. The stone is displaced downwards from its equilibrium position by a distance of 25 mm then released. The time for 20 oscillations is measured at 22.0 s. Calculate **(a)** its time period, **(b)** its angular frequency, **(c)** its maximum speed, **(d)** the maximum tension in the spring.
2. **(a)** Calculate the time period of a simple pendulum of length 2.0 m.

   (b) When a simple pendulum of length 2.0 m is in oscillating motion, its thread makes a maximum angle with its equilibrium position of 5°. Calculate **(i)** the maximum height of the bob above its equilibrium position, **(ii)** the maximum speed of the bob.

28.4 Resonance

The energy of an oscillating system

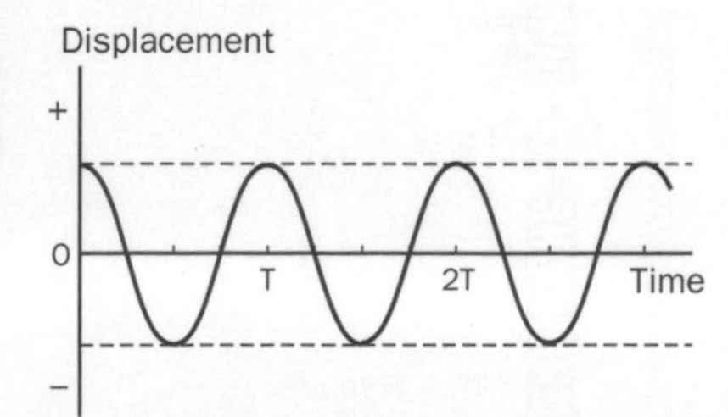

Fig. 28.13 Free oscillations.

The energy of an oscillating system changes every half cycle from kinetic energy at maximum speed to potential energy at maximum displacement and back to kinetic energy. At any instant, the total energy of the system is the sum of its kinetic energy and its potential energy.

Free oscillations

If there are no frictional forces present, the total energy remains constant and the amplitude is unchanged. The oscillations are described as **free** because there are no forces present that would cause the energy of the system to be dissipated.

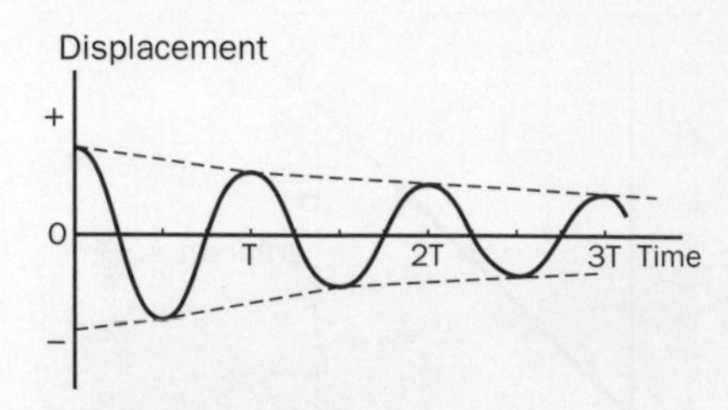

Fig. 28.14 Light damping.

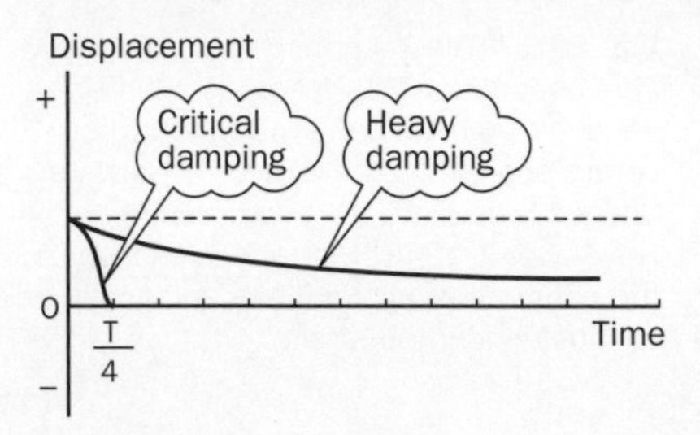

Fig. 28.15 More damping.

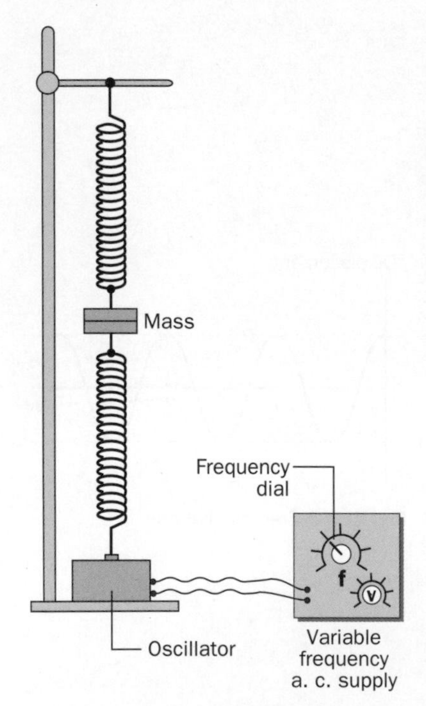

Fig. 28.16 Forced oscillations.

Damped oscillations

The oscillations of a simple pendulum gradually die away as air resistance causes the energy of the simple pendulum to be gradually reduced to zero. The amplitude becomes smaller and smaller, as shown in Fig. 28.14. The presence of air resistance 'damps' the oscillations. If it were possible to remove air resistance, the amplitude would remain constant as the pendulum would not lose its energy. The degree of damping depends on the maximum damping force in relation to the maximum restoring force.

- **Light damping** is where the damping force is much smaller than the maximum restoring force. The amplitude gradually becomes smaller and smaller as the damping force gradually dissipates the energy of the system, as shown in Fig. 28.14.
- **Heavy damping** is where the system returns gradually to equilibrium without oscillating, after being displaced then released. For example, a mass on a spring in a tube of viscous oil would return slowly to equilibrium without oscillating if released from a non-equilibrium position. Fig. 28.15 shows how the displacement of a heavily damped object changes after it has been released.
- **Critical damping** is where the system returns to equilibrium in as little time as possible without oscillating, as shown in Fig. 28.15. This type of damping is an important design feature of an analogue meter, to ensure the pointer of such a meter does not oscillate and does not take too long to reach its correct position when the current changes.

Forced oscillations

If a child on a swing is given a push every time the child reaches maximum height, the amplitude of the swing increases and increases. The force of the push supplies energy to the child so the total energy of the child increases. The push force needs to be applied at the same frequency as the natural frequency of oscillation of the spring.

The amplitude of any oscillating system subjected to a periodic force becomes very large if the periodic force supplies energy at a certain point in every cycle of oscillation. This effect is known as **resonance** and the frequency of the periodic force at which it occurs is called the **resonant frequency**. At any other frequency, the system is forced to oscillate with a small variable amplitude without resonating.

Fig. 28.16 shows how an oscillator may be used to apply forced oscillations to a mass held between two vertical springs. The frequency of the oscillator is altered gradually until it is equal to the resonant frequency of the system. When this occurs, the amplitude of oscillations increases until energy losses due to damping equal the energy supplied by the oscillator. The lighter the damping, the greater the amplitude.

The resonant frequency is equal to the natural frequency of oscillation of the system, only if the damping is very light.

Further examples of mechanical resonance

Fig. 28.17 The Tacoma bridge collapse. This photograph was taken as a large section of the concrete roadway in the centre span of the new Tacoma Narrows bridge hurtled into the Puget Sound on November 7 1940. High winds caused the bridge to sway, undulate and finally collapse under the strain.

1. The Tacoma bridge collapse was due to forced oscillations caused by periodic eddies created at a certain wind speed. Each eddy caused the bridge to sway. On the day of its destruction, the wind speed was such that the frequency of the eddies was the same as the natural frequency of the bridge span twisting and swinging from side to side. The twisting motion caused more eddies, which caused more twisting until the bridge destroyed itself. Since then, bridge engineers have incorporated damping mechanisms to prevent resonance.
2. A wine glass shatters when subjected to sound at a certain frequency. The sound waves make the wine glass oscillate and if the sound frequency is matched to the natural frequency of the wine glass, the glass shatters.
3. A vehicle panel that vibrates at a certain engine frequency has its own natural frequency of oscillation. The panel is forced to vibrate by the engine's motion. When the frequency of the engine is equal to the natural frequency of vibration of a panel, the panel vibrates at resonance.

Questions 28.4

1. A washing machine vibrates excessively at a certain speed.
   (a) Explain why this occurs at a certain speed.
   (b) Describe how such excessive vibrations could be stopped.
2. The chassis of a car of mass 1200 kg is 20 cm clear of the ground when the car is unloaded. When the car carries four people of combined mass 300 kg, the chassis is only 15 cm clear of the ground.
   (a) Calculate **(i)** the additional weight on each of the four wheels of the car when it carries four people, **(ii)** the spring constant of the suspension spring of each wheel.
   (b) (i) Show that the time period of the oscillations of the suspension system is 1.0 s.
   (ii) Explain why the car oscillates uncomfortably when it travels at a speed of 10 m s^{-1} over speed bumps spaced 10 m apart.

See www.palgrave.com/foundations/breithaupt to model oscillating motion using a spreadsheet.

Summary

- **Amplitude** is the maximum displacement from equilibrium.
- **Time period,** T, is the time for one complete cycle of oscillation.
- **Frequency,** $f = \frac{1}{T}$
- **Acceleration** $= -\omega^2 \times$ displacement,

  where ω = angular frequency $= \frac{2\pi}{T}$

◆ **Maximum speed, $\upsilon_{MAX} = \omega \times$ amplitude.**

◆ **Time period formulae**

1. Mass, m, on a spring, $T = 2\pi\left(\frac{m}{k}\right)^{\frac{1}{2}}$, where k is the spring constant.

2. Simple pendulum of length L, $T = 2\pi\left(\frac{L}{g}\right)^{\frac{1}{2}}$

◆ **Resonance** for light damping occurs when the frequency of the applied force is equal to the natural frequency of oscillation of the system.

Revision questions

$g = 9.8\ \mathrm{m\ s^{-2}}$

28.1. An object oscillates vertically with an amplitude of 100 mm in simple harmonic motion.
(a) The object takes 15.0 s to undertake 20 oscillations. Calculate **(i)** its time period of oscillation, **(ii)** its maximum velocity, **(iii)** its maximum acceleration.
(b) The displacement, s, of the object varies with time in accordance with the equation, $s = r \sin \omega t$, where r is the amplitude of oscillations and T is the time period.
(i) Sketch a graph to show how the displacement varies with time in accordance with the equation, $s = r \sin \omega t$.
(ii) Indicate on your graph a point where the speed is a maximum. Label this point X.
(iii) Indicate on your graph a point where the acceleration is a maximum. Label this point Y.

28.2. A vertical spring is suspended from a fixed point. When an object of mass 0.30 kg is suspended from the lower end of the spring, the spring stretches by 90 mm.
(a) Calculate the spring constant of the spring.
(b) The object on the spring is pulled vertically down by a distance of 0.060 m from its equilibrium position, then released. Calculate **(i)** the time period of oscillations of the spring, **(ii)** the maximum speed of the mass.

28.3. In a vehicle test, a loaded vehicle of total mass 900 kg travelled over a bump on the road surface at constant speed. A passenger in the car counted two complete cycles of oscillation in 3.0 s after passing over the bump.
(a) Calculate **(i)** the time period of the oscillations, **(ii)** the spring constant of the suspension spring at each wheel.
(b) The mass of the unloaded vehicle and driver is 750 kg. Calculate the time period of oscillations of the unloaded vehicle and driver.

28.4. A simple pendulum consists of a small sphere of diameter 2.0 cm attached to a thread of length 1.18 m.
(a) Calculate the time period of small oscillations of this pendulum.
(b)(i) The pendulum bob is displaced from equilibrium with the string taut until the thread makes an angle of 10° with a vertical line. Calculate the displacement of the bob at this position.
(ii) The bob is then released from this position. Calculate its maximum speed as it passes through its lowest position.

28.5. A trailer body has a mass of 400 kg and is mounted on four springs, one at each corner. The floor of the trailer body is lowered by a distance of 20 mm when four 30 kg bags of sand are evenly placed in the trailer.
(a) Calculate **(i)** the total weight of the bags of sand, **(ii)** the spring constant of each spring, **(iii)** the time period of oscillations of the loaded trailer.
(b) Explain why the trailer oscillates considerably when it is driven at a certain speed over a series of speed bumps.

Further Questions

(Numerical answers only are supplied on p. 473)

Unit 1

Q1.1 (a) List the following types of electromagnetic waves in order of increasing frequency.

infra-red radiation microwaves visible light X-rays

(b) The speed of electromagnetic waves in free space is 3.00×10^8 m s^{-1}. Calculate **(i)** the frequency of light of wavelength 580 nm, **(ii)** the wavelength of radio waves of frequency 95 MHz.

Q1.2. (a) (i) State which of the following types of waves are transverse waves.

electromagnetic waves sound waves waves on a string water waves

(ii) Which one of the following properties is shown by transverse waves and not by longitudinal waves?

diffraction polarisation refraction reflection

(b) Sound waves of frequency 500 Hz travel through a substance at a speed of 1500 m s^{-1}.

(i) Calculate the wavelength of these waves.

(ii) Calculate the phase difference between two points that lie along the direction of propagation of the wave at a distance of 1.0 m apart.

Q1.3 (a) Explain why the wider a satellite dish is **(i)** the stronger the signal it is capable of collecting, **(ii)** the more difficult it is to align it to obtain maximum signal strength.

(b) Straight waves on the surface of a pool cross a straight boundary from shallow to deep water at an angle of 45° to the boundary.

(i) Water waves travel faster in deep water than in shallow water. What change occurs to the wavelength of the water waves on crossing the boundary?

(ii) Sketch a top view of a wavefront as it is crossing the boundary, indicating its direction of propagation on each side of the boundary.

Q1.4 (a) Describe with the aid of a diagram what is meant by a polarised wave on a vibrating string.

(b) Describe an experimental test that could be carried out using two polaroid squares to find out if light from a point light source is polarised.

Unit 2

Q2.1 (a) (i) State the approximate frequency range of the normal human ear.

(ii) State an approximate value of frequency of sound which the ear is most sensitive to.

(b) A motor cycle produces a maximum loudness level of 80 dB as it passes by a roadside monitor. Calculate the loudness level produced by four such motor cycles passing by the same roadside monitor.

Q2.2 (a) With the aid of a diagram, describe and explain the principle of operation of an ultrasonic transducer.

(b) An ultrasonic transducer on the surface of a patient emits short pulses at a rate of 1000 Hz into the patient.

(i) Calculate the time taken for an ultrasonic pulse travelling at a speed of 1500 m s^{-1} to travel from the transducer to the other side of the patient and back, a total distance of 0.60 m.

(ii) Calculate the time between such a reflected pulse returning to the patient and the next pulse being transmitted.

Q2.3 (a) What change occurs in the note produced by a guitar string if the string **(i)** is slackened without change of length, **(ii)** made shorter without change of tension?

(b) A string of length 1.80 m is made to vibrate in its fundamental mode at a frequency of 60 Hz.

(i) Calculate the wavelength and speed of the waves on the string in this condition.

(ii) The frequency is raised without changing the length or the tension of the string until two equally spaced nodes are observed between its ends. Sketch this pattern of vibration and calculate the frequency at which it occurs.

Q2.4 A pipe of length 0.70 m is closed at one end. A small speaker, connected to a variable frequency a.c. supply, is held near the open end of the pipe and the frequency is adjusted until the pipe resonates. The lowest frequency at which this occurs is 120 Hz.

(a) Explain why the pipe resonates at this frequency.

(b) (i) Estimate the speed of sound in the pipe.

(ii) What is the next highest frequency at which this pipe resonates?

Unit 3

Q3.1 (a) A light ray is directed at the flat surface of a glass block of refractive index 1.5 at an angle of incidence of **(i)** 20°, **(ii)** 50°. Calculate the angle of refraction in each case.

(b) (i) What is meant by total internal reflection of light?

(ii) Calculate the critical angle of glass of refractive index 1.5.

(iii) With the aid of a diagram, explain why light entering an optical fibre emerges from the other end even if the fibre is bent.

Q3.2 (a) In an experiment to demonstrate interference of light, a laser beam of wavelength 650 nm was directed at a pair of closely spaced slits. An interference pattern consisting of alternate bright and dark fringes was observed on a screen 1.5 m away from the slits, placed perpendicular to the direction of the beam.

(i) The distance across four fringe spaces was 6.0 mm. Calculate the fringe spacing.

(ii) Use your answer to **(i)** and the other data to calculate the slit spacing.

(b) The pair of slits was moved 0.5 m away from the screen. Calculate the distance across four fringe spaces for this arrangement.

Q3.3 (a) Determine the position and nature of the image of an object formed by a 150 mm focal length convex lens for an object distance of **(i)** 100 mm, **(ii)** 300 mm.

(b) A concave lens of focal length 0.20 m is used to view an object which is held 0.20 m from the lens. Calculate the position of the image formed and draw a ray diagram to show how the image is formed.

Q3.4 (a) Draw a ray diagram to show how a refracting telescope forms at infinity an image of a distant point object which is off the axis of the telescope.

(b) A telescope which has a magnifying power of ×8 is used to observe two stars which are 0.5° apart. Calculate how far apart the two stars appear when viewed through the telescope.

(c) The objective lens of this telescope has a focal length of 0.60 m. What is the focal length of the eyepiece for a magnifying power of ×8?

Unit 4

Q4.1 (a) (i) State the three primary colours in order of increasing wavelength.

(ii) State an approximate value of the wavelength of blue light.

(b) A parallel beam of white light directed into a glass prism emerges to form a spectrum on a white screen.

(i) What type of spectrum is formed?

(ii) Which colour of light on the screen is refracted least?

Q4.2 A parallel beam of monochromatic light is directed normally at a diffraction grating which has 600 lines per millimetre.

(a) Calculate the spacing between adjacent lines of this grating.

(b) Use the diffraction grating equation to calculate the angle of diffraction of the first order diffracted beam.

(c) Calculate the maximum order number.

Q4.3 A light source emits light of wavelength 430 nm and 550 nm. Parallel light from this source was directed at a diffraction grating which had 500 lines per millimetre.

(a) Calculate the angle between these two wavelengths in the second order spectrum.

(b) Calculate the maximum angle of diffraction and state which wavelength corresponds to this angle.

Q4.4 (a) State which type of electromagnetic radiation is used for each purpose below and which of its properties makes it suitable for that purpose.

(i) Communication from a satellite to the Earth.

(ii) Detection of movement of people and animals in darkness.

(iii) Image formation of a damaged bone.

(b) (i) Describe how amplitude modulation and frequency modulation differ.

(ii) Why is an infra-red signal capable of carrying more pulses per second than a microwave signal?

Unit 5

The Avogadro constant $= 6.02 \times 10^{23}\ \text{mol}^{-1}$

Q5.1 The relative atomic masses of hydrogen, carbon and oxygen are 1, 12 and 16 repectively.

(a) Calculate the molar mass of **(i)** water, H_2O, **(ii)** carbon dioxide, CO_2.

(b) (i) Calculate the mass of a water molecule.

(ii) Estimate the linear size of a water molecule. The density of water $= 1000\ \text{kg m}^{-3}$.

Q5.2 (a) What type of bond holds the atoms together in **(i)** a molecule of carbon dioxide, **(ii)** a sodium chloride crystal?

(b) (i) Describe how the arrangement of molecules in a solid differs from the arrangement in a liquid.

(ii) When a pure substance melts, its temperature does not change. Explain why thermal energy must be supplied to melt it when there is no change of temperature.

Q5.3 (a) The isotopes of an element have different physical properties such as atomic mass.

(i) What is meant by the phrase 'isotopes of an element'?

(ii) Why do the isotopes of an element form identical bonds when they react with other elements?

(b) Every neon atom in neon gas contains 10 protons and either 10 or 12 neutrons.

(i) Write down the symbol for each of these two isotopes of neon.

(ii) The relative atomic mass of neon gas is 20.2. Calculate the proportion of each isotope in neon gas.

Q5.4 Solid aluminium has a density of 2700 kg m^{-3} and its relative atomic mass is 27.

(a) Calculate **(i)** the molar volume of solid aluminium, **(ii)** the number of atoms in a volume of 1.0 cm^3 of aluminium.

(b) Estimate the distance between the centres of adjacent aluminium atoms in solid aluminium.

Unit 6

Q6.1 (a) Calculate the expansion of a brass bar of length 2.50 m when its temperature is raised from 20°C to 50°C.

(b) An elevator in a vertical shaft is supported by several steel cables, each of length 200 m. Calculate the extension of each cable due to a 20 K change of temperature.

Values of linear expansivity: brass 1.9×10^{-5}, steel 1.1×10^{-5}.

Q6.2 (a) 52 kg of water at 15°C in an insulated copper tank of mass 15 kg is heated by a 4.0 kW immersion heater for 30 minutes. Calculate **(i)** the electrical energy supplied in this time, **(ii)** the final temperature of the water, assuming no heat loss occurs.

(b) Estimate the length of time taken to boil away 0.5 kg of water in a 3.0 kW electric kettle.

Specific heat capacities: water 4200 J kg^{-1} K^{-1}, copper 390 J kg^{-1} K^{-1}.

Specific latent heat of vaporisation of water = 2.3 MJ kg^{-1}.

Q6.3 (a) Describe two methods of reducing heat loss from a house in winter. For each method, state which physical process is responsible for this reduction.

(b) Calculate the heat loss per second through a glass window pane of area 0.80 m^2 and thickness 6.0 mm when the temperature difference across the window pane is 8.0°C.

The thermal conductivity of glass = 0.12 W m^{-2} K^{-1}.

Q6.4 (a) Use Wien's Law to estimate the temperature of the light-emitting surface of the Sun, assuming the wavelength of light from the Sun at peak intensity is 500 nm.

(b) (i) The intensity of solar radiation at the Earth, which is 1.5×10^{11} m from the Sun, is 1.4 kW m^{-2}. Use this data to estimate the energy per second radiated from the Sun.

(ii) The Sun's diameter is 1.4×10^9 m. Use this date and your estimate in **(a)** of the Sun's temperature to calculate the emissivity of the Sun. The Stefan constant = 5.67×10^{-8} W m^{-2} K^{-4}.

Unit 7

Q7.1 (a) Sketch curves to show how the extension varies with tension for **(i)** a steel spring when it is stretched, **(ii)** a rubber band when it is stretched.

(b) A vertical steel spring is fixed at its upper end and supports a weight of 2.0 N at its lower end. When an extra 1.0 N weight is added to its lower end, its length increases from 405 mm to 445 mm.

(i) Calculate its length when it is unloaded.

(ii) The spring is unloaded and an object of unknown weight is hung from its lower end, causing its length to increase to 500 mm. Calculate the weight of this object.

Q7.2 (a) What is meant by the breaking stress of a material?

(b) A wire of diameter 0.25 mm snapped when the tension in it reached 49 N.

(i) Calculate the breaking stress of the material of this wire.

(ii) Calculate the tension at which a wire of identical material of diameter 0.36 mm would snap.

Q7.3 A vertical steel wire of diameter 0.28 mm and of length 1.8 m is fixed at its upper end.

(a) Calculate the stress in the wire when it supports a 20 N weight.

(b) Calculate the increase in the length of this wire if a weight of 50 N is attached to its lower end. The Young modulus of steel = 2.0×10^9Pa.

Q7.4 A wire of diameter 0.26 mm and of length 5.00 m extends by 30 mm when the tension in it is increased from zero to 40 N.

(a) Calculate the Young modulus of the metal.

(b) Calculate the energy stored at a tension of 40 N.

Unit 8

Q8.1 (a) Calculate the pressure at the base of an upright concrete post of weight 400 N and of dimensions 1.6 m $\times$ 0.10 m $\times$ 0.10 m.

(b) Calculate the pressure the post in **(a)** would exert on level ground if it was laid flat.

Q8.2 (a) Calculate the pressure due to sea water of density 1050 kg m^{-3} at a depth of 5.0 m below the surface.

$g = 9.8$ m s^{-2}

(b) Explain why a deep sea diver could not inhale air from the surface via a long tube.

Q8.3 (a) Explain why a small amount of air in a vehicle brake system reduces the effectiveness of the brakes.

(b) In a hydraulic brake system, a force of 200 N is applied to the master piston which has a cross-sectional area of 4.0 cm^2.

(i) Calculate the pressure, in pascals, on the brake fluid in the piston due to this force.

(ii) Each slave piston has a cross-sectional area of 60 cm^2. Calculate the force exerted by each slave piston.

Q8.4 A flat barge floats with its deck 1.20 m above the water when unloaded. When it is loaded with material of total weight 18 kN, it floats 50 mm lower in the water.

(a) Calculate the pressure due to 50 mm of water. The density of water = 1000 kg m^{-3}. $g = 9.8$ m s^{-2}.

(b) Estimate the area of the barge hull.

(c) If the displacement of the boat is not allowed to exceed 200 mm, calculate the maximum load it can carry.

Unit 9

Q9.1 (a) Calculate the components parallel and perpendicular to a horizontal line from North to South of a horizontal force of magnitude 8N acting **(i)** 30°E of due North, **(ii)** 70° W of due North.

(b) Calculate the magnitude and direction of the resultant force on a point object due to a force of 10 N acting due North and a force of 8 N acting **(i)** due South, **(ii)** due East, **(iii)** 60° E of due North.

Q9.2 Calculate the force necessary to maintain a point object of weight 6.0 N in equilibrium when it is also acted on by a **(a)** an upward vertical force of magnitude 5.0 N, **(b)** a horizontal force of magnitude 5.0 N.

Q9.3 A uniform wooden plank of length 1.60 m and weight 80 N rests horizontally on two bricks X and Y, where X is at a distance of 0.40 m from one end and Y is at a distance of 0.20 m from the other end.

(a) Sketch a free body diagram showing the forces acting on the plank.

(b) Calculate the support force acting on the plank due to each brick.

(c) A child of weight 200 N stands on the plank at its mid-point and walks along the plank towards the end beyond X. Calculate the distance of the child from the mid-point at which the plank lifts off the brick at Y.

Q9.4 (a) A garden rake consists of a 8.0 N handle of length 1.50 m and a metal 'comb' of weight 12.0 N and of width 0.30 m attached to the handle at its midpoint. Calculate the distance from the free end of the rake to its centre of gravity.

(b) A wheelbarrow is designed so that its handles are 1.2 m from its wheel base. The wheelbarrow is loaded with a 280 N bag of sand which is placed 0.30 m from its wheel base. Sketch the arrangement and calculate the force that must be applied to its handles to raise it.

Unit 10

$g = 9.8$ m s^{-2}

Q10.1 A vehicle was travelling at 25 m s^{-1} when the driver saw traffic lights, 65 m away, turn red. The driver applied the brakes and the car decelerated uniformly to stop at the lights.

(a) The driver's reaction time before applying the brakes was 0.6 s. Calculate the distance travelled by the car in this time moving at 25 m s^{-1}.

(b) Calculate the deceleration of the vehicle during the period when the brakes were applied.

Q10.2 A train travelling between two stations accelerated from rest to a speed of 15 m s^{-1} for 75 s then moved at constant speed for 240 s before decelerating to a standstill in 60 s.

(a) Sketch a speed v. time graph for this journey.

(b) Calculate **(i)** the acceleration, **(ii)** the distance moved, in each of the three parts of the journey.

(c) Calculate the distance apart of the two stations and the average speed of the train for the journey beween the two stations.

Q10.3 A stone released from rest at the top of a well hits the water in the well 1.4 s later.

(a) Calculate **(i)** the speed, **(ii)** the distance fallen by the stone just before impact with the water surface.

(b) Sketch a speed–time graph to represent the motion of the stone from when it was released to when it hit the water surface.

Q10.4 A tennis ball is thrown vertically upwards and returns to the thrower 3.2 s later.

(a) Calculate **(i)** the time it takes to reach maximum height, **(ii)** the maximum height reached, **(iii)** its speed of projection, **(iv)** its speed just before returning to the thrower.

(b) Sketch a velocity–time graph to represent the motion of the tennis ball.

Unit 11

$g = 9.8$ m s^{-2}

Q11.1 (a) An object of mass 5.0 kg, initially at rest, is acted on by a resultant force of 15 N for 10 s. Calculate **(i)** its acceleration, **(ii)** its speed after 10 s, **(iii)** its momentum after 10 s.

(b) A ball of mass 0.15 kg, moving at right angles to a vertical wall at a speed of 14 m s^{-1}, rebounds from the wall at a speed of 10 m s^{-1}.

(i) Calculate the change of momentum of the ball.

(ii) The ball was in contact with the wall for 25 ms. Calculate the force of impact of the ball on the wall.

Q11.2 (a) A brick of mass 4.7 kg, dropped from rest at a height of 8.4 m, falls onto a bed of sand, causing an indentation in the sand of 42 mm. Calculate **(i)** the speed of the brick just before it hits the sand, **(ii)** its deceleration in the sand, **(iii)** the impact force.

(b) A lorry of total mass 38 000 kg, travelling at a speed of 31 m s^{-1} is brought to rest in a distance of 185 m when its brakes are applied. Calculate **(i)** the deceleration of the lorry, **(ii)** the braking force.

Q11.3 (a) Water emerges from a hosepipe at a speed of 11 m s^{-1} at a flow rate of 0.17 kg s^{-1}. Calculate **(i)** the momentum per second carried away by the water from the hosepipe, **(ii)** the force on the hosepipe due to this rate of loss of water.

(b) A cannon of mass 340 kg is designed to fire a cannon ball of mass 12.5 kg at a speed of 74 m s^{-1} from a barrel of length 1.35 m. Calculate **(i)** the acceleration of the ball in the barrel, **(ii)** the force on the ball, **(iii)** the recoil velocity of the cannon.

Q11.4 A rail wagon of mass 1200 kg, moving at a steady speed of 1.8 m s^{-1}, collides with another wagon of mass 900 kg and couples to it. Calculate the velocity of the two wagons immediately after the impact if the lighter wagon was initially **(a)** stationary, **(b)** moving at a speed of 2.5 m s^{-1} in the opposite direction to the other wagon.

Unit 12

$g = 9.8$ m s^{-2}

Q12.1 (a) Calculate the change of potential energy and kinetic energy of a ball of mass 0.30 kg, released from rest at a height of 2.4 m above the floor, when it is **(i)** 1.2 m above the floor, **(ii)** just above the floor.

(b) A ball of mass 0.080 kg, released from a height of 1.50 m above a tiled floor, rebounds to a height of 1.20 m above the floor. Calculate **(i)** its kinetic energy and speed immediately before the impact, **(ii)** its kinetic energy and speed immediately after the impact, **(iii)** its loss of kinetic energy due to the impact.

Q12.2 A roller coaster car of total mass (including passengers) 2500 kg descends from a height of 55 m along a steep incline onto a level section of track.

(a) Calculate its loss of potential energy due to this descent.

(b) Its speed before descent was 3 m s^{-1} and its speed immediately after the descent was 25 m s^{-1}. Calculate its gain of kinetic energy as a result of the descent.

(c) Calculate the loss of total energy of the roller coaster in this descent.

Q12.3 (a) Estimate the average power from a tidal power station operating at an efficiency of 0.40, that traps sea water over a level area of 15 km^2 when high tide is 3.0 m above low tide. The density of sea water = 1050 kg m^{-3}.

(b) A certain type of fuel is capable of releasing 30 MJ of energy per kilogram. A vehicle travelling at a steady speed of 25 m s^{-1} uses this type of fuel at a rate of 12 km per kilogram. Estimate the input power from the fuel to the engine.

Q12.4 A vehicle of mass 1100 kg, moving at a speed of 20 m s^{-1}, collided with a stationary vehicle of mass 900 kg. The two vehicles locked together as a result of the impact.

(a) Calculate **(i)** the speed of the two vehicles immediately after the impact, **(ii)** the loss of kinetic energy due to the impact.

(b) The impact caused the two vehicles to shorten by a combined length of 1.24 m. Calculate the loss of kinetic energy per metre during the impact and hence determine the impact force.

Unit 13

Q13.1 A plastic rod is charged by friction. When it is held over a negatively charged electroscope, the electroscope leaf rises.

(a) Use this observation to deduce the type of charge on the rod.

(b) Explain, in terms of electron transfer, how the rod became charged.

Q13.2 (a) Calculate the combined resistance of a 3.0 Ω resistor and a 6.0 Ω resistor **(i)** in series with each other, **(ii)** in parallel with each other.

(b) A 4.0 Ω resistor and a 6.0 V battery of negligible internal resistance are connected in series with the combination in **(a) (ii)**. Sketch the circuit diagram and calculate the current, p.d. and power dissipated for each resistor.

Q13.3 An electric kettle is designed to use 2.5 kW of electrical power at 240 V.

(a) Calculate **(i)** the current through the kettle element when it operates normally, **(ii)** the charge passing through the kettle in 3 minutes, **(iii)** the electrical energy delivered to the kettle in 3 minutes.

(b) Given a 3 A fuse, a 5 A fuse and a 13 A fuse, which one of these fuses would you select for this kettle?

Q13.4 (a) Sketch a graph to show how the current varies with p.d. for a silicon diode.

(b) A silicon diode is connected in its 'forward direction' in series with a 1.5 kΩ resistor and a 1.5 V cell.

(i) Sketch the circuit diagram for this arrangement.

(ii) Assuming the p.d. across the silicon diode remains at 0.6 V when it conducts, show that the current through the diode in this circuit is 0.60 mA.

(iii) Calculate the power dissipated in the diode and in the resistor and the power supplied by the cell when the current through the diode is 0.60 mA.

Unit 14

Q14.1 A battery of e.m.f. 12.0 V and internal resistance 1.5 Ω is connected to two 5.0 Ω resistors in parallel with each other.

(a) Sketch the circuit diagram and calculate the current through the battery.

(b) Calculate the p.d. across the parallel combination and the current through each of the resistors in the combination.

(c) (i) Calculate the power supplied by the battery and the power delivered to each 5.0 Ω resistor.

(ii) Account for the difference between the power supplied by the battery and the total power delivered to the two 5.0 Ω resistors.

Q14.2 (a) A Wheatstone bridge circuit consists of a metre length of uniform wire, a 5.0 Ω standard resistance, S, and an unknown resistor, X, as in Fig Q14.2. When the bridge is balanced, the balance point is 450 mm from the end of the wire connected to S. Calculate the resistance of X.

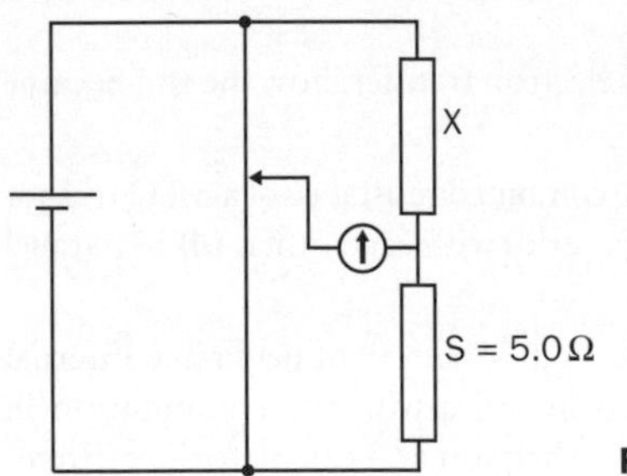

Fig. Q14.2.

(b) A second 5.0 Ω resistance is connected in parallel with S. Calculate the new balance position.

Q14.3 A potential divider consisting of a 10 kΩ resistor and an LDR is connected to a 5.0 V battery. A high-resistance voltmeter is connected across the 10 kΩ resistor.

(a) (i) Sketch the circuit diagram for this arrangement.

(ii) When the LDR is in darkness, the voltmeter reading is 0.20 V. Calculate the resistance of the LDR in darkness.

(b) Describe and explain how the voltmeter reading would change when the LDR is exposed to light.

Q14.4 (a) Calculate the resistivity of a 2.5 m length of metal wire of diameter 0.35 mm which has a resistance of 12.5 Ω.

(b) Calculate the length of wire necessary to make a 5.0 Ω resistor using wire of diameter 0.28 mm and resistivity 5.2×10^{-7} Ω m.

Unit 15

$\varepsilon_0 = 8.85 \times 10^{-12}$ F m^{-1}

Q15.1 A capacitor is connected in series with a battery, a switch, a milliammeter and a variable resistor. The switch is closed and the variable resistor is used to keep the current at a constant value of 1.2 mA. The switch is opened 60 s after it has been closed.

(a) Calculate the charge stored by the capacitor in this time.

(b) The p.d. across the capacitor plates after 60 s is measured at 3.3 V using a digital voltmeter. Calculate the capacitance of the capacitor.

Q15.2 (a) Calculate the combined capacitance of a 6 μF capacitor and a 12 μF capacitor **(i)** in parallel, **(ii)** in series.

(b) The series combination in **(a)** is connected to a 6.0 V battery. Calculate **(i)** the charge and energy stored in each capacitor, **(ii)** the total charge and energy stored.

Q15.3 (a) (i) Calculate the capacitance of a parallel plate capacitor consisting of two square, metal plates of length 0.26 m, separated by an air gap of width 2.0 mm.

(ii) The capacitor in (i) was connected to a 12 V battery. Calculate the charge and energy stored by the capacitor in this arrangement.

(b) In (a)(ii), the capacitor was disconnected from the battery and the plates moved together to a separation of 1.0 mm. Calculate **(i)** the new capacitance, **(ii)** the p.d. between the plates at 1.0 mm apart.

Q15.4 A 4.7 μF capacitor was connected to a 9.0 V battery and then discharged through a 1.5 MΩ resistor. Calculate **(a)** the initial charge and energy stored by the capacitor, **(b)** the initial current, **(c)** the charge and energy stored after 5.0 s.

Unit 16

Q16.1 (a) Sketch the symbol and write down the truth table for **(i)** an OR gate, **(ii)** a NAND gate.

(b) Design a logic gate system that will switch an LED on when the light intensity in a room falls below a certain level and the temperature of the room falls below a certain value. Include a test switch in your system which will make the LED switch on when the switch is closed.

Q16.2 (a) An a.c. signal generator is connected between 0 V and the non-inverting input of an op-amp on open-loop. The inverting input is connected to 0 V. Sketch the circuit diagram and the output waveform, assuming the output voltage is limited to the range +15 V when the signal generator is adjusted to produce a sine wave voltage of peak value 1.0 V at a frequency of 1.0 kHz.

(b) In the circuit in (a), a 1.0 MΩ resistor is connected between the output terminal of the op-amp and the inverting input. Also, the inverting input is disconnected from 0 V and reconnected via a 0.5 MΩ resistor. Sketch the circuit diagram now and calculate the peak value of the output voltage for the same a.c. voltage as before applied between the non-inverting input and 0 V.

Q16.3 (a) Sketch a circuit diagram to show how an op-amp can be used as a voltage follower, to convert a low-resistance voltmeter to be a high-resistance voltmeter.

(b) (i) Sketch the output voltage waveform of an astable multivibrator which operates at a frequency of 5 kHz.

(ii) Estimate a suitable resistance value for a 5 kHz multivibrator which is to be constructed using 0.1 μF capacitors.

Q16.4 Design an electronic thermometer capable of giving an output voltage that increases smoothly as the temperature of a thermistor increases over a limited range.

Unit 17

$\varepsilon_0 = 8.85 \times 10^{-12}$ F m^{-1}, e = 1.6×10^{-19} C

Q17.1 (a) Calculate the electric field strength and the electric potential at a distance of 3.0×10^{-10} m from a nucleus carrying a charge of 9.6×10^{-19}C.

(b) Calculate the work done to remove an electron at the above position away from the nucleus to infinity.

Q17.2 (a) Calculate the force between a +1.5 nC point charge and a +3.6 nC point charge which are at a distance of 50 mm apart.

(b) Calculate the electric field strength and the potential at the mid-point of these two point charges.

Q17.3 (a) (i) Calculate the electric field strength between two horizontal, parallel plates spaced 40 mm apart when one of the plates is earthed and the other is at a positive potential of 5.0 kV.

(ii) Sketch a graph to show how the potential varies from the earthed plate to the positive plate.

(b) A droplet carrying a charge of $+2.7 \times 10^{-15}$ C falls from the top plate to the bottom plate. Calculate the work done by the electric field on this droplet.

Q17.4 (a) Sketch the pattern of lines of force near a negatively charged conducting sphere.

(b) A conducting sphere of radius 0.15 m is negatively charged to a potential of 4.6 kV. Calculate **(i)** the charge on the sphere, **(ii)** the potential at a distance of 1.0 m from the sphere's surface.

Unit 18

$\mu_0 = 4\pi \times 10^{-7}$ H m^{-1}

Q18.1 (a) Sketch the pattern of magnetic field lines near **(i)** a bar magnet, **(ii)** a long straight wire carrying a direct current.

(b) The magnetic flux density at a distance of 80 mm from a long straight wire of diameter 0.80 mm was 38 μT. Calculate **(i)** the current in the wire, **(ii)** the magnetic flux density at the surface of the wire.

Q18.2 (a) A wire of length 0.065 m was placed perpendicular to the lines of force of a uniform magnetic field. When a current of 0.42 A was passed through the wire, a force of 1.50 mN was exerted on the wire. Calculate the magnetic flux density.

(b) The wire was turned so it made an angle of 40° with the lines of force of the magnetic field. Calculate the force on the wire in this position for the same current.

Q18.3 In an experiment to measure the horizontal component of the Earth's magnetic flux density, B_H, a 500 turn solenoid of length 0.20 m was aligned with its axis horizontal along a line from North to South. A plotting compass was placed at one end of the solenoid. An increasing current was passed through the solenoid until the direction of the plotting compass reversed.

(a) The current at the point of reversal was 5.7 mA. Calculate the horizontal component of the Earth's magnetic flux density, B_H.

(b) Sketch an end-view of the end of the solenoid where the plotting compass was positioned, indicating the direction of current around the solenoid.

Q18.4 (a) Sketch a graph to show how the magnetisation of a ferromagnet in a solenoid varies with the solenoid current.

(b) Explain each of the following:

(i) An iron bar is magnetised when it is placed in a direct current solenoid.

(ii) An iron bar can be demagnetised by removing it gradually from an a.c. solenoid.

Unit 19

Q19.1 (a) A coil of wire was connected to a centre-reading microammeter.

(i) When a bar magnet with its north pole nearest the coil was moved towards the coil, the needle of the microammeter deflected briefly to the left of centre. Explain why this happened and describe what would have happened if the south pole of the bar magnet had been nearest the coil.

(ii) Explain why the needle did not deflect when the magnet was held at rest close to the coil.

(b) A rod of length 0.20 m, moving at a speed of 25 m s^{-1}, cuts the lines of force of a uniform magnetic field of flux density 38 mT at right angles to the lines of force. Calculate the e.m.f. induced in the rod.

Q19.2 (a) (i) With the aid of a diagram, explain the operation of an a.c. generator.

(ii) Explain why the peak voltage from an a.c. generator occurs when the coil is parallel to the lines of force of the magnetic field.

(b) A 120 turn coil of length 0.15 m and width 0.04 m spins at a constant frequency of 20 Hz in a uniform magnetic field of flux density 85 mT. Calculate **(i)** the peak magnetic flux through the coil, **(ii)** the peak voltage produced by the coil.

Q19.3 (a) Explain why the current passing through the armature of a d.c. motor connected to a battery increases when the motor speed is reduced.

(b) A d.c. electric motor with an armature of resistance 2.5 Ω is connected to a 12 V battery of internal resistance 0.5 Ω. When the motor rotates at a frequency of 25 Hz, the current in the circuit is 0.20 A. Calculate **(i)** the back e.m.f. at this frequency, **(ii)** the current when the frequency is reduced to 15 Hz.

Q19.4 A transformer consists of a 60 turn primary coil and a 1200 turn secondary coil. A 240 V, 60 W light bulb is connected to the secondary coil.

(a) Calculate the primary voltage which would light the bulb normally.

(b) Calculate the maximum primary current to light this bulb normally.

Unit 20

Q20.1 The current in an a.c. circuit changes with time according to the equation I (in amperes) = 1.60 sin (100 πt).

(a) Calculate the peak current and the frequency of this alternating current.

(b) Calculate the current in this circuit at t = **(i)** 2.50 ms, **(ii)** 5.00 ms, **(iii)** 12.0 ms.

Q20.2 (a) Calculate **(i)** the r.m.s. value of a sinusoidal current of peak value 2.4 A, **(ii)** the peak voltage of a sinusoidal voltage of r.m.s. value 12.0 V.

(b) A 5.0 Ω resistor is connected to a 50 Hz, 6.0 V r.m.s. alternating voltage supply unit. Calculate **(i)** the r.m.s. current through the resistor, **(ii)** the mean power dissipated in the resistor.

Q20.3 (a) A 2.2 μF capacitor is connected in series with a 2.0 kΩ resistor and a 50 Hz, 6.0 V r.m.s. alternating voltage supply unit. Sketch the circuit and calculate **(i)** the r.m.s. current in the circuit, **(ii)** the r.m.s. voltage across each component, **(iii)** the power dissipated in the circuit.

(b) The capacitor in **(a)** is replaced by a 1.2 H inductor which has negligible resistance. Sketch this circuit and calculate **(i)** the r.m.s. current in the circuit, **(ii)** the r.m.s. voltage across each component, **(iii)** the power dissipated in the circuit.

Q20.4 A 4.7 μF capacitor is connected in series with a coil of inductance 7.5 mH and resistance 25 Ω and a variable frequency a.c. supply unit.

(a) Calculate the resonant frequency of this circuit.

(b) The supply unit is adjusted to give an output at 1.5 kHz and 6.0 V r.m.s. Calculate **(i)** the circuit impedance, **(ii)** the r.m.s. current, **(iii)** the r.m.s. voltage across the coil.

Unit 21

$e = 1.6 \times 10^{-19}$ C, $h = 6.6 \times 10^{-34}$ J s, $m_e = 9.1 \times 10^{-31}$ kg

Q21.1 (a) With the aid of a diagram, explain how a beam of electrons is produced from a hot wire filament.

(b) Calculate the kinetic energy and speed of electrons produced in an electron tube operating at an anode voltage of 4.5 kV.

Q21.2 A narrow beam of electrons is directed into a uniform magnetic field of magnetic flux density 3.2 mT at 90° to the lines of force of the magnetic field. The beam forms a circle of diameter 78 mm.

(a) Calculate **(i)** the speed and **(ii)** the kinetic energy of the electrons in this beam.

(b) (i) Sketch the path of the beam in the field and indicate the direction of the magnetic field.

(ii) Mark a point on the path of the beam and indicate the direction of the force due to the magnetic field on an electron in the beam at that point.

(iii) Explain why the kinetic energy of each electron in the beam is constant, even though a force acts on each electron.

Q21.3 In an experiment to measure the charge on an oil droplet, an oil droplet was held stationary in a uniform electric field of strength 252 kV m^{-1}. When the field was switched off, the droplet descended at steady speed, taking 16.2 s to fall through a distance of 2.9 mm. Calculate **(a)** the terminal speed of the droplet, **(b)** the droplet's radius and mass, **(c)** the charge carried by the droplet:

The viscosity of air = 1.8×10^{-5} N s m^{-2}, the density of the oil used = 850 kg m^{-3}.

Q21.4 (a) Calculate the maximum frequency of X-rays produced by an X-ray tube operating at 55 kV.

(b) Light of wavelength 550 nm is directed at an earthed metal plate, causing photoelectric emission. This emission ceases if the plate potential is raised to +0.52 V. Calculate the work function and the threshold frequency of the metal surface.

Unit 22

Q22.1 (a) Explain, in terms of protons and neutrons, the change that takes place in an unstable nucleus when it emits **(i)** an α-particle, **(ii)** a β-particle.

(b) A sample of a radioactive isotope of half-life 15 hours has an initial activity of 0.64 MBq. Calculate its activity after **(i)** 30 hours, **(ii)** 7 days.

Q22.2 A Geiger counter was placed at a distance of 0.20 m from a radioactive source. The average count rate was measured at 25.20 s^{-1}. When a metal plate of thickness 2.00 mm was placed between the source and the counter, the count rate decreased to 9.40 s^{-1}. When the source was removed, the count rate was 0.40 s^{-1}.

(a) What type of radiation was emitted by the source? Explain your answer.

(b) Calculate the half value thickness of the metal.

Q22.3 Carbon 14 is a radioactive isotope which has a half-life of 5570 years. It decays to form nitrogen 14.

(a) Write down the equation for this radioactive change.

(b) Living wood has an activity of 270 Bq kg^{-1} due to the decay of carbon 14. A 1.2 g sample of wood from the hull of an ancient ship has an activity of 0.26 Bq. Estimate the age of the sample.

Q22.4 A point source of gamma radiation is at a distance of 0.50 m from the end of a GM tube of cross-sectional area 3.1×10^{-4} m^2. The corrected count rate is 35.2 counts per second.

(a) Estimate the activity of the source.

(b) Calculate the corrected count rate if the source-tube distance is reduced to 0.20 m.

Unit 23

$\varepsilon_0 = 8.85 \times 10^{-12}$ F m^{-1}, e = 1.6×10^{-19} C 1, u = 931 MeV

The Avogadro constant = 6.02×10^{23} mol^{-1}

Q23.1 (a) Estimate the least distance of approach of a 5.0 MeV α-particle to a nitrogen nucleus (Z = 7).

(b) Use your estimate in **(a)** to show that the density of the nucleus is much greater than the density of ordinary matter.

Q23.2 (a) (i) Explain what is meant by the binding energy of a nucleus.

(ii) Sketch a graph to show how the binding energy per nucleon of a nucleus varies with the mass number of the nucleus.

(b) Calculate the binding energy per nucleon of **(i)** a helium 4 nucleus (Z = 2), **(ii)** a uranium 238 nucleus (Z = 92).

(Atomic masses: helium 4 = 4.00260 u, uranium 238 = 238.05076 u, proton mass = 1.00728 u, neutron mass = 1.00867 u, electron mass = 0.00055 u)

Q23.3 Cobalt 60 (Z = 27) is a radioactive isotope that emits β-particles to form an isotope of nickel (Z = 28).

(a) Write down the equation representing the decay of this isotope.

(b) Calculate the Q-value for this decay. (Atomic masses: cobalt 60 = 59.93381 u, nickel 60 = 59.93078 u)

(c) Cobalt 60 has a half-life of 5.3 years. Calculate the activity of 1 gram of cobalt 60.

Q23.4 (a) In a nuclear reactor, uranium 235 nuclei undergo a controlled chain reaction of neutron-induced fission reactions. Explain what is meant by **(i)** neutron-induced fission, **(ii)** a controlled chain reaction of fission reactions.

(b) Explain the function of each of the following in a nuclear reactor:

(i) the moderator, **(ii)** the control rods, **(iii)** the coolant.

Unit 24

The Avogadro constant = 6.02×10^{23} mol^{-1}, R = 8.31 J mol^{-1} K^{-1}

Q24.1 (a) A piston contains 0.12 m^3 of dry air at 300 K and a pressure of 102 kPa. Calculate **(i)** the number of moles of air in the piston, **(ii)** the pressure in the piston if the volume is reduced to 0.02 m^3 and the temperature is changed to 400 K.

(b) In a chemistry experiment, a volume of 18.7 cm^3 of gas was collected at 14°C and at a pressure of 105 kPa. Calculate the volume of this amount of gas at 0°C and 102 kPa pressure.

Q24.2 (a) A hand-operated pump of volume 100 cm^3 is designed to reduce the air pressure in a flask of volume 500 cm^3. The air in the pump is initially at a pressure of 102 kPa. Calculate the pressure of the air in the flask after **(i)** a single stroke of the pump which increases the volume of air from 500 to 600 cm^3, **(ii)** 10 strokes of the pump.

(b) Calculate the number of molecules per unit volume in a gas at a pressure of 10 kPa and a temperature of 300 K.

Q24.3 (a) (i) State the kinetic theory equation for an ideal gas, defining the symbols you use.

(ii) State the assumptions of the kinetic theory of gases.

(b) A sealed cylinder of volume 0.25 m^3 contains nitrogen gas at a pressure of 120 kPa and a temperature of 290 K. Calculate **(i)** the number of moles of gas present, **(ii)** the mass of gas present, **(iii)** the r.m.s. speed of the gas molecules, **(iv)** the mean kinetic energy of the gas molecules.

The molar mass of nitrogen = 0.028 kg

Q24.4 (a) Define the root mean square speed of the molecules of a gas.

(b) Calculate the root mean square speed of the molecules of **(i)** oxygen (molar mass = 0.032 kg) at 15°C, **(ii)** oxygen at 0°C, **(iii)** hydrogen (molar mass = 0.002 kg) at 0°C.

Unit 25

Q25.1 (a) State what physical property is measured in **(i)** a liquid-in-glass thermometer, **(ii)** a thermocouple thermometer.

(b) The pressure of a constant volume gas thermometer relative to atmospheric pressure was −9.2 kPa at 0°C and 27.4 kPa at 100°C. Calculate the temperature when the pressure was 5.1 kPa in **(i)** °C, **(ii)** K.

Q25.2 (a) A gas cylinder of volume 0.058 m^3 contained an ideal gas at 20°C at a pressure of 101 kPa. Calculate **(i)** the number of moles of gas in the cylinder, **(ii)** the internal energy of the gas.

(b) The temperature of the cylinder was raised to 100°C. Calculate **(i)** the pressure of the gas, **(ii)** the increase of internal energy of the gas.

Q25.3 Four moles of an ideal gas at 300 K were heated at a constant pressure of 100 kPa until its temperature was 350 K.

(a) Calculate **(i)** the initial volume of the gas, **(ii)** the final volume of the gas, **(iii)** the work done on the gas.

(b) (i) Show that the increase of internal energy of the gas was 1.66 kJ.

(ii) Calculate the heat supplied to the gas.

Q25.4 (a) A piston containing 0.0301 m^3 of air at 102 kPa and 20°C was compressed adiabatically to a volume of 0.0040 m^3. Calculate the pressure of the gas after the compression and show that its temperature was 384 °C. (Assume γ for air = 1.4)

(b) Calculate the maximum efficiency of a heat engine operating between 20°C and 384°C.

Unit 26

Q26.1 A spin drier has a drum of diameter 0.35 m which is designed to spin at a speed of 600 turns per minute. Calculate **(a)** the time taken for one turn at this frequency, **(b)** the speed and centripetal acceleration of a point on the drum.

Q26.2 A vehicle fitted with tyres of diameter 0.45 m is travelling at a steady speed of 31 m s^{-1}. Calculate **(a)** the time taken for each wheel to rotate once, **(b)** the frequency of rotation of the wheels, **(c)** the centripetal acceleration of a point on one of the tyres.

Q26.3 A vehicle of mass 950 kg travels over a bridge of radius of curvature 128 m at a steady speed of 24 m s^{-1}. Calculate **(a)** the weight of the vehicle, **(b)** the centripetal acceleration of the vehicle on the bridge, **(c)** the force of the vehicle on the bridge as it passes over the highest point of the bridge.

Q26.4 A pendulum consists of a bob of mass 0.12 kg on the end of a string of length 0.95 m. The bob is released from a stationary position with the string taut and horizontal. Calculate **(a)** the loss of potential energy of the bob between its initial position and its lowest position, **(b)** the speed of the bob as it passes through its lowest position, **(c)** the centripetal acceleration of the bob at its lowest position, **(d)** the tension in the string as the bob passes through its lowest position.

Unit 27

See the question box on p. 400 for data for the questions below.

Q27.1 (a) Calculate the force of gravity on a 100 kg satellite in a circular orbit 3000 km above the Earth.

(b) Calculate the speed and time period of the satellite in **(a)**.

Q27.2 (a) The gravitational field strength at the surface of the Earth is 9.8 N kg^{-1}. Calculate the height above the Earth at which the gravitational field strength of the Earth is 9.0 N kg^{-1}.

(b) Show that the escape speed from the Earth's surface is 11.2 km s^{-1}.

Q27.3 (a) Define gravitational potential.

(b) (i) Show that the gravitational potential at the surface of the Moon is −2.8 MJ kg^{-1}.

(ii) Calculate the escape speed off the lunar surface.

Q27.4 (a) Calculate the gravitational field strength of the Sun at the Earth's orbit.

(b) (i) Calculate the distance from the centre of the Sun to the point at which its gravitational field strength is equal to 9.8 N kg^{-1}.

(ii) Calculate the ratio of the distance in **(i)** to the mean distance from the Earth to the Sun.

Unit 28

$g = 9.8$ m s^{-2}

Q28.1 (a) Define simple harmonic motion.

(b) A mass on a spring oscillates vertically, taking 25.2 s to complete 10 complete cycles of oscillation. Its height above the floor changes from 50 mm to 180 mm during each half cycle.

Calculate **(i)** its time period, **(ii)** its amplitude of oscillation, **(iii)** its maximum speed.

Q28.2 A simple pendulum consists of a spherical bob of mass 0.052 kg and diameter 1.1 cm attached to a thread of length 920 mm.

(a) Calculate the time period of this pendulum.

(b) The bob is released at rest with the thread taut at an angle of 5° to the vertical. Calculate **(i)** the amplitude of oscillation, **(ii)** the maximum speed of the bob, **(iii)** the maximum acceleration of the bob.

Q28.3 (a) A 300 mm steel spring fixed at its upper end is used to support a 0.25 kg mass. Its equilibrium length when supporting this mass is 348 mm. Calculate **(i)** the stiffness constant of this spring, **(ii)** the time period of vertical oscillations of the 0.25 kg mass on the spring.

(b) The 0.25 kg mass is displaced vertically by 20 mm from its equilibrium position then released. Calculate **(i)** the maximum speed of the mass, **(ii)** the minimum tension in the spring.

Q28.4 (a) What is meant by **(i)** free oscillations, **(ii)** damped oscillations?

(b) (i) Explain what is meant by resonance.

(ii) Describe an example of a mechanical system in resonance.

Using Spreadsheets

A spreadsheet may be used to carry out repeatedly a sequence of calculations. The simple example to the right is designed to illustrate how to set a spreadsheet up.

A general dynamics spreadsheet

The spreadsheet instructions that follow are designed to enable you to calculate the displacement, velocity and acceleration of any object at successive intervals, given its acceleration in terms of displacement and velocity.

The acceleration, a, is expressed as a general formula $a = \alpha + bv + cs$, where α, b and c are constants. The values of these constants are inserted into the spreadsheet as well as the time interval, dt. The shorter the time interval, the more accurate the calculations. However, more rows are then needed to reach a particular point in time.

The initial values of s, v and time, t, are inserted. The spreadsheet then automatically calculates s, v and a at subsequent times dt, 2dt, 3dt, 4dt, etc. The results may be plotted as a graph using the related chart display.

The instructions to set up this spreadsheet using LOTUS 123 are given below. Some changes may be necessary for other spreadsheet packages.

Example 1

Calculate the displacement every second from O, of an object moving with a constant velocity of 5 m s^{-1}. The object is initially at O.

The displacement, s, at time, t, is given by $s = vt$, where $v = 5$ m s^{-1}.

Step 1 Insert the velocity value in cell A1 and the time interval (dt) in cell B1.

Step 2 Label cell A2 ‘t’ and cell B2 ‘s’; use “ before t or s to insert the letters.

Step 3 Insert 0 into cells A3 and B3 for the initial values of the time and displacement.

Step 4 Insert the formula +A3+$B$1 into cell A4 to display the time at the end of the first time interval. The $ sign is necessary to give an absolute reference to cell B1.

Step 5 Insert the formula +$A$1*A4 into cell B4 to calculate and display the displacement.

Step 6 Copy cells A4 and B4 down as far as desired. The spreadsheet should then show the time and displacement on each row. Note that in each cell below A4 or B4, the formulae retain addresses labelled with the $ sign (absolute addresses) and index addresses without the $ sign.

LOTUS 123 set up

Acceleration = α +bv + cs; insert values of constants α, b and c into cells A1, B1 and C1 respectively.

Insert the column titles t, t+dt, v, s, a, dv, v+dv, ds, s+ds into cells A2 to I2; use “ to insert these labels.

Initial time, t, and time interval, dt: insert values into cells A3 and B3 respectively.

Initial displacement and velocity: insert values into cells D3 and C3 respectively.

Calculation of acceleration: insert the formula +$A$1+$B$1*C3+$C$1*D3 into cell E3 to calculate α +bv + cs.

Calculation of change of velocity, dv, and final velocity, v+ dv: insert +E3*$B$3 into cell F3 to calculate dv; insert + F3+C3 into cell G3 to calculate v + dv.

Calculation of change of displacement, ds, and s: insert +0.5*(G3+C3)*$B$3 into cell H3 to calculate ds; insert +H3+D3 into cell I3 to calculate s+ds.

Calculation of time at the start of the next interval: insert +A3+$B$3 into cell A4.

Calculation of time at the end of the interval (t+dt): insert +A4+$B$3 into cell B4.

Next values of displacement and velocity: insert +I3 into cell D4; insert +G3 into cell C4.

Next values of a, dv, v+dv, ds and s + ds: copy cells E3 to I3 into cells E4 to I4 respectively. Use the ‘copy down’ facility for this.

Next values: copy down the row of cells A4 to I4 as far as required.

Example 2

An object released from rest

Acceleration, a = -9.8 m s^{-2}, hence $\alpha = -9.8$, b = c = 0. Also, the initial values of displacement, velocity and time are zero. In the spreadsheet below, dt = 0.200 s.

−9.800	**0.000**	**0.000**						
t	t+dt	v	s	a	dv=a*dt	v + dv	ds	s+ds
0.000	0.200	0.000	0.000	−9.800	−1.960	−1.960	−0.196	−0.196
0.200	0.400	−1.960	−0.196	−9.800	−1.960	−3.920	−0.588	−0.784
0.400	0.600	−3.920	−0.784	−9.800	−1.960	−5.880	−0.980	−1.764
0.600	0.800	−5.880	−1.764	−9.800	−1.960	−7.840	−1.372	−3.136
0.800	1.000	−7.840	−3.136	−9.800	−1.960	−9.800	−1.764	−4.900
1.000	1.200	−9.800	−4.900	−9.800	−1.960	−11.760	−2.156	−7.056
1.200	1.400	−11.760	−7.056	−9.800	−1.960	−13.720	−2.548	−9.604
1.400	1.600	−13.720	−9.604	−9.800	−1.960	−15.680	−2.940	−12.544
1.600	1.800	−15.680	−12.544	−9.800	−1.960	−17.640	−3.332	−15.876
1.800	2.000	−17.640	−15.876	−9.800	−1.960	−19.600	−3.724	−19.600
2.000	2.200	−19.600	−19.600	−9.800	−1.960	−21.560	−4.116	−23.716
2.200	2.400	−21.560	−23.716	−9.800	−1.960	−23.520	−4.508	−28.224

Note

Presentation of the spreadsheet may be improved by reducing the column widths to 8 characters per column, setting the number of decimal places to 3 and centering all the entries.

Example 3

An object released from rest in a fluid

Acceleration, a = -9.8 m s^{-2}, hence $\alpha = -9.8$, b = -0.10, c = 0. Also, the initial values of displacement, velocity and time are zero. In the spreadsheet below, dt = 0.500 s.

−9.800	**−0.100**	**0.000**						
t	t+dt	v	s	a	dv	v+dv	ds	s+ds
0.000	0.200	0.000	0.000	−9.800	−1.960	−1.960	−0.196	−0.196
0.200	0.400	−1.960	−0.196	−9.604	−1.921	−3.881	−0.584	−0.780
0.400	0.600	−3.881	−0.780	−9.412	−1.882	−5.763	−0.964	−1.744
0.600	0.800	−5.763	−1.744	−9.224	−1.845	−7.608	−1.337	−3.082
0.800	1.000	−7.608	−3.082	−9.039	−1.808	−9.416	−1.702	−4.784
1.000	1.200	−9.416	−4.784	−8.858	−1.772	−11.187	−2.060	−6.844
1.200	1.400	−11.187	−6.844	−8.681	−1.736	−12.924	−2.411	−9.255
1.400	1.600	−12.924	−9.255	−8.508	−1.702	−14.625	−2.755	−12.010
1.600	1.800	−14.625	−12.010	−8.337	−1.667	−16.293	−3.092	−15.102
1.800	2.000	−16.293	−15.102	−8.171	−1.634	−17.927	−3.422	−18.524
2.000	2.200	−17.927	−18.524	−8.007	−1.601	−19.528	−3.746	−22.270
2.200	2.400	−19.528	−22.270	−7.847	−1.569	−21.098	−4.063	−26.332

See www.palgrave. com/foundations/breithaupt to investigate motion using a spreadsheet.

List of Experiments in the Book

Key

D short experiments that **demonstrate** a specific point
E standard **experiments** that can be completed in a single session
I short **investigations** or tests that develop or reinforce a topic
P longer experiments that could be developed into **projects**
* Website only (see page reference and www.palgrave.com/foundations/breithaupt)

Safety note
A risk assessment must be carried out before any practical work (including demonstrations) is carried out. All necessary safety information should be supplied to students.

Location Guide to Mathematical Skills

See www.palgrave.com/foundations/breithaupt for exercises on these mathematical skills.

Algebra

1. Rearranging an equation ***Units and measurements***
2. Using the dynamics equations **10.3**
3. Pythagoras theorem **Fig. 9.6**
4. Simultaneous equations **Fig. 14.7**

Calculus

1. Rates of change in dynamics **10.6, 11.2**
2. Rates of change in exponential decay
capacitor discharge **15.5**
radioactive decay **22.5**
3. Differentiation of x^n and e^x **15.5**
4. Differentiating a sine function **Fig. 20.21**

Graphs

1. Straight line graphs
through the origin **Fig. 7.2 and Fig. 7.9**
$y = mx + c$ **Fig. 10.15**
2. Inverse square law
Coulomb's law **17.1**
gamma radiation **22.3**
Newton's theory of gravity **27.1**
3. Rates of change
dynamics **10.6, 11.2**
exponential decay:
capacitor discharge **15.5**
radioactive decay **22.5**
4. Area under a curve
energy stored in a spring **7.1**
energy stored in a capacitor **15.3**
energy to escape from a planet **27.3**

Trigonometry

1. Sin, cos and tan **3.2, 9.1**
2. Sine wave formula **Fig. 19.15, Fig. 20.4, Fig. 28.7**
3. Circular measures **26.1**

Using a calculator

1. Standard notation and powers of ten ***Units and measurements***
2. Raising a number to a given power using the x^y key of a calculator
finding a cube root **5.3**
adiabatic equation **25.3**
3. Using the 'ln' key and 'e' key of a calculator
capacitor discharge **15.5**
radioactive decay **22.5**

Vectors

1. Representing vectors **p. 117, Fig. 9.1**
2. Parallelogram rule **Fig. 9.3**
3. Resolving a force **Fig. 9.6**
4. Calculating the resultant **Fig. 9.7**
5. i and j vectors **Fig. 9.1**
6. Pythagoras theorem **Fig. 9.6**

Summary of Equations

General

Density

$\rho = \frac{m}{V}$ where m = mass, V = volume

Weight

Weight = mg

Circle of radius, r

area, $A = \pi r^2$

circumference, $C = 2\pi r$

arc length, $s = r\theta$, where θ = angle subtended by the arc in radians

Sphere of radius, r

surface area = $4\pi r^2$

volume, $V = \frac{4}{3}\pi r^3$

Cylinder of radius, r and length, L

volume = $\pi r^2 L$

Part 1: Waves

Frequency

$f = \frac{1}{T}$ where T = time period

Wave speed

$\upsilon = f\lambda$ where f = frequency and λ = wavelength

Phase difference

$$\Delta\Phi = \frac{2\pi x}{\lambda}$$

Stationary waves on a string of length, L

$2L = m\lambda$

Stationary waves in a pipe of length, L, open at one end

$$L + e = \frac{(2m+1)\lambda}{4}$$

Stationary waves in a pipe of length, L, open at both ends

$$L + 2e = \frac{m\lambda}{2}$$

Young's slits

$$\lambda = \frac{yd}{X}$$

where y = fringe spacing, d = slit spacing, X = slit-screen distance

Snell's law

$$\frac{\sin i}{\sin r} = \frac{\lambda_i}{\lambda_r} = \frac{\upsilon_i}{\upsilon_r} = \text{refractive index, } n$$

Lens formula

$$\frac{1}{u} + \frac{1}{\upsilon} = \frac{1}{f}$$

Diffraction grating equation

$d \sin\theta_m = m\lambda$ where d = slit spacing

Part 2: Properties of Materials

Thermal expansion

$\Delta L = \alpha L \Delta T$

Energy transferred $= mc\,(T_2 - T_1)$

for change of temperature $(T_2 - T_1)$ of mass, m

Energy transferred $= ml$

for change of state of mass, m

Power radiated

$W = e\sigma AT^4$ from a surface at temperature, T

Heat transfer

$\frac{Q}{t} = \frac{kA(T_2 - T_1)}{L}$ through a thermal conductor of thermalconductivity, k

Hooke's law

$T = ke$ where T = tension and k = spring constant

Young modulus of elasticity

$$\frac{\text{stress}}{\text{strain}} = \frac{(T \div A)}{(e \div L)} = \frac{TL}{Ae}$$

Energy stored in a stretched spring

$= \frac{1}{2}ke^2$ [$= \frac{1}{2}(AE/L)e^2$ for a stretched wire]

Pressure

$p = \frac{F}{A}$ where F = force applied at right angles to a surface of area, A

Pressure of a liquid column of depth, H

$p = H\rho g$

Part 3: Mechanics

Component of a force, F, along a line at angle, θ
$= F\cos\theta$

Moment of a force $= Fd$

Coefficient of friction
$$\mu = \frac{F}{N}$$

Equations for uniform acceleration
$v = u + at$; $s = \frac{1}{2}(u + v)t$; $s = ut + \frac{1}{2}at^2$; $v^2 = u^2 + 2as$

Momentum $= mv$

Newton's second law
$F = \frac{d}{dt}(mv)$ becomes $F = ma$ for constant mass, m

and $F = v\frac{dm}{dt}$ for constant rate of change of mass

Work = force × distance

Power $= \frac{\text{energy transferred}}{\text{time taken}}$

Kinetic energy $= \frac{1}{2}mv^2$

Change of potential energy $= mgh$

Part 4: Electricity

Charge
$Q = It$ for constant current and
$I = \frac{dQ}{dt}$ for changing current

Potential difference
$V = \frac{E}{Q}$ where E = electrical energy transferred

Resistance
$$R = \frac{V}{I}$$

Resistor combination rules
in series, $R = R_1 + R_2$; in parallel, $\frac{1}{R} = \frac{1}{R_1} + \frac{1}{R_2}$

Potential divider equation
p.d. across $R_1 = \frac{R_1}{R_1 + R_2}V$

Potentiometer equation
$$\frac{E_1}{E_2} = \frac{l_1}{l_2}$$

Wheatstone bridge equation
$$\frac{P}{R} = \frac{Q}{S}$$

Resistivity
$$\rho = \frac{RA}{L}$$

Capacitance
$$C = \frac{Q}{V}$$

Capacitor combination rules
in series, $\frac{1}{C} = \frac{1}{C_1} + \frac{1}{C_2}$; in parallel, $C = C_1 + C_2$

Energy stored in a capacitor $= \frac{1}{2}CV^2$

Capacitor discharge
$Q = Q_0e^{-t/RC}$

Time constant $= CR$

Capacitance of parallel plates
$$C = \frac{A\varepsilon_o\varepsilon_r}{d}$$

Voltage gain
$\frac{V_{OUT}}{V_{IN}} = -\frac{R_F}{R_1}$ for an inverting amplifier

Voltage gain
$\frac{V_{OUT}}{V_{IN}} = -\frac{R_F + R_1}{R_1}$ for an non-inverting amplifier

Part 5: Fields

Coulomb's law
$$F = \frac{Q_1Q_2}{4\pi\varepsilon_o r^2}$$

Electric field strength due to two parallel plates
$$E = \frac{V}{d} = \frac{Q}{A\varepsilon_o}$$

Electric field near a point charge, Q
Electric field strength, $E = \frac{Q}{4\pi\varepsilon_o r^2}$

Electric potential, $V = \frac{Q}{4\pi\varepsilon_o r}$

Force on a current-carrying conductor
$F = BIL\sin\theta$

Force on a moving charge
$F = Bq\upsilon \sin\theta$

Hall voltage
$V_H = B\upsilon d$

Magnetic flux density formulae

at distance, r, from a long straight wire, $B = \frac{\mu_0 I}{2\pi r}$

inside a long solenoid, $B = \frac{\mu_0 NI}{l}$

Magnetic flux linkage
$\Phi = BAN$

Faraday's law of electromagnetic induction

induced e.m.f., $V = -\frac{d\Phi}{dt}$

Induced e.m.f. formulae
for a moving rod: $V = BL\upsilon$
for a generator coil: $V = 2\pi fBAN \sin(2\pi ft)$

Transformer equation
$\frac{V_S}{V_P} = \frac{N_S}{N_P}$

Self-inductance
$L = \frac{\Phi}{I}$

Energy stored in a self-inductance $= \frac{1}{2}LI^2$

R.m.s. value

R.m.s. value of an alternating current $= \frac{\text{peak value}}{\sqrt{2}}$

Reactance of a capacitor
$\frac{V_o}{I_O} = \frac{1}{2\pi fC}$

Reactance of an inductor
$\frac{V_o}{I_O} = 2\pi fL$

Impedance of a series LCR circuit
$Z = [R^2 + (2\pi fL - \frac{1}{2}\pi fC)^2]^{1/2}$

Resonant frequency

Resonant frequency $= \frac{1}{2\pi (LC)^{1/2}}$

Part 6: Atomic and Nuclear Physics

Electron gun equation
$\frac{1}{2}m\upsilon^2 = eV_A$

Force on an electron between oppositely charged parallel plates

$F = \frac{eV}{d}$

Force on an electron moving across a magnetic field
$F = Be\upsilon$

Millikan's experiment

$\frac{QV}{d} = mg$

Stokes Law
$F = 6\pi\eta r\upsilon$

Photon energy
$E = hf$

Maximum k.e. of emitted photoelectron $= hf - \phi$

Maximum frequency of X-rays from an X-ray tube

$f_{max} = \frac{eV_A}{h}$

Radioactive decay equations

activity, $A = \frac{dN}{dt} = -\lambda N$

$N = N_o e^{-\lambda t}$

decay constant, $\lambda = \frac{\log_e 2}{T_{1/2}}$

Energy
$E = mc^2$

Part 7: Further Physics

Boyle's law
$pV = \text{constant}$

Charles' Law
$\frac{V}{T} = \text{constant}$

Pressure law
$\frac{p}{T} = \text{constant}$

Ideal gas equation
$pV = nRT$

Kinetic theory equation
$pV = \frac{1}{3}Nm\upsilon^2_{RMS}$ or $p = \frac{1}{3}\rho\upsilon^2_{RMS}$

Mean kinetic energy of a gas molecule $= \frac{3}{2}kT$

Absolute temperature
$$T = \frac{pV}{(pV)_{Tr}} \times 273.16$$

Celsius temperature
t = absolute temperature, T − 273.15

1st Law of Thermodynamics
$\Delta Q = \Delta U + \Delta W$

Work done by a gas $= p\Delta V$

Molar heat capacities of a gas
$C_P - C_V = R$

monoatomic gas: $C_P = \frac{5}{2}RT$, $C_V = \frac{3}{2}RT$, $\gamma = \frac{5}{3}$
diatomic gas: $C_P = \frac{7}{2}RT$, $C_V = \frac{5}{2}RT$, $\gamma = \frac{7}{5}$
polyatomic gas: $C_P = \frac{8}{2}RT$, $C_V = \frac{6}{2}RT$, $\gamma = \frac{4}{3}$

Equation for an adiabatic change of a gas
pV^γ = constant

Efficiency of a heat engine
$$= \frac{W}{Q_1} = \frac{Q_1 - Q_2}{Q_1}$$

Maximum efficiency of a heat engine
$$= \frac{T_1 - T_2}{T_1}$$

Angular speed
$$\omega = \frac{2\pi}{T} = 2\pi f$$

Centripetal acceleration
$$a = \frac{-\upsilon^2}{r} = -\omega^2 r$$

Newton's law of gravitation
$$F = \frac{Gm_1m_2}{r^2}$$

Gravitational field strength
$$g = \frac{F}{m}$$

Potential energy of two masses
$$= -\frac{Gm_1m_2}{r}$$

At distance, r, from the centre of a spherical planet of mass, M
$$g = \frac{GM}{r^2}$$

Escape speed $= \sqrt{(2gr)}$

Satellite equation
$$r^3 = \frac{GM}{4\pi^2}T^2$$

Simple harmonic motion equations
acceleration, $a = -\omega^2 s$ where s = displacement
maximum speed, $\upsilon_{max} = \omega r$ where r = amplitude
time period, $T = 2\pi\left(\frac{m}{k}\right)^{½}$ for mass, m, on a spring
time period, $T = 2\pi\left(\frac{L}{g}\right)^{½}$ for a simple pendulum of length, L

Useful Data

Table of physical constants

The Avogadro constant	$N_A = 6.02 \times 10^{23}\ mol^{-1}$
1 atomic mass unit	$= 1.666 \times 10^{-27}$ kg
The Stefan constant	$\sigma = 5.67 \times 10^{-8}\ W\ m^{-2}K^{-4}$
Absolute permittivity of free space	$\varepsilon_o = 8.85 \times 10^{-12}\ F\ m^{-1}$
Absolute permeability of free space	$\mu_o = 4\pi \times 10^{-7}\ H\ m^{-1}$
Speed of light in a vacuum	$= 3.00 \times 10^{8}\ m\ s^{-1}$
The Planck constant	$h = 6.63 \times 10^{-34}$ J s
Electronic charge	$e = 1.60 \times 10^{-19}$ C
Specific charge of the electron	$e/m = 1.76 \times 10^{11}\ C\ kg^{-1}$
Mass of the electron	$= 9.11 \times 10^{-31}$ kg
Mass of the proton	$m_p = 1.67 \times 10^{-27}$ kg
Molar gas constant	$R = 8.31\ J\ K^{-1}\ mol^{-1}$
The Boltzmann constant	$k = 1.38 \times 10^{-23}\ J\ K^{-1}$
Gravitational constant	$G = 6.67 \times 10^{-11}\ Nm^2\ kg^{-2}$

Useful numerical data

The weight of 1 kg at the surface of the Earth =	9.8 N
2π radians =	360°
1 electron volt =	1.60×10^{-19} J
1 u =	931 MeV

Answers to In-text Questions and Revision Questions

The numerical answers for each question are usually given to the same number of significant figures as the data in the question.

Units and Measurements

In-text questions

U1

1. (a) (i) 0.500 m **(ii)** 320 cm **(iii)** 95.60 m.
(b) (i) 450 g **(ii)** 1.997 kg **(iii)** 5.4×10^7 g.
2. (i) 1.50×10^{11} m **(ii)** 3.15×10^7 s **(iii)** 6.30×10^{-7} m
(iv) 2.58×10^{-8} kg **(v)** 1.50×10^6 m **(vi)** 1.25×10^{-6} m.

U2

1. (a) 1.90 kg **(b)** 2.40×10^{-4} m^3 **(c)** 7920 kg m^{-3}.
2. (a) 0.048 kg **(b)** 1.77×10^{-5} m^3 **(c)** 2710 kg m^{-3}.
3. (a) (i) 0.960 kg **(ii)** 9.60×10^{-4} m^3 **(iii)** 0.101 m
(b) (i) 0.105 kg **(ii)** 4.0×10^{-5} m^3 **(iii)** 2630 kg m^{-3}.

Revision questions

1. (a) C **(b)** E **(c)** B **(d)** E.
2. (a) (i) 54.0 g **(ii)** 0.1 g **(b) (i)** 103.4 g **(ii)** 0.1 g **(c) (i)** 107 cm^3
(ii) 2 cm^3

(d) (i) $\text{density} = \dfrac{\text{mass}}{\text{volume}} = \dfrac{103.4 \times 10^{-3}\,\text{kg}}{107 \times 10^{-6}\,\text{m}^3} = 966\ \text{kg m}^{-3}$

(ii) $\text{maximum density} = \dfrac{\text{maximum mass}}{\text{minimum volume}} = \dfrac{103.5 \times 10^{-3}\,\text{kg}}{105 \times 10^{-6}\,\text{m}^3} = 986\ \text{kg m}^{-3}$

Hence the uncertainty in the density = 20 kg m^{-3}. The density is therefore between 946 and 986 kg m^{-3}, i.e. density = 966 ± 20 kg m^{-3}. The percentage uncertainty in the density is therefore $\dfrac{20}{966} \times 100\% = 2\%$.

3. (a) 0.330 m **(b)** 6.5×10^{-3} m^2 **(c)** 1.63×10^{-5} m^3 (= 0.100m × 0.065 m × 0.0025 m) **(d)** 7380 kg m^{-3} (= 0.120 kg ÷ 1.63×10^{-5} m^3).

4. (a) volume = $\pi r^2 L$, where r, the radius, = 26 mm = 0.026 m and L, the length, = 1.2 m
$\therefore$ volume = $\pi \times (0.026)^2 \times 1.2 = 2.55 \times 10^{-3}$ m^3
(b) mass = volume × density = $2.55 \times 10^{-3} \times 2700 = 6.88$ kg.

5. (a) Volume V = $\frac{4}{3}\pi r^3$, where r, the radius, = 12 mm = 0.012 m.
Hence the volume = $\frac{4}{3}\pi \times (0.012)^3 = 7.24 \times 10^{-6}$ m^3
(b) Rearranging the volume equation gives

$r^3 = \dfrac{3V}{4\pi} = \dfrac{3 \times 0.60 \times 10^{-6}}{4\pi} = 1.43 \times 10^{-7}\ \text{m}^3.$

Hence r = $(1.43 \times 10^{-7})^{1/3} = 5.2 \times 10^{-3}$ m = 5.2 mm.
$\therefore$ diameter = 10.4 mm.

Unit 1

In-text questions

1.1

1. (a) 12 mm **(b)** 50 mm.
2. (a) 0.2 s **(b)** 0.25 m s^{-1}.
3. (a) 3400 Hz **(b)** 0.11 m.
4. (a) 200 kHz **(b)** 250 m.
5. (a) (i) 180° **(ii)** 180° **(iii)** 360° **(iv)** 270°
(b) (i) π **(ii)** π **(iii)** 2π **(iv)** $\frac{3}{2}\pi$.

1.3

1. See Fig. A1.1.

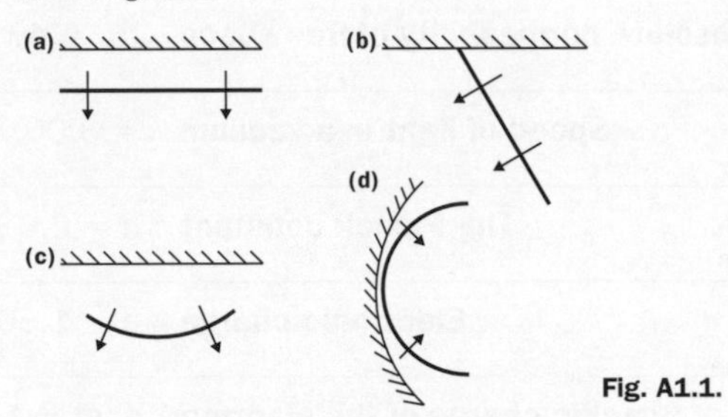

Fig. A1.1.

2. (a) refraction **(b)** reflection **(c)** diffraction.

1.4

1. (a) Move one end of the coil from side to side rapidly.
(b) (i) longitudinal **(ii)** and **(iii)** transverse.
2. (a) The pressure of the sound waves forces the ear drum to and fro repeatedly.
(b) The plane of the aerial is at right angles to the plane of polarisation of the radio waves so the radio waves cannot be detected.

Revision questions

1.1. (a) gamma, X, UV, visible, IR, microwave, radio.
(b) (i) radio **(ii)** visible **(iii)** microwave **(iv)** gamma.
1.2. (a) Transverse: vibration at 90° to direction of propagation; longitudinal: vibrations parallel to direction of propagation.
(b) transverse: microwaves, light; longitudinal: sound waves, ultrasonics, primary seismic waves.
1.3.(a) (i) 15 mm **(ii)** 30 mm
(b) (i) π **(ii)** $\frac{3}{2}\pi$ **(iii)** $\pi/2$
(c) (i) −15 mm **(ii)** 0 **(iii)** 0 **(iv)** +15 mm.
1.4. (a) (i) 6.0×10^{14} Hz **(ii)** 600 nm
(b) (i) 6800 Hz **(ii)** 28 mm.
1.5. (a) See Fig. A1.2.

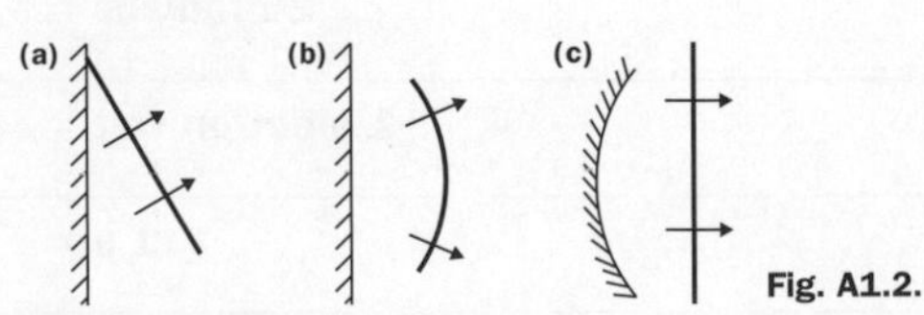

Fig. A1.2.

(b) (i) To focus the microwaves to a point, **(ii)** The detector is placed at the focal point of the dish.
(c) The waves slow down and lose energy as the water becomes shallower.
1.6. (a) (i) See Fig. A1.3.

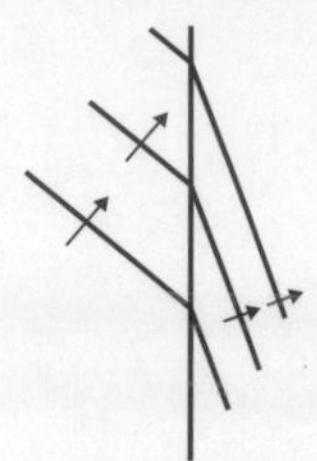

Fig. A1.3.

(ii) The wavelength is less in the shallow water than in the deep water. The direction of the waves is closer to the normal (i.e. closer to the line perpendicular to the boundary) in the shallow water.
(iii) Waves at an angle to the beach turn towards the beach because the part of each wavefront nearer the beach slows down, so the rest of the wavefront catches up and the wavefront turns towards the beach.
1.7. (a) See Fig. A1.4.

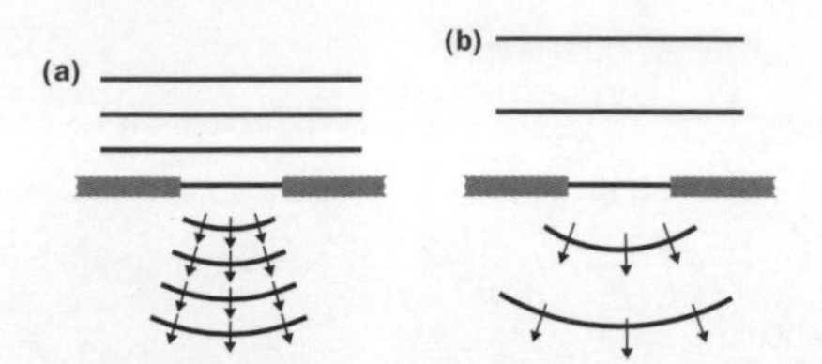

Fig. A1.4.

(b) Diffraction is less for a larger dish so the focus is more difficult to find.
1.8. (a) See p. 20.
(b) (i) Polarised light from the filter nearest the light source cannot pass through the other filter when the two filters are 'crossed'.
(ii) The intensity of light changes from minimum to maximum at 90° then minimum at 180° then maximum at 270° and back to minimum at 360°.
(iii) The same as in (i).

Unit 2

In-text questions

2.1

1. (a) 50 Hz **(b) (i)** and **(ii)** See Fig. A2.1(i) and (ii).
2. (a) 0.05 s **(b)** 75 m.

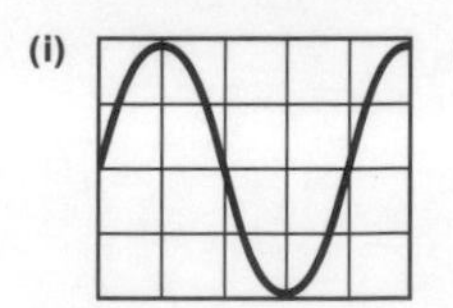

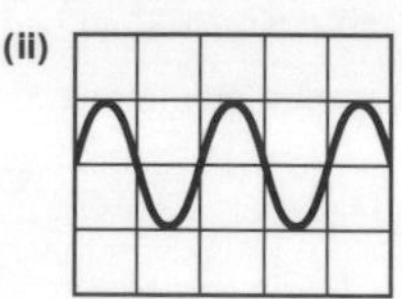

Fig. A2.1.

2.2

1. (a) To amplify the force of the vibrations, to filter out unwanted sounds, to protect the inner ear from excessive loudness.
(b) A recorder does not record higher frequencies as effectively as lower frequencies. Therefore, a recording sounds deeper than your direct speech. Also, the speed of sound through bone depends on frequency, so direct sounds are distorted.
(c) (i) 30 dB **(ii)** 1000 times as much ($= 10 \times 10 \times 10$).
2. (a) About 3000 Hz, **(b)** about 25 dB, **(c)** Hearing loss at the upper end of the frequency range is greater than at the other end. The upper frequency limit is reduced to about 15 kHz.

2.3

1. (a) If the pulses last too long, reflected pulses might return before the end of a transmitted pulse and would therefore not be detected.
(b) The time for one cycle, $T = \frac{1}{f} = \frac{1}{(1.5 \times 10^6)} = 0.67 \times 10^{-6}$ s.
Hence the time for 10 pulses = 6.7 μs.
(c) Ratio $= \frac{6.7\ \mu s}{1\ ms} = \frac{6.7 \times 10^{-6}}{1 \times 10^{-3}} = 6.7 \times 10^{-3}$.
2. (a) $\lambda = \frac{v}{f} = \frac{1500\ m\ s^{-1}}{40\ kHz} = \frac{1500}{40\,000} = 3.75 \times 10^{-2}$ m.
(b) (i) The wavelength is greater than the width of the source of waves so diffraction is significant. **(ii)** This is desirable so that waves spread out in all directions from the transducer.

2.4

1. (a) See Fig. A2.2.

Fig. A2.2.

(b) The fundamental frequency $f_o = \frac{v}{2L}$.
Hence the frequency is inversely proportional to the length.
(i) Doubling the length halves the frequency, hence the frequency becomes 60 Hz.
(ii) Halving the length doubles the frequency, hence the frequency becomes 240 Hz.
2. (a) See Fig. A2.3.

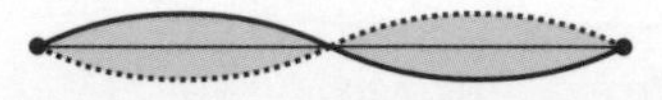

Fig. A2.3.

(b) (i) The amplitude is zero at the ends and the middle, and greatest one quarter and three quarters of the way along the wire.
(ii) The phase difference is zero for any two points either side of the centre and 180° for any two points separated by the node in the middle.

2.5

1. (a) $f_1 = 3f_o = 312$ Hz
(b) (i) $\lambda_o = \frac{v}{f_o} = \frac{340}{104} = 3.27$ m
(ii) $e = \frac{1}{4}\lambda_o - L = 0.817 - 0.815 = 0.002$ m.
2. (i) $\lambda_o = 2L = 2 \times 0.60 = 1.20$ m
(ii) $f_o = \frac{v}{\lambda_o} = \frac{340}{1.2} = 283$ Hz.

Revision questions

2.1. (a) (i) 0.025 s **(ii)** 40 Hz.
(b) 8.5 m ($= 340\ m\ s^{-1} \div 40$ Hz).
(c) See Fig. A2.4.

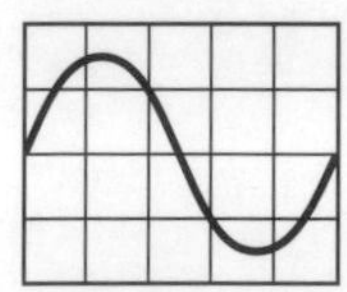

Fig. A2.4.

2.2 (a) (i) D **(ii)** B.
(b) (i) 204 m ($= 340\ m\ s^{-1} \times 1.2\ s \div 2$), **(ii)** Time a further short blast. If the timing is less than the first timing, the ship has moved closer to the cliffs.
2.3. (a) (i) 18 kHz, **(ii)** Worn bones in the middle ear, inner ear nerve damage.
(b) (i) ×10 **(ii)** ×10 **(iii)** ×10 000.
(c) 100.
2.4. (a) (i) Each reflected pulse is due to an internal boundary. Each boundary both transmits and reflects some ultrasonic energy. Transmitted ultrasonics from one boundary will partially reflect at the next boundary.
(ii) Ultrasonic energy is partly absorbed by the medium it passes through. The amount of absorption depends on the distance travelled, which depends on how far the tissue boundary is. Also, the strength of a reflection depends on the type of boundary.
(b) (i) The slice will not vibrate at resonance unless the frequency matches.

(ii) The wavelength is small enough at frequencies of the order of MHz for diffraction to be insignificant. If diffraction was significant, the ultrasonic pulses would spread out and weaken.

2.5. (a) (i) increased **(ii)** increased.

(b) (i) These two frequencies are the fundamental and the second overtone.

(ii) Both the frequencies in (i) have an antinode at the centre of the wire. The wire was plucked at the middle so a node is not likely there. Hence the first overtone at 512 Hz is missing.

2.6. (a) Since the frequency is inversely proportional to the length (i.e. $f = \frac{\text{constant}}{L}$), then f L is constant

(i) $f \times 0.4 = 384 \times 0.80$ gives $f = 384 \times 2 = 768$ Hz

(ii) $f \times 1.2 = 384 \times 0.80$ gives $f = 384 \times 0.80 \div 1.20 = 768 \div 3 = 256$ Hz.

(b) 5% of 384 Hz = 19 Hz. To raise the frequency to 403 Hz, the length must be reduced to 0.76 m (from $384 \times 0.8 = 403 \times L$). To reduce the frequency to 365 Hz, the length must be raised to 0.84 m ($= 384 \times 0.8 = 365 \times L$).

2.7. (a) (i) $\lambda = \frac{340}{150} = 2.27$ m

(ii) See Fig. A2.5

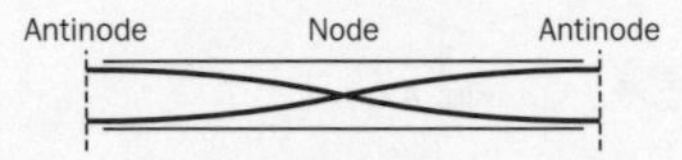

Fig. A2.5.

(iii) One half wavelength at 150 Hz = 1.13(3) m. Since $L + 2e = \frac{1}{2}\lambda$, where L, the pipe length, = 1.10 m and e is the end correction, then 2e = 1.133 m − 1.10 m = 0.033 m. Hence e = 0.017 m.

(b) The first overtone frequency = 2 × 150 = 300 Hz.

(c) For a closed pipe vibrating at its fundamental frequency, $\frac{\lambda}{4} = L + e$.

Hence $\lambda = 4 \times 1.117 = 4.47$ m $\therefore f = \frac{340}{4.47} = 76$ Hz.

8. (a) The fundamental wavelength = 2 × the pipe length, ignoring the end-corrections. Hence the wavelengths range from 80 mm to 8 m. The fundamental frequency range is therefore from

42.5 Hz ($= \frac{340}{8}$) to 4250 Hz ($= \frac{340}{0.080}$).

(b) 8.5 kHz, 12.75 kHz, 17.0 kHz.

Unit 3

In-text questions

3.1

1. (a) (i) d = 0.50 mm, y = 0.90 mm, X = 0.80 m

$$\lambda = \frac{yd}{X} = \frac{0.90 \times 10^{-3} \times 0.50 \times 10^{-3}}{0.80} = 563 \times 10^{-9}\ \text{m} = 563\ \text{nm}$$

(ii) green.

2. (a) The fringe pattern would disappear.

(b) (i) The fringes would be closer together. This is because the fringe spacing × the slit spacing is constant, so if the slit spacing is made larger, the fringe spacing is decreased. **(ii)** The fringe spacing would be unchanged but there would be fewer fringes visible because there is less overlap. This is because diffraction is less from a wider slit.

3.2

1. (a) See Fig. 3.6 **(b)** 3.0 m.

2. (a) See Fig. 3.11.

(b) $\sin c = \frac{n_r}{n_i} = \frac{\text{refractive index of cladding}}{\text{refractive index of the core}} = \frac{1.2}{1.5} = 0.80 \quad \therefore c = 53°$.

(c) The lower the critical angle, the more the fibre can be bent without light leaving the core. A higher refractive index for the cladding would mean a larger critical angle at the core-cladding boundary.

3.3

1. (a) See Figs 3.17–3.20.

(b) (i) $\frac{1}{0.5} + \frac{1}{v} = \frac{1}{0.2} \quad \therefore \frac{1}{v} = \frac{1}{0.2} - \frac{1}{0.5} = 5.0 - 2.0 = 3.0 \quad \therefore v = 0.33$ m

$m = \frac{v}{u} = \frac{0.333}{0.50} = 0.67$

(ii) $\frac{1}{0.25} + \frac{1}{v} = \frac{1}{0.2} \quad \therefore \frac{1}{v} = \frac{1}{0.2} - \frac{1}{0.25} = 5.0 - 4.0 = 1.0 \quad \therefore v = 1.00$ m

$m = \frac{v}{u} = \frac{1.00}{0.25} = 4.0$

(iii) $\frac{1}{0.15} + \frac{1}{v} = \frac{1}{0.2} \quad \therefore \frac{1}{v} = \frac{1}{0.2} - \frac{1}{0.15} = 5.0 - 6.7 = -1.7 \quad \therefore v = -0.60$ m

$m = \frac{v}{u} = \frac{-0.60}{0.15} = -4.0.$

2. (a) See Fig. 3.20. **(b)** See Fig. 3.22.

Revision questions

3.1. (a) See Fig, 3.2 and pp. 37–38.

(b) (i) λ = 590 nm, X = 0.90 m, y = 0.50 mm

Rearranging $\frac{\lambda}{d} = \frac{y}{X}$ gives $d = \frac{\lambda X}{y} = \frac{590 \times 10^{-9} \times 0.90}{0.5 \times 10^{-3}} = 1.1 \times 10^{-3}$ m

(ii) X = 0.90 m, y = 0.50 mm, d = 0.80 mm

Rearranging $\frac{\lambda}{d} = \frac{y}{X}$ gives $\lambda = \frac{yd}{X} = \frac{0.5 \times 10^{-3} \times 0.8 \times 10^{-3}}{0.90} = 440 \times 10^{-9}$ m.

3.2. (a) (i) 650 nm **(ii)** 450 nm.

(b) (i) refractive index, n = wavelength in air ÷ wavelength in glass

$\therefore$ wavelength in glass = wavelength in air ÷ n = $\frac{590\ \text{nm}}{1.55} = 381$ nm

(ii) The speed is slower in glass.

3.3. (a) Towards. **(b)** For light passing from air into glass, $n = \frac{\sin i}{\sin r}$

$\therefore \sin r = \frac{\sin i}{n}$

(i) For i = 25° and n = 1.50, $\sin r = \frac{\sin 25}{1.50} = 0.28, \quad \therefore r = 16°$

(ii) For i = 50° and n = 1.50, $\sin r = \frac{\sin 50}{1.50} = 0.51, \quad \therefore r = 31°$.

(c) (i) For light passing from glass to air, $\frac{\sin i}{\sin r} = \frac{1}{n}$

$\therefore \sin r = n \sin i = 1.5 \sin 40 = 0.96 \quad \therefore r = 75°$

(ii) $\sin c = \frac{1}{n} = \frac{1}{1.50} = 0.67 \quad \therefore c = 42°$

(iii) The light ray undergoes total internal reflection. See Fig. A3.1.

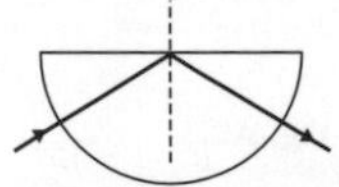

Fig. A3.1.

3.4. (a) (I) The image is real, inverted, 0.18 m from the lens on the other side, and 0.18 × the size of the object.

(ii) The image is real, inverted, 0.60 m from the lens on the other side, and 3.0 × the size of the object.

(iii) The image is virtual, upright, 0.3 m from the lens on the same side, and 3.0 × the size of the object.

(b) Rearranging $\frac{1}{u} + \frac{1}{v} = \frac{1}{f}$ gives $\frac{1}{v} = \frac{1}{f} - \frac{1}{u}$

(i) u = 1.00 m, f = 0.15 m $\therefore \frac{1}{v} = \frac{1}{0.15} - \frac{1}{1.00} = 6.7 - 1.0 = 5.7$

$\therefore v = 0.18$ m

Linear magnification $= \frac{v}{u} = \frac{0.18}{1.0} = 0.18$

(ii) u = 0.20 m, f = 0.15 m $\therefore \frac{1}{v} = \frac{1}{0.15} - \frac{1}{0.20} = 6.7 - 5.0 = 1.7$

$\therefore v = 0.60$ m

Linear magnification $= \frac{v}{u} = \frac{0.60}{0.20} = 3$

(i) $u = 0.10$ m, $f = 0.15$ m $\therefore \frac{1}{v} = \frac{1}{0.15} - \frac{1}{0.10} = 6.7 - 10 = -3.3$

$\therefore v = -0.30$ m

Linear magnification $= \frac{v}{u} = \frac{-0.30}{0.10} = -3.0.$

3.5. (a) See Fig. 3.26 and p. 46.

(b) The lens should be moved away from the film to focus the image on the film because the object distance is less than before. The aperture should be made wider if the light level indoors is poor.

3.6. (a) See Fig. 3.27 and p. 47.

(b) Diffraction is reduced using blue light because it has a smaller wavelength than all the other components of white light. Consequently, points closer together on the object can be seen separately.

(c) Spherical aberration of one or both lenses is the likely cause. See p. 46.

3.7. (a) See Fig. 3.28 and p. 47.

(b) A wide objective would collect more light and make the images of point objects brighter. Also, there would be less diffraction and more detail could be seen in extended images.

Unit 4

In-text questions

4.1

1. (a) See Fig. 4.1

(b) Only the blue part of the spectrum would be seen. All the other colours would not be reflected from the screen.

2. (a) Red + cyan; green + magenta; blue + yellow, **(b) (i)** red **(ii)** black.

4.2

1. (a) $m = 1$, $d = 1/600$ mm $= 1.67 \times 10^{-6}$ metres; $\theta_1 = 122° 06' - 101° 22' = 20° 44' = 20.73°$

Using $m\lambda = d\sin\theta_m$ gives $\lambda = d\sin\theta_1 = 1.67 \times 10^{-6} \times \sin 20.73 = 5.91 \times 10^{-7}$ metres.

(b) For $m = 2$, $\sin\theta_2 = \frac{2\lambda}{d} = 2\sin 20.73 = 0.708 \quad \therefore \theta_2 = 45.07°$;

for $m = 3$, $\sin\theta_3 = \frac{3\lambda}{d} = 3\sin 20.73 = 1.062$; no third order is possible.

2. (a) Order number $m = 1$, $\lambda = 629$ nm, $\theta_1 = 10° 52' = 10.87°$

Rearranging $m\lambda = d\sin\theta_m$ for $m = 1$ gives

$d = \frac{\lambda}{\sin\theta_1} = \frac{629 \times 10^{-9}}{\sin 10.87} = 3.33 \times 10^{-6}$ metres

No. of lines per mm $= \frac{1}{3.33 \times 10^{-3}} = 300.$

(b) Maximum order number $= \frac{d}{\lambda}$ rounded down $= \frac{3.33 \times 10^{-6}}{629 \times 10^{-9}} = 5.29$

rounded down = 5.

4.3A

1. (a) The sea is usually colder than the land so emits less infra-red radiation. **(b)** Urban areas are usually warmer than rural areas so emit more infra-red radiation.

2. (a) More infra-red radiation is received than is emitted due to the presence of the hot plate. **(b)** Less infra-red radiation is received than is emitted due to the presence of the beaker of ice.

4.3B

1. (a) Microwaves penetrate food and heat water molecules inside the food. **(b)** Microwaves have a very high frequency so a beam can carry many more digital signals than radio waves. Also, they are not absorbed by the atmosphere at this frequency.

2. (a) To ensure there are no cold spots in the food due to microwave nodes. **(b)** Microwaves at such high power are harmful to humans and animals.

4.3C

1. (a) (i) $c = 300\,000$ km s^{-1}, $\lambda = 1500$ m $f = \frac{c}{\lambda} = \frac{300\,000 \times 10^3}{1500} = 200$ kHz.

(ii) These radio waves travel round the Earth following the Earth's curvature to reach distant countries.

(b) Absorption by the atmosphere makes the signal too weak beyond a certain range. Radio waves above 30 MHz pass through the ionosphere into space.

2. (a) Polarisation. **(b)** The microwave beam from the satellite has further to travel through the atmosphere to reach N. Europe and therefore it becomes weaker and a wider dish is needed to compensate. A wider dish produces less diffraction so it is more difficult to focus the microwaves onto the aerial.

4.4A

1. Produce a visible spectrum in a dark room as in Fig. 4.1. Mark some white paper using a fluorescent marker pen. Place the marks on the paper beyond the violet part of the visible spectrum. The marks should glow if ultraviolet radiation is present.

2. (a) The fabric of the white shirt contains chemicals from washing powder that glow in ultraviolet radiation. **(b)** The skin cream absorbs the ultraviolet radiation and prevents it from reaching the skin.

4.4B

1. (a) The image on the film would be blurred otherwise as the shadow would be ill-defined.

(b) Light would blacken the film and the X-rays would make no difference if the film was not in a light-proof wrapper.

2. (a) $f = \frac{c}{\lambda} = \frac{3.0 \times 10^8}{1 \times 10^{-14}} = 3.0 \times 10^{22}$ Hz.

(b) (i) It kills living cells **(ii)** It kills living cells.

Revision questions

4.1. (a) See Fig. 4.1. **(b)** Direct a narrow beam of white light into a semi-circular glass block, as in Fig. 3.10. The beam is split into colours where it emerges from the block. Adjust the incident angle so the red component emerges along the boundary. The blue component is totally internally reflected at this position because the critical angle for blue light is less than for red light.

4.2. (a) Black, since it absorbs blue light. **(b) (i)** Blue print on a black background, as the background absorbs blue light. **(ii)** Black print on a red background, since the print absorbs yellow light and the background absorbs the green component of yellow light.

4.3. (a) (i) $\tan\theta_1 = \frac{0.29}{1.5} = 0.193 \quad \therefore \theta_1 = 10.9°$

(ii) $d = \frac{1 \text{ mm}}{300 \text{ mm}} = 3.33 \times 10^{-6}$ metres. Using $d\sin\theta_m = m\lambda$ gives $\lambda = d\sin\theta_1 = 3.33 \times 10^{-6} \times \sin 10.9 = 6.3 \times 10^{-7}$ m $= 630$ nm.

(b) $\frac{d}{\lambda} = \frac{3.33 \times 10^{-6}}{6.3 \times 10^{-7}} = 5.3 \quad \therefore$ five orders are produced.

4.4. (a) (i) Blue: $\theta_1 = (125° 33' - 95° 39') \div 2 = 14° 57'$;

Yellow: $\theta_1 = (131° 20' - 89° 52') \div 2 = 20° 44'$

(ii) For the first order yellow line, $\lambda = 590$ nm, $\theta_1 = 20°44'$. Using $d\sin\theta_m = m\lambda$ gives $d = \frac{\lambda}{\sin\theta_1} = \frac{590 \times 10^{-9}}{\sin 20° 44'} = 1.67 \times 10^{-6}$ metres

$\therefore$ the number of lines per mm $= \frac{1}{1.67 \times 10^{-3}} = 600$

(iii) For the first order blue line, $\theta_1 = 14° 57' \quad \therefore \lambda = d\sin\theta_1 = 1.67 \times 10^{-6} \times \sin 14° 57' = 4.3 \times 10^{-7}$ metres $= 430$ nm.

(b) (i) Blue: $\sin\theta_2 = \frac{2m\lambda}{d} = \frac{2 \times 4.3 \times 10^{-7}}{1.67 \times 10^{-6}} = 2 \times 0.257 = 0.514$

$\therefore \theta_2 = 31.0°$; $\sin\theta_3 = \frac{3m\lambda}{d} = \frac{3 \times 4.3 \times 10^{-7}}{1.67 \times 10^{-6}} = 3 \times 0.257 = 0.77$

$\therefore \theta_2 = 50.4°$; no fourth order is formed.

(ii) Yellow: $\sin\theta_2 = \frac{2m\lambda}{d} = \frac{2 \times 5.9 \times 10^{-7}}{1.67 \times 10^{-6};} = 2 \times 0.353 = 0.706$

$\therefore \theta_2 = 45.0°$; no third order is formed.

4.5. **(a)** **(i)** radio **(ii)** infra-red **(iii)** ultraviolet **(b)** **(i)** 300 kHz **(ii)** 10 Ghz **(iii)** 5×10^{14} Hz.
4.6. **(a)** Humans and animals are hotter and therefore emit more infra-red radiation than their surroundings. **(b)** Invisible ink glows in ultraviolet radiation. **(c)** Microwaves are not absorbed by the atmosphere and are at a high enough frequency to carry several TV channels. **(d)** X-rays pass through soft tissue and are absorbed by bone. **(e)** Gamma radiation kills living cells due to its ionising properties. **(f)** Line of sight communication; high carrier frequency allows many channels.

Unit 5

In-text questions

5.1

1. **(a)** Solid: it keeps its own shape after removal from its dish. **(b)** Liquid: the spray consists of tiny droplets which gradually fall unless they adhere to a surface.
2. **(a)** Ice is a solid which turns to water in the liquid state at 0°C, which then boils and turns to steam at 100°C.
(b) Water droplets in the fabric evaporate and the vapour is removed by the wind.

5.2A

1. **(a)** 1p + 2n **(b)** 8p + 8n **(c)** 11p + 12n **(d)** 82p + 124n **(e)** 92p + 146n.
2. **(a)** ${}^{4}_{2}$He **(b)** ${}^{12}_{6}$C.

5.2B

1. **(a)** 18 **(b)** 28 **(c)** 16 **(d)** 17 **(e)** 80.
2. Radius of oil droplet, r = 0.40 mm; diameter of oil patch, D = 350 mm.

$\therefore$ volume of oil patch, $\frac{\pi D^2 t}{4}$ = volume of oil droplet, $\frac{4\pi r^3}{3}$

where t = patch thickness.

$\therefore$ estimate of length of oil molecule $= t = \frac{4\pi (0.40 \times 10^{-3})^3 \times 4}{3\pi (0.35)^2} = 2.8 \times 10^{-9}$ m

5.3

1. **(a)** **(i)** 12 g of carbon contains 6.02×10^{23} atoms $\therefore$ 1 kg of carbon contains 5.0×10^{25} atoms ($6.02 \times 10^{23}/0.012$).
(ii) 235 g of Uranium contains 6.02×10^{23} atoms $\therefore$ 1 kg of uranium contains 2.5×10^{24} atoms ($6.02 \times 10^{23}/0.235$).
(b) **(i)** 44 g of CO_2 contains 6.02×10^{23} molecules $\therefore$ 1 kg of carbon dioxide contains 1.4×10^{25} molecules.
(ii) 16 g of methane contains 6.02×10^{23} molecules $\therefore$ 1 kg of methane contains 3.8×10^{25} molecules.
2. **(a)** 27 g of aluminium contains 6.02×10^{23} atoms $\therefore$ the mass of 1 atom of aluminium = 27 g/6.02×10^{23} = 4.5×10^{-23} grams = 4.5×10^{-26} kg.
(b) The volume of 1 aluminium atom = mass of 1 atom/density = 1.67×10^{-29} m^3
$\therefore$ diameter = $(1.67 \times 10^{-29})^{1/3} = 2.6 \times 10^{-10}$ m.

5.4

1. **(a)** **(i)** covalent **(ii)** metallic **(iii)** ionic **(iv)** molecular **(v)** none. **(b)** **(i)** 2, 4 **(ii)** See Fig. 5.14.
2. **(a)** no, yes **(b)** yes, yes **(c)** yes, no **(d)** yes, no.

Revision questions

5.1. **(a)** The atoms and molecules are locked together. **(b)** The molecules of a liquid move about at random and therefore the liquid flows to take the shape of its container. **(c)** Carbon dioxide molecules are heavier than air molecules so they do not move away as quickly as the molecules of a lighter gas would.
5.2. **(a)** Water droplets freeze. **(b)** Water vapour condenses to form a film of water droplets on the windscreen. **(c)** Solid ice turns directly to vapour.
5.3. **(a)** 17p + 20n, 17p + 18n. **(b)** 35.5 g (= (3 × 35 + 1 × 37) ÷ 4).
5.4. **(a)** **(i)** 2, 4 **(ii)** 1.
(b) **(i)** 16 g (= 12 + 4) **(ii)** 2.7×10^{-26} kg (= 0.016 kg ÷ N_A)
(iii) See Fig. A5.1.
5.5. **(a)** 92p + 143n **(b)** 2.6×10^{21} (= N_A ÷ 235).
5.6. **(a)** 3.0×10^{-26} kg (= 0.018 ÷ N_A). **(b)** 3.1×10^{-10} m (= $(3.0 \times 10^{-26} \div 1000)^{1/3}$).
5.7. **(a)** 3.4×10^{-25} kg (= 0.207 ÷ N_A) **(b)** 3.1×10^{-10} m (= $(3.4 \times 10^{-25} \div 11340)^{1/3}$).
5.8. **(a)** 0.80 × 28 g + 0.20 × 32 g = 28.8 g **(b)** **(i)** 3.4 nm {= $[0.029 \div (N_A \times 1.2)]^{1/3}$} **(ii)** The average spacing is about 10× the diameter of an air molecule.

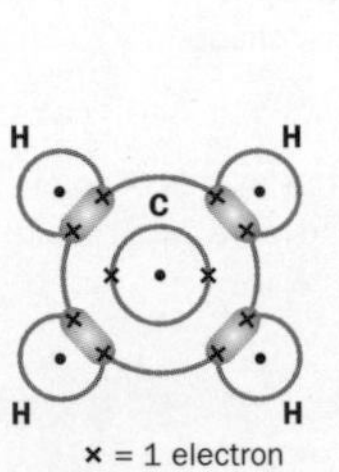

Fig. A5.1.

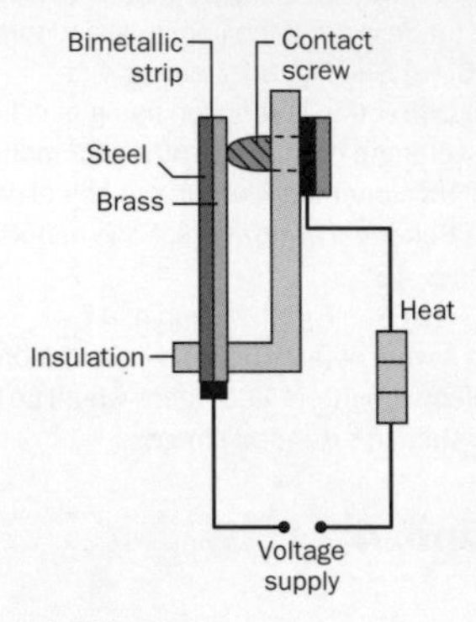

Fig. A6.1.

Unit 6

In-text questions

6.1

1. See Fig. A6.1.
2. **(a)** $\Delta L = \alpha L \Delta T = 1.9 \times 10^{-5} \times 0.80 \times 5 = 7.6 \times 10^{-5}$ m.
(b) $\Delta L = \alpha L \Delta T = 2.6 \times 10^{-5} \times 1.80 \times 25 = 9.8 \times 10^{-4}$ m.

6.2

1. **(a)** Energy needed = 5.0 × 2100 × (50 − 10) = 420 000 J.
(b) Energy needed = 1500 × 850 × (35 − 5) = 3.8×10^7 J = 38 MJ.
(c) Energy needed = (15 × 900 × 60) + (95 × 4200 × 60) = 0.8×10^6 + 23.9×10^6 = 24.7 MJ.
2. Energy needed to heat the calorimeter = 0.055 × 390 × (62 − 15) = 1010 J $\therefore$ energy available to heat the liquid = 9340 − 1010 = 8330 J. Mass of liquid = 143 − 55 = 88 g = 0.088 kg; temperature rise of liquid = 47 K.
Since energy needed to heat the liquid = mass of liquid × specific heat capacity (c) × temperature rise, then 0.088 × c × 47 = 8330

$\therefore c = \frac{8330}{0.088 \times 47} = 2010$ J kg^{-1} K^{-1}.

6.3

1. Energy to heat 0.20 kg of ice from −10°C to 0°C = 0.20 × 2100 × 10 = 4200 J. Energy needed to melt 0.20 kg of ice = 0.20 × 340 000 = 68 000 J. Energy needed to heat 0.20 kg of water from 0°C to 100°C = 0.20 × 4200 × (100 − 0) = 84 000 J. Energy needed to boil away 0.20 kg of water = 0.20 × 2.3 MJ = 460 000 J. Total energy needed = 4.2 kJ + 68 kJ + 84 kJ + 460 kJ = 616(.2) kJ.
2. **(a)** Energy needed = 0.12 × 4200 × (52 − 20) = 2.6×10^4 J.
(b) Energy released by mass, m, of steam when it condenses = ml, where l = 2.3×10^6 J kg^{-1}. Energy released when mass, m, of water cools from 100°C to 52°C = m × 4200 × (100 − 52) = 2.18×10^5 m $\therefore$ total energy released by mass m = 2.18×10^5 m + 2.3×10^6 m = 2.32×10^6 m. Assuming energy released by the steam = energy needed to heat the water, 2.32×10^6 m = 2.6×10^4

$\therefore m = \frac{2.6 \times 10^4}{2.32 \times 10^6} = 1.13 \times 10^{-2}$ kg = 11.3 g.

6.4

1. (a) To minimise heat loss due to thermal radiation. **(b)** To reduce heat loss due to thermal conduction. **(c)** To prevent heat loss due to convection and evaporation. **(d)** Cork, polystyrene or any other material which is a good thermal insulator.

2. The surface area of the hot plate $\frac{\pi d^2}{4} = \frac{\pi(0.10)^2}{4} = 7.9 \times 10^{-3}\,\text{m}^2$,

Using Stefan's Law, energy per second radiated from the surface $= e\sigma AT^4$
$= 1 \times 7.9 \times 10^{-3} \times 5.67 \times 10^{-8} \times (1200)^4 = 920\,\text{W}$.

6.5

1. Using $\frac{Q}{t} = kA\frac{(T_1 - T_2)}{L}$ with $A = 1.0\,\text{m}^2$, $(T_1 - T_2) = 15°\text{C}$ and $L = 0.120\,\text{m}$

$\therefore \frac{Q}{t} = 0.40 \times 1 \times \frac{15}{0.12} = 50\,\text{W}$.

2. (a) The mass of water boiled away in 1 second $= \frac{0.10\,\text{kg}}{120\,\text{s}}$
$= 8.3 \times 10^{-4}\,\text{kg s}^{-1}$
$\therefore$ energy needed per second $= ml = 8.3 \times 10^{-4} \times 2.3 \times 10^6$
$= 1.9 \times 10^3\,\text{J s}^{-1}$.

(b) Base area, $A = \frac{\pi d^2}{4} = \frac{\pi(0.120)^2}{4} = 1.13 \times 10^{-2}\,\text{m}^2$.

Rearranging $\frac{Q}{t} = kA\frac{(T_1 - T_2)}{L}$ gives $(T_1 - T_2) = \frac{LQ/t}{kA}$

$\therefore (T_1 - T_2) = \frac{0.120 \times 1.9 \times 10^3}{210 \times 1.13 \times 10^{-2}} = 96\,\text{K} = 96°\text{C}$.

Since the top side of the base is at 100°C, the underside must therefore be at a temperature of 196°C.

3. (a) Total window area = 4 x 1 x 1.5 = 6 m².

(b) (i) Total heat loss per second = 6 x 4.3 x 10 = 258 W (ii) Total heat loss per second = 6 x 3.2 x 10 = 192 W

(c) For (b)(i), heat loss per day = 258 x 24 x 3600 = 2.2×10^7 J
.'. cost per day = 22 p

For (b)(ii), heat loss per day = 192 x 24 x 3600 = 1.7×10^7 J
.'. cost per day = 17 p

Revision questions

6.1. (a) (i) 373 K **(ii)** 243 K **(b)(i)** −173 K **(ii)** 727 K.

6.2. (a) (i) To allow for thermal expansion in summer. Without the gaps, the road spans would press against each other and crack.
(ii) The tyre expands when heated. It can then be fitted on the wheel and when it cools it tightens on the wheel because it contracts.
(b) Change of length $= \alpha L\Delta T = 1.1 \times 10^{-5} \times 120 \times (35 - -5)$
$= 5.3 \times 10^{-2}\,\text{m}$.

6.3. (a) (i) Energy to heat water $= 0.185 \times 4200 \times (27 - 15) = 9320\,\text{J}$
(ii) Energy needed to heat the calorimeter $= 0.065 \times 390 \times (27 - 15) = 304\,\text{J}$.
(b) Total energy released by metal $= 9320 + 304 = 9624\,\text{J}$

$\therefore 0.235 \times c \times (100 - 27) = 9624$. Hence $c = \frac{9624}{0.235 \times 73}$
$= 560\,\text{J kg}^{-1}\text{K}^{-1}$.

6.4. (a) (i) Energy removed $= (0.080 \times 390 \times 16) + (0.120 \times 4200 \times 16) = 8560\,\text{J}$. **(ii)** Energy removed on freezing $= 0.120 \times 340\,000 = 40\,800\,\text{J}$.
(b) Total energy removed in 35 minutes $= 40\,800 + 8560 = 49\,360\,\text{J}$

$\therefore$ energy removed per second $= \frac{49360}{35 \times 60} = 24\,\text{J s}^{-1}$.

6.5 (a) Net rate of heat transfer $= [0.6 \times 5.67 \times 10^{-8} \times 1.6 \times (273 + 45)^4] - [0.6 \times 5.67 \times 10^{-8} \times 1.6 \times (273 + 25)^4] = 557 - 429 = 128\,\text{W}$.
(b) For $A = 1\,\text{m}^2$, net rate of heat loss $= [1 \times 5.67 \times 10^{-8} \times 1 \times (273 + 10)^4] - [1 \times 5.67 \times 10^{-8} \times 1 \times (273 + 0)^4] = 364 - 315 = 49\,\text{W}$.

6.6. (a) (i) Temperature gradient $= \frac{(54 - 26)}{15 \times 10^{-3}} = 1870\,\text{K m}^{-1}$
(ii) Heat conducted per second $= 0.037 \times 0.95 \times 1870 = 66\,\text{J s}^{-1}$.

(b) Heat conducted per second $= 0.2 \times 0.12 \times \frac{(22 - 18)}{10 \times 10^{-3}} = 9.6\,\text{W}$.

6.7 (i) Heat loss/second = 2.0 x 80 x 20 = 3200 W , (ii) Heat loss/second = 0.5 x 80 x 20 = 800 W

6.8 (a)(i) Area of sides − windows area = (24 x 2.5) − 4 = 56 m²
(ii) Floor area = roof area = 9 x 3 = 27 m²,
(b) (i) Heat loss/second = (2.5 x 56 x 5) for the sides + (2.0 x 27 x 5) for the roof + (2.0 x 27 x 5) for the floor + (4.0 x 4 x 5) for the windows = 1390 W
(ii) Heat loss in 1 week = 1390 x 3600 x 24 x 7 = 8.4×10^8 J.
Cost = 2p x 8.4×10^8 / 1×10^6 = £16.78.

Unit 7

In-text questions

7.1

1. (a) (i) Weight $= 0.40 \times 9.8 = 3.9\,\text{N}$ **(ii)** Extension = 120 mm
(iii) $k = 3.9\,\text{N}/0.120\,\text{m} = 32.5\,\text{N m}^{-1}$.
(b) (i) Extension, $e = 390 - 300 = 90\,\text{mm}$
$\therefore$ tension $= ke = 32.5 \times 0.090 = 2.93\,\text{N}$
$\therefore$ weight $= 2.93\,\text{N}$.
(ii) Mass $= \text{weight}/g = 0.30\,\text{kg}$.

2. (a)

Tension/N	0	2.0	4.0	6.0	8.0
Extension/m	0	0.052	0.100	0.152	0.198

(b) (ii) Spring constant = gradient of line $= 40\,\text{N m}^{-1}$.

7.2

1. (a) Brittle, not too stiff or strong. **(b)** Tough, strong, not too stiff, lightweight. **(c)** Tough, strong, stiff, lightweight.

2. No, the extension is not proportional to the length so equal weights would not extend the rubber band by equal increases of length. Also, the unloading curve differs to the loading curve so the extension depends on whether the rubber band is being loaded or unloaded.

7.3

1. (a) (i) Area of cross-section, $A = \pi(0.38 \times 10^{-3})^2/4 = 1.13 \times 10^{-7}\,\text{m}^2$

$\therefore \text{stress} = \frac{\text{tension}}{\text{area of cross-section}} = 1.0 \times 10^9\,\text{Pa}$

(ii) $\text{Strain} = \frac{\text{stress}}{E} = \frac{1.0 \times 10^9}{2.0 \times 10^{11}} = 5.0 \times 10^{-3}$

$\therefore$ extension = strain × length $= 5.0 \times 10^{-3} \times 2.37 = 1.2 \times 10^{-2}\,\text{m}$.
(b) Maximum weight = breaking stress × area of cross-section
$= 1.1 \times 10^9 \times 1.13 \times 10^{-7} = 120\,\text{N}$.

2. (a) See Fig. A7.1.
(b) (i) Approx. 2.5 mm

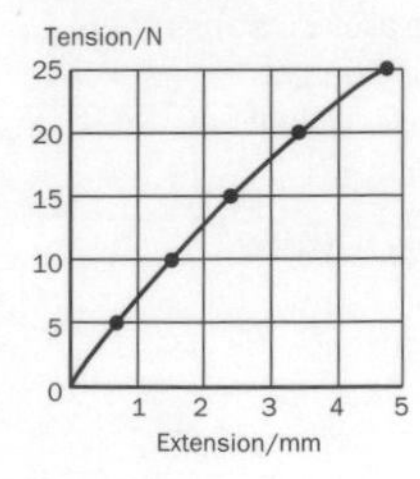

Fig. A7.1.

(ii) Gradient of straight section of line $= \frac{10}{1.6 \times 10^{-3}} = 6.25 \times 10^3\,\text{N m}^{-1}$

$\therefore \text{Young modulus} = \frac{\text{gradient} \times \text{length}}{\text{area of cross-section}} = \frac{6.25 \times 10^3 \times 1.28}{\pi(0.28 \times 10^{-3})^2/4}$
$= 1.3 \times 10^{11}\,\text{Pa}$.

7.4

1. (a) (i) See Fig. 7.14. **(ii)** The force between two atoms is proportional to the change of separation, up to a limit. **(b)** See p. 99.

2. (a) B is stronger because its UTS is greater. **(b)** A is stiffer because its Young modulus (= initial gradient) is greater.

7.5

1. (a) (i) 0.195 m (= 515 − 320 mm) **(ii)** $k = \frac{T}{e} = \frac{5.0}{0.195} = 25.6\ N\ m^{-1}$.

(b) Energy stored $= \frac{1}{2}Te = \frac{1}{2} \times 5.0 \times 0.195 = 0.49$ J.

2. (a) (i) 0.035 m (=785 − 750 mm) **(ii)** Area of cross-section $A = \pi d^2/4 = \frac{1}{4}\pi(5.0 \times 10^{-4})^2 = 1.96 \times 10^{-7}\ m^2$

Rearranging $E = \frac{TL}{Ae}$ gives

$$T = \frac{AEe}{L} = \frac{1.96 \times 10^{-7} \times 3.0 \times 10^9 \times 0.035}{0.750} = 27.5\ N.$$

(b) Volume $= AL = 1.96 \times 10^{-7} \times 0.75 = 1.47 \times 10^{-7}\ m^3$. Energy stored $= \frac{1}{2}Te = \frac{1}{2} \times 27.5 \times 0.035 = 0.48$ J

∴ energy stored per unit volume $= \frac{0.48}{1.47 \times 10^{-7}} = 3.3 \times 10^6\ J\ m^{-3}$.

Revision questions

7.1. (a) (i) Extension = 345 mm = 0.345 m ∴ spring constant, $k = T/e = 6.0\ N/0.345\ m = 17.4\ N\ m^{-1}$

(ii) Energy stored $= \frac{1}{2}Te = 0.5 \times 6.0 \times 0.345 = 1.04$ J.

(b) (i) Using $T = ke$, weight = tension = 17.4 × (0.728 − 0.500) = 4.0 N

(ii) Mass = weight/g = 4.0/9.8 = 0.40 kg.

7.2. (a) (i) B **(ii)** B.

(b) (i) See Fig. 7.5. **(ii)** Elasticity is the ability to regain shape after being distorted. The elastic band regains its original length when it is unloaded. The polythene strip does not regain its original length and it therefore undergoes plastic behaviour.

7.3. (a) Extension = 1.414 − 1.393 = 0.021 m due to 60 N of extra tension.

(b) Young modulus $= \frac{TL}{eA} = \frac{60 \times 1.393}{0.021 \times \pi(0.20 \times 10^{-3})^2/4} = 1.3 \times 10^{11}$ Pa.

(c) Energy stored $= \frac{1}{2}Te = 0.5 \times 60 \times 0.021 = 0.63$ J.

7.4. (a) Maximum weight = maximum allowed stress × cross-sectional area of the cable $= 2.0 \times 10^8 \times \pi(0.030)^2 / 4 = 1.4 \times 10^5$ N.

(b) (i) $E = \frac{TL}{Ae}$ $\therefore e = \frac{TL}{AE} = \frac{1.4 \times 10^5 \times 55}{\pi(0.030)^2/4 \times 2.0 \times 10^{11}} = 5.5 \times 10^{-2}$ m.

(ii) Energy stored $= \frac{1}{2}Te = 0.5 \times 1.4 \times 10^5 \times 5.5 \times 10^{-2} = 3.9 \times 10^3$ J. Volume of wire $= LA = 55 \times \pi(0.030)^2/4 = 3.9 \times 10^{-2}\ m^3$

∴ energy stored per unit volume $= \frac{3.9 \times 10^3\ J}{3.9 \times 10^{-2}\ m^3} = 1.0 \times 10^5\ J\ m^{-3}$.

7.5. (a) 0.25 mm. **(b)** Rearranging $E = \frac{TL}{Ae}$ gives

$$T = \frac{Eae}{L} = \frac{1.3 \times 10^{11} \times \pi(0.010)^2 \times 0.25 \times 10^{-3}}{60 \times 10^{-3} \times 4} = 4.3 \times 10^4\ N.$$

7.6. (a) Area of cross-section of each cable $= \pi(0.005)^2/4 = 2.0 \times 10^{-5}\ m^2$ ∴ maximum safe tension in each cable = maximum allowed stress × area of cross-section $= 2.0 \times 10^3$ N ∴ maximum allowed weight $= 2.0 \times 10^3 \times 8 = 1.6 \times 10^4$ N. Maximum passenger weight = 16.0 kN − 1.80 kN = 14.2 kN.

(b) Rearrange $E = \frac{TL}{Ae}$ to give $e = \frac{TL}{AE}$ ∴ extension for tension of

$$2.0 \times 10^3\ N = \frac{2.0 \times 10^3 \times 65}{2.0 \times 10^{-5} \times 2.0 \times 10^{11}} = 3.2 \times 10^{-2}\ m.$$

Unit 8

In-text questions

8.1

1. (a) The sharper the needle, the less the area of contact between its tip and the cloth when it is pushed into the cloth. The pressure of a sharp needle on the cloth is therefore greater than that of a blunt needle for the same force.

(b) The contact area between the tyre and the earth is larger for a wide tyre in comparison with a narrow tyre. Hence the pressure of the tractor on the earth is less and it is less likely to sink into the earth.

2. (a) $p = \frac{F}{A} = \frac{750}{8.5 \times 10^{-3}} = 8.8 \times 10^4$ Pa.

(b) Total area, $A = \frac{F}{p} = \frac{12\,000}{250\,000} = 4.8 \times 10^{-2}\ m^2$

∴ area of each tyre in contact $= 1.2 \times 10^{-2}\ m^2$.

8.2

1. The effort is applied to one end of the lever which causes a greater force to be applied to the piston in the narrow cylinder. The force on the piston in the master cylinder is greater than the force on the piston in the narrow cylinder. The force of the master cylinder on the load acts via another lever system which enables the piston in the master cylinder to exert a larger force on the load.

2. Maximum force = maximum pressure × total area of all four pistons $= 500 \times 10^3 \times 4 \times 0.012 = 2.4 \times 10^4$ N ∴ maximum additional weight = 24 000 − 2500 N = 21 500 N.

8.3

1. $p = H\rho g = 20.0 \times 1030 \times 9.8 = 2.0 \times 10^5$ Pa.

2. Rearrange $p = H\rho g$ to give $H = \frac{p}{\rho g} = \frac{101 \times 10^3}{1030 \times 9.8} = 10.1$ m.

8.4

1. (a) (i) $p = H\rho g = 0.25 \times 1000 \times 9.8 = 2.5$ kPa.

(ii) 2.4% $(= \frac{2.5\ kPa}{101\ kPa} \times 100\%)$.

(b) (i) It would be insufficient to drive gas through the pipes. **(ii)** It would cause gas to leak from the pipes.

2. (a) The pressure in the space at the top is negligible. The column of mercury is supported by atmospheric pressure acting on the open surface of mercury in the reservoir.

(b) (i) For H = 80 mm, $p = H\rho g = 0.080 \times 13\,600 \times 9.8 = 1.1 \times 10^4$ Pa = 11 kPa. For H = 120 mm, $p = H\rho g = 0.120 \times 13600 \times 9.8 = 1.6 \times 10^4$ Pa = 16 kPa.

(ii) To produce the same pressure with water rather than mercury, the manometer tube would need to be about 13 times longer, which is unrealistic.

8.5

1. (a) $p = \frac{F}{A} = \frac{700}{2.0 \times 0.060} = 580$ Pa.

(b) Rearrange $p = H\rho g$ to give $H = \frac{p}{\rho g} = \frac{580}{1000 \times 9.8} = 0.060$ m.

2. (a) Weight of water displaced when it carries maximum load = 0.05 × $A\rho g$, where A is the area of the base of the hull. This is equal to the extra upthrust to keep it floating and the extra upthrust is equal to the extra weight, i.e. 0.050 $A\rho g$ = 40 000. When empty, the upthrust = weight of water displaced = 1.20 $A\rho g$. This is equal to the weight of the empty boat.

∴ weight of empty boat $= 1.20 A\rho g = \frac{1.20 \times 40\,000}{0.050} = 960\,000$ N.

(b) The total weight when fully loaded = 1 000 000 N ∴ the weight of water displaced = upthrust = 1 000 000 N. Hence $H'A\rho' g = 1\,000\,000$ where H′ = depth of the base of the hull below the water line, and $\rho' = 1000\ kg\ m^{-3}$.

Since $0.050 A\rho g = 40\,000$, then $\frac{H'}{0.050} = \frac{1\,000\,000\rho}{40\,000\rho'} = 25 \times \frac{1050}{1000} = 26.25$.

∴ H′ = 26.3 × 0.050 = 1.31 m.

Revision questions

8.1. (a) Weight = mass × g = volume × density × g = (0.450 × 0.300 × 0.025) × 2600 × 9.8 = 86 N.

(b) (i) Face area $= 0.45 \times 0.30 = 0.135\ m^2$

∴ pressure $= \frac{weight}{area} = \frac{86}{0.135} = 640$ Pa.

(ii) Edge area $= 0.30 \times 0.025 = 0.0075\ m^2$

$\therefore$ pressure $= \dfrac{\text{weight}}{\text{area}} = \dfrac{86}{0.0075} = 11500$ Pa.

8.2. (a) Area of piston $= \pi(0.45)^2/4 = 0.16\ m^2$

$\therefore p = \dfrac{F}{A} = \dfrac{120\,000}{0.16} = 755\,000$ Pa.

(b) Effort force = pressure × area of narrow piston $= 755\,000 \times \pi(0.025)^2/4 = 370$ N.

8.3. (a) Rearrange $p = H\rho g$ to give $H = \dfrac{p}{\rho g} = \dfrac{101 \times 10^3}{1.2 \times 9.8} = 8600$ m.

(b) The density of the atmosphere decreases with increased height.

8.4. (a) $p = H\rho g = 2.0 \times 1000 \times 9.8 = 2.0 \times 10^4$ Pa.

(b) The maximum lung pressure of a human is no more than about 20 kPa. At a depth of more than 2.0 m, the pressure due to the water would be greater than maximum lung pressure so it would not be possible to inflate the lungs naturally at this depth or below.

8.5. (a) See p. 108.

(b) If the arm is raised, the blood pressure in it is lowered so the reading would be less.

8.6. (a) The submarine weighs less when the water is forced from its ballast tanks. The upthrust is unchanged because the volume of the boat and hence the weight of water displaced is the same. Therefore the submarine rises because the upthrust on it exceeds its weight.

(b) Weight of test tube = upthrust for flotation. Upthrust = weight of water displaced $= HA\rho g$, where A is the area of cross-section of the test tube and H is its depth in the water.

$\therefore HA\rho g$ is the same in both liquids (= weight of test tube) $\therefore H_1\rho_1 = H_2\rho_2$ which gives $\rho_2 = \dfrac{H_1\rho_1}{H_2} = \dfrac{0.060 \times 1000}{0.068} = 880\ kg\ m^{-3}$.

Unit 9

In-text questions

9.1A

1. (a) 17.0 N due East **(b)** 7.0 N due West **(c)** 13.0 N at an angle of 67.4° North of due E **(d)** 15.1 N at 43.3° North of due E **(e)** 10.4 N at 84.5° North of due West.

(c)

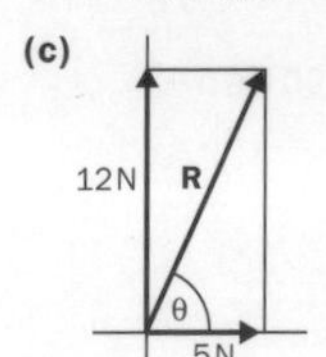

(d)

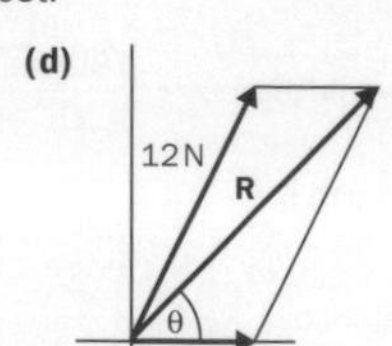

(e)

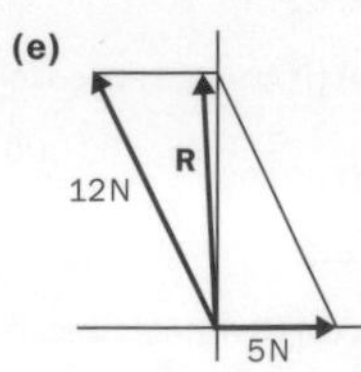

Fig. A9.1.

2. (a) 10.0 N **(b)** 13.0 N.

(a)

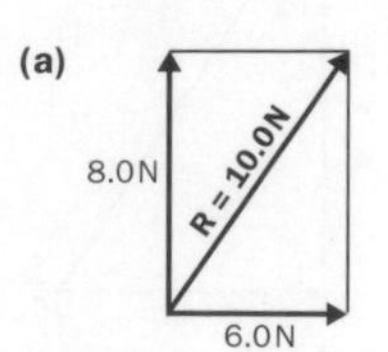

(b)

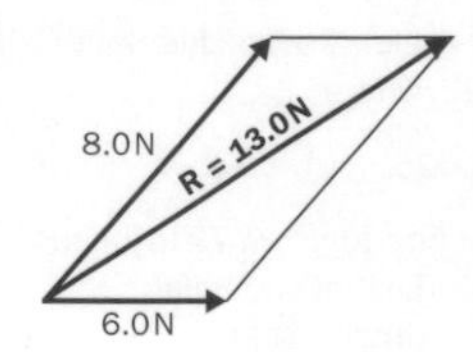

Fig. A9.2.

9.1B

1. (a) $A_x = +3\cos 45$ N, $A_y = +3\sin 45$ N, $B_x = +2.0$ N, $B_y = 0$

$\therefore R_x = 3\cos 45 + 2 = 4.1$ N, $R_y = 2.0$ N, so $R = (4.1^2 + 2^2)^{1/2} = 4.6$ N at $\theta = 27°$ above the x-axis (since $\tan\theta = 2.0/4.1$).

(b) $A_x = +4\cos 30$ N, $A_y = +4\sin 30$ N; $B_x = -2.5$ N, $B_y = 0$

$\therefore R_x = 4\cos 30 - 2.5 = +1.0$ N, $R_y = +2.0$ N, so $R = (1.0^2 + 2.0^2)^{1/2} = 2.2$ N at $\theta = 26°$ above the x-axis (since $\tan\theta = 1.0/2.0$).

2. In each case, C is equal and opposite to R in question 1.

9.2

1. Use $W_1d_1 + W_2d_2 = W_3d_3$. **(a)** 2.0 m **(b)** 1.5 m **(c)** 0.5 m **(d)** 2.0 N.

2. (a) $W \times 0.50 = 2.0 \times 0.15$ $\therefore W = 2.0 \times 0.15/0.50 = 0.6$ N.

(b) $(W_x + 0.60) \times 0.50 = 2.0 \times 0.36$ $\therefore W_x + 0.60 = 2.0 \times 0.36/0.50 = 1.44$ $\therefore W_x = 0.84$ N.

9.3

1. Let d represent the distance from the centre of gravity of the whole placard to the free end of the pole. Consider the pole horizontal, as in Fig. A9.3. The moment of the pole about its free end = 30 N × 2.0 m = 60 N m anticlockwise. The moment of the square board about the free end of the pole = 20 N × 4.0 m = 80 N m anticlockwise

$\therefore$ total moment = 60 + 80 = 140 Nm anticlockwise.

The total weight of 50 N acting at distance d from the free end of the pole gives the same moment.

Hence 50 d = 140 $\therefore$ d = 140/50 = 2.8 m.

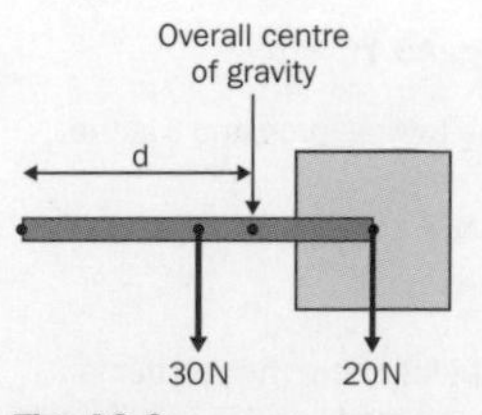

Fig. A9.3.

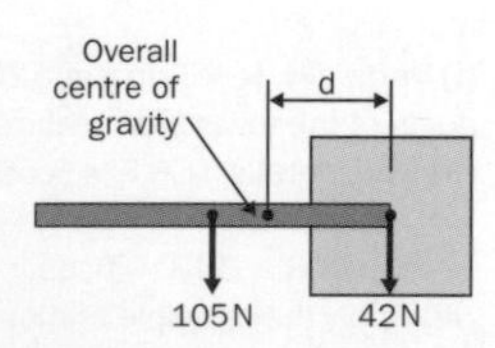

A9.4.

2. (a) Area of rectangle $= 0.4\ m^2$; area of square $= 0.16\ m^2$ $\therefore$ weight of rectangle $= 15g \times 0.40/0.56 = 105$ N.

(b) Weight of square $= 15g - 105 = 42$ N.

(c) Consider the rectangular section horizontal as in Fig. A9.4. The moment of the rectangular section about the centre of the square = 105 N × 1.2 m = 126 N m. This is the same as the moment of the whole board about the centre of the square, i.e. $15gd$, where d is the distance from the centre of the square to the centre of gravity of the whole board.

$\therefore 15gd = 126$ $\therefore d = 126/15g = 0.86$ m $\therefore$ the height of the centre of gravity $= 2.20 - 0.86 = 1.34$ m.

9.4

1. (a) $F = W\sin\theta = 12\sin 58 = 10.2$ N. **(b)** $N = W\cos\theta = 12\cos 58 = 6.4$ N. **(c)** $\mu = \tan 58° = 1.6$.

2. Fig. A9.5 shows a free-body diagram of the forces on the ladder. Let θ represent the angle between the ladder and the floor at the point of slipping. The force on the ladder due to the floor has been resolved into a normal component, N_1, and a friction component, F. Considering all the forces acting on the ladder:

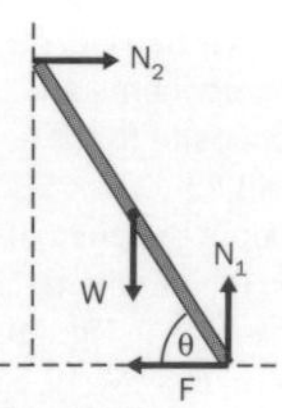

Fig. A9.5.

- The upward vertical forces = downward vertical components $\therefore N_1 = W$.
- The horizontal components acting leftwards = the horizontal components acting rightwards $\therefore N_2 = F$.

Taking moments about the lower end of the ladder at the point of contact with the floor:

- Clockwise moment $= N_2h = F \times 8.0\sin\theta$, since h, the height of the point of contact of the ladder on the wall $= 8.0\sin\theta$, and $N_2 = F$.
- Anticlockwise moment $= Wd/2 = N_1 \times 8.0\cos\theta/2$ since d, the distance from the wall of the point of contact of the ladder on the floor $= 8.0\cos\theta$ and $N_1 = W$.

• In equilibrium, $F \times 8.0\sin\theta = N_1 \times 4.0\cos\theta$.
Since $F = \mu N_1 = 0.4N_1$, then $0.4\,N_1 \times 8.0\sin\theta = N_1 \times 4.0\cos\theta$.
Simplifying this expression gives $0.8\sin\theta = \cos\theta$

$\therefore \tan\theta = \dfrac{\sin\theta}{\cos\theta} = \dfrac{1}{0.8} = 1.25 \therefore \theta = 51°$.

9.5

1. Consider the free-body force diagram for the shelf, as in Fig. A9.6.
Let X and Y represent the two support forces as in the diagram
$\therefore X + Y = 25 + 40 = 65$ N.

Fig. A9.6.

Take moments about the support at X:
• Clockwise moments = $(40 \times 0.5) + (25 \times 1.0) = 45$ N m.
• Anticlockwise moment = 2.0Y.
• For equilibrium, $2.0Y = 45 \quad \therefore Y = 22.5$ N.
Hence $X = 65 - 22.5 = 42.5$ N.

2. Consider the free-body force diagram for the caravan, as shown in Fig. A9.7.

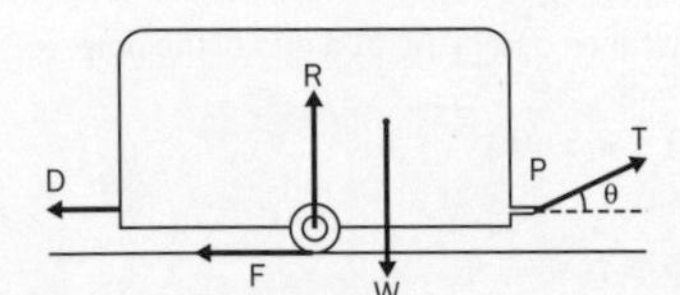

Fig. A9.7.

(i) Vertically, $R + T\sin\theta = 3200$, where T is the towing force and θ is the angle of the towing force above the horizontal.
(ii) Horizontally, $D + F = T\cos\theta$, where $D = 0.5 \times 1086 = 543$ N and $F = \mu R = 0.4R$.
Hence $0.4R + 543 = T\cos\theta$.
(iii) Apply the principle of moments about the point where the tow bar is attached to the caravan, which is 0.5 m above the road and 1.5 m in front of the front wheels, as in Fig. 9.37.
• Total clockwise moment = $3.0R + 0.5F = 3.2R$, since $F = 0.4R$.
• Total anticlockwise moment = $2.0W + (1.3 - 0.5)D = 6400 + (0.8 \times 1086) = 7270$ N m.
Hence $3.2R = 7270 \quad \therefore R = 7270/3.2 = 2270$ N $\quad \therefore T\sin\theta = 3200 - R = 3200 - 2270 = 930$ N.
Also, $T\cos\theta = 0.4R + 543 = (0.4 \times 2270) + 543 = 1450$ N

$\therefore \tan\theta = \dfrac{T\sin\theta}{T\cos\theta} = \dfrac{930}{1450} = 0.64\ \theta = 33°$.

$T = 930/\sin\theta = 930/\sin 33 = 1720$ N.

Revision questions

9.1. (a) (i) mass, m = volume × density = $0.050 \times 0.500 \times 0.800 \times 2700 = 54$ kg. **(ii)** weight = $mg = 54 \times 9.8 = 530$ N.
(b) (i) See Fig. A9.8. **(ii)** There are no other vertical components and therefore the weight is equal and opposite to the normal reaction at the floor.
(iii) $F = 0.4 \times 530 = 212$ N. **(iv)** Take moments about the point of contact at the floor; $W \times 0.5b = R \times h$, where R is the reaction at the wall, b and h are as in Fig. A9.8. $\therefore b/h = 2R/W$. Hence $\tan\theta = b/h = 2R/W$. Since reaction R is equal and opposite to the frictional force, R = 212 N.
$\tan\theta = 2 \times 212/530 = 0.8 \quad \therefore \theta = 39°$.

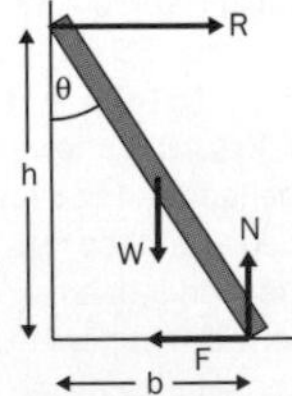

Fig. A9.8.

9.2. (a) Initially, the front wheels bear less weight than the rear wheels. As the skip is raised and moved onto the lorry, an increasing fraction of the weight is carried by the front wheels.
(b) Let X = reaction on the front wheels, Y = reaction at the rear wheels $\therefore X + Y = 130$ kN. Take moments about the rear wheels to give $4.5X + 10 \times 2.5 = 120 \times 3.2$. $4.5X = 384 - 25 = 359 \quad \therefore X = 359/4.5 = 80$ kN
$\therefore Y = 130 - 80 = 50$ kN.
9.3. (i) Take moments about the pivot: $W \times (250 - 10) = 6 \times (50 - 10)$
$\therefore W \times 240 = 6 \times 40 = 240$. Hence $W = 240/240 = 1.0$ N. **(ii)** Force on the rule due to the pivot = $6 - 1 = 5$ N downwards.
(b) Take moments about the pivot: $S \times (50 - 10) = 2 \times (410 - 10) + 1.0 \times (250 - 10) \therefore 40s = 800 + 240 = 1040$. Hence $S = 1040/40 = 26$ N.
9.4. (a) Downward pull = $80 + 50g = 570$ N.
(b) Let X and Y represent the support forces, so $X + Y = 2400 + 570 = 2970$ N.
Take moments about Y: $8.0X = 2400 \times 4.0 + 570 \times 6.0 = 9600 + 3420 = 13020 \therefore X = 13020/8 = 1630$ N at the end near the hose.
Hence $Y = 2970 - 1630 = 1340$ N.
9.5. (a) Take moments about the front wheels: $F \times 1.5 = 1500 \times (0.40 - 0.25) \quad \therefore F \times 1.5 = 1500 \times 0.15$. Hence $F = 1500 \times 0.15/1.5 = 150$ N.
(b) Take moments about the front wheels: $300\,h - 1500 \times (0.40 - 0.25)$
$\therefore h = 1500 \times 0.15/300 = 0.75$ m.

Unit 10

In-text questions

10.1

1. 110 km.
2. Speed = $\dfrac{\text{distance}}{\text{time}} = \dfrac{20\text{ km}}{0.5\text{ h}} = 40\text{ km h}^{-1}$.
3. (a) 50 km (= 180 − 110 − 20 km). **(b)** 30 minutes.
(c) Average speed for final section = $\dfrac{\text{distance}}{\text{time}} = \dfrac{50\text{ km}}{0.5\text{ h}} = 100\text{ km h}^{-1}$.
4. (a) See Fig. A10.1.

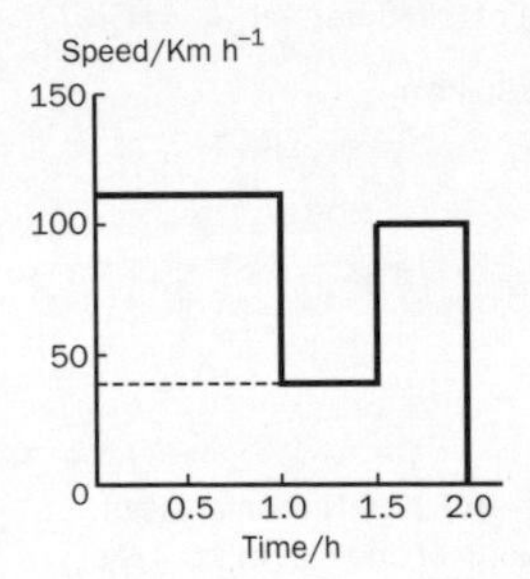

Fig. A10.1.

(b) (i) Average speed for the whole journey = $\dfrac{180\text{ km}}{2\text{ h}} = 90\text{ km h}^{-1}$.

(ii) $90\text{ km h}^{-1} = \dfrac{90\,000\text{ m}}{3600\text{ s}} = 25\text{ m s}^{-1}$.

10.2

1. (a) (i) Average speed = $\dfrac{\text{distance}}{\text{time}} = \dfrac{84\text{ km}}{(70/60\text{ h})} = 72\text{ km h}^{-1}$.

(ii) $72\text{ km h}^{-1} = \dfrac{72000\text{ m}}{3600\text{ s}} = 20\text{ m s}^{-1}$.
(b) Direct distance from junction 22 to junction 5 = $(48^2 + 30^2)^{½} = 57$ km.
Junction 5 is at angle θ East of due South from Junction 22, where

$\tan\theta = \dfrac{30}{48} = 0.625 \quad \therefore \theta = 32°$

$\therefore$ displacement from junction 22 to junction 5 is 57 km at 32° East of due South.
(c) (i) due East **(ii)** due South.

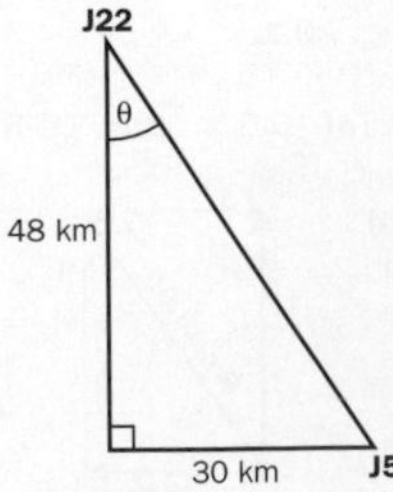

Fig. A10.2.

10.3

1. (a) The vehicle slows down from an initial speed, u, of 30 m s⁻¹ to rest (i.e. $v = 0$) in 20 s $\therefore$ acceleration, $a = \dfrac{(v-u)}{t} = \dfrac{(0-30)}{20} = -1.5\text{ m s}^{-2}$.
(b) Distance travelled = area under the line = $\frac{1}{2}(u+v)t = 0.5 \times 30 \times 20 = 300$ m.

2. (a) Acceleration $= \frac{(v - u)}{t} = \frac{(12 - 0)}{30} = 0.4\ \text{m s}^{-2}$. Distance travelled $= \frac{1}{2}(u + v)t = 0.5 \times 12 \times 30 = 180$ m.

(b) (i) The speed is constant for this period. **(ii)** Distance travelled = speed × time = 12 × 100 = 1200 m.

(c) (i) Acceleration $= \frac{(v - u)}{t} = \frac{(0 - 12)}{20} = -0.6\ \text{m s}^{-2}$.

(ii) Distance travelled $= \frac{1}{2}(u + v)t = 0.5 \times 12 \times 20 = 120$ m.

(d) (i) Total distance travelled = 180 + 1200 + 120 m = 1500 m.

(ii) Average speed $= \frac{1500\ \text{m}}{150\ \text{s}} = 10\ \text{m s}^{-1}$.

10.4

1. $u = 0$, $s = 210$ m, $v = 85\ \text{m s}^{-1}$.

(a) To calculate t, use $s = \frac{(u + v)t}{2}$ which becomes $s = \frac{vt}{2}$ since $v = 0$.

Rearranging this equation gives $t = \frac{2s}{v} = \frac{2 \times 210}{85} = 4.9$ s.

(b) To calculate a, use $a = \frac{(v - u)}{t} = \frac{(85 - 0)}{4.9} = 17\ \text{m s}^{-2}$.

2. $u = 115\ \text{m s}^{-1}$, $v = 0$, $s = 0.045$ m.

(a) To calculate t, rearrange $s = \frac{(u + v)t}{2}$ with $v = 0$ to give $t = \frac{2s}{v}$

$= \frac{2 \times 0.045}{115} = 0.00078$ s.

(b) To calculate a, use $a = \frac{(v - u)}{t} = \frac{(0 - 115)}{0.00078} = -147\ 000\ \text{m s}^{-2}$

$= -1.47 \times 10^5\ \text{m s}^{-2}$.

3. $u = 28\ \text{m s}^{-1}$, $v = 0$, $s = 1.2$ m.

To calculate a, rearrange $v^2 = u^2 + 2$ as with $v = 0$ to give

$a = \frac{-v^2}{2s} = \frac{-28 \times 28}{2 \times 1.2} = -330\ \text{m s}^{-2}$.

10.5A

1. $u = 0$, $t = 1.8$ s, $a = g$.

(a) To calculate the depth of the well, use $s = ut + \frac{1}{2}at^2 = (0 \times 1.8) + (\frac{1}{2} \times 9.8 \times 1.8^2) = 16$ m.

(b) To calculate the speed just before impact, use $v = u + at = 0 + (9.8 \times 1.8) = 18\ \text{m s}^{-1}$.

2. $u = 4.0\ \text{m s}^{-1}$ downwards, $s = 36$ m, $a = g$.

(a) To calculate the speed of the package just before impact, use $v^2 = u^2 + 2as = 4.0^2 + (2 \times 9.8 \times 36) = 722\ \text{m}^2\ \text{s}^{-2}$ $\therefore v = \sqrt{722} = 27\ \text{m s}^{-1}$.

(b) To calculate the time taken, rearrange $v = u + at$ to give

$t = \frac{(v - u)}{a} = \frac{(27 - 4)}{9.8} = 2.3$ s.

10.5B

1. (a) The ball takes 2.2 s (= 0.5 × 4.4 s) to reach maximum height.

(b) Consider the ball from the point of projection to maximum height: $t = 2.2$ s, $v = 0$, $a = -9.8\ \text{m s}^{-2}$.

To calculate u, rearrange $v = u + at$ with $v = 0$ to give $u = -at$ $= -(-9.8 \times 2.2) = 22\ \text{m s}^{-1}$.

(c) To calculate the maximum height, use $s = ut + \frac{1}{2}at^2$ $= (22 \times 2.2) + (0.5 \times -9.8 \times 2.2^2) = 24$ m.

(d) Its speed on returning to the same height is the same as its speed of projection =.22 m s^{-1}.

2. $u = +15\ \text{m s}^{-1}$, $a = -9.8\ \text{m s}^{-2}$, $s = -90$ m.

(a) To calculate v, use $v^2 = u^2 + 2as = 15^2 + (2 \times -9.8 \times -90)$ $= 1989\ \text{m}^2\ \text{s}^{-2}$.

$\therefore v = \sqrt{1989} = 45\ \text{m s}^{-1}$.

(b) To calculate the time taken, rearrange $v = u + at$ to give $t = \frac{(v - u)}{a}$.

Note that $v = -45\ \text{m s}^{-1}$ since the package is moving downwards before impact.

Hence $t = \frac{(-45 - 15)}{-9.8} = 6.1$ s.

10.6

1.

t/s	s/m	v/m s^{-1}	t/s	s/m	v/m s^{-1}
0.000	0.000	24.500	3.000	29.400	−4.900
0.500	11.020	19.600	3.500	25.730	−9.800
1.000	19.600	14.700	4.000	19.600	−14.700
1.500	25.730	9.800	4.500	11.020	−19.600
2.000	29.400	4.900	5.000	0.000	24.500
2.500	30.630	0.000			

Revision questions

10.1. (a) $a = 1.5\ \text{m s}^{-2}$, $u = 0$, $t = 20$ s.

To calculate v, use $v = u + at = 0 + 1.5 \times 20 = 30\ \text{m s}^{-1}$.

(b) See Fig. A10.3.

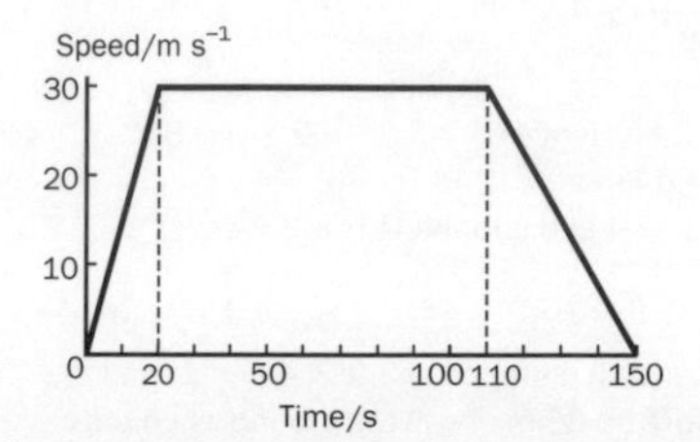

Fig. A10.3.

(c) $t = 40$ s, $u = 30\ \text{m s}^{-1}$, $v = 0$.

To calculate a, use $v = u + at$; $0 = 30 + 40a$. Hence $40a = -30$

$\therefore a = -30/40 = -0.75\ \text{m s}^{-2}$.

(d) (i) 0→20 s: $s = \frac{1}{2}(u + v)t = \frac{1}{2}(0 + 30) \times 20 = 300$ m.

Next 90 s: $s = \frac{1}{2}(u + v)t = 30 \times 90 = 2700$ m (average speed $= \frac{1}{2}(u + v) = 30\ \text{m s}^{-1}$).

Final 40 s: $s = \frac{1}{2}(u + v)t = \frac{1}{2}(30 + 0) \times 40 = 600$ m.

(ii) Average speed $= \frac{\text{total distance}}{\text{time taken}} = \frac{(300 + 2700 + 600)\ \text{m}}{150\ \text{s}} = 24\ \text{m s}^{-1}$.

10.2. (a) $u = 0$, $v = 6.5\ \text{m s}^{-1}$, $t = 30$ s.

To calculate a, use $v = u + at$ to give $6.5 = 0 + 30a$ so $30a = 6.5$.

$\therefore a = 6.5/30 = 0.22\ \text{m s}^{-2}$.

To calculate s, use $s = \frac{1}{2}(u + v)t = \frac{1}{2}(0 + 6.5) \times 30 = 97.5$ m.

(b) $u = 6.5\ \text{m s}^{-1}$, $v = 0$, $s = 80$ m.

To calculate a, use $v^2 = u^2 + 2as$ to give $0 = u^2 + 2as$ or $2as = -u^2$

$\therefore a = \frac{-u^2}{2s} = \frac{-6.5^2}{(2 \times 80)} = -0.26\ \text{m s}^{-2}$.

To calculate t, use $s = \frac{1}{2}(u + v)t$ to give $s = ut/2$ or $t = 2s/u$

$\therefore t = 2 \times 80/6.5 = 25$ s.

(c) (i) See Fig. A10.4.

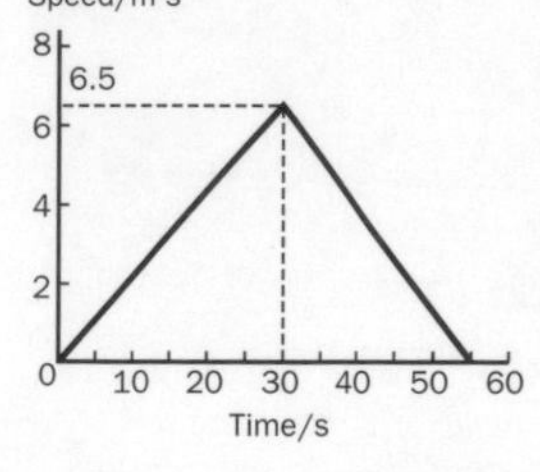

Fig. A10.4.

(ii) Average speed $= \frac{\text{total distance}}{\text{time taken}} = \frac{97.5 + 80}{(30 + 25)} = \frac{177.5}{55} = 3.25\ \text{m s}^{-1}$.

10.3. (a) $u = +15\ \text{m s}^{-1}$, $v = 0$ at max height, $a = -9.8\ \text{m s}^{-2}$.

To calculate s (its maximum height), use $v^2 = u^2 + 2as$ to give $0 = u^2 + 2as$ or $2as = -u^2$. Hence $s = \frac{-u^2}{2a} = \frac{-15^2}{2 \times -9.8} = 11.5$ m.

To calculate t, use $v = u + at$ to give $0 = u + at$ or $at = -u$.

Hence $t = -u/a = -15/-9.8 = 1.5$ s.

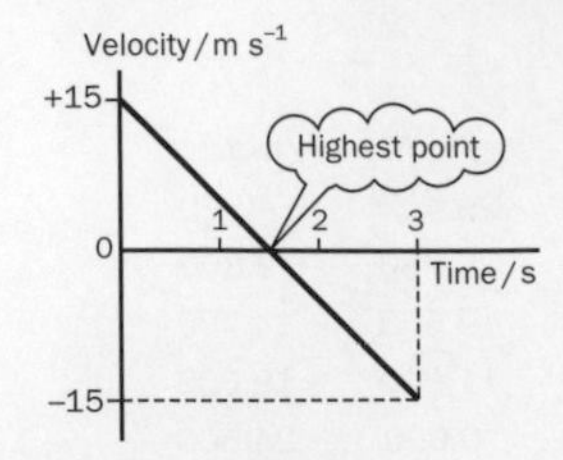

Fig. A10.5.

(b) (i) $u = +15\ m\ s^{-1}$, $a = -9.8\ m\ s^{-2}$, $s = +5.0\ m$.
To calculate v, use $v^2 = u^2 + 2as$. Hence $v^2 = 15^2 + 2 \times -9.8 \times 5.0 = 225 - 98 = 127$.
$\therefore v = \sqrt{127} = -11(.3)\ m\ s^{-1}$, (– for downwards).
To calculate t, use $v = u + at$ to give $at = v - u$ or $t = (v - u)/a$.
Hence $t = \frac{(-11.3 - 15)}{-9.8} = \frac{-26.3}{-9.8} = 2.7\ s$.

10.4. $u = -3.2\ m\ s^{-1}$, $a = -9.8\ m\ s^{-2}$, $s = -100\ m$.
(a) To calculate v, use $v^2 = u^2 + 2as$. Hence $v^2 = 3.2^2 + 2 \times -9.8 \times -100 = 1970$ $\therefore v = \sqrt{1970} = 44\ m\ s^{-1}$.
(b) To calculate t, rearrange $v = u + at$ and substitute $v = -44\ m\ s^{-1}$.
Hence $t = \frac{(v - u)}{a} = \frac{(-44 - 3.2)}{-9.8} = 4.8\ s$.
(c) In 4.8 s, the parachutist descends a distance of $3.2 \times 4.8 = 15\ m$ (= speed × time). She will therefore be 85 m above the ground when the object hits the ground.

10.5. (a) (i) $u = 0$, $a = 8.0\ m\ s^{-2}$, $t = 40\ s$.
To calculate v, use $v = u + at = 0 + 8.0 \times 40 = 320\ m\ s^{-1}$. To calculate its height at cut-off, use $s = ut + \frac{1}{2}at^2 = \frac{1}{2} \times 8.0 \times 40^2 = 6400\ m$.
(ii) To calculate its maximum height, calculate the distance moved upwards after cut-off, i.e. $u = +320\ m\ s^{-1}$, $a = -9.8\ m\ s^{-2}$, $v = 0$.
To calculate s, use $v^2 = u^2 + 2as$ to give $0 = u^2 + 2as$ or $2as = -u^2$
Hence $s = \frac{-u^2}{2a} = \frac{-320^2}{(2 \times -9.8)} = 5200\ m$ $\therefore$ maximum height $= 6400 + 5200 = 11600\ m$.
(iii) To calculate its velocity, v, just before impact, consider its motion under gravity from cut-off to impact:
$a = -9.8\ m\ s^{-2}$, $u = +320\ m\ s^{-1}$, $s = -6400\ m$.
Use $v^2 = u^2 + 2as = 320^2 + (2 \times -9.8 \times -6400) = 2.28 \times 10^5\ m^2\ s^{-2}$.
$\therefore v = -480\ m\ s^{-1}$ (– for downwards).
(iv) To calculate the time, t, taken from cut-off to impact, use $v = u + at$ or $at = v - u$.
Hence $t = \frac{(v - u)}{a} = \frac{(-480 - 320)}{-9.8} = 82\ s$.
$\therefore$ total time taken = 82 + 40 = 122 s.
(b) See Fig. A10.6.

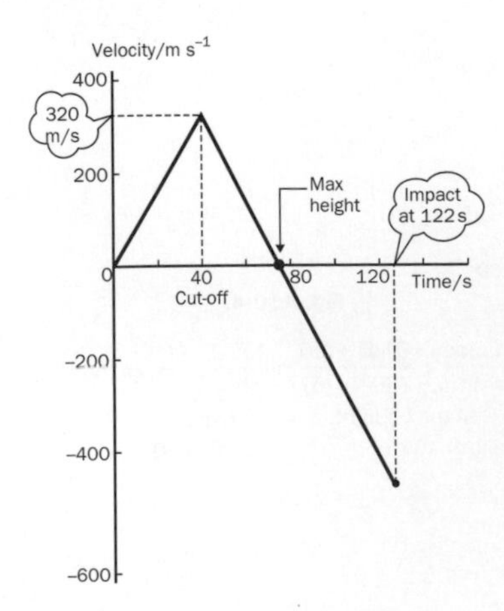

Fig. A10.6.

Unit 11

In-text questions

11.1

1. (a) $u = 25\ m\ s^{-1}$, $v = 0$, $t = 8.0\ s$.
To calculate a, rearrange $v = u + at$ to give $a = \frac{(v-u)}{t} = \frac{(0-25)}{8.0} = 3.1\ m\ s^{-2}$.
(b) Force needed = mass × acceleration = 1200 × 3.1 = 3700 N.
2. (a) Weight = mg = 8000 × 9.8 = 78 400 N. **(b)** $T - mg = ma$, where T is the thrust and a is the acceleration $\therefore T = mg + ma$ = 78 400 + (8000 × 6.0) = 126 400 N.
3. (a) $T = mg$ = 1500 × 9.8 = 14 700 N.
(b) $a = +1.2\ m\ s^{-2}$. Using $T - mg = ma$ gives $T = mg + ma$ = 14 700 + (1500 × 1.2) = 16 500 N.
(c) $a = +1.2\ m\ s^{-2}$ since the deceleration direction is upwards, in the opposite direction to the velocity.
Using $T - mg = ma$ gives $T = mg + ma$ = 14 700 + (1500 × 1.2) = 16 500 N.

11.2

1. (a) Change of momentum = final momentum – initial momentum = (–0.12 × 25) – (0.12 × 25) = –6.0 kg m s^{-1}.
(b) Impact force = $\frac{\text{change of momentum}}{\text{time taken}} = \frac{-6.0}{0.004} = -1500$ N.
2. (a) Weight = mg = 2000 × 9.8 = 19 600 kg.
(b) $T - mg = ma$, where T is the thrust $\therefore T = mg + ma$ = 19 600 + (2000 × 6.5) = 32 600 N.

11.3

1. (a) Loss of momentum of first vehicle = (800 × 25) – (800 × 5) = 16 000 kg m s^{-1}.
(b) Gain of momentum of second vehicle = 1600v, where v is its speed after the impact $\therefore$ 1600 v = 16 000, which gives v = 10 m s^{-1}.
(c) Impact force = $\frac{\text{change of momentum of either vehicle}}{\text{time taken}} = \frac{16\,000}{0.2}$ = 80 000 N.
2. Let v = recoil speed of gun $\therefore$ 500 v = 4.0 × 115, to give v = 0.86 m s^{-1}.

Revision questions

11.1. (a)(i) $a = \frac{v - u}{t} = \frac{12 - 0}{10} = 1.2\ m\ s^{-2}$.
(ii) $F = ma$ = 1500 × 1.2 = 1800 N.
(b) (i) $a = 0$ $\therefore 4T = mg$ = 850 × 9.8 = 8300 N. T = 2075 N.
(ii) $a = +0.5\ m\ s^{-2}$, $4T - mg = ma$ $\therefore 4T = mg + ma$ = 8300 + (850 × 0.5) = 8700 N. (to 2 sig. figs) $\therefore$ T = 2175 N.
(c) (i) Fig. A11.1 shows the forces acting on the block. Consider the forces and their components parallel and perpendicular to the surface:
- Perpendicular: $N = mg$ = 12 × 9.8 = 118 N.
- Parallel: $F - F_o = ma$, where $F_o = \mu N = \mu mg$ = 0.4 × 12 × 9.8 = 47 N.

$\therefore$ 12a = 50 – 47 = 3. Hence a = 3/12 = 0.25 m s^{-2}.

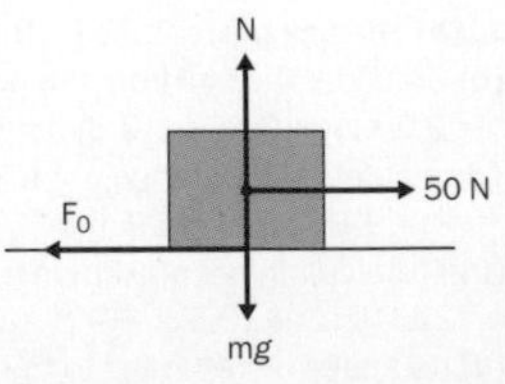

Fig. A11.1.

(ii) Fig. A11.2 shows the forces acting on the block. Consider the forces and their components parallel and perpendicular to the surface:
- Perpendicular: $N = mg\cos 30$ = 12 × 9.8 × cos 30 = 102 N.
- Parallel: $F + mg\sin 30 - F_o = ma$, where $F_o = \mu N$ = 0.4 × 102 = 41 N.

$\therefore$ 12 a = 50 + (12 × 9.8 × sin 30) – 41 = 68 $\therefore$ a = 68/12 = 5.7 m s^{-2}.

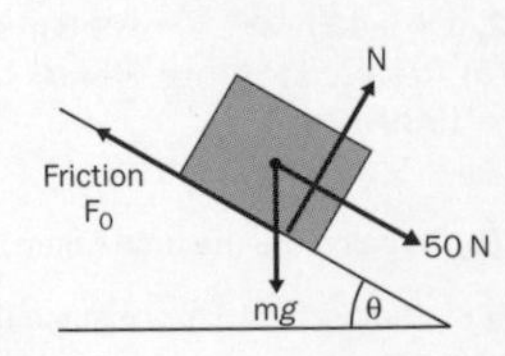

Fig. A11.2.

11.2. (a) Loss of momentum = 0.001 × 100 = 0.10 kg m s^{-1}.

(b) Rearrange $s = \frac{1}{2}(u+v)t$ with $v = 0$ to give $t = \frac{2s}{(u+v)} = \frac{2 \times 0.050}{100} = 0.0010$ s.

(c) Force = $\frac{\text{change of momentum}}{\text{time taken}} = \frac{0.10}{0.0010} = 100$ N.

11.3. (a) Weight = mg = 11800 N;

acceleration = $\frac{\text{resultant force}}{\text{mass}} = \frac{16000 - 11800}{1200} = 3.5$ m s^{-2}.

(b) Thrust force, $T = \frac{v\,dm}{dt}$ $\therefore \frac{dm}{dt} = \frac{T}{v} = \frac{16\,000}{1200} = 13.3$ kg s^{-1}.

(c) Time taken to use 800 kg of fuel at 13.3 kg s^{-1} = $\frac{800}{13.3}$ = 60 s.

11.4. (a) Total initial momentum = 3.0 × 1000 = 3000 kg m s^{-1}. Total final momentum = (1000 + 2000)V, where V is the final speed of the two wagons ∴ 3000V = 3000. Hence V = 1 m s^{-1}.

(b) Total initial momentum = (3.0 × 1000) + (2.0 × 2000) = 7000 kg m s^{-1}. Total final momentum = (1000 + 2000)V, where V is the final speed of the two wagons ∴ 3000V = 7000. Hence V = 7/3 = 2.3 m s^{-1}.

(c) Total initial momentum = (3.0 × 1000) − (2.0 × 2000) = −1000 kg m s^{-1} where the minus sign indicates the direction of the 2000 kg wagon. Total final momentum = (1000 + 2000)V, where V is the final speed of the two wagons ∴ 3000V = −1000. Hence V = −1/3 = −0.3 m s^{-1}.

11.5. Let m represent the mass of the unknown object ∴ (m + 0.5) × 0.22 = 0.60 × 0.35

$m + 0.5 = \frac{0.60 \times 0.35}{0.22} = 0.95$ kg ∴ m = 0.95 − 0.50 = 0.45 kg.

Unit 12

In-text questions

12.1

1. (a) (i) Initial k.e. = $\frac{1}{2}mv^2$ = 0.5 × 0.12 × 22^2 = 29 J. **(ii)** At maximum height, the ball is stationary for an instant and has zero k.e. **(iii)** Gain of p.e. = loss of k.e. = 29 J. **(iv)** mgh = 29; h = 29/0.12 g = 25 m.

(b) (i) Loss of p.e. = mgh = 150 × 9.8 × 200 = 290 kJ. **(ii)** Its gain of k.e. = $\frac{1}{2}mv^2$ = 0.5 × 150 × 55^2 = 230 kJ. **(iii)** Work done against friction, etc. = 290 − 230 = 60 kJ. **(iv)** Friction force, F × distance = 60 kJ ∴ F = 60 000/1500 = 40 N.

12.2

1. (a) (i) k.e. = $\frac{1}{2}mv^2$ = 0.5 × 70 × 10^2 = 3500 J.

(ii) power output = $\frac{\text{k.e. gain}}{\text{time taken}} = \frac{3500\text{ J}}{3.5\text{ s}}$ = 1000 W.

(b) (i) p.e. gain = mgh = 55 × 9.8 × 0.40 = 220 J. **(ii)** Time taken for 1 step = 3.0 s ∴ power output = $\frac{\text{p.e. gain}}{\text{time taken}} = \frac{220\text{ J}}{3.0\text{ s}}$ = 73 W.

2. (a) Use $P = Fv$ to give $F = \frac{P}{v} = \frac{175\,000}{55}$ = 3200 N.

(b) At constant speed, resistive force = output force = 3200 N.

(c) Method 1: rearrange $v^2 = u^2 + 2as$ with v = 0, u = 55 m s^{-1}, s = 1000 m, to obtain $a = \frac{(v^2 - u^2)}{2s} = \frac{3025}{(2 \times 1000)}$ = 1.5 m s^{-2}.

Hence F = ma = 60 000 × 1.5 = 90 000 N.

Method 2: Calculate the loss of k.e. and use work done, Fs = loss of k.e. to calculate F.

12.3

1. (a) Electrical energy supplied to motor = power × time = 300 × 25 = 7500 J.

(b) Useful energy from motor = mgh = 320 × 9.5 = 3000 J.

(c) Energy wasted = 7500 − 3000 = 4500 J.

(d) Efficiency = $\frac{3000}{7500}$ = 0.40.

2. (a) p.e. gain = 150 N × 2.5 m = 375 J.

(b) Work done by effort = effort × distance moved = 40 N × (4 × 2.5 m) = 400 J.

(c) Efficiency = $\frac{375}{400}$ = 0.94.

12.4

1. (a) 10 m^2 (= 5000 W/500 W).

(b) Energy needed to heat 0.012 kg from 15°C to 30°C = 0.012 × 4200 × (30 − 15) = 760 J ∴ power = 760 W.

2. (a) $P = \frac{1}{2} \times 1.2 \times \pi \times 15^2 \times 5^3$ = 50 000 W. **(b)** $P = \frac{1}{2} \times 1.2 \times \pi \times 15^2 \times 15^3 = 1.4 \times 10^6$ W.

Revision questions

12.1. (a) (i) h = 40 sin 5° = 3.5 m. **(ii)** P.e. gain = mgh = 120 × 9.8 × 3.5 = 4120 J.

(b) (i) F = $\frac{\text{p.e. gain}}{\text{distance}} = \frac{4120}{40}$ = 103 N.

(ii) Friction between the wheels and the wheel bearings must be overcome.

(c) Work done/sec = $\frac{\text{p.e. gain}}{\text{time taken}} = \frac{4120}{45}$ = 91 W.

12.2. (a) (i) h = $\frac{s}{20} = \frac{21}{20}$ = 1.05 m, since s = 21 m in 1 second.

(ii) P.e. gain = mgh = 600 × 9.8 × 1.05 = 6.2 kJ.

(b) (i) Energy delivered by engine in 1 s = 25 kJ. Work done against resistive forces = energy delivered − p.e. gain = 25 − 6.2 = 18.8 kJ.

(ii) Resistive force = $\frac{\text{work done against friction}}{\text{distance}} = \frac{1880}{21}$ = 895 N.

12.3. (a) K.e. gain of pile driver = loss of p.e. $\frac{1}{2}mv^2 = mgh$ ∴ $v = \sqrt{2gh} = \sqrt{(2 \times 9.8 \times 3.0)}$ = 7.7 m s^{-1}.

(b) (4000 + 6000) V = 4000 × 7.7, where V is the speed of the pile driver and the girder immediately after impact. Thus V = $\frac{4000 \times 7.7}{10\,000}$ = 3.1 m s^{-1}.

(c) K.e. of girder and block after impact = $\frac{1}{2}$(4000 + 6000) × 3.1^2 = 48 050 J.

Average resistive force = $\frac{\text{loss of k.e. of girder and block}}{\text{distance moved}} = \frac{48\,050}{1.5}$ = 32 000 N.

12.4. (a) (i) P.e. gain = $m_x g h_0$ = 0.25 × 9.8 × 0.10 = 0.245 J. **(ii)** K.e. gain = loss of p.e. on downswing ∴ $\frac{1}{2}m_x u_0{}^2 = m_x g h_0$ hence $u_0 = \sqrt{(2gh_0)} = \sqrt{(2 \times 9.8 \times 0.10)}$ = 1.4 m s^{-1}.

(b) (i) P.e. gain = $m_y g h_2$ = 0.15 × 9.8 × 0.05 = 0.27 J. **(ii)** K.e. loss = p.e. gain on upswing ∴ $\frac{1}{2}m_y v_2{}^2 = m_y g h_2$ hence $v_2 = \sqrt{(2gh_2)} = \sqrt{(2 \times 9.8 \times 0.05)}$ = 0.99 m s^{-1}.

(c) (i) $m_x u_0 = m_x v_1 + m_y v_2$. Hence 0.25 × 1.4 = 0.25v_1 + 0.15 × 0.99 ∴ 0.25v_1 = 0.20. Hence v_1 = 0.20/0.25 = 0.80 m s^{-1}.

(ii) Using p.e. gain = k.e. loss on the upswing, $mgh_1 = \frac{1}{2}m_x v_1{}^2$ ∴ $h_1 = \frac{v_1{}^2}{2g} = \frac{0.8^2}{2 \times 9.8}$ = 0.032 m.

12.5. (a) Using conservation of momentum, (1200 + 800)V = 1200 × 30, where V is the speed after the impact ∴ V = $\frac{1200 \times 30}{2000}$ = 18 m s^{-1}.

(b) Loss of k.e. = initial k.e. − final k.e. = ($\frac{1}{2}$ × 1200 × 30^2) − ($\frac{1}{2}$ × 2000 × 18^2) = 220 000 J.

(c) Force = $\frac{\text{change of momentum}}{\text{time taken}} = \frac{(1200 \times 18) - (1200 \times 30)}{0.12}$ = 120 000 N.

Unit 13

In-text questions

13.1

1. (a) Electrons from the cloth transfer to the rod. **(b)** Negative. **(c)** Charge the electroscope negatively by direct contact with the charged polythene rod. Charge the unknown rod and hold it directly above the cap of the electroscope. If the leaf rises further, the unknown rod is negative. If the leaf falls, the rod could be positive or uncharged. To decide which, earth the electroscope and repeat the test with the electroscope charged positively from the charged perspex rod.

2. (a) See Fig. A13.1.

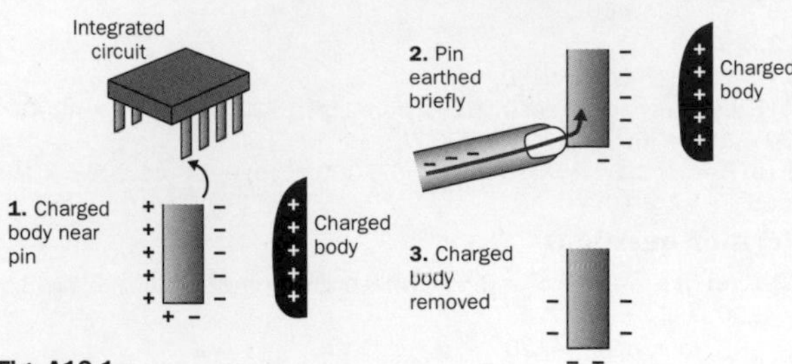

Fig. A13.1.

(b) Friction between the flowing oil and the pipe generates static electricity which would build up on the pipe if the pipe was not earthed. A spark due to static electricity could cause an explosion as oil is flammable.

13.2

1. (a) air, glass, germanium **(b)** 5 A, 0.3 C, 3 s, 2.5 A.

2. (a) 2.0 A **(b)** 0.35 A **(c)** 4.2 A leaving the junction.

13.3

1. (a) 3 A (= 36 W/12 V). **(b)** $Q = It = 3 \times 10 \times 60 = 1800$ C.

(c) Energy supplied = 36 W × 600 s = 21 600 J.

2. (a) 2.8 kW h = 2800 W for 3600 s = 10 MJ.

(b) Rearrange $V = \frac{E}{Q}$ to give $Q = \frac{E}{V} = \frac{10\text{ MJ}}{230\text{ V}} = 44$ kC.

(c) Current, $I = \frac{Q}{t} = \frac{44\text{ kC}}{30 \times 60\text{ s}} = 24$ A.

13.4A

1. 7.5 Ω, 200 V, 0.68 mA, 2.4 kΩ, 0.50 mA.

2. (a) 3900 ohms **(b)** 1.5 mA.

13.4B

1. The circuit diagram is shown in Fig. A13.2.

(a) (i) 9 Ω. **(ii)** $I = V/R = 9\text{ V}/9\ \Omega = 1.0$ A.

(iii) 3 Ω resistor: $V = IR = 1 \times 3 = 3$ V, $P = IV = 1 \times 3 = 3$ W. 6 Ω resistor: $V = IR = 1 \times 6 = 6$ V, $P = IV = 1 \times 6 = 6$ W.

Fig. A13.2.

(b) The circuit diagram is shown in Fig. A13.3.

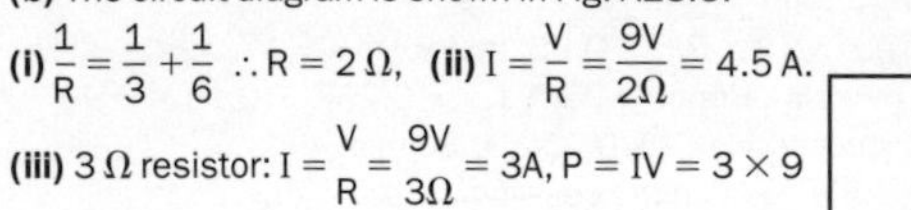

(i) $\frac{1}{R} = \frac{1}{3} + \frac{1}{6}$ $\therefore R = 2\ \Omega$, **(ii)** $I = \frac{V}{R} = \frac{9\text{V}}{2\Omega} = 4.5$ A.

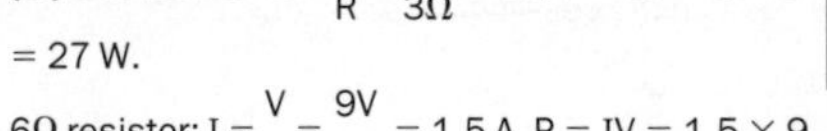

(iii) 3 Ω resistor: $I = \frac{V}{R} = \frac{9\text{V}}{3\Omega} = 3\text{A}$, $P = IV = 3 \times 9 = 27$ W.

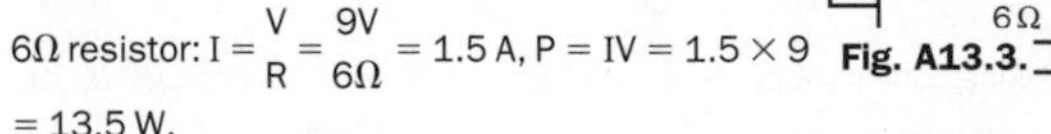

6Ω resistor: $I = \frac{V}{R} = \frac{9\text{V}}{6\Omega} = 1.5$ A, $P = IV = 1.5 \times 9 = 13.5$ W.

Fig. A13.3.

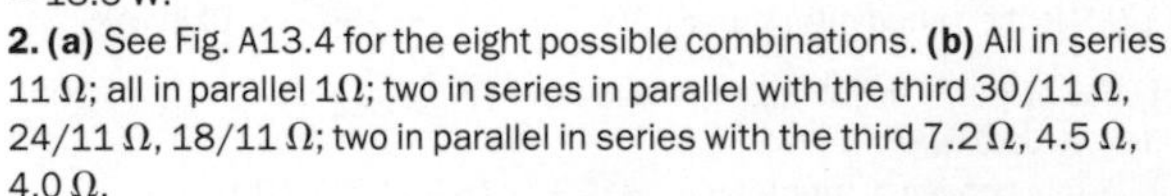

2. (a) See Fig. A13.4 for the eight possible combinations. **(b)** All in series 11 Ω; all in parallel 1Ω; two in series in parallel with the third 30/11 Ω, 24/11 Ω, 18/11 Ω; two in parallel in series with the third 7.2 Ω, 4.5 Ω, 4.0 Ω.

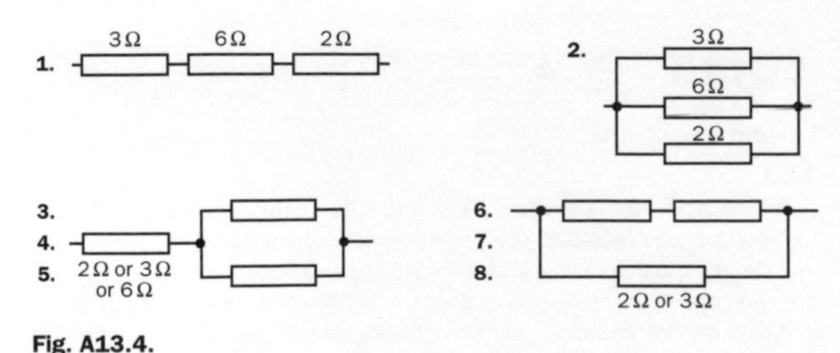

Fig. A13.4.

Revision questions

13.1. (a) Static electric charge is created on the balloon when it is rubbed. This charge attracts opposite charges in the ceiling which holds the balloon on the ceiling.

(b) The plastic ruler becomes positively charged, the same as the perspex rod, when it is rubbed. The charge on the ruler forces the positive charge on the cap of the electroscope onto the leaf and the stem. Hence the leaf rises more.

(c) Friction between clothing and the seat creates static electricity. The person becomes charged and discharges on touching the door handle.

13.2. (a) More conduction electrons are created when the temperature is increased. The material therefore conducts more easily.

(b) The atoms in the metal vibrate more when the temperature of the metal is raised. The vibrating atoms impede the flow of electrons through the metal. The material therefore conducts less easily.

13.3. (a) (i) $Q = It = 0.25\text{ A} \times 600\text{ s} = 150$ C. **(ii)** $P = IV = 0.25\text{ A} \times 1.5\text{ V} = 0.375$ W. **(iii)** $E = Pt = 0.375 \times 600 = 225$ J.

(b) (i) $Q = It = 0.25\text{ A} \times 3600\text{ s} = 900$ C. **(ii)** $E = QV = 900\text{ C} \times 1.5\text{ V} = 1350$ J.

13.4. (a) (i) $I = 0.1 + 0.1 = 0.2$ A. **(ii)** $V_{batt} = V_{bulb} + V_{res} \therefore V_{res} = 4.5 - 3.0 = 1.5$ V.

(iii) $P_x = P_y = IV = 0.1\text{ A} \times 3.0\text{ V} = 0.30$ W. **(iv)** $P_{batt} = I_{batt} V_{batt} = 0.2\text{ A} \times 4.5\text{ V} = 0.90$ W.

(b) The power dissipated in the variable resistor = 0.2 A × 1.5 V = 0.3 W. This is the difference between the power supplied by the battery and the power delivered to each torch bulb.

13.5. (a) (i) $I = \frac{P}{V} = \frac{750\text{ W}}{230\text{ V}} = 3.3$ A. **(ii)** Number of units used = 0.75 kW × 0.5 h = 0.375 kW h.

(b) 5 A.

13.6 (a) See Fig. A 13.5. **(b) (i)** $R = \frac{V}{I} = \frac{5\text{ V}}{0.44\text{ A}} = 11.4\ \Omega$.

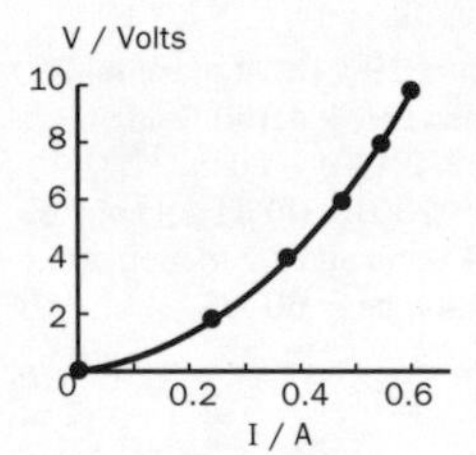

Fig. A13.5.

(ii) $R = \frac{10\text{ V}}{0.66\text{ A}} = 15.1\ \Omega$.

(c) The element becomes hotter as the current increases.

13.7. (a) See Fig. A13.6.

Fig. A13.6.

(b) (i) $V_R = 9.0 - 0.6 = 8.4$ V. **(ii)** $R = \frac{V_R}{I} = \frac{8.4\text{ V}}{2.0\text{ A}} = 4.2\ \Omega$.

13.8. (a) See Fig. A13.7.

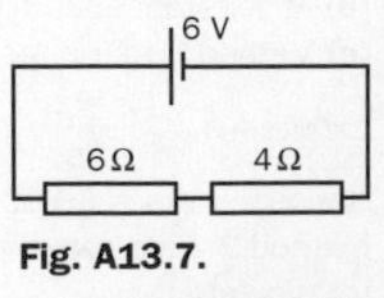

Fig. A13.7.

(b) (i) $R = 4 + 6 = 10\ \Omega$.

(ii) $I_{batt} = \frac{V_{batt}}{R} = \frac{6.0\text{ V}}{10\ \Omega} = 0.60$ A.

(iii) 4 Ω: $V_4 = IR_4 = 0.60 \times 4.0 = 2.4$ V: 6 Ω: $V_6 = IR_6 = 0.60 \times 6.0 = 3.6$ V.

(iv) 4 Ω: $P = IV_4 = 0.60 \times 2.4 = 1.44$ W: 6 Ω: $P = IV_6 = 0.60 \times 3.6 = 2.16$ W.

13.9. (a) See Fig. A13.8.

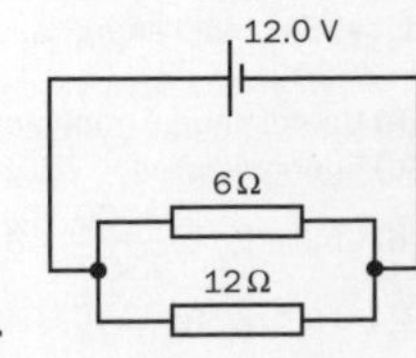

Fig. A13.8.

(b) (i) $\frac{1}{R} = \frac{1}{6} + \frac{1}{12} = \frac{12+6}{12 \times 6}$ $\therefore R = \frac{72}{18} = 4\,\Omega$.

(ii) $I = \frac{V_{batt}}{R} = \frac{12}{4} = 3\,A$. **(iii)** $I_6 = \frac{V_{batt}}{6} = \frac{12}{6} = 2\,A$, $I_{12} = \frac{V_{bat}}{12} = \frac{12}{12} = 1\,A$.

(iv) $P_6 = I_6 V_{batt} = 2 \times 12 = 24\,W$; $P_{12} = I_{12} V_{batt} = 1 \times 12 = 12\,W$.

13.10. (a) See Fig. A13.9.

(b) (i) $R = 4 + \left(\frac{1}{3} + \frac{1}{6}\right)^{-1}$

$= 4 + 2 = 6\,\Omega$.

(ii) $I_{batt} = \frac{V_{batt}}{R} = \frac{6}{6} = 1.0\,A$.

(iii) $V_4 = I_{batt} \times 4 = 1.0 \times 4 = 4.0\,V$;

$V_3 = V_6 = V_{batt} - V_4 = 6 - 4 = 2.0\,V$.

Fig. A13.9.

(iv) $I_4 = I_{batt} = 1.0\,A$;

$I_3 = \frac{V_3}{R_3} = \frac{2.0}{3} = 0.67\,A$; $I_6 = \frac{V_6}{R_6} = \frac{2.0}{6} = 0.33\,A$.

(v) $P_4 = I_4 V_4 = 1.0 \times 4.0 = 4.0\,W$; $P_6 = I_6 V_6 = 0.33 \times 2.0 = 0.7\,W$;

$P_3 = I_3 V_3 = 0.67 \times 2.0 = 1.3W$.

Unit 14

In-text questions

14.1

1. (a) See Fig. A14.1.

(b) Use $E = V + Ir$ to find V, the p.d. across the torch bulb (= IR). Hence $2.0 = V + (0.25 \times 0.5)$

$\therefore V = 2.0 - (0.25 \times 0.5) = 1.875\,V$. Power delivered to torch bulb $= IV = 0.25 \times 1.875 = 0.47\,W$.

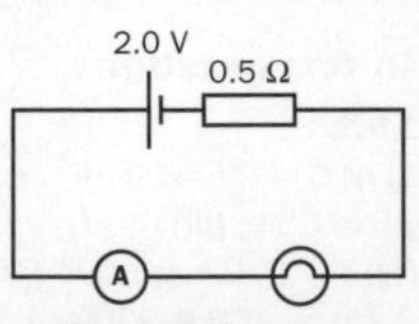

Fig. A14.1.

(c) Power supplied by cell $= IE = 0.25 \times 2.0 = 0.50\,W$.

(d) Power dissipated in the cell due to its internal resistance $= I^2 r = 0.25^2 \times 0.5 = 0.3\,W$. The cell supplies 5.0 W of electrical power, the bulb uses 0.47 W of electrical power, the internal resistance of the cell causes 0.3 W of electrical power to be dissipated.

2. (a) See Fig. A14.2.

The voltmeter reading decreases because the reduced resistance allows more current so there is a bigger lost voltage. Hence the cell p.d. (= e.m.f. − lost voltage) falls.

(b) Use $E = V + Ir$ where $V = IR$, for each pair of values of I and R to give

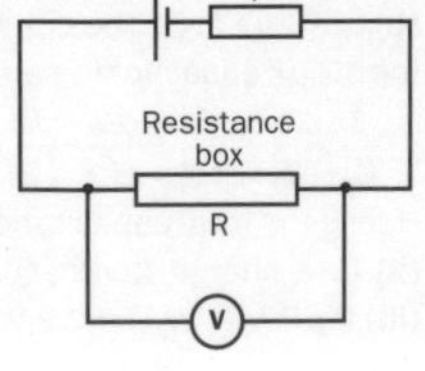

Fig. A14.2.

$E = 1.20 + I_1 r$, where $I_1 = \frac{V}{R} = \frac{1.2}{16.0} = 0.075\,A$. Hence $E = 1.20 + 0.075r$.

Also, $E = 1.00 + I_2 r$ where $I_2 = \frac{V}{R} = \frac{1.00}{8.0} = 0.125\,A$. Hence $E = 1.00 + 0.125\,r$

$\therefore E = 1.20 + 0.075\,r = 1.00 + 0.125r$ which gives $0.125\,r - 0.075\,r$

$= 1.20 - 1.00$ $\therefore 0.050\,r = 0.20$ to give $r = \frac{0.20}{0.050} = 4.0\,\Omega$ and $E = 1.00 + (0.125 \times 4.0) = 1.50\,V$.

14.2

1. Using $V_y = \frac{R_y}{(R_x + R_y)} V_{batt}$ gives **(a)** 1.0 mA, 1.0 V **(b)** 0.25 mA, 0.25 V **(c)** 0.11 mA, 3.45V **(d)** 0.030 A, 6.0 V.

2. (a) A current passes through cell X and the 5 ohm resistor at balance. The cell p.d. is therefore less than the cell's e.m.f. because there is a 'lost voltage' across the internal resistance of the cell when current passes through the cell. Note that no current passes through the centre-reading meter because the sliding contact is at the balance position.

(b) (i) Using $E_x = \frac{L_x}{L_s} E_s = \frac{0.775}{0.558} \times 1.08 = 1.50\,V$.

(ii) Cell p.d. at balance with the 5 ohm resistor in circuit,

$V = E_s L'_x = \frac{1.08}{0.558} \times 0.692 = 1.34\,V$ $\therefore$ current $I = \frac{V}{R} = \frac{1.34}{5.0} = 0.27\,A$.

Hence $r = \frac{(E - V)}{I} = \frac{(1.50 - 1.34)}{0.27} = 0.6\,\Omega$.

14.3

1. (a) $\frac{X}{50} = \frac{0.452}{0.548} = 0.82$ $\therefore X = 0.82 \times 50 = 41\,\Omega$.

(b) Let L_w = the length of wire from the balance point to the end of the wire at X. The effective resistance of two 50 Ω resistors in parallel = 25 Ω

$\therefore \frac{X}{25} = \frac{L}{(L_w - L)}$ where L_w, the length of the wire = 1.00 m.

Hence $\frac{L}{(L_w - L)} = \frac{41}{25} = 1.64$ $\therefore L = 1.64\,(L_w - L)$ to give $2.64\,L = 1.64\,L_w$.

Hence $L = \frac{1.64\,L_w}{2.64} = 0.621\,m$ since $L_w = 1.00\,m$.

2. (a) The LDR's resistance is greater in darkness than in daylight. The potential difference across the LDR therefore increases, thus reducing the p.d. across the variable resistor. Consequently, the potential difference from X to Z (which is unchanged) exceeds the potential difference from Y to Z. Therefore X becomes positive relative to Y.

(b) The variable resistance needs to be increased until it equals that of the LDR in darkness. The p.d. between Y and Z will then equal the p.d. between X and Z.

14.4

1. (a) $\rho = \frac{RA}{L} = \frac{58 \times \pi (0.24 \times 10^{-3})^2/4}{5.5} = 4.8 \times 10^{-7}\,\Omega\,m$.

(b) $R = \frac{\rho L}{A} = \frac{4.8 \times 10^{-7} \times 1.0}{\pi (0.35 \times 10^{-3})^2/4} = 5.0\,\Omega$

2. Area of cross-section, A = width, w × thickness, t

$\therefore$ resistivity $\rho = \frac{RA}{L} = \frac{Rwt}{L}$.

Hence $t = \frac{\rho L}{Rw} = \frac{3.0 \times 10^{-5} \times 10 \times 10^{-3}}{15 \times 1.0 \times 10^{-3}} = 2.0 \times 10^{-5}\,m$.

14.5

1. (a) $V_{max} = 1.0\,mA \times 100\,\Omega = 0.10\,V$.

(i) Current through shunt resistor, i = 99 mA

$\therefore$ shunt resistance $= \frac{0.10\,V}{0.099\,A} = 1.01\,\Omega$.

(ii) Current through shunt resistor, i = 9.999 A

$\therefore$ shunt resistance $= \frac{0.10\,V}{9.999\,A} = 0.010\,\Omega$.

(b) Current through series resistor = 1.0 mA.

(i) Voltage across series resistor = 10.0 − 0.10 = 9.9 V

$\therefore$ series resistance $= \frac{9.9\,V}{0.001\,A} = 9900\,\Omega$.

(ii) Voltage across series resistor = 30.0 − 0.10 = 29.9 V

$\therefore$ series resistance $= \frac{29.9\,V}{0.001\,A} = 29\,900\,\Omega$.

2. Maximum meter current $= \frac{200\,mV}{5.0\,M\Omega} = 40.0\,nA$.

(a) Maximum current through series resistor = 40.0 nA.

(i) Maximum voltage across series resistor = 2.0 − 0.2 = 1.8 V

$\therefore$ series resistance $= \frac{1.8\,V}{40.0\,nA} = 45\,M\Omega$.

(ii) Maximum voltage across series resistor = 100.0 − 0.2 = 99.8 V

$\therefore$ series resistance $= \frac{99.8\,V}{40.0\,nA} = 2495\,M\Omega$.

(b) Maximum voltage across shunt resistor = 200 mV.

(i) Maximum current through shunt resistor = 200 mA (−40 nA negligible)

$\therefore$ shunt resistance $= \frac{200\,mV}{200\,mA} = 1.0\,\Omega$.

(ii) Maximum current through shunt resistor = 10 A (−40 nA negligible)

$\therefore$ shunt resistance $= \dfrac{200\text{ mV}}{10\text{ A}} = 20\text{ m}\Omega$.

Revision questions

14.1. (a) (i) See Fig. A14.3.

(ii) $E = I(R + r) \therefore I = \dfrac{2.0}{(4.2 + 0.8)} = 0.40\text{ A}$,

$V = IR = 0.40 \times 4.2 = 1.68\text{ V}$.

(b) (i) The lost voltage across the internal resistance of the cell = Ir $= 0.40 \times 0.80 = 0.32\text{ V}$, which is the difference between the cell's e.m.f. and the p.d. across R.

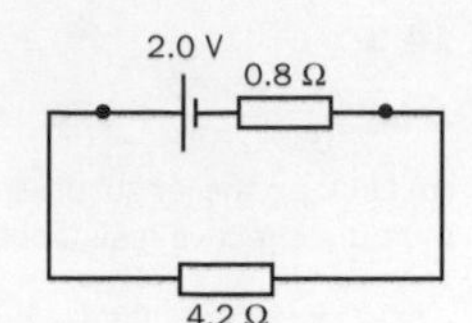

Fig. A14.3.

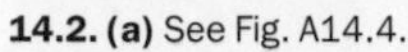

14.2. (a) See Fig. A14.4.

(b) (i) $P = IV = 0.25 \times 2.5 = 0.625\text{ W}$.

(ii) Electric energy/sec produced = IE $= 0.25 \times 3.0 = 0.75\text{ W}$.

(c) Difference $= 0.75 - 0.625 = 0.175\text{ W}$; power dissipated in the cells due to internal resistance = I × lost voltage (Ir) $= I^2r = 0.25^2 \times 1.0 = 0.0625\text{ W}$; power dissipated in variable resistor accounts for $(0.175 - 0.0625)\text{ W} = 0.1125\text{ W}$.

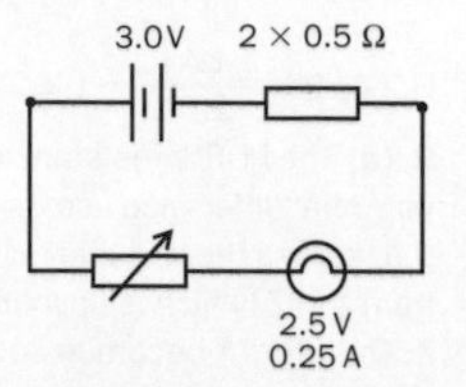

Fig. A14.4.

14.3. (a) $E = IR + Ir \therefore E = (0.1 \times 15) + 0.1r = 1.5 + 0.1r$; also, $E = (0.2 \times 5.0) + 0.2r = 1.0 + 0.2r \quad \therefore 1.0 + 0.2r = 1.5 + 0.1r$ so $0.1r = 0.5$ $\therefore r = 5.0\ \Omega$; $E = 1.0 + (0.2 \times 5.0) = 2.0\text{ V}$.

(b) $I = \dfrac{E}{R + r} = \dfrac{2.0}{5.0 + 3.0} = 0.25\text{ A}$.

14.4. (a) Thermistor resistance = 20 kΩ at 20°C

$\therefore V_R = \dfrac{10}{(20 + 10)} \times 12.0\text{ V} = 4.0\text{ V}$.

(b) $V_R = \dfrac{10}{(10 + R)} \times 12.0 = 5.0 \quad \therefore 5(R + 10) = 12 \times 10$

so $5R = 120 - 50 = 70 \quad \therefore R = 14\text{ k}\Omega$.

$\therefore$ temperature = 35°C.

14.5. (a) See Fig. A14.5.

(b) (i) 1.0 V. **(ii)** Total resistance = 1.5 kΩ since voltmeter and 1 kΩ resistor in parallel are equal to 0.5 kΩ; hence cell current $= \dfrac{2.0\text{ V}}{1.5\text{ k}\Omega} = 1.33$, mA so voltmeter reading = p.d. across 0.5 kΩ combination $= 0.5\text{ k}\Omega \times 1.33\text{ mA} = 0.67\text{ V}$.

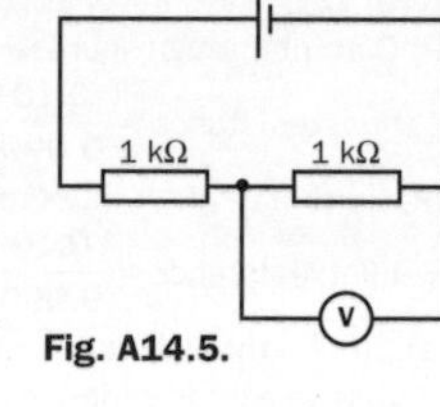

Fig. A14.5.

14.6. (a) $E_x = \dfrac{L_x}{L_s} E_s = \dfrac{0.651}{0.885} \times 1.5 = 1.1\text{ V}$.

(b) (i) $V = \dfrac{L'_x}{L_s} E_s = \dfrac{0.542}{0.885} \times 1.5 = 0.92\text{ V}$.

(ii) $I = \dfrac{V}{R} = \dfrac{0.92}{2.0} = 0.46\text{ A} \quad \therefore r = \dfrac{(E - V)}{I} = \dfrac{(1.1 - 0.92)}{0.46} = 0.4\ \Omega$.

14.7. Use $\dfrac{P}{Q} = \dfrac{R}{S}$ **(a)** $R = \dfrac{PS}{Q} = \dfrac{5.0 \times 50.0}{12.0} = 20.8\ \Omega$.

(b) $Q = \dfrac{PS}{R} = \dfrac{5.0 \times 50.0}{20.0} = 12.5\ \Omega$.

(c) $P = \dfrac{QR}{S} = \dfrac{12.0 \times 20.0}{50.0} = 4.8\ \Omega$.

14.8. (a) (i) $\dfrac{R}{S} = \dfrac{(1000 - 438)}{438} \quad \therefore R = 1.28\ S = 15.4\ \Omega$.

(ii) For $L = 438 + 2 = 440$ mm, $R = \dfrac{(1000 - 440)S}{440} = 15.3\ \Omega$

$\therefore$ probable error in $R = 0.1\ \Omega$.

(iii) $\dfrac{1}{S'} = \dfrac{1}{6} + \dfrac{1}{12} \quad \therefore S' = 4.0\ \Omega \quad \therefore \dfrac{(1000 - L)}{L} = \dfrac{15.4}{4.0}$

$4000 - 4.0\ L = 15.4\ L$; hence $19.4\ L = 4000 \quad \therefore L = \dfrac{4000}{19.4} = 206$ mm.

14.9. (a) $R = \dfrac{\rho L}{A} = \dfrac{4.8 \times 10^{-7} \times 1.50}{\pi(0.36 \times 10^{-3})^2/4} = 7.1\ \Omega$.

(b) $A = \dfrac{\rho L}{R} = \dfrac{4.8 \times 10^{-7} \times 10}{105} = 4.57 \times 10^{-8}\text{ m}^2 \quad \therefore \dfrac{\pi d^2}{4} = 4.57 \times 10^{-8}$;

hence $d^2 = 4.57 \times 10^{-8} \times 4/\pi = 5.82 \times 10^{-8}\text{ m}^2 \quad \therefore d = 2.4 \times 10^{-4}$ m.

(c) $L = \dfrac{RA}{\rho} = \dfrac{0.040 \times \pi(2.0 \times 10^{-3})^2/4}{1.7 \times 10^{-8}} = 7.4$ m.

14.10. (a) Fit a shunt resistor, S, in parallel with the meter, where $S = \dfrac{ir}{(I - i)}$.

(i) $S = \dfrac{0.10\text{ mA} \times 500\ \Omega}{(100 - 0.1)\text{ mA}} = 0.50\ \Omega$. **(ii)** $S = \dfrac{0.10\text{ mA} \times 500\ \Omega}{(5.0 - 0.0001)\text{ A}} = 10\text{ m}\Omega$.

(b) Fit a multiplier resistor, R, in series with the meter where $R = \dfrac{(V - v)\,r}{v}$ where $v = ir$.

(i) $R = \dfrac{(1.0 - 0.05) \times 500}{0.05} = 9500\ \Omega$.

(ii) $R = \dfrac{(15.0 - 0.05) \times 500}{0.05} = 149\,500\ \Omega$.

Unit 15

In-text questions

15.1

1. (i) $Q = CV = 10\ \mu F \times 6.0\text{ V} = 60.0\ \mu C$. **(ii)** $V = Q/C = 0.33\ \mu C/0.22\ \mu F = 1.5\text{ V}$. **(iii)** $C = Q/V = 9.90 \times 10^4\ \mu C/4.50\text{ V} = 22\,000\ \mu F$.

(iv) $V = Q/C = 5.00 \times 10^{-2}\ \mu C / 1.00 \times 10^{-3}\ \mu F = 50\text{ V}$.

2. (a) $Q = CV = 2200\ \mu F \times 9.0\text{ V} = 19800\ \mu C$.

(b) Rearrange Q = It to give $t = Q/I = 19800\ \mu C/0.25\text{ mA} = 79.2$ s.

15.2

1. (a) (i) $\dfrac{1}{C} = \dfrac{1}{3} + \dfrac{1}{6} = \dfrac{6 + 3}{3 \times 6} = \dfrac{9}{18} = \dfrac{1}{2} \quad \therefore C = 2\ \mu F$. **(ii)** $C = 3 + 6 = 9\ \mu F$.

(b) CIRCUIT A: **(i)** The combined capacitance, C_1, of the 6 μF capacitor and the 12 μF capacitor in series is given by

$\dfrac{1}{C_1} = \dfrac{1}{6} + \dfrac{1}{12} = \dfrac{12 + 6}{6 \times 12} = \dfrac{18}{72} \quad \therefore C_1 = \dfrac{72}{18} = 4\ \mu F$.

Hence the total capacitance, $C = 4 + 2 = 6\ \mu F$.

(ii) Total charge stored, $Q = CV_{batt} = 6\ \mu F \times 12\text{ V} = 72\ \mu C$.

(iii) 2 μF: $V = V_{batt} = 12\text{ V} \therefore Q = CV = 2\ \mu F \times 12\text{ V} = 24\ \mu C$;

6 μF; $Q = 72 - 24 = 48\ \mu C \quad \therefore V = \dfrac{Q}{C} = \dfrac{48\ \mu C}{6} = 8\text{ V}$; 12 μF: $Q = 48\ \mu C$,

$V = 12 - 8 = 4\text{ V}$.

(c) CIRCUIT B: **(i)** The combined capacitance of 1 μF and 2 μF in parallel = 3 μF. Hence the combined capacitance of 3 μF and 6 μF in series = 2 μF.

(ii) $Q = CV = 2\ \mu F \times 5.0\text{ V} = 10\ \mu C$.

(iii) 6 μF; Q = 10 μC (same as the total charge stored)

$\therefore V = \dfrac{Q}{C} = \dfrac{10\ \mu C}{6\ \mu F} = 1.67\text{ V}$; 1 μF: $V = 5.0 - 1.67 = 3.33\text{ V}$

$\therefore Q = CV = 1\ \mu F \times 3.33\text{ V} = 3.33\ \mu C$; 2 μF: $V = 3.33\text{ V}$

$\therefore Q = CV = 2\ \mu F \times 3.33(3)\text{ V} = 6.67\ \mu C$.

2. (a) Total charge stored, $Q = CV = 4.7\ \mu F \times 6.0\text{ V} = 28\ \mu C$.

(b) Combined capacitance $= 10 + 4.7 = 14.7\ \mu F$

$\therefore$ final p.d. $= \dfrac{\text{total charge stored, Q}}{\text{combined capacitance}} = \dfrac{28\ \mu C}{14.7\ \mu F} = 1.90\text{ V}$.

(c) 4.7 μF: $Q = CV = 4.7\ \mu F \times 1.90\text{ V} = 9\ \mu C$; 10 μF; $Q = CV = 10\ \mu F \times 1.90\text{ V} = 19\ \mu C$.

15.3

1. (a) (i) $E = \tfrac{1}{2}CV^2 = \tfrac{1}{2} \times 5.0 \times 10^{-6} \times 12^2 = 360 \times 10^{-6}\text{ J} = 360\ \mu J$.

(ii) $V = Q/C = 1.8\text{ mC}/100\ \mu F = 18\text{ V} \quad \therefore E = \tfrac{1}{2}CV^2 = \tfrac{1}{2} \times 100 \times 10^{-6} \times 18^2 = 1.62 \times 10^{-2}\text{ J}$.

(iii) $E = \frac{1}{2}QV = 0.5 \times 30 \times 10^{-3} \times 6.0 = 9.0 \times 10^{-2}$ J.
(b) (i) Rearrange $E = \frac{1}{2}CV^2$ to give $V^2 = 2E/C = 2 \times 250 \times 10^{-6}/20 \times 10^{-6} = 25 \quad \therefore V = 5.0$ V.
(ii) $E = \frac{1}{2}CV^2 = \frac{1}{2} \times 50\,000 \times 10^{-6} \times 6.0^2 = 0.90$ J

$$\therefore \text{time} = \frac{\text{energy}}{\text{power}} = \frac{0.90\text{ J}}{0.2\text{ W}} = 4.5\text{ s.}$$

2. (a) (i) $Q = CV = 100\ \mu F \times 6.0\ V = 600\ \mu C$; $E = \frac{1}{2}CV^2 = \frac{1}{2} \times 100 \times 10^{-6} \times 6.0^2 = 1.8 \times 10^{-3}$ J.
(ii) Combined capacitance = 100 + 50 = 150 μF

$$\therefore \text{final p.d.} = \frac{\text{initial charge}}{\text{combined capacitance}} = \frac{600\ \mu C}{150\ \mu F} = 4.0\text{ V.}$$

(iii) 100 μF: $Q = CV = 100\ \mu F \times 4.0\ V = 400\ \mu C$; $E = \frac{1}{2}CV^2 = \frac{1}{2} \times 100 \times 10^{-6} \times 4.0^2 = 8.0 \times 10^{-4}$ J; 50 μF: $Q = CV = 50\ \mu F \times 4.0\ V = 200\ \mu C$; $E = \frac{1}{2}CV^2 = \frac{1}{2} \times 50 \times 10^{-6} \times 4.0^2 = 4.0 \times 10^{-4}$ J.
(b) Energy dissipated = 1.8 mJ − (0.8 + 0.4) mJ = 0.6 mJ.

15.4

1. (a) Using $C = \frac{A\varepsilon_0\varepsilon_r}{d}$ gives

$$C = \frac{1.5 \times 0.04 \times 8.85 \times 10^{-12} \times 2.5}{0.010 \times 10^{-3}} = 1.3 \times 10^{-7}\text{ F.}$$

(b) (i) For 0.10 mm thickness, $V = 700\text{ kV mm}^{-1} \times 0.01\text{ mm} = 7$ kV.
(ii) $E = \frac{1}{2}CV^2 = 0.5 \times 1.3 \times 10^{-7} \times 7000^2 = 3.2$ J.

2. (a) Using $C = \frac{A\varepsilon_0}{d}$ gives $C = \frac{6.0 \times 8.85 \times 10^{-12}}{0.10} = 5.3 \times 10^{-10}$ F.
(b) $Q = CV = 5.3 \times 10^{-10} \times 1000 = 5.3 \times 10^{-7}$ C; $E = \frac{1}{2}CV^2 = 0.5 \times 5.3 \times 10^{-10} \times 1000^2 = 2.7 \times 10^{-4}$ J.

15.5

1. (a) $Q_0 = CV_0 = 5.0\ \mu F \times 12.0\ V = 60\ \mu C$; $E = \frac{1}{2}CV^2 = \frac{1}{2} \times 5.0 \times 10^{-6} \times 12.0^2 = 360\ \mu J$.

(b) $\frac{t}{CR} = \frac{5.0}{5.0 \times 10^{-6} \times 0.5 \times 10^6} = 2.0 \quad \therefore Q = Q_0\, e^{-t/CR} = 60\, e^{-2}$

$= 60 \times 0.135 = 8.1\ \mu C \quad \therefore V = \frac{Q}{C} = \frac{8.1 \times 10^{-6}}{5.0 \times 10^{-6}} = 1.6$ V.

2. (a) $Q_0 = CV_0 = 2200 \times 10^{-6} \times 6.0 = 0.013$ C.
(b) $CR = 2200 \times 10^{-6} \times 100 \times 10^3 = 220$ s.
(c) At $t = CR$, $Q = Q_0\, e^{-1} = 0.37\, Q_0 = 0.0048$ C;

$$V = \frac{Q}{C} = \frac{0.0048}{2200 \times 10^{-6}} = 2.2\text{ V.}$$

(d) $I = \frac{V}{R} = \frac{2.2}{100 \times 10^3} = 2.2 \times 10^{-5}$ A.

Revision questions

15.1. (a) (i) The charging current is sufficent to light the LED. **(ii)** The charging current decreases and becomes too small to light the LED.
(b) (i) $Q = CV = 5000 \times 10^{-6}\text{ F} \times 3.0\text{ V} = 1.5 \times 10^{-2}$ C.
(ii) $t = \frac{Q}{I} = \frac{1.5 \times 10^{-2}}{0.15 \times 10^{-3}} = 100$ s.

15.2. (a) (i) $C = 3 + 6 = 9\ \mu F$.
(ii) $\frac{1}{C} = \frac{1}{3} + \frac{1}{6} = \frac{6+3}{3 \times 6} = \frac{9}{18} \quad \therefore C = \frac{18}{9} = 2\ \mu F$.
(b) (i) The combined capacitance of the 3 μF and the 6 μF capacitor in parallel is 9μF; therefore the total capacitance is given by

$$\frac{1}{C} = \frac{1}{9} + \frac{1}{11} = \frac{11+9}{9 \times 11} = \frac{20}{99} \quad \therefore C = \frac{99}{20} = 4.95\ \mu F.$$

(ii) $Q = CV = 4.95 \times 10^{-6} \times 6.0 = 2.97 \times 10^{-5}$ C.
(iii) 11 μF: $Q = 2.97 \times 10^{-5}\text{ C} \quad \therefore V = \frac{Q}{C} = \frac{2.97 \times 10^{-5}}{11 \times 10^{-6}} = 2.7$ V;
$E = \frac{1}{2}CV^2 = 0.5 \times 11 \times 10^{-6} \times 2.7^2 = 4.0 \times 10^{-5}$ J; 3 μF; $V = 6.0 - 2.7 = 3.3\text{ V} \quad \therefore Q = CV = 3 \times 10^{-6} \times 3.3 = 9.9 \times 10^{-6}$ C; $E = \frac{1}{2}CV^2 = 0.5 \times 3 \times 10^{-6} \times 3.3^2 = 1.6 \times 10^{-5}$ J; 6 μF: $V = 6.0 - 2.7 = 3.3\text{ V} \quad \therefore Q = CV = 6 \times 10^{-6} \times 3.3 = 19.8 \times 10^{-6}$ C; $E = \frac{1}{2}CV^2 = 0.5 \times 6 \times 10^{-6} \times 3.3^2 = 3.2 \times 10^{-5}$ J.

15.3. (a) $C = \frac{A\varepsilon_0\varepsilon_r}{d} = \frac{0.30 \times 0.25 \times 8.85 \times 10^{-12} \times 3.5}{1.5 \times 10^{-3}} = 1.55 \times 10^{-9}$ F.

(b) $Q = CV = 1.55 \times 10^{-9} \times 50 = 7.7 \times 10^{-8}$ C; $E = \frac{1}{2}CV^2 = 0.5 \times 1.55 \times 10^{-9} \times 50^2 = 1.9 \times 10^{-6}$ J.

15.4. (a) (i) $Q_0 = CV_{batt} = 4.7 \times 10^{-6} \times 9.0 = 4.2 \times 10^{-5}$ C, $E = \frac{1}{2}CV^2 = 0.5 \times 4.7 \times 10^{-6} \times 9.0^2 = 1.9 \times 10^{-4}$ J.
(ii) Time constant, $CR = 4.7 \times 10^{-6} \times 10 \times 10^6 = 47$ s. **(iii)** At t = 100 s, $t/CR = 100/47 = 2.12 \therefore Q = Q_0 e^{-t/CR} = 4.2 \times 10^{-5} \times e^{-2.12} = 5.0 \times 10^{-6}$ C;

$$V = \frac{Q}{C} = \frac{5.0 \times 10^{-6}}{4.7 \times 10^{-6}} = 1.06\text{ V;}$$

$E = \frac{1}{2}CV^2 = 0.5 \times 4.7 \times 10^{-6} \times 1.06^2 = 2.6 \times 10^{-6}$ J.
(b) (i) Initial charge = 4.2×10^{-5} C; total capacitance = 4.7 + 2.2 = 6.9 μF

$$\therefore V = \frac{Q}{C} = \frac{4.2 \times 10^{-5}}{6.9 \times 10^{-6}} = 6.1\text{ V.}$$

(ii) 4.7 μF: $E = \frac{1}{2}CV^2 = 0.5 \times 4.7 \times 10^{-6} \times 6.1^2 = 8.7 \times 10^{-5}$ J;
2.2 μF: $E = \frac{1}{2}CV^2 = 0.5 \times 2.2 \times 10^{-6} \times 6.1^2 = 4.1 \times 10^{-5}$ J.
(iii) Loss of stored energy = $1.9 \times 10^{-4}\text{ J} - (8.7 \times 10^{-5}\text{ J} + 4.1 \times 10^{-5}\text{ J}) = 7.2 \times 10^{-5}$ J.

15.5. (a) Input resistance = $\frac{6.0\text{ V}}{0.5\text{ mA}} = 12\,000\ \Omega$.
(b) (i) $CR = 50\,000 \times 10^{-6} \times 12\,000 = 600$ s.
(ii) $\frac{t}{CR} = \frac{100}{600} = 0.167 \quad \therefore Q = Q_0\, e^{-t/CR} = Q_0\, e^{-0.167} = 0.85\, Q_0$

$$\therefore V = \frac{Q}{C} = 0.85\frac{Q_0}{C} = 0.85\, V_0 = 5.1\text{ V.}$$

(iii) $Q = Q_0\, e^{-t/CR} \quad \therefore V = \frac{Q}{C} = \frac{Q_0}{C}e^{-t/CR} = V_0\, e^{-t/CR} \quad \therefore 5.0 = 6.0\, e^{-t/CR}$

$$\therefore e^{-t/CR} = \frac{5.0}{6.0} = 0.83$$

$$\therefore \frac{-t}{CR} = \ln 0.83 = -0.186,\ t = 0.186\,CR = 0.186 \times 600 = 112\text{ s.}$$

Unit 16

In-text questions

16.1

1. Digital **(a)** and **(b)**; analogue **(c)** and **(d)**.
2. See Fig. A16.1.

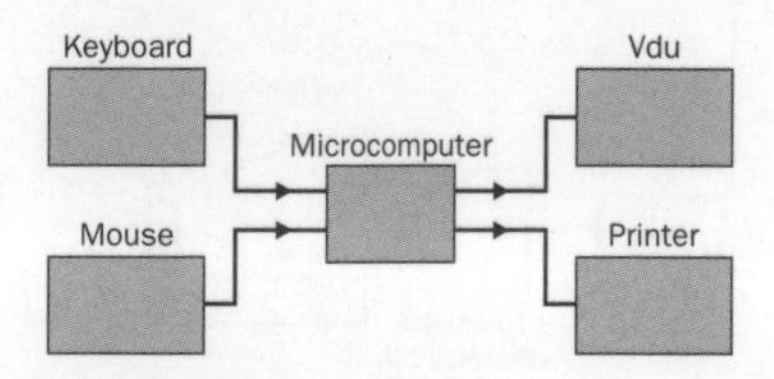

Fig. A16.1.

16.2

1.

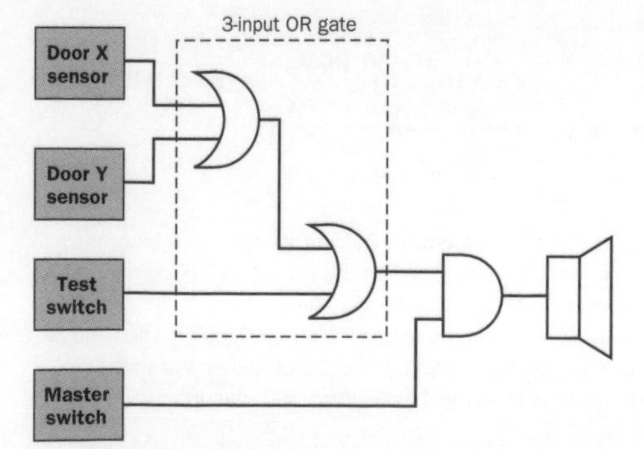

Fig. A16.2.

A	B	Output for (a)	Output for (b)	Output for (c)
0	0	1	1	0
0	1	1	0	1
1	0	0	1	1
1	1	1	1	0

2. (a) ALARM = 1 if master switch = 1 AND door X sensor = 1 OR door Y sensor = 1 OR test switch = 1.
(b) ALARM = 1 if master switch = 1 AND smoke sensor OR panic button OR test switch = 1.

16.3

1. (a) The test switch supplies logic 1 to the OR gate, which therefore switches the transistor and relay on.

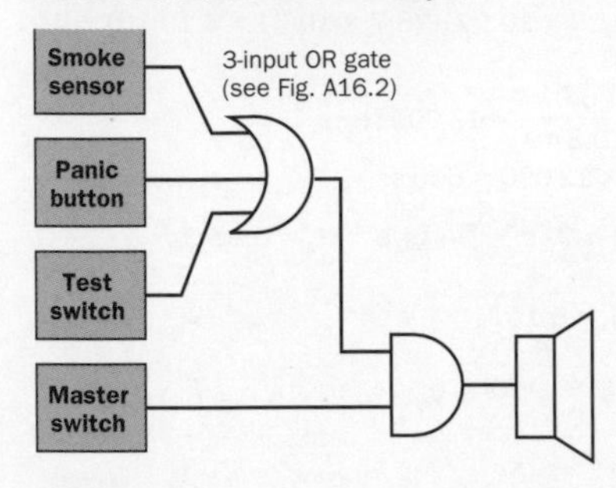

Fig. A16.3.

(b) (i) The output voltage from the temperature sensor rises as the thermistor warms up. At a certain temperature, the output voltage is high enough to count as logic 1 to the OR gate input. Hence the OR gate switches the fan on. **(ii)** The temperature at which the thermistor is switched on is raised. Reducing this resistance lowers the voltage from the potential divider output, so the thermistor must warm up more to raise the voltage to the level necessary to switch the output state of the OR gate.
2. See Fig. A16.4.

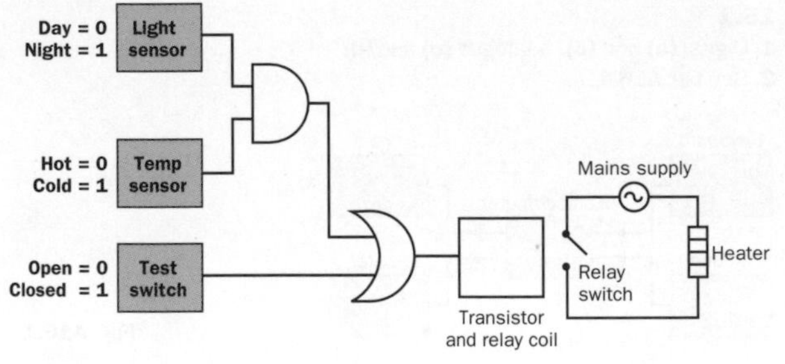

Fig. A16.4.

16.4

1. Circuit A:

(a) Voltage gain $= \dfrac{-R_F}{R_1} = \dfrac{-15\ \text{M}\Omega}{0.5\ \text{M}\Omega} = -30$. **(b)** $V_{IN} = \dfrac{-V_{OUT}}{-30} = \dfrac{15}{30} = 0.5$ V.

Circuit B: **(a)** Voltage gain $= \dfrac{R_F + R_1}{R_1} = \dfrac{(10+1)}{1} = 11$.

(b) $V_{IN} = \dfrac{V_{OUT}}{11} = \dfrac{15}{11} = 1.4$ V.

2. (a) The voltage to input Q is constant. The voltage to input P increases as the variable resistor is increased. When the voltage at P becomes greater than the voltage at Q, the op-amp switches polarity.
(b) The thermistor resistance increases as it becomes cooler. The voltage to input P therefore falls as the thermistor becomes cooler. When the voltage at P becomes less than the fixed voltage at Q, the op-amp output switches polarity to +15 V.

16.5

1. (a) The time delay would be lengthened because the time constant, CR, would be increased.
(b) The time delay would be reduced because the time constant, CR, would be reduced.
2. (a) (i) $T = CR = 1 \times 10^{-6}\ \text{F} \times 1000\ \Omega = 1 \times 10^{-3}$ s. **(ii)** Each pulse would be 2×10^{-3} ms in duration so the frequency of the pulses would be 500 Hz ($= \frac{1}{2} \times 10^{-3}$s).
(b) One state would last 4.7× longer and the other state would last 10× longer. The pulse would therefore be 14.7× longer and the pulse frequency would be reduced to about 34 Hz (= 500 Hz/14.7).

Revision questions

16.1.

Input A	Input B	Output for (a)	Output for (b)	Output for (c)
0	0	0	1	1
0	1	0	1	1
1	0	0	1	1
1	1	1	0	0

16.2. (a) (i) To test the system **(ii)** Increase the variable resistance **(iii)** OR.
(b) High voltage at the output of the logic circuit makes a current enter the base of the transistor, so allowing a current to pass through the relay coil and the transistor via its collector. The relay coil is therefore energised and the fan motor is switched on.
16.3. (a) See Fig. A16.5.

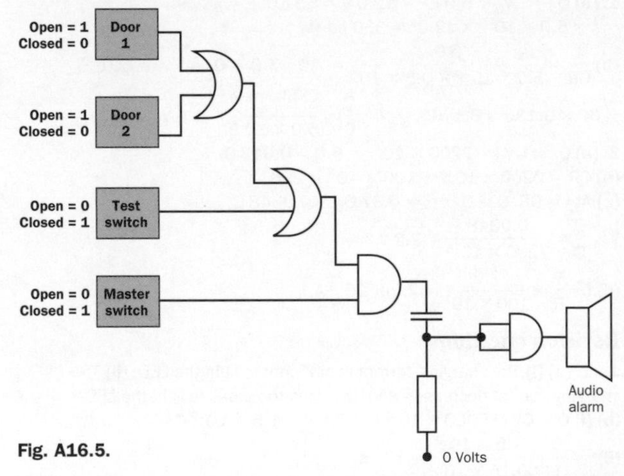

Fig. A16.5.

(b) The time delay is increased because the capacitor takes longer to charge up, so the voltage at the input of the first NOT gate takes longer to reach logic level 1.
16.4. See Fig. A16.6.

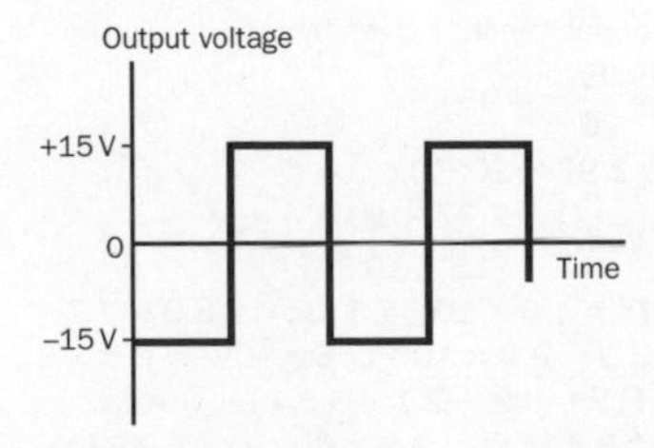

Fig. A16.6.

(ii) The output waveform is at positive saturation when the input voltage at Q is more than the voltage at P, and at negative saturation when the input voltage at P is more than the voltage at Q. Since the voltage at Q is +0.5 V and the input voltage is an alternating voltage of peak voltage 1.0 V, the output is at positive saturation more than it is at negative saturation.

(b) (i) A non-inverting amplifier. **(ii)** 6 $[= \frac{(2.0 + 0.4)}{0.4}]$. **(iii)** 2.5 V (= 15V/6).

16.5. (a) (i) $-2.0\,V\,(= \frac{-R_F}{R_1}V_1 = \frac{-1.0}{1.0} \times 2.0)$.

(ii) $-3.0\,V\,(= \frac{-R_F}{R_2}V_2 = \frac{-1.0}{0.5} \times 1.5)$. **(iii)** $-5.0\,V\,(= -2.0\,V - 3.0\,V)$.

(b) See Table A16.1 and Fig. A16.7.

Table A16.1

Binary number			Output voltage/ volts
V_C	V_B	V_A	
0	0	0	0
0	0	1	1
0	1	0	2
0	1	1	3
1	0	0	4
1	0	1	5
1	1	0	6
1	1	1	7

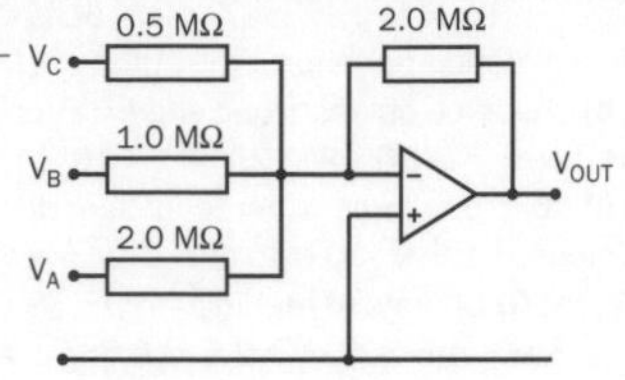

Fig. A16.7.

Unit 17

In-text questions

17.1

1. $F = \frac{1}{4\pi\varepsilon_o} \times \frac{Q_1 Q_2}{r^2} = \frac{+1.6 \times 10^{-19} \times -1.6 \times 10^{-19}}{4\pi \times 8.85 \times 10^{-12} \times (0.10 \times 10^{-9})^2} = -2.3 \times 10^{-8}\,N.$

2. The force between the +2.5 nC charge and the +1.5 μC charge,

$F_1 = \frac{+2.5 \times 10^{-9} \times +1.5 \times 10^{-6}}{4\pi \times 8.85 \times 10^{-12} \times (0.10)^2} = 3.4\,mN$ (repulsion).

The force between the +2.5 nC charge and the −3.5 μC charge,

$F_1 = \frac{+2.5 \times 10^{-9} \times -3.5 \times 10^{-6}}{4\pi \times 8.85 \times 10^{-12} \times (0.10)^2} = -7.9\,mN$ (attraction).

The +2.5 nC charge therefore experiences a force 3.4 mN away from the +1.5 μC charge and a force of 7.9 mN towards the −3.5 μC charge. These two forces on the +2.5 nC charge are in the same direction. Therefore, the total force on the +2.5 nC charge is 11.3 mN (= 7.9 mN + 3.4 mN).

17.2

1. (a) Change of p.e., $\Delta PE = q\,(V_2 - V_1)$. **(i)** $\Delta PE = +2.5\,nC \times (6.0 - 4.0) = 5.0\,nJ$. **(ii)** $\Delta PE = 0$. **(iii)** $\Delta PE = -5.0\,nJ$.

(b) $Fd = q(V_2 - V_1)$ $\therefore F = \frac{5.0 \times 10^{-9}\,J}{5.0 \times 10^{-3}\,m} = 1.0 \times 10^{-6}\,N.$

2. (a) (i) Negative. **(ii)** Rearranging $\frac{qV}{d} = mg$ gives

$q = \frac{mgd}{V} = \frac{2.5 \times 10^{-15} \times 9.8 \times 12.0 \times 10^{-3}}{600} = 4.9 \times 10^{-19}\,C.$

(b) See Fig. A17.1.

(c) (i) V = 100 V at 2.0 mm above the earthed plate $\therefore$ p.e. = qV = $+4.9 \times 10^{-17}$ J.

(ii) V = 300 V at 6.0 mm above the earthed plate $\therefore$ p.e. = qV = $+1.47 \times 10^{-16}$ J.

(iii) V = 500 V at 10.0 mm above the earthed plate $\therefore$ p.e. = qV = $+2.45 \times 10^{-16}$ J.

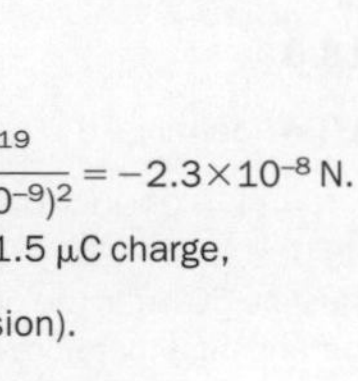

Fig. A17.1.

17.3

1. (a) $F = qE = 3.2 \times 10^{-19} \times 120\,000 = 3.8 \times 10^{-14}\,N.$

(b) $\Delta E_P = q\Delta V = qE\Delta x = 3.8 \times 10^{-14} \times 5.0 \times 10^{-6} = 1.9 \times 10^{-19}\,J.$

2. (a) $E = \frac{V}{d} = \frac{300}{0.010} = 30\,000\,V\,m^{-1}.$

(b) $Q/A = \varepsilon_o E = 8.85 \times 10^{-12} \times 30\,000 = 2.7 \times 10^{-7}\,C\,m^{-2}.$

(c) No. of electrons per $m^2 = \frac{2.7 \times 10^{-7}}{1.6 \times 10^{-19}} = 1.7 \times 10^{12}$

$\therefore$ No. of electrons per $mm^2 = 1.7 \times 10^{12} \times 10^{-6} = 1.7 \times 10^{6}.$

17.4

1. (a) $E = \frac{Q}{4\pi\varepsilon_o r^2} = \frac{+8.5 \times 10^{-12}}{4\pi \times 8.85 \times 10^{-12} \times (5.0 \times 10^{-3})^2} = 3.1 \times 10^{3}\,V\,m^{-1}.$

$V = \frac{Q}{4\pi\varepsilon_o r} = \frac{+8.5 \times 10^{-12}}{4\pi \times 8.85 \times 10^{-12} \times (5.0 \times 10^{-3})} = 15.3\,V.$

(b) $E = \frac{V_P}{d} = \frac{200}{40 \times 10^{-3}} = 5000\,V\,m^{-1}$. At the mid-point, V = 100 V.

2. (a) Rearrange $V_S = \frac{Q}{4\pi\varepsilon_o R_S}$ to give $Q = 4\pi\varepsilon_o R_S V_S = 4\pi \times (8.85 \times 10^{-12}) \times 0.20 \times 100\,000 = 2.2 \times 10^{-6}\,C.$

(b) $E = \frac{Q}{4\pi\varepsilon_o R_S{}^2} = \frac{4\pi\varepsilon_o R_S V_S}{4\pi\varepsilon_o R_S{}^2} = \frac{V_S}{R_S} = \frac{100\,000}{0.20} = 500\,000\,V\,m^{-1}.$

(c) At r = 2.0 m, $E = \frac{Q}{4\pi\varepsilon_o r^2} = \frac{+2.2 \times 10^{-6}}{4\pi \times 8.85 \times 10^{-12} \times (2.0)^2} = 5000\,V\,m^{-1}.$

$V = \frac{Q}{4\pi\varepsilon_o r} = \frac{+2.2 \times 10^{-6}}{4\pi \times 8.85 \times 10^{-12} \times (2.0)} = 10\,000\,V.$

Revision questions

17.1. (a) See Fig. A17.2.

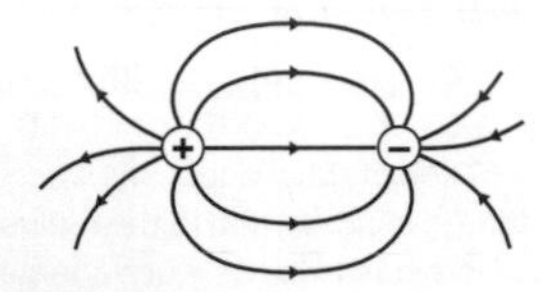

Fig. A17.2.

(b) $F_1 = \frac{+2.5 \times 10^{-12} \times -7.2 \times 10^{-12}}{4\pi \times 8.85 \times 10^{-12} \times (0.020)^2} = 4.0 \times 10^{-10}\,N$ (attraction).

(c) Let Q_1 represent the positive charge and Q_2 the negative charge.

The electric field strength due to Q_1,

$E_1 = \frac{Q_1}{4\pi\varepsilon_o r^2} = \frac{+2.5 \times 10^{-12}}{4\pi \times 8.85 \times 10^{-12} \times (0.010)^2} = 225\,V\,m^{-1}$ away from Q_1.

The electric field strength due to Q_2,

$E_2 = \frac{Q_2}{4\pi\varepsilon_o r^2} = \frac{-7.2 \times 10^{-12}}{4\pi \times 8.85 \times 10^{-12} \times (0.010)^2} = -648\,V\,m^{-1}$ towards Q_2.

The total electric field, $E = E_1 + E_2$ because the electric field contributions are in the same direction at the midpoint. Hence $E = 225 + 648 = 873\,V\,m^{-1}$. The electric potential, V_1, at the midpoint, due to

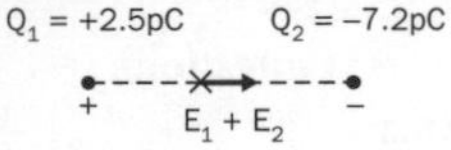

Fig. A17.3.

$Q_1 = \frac{Q_1}{4\pi\varepsilon_o r} = \frac{+2.5 \times 10^{-12}}{4\pi \times 8.85 \times 10^{-12} \times (0.010)} = 2.25\,V.$

The electric potential, V_2, at the midpoint, due to

$Q_2 = \frac{Q_2}{4\pi\varepsilon_o r} = \frac{-7.2 \times 10^{-12}}{4\pi \times 8.85 \times 10^{-12} \times (0.010)} = -6.48\,V.$

The total potential, $V = V_1 + V_2 = (2.25) + (-6.48) = -4.23\,V.$

17.2. (a) (i) $E = \frac{V}{d} = \frac{430}{5.0 \times 10^{-3}} = 8.6 \times 10^{4}\,V\,m^{-1}.$

(ii) Rearrange $\frac{qV}{d} = mg$ to obtain

$q = \frac{mgd}{V} = \frac{5.6 \times 10^{-15} \times 9.8 \times 5.0 \times 10^{-3}}{430} = 6.4 \times 10^{-19}\,C.$

(b) For a balanced droplet with constant charge, its mass, m, is proportional to the holding voltage, V $\therefore$ the new mass of the droplet

$m' = \frac{mV'}{V} = \frac{5.6 \times 10^{-15} \times 620}{430} = 8.1 \times 10^{-15}\,kg$ $\therefore$ the mass of the uncharged droplet $= m' - m = 2.5 \times 10^{-15}\,kg.$

17.3. (a) $F = \frac{-e^2}{4\pi\varepsilon_o r^2} = \frac{-(1.6 \times 10^{-19})^2}{4\pi \times 8.85 \times 10^{-12} \times (1.0 \times 10^{-10})^2} = 2.3 \times 10^{-8}\,N.$

(b) Potential due to +e at distance r

$= \frac{e}{4\pi\varepsilon_o r} = \frac{1.6 \times 10^{-19}}{4\pi \times 8.85 \times 10^{-12} \times (1.0 \times 10^{-10})} = 14.4\,V$

$\therefore$ potential energy of an electron at 14.4 V = −14.4 e = $-14.4 \times 1.6 \times 10^{-19} = -2.3 \times 10^{-18}$ J $\therefore$ work done to separate electron and proton to infinity = 2.3×10^{-18} J.

17.4. (a) A small test charge, q, placed at the mid-point would experience a force towards one of the two charges and an equal and opposite force towards the other charge. Hence the resultant force on the test charge is zero.

(b) (i) Potential at 20 mm from 1.6 nC charge

$$= \frac{Q}{4\pi\varepsilon_0 r} = \frac{+1.6 \times 10^{-9}}{4\pi \times 8.85 \times 10^{-12} \times (0.020)} = +720\text{ V}.$$

Potential due to the other charge at the same distance is also +720 V

∴ total potential = 720 + 720 = 1440 V.

(ii) See Fig. A17.4.

(c) $W = qV = 1.5 \times 10^{-17} \times 1440 = 2.2 \times 10^{-14}$ J.

17.5. (a) $E_S = \frac{Q}{4\pi\varepsilon_0 R_S^2} = \frac{V_S}{R_S}$ since $V_S = \frac{Q}{4\pi\varepsilon_0 R_S}$.

(i) $E_S = \frac{10\,000}{100} = 100\text{ V m}^{-1}$.

(ii) $E_S = \frac{10\,000}{1.0 \times 10^{-3}} = 1.0 \times 10^{7}\text{ V m}^{-1}$.

(b) Charge gathers at sharp points. Because the radius of curvature is very small at a sharp point, the electric field is very strong.

(c) (i) Rearrange $V_S = \frac{Q}{4\pi\varepsilon_0 R_S}$ to give $Q = 4\pi\varepsilon_0 R_S V = 4\pi \times 8.85 \times 10^{-12} \times (0.100) \times 5000 = 5.6 \times 10^{-8}$ C.

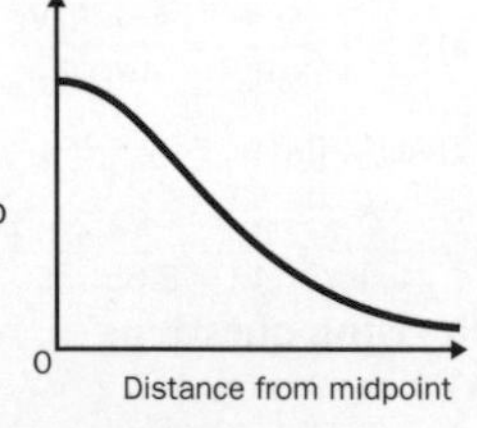

Fig. A17.4.

(ii) The potential equalises when contact is made. Hence the charge is shared in proportion to the radius since $Q = 4\pi\varepsilon_0 R_S V$. Therefore the larger sphere has twice as much charge as the smaller sphere since its radius is twice as large. The charge on the larger sphere therefore becomes 3.7×10^{-8} C ($= \frac{2}{3} \times 5.6 \times 10^{-8}$ C) and the charge on the smaller sphere becomes 1.9×10^{-8} C ($= \frac{1}{3} \times 5.6 \times 10^{-8}$ C).

(iii) $V_S = \frac{Q}{4\pi\varepsilon_0 R_S} = \frac{3.7 \times 10^{-8}}{4\pi \times (8.85 \times 10^{-12}) \times 0.100} = 3330$ V for both spheres.

Unit 18

In-text questions

18.1

1. (a) Steel does not lose its magnetism whereas iron does.

(b) (i) Position the two magnets so they are equidistant from the plotting compass with the N-pole of each magnet pointing towards the plotting compass. The plotting compass needle is repelled more by the stronger N-pole.

(ii) With the magnets as in Fig. 18.6 and the N-pole of each magnet pointing towards the plotting compass, adjust the distances so the plotting compass needle is at 45° to the axis of each magnet. Turn one of the magnets around. If the plotting compass needle turns through more than 90°, the S-pole of that magnet is stronger than its N-pole.

2. See Fig. A18.1.

(a)

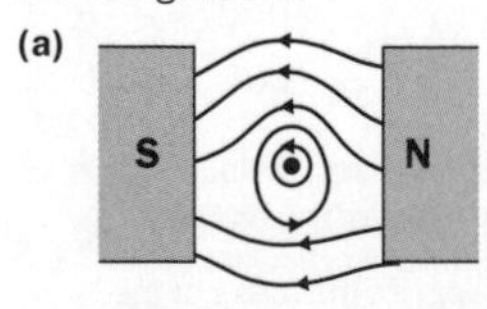

(b)

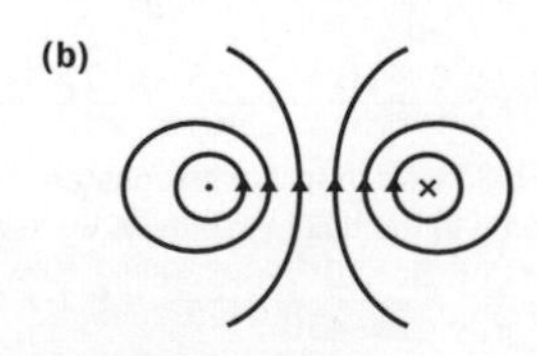

Fig. A18.1.

18.2

1. (a) Iron magnetises and demagnetises easily.

(b) The resistivity of copper is very low, so the resistance of the windings is very small.

(c) It does not oxidise and it has a high melting point, so it withstands sparks.

2. (a) See p. 255.

(b) A fuse melts and needs to be replaced. A circuit breaker does not need to be replaced and is reset manually.

18.3

1. (a) Anticlockwise; see Fig. 18.14 if necessary.

(b) The split-ring commutator reverses the current direction around the coil every half-turn. This is necessary to ensure that when each side is adjacent to a certain pole, the current is always in the same direction, so the force is always in the same direction.

2. (a) X deflects less because the forces on the coil due to the magnet are less.

(b) X deflects more than Y because the spiral spring can be tightened more before it stops the coil turning.

18.4

1. (a) (i) $F = BIL = 0.080 \times 5.2 \times 10^{-2} \times 3.2 = 13.3$ mN due North.

(ii) Rearrange $F = BIL$ to give $I = \frac{F}{BL} = \frac{0.015}{0.25 \times 0.040} = 1.5$ A in a direction from north to south.

(b) The force on each long edge = BILn, where n is the number of turns. Hence $F = 0.065 \times 7.2 \times 40 \times 10^{-3} \times 80 = 1.5$ N.

(ii) Couple = force × perpendicular distance between lines of action of the forces = 1.5 N × 0.030 m = 0.045 N m.

2. (a) (i) Change of reading = 105.38 − 104.92 = 0.46 g = 4.6×10^{-4} kg

∴ force = $4.6 \times 10^{-4} \times 9.8 = 4.5 \times 10^{-3}$ N.

(ii) $B = \frac{F}{IL} = \frac{4.5 \times 10^{-3}}{6.5 \times 0.035} = 2.0 \times 10^{-2}$ T.

(b) The angle between the lines of force and the wire is 60°

∴ $F = BIL \sin 60 = 2.0 \times 10^{-2} \times 6.5 \times 0.035 \times \sin 60 = 3.9 \times 10^{-3}$ N

∴ new reading $= 105.38 - \left(\frac{3.9 \times 10^{-3}}{g} \times 1000\right) = 104.98$ g.

18.5

1. (a) Rearrange $B = \mu_0 nI$ to give $n = \frac{B}{\mu_0 I} = \frac{25 \times 10^{-3}}{4\pi \times 10^{-7} \times 8.0} = 2490$

∴ N = nL = 2490 × 0.800 = 1990 turns.

(b) (i) B = 50 + 25 = 75 mT. **(ii)** For zero field, the solenoid's own field must be 50 mT in the opposite direction to the external field. The solenoid current must therefore be 16.0 A, since a current of 8.0 A is needed to produce a 25 mT field. If the external field is left to right, the solenoid field must be right to left. The solenoid current must therefore be anticlockwise around the solenoid, viewing the left-hand end of the solenoid.

2. (a) (i) For r = 12.5 mm (= 25 mm/2),

$$B = \frac{\mu_0 I}{2\pi r} = \frac{2.0 \times 10^{-7} \times 1000}{12.5 \times 10^{-3}} = 1.6 \times 10^{-2}\text{ T}.$$

(ii) For r = 10 m, $B = \frac{\mu_0 I}{2\pi r} = \frac{2.0 \times 10^{-7} \times 1000}{10} = 2.0 \times 10^{-5}$ T.

(b) The force per unit length on the second cable = $BI_2 = 2.0 \times 10^{-5} \times 1000 = 0.020\text{ N m}^{-1}$.

18.6

1. (a) The magnetic field of the solenoid aligns all the domains in the ferromagnet to magnetise the ferromagnet fully.

(b) The magnetic field of the solenoid alternates, making the domains repeatedly change direction. As the ferromagnet is withdrawn, the solenoid field has less effect on the ferromagnet and the domains gradually lose their alignment, which becomes disordered.

2. (a) (i) Steel remains more strongly magnetised than iron in a direct current solenoid after the current is switched off. **(ii)** Steel has a larger loop area as it is harder to magnetise and demagnetise.

(b) Number of turns per unit length, $n = \frac{200}{\pi \times 0.040}$

∴ $B = \mu_r \mu_0 nI = 2000 \times 4\pi \times 10^{-7} \times \frac{200}{\pi \times 0.040} \times 0.05 = 0.20$ T.

Revision questions

18.1. (a) See Fig. A18.2.

(b) (i) The bar magnet makes the domains align along the lines of force of the bar magnet's field. After the bar magnet is removed, the domains remain partly aligned.

(ii) The bar magnet attracts the opposite pole of each domain so it leaves the end of the nail with the opposite polarity.

18.2. (a) See p. 256.

(b) (i) $F = BIL\,N = 0.120 \times 5.0 \times 0.040 \times 200 = 4.8$ N. **(ii)** The force on each long side is unchanged (the perpendicular distance between the two forces is zero so the forces have no turning effect at this position).

18.3. (a) (i) Into the plane of the diagram.

(ii) rearrange $\frac{qV}{d} = Bq\upsilon$ to give $B = \frac{V}{\upsilon d}$.

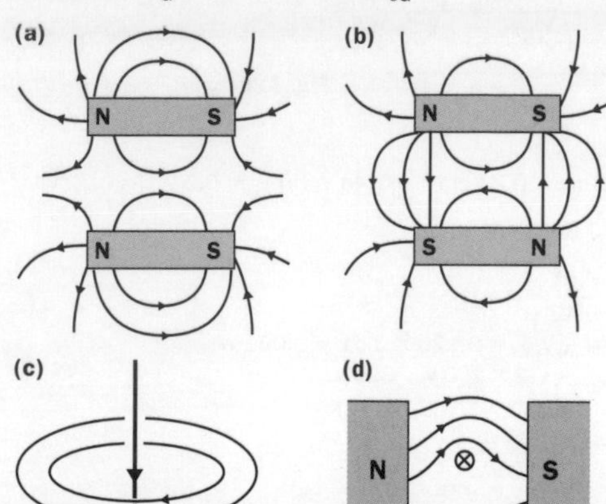

Fig. A18.2.

(iii) $B = \frac{V}{\upsilon d} = \frac{3500}{2.8 \times 10^{7} \times 50 \times 10^{-3}} = 2.5 \times 10^{-3}$ T.

(b) V needs to be doubled if B is doubled. Hence V = 7000 V.

18.4. (a) (i) $B = \mu_0 nI = 4\pi \times 10^{-7} \times \frac{500}{0.250} \times 6.5 = 1.63 \times 10^{-2}$ T.

(ii) At the end, $B = 0.5\,\mu_0 nI = 8.2 \times 10^{-4}$ T.

(b) Rearrange $B = \mu_0 nI$ to give

$$I = \frac{B}{\mu_0 n} = \frac{60 \times 10^{-6}}{4\pi \times 10^{-7} \times (500/0.250)} = 2.4 \times 10^{-2}\text{ T.}$$

18.5. (a) At the nearer edge, the magnetic flux density due to the wire,

$B_1 = \frac{\mu_0 I}{2\pi r} = \frac{2.0 \times 10^{-7} \times 8.5}{0.040} = 4.25 \times 10^{-5}$ T $\therefore$ force on the nearer edge $= B_1 I_{coil} L = 4.25 \times 10^{-5} \times 2.0 \times 0.060 = 5.1 \times 10^{-6}$ N.

At the further edge, which is twice as far from the wire, $B = 0.5\,B_1$ so the force is half the force on the nearer edge and therefore equals 2.55×10^{-6} N.

(b) The nearer edge is attracted to the wire and the further edge is repelled. Therefore the resultant force is $(5.1 - 2.55) \times 10^{-6}$ N $= 2.55 \times 10^{-6}$ N towards the wire.

Unit 19

In-text questions

19.1

1. See Fig. A19.1.

2. The induced e.m.f. is due to the wingspan cutting the vertical component, B_V, of the Earth's field. Since $B_V = 60\sin 70 = 56\ \mu$T, then the induced voltage, $V = B_V L\upsilon = 56 \times 10^{-6} \times 22 \times 180 = 0.22$ V.

19.2

1. (a) $\upsilon = u + at = 0 + (9.8 \times 0.5) = 4.9$ m s^{-1}.

(b) The rod moves vertically and therefore cuts the horizontal component, B_H, of the Earth's field. $B_H = 60\cos 70 = 20.5\ \mu$T $\therefore V = B_H L\upsilon = 20.5 \times 10^{-6} \times 0.750 \times 4.9 = 7.5 \times 10^{-5}$ V.

2. (a) Total magnetic flux $= BAN = 90 \times 10^{-3} \times (0.040 \times 0.040) \times 50 = 7.2 \times 10^{-3}$ Wb.

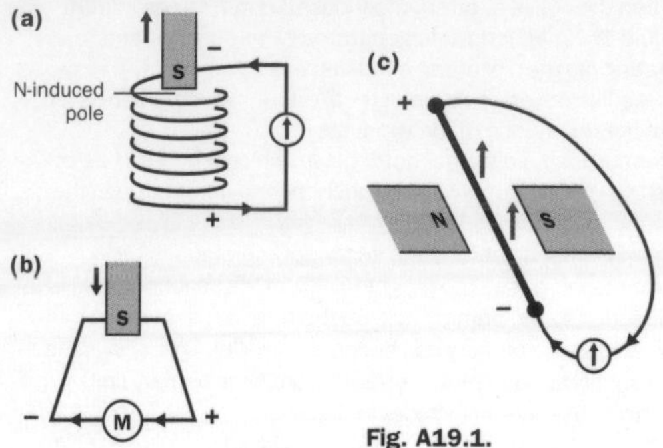

Fig. A19.1.

(b) Time taken, $t = \frac{\text{distance}}{\text{speed}} = \frac{0.040}{0.16} = 0.25$ s.

(ii) $V = \frac{\text{change of total flux}}{\text{time taken}} = \frac{7.2 \times 10^{-3}}{0.25} = 2.9 \times 10^{-2}$ V.

19.3

1. (a) See Fig. 19.15b.

(b) (i) any zero **(ii)** any peak.

(c) The peak voltage is halved and the time for each cycle is doubled, so the waveform is half the height and twice as long as in (a).

2. (a) $V_{MAX} = 2\pi f\,BAN = 2\pi \times 50 \times 0.150 \times (0.500 \times 0.300) \times 1500 = 10600$ V.

(b) See Fig. 19.15b.

(c) The force due to the magnetic field on each side = BILN $= 0.150 \times 10.0 \times 0.50 \times 1500 = 1125$ N. The couple at peak current = force × width of the coil $= 1125 \times 0.30 = 340$ Nm.

19.4

1. (a) Rearrange $\frac{V_S}{V_P} = \frac{N_S}{N_P}$ to give $V_P = \frac{N_P \times V_S}{N_S} = \frac{60 \times 230}{1200} = 11.5$ V.

(b) Power delivered by the secondary coil

$I_S V_S = 100$ W, hence $I_S = \frac{100\text{ W}}{230\text{ V}} = 0.43$ A.

Assuming 100% efficiency, $I_P V_P = I_S V_S$.

Rearranging this equation gives $I_P = \frac{I_S V_S}{V_P} = \frac{0.43 \times 230}{11.5} = 8.7$ A.

2. (a) Turns ratio $\frac{N_S}{N_P} = \frac{V_S}{V_P} = \frac{230}{110} = 2.1$.

(b) Maximum generator current $= \frac{20\text{ kW}}{110\text{ V}} = 182$ A.

(c) Maximum current from the cables $= I_S = \frac{I_P V_P}{V_S} = \frac{182 \times 110}{230} = 87$ A.

19.5

1. (a) (i) The growth of current through the coil causes an induced e.m.f. which opposes the battery and slows the growth of current.

(ii) Estimated rate of growth of current = 0.25 A/1 s = 0.25 A s^{-1} approximately. Hence L = 3 V/0.25 A s^{-1} = 12 H approximately for the two coils.

(b) The magnetic fields of the two coils were in opposite directions as a result of the reversal of the connections to one coil. Hence the magnetic flux in the core is zero, regardless of the current in the coils. Hence no induced e.m.f. can be produced.

2. (a) Initial rate of change of current, $\frac{di}{dt} = \frac{V_{BATT}}{L} = \frac{12}{25} = 0.48$ A s^{-1}.

(b) (i) Total resistance, $R = 2.0 + 1.0 = 3.0\ \Omega$ $\therefore I_0 = \frac{V_{BATT}}{R} = \frac{12}{3.0} = 4.0$ A.

(ii) Energy stored $= \frac{1}{2}LI^2 = 0.5 \times 25 \times 4.0^2 = 200$ J.

Revision questions

19.1. (a) (i) North **(ii)** left to right.

(b) (i) Smaller and in the same direction. **(ii)** Larger and in the opposite direction.

19.2. (a) (i) When the coil is shorted, the induced e.m.f. forces current around the circuit. The magnetic field generated by this current acts against the rotating magnet, exerting a braking effect.
(ii) The loss of k.e. is converted into electrical energy, which is converted into heat due to the resistance of the windings.
(b) (i) The increased load slows the motor down, which reduces the back e.m.f. enabling the voltage supply unit to push more current through the motor.
(ii) Connecting an extra load across the secondary coil causes the secondary current to increase. The magnetic flux of the secondary coil is in the opposite direction to the primary flux, so the total flux decreases and the back e.m.f. across the primary coil becomes smaller. This allows the primary voltage supply to push more current through the primary coil, increasing the flux in the core back to its initial level.
19.3. (a) $V = BL\upsilon = 80 \times 10^{-6} \times 35 \times 550 = 1.54$ V.
(b) (i) Each block represents a change of flux of 1.0×10^{-4} Wb (= 5 mV × 0.02 s). There are 36 blocks under the curve. Hence the change of flux linkage is 3.6×10^{-3} Wb (= $36 \times 1.0 \times 10^{-4}$ Wb).
(ii) $BAN = 3.6 \times 10^{-4}$ Wb $\quad \therefore B = \dfrac{3.6 \times 10^{-3}}{\pi(12.5 \times 10^{-3})^2 \times 120} = 6.1 \times 10^{-2}$ T.
19.4. (a) (i) The sudden growth of current through the primary coil causes a sudden growth of magnetic flux in the core which induces a large e.m.f. in the secondary coil. This e.m.f. is large enough to cause a spark across the gap between the rods.
(ii) The magnetic flux in the core suddenly collapses, generating a large reverse e.m.f. in the secondary coil which also creates a spark.
(iii) No change in the primary current occurs so there is no change in the magnetic flux in the core and hence no induced e.m.f.
(b) (i) Rearrange $\dfrac{V_S}{V_P} = \dfrac{N_S}{N_P}$ to give $N_S = \dfrac{N_P \times V_S}{V_P} = \dfrac{1200 \times 9}{230} = 47$.
(II) $I_S = \dfrac{36\text{ W}}{9.0\text{ V}} = 4.0$ A. **(iii)** $I_P = \dfrac{36\text{ W}}{230\text{ V}} = 0.16$ A.
19.5. (a) (i) $B = 2000 \times 4\pi \times 10^{-7} \times \dfrac{80}{\pi \times 0.035} \times 0.06 = 0.11$ T.
(ii) $\Phi = BAN = 0.11 \times 6.0 \times 10^{-5} \times 80 = 5.3 \times 10^{-4}$ Wb
$\therefore L = \dfrac{\Phi}{I} = \dfrac{5.3 \times 10^{-4}}{0.06} = 8.8 \times 10^{-3}$ H.
(b) (i) Initial rate of growth of current $= \dfrac{V_{BATT}}{L} = 170$ A s^{-1}.
(ii) $E = \frac{1}{2}LI^2 = 0.5 \times 8.8 \times 10^{-3} \times (0.06)^2 = 1.6 \times 10^{-5}$ J.

Unit 20

In-text questions

20.1

1. The trace measures 58 mm from the bottom to the top. Also, one complete cycle covers a horizontal distance of 32 mm.
(a) $2V_0 = 5.8$ cm × 0.2 V cm^{-1} = 1.16 V $\quad \therefore V_0 = 0.58$ V; T = 3.2 cm × 10 μs cm^{-1} = 32 μs
$\therefore f = \dfrac{1}{32 \times 10^{-6}\text{ s}} = 3.13 \times 10^4$ Hz.
(b) $2V_0 = 5.8$ cm × 5.0 V cm^{-1} = 29 V $\quad \therefore V_0 = 14.5$ V;
T = 3.2 cm × 5 ms cm^{-1} = 16 ms
$\therefore f = \dfrac{1}{16 \times 10^{-3}\text{ s}} = 62.5$ Hz.
2. (a) See Fig. A20.1.
(b) (i) $V = V_0 \sin(2\pi ft) = 4.0 \sin(2\pi \times 200 \times 1.0 \times 10^{-3}$ radians$) = 3.8$ V.
(ii) $V = V_0 \sin(2\pi ft) = 4.0 \sin(2\pi \times 200 \times 3.0 \times 10^{-3}$ radians$) = -2.4$ V.

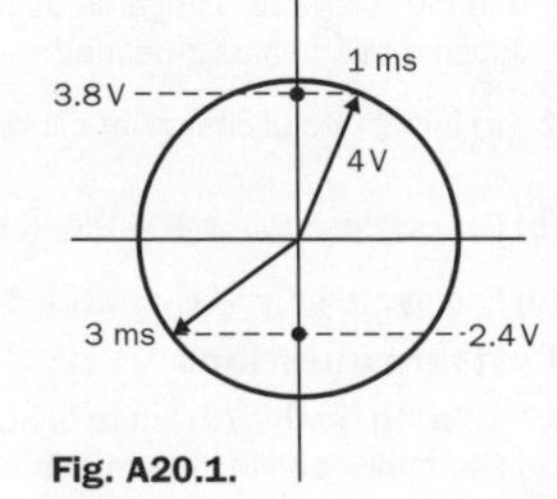

Fig. A20.1.

20.2

1. (a) (i) Right to left **(ii)** A and C.
(b) A half-wave rectified waveform is produced. See Fig. 20.7.
2. (a) Right to left. **(b)** The peak current is greater in alternate half-cycles when diode A conducts, because current passes through the lesser resistance when A conducts and through the higher resistance when B conducts.
See Fig. A20.2.

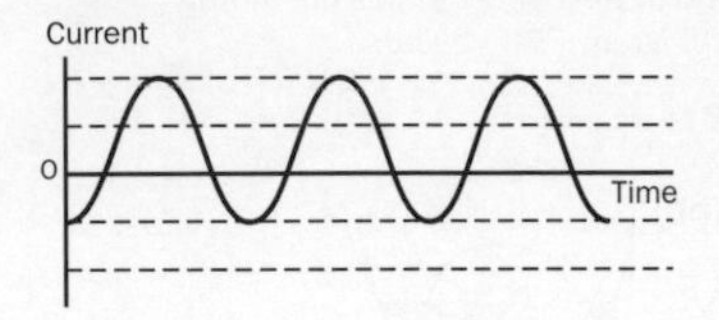

Fig. A20.2.

20.3

1. (a) (i) $V_{pp} = 7.2$ cm × 0.2 V cm^{-1} = 1.44 V. **(ii)** $V_o = 0.5\,V_{pp} = 0.72$ V.
(iii) $V_{rms} = \dfrac{1V_0}{\sqrt{2}} = 0.51$ V.
(b) (i) $R = \dfrac{V_{rms}}{I_{rms}} = \dfrac{0.51}{0.12} = 4.2\ \Omega$.
(ii) Mean power $= I_{rms}V_{rms} = 0.120 \times 0.51 = 0.061$ W.
2. (a) $V_o = \sqrt{2}\,V_{rms} = \sqrt{2} \times 230$ V = 325 V.
(b) (i) Rearrange mean power $= I_{rms}V_{rms}$ to give
$I_{rms} = \dfrac{\text{mean power}}{V_{rms}} = \dfrac{1000\text{ W}}{230\text{ V}} = 4.3$ A.
(ii) $I_o = \sqrt{2}I_{rms} = \sqrt{2} \times 4.3 = 6.1$ A. **(iii)** $R = \dfrac{V_{rms}}{I_{rms}} = \dfrac{230}{4.3} = 53\ \Omega$.

20.4

1. (a) See Fig. A20.3.

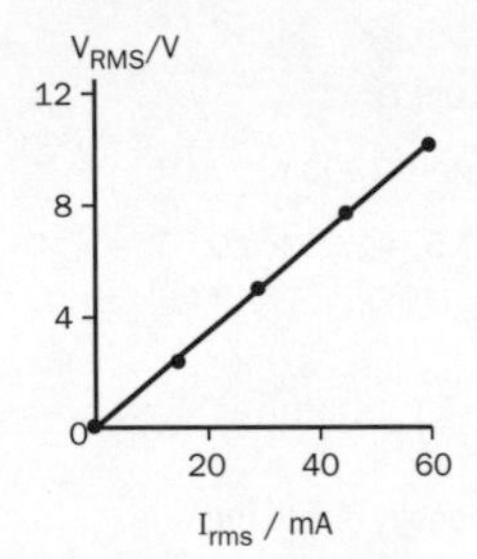

Fig. A20.3.

(b) Reactance = gradient of graph $= \dfrac{V_{rms}}{I_{rms}} = \dfrac{10.0}{60.0 \times 10^{-3}} = 167\ \Omega$.
(c) Rearrange $X_C = \dfrac{1}{2\pi fC}$ to give $C = \dfrac{1}{2\pi fX_C} = \dfrac{1}{2\pi \times 2000 \times 167} = 4.8 \times 10^{-7}$ F.
2. (a) (i) Trace height (top to bottom) = 60 mm
∴ peak-to-peak voltage = 2.0 × 6.0 = 12.0 V $\quad \therefore V_o = 6.0$ V.
(ii) $V_{rms} = \dfrac{V_0}{\sqrt{2}} = 4.24$ V.
(b) One full cycle on the screen = 40 mm horizontally
∴ time period = 0.5 × 4.0 ms = 2.0 ms
∴ frequency, $f = \dfrac{1}{T} = \dfrac{1}{2.0 \times 10^{-3}} = 500$ Hz.
(c) (i) Reactance $= \dfrac{V_{rms}}{I_{rms}} = \dfrac{4.24}{0.48 \times 10^{-3}} = 8830\ \Omega$.
(ii) Rearrange $X_C = \dfrac{1}{2\pi fC}$ to give
$C = \dfrac{1}{2\pi fX_C} = \dfrac{1}{2\pi \times 500 \times 8830} = 3.6 \times 10^{-8}$ F.

20.5

1. (a) (i) $X_L = \dfrac{V_{rms}}{I_{rms}} = \dfrac{12.0}{0.110} = 109\ \Omega$.
(ii) $X_L = 2\pi fL \quad \therefore L = \dfrac{X_L}{2\pi f} = \dfrac{109}{2\pi \times 1500} = 1.16 \times 10^{-2}$ H

(b) $X_L = 2\pi fL = 2\pi \times 20\,000 \times 1.16 \times 10^{-2} = 1460\,\Omega$

$\therefore I_{rm,s} = \dfrac{V_{rms}}{X_L} = \dfrac{12.0}{1460} = 8.2 \times 10^{-3}$ A.

2. (a)(i) $X_L = \dfrac{V_{rms}}{I_{rms}} = \dfrac{6.0}{0.028} = 214\,\Omega$.

(ii) $X_L = 2\pi fL \quad \therefore f = \dfrac{X_L}{2\pi L} = \dfrac{214}{2\pi \times 0.045} = 757$ Hz.

(b) $X_C = X_L$ for the same current

$\therefore \dfrac{1}{2\pi fC} = X_L$, hence $C = \dfrac{1}{2\pi fX_L} = \dfrac{1}{2\pi \times 757 \times 214} = 9.8 \times 10^{-7}$ F.

20.6

1. (a) (i) $X_C = \dfrac{1}{2\pi fC} = \dfrac{1}{2\pi \times 50 \times 10 \times 10^{-6}} = 318\,\Omega$.

(ii) $X_L = 2\pi fL = 2\pi \times 50 \times 0.48 = 151\,\Omega$.

(iii) $Z = [R^2 + (X_C - X_L)^2]^{1/2} = [55^2 + (318 - 151)^2]^{1/2} = 176\,\Omega$.

(iv) $\therefore I_{rms} = \dfrac{V_{rms}}{Z} = \dfrac{6.0}{176} = 0.034$ A.

(b) (i) r.m.s. value of $V_R = I_{rms} R = 0.034 \times 55 = 1.9$ V.

(ii) r.m.s. value of $V_L = I_{rms} X_L = 0.034 \times 151 = 5.1$ V.

(iii) r.m.s. value of $V_C = I_{rms} X_C = 0.034 \times 318 = 10.8$ V.

(c) (i) See Fig. A20.4.

(a) Phasor diagram **(b)** Simplified diagram

Fig. A20.4.

(ii) $\tan\phi = \dfrac{(V_C - V_L)}{V_R} = \dfrac{10.8 - 5.1}{1.9} = 3.0 \quad \therefore \phi = 72°$.

2. (a) (i) $f_O = \dfrac{1}{2\pi\sqrt{(LC)}} = \dfrac{1}{2\pi\sqrt{(0.15 \times 10^{-3} \times 0.47 \times 10^{-6})}} = 1.9 \times 10^4$ Hz.

(ii) $X_C = \dfrac{1}{2\pi fC} = \dfrac{1}{2\pi \times 1.9 \times 10^4 \times 0.47 \times 10^{-6}} = 17.8\,\Omega$.

(b) (i) $R = \dfrac{V_{rms}}{I_{rms}} = \dfrac{5.0}{0.350} = 14.3\,\Omega$.

(ii) r.m.s. voltage across the capacitor $= I_{rms}X_C = 0.350 \times 17.8 = 6.2$ V.

(iii) The capacitor voltage and the inductor voltage are equal and opposite. See Fig. A20.5 for the phasor diagram.

Fig. A20.5.

Revision questions

20.1. (a) and **(b)** See p. 288.

20.2. (a) (i) $I_O = \sqrt{2}I_{rms} = 1.06$ A.
(ii) $V_O = I_O R = 1.06 \times 47 = 50$ V.
(iii) Peak power $= I_O V_O = 1.06 \times 50 = 53$ W. **(iv)** Mean power $= 0.5 \times$ peak power $= 26$ W.
(b) See Fig. A20.6 **(ii)** 5.1 V **(iii)** 11.2 V.

Fig. A20.6.

20.3. (a) (i) Time period, $T = 4.0$ cm $\times$ 5 ms cm^{-1} = 20 ms

$\therefore f = \dfrac{1}{0.020\text{ s}} = 50$ Hz.

(ii) $V_O = 3.0$ cm $\times$ 2.0 V cm^{-1} = 6.0 V.

(iii) $V_{rms} = \dfrac{V_O}{\sqrt{2}} = 4.2$ V.

(b) (i) $X_C = \dfrac{V_{rms}}{I_{rms}} = \dfrac{4.2\text{ V}}{2.8 \times 10^{-3}\text{ A}} = 1.5 \times 10^3\,\Omega$.

(ii) Rearrange $X_C = \dfrac{1}{2\pi fC}$ to give

$C = \dfrac{1}{2\pi f X_C} = \dfrac{1}{2\pi \times 50 \times 1500} = 2.1 \times 10^{-6}$ F.

20.4. (a) (i) $X_L = \dfrac{V_{rms}}{I_{rms}} = \dfrac{6.0\text{V}}{0.35\text{ A}} = 17.1\,\Omega$.

(ii) Rearrange $X_L = 2\pi fL$ to give $L = \dfrac{X_L}{2\pi f} = \dfrac{17.1}{2\pi \times 50} = 5.4 \times 10^{-2}$ H.

(b) $X_L = X_C$ and $X_C = \dfrac{1}{2\pi fC} \quad \therefore \dfrac{1}{2\pi fC} = 17.1\,\Omega$

$\therefore C = \dfrac{1}{2\pi \times 50 \times 17.1} = 1.9 \times 10^{-4}$ F.

(c) (i) X_L would double and therefore the r.m.s. current would halve to 0.175 A.
(ii) X_C would halve and therefore the r.m.s. current would double to 0.70 A.

20.5. (a) The circuit is in resonance at 750 Hz. The capacitor reactance is exactly equal to the inductor reactance at this frequency so the circuit impedance is a minimum at 750 Hz. Hence the current is at a maximum.

(b) (i) $X_C = \dfrac{1}{2\pi fC} = \dfrac{1}{2\pi \times 750 \times 2.2 \times 10^{-6}} = 97\,\Omega$.

(ii) $X_L = X_C$ and $X_L = 2\pi fL \quad \therefore 2\pi fL = X_C$

$\therefore L = \dfrac{X_C}{2\pi f} = \dfrac{96.5}{2\pi \times 750} = 2.1 \times 10^{-2}$ H.

(iii) $R = \dfrac{\text{r.m.s. supply voltage}}{I_{rms}} = \dfrac{6.0\text{ V}}{65 \times 10^{-3}\text{ A}} = 92\,\Omega$.

(c) (i) r.m.s. voltage across R = 6.0 V.
(ii) r.m.s. voltage across C $= I_{rms}X_C = 65 \times 10^{-3} \times 97 = 6.3$ V. **(iii)** r.m.s. voltage across L = 6.3 V.
(d) See Fig. A 20.7: at resonance, $V_C - V_L = 0$ so the supply voltage is equal to V_R which is in phase with I.

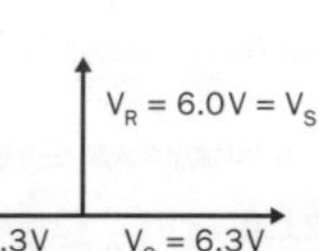

Fig. A20.7.

Unit 21

In-text questions

21.1

1. Rearranging $\frac{1}{2}mv^2 = eV_A$ gives $v^2 = \dfrac{2eV_A}{m}$.

(a) $v^2 = \dfrac{2eV_A}{m} = 2 \times 1.76 \times 10^{11} \times 100 = 3.52 \times 10^{13}$

$\therefore v = 5.9 \times 10^6$ m s^{-1}.

(b) $v^2 = \dfrac{2eV_A}{m} = 2 \times 1.76 \times 10^{11} \times 4000 = 1.41 \times 10^{15}$

$\therefore v = 3.8 \times 10^7$ m s^{-1}.

2. (a) Rearrange $\frac{1}{2}mv^2 = eV_A$ to calculate v.

Hence $v^2 = \dfrac{2eV_A}{m} = 2 \times 1.76 \times 10^{11} \times 3200 = 1.1 \times 10^{15}$

$\therefore v = 3.4 \times 10^7$ m s^{-1}.

(b) Rearrange $Bev = \dfrac{(eV_P)}{d}$ to give

$B = \dfrac{V_P}{vd} = \dfrac{4200}{3.4 \times 10^7 \times 40 \times 10^{-3}} = 3.1$ mT.

21.2

1. (a) Positive, as the top plate is negative.

(b) Rearranging $\dfrac{QV_P}{d} = mg$ gives

$Q = \dfrac{mgd}{V_P} = \dfrac{3.8 \times 10^{-15} \times 9.8 \times 5.0 \times 10^{-3}}{595} = 3.2 \times 10^{-19}$ C.

2. (a) (i) $v = \dfrac{s}{t} = \dfrac{1.0 \times 10^{-3}}{16.5} = 6.1 \times 10^{-5}$ m s^{-1}.

(ii) $r^2 = \dfrac{9\eta v}{2\rho g} = \dfrac{9 \times 1.8 \times 10^{-5} \times 6.1 \times 10^{-5}}{2 \times 960 \times 9.8} = 5.2 \times 10^{-13}$ m^2

$\therefore r = 7.2 \times 10^{-7}$ m.

(iii) $m = \frac{4}{3}\pi r^3\rho = \frac{4}{3}\pi (7.2 \times 10^{-7})^3 \times 960 = 1.5 \times 10^{-15}$ kg.

(iv) $Q = \dfrac{mgd}{V_P} = \dfrac{1.5 \times 10^{-15} \times 9.8 \times 4.0 \times 10^{-3}}{375} = 1.6 \times 10^{-19}$ C.

(b) (i) The drag force increases with speed as the droplet falls. Hence the resultant force (= weight − drag force) and therefore the acceleration

decreases to zero. At the terminal speed, the drag force is equal and opposite to the weight.
(ii) The electrostatic force on the droplet would reverse direction and the droplet would be attracted to the lower plate. It would therefore accelerate towards the lower plate, reaching a constant speed when the drag force became equal and opposite to the electrostatic force + the weight.

21.3

1. (a) (i) $E = hf = \frac{hc}{\lambda} = \frac{6.6 \times 10^{-34} \times 3.0 \times 10^{8}}{600 \times 10^{-9}} = 3.3 \times 10^{-19}$ J.

(ii) $E = hf = \frac{hc}{\lambda} = \frac{6.6 \times 10^{-34} \times 3.0 \times 10^{8}}{100 \times 10^{-9}} = 2.0 \times 10^{-18}$ J.

(b) (i) $\phi = 0.64$ eV $= 0.64 \times 1.6 \times 10^{-19}$ J $= 1.0 \times 10^{-19}$ J
$\therefore$ max. k.e. $= hf - \phi = 3.3 \times 10^{-19} - 1.0 \times 10^{-19} = 2.3 \times 10^{-19}$ J.
(ii) Max. k.e. $= hf - \phi = 2.0 \times 10^{-18} - 1.0 \times 10^{-19} = 1.9 \times 10^{-18}$ J.

2. (a) (i) $E = hf = \frac{hc}{\lambda} = \frac{6.6 \times 10^{-34} \times 3.0 \times 10^{8}}{550 \times 10^{-9}} = 3.6 \times 10^{-19}$ J.
(ii) Work done $= eV = 1.6 \times 10^{-19} \times 0.58 = 9.3 \times 10^{-20}$ J.
(iii) Rearrange $eV_S = hf - \phi$ to give $\phi = hf - eV_S = 3.6 \times 10^{-19} - 9.3 \times 10^{-20} = 2.7 \times 10^{-19}$ J.

(b) The threshold frequency, $f_0 = \frac{\phi}{h} = \frac{2.7 \times 10^{-19}}{6.6 \times 10^{-34}} = 4.1 \times 10^{14}$ Hz

$\therefore$ minimum wavelength $= \frac{c}{f_o} = \frac{3.0 \times 10^{8}}{4.1 \times 10^{14}} = 7.3 \times 10^{-7}$ m.

21.4

1. (a) Photon energy $= E_1 - E_2 = 7.2 - 0 = 7.2$ eV $= 7.2 \times 1.6 \times 10^{-19}$ $= 11.5 \times 10^{-19}$ J. Rearrange photon energy, $E = hf = \frac{hc}{\lambda}$ to give

$\lambda = \frac{hc}{E} = \frac{6.6 \times 10^{-34} \times 3.0 \times 10^{8}}{11.5 \times 10^{-19}} = 1.7 \times 10^{-7}$ m.

(b) (i) Photon energy $= E_1 - E_2 = 8.9 - 7.6 = 1.3$ eV $= 1.3 \times 1.6 \times 10^{-19} = 2.1 \times 10^{-19}$ J. Rearrange photon energy, $E = hf = \frac{hc}{\lambda}$ to give

$\lambda = \frac{hc}{E} = \frac{6.6 \times 10^{-34} \times 3.0 \times 10^{8}}{2.1 \times 10^{-19}} = 9.4 \times 10^{-7}$ m.

(ii) Photon energy $= E_1 - E_2 = 8.9 - 4.9 = 4.0$ eV $= 4.0 \times 1.6 \times 10^{-19}$ $= 6.4 \times 10^{-19}$ J. Rearrange photon energy, $E = hf = \frac{hc}{\lambda}$ to give

$\lambda = \frac{hc}{E} = \frac{6.6 \times 10^{-34} \times 3.0 \times 10^{8}}{6.4 \times 10^{-19}} = 3.1 \times 10^{-7}$ m.

2. (a) $E_2 = \frac{-13.6}{4} = -3.4$ eV; $E_3 = \frac{-13.6}{9} = -1.5$ eV;

$E_4 = \frac{-13.6}{16} = -0.85$ eV; $E_5 = \frac{-13.6}{25} = -0.54$ eV.

(b)(i) See Fig. A21.1.

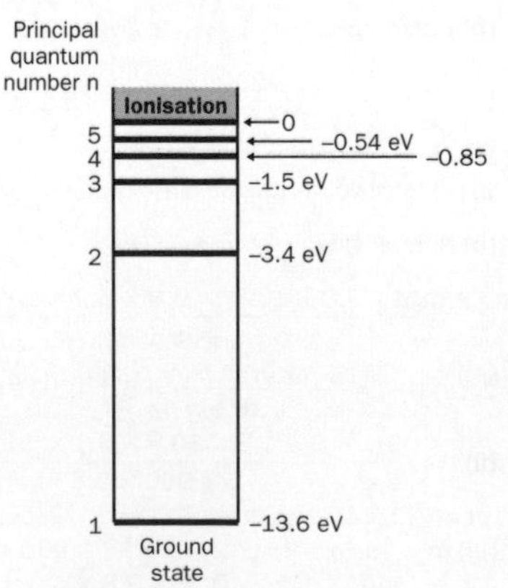

Fig. A21.1.

(ii) Photon energy, $E = (-0.54) - (-0.85) = 0.31$ eV $= 0.31 \times 1.6 \times 10^{-19}$ J $= 5.0 \times 10^{-20}$ J. Rearrange photon energy, $E = hf = \frac{hc}{\lambda}$ to give

$\lambda = \frac{hc}{E} = \frac{6.6 \times 10^{-34} \times 3.0 \times 10^{8}}{5.0 \times 10^{-20}} = 4.0 \times 10^{-6}$ m.

(iii) Infra-red.

21.5

1. (a) $\lambda_{MIN} = \frac{hc}{eV_A} = \frac{6.6 \times 10^{-34} \times 3.0 \times 10^{8}}{1.6 \times 10^{-19} \times 25\,000} = 5.0 \times 10^{-11}$ m.

(b) Electrical power used by the tube $= IV_A = 30 \times 10^{-3} \times 25\,000 = 750$ W.
(c) Heat produced per second $= 98\% \times 750$ W $= 735$ W.
2. See p. 320.

Revision questions

21.1. (a) (i) K.e. $= eV_A = 1.6 \times 10^{-19} \times 4200 = 6.7 \times 10^{-16}$ J.

(ii) $\frac{1}{2}mv^2 = 6.7 \times 10^{-16}$ $\therefore v^2 = \frac{2 \times 6.7 \times 10^{-16}}{9.1 \times 10^{-31}} = 1.48 \times 10^{15}$

$\therefore v = 3.8 \times 10^{7}$ m s^{-1}.

(b) (i) $F = \frac{eV_P}{d} = \frac{1.6 \times 10^{-19} \times 5000}{50 \times 10^{-3}} = 1.6 \times 10^{-14}$ N.

(ii) Using $F = Bev$ gives $B = \frac{F}{ev} = \frac{1.6 \times 10^{-14}}{1.6 \times 10^{-19} \times 3.8 \times 10^{7}} = 2.6 \times 10^{-3}$ T into the plane of the diagram.

21.2. (a) (i) Negative charge. **(ii)** The viscous drag increases as the speed increases until it becomes equal and opposite to the weight when the droplet is falling at its terminal speed.

(b) (i) $v = \frac{s}{t} = \frac{1.2 \times 10^{-3}}{14.6} = 8.2 \times 10^{-5}$ m s^{-1}.

(ii) $r^2 = \frac{9\eta v}{2\rho g} = \frac{9 \times 1.8 \times 10^{-5} \times 8.2 \times 10^{-5}}{2 \times 960 \times 9.8} = 7.0 \times 10^{-13}$

$\therefore r = 8.4 \times 10^{-7}$ m.

(iii) $m = \frac{4}{3}\pi r^3 \rho = \frac{4}{3}\pi(8.4 \times 10^{-7})^3 \times 960 = 2.4 \times 10^{-15}$ kg.

(iv) $Q = \frac{mgd}{V} = \frac{2.4 \times 10^{-15} \times 9.8 \times 4.0 \times 10^{-3}}{590} = 1.6 \times 10^{-19}$ C.

21.3. (a) (i) $f = \frac{c}{\lambda} = \frac{3.0 \times 10^{8}}{500 \times 10^{-9}} = 6.0 \times 10^{14}$ Hz;
$E = hf = 6.6 \times 10^{-34} \times 6.0 \times 10^{14} = 4.0 \times 10^{-19}$ J.

(ii) $f = \frac{c}{\lambda} = \frac{3.0 \times 10^{8}}{50 \times 10^{-9}} = 6.0 \times 10^{15}$ Hz;
$E = hf = 6.6 \times 10^{-34} \times 6.0 \times 10^{15} = 4.0 \times 10^{-18}$ J.

(b) (i) Work done $= eV = 1.6 \times 10^{-19} \times 0.36 = 5.8 \times 10^{-20}$ J.
(ii) Rearranging $eV_S = hf - \phi$ gives $\phi = hf - eV_S = 4.0 \times 10^{-19} - 5.8 \times 10^{-20} = 3.4 \times 10^{-19}$ J $= \frac{3.4 \times 10^{-19}}{1.6 \times 10^{-19}} = 2.1$ eV.

(iii) Threshold frequency, $f_o = \frac{\phi}{h} = 5.2 \times 10^{14}$ Hz

$\therefore \lambda = \frac{c}{f_o} = \frac{3.0 \times 10^{8}}{5.2 \times 10^{14}} = 5.8 \times 10^{-7}$ m.

21.4. (a) (i) $hf = E_4 - E_3 = \frac{(-13.6)}{16} - \frac{(-13.6)}{9}$ eV $= 0.66$ eV $= 0.66 \times 1.6 \times 10^{-19}$ J $\therefore f = \frac{E_4 - E_3}{h} = \frac{0.66 \times 1.6 \times 10^{-19}}{6.6 \times 10^{-34}} = 1.6 \times 10^{14}$ Hz

$\therefore \lambda = \frac{c}{f} = \frac{3.0 \times 10^{8}}{1.6 \times 10^{14}} = 1.9 \times 10^{-6}$ m.

(ii) $hf = E_2 - E_1 = \frac{(-13.6)}{4} - \frac{(-13.6)}{1}$ eV $= 10.2$ eV $= 10.2 \times 1.6 \times 10^{-19}$ J

$\therefore f = \frac{E_2 - E_1}{h} = \frac{10.2 \times 1.6 \times 10^{-19}}{6.6 \times 10^{-34}} = 2.5 \times 10^{15}$ Hz

$\therefore \lambda = \frac{c}{f} = \frac{3.0 \times 10^{8}}{2.5 \times 10^{15}} = 1.2 \times 10^{-7}$ m.

(b) 1. For $\lambda = 565$ nm,

$E = hf = \frac{hc}{\lambda} = \frac{6.6 \times 10^{-34} \times 3.0 \times 10^{8}}{565 \times 10^{-9}} = 3.5 \times 10^{-19}$ J $= 2.2$ eV which corresponds to a transition from the energy level at -9.8 eV to the ground state at -12.0 eV.

2. For $\lambda = 430$ nm,

$E = hf = \frac{hc}{\lambda} = \frac{6.6 \times 10^{-34} \times 3.0 \times 10^{8}}{430 \times 10^{-9}} = 4.6 \times 10^{-19}$ J = 2.9 eV which corresponds to a transition from the energy level at −6.9 eV to the energy level at −9.8 eV.

21.5. (a) (i) See Fig. 21.22. **(ii)** The beam electrons knock electrons out of the shells of the target atoms. When a vacancy in the shell is refilled, a photon of a certain energy is emitted. Since the frequency of a photon is proportional to its energy, photons of certain frequencies only are produced by this mechanism. The intensity distribution is therefore much more intense at these frequencies.

(b) Combine $\lambda = \frac{c}{f}$ and $eV_A = hf$ to give $\lambda = \frac{hc}{eV_A}$.

(i) $\lambda = \frac{hc}{eV_A} = \frac{6.6 \times 10^{-34} \times 3.0 \times 10^{8}}{1.6 \times 10^{-19} \times 25\,000} = 5.0 \times 10^{-11}$ m.

(ii) $\lambda = \frac{hc}{eV_A} = \frac{6.6 \times 10^{-34} \times 3.0 \times 10^{8}}{1.6 \times 10^{-19} \times 100\,000} = 1.25 \times 10^{-11}$ m.

Unit 22

In-text questions

22.1

1. (a) 92 p + 143 n **(b)** 6 p + 6 n **(c)** 17 p + 20 n.

2. (a) 235 grams of U 235 contains 6.02×10^{23} atoms, hence 1.0 kg of U 235 contains 2.6×10^{24} atoms (= $6.02 \times 10^{23}/0.235$).

(b) 4.0 grams of He 4 contains 6.02×10^{23} atoms, hence 1.0 g of He 4 contains 1.5×10^{23} atoms (= $6.02 \times 10^{23}/4$).

22.2

1. (a) 2, 234 **(b)** 27, 0, 60.

2. (a) Similarity; both cause ionisation or both emitted by the nucleus.

(b) (i) The nucleus loses two protons and two neutrons. **(ii)** A neutron in the nucleus changes into a proton, creating and emitting a beta particle at the instant of change.

22.3

1. (a) (i) Beta radiation; the paper made no difference so there was no alpha radiation emitted; also, the 4 mm plate prevented the radiation from the source reaching the tube so no gamma radiation was present.

(ii) With no source present, the count rate should decrease to the background count rate if the tube is moved more than about 50 cm from the source.

(b) (i) Corrected count rate without the metal plate present = 602 counts per minute; corrected count rate with the metal plate present = 456 counts per minute.

(ii) Percentage transmitted $= \frac{456}{602} \times 100\% = 76\%$.

(iii) Corrected count rate with two metal plates present = 76% × 76% of 602 counts per minute = 348 counts per minute. Hence the corrected count rate = 374 counts per minute (= 348 + 26 counts per minute).

2. (a) Average background count rate $= \frac{(143 + 124 + 136)}{3} \div 5$ minutes = 27 counts per minute. Average count rate with source present $= \frac{(725 + 746 + 738)}{3} \div 3$ minutes = 245 counts per minute. Corrected count rate due to source = 245 − 27 = 218 counts per minute.

(b) The corrected count rate varies with distance from the source according to the inverse square law.

(i) At 0.20 m, the distance is halved so the corrected count rate increases by × 4 to 872 counts per minute (= 4 × 218). Hence the count rate would be 899 counts per minute (= 872 + 27).

(ii) At 0.15 m, the distance is reduced by a factor of 0.375 (= 0.15/0.40). Hence the corrected count rate would increase by a factor of $\frac{1}{0.375^2} = \times 7.1$ to approximately 1550 counts per minute (= 7.1 × 218). Hence the count rate would be about 1580 counts per minute (= 1550 + 27 to the nearest 10).

22.4

1. (a) (i) 105 counts per minute (c.p.m.) **(ii)** 52 c.p.m.

(b) Approximately four half-lives (= 270 s to the nearest 10 s).

2. (a) (i) 80 kBq **(ii)** 40 kBq.

(b) 100 years is approx. four half-lives so the activity would be about 10 kBq.

22.5

1. (a) $\lambda = \frac{\log_e 2}{T_{1/2}} = \frac{0.693}{140 \times 24 \times 3600 \text{ s}} = 5.7 \times 10^{-8}\text{ s}^{-1}$.

(b) 210 g of Po 210 contains 6.02×10^{23} atoms $\therefore$ 1.0 mg contains 2.87×10^{18} atoms $\left(= 6.02 \times 10^{23} \times \frac{0.001}{210}\right)$.

(c) Activity $= \lambda N = 5.7 \times 10^{-8} \times 2.87 \times 10^{18} = 1.6 \times 10^{11}$ Bq.

(d) $A = A_0 e^{-\lambda t} = 1.6 \times 10^{11} \times \exp - (5.7 \times 10^{-8} \times 365\tfrac{1}{4} \times 24 \times 3600) = 2.6 \times 10^{10}$ Bq.

(Note: $\exp - (5.7 \times 10^{-8} \times 365\tfrac{1}{4} \times 24 \times 3600) = 0.166$)

2. (a) $\lambda = \frac{\log_e 2}{T_{1/2}} = \frac{0.693}{1940 \times 24 \times 3600} = 4.1 \times 10^{-9}\text{ s}^{-1}$.

(b) $N = \frac{A}{\lambda} = \frac{2.0 \times 10^{6}}{4.1 \times 10^{-9}} = 4.8 \times 10^{14}$.

(c) $A = A_0 e^{-\lambda t}$ gives $0.1 = 2.0 e^{-\lambda t}$ $\therefore e^{-\lambda t} = \frac{0.1}{2.0} = 5.0 \times 10^{-2}$.

Hence $-\lambda t = \log_e 5.0 \times 10^{-2} = -3.0$

$\therefore t = \frac{-3.0}{-\lambda} = \frac{-3.0 \text{ s}}{-4.1 \times 10^{-9}} = 0.73 \times 10^{9}$ s = 23.2 years.

Revision questions

22.1. (a) 28, 63, −1 **(b)** 214, 2, 82.

22.2. (a) Average background count rate $= \frac{(128 + 136 + 138)}{3 \times 5} = 26.8$ counts per minute.

(b) Beta radiation; alpha radiation would not reach the tube at 10 cm; gamma radiation would not be affected as much by a metal plate of thickness 0.8 mm.

(c) Corrected count rate without the absorber $= \frac{(565 + 572 + 552)}{3 \times 3} - 26.8 = 35.8$ c.p.m. Corrected count rate with the absorber present $= \frac{(384 + 368 + 372)}{3 \times 3} - 26.8 = 14.9$ c.p.m. Percentage transmitted $= \frac{14.9}{35.8} \times 100\% = 42\%$.

22.3. (a) Let d = distance of the source inside the container. Using the inverse square law in the form, corrected count rate C $= \frac{k}{r^2}$, then $C_1 = \frac{k}{(0.090 + d)^2}$ and $C_2 = \frac{k}{(0.190 + d)^2}$.

Hence $\frac{C_1}{C_2} = \frac{(0.190 + d)^2}{(0.090 + d)^2}$. As $\frac{C_1}{C_2} = \frac{1284}{330} = 4$ then $\frac{(0.190 + d)^2}{(0.090 + d)^2} = 4$

so $\frac{(0.190 + d)}{(0.090 + d)} = 2$ $\therefore 0.190 + d = 2\,(0.090 + d) = 0.180 + 2d$

$\therefore d = 0.190 - 0.180 = 0.010$ m = 10 mm.

(b) Measure the count rate with paper and with metal plates of different thicknesses between the source and the tube. If the paper absorbs the radiation, it is alpha radiation. If the count rate is scarcely affected by several millimetres of aluminium, the radiation is gamma radiation. Otherwise, it is beta radiation.

22.4. (a) $\lambda = \frac{\log_e 2}{T_{1/2}} = \frac{0.693}{14.8 \text{ h}} = 0.047\text{ h}^{-1} = 1.3 \times 10^{-5}\text{ s}^{-1}$.

(b) 24 g of Na 24 contains 6.0×10^{23} atoms

$\therefore$ 1.0 mg contains 2.5×10^{19} atoms $\left(= \frac{1.0 \times 10^{-3} \times 6.0 \times 10^{23}}{24}\right)$.

(c) Activity $= \lambda N = 3.3 \times 10^{14}$ Bq.

(d) $A = A_0 e^{-\lambda t} = 3.3 \times 10^{14} \times \exp(-1.3 \times 10^{-5} \times 24 \times 3600) = 1.1 \times 10^{14}$ Bq.

22.5. (a) (i) $\lambda = \frac{\log_e 2}{T_{½}} = \frac{0.693}{100 \times 365.25 \times 24 \times 3600 \text{ s}} = 2.2 \times 10^{-10} \text{ s}^{-1}$.
(ii) 239 g contains 6.0×10^{23} atoms
$\therefore$ 1.0 g contains 2.5×10^{21} atoms $\left(= \frac{6.0 \times 10^{23}}{239}\right)$.
(b) (i) Activity of 1.0 g = $\lambda N = 2.2 \times 10^{-10} \times 2.5 \times 10^{21} = 5.5 \times 10^{11}$ Bq
$\therefore$ mass of sample of activity 1000 MBq = $\frac{1.0 \times 10^9}{5.5 \times 10^{11}} = 1.8 \times 10^{-3}$ g.
(ii) $A = A_0 e^{-\lambda t} = 1000 \times \exp(-2.2 \times 10^{-10} \times 10 \times 365.25 \times 24 \times 3600) = 933$ MBq.

Unit 23

In-text questions

23.1

1. (a) Gravity. **(b) (i)** The strong nuclear force. **(ii)** The weak nuclear force. **(c)** The electromagnetic force.
2. (a) (i) $F = \frac{Q_1 Q_2}{4\pi\varepsilon_0 r^2} = \frac{2 \times 1.6 \times 10^{-19} \times 7 \times 1.6 \times 10^{-19}}{4\pi\varepsilon_0 \times (10 \times 10^{-15})^2} = 32$ N.
(ii) $F = \frac{Q_1 Q_2}{4\pi\varepsilon_0 r^2} = \frac{2 \times 1.6 \times 10^{-19} \times 7 \times 1.6 \times 10^{-19}}{4\pi\varepsilon_0 \times (2 \times 10^{-15})^2} = 800$ N.
(b) Attraction due to the strong nuclear force exceeds the repulsion due to the electrostatic force only if the alpha particle has enough kinetic energy to overcome the electrostatic force and come within range of the strong nuclear force.

23.2

1. (a) (i) Mass defect, $\Delta m = Zm_p + (A - Z) m_n - M = 8 \times 1.00728 + 8 \times 1.00867 - (15.99492 - 8 \times 0.00055) = 0.137$ u $\therefore$ B.E. of O–16 nucleus = 0.137 u × 931 MeV/u = 128 MeV
$\therefore$ B.E. per nucleon = $\frac{\text{B.E.}}{A} = \frac{128}{16} = 8.0$ MeV per nucleon.
(ii) Mass defect, $\Delta m = Zm_p + (A - Z) m_n - M = 38 \times 1.00728 + 52 \times 1.00867 - (89.90730 - 90 \times 0.00055) = 0.870$ u $\therefore$ B.E. of Sr-90 nucleus = 0.870 u × 931 MeV/u = 810 MeV
$\therefore$ B.E. per nucleon = $\frac{\text{B.E.}}{A} = \frac{870}{90} = 9.0$ MeV per nucleon.
2. (a) (i) ${}^{238}_{92}\text{U} \rightarrow {}^{234}_{90}\text{Th} \rightarrow {}^{4}_{2}\alpha$
(ii) Δm = atomic mass of U 238 – [atomic mass of Th 234 + atomic mass of He 4] = 238.05076 – [234.04357 + 4.00260] = 0.00459 u
$\therefore Q = \Delta m c^2 = 0.00459 \text{ u} \times 931 \text{ MeV/u} = 4.3$ MeV.
(b)(i) ${}^{60}_{27}\text{Co} \rightarrow {}^{0}_{-1}\alpha + {}^{60}_{28}\text{Ni}$
(ii) Δm = atomic mass of Co 60 – atomic mass of Ni 60 = 59.93381 – 59.93078 = 0.00303 u $\therefore Q = \Delta mc^2 = 0.00303 \text{ u} \times 931$ MeV/u = 2.8 MeV.

23.3

1. (a) (i) Induced fission is the splitting of a heavy nucleus into two approximately equal fragments as a result of a neutron colliding with and being absorbed by the nucleus.
(ii) Two or three neutrons are released when a heavy nucleus is fissioned. These neutrons can then go on to produce further fission, leading to more released neutrons and more fission, etc.
(b) (i) Below a certain mass, too many neutrons would escape and not produce further fission. The probability of escape is proportional to the surface area and the probability of fission is proportional to the volume of the material. If the material is below a certain mass, its surface to volume ratio is too low for a chain reaction to occur.
(ii) For a steady rate of fission, exactly one neutron per fission must go on to produce further fission. Neutron-absorbing control rods are inserted into the core to keep the fission rate constant by absorbing surplus neutrons. Energy is released at a steady rate when the fission rate is constant.
(c) (i) The moderator reduces the kinetic energy of fission neutrons so they can produce further fission of U 235. The fission neutrons collide with the moderator atoms and transfer kinetic energy to the atoms.
(ii) For each successive collision, the k.e. of a neutron is reduced to 75% (i.e. to 0.75 × the k.e. before impact). Hence for n successive collisions, the k.e. is reduced to $(0.75)^n$ $\therefore$ 0.01 eV = $(0.75)^n \times 1$ MeV,
hence $(0.75)^n = \frac{0.01\text{eV}}{1 \text{ MeV}} = 10^{-8}$.
Taking logs to the base 10 either side therefore gives $n\log 0.75 = -8$
$\therefore n = 64 (= -8/\log 0.75)$.
2. (a) Energy output in 1 day = 1000 MJ × 24 h × 3600 s = 8.6×10^{13} J
$\therefore$ energy needed from fuel per day = $\frac{8.6 \times 10^{13}}{0.25} = 3.4 \times 10^{14}$ J.
(i) Mass of oil needed per day = $\frac{3.4 \times 10^{14}}{30 \times 10^6}$ = 10 million kg approx.
(ii) Mass of uranium needed per day = $\frac{\text{10 million kg approx}}{50\,000}$ = 200 kg approx.
(b) (i) See p. 352. **(ii)** To recover uranium 238 and plutonium 239. **(ii)** The fuel is plutonium 239 which is fissioned by fast neutrons. Fission neutrons do not therefore need to be slowed down to produce further fission of plutonium.

23.4

1. (a) ${}^{9}_{4}\text{Be} + {}^{4}_{2}\alpha \rightarrow {}^{1}_{0}\text{n} + {}^{12}_{6}\text{C}$.
(b)(i) Charge of uud = $+\frac{2}{3}e + \frac{2}{3}e - \frac{1}{3}e = +e$; charge of udd = $+\frac{2}{3}e - \frac{1}{3}e - \frac{1}{3}e = 0$.
(ii) Charge of uus = $+\frac{2}{3}e + \frac{2}{3}e - \frac{1}{3}e = +e$; strangeness = 0 + 0 + 1.
2. (a) (i) The particle and antiparticle carry opposite charges.
(ii) The particle and antiparticle lose kinetic energy and slow down. Each is forced into a spiral as it slows down gradually.
(b) Total energy available = 2 + 2 + 1 + 1 = 6 GeV $\therefore$ maximum gamma energy = 3 GeV.

Revision questions

23.1. (a) $Q_1 = 2e$, $Q_2 = 4e$ $\therefore$ assuming initial KE = PE at closest approach, $5 \times 10^6 e = \frac{2e \times 4e}{4\pi\varepsilon_0 r}$
$\therefore r = \frac{8e^2}{4\pi\varepsilon_0 \times 5 \times 10^6 e} = \frac{8 \times 1.6 \times 10^{-19}}{4\pi \times 8.85 \times 10^{-12} \times 5 \times 10^6} = 2.3 \times 10^{-15}$m.
(b) The alpha particle will be attracted into the nucleus at a separation of 2.3 fm, since the strong nuclear force extends to this distance. A nuclear reaction is therefore likely.
23.2. (a) (i) $\Delta m = (6 \times 1.00728 + 6 \times 1.00867) - (12.0000 - 6 \times 0.00055) = 0.092$ u $\therefore$ B.E. = 0.092 × 931 MeV = 86.0 MeV; B.E./nucleon = 86.0/12 = 7.2 MeV.
(ii) $\Delta m = (82 \times 1.00728 + 124 \times 1.00867) - (205.97446 - 82 \times 0.00055) = 1.652$ u $\therefore$ B.E. = 1.652 × 931 MeV = 1540 MeV; B.E./nucleon = 1540/206 = 7.5 MeV.
(b) (i) See Fig. 23.5.
(ii) When a heavy nucleus fissions, the binding energy per nucleon rises as the fission fragments are more stable. The energy released is equal to the rise of the total binding energy.
23.3. (a) (i) ${}^{210}_{84}\text{Po} \rightarrow {}^{4}_{2}\alpha + {}^{206}_{82}\text{Pb}$.
(ii) Q = 931 × (209.98287 – 205.97446 – 4.00260) = 5.4 MeV.
(b) (i) ${}^{24}_{11}\text{Na} \rightarrow {}_{-1}^{0}\beta + {}^{24}_{12}\text{Mg}$.
(ii) Q = 931 × (23.99097 – 23.98505) = 5.5 MeV.
23.4. (a) (i) Fission neutrons go on to produce further fission which releases more neutrons which produce further fission, etc.
(ii) Control rods are inserted into the reactor core to absorb surplus neutrons so that exactly 1 neutron per fission goes on to produce further fission.
(b) (i) The moderator reduces the k.e. of the fission neutrons so they produce further fission of U 235.
(ii) Low neutron absorption, high melting point if solid, high specific heat capacity.
(c) (i) Spent fuel rods contain radioactive fission fragments, and plutonium 239 and uranium 238 which are highly radioactive.

(ii) Decay constant, $\lambda = \frac{\log_e 2}{T_{1/2}} = \frac{0.693}{100 \text{ yrs}}$;

number of atoms in 1 kg, $N = \frac{6.02 \times 10^{23}}{0.235}$;

activity $= \lambda N = \frac{0.693}{100 \times 365.25 \times 24 \times 3600 \text{ s}} \times \frac{6.02 \times 10^{23}}{0.235} = 5.6 \times 10^{14}$ Bq.

23.5. (a) (i) A ring of magnets forces the particles around a circular path. **(ii)** The speed increases until the particle radiates energy at the same rate as it gains it from the accelerator. The speed is then constant.

(b) (i) $E = mc^2 \quad \therefore m = \frac{E}{c^2} = \frac{20\,000 \text{ MeV}}{931} = 21.5$ u.

(ii) As it gains kinetic energy, its mass increases and its speed approaches closer to the speed of light. Its mass tends to infinity as it approaches the speed of light.

(c) (i) $Q = +\frac{2}{3}e - \frac{1}{3}e - \frac{1}{3}e = 0$; $S = 0 + 0 - 1 = -1$. **(ii)** $Q = +\frac{2}{3}e + \frac{1}{3}e = +e$; $S = 0 + 1$.

(d) (i) $S = 0$ so there are no strange quarks present. Since $Q = +2e$, there must be 3 up quarks present ($Q = 3 \times +\frac{2}{3}e$).

(ii) $S = +1$, so the meson contains the strange antiquark. The charge of this antiquark is $+\frac{1}{3}e$ so the meson must also include a down quark (charge $= -\frac{1}{3}e$) so the total charge is 0.

Unit 24

In-text questions

24.1A

1. (a) Initial volume of air = volume of pump + tyre = 1545 cm^3. Final volume of air in tyre = 1500 cm^3 $\therefore p \times 1500 = 110 \times 1545$,

so $p = \frac{110 \times 1545}{1500} = 113.3$ kPa.

(b) Pressure increase = 3.3 kPa $\therefore$ % increase of pressure $= \frac{3.3}{110} \times 100\% = 3\%$.

(c) Pressure after 20 strokes $= 103\%^{20} \times 110$ kPa $= 1.03^{20} \times 110$ kPa = 199 kPa.

2. (a) Initial volume of air $= V_F + V_P$ where V_F = volume of flask and V_P = volume of pump. Final volume of air $= V_F$ $\therefore 100\,(V_F + V_P) = 118\,V_F$ so $100\,V_P = 18\,V_F$.

$V_P = 0.18 V_F = 0.18 \times 250 \text{ cm}^3 = 45 \text{ cm}^3$.

(b) Initial volume of air = 295 cm^3 $- \upsilon$, where υ is the powder volume. Final volume of air = 250 cm^3 $- \upsilon$ $\therefore 100\,(295 - \upsilon) = 140\,(250 - \upsilon)$

$\therefore 29500 - 100\upsilon = 35000 - 140\upsilon$. Hence $40\upsilon = 5500$

so $\upsilon = \frac{5500}{40} = 138 \text{ cm}^3$.

24.1B

1. Use $\frac{p_2V_2}{T_2} = \frac{p_1V_1}{T_1}$ to give

(a) $V_2 = \frac{p_1V_1T_2}{p_2T_1} = \frac{100 \times 0.05 \times 350}{110 \times 300} = 0.054 \text{ m}^3$.

(b) $T_2 = \frac{p_2V_2T_1}{p_1V_1} = \frac{101 \times 0.12 \times 400}{105 \times 0.24} = 192$ K.

(c) $p_2 = \frac{p_1V_1T_2}{V_2T_1} = \frac{0.35 \times 0.85 \times 250}{0.58 \times 350} = 0.37$ kPa.

(d) $T_1 = \frac{p_1V_1T_2}{p_2V_2} = \frac{101 \times 0.42 \times 300}{101 \times 0.38} = 332$ K.

(e) $V_1 = \frac{p_2V_2T_1}{p_1T_2} = \frac{101 \times 0.161 \times 290}{110 \times 273} = 0.157 \text{ m}^3$.

2. (a) $p_1 = 105$ kPa; $V_1 = 20 \text{ cm}^3 = 20 \times 10^{-6} \text{ m}^3$, $T_1 = 288$ K; $p_2 = 101$ kPa; V_2 = to be calculated; $T_2 = 273$ K.

Using $\frac{p_2V_2}{T_2} = \frac{p_1V_1}{T_1}$ gives $\frac{101 \times 10^3 \times V_2}{273} = \frac{105 \times 10^3 \times 2.0 \times 10^{-5}}{288}$.

Hence $V_2 = \frac{105 \times 10^3 \times 2.0 \times 10^{-5} \times 273}{288 \times 101 \times 10^3} = 1.97 \times 10^{-5} \text{ m}^3 = 19.7 \text{ cm}^3$.

(b) $p_1 = 101$ kPa; $V_1 = 30 \text{ cm}^3 = 30 \times 10^{-6} \text{ m}^3$; $T_1 = 20 + 273 = 293$ K; p_2 = to be calculated; $V_2 = 31 \text{ cm}^3$; $T_2 = 100 + 273 = 373$ K.

Using $\frac{p_2V_2}{T_2} = \frac{p_1V_1}{T_1}$ gives $\frac{p_2 \times 3.1 \times 10^{-5}}{273} = \frac{101 \times 10^3 \times 3.0 \times 10^{-5}}{293}$.

Hence $p_2 = \frac{101 \times 10^3 \times 3.0 \times 10^{-5} \times 373}{293 \times 3.1 \times 10^{-5}} = 124$ kPa.

24.2

1. (a) Rearrange $pV = nRT$ to give

$n = \frac{pV}{RT} = \frac{120 \times 10^3 \times 200 \times 10^{-6}}{8.31 \times (273 + 15)} = 1.00 \times 10^{-2}$ mol.

(b) V is constant $\therefore \frac{p_1}{T_1} = \frac{p_2}{T_2}$ where $p_1 = 120$ kPa, $p_2 = 150$ kPa, $T_1 = 288$ K, $T_2 = ?$

Rearranging this equation gives $T_2 = \frac{p_2T_1}{p_1} = \frac{150 \times 10^3 \times 288}{120 \times 10^3} = 360$ K.

(c) (i) At 100°C, T = 373 K, V = 200 cm^3 $= 200 \times 10^{-6} \text{ m}^3$, p = 150 kPa

$\therefore n = \frac{pV}{RT} = \frac{150 \times 10^3 \times 200 \times 10^{-6}}{8.31 \times 373} = 9.7 \times 10^{-3}$ mol.

Number of moles lost $= 1.00 \times 10^{-2} - 9.7 \times 10^{-3} = 3 \times 10^{-4}$ mol.

(ii) Fraction of gas lost $= \frac{\text{number of moles lost}}{\text{initial number of moles}} = \frac{3 \times 10^{-4}}{1.00 \times 10^{-2}} = 0.03$.

2. (i) $V_m = \frac{RT}{p} = \frac{8.31 \times 273}{101 \times 10^3} = 0.0224 \text{ m}^3$;

density, $\rho = \frac{\text{mass}}{\text{volume}} = \frac{0.028}{0.0224} = 1.25 \text{ kg m}^{-3}$.

(ii) $V_m = \frac{RT}{p} = \frac{8.31 \times 373}{101 \times 10^3} = 0.0306 \text{ m}^3$;

density, $\rho = \frac{\text{mass}}{\text{volume}} = \frac{0.028}{0.0306} = 0.91 \text{ kg m}^{-3}$.

24.3

1. (a) (i) Use $pV_m = RT$ to calculate the volume of 1 mole at 0°C and 101 kPa pressure.

Hence $V_m = \frac{RT}{p} = \frac{8.31 \times (273 + 100)}{150 \times 10^3} = 2.07 \times 10^{-2} \text{ m}^3$

$\therefore$ density, $\rho = \frac{\text{mass}}{\text{volume}} = \frac{\text{molar mass}}{\text{molar volume}} = \frac{0.028 \text{ kg}}{2.07 \times 10^{-2} \text{ m}^3} = 1.36 \text{ kg m}^{-3}$.

(ii) Rearrange $p = \frac{1}{3}\rho \upsilon_{RMS}^2$ to give

$\upsilon_{RMS}^2 = \frac{3p}{\rho} = \frac{3 \times 150 \times 10^3}{1.36} = 3.31 \times 10^5 \text{ m}^2 \text{ s}^{-2}$ $\therefore \upsilon_{RMS} = 575 \text{ m s}^{-1}$.

(b) (i) Mean k.e. $= \frac{3}{2}kT = 1.5 \times 1.38 \times 10^{-23} \times (100 + 273) = 7.72 \times 10^{-21}$ J.

(ii) Since the mean k.e. is proportional to T, then the absolute temperature needs to be doubled to double the mean k.e., i.e. T needs to be raised to 746 K (= 2 × 373 K).

2. (a) (i) At 0°C, mean k.e. $= \frac{3}{2}kT = 1.5 \times 1.38 \times 10^{-23} \times 273 = 5.65 \times 10^{-21}$ J.

(ii) For 1 mole, $\frac{1}{2}N_A m \upsilon_{RMS}^2 = \frac{3}{2}RT$ and $N_A m$ = molar mass, M

$\therefore \frac{1}{2}M\upsilon_{RMS}^2 = \frac{3}{2}RT$ so $\upsilon_{RMS}^2 = \frac{3RT}{M} = \frac{3 \times 8.31 \times 273}{0.002}$

$= 3.40 \times 10^6 \text{ m}^2 \text{ s}^{-2}$. $\upsilon_{RMS} = 1.84 \times 10^3 \text{ m s}^{-1}$.

(b) (i) At 100°C, mean k.e. $= \frac{3}{2}kT = 1.5 \times 1.38 \times 10^{-23} \times (100 + 273) = 7.72 \times 10^{-21}$ J.

(ii) $\upsilon_{RMS}^2 = \frac{3RT}{M} = \frac{3 \times 8.31 \times 373}{0.002} = 4.65 \times 10^6 \text{ m}^2 \text{ s}^{-2}$.

$\upsilon_{RMS} = 2.16 \times 10^3 \text{ m s}^{-1}$.

Revision questions

24.1 (a) (i) Rearrange $\frac{p_1V_1}{T_1} = \frac{p_2V_2}{T_2}$ with $p_1 = 103$ kPa, $V_1 = 40 \text{ cm}^3 = 4.0 \times 10^{-5} \text{ m}^3$, $T_1 = 273 + 25 = 298$ K, $p_2 = 101$ kPa, $T_2 = 273$ K, V_2 to be calculated.

$\therefore V_2 = \frac{p_1V_1T_2}{p_2T_1} = \frac{103 \times 10^3 \times 4.0 \times 10^{-5} \times 273}{101 \times 10^3 \times 298} = 3.7 \times 10^{-5} \text{ m}^3$.

(ii) Rearrange $pV = nRT$ to give
$n = \frac{pV}{RT} = \frac{103 \times 10^3 \times 4.0 \times 10^{-5}}{8.31 \times 298} = 1.66 \times 10^{-3}$ moles.
Mass = nM, where M = molar mass $\therefore$ mass $= 1.66 \times 10^{-3} \times 0.032$
$= 5.3 \times 10^{-5}$ kg.

(b) (i) Rearrange $\frac{p_1V_1}{T_1} = \frac{p_2V_2}{T_2}$ with $p_1 = 101$ kPa, $V_1 = 1.2\ cm^3$
$= 1.2 \times 10^{-6}\ m^3$, $T_1 = 273 + 10 = 283$ K, $p_2 = 200$ kPa, $T_2 = 273 + 20$
$= 293$ K, V_2 to be calculated.
$\therefore V_2 = \frac{p_1V_1T_2}{p_2T_1} = \frac{101 \times 10^3 \times 1.2 \times 10^{-6} \times 293}{200 \times 10^3 \times 283} = 6.3 \times 10^{-7}\ m^3$.

(ii) Rearrange $pV = nRT$ to give $n = \frac{pV}{RT} = \frac{101 \times 10^3 \times 1.2 \times 10^{-6}}{8.31 \times 283} = 5.2 \times 10^{-5}$.
Mass = nM, where M = molar mass $\therefore$ mass $= 5.2 \times 10^{-5} \times 0.029$
$= 1.5 \times 10^{-6}$ kg.

24.2. (a) (i) Rearrange $\frac{p_1V_1}{T_1} = \frac{p_2V_2}{T_2}$ with $p_1 = 150$ kPa, $V_1 = 1600\ cm^3$
$= 1.60 \times 10^{-3}\ m^3$, $T_1 = 273 + 25 = 298$ K, $V_2 = 1500\ cm^3$
$= 1.5 \times 10^{-3}\ m^3$, $T_2 = 273$ K, p_2 to be calculated.
$\therefore p_2 = \frac{p_1V_1T_2}{V_2T_1} = \frac{150 \times 10^3 \times 1.6 \times 10^{-3} \times 273}{1.5 \times 10^{-3} \times 298} = 1.47 \times 10^5$ Pa.

(ii) Rearrange $pV = nRT$ to give
$n = \frac{pV}{RT} = \frac{150 \times 10^3 \times 1.6 \times 10^{-3}}{8.31 \times 298} = 9.69 \times 10^{-2}$ moles.
Mass = nM, where M = molar mass $\therefore$ mass $= 9.69 \times 10^{-2} \times 0.029$
$= 2.8 \times 10^{-3}$ kg.

(b) Sealed end lower than the upper end: the pressure in the trapped air, in millimetres of mercury, $p_1 = p_o + L$, where L is the length of the mercury thread, and p_o is the atmospheric pressure in millimetres of mercury. The volume of the trapped air column, V_1 = length of air column × area of cross-section, A = 0.12A.
Sealed end above the lower end: the pressure in the trapped air, in millimetres of mercury, $p_2 = p_o - L$, where L is the length of the mercury thread, and p_o is the atmospheric pressure in millimetres of mercury. The volume of the trapped air column, V_2 = length of air column × area of cross-section, A = 0.18A.
Using $p_1V_1 = p_2V_2$ gives $(p_o + L) \times 0.12A = (p_o - L) \times 0.18A$
$\therefore 0.18p_o - 0.12p_o = 0.18L + 0.12L$
$0.06p_o = 0.30L \quad \therefore p_o = \frac{0.30}{0.06}L = 5L = 5 \times 0.150\ m = 0.750\ m$
= 750 mm of mercury.

24.3. (a) and **(b)** See Fig. A24.1.

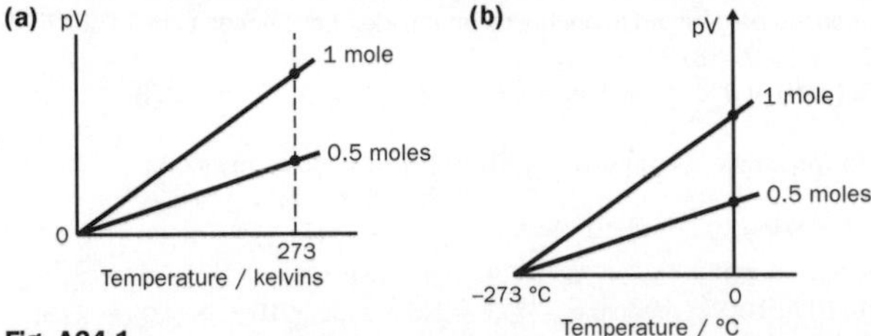

Fig. A24.1.

24.4. (a) (i) Rearrange $pV = nRT$ to give
$n = \frac{pV}{RT} = \frac{140 \times 10^3 \times 500 \times 10^{-6}}{8.31 \times (20 + 273)} = 2.87 \times 10^{-2}$ moles $\therefore$ mass = nM,
where M = molar mass, hence mass $= 2.87 \times 10^{-2} \times 0.028$ kg
$= 8.0 \times 10^{-4}$ kg.

(ii) Number of molecules $= n \times N_A = 2.87 \times 10^{-2} \times 6.02 \times 10^{23}$
$= 1.73 \times 10^{22}$.

(iii) Rearrange $pV = \frac{1}{3}Nm{v_{RMS}}^2$ to give
${v_{RMS}}^2 = \frac{3pV}{Nm}$, where Nm = the total mass of gas
$\therefore {v_{RMS}}^2 = \frac{3 \times 140 \times 10^3 \times 500 \times 10^{-6}}{8.0 \times 10^{-4}} = 2.63 \times 10^5$,
hence $v_{RMS} = 512\ m\ s^{-1}$.

(b) (i) Use $\frac{p_2}{T_2} = \frac{p_1}{T_1}$ where $p_1 = 140$ kPa, $p_2 = 170$ kPa, $T_1 = (20 + 273)$
$= 293$ K and T_2 is to be calculated.
$T_2 = \frac{p_2T_1}{p_1} = \frac{170 \times 293}{140} = 356\ K = 83°\ C$.

(ii) With the valve open, the pressure and volume remain constant and the number of moles falls. Using $pV = nRT$, $nT = \frac{pV}{R}$ = constant, hence $n_2T_2 = n_1T_1$
$\therefore n_2 = \frac{n_1T_1}{T_2} = \frac{2.87 \times 10^{-2} \times 356}{(273 + 100)} = 2.74 \times 10^{-2}$ moles $\therefore$ number of moles lost $= 2.87 \times 10^{-2} - 2.74 \times 10^{-2} = 1.3 \times 10^{-3}$ moles
$\therefore$ mass loss $= 1.3 \times 10^{-3} \times 0.028$ kg $= 3.6 \times 10^{-5}$ kg

24.5. (a) (i) The molecules hit the sides less often because they have more distance to travel between successive impacts on average, so the rate of impacts is less and hence the pressure is less.

(ii) The molecules hit the sides less often because they are moving more slowly and the average impact force is less because the molecules move more slowly. Hence the pressure is less because the rate of impacts is smaller and the force of each impact is less.

(b) Rearrange $pV = \frac{1}{3}Nm{v_{RMS}}^2$ to give ${v_{RMS}}^2 = \frac{3pV}{Nm} = \frac{3nRT}{Nm} = \frac{3RT}{N_Am} = \frac{3RT}{M}$
where $M = N_Am$ = the molar mass.

(i) M = 0.002 kg $\therefore {v_{RMS}}^2 = \frac{3RT}{M} = \frac{3 \times 8.31 \times (273 + 20)}{0.002}$
$= 3.65 \times 10^6\ m^2\ s^{-2}$ $\therefore v_{RMS} = 1910\ m\ s^{-1}$.

(ii) M = 0.032 kg $\therefore {v_{RMS}}^2 = \frac{3RT}{M} = \frac{3 \times 8.31 \times (273 + 20)}{0.032}$
$= 2.28 \times 10^5\ m^2\ s^{-2}$ $\therefore v_{RMS} = 478\ m\ s^{-1}$.

(iii) The rms speed of hydrogen molecules is over 4× more than the rms speed of oxygen molecules at the same temperature. Hydrogen molecules are moving fast enough to escape from the Earth's gravity whereas oxygen molecules are not.

Unit 25

In-text questions

25.1

1. (a) $t = \frac{h - h_o}{h_{100} - h_o} \times 100\ °C = \frac{105 - -50}{220 - -50} \times 100°C = \frac{155}{270} \times 100 = 57.4°C$.

(b) (i) Midway = 50 degrees.

(ii) $t = \frac{h - h_o}{h_{100} - h_o} \times 100\ °C = \frac{80 - -50}{220 - -50} \times 100°C = \frac{130}{270} \times 100 = 48.1°C$.

2. $E = at - bt^2$, where $a = 40.5\ \mu V\ °C^{-1}$ and $b = 0.065\ \mu V\ °C^{-2}$.

(a)

t/°C	0	20	40	60	80	100
E/μV	0	784	1516	2196	2824	3400

(b) At E = 1700 μV, t = 45°C (to within 0.5°C).

25.2

1. (a) (i) $\Delta Q = \Delta U + \Delta W = +1000 + 300 = +1300$ J; i.e. 1300 J of heat transfer to the body.

(ii) $\Delta Q = \Delta U + \Delta W = +1000 - 300 = +700$ J; i.e. 700 J of heat transfer to the body.

(b) (i) $\Delta Q = \Delta U + \Delta W = -500 + 200 = -300$ J; i.e. 300 J of heat transfer from the body.

(ii) $\Delta Q = \Delta U + \Delta W = -500 - 200 = -700$ J; i.e. 700 J of heat transfer from the body.

2. (a) Rearrange $\Delta Q = \Delta U + \Delta W$ to give $\Delta W = \Delta Q - \Delta U = 400 - 2000$
$= -1600$ J (ie. 1600 J of work is done on the body).

(b) $\Delta Q = \Delta U + \Delta W = -2000 + 2400 = +400$ J (i.e. heat transfer = 400 J into the body).

(c) Rearrange $\Delta Q = \Delta U + \Delta W$ to give $\Delta U = \Delta Q - \Delta W = -800 - -2400 = 1600$ J (i.e. heat transfer into body = 1600 J).
(d) Rearrange $\Delta Q = \Delta U + \Delta W$ to give $\Delta W = \Delta Q - \Delta U = -400 - -1000 = 600$ J (i.e. 600 J of work is done by the body).

25.3

1. (a) (i) $\frac{V}{T}$ = constant $\therefore T_2 = \frac{V_2}{V_1}T_1 = \frac{360}{240} \times (273 + 17) = 435$ K.
(ii) $\Delta W = p\Delta V = 80 \times 10^3 (360 \times 10^{-6} - 240 \times 10^{-6}) = 9.6$ J
$\Delta U = \frac{3}{2}nR\,\Delta T = \frac{3}{2}p\Delta V = 1.5 \times 9.6 = 14.4$ J
$\Delta Q = \Delta U + \Delta W = 24.0$ J.
(iii) See Fig. A25.1.

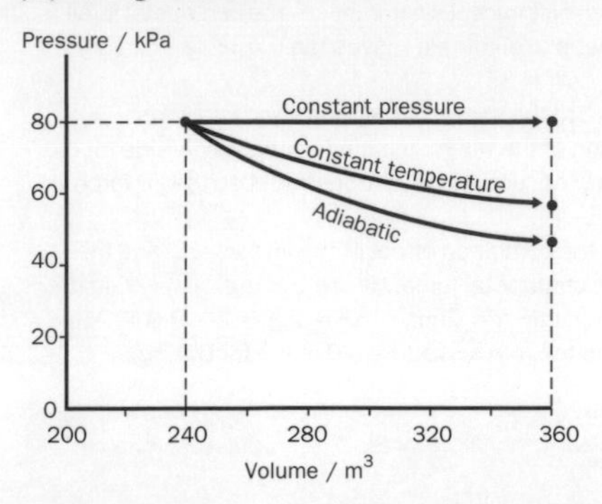

Fig. A25.1.

(b) (i) pV = constant $\therefore p_2 = \frac{p_1V_1}{V_2} = \frac{80 \times 10^3 \times 240 \times 10^{-6}}{360 \times 10^{-6}} = 53.3$ kPa.
(ii) See Fig. A25.1.
(iii) Estimate the area under the curve to give $\Delta W = 7.8$ J, $\Delta U = 0$
$\therefore \Delta Q = 7.8$ J.
(c) (i) Use pV^γ = constant, hence $p \times 360^{1.4} = 80 \times 240^{1.4}$

$\therefore p \times 3790 = 80 \times 2149 \quad \therefore p = \frac{80 \times 2149}{3790} = 45$ kPa.

To find T, use pV = nRT $\therefore \frac{pV}{T}$ = constant so $\frac{45 \times 360}{T} = \frac{80 \times 240}{(273 + 17)}$;

$T = \frac{45 \times 360 \times 290}{80 \times 240} = 245\text{ K} = -28°\text{C}$.
(ii) See Fig. A25.1.
(iii) $\Delta U = \frac{3}{2}nR\Delta T = \frac{3}{2}\frac{p_1V_1}{T_1}\Delta T = 1.5 \times \frac{80 \times 10^3 \times 240 \times 10^{-6}}{290} \times$
$(245 - 290) = -4.5$ J; $\Delta Q = 0$ so $\Delta W = -\Delta U = 4.5$ J.

2. (a) (i) Rearrange pV = nRT to give $n = \frac{pV}{RT} = \frac{110 \times 10^3 \times 1200 \times 10^{-6}}{8.31 \times (273 + 15)}$
= 0.055 moles.
(ii) Mass of air = n × molar mass = $0.055 \times 0.029\text{ kg} = 1.6 \times 10^{-3}$ kg.
(iii) Internal energy, $U = \frac{5}{2}nRT = \frac{5}{2}pV = 2.5 \times 110 \times 10^3 \times 1200 \times 10^{-6} = 330$ J.
(b) (i) Use pV^γ = constant, hence $p \times 150^{1.4} = 110 \times 1200^{1.4}$

$\therefore p \times 1110 = 110 \times 20460 \quad \therefore p = \frac{110 \times 20460}{1110} = 2030$ kPa.

To find the temperature, use pV = nRT
$\therefore \frac{pV}{T}$ = constant, so $\frac{2030 \times 150}{T} = \frac{110 \times 1200}{(273 + 15)}$;

$T = \frac{2030 \times 150 \times 288}{110 \times 1200} = 664\text{ K} = 391°\text{C}$.
(ii) $\Delta U = \frac{3}{2}nR\,\Delta T = \frac{3}{2} \times 0.055 \times 8.31 \times (664 - 288) = 258$ J.
(iii) $\Delta Q = 0$, so $\Delta W = -\Delta U = -258$ J.

25.4

1. (a) (i) 500 s (= 15 000 m/15 m s^{-1}).
(ii) 20 MJ/500 s = 40 kW.
(b) Efficiency = $\frac{\text{output power}}{\text{energy supplied per second}} = \frac{6\text{ kW}}{40\text{ kW}} = 0.15$.
2. (a) $Q_1 = 160\text{ W} + 40\text{ W} = 200$ W.
(b) (i) Efficiency = $\frac{40\text{ W}}{200\text{ W}} = 0.20$.
(ii) Maximum possible efficiency = $\frac{400 - 300}{400} = 0.25$.

Revision questions

25.1. (a) (i) $t = \frac{20 - -6}{29 - -6} \times 100 = 74°$C. **(ii)** $T = t + 273 = 347$ K.
(b) (i) $V = 1.25 \times 50 \times (1 - 50 \times 20 \times 10^{-4}) = 56.3$ mV.
(ii) Centigrade temperature = 56.3 degrees.

25.2. (a) $\Delta Q = \Delta U + p\Delta V = 1000 + (100 \times 10^3 \times 2.0 \times 10^{-3}) = 1200$ J.
(b) $\Delta U = \Delta Q - p\Delta V = -600 - (150 \times 10^3 \times -2.0 \times 10^{-3}) = -600 + 300 = -300$ J.
(c) $p\Delta V = \Delta Q - \Delta U \quad \therefore 100 \times 10^3\,\Delta V = 200 - -800 = 1000$
$\therefore \Delta V = 10 \times 10^{-3}\text{ m}^3$.
(d) $p\Delta V = \Delta Q - \Delta U \quad \therefore p \times -5.0 \times 10^{-3} = 400 - 1200 = -800$
$\therefore p = 160$ kPa.

25.3. (a) (i) Rearrange pV = nRT to give
$n = \frac{pV}{RT} = \frac{10 \times 10^3 \times 60 \times 10^{-6}}{8.31 \times 290} = 2.5 \times 10^{-4}$.
(ii) $U = \frac{3}{2}nRT = 1.5 \times 2.5 \times 10^{-4} \times 8.31 \times 290 = 0.90$ J.
(b) (i) pV = nRT for constant mass and volume becomes $\frac{p}{T}$ = constant

$\therefore p_2 = \frac{T_2}{T_1}p_1 = \frac{380}{290} \times 10\text{ kPa} = 13.1$ kPa.
(ii) $U_1 = 0.90$ J; $U_2 = \frac{3}{2}nRT = 1.5 \times 2.5 \times 10^{-4} \times 8.31 \times 380 = 1.18$ J
$\therefore \Delta U = 0.28$ J.

25.4. (a) (i) Rearrange pV = nRT to give
$n = \frac{pV}{RT} = \frac{101 \times 10^3 \times 1.00 \times 10^{-4}}{8.31 \times (273 + 20)} = 4.1 \times 10^{-3}$ mol.
(ii) For a diatomic gas, $U = \frac{5}{2}nRT = 2.5 \times 4.1 \times 10^{-3} \times 8.31 \times 293 = 25.3$ J.
(b) (i) Use pV^γ = constant where $\gamma = 1.4 \quad \therefore p \times 20^{1.4} = 101 \times 100^{1.4}$
$\therefore 66.3\,p = 101 \times 631$, so $p = \frac{101 \times 631}{66.3} = 961$ kPa.

To calculate T, use pV = nRT, hence
$T = \frac{pV}{nR} = \frac{961 \times 10^3 \times 20 \times 10^{-6}}{4.1 \times 10^{-3} \times 8.31} = 564$ K.
(ii) $U = \frac{5}{2}nRT = 2.5 \times 4.1 \times 10^{-3} \times 8.31 \times 564 = 48.7$ J.
(iii) $\Delta W = -\Delta U$ as $\Delta Q = 0 \quad \therefore \Delta W = -(48.7 - 25.3) = -23.4$ J
$\therefore$ work done on air = 23.4 J.

25.5. (a) (i) $Q_2 = 2000 - 600 = 1400$ J in 1 s.
(ii) Efficiency = $\frac{600}{2000} = 0.30$.
(iii) Maximum possible efficiency = $\frac{450 - 300}{450} = 0.33$.
(b) (i) Consider A to B; $T_B = 1000$ K,
$p_B = p_A \times \frac{T_B}{T_A} = 100\text{ kPa} \times \frac{1000}{300} = 333$ kPa.
Consider B to C; Using pV^γ = constant $\therefore 333 \times 0.0020^\gamma = 100 \times V_c^\gamma$,
hence $V_c^{1.4} = 3.33 \times 0.0020^{1.4} = 5.54 \times 10^{-4} \quad \therefore 1.4 \ln V_C = \ln 5.54 \times 10^{-4} = -7.50 \quad \therefore \ln V_C = -5.36 \quad \therefore V_C = 4.7 \times 10^{-3}\text{ m}^3$.
Using pV = nRT, $T_C = \frac{p_cV_c}{nR} = \frac{p_cV_c T_B}{p_BV_B} = \frac{100 \times 4.7 \times 10^{-3} \times 1000}{333 \times 0.0020} = 706$ K.
(ii) Maximum possible efficiency = $\frac{1000 - 300}{1000} = 0.70$.
(iii) Estimating the work done from the 'area' of the loop gives about 220 J.
(iv) Heat gain occurs between A and B; as no work is done, then
$\Delta Q_1 = \Delta U = \frac{5}{2}nR\,(T_B - T_A) = \frac{\frac{5}{2}p_AV_A(T_B - T_A)}{T_A}$
$= \frac{2.5 \times 100 \times 10^3 \times 0.0020 \times 700}{300} = 1170$ J.

(v) Efficiency $= \frac{W}{Q_1} = \frac{220}{1170} = 0.19$.

Unit 26

In-text questions

26.1

1. (a) $T = \frac{1}{f} = \frac{60}{800} = 0.075$ s.

(b) $\omega = 2\pi f = 2\pi \times \frac{800}{60} = 83.8$ rad s^{-1}.

(c) $\upsilon = \omega r = 83.8 \times 0.20 = 16.8$ m s^{-1}.

2. (a) $f = \frac{3000}{60} = 50$ Hz, $T = \frac{1}{f} = 0.02$ s.

(b) $\omega = 2\pi f = 2\pi \times 50 = 314$ rad s^{-1}.

3. (a) (i) $C = 2\pi r = \pi \times 0.040 = 0.126$ m.

(ii) $n = \frac{\text{tape length}}{C} = \frac{2.0}{0.126} = 15.9$.

(b) (i) $\upsilon = 2\pi f r = 2\pi \times 2 \times 0.020 = 0.25$ m s^{-1}. **(ii)** $t = \frac{2.0\text{ m}}{0.25\text{ m s}^{-1}} = 8.0$ s.

(c) (i) $\theta = 0.9$ of a turn $\times 2\pi = 5.75$ rad. **(ii)** $\theta = 0.9 \times 360° = 330°$

$\therefore$ angle between ends = 30°.

26.2

1. (a) $a = \frac{\upsilon^2}{r} = \frac{20 \times 20}{30} = 13.3$ m s^{-2}.

(b) $F = ma = 800 \times 13.3 = 10\,700$ N.

2. $F = ma = \frac{m\upsilon^2}{r}$ $\therefore \upsilon^2 = \frac{Fr}{m} = 0.5mgr/m = 0.5\,gr = 0.5 \times 9.8 \times 30 = 147$

$\therefore \upsilon = 12.1$ m s^{-1}.

3. (b) $\upsilon = \sqrt{gr} = \sqrt{(9.8 \times 6400\,000)} = 7920$ m s^{-1}.

26.3

1. (a) $F = mg - \frac{m\upsilon^2}{r} = 2000 \times 9.8 - \frac{2000 \times 15^2}{25} = 1600$ N.

(b) $F = mg - \frac{m\upsilon^2}{r} = 0$ at the maximum speed. $\upsilon_{MAX} = \sqrt{gr} = \sqrt{9.8 \times 25}$

$= 15.6$ m s^{-1}.

2. Centripetal acceleration $= \frac{\upsilon^2}{r} = \frac{95^2}{450} = 20.0$ m s^{-2}

$\therefore$ extra '*g* force' $= (20/9.8)\,g = 2.0\,g$

Revision questions

26.1. (a) Time for one rotation, $T = 2 \times 60 \times 60 = 7200$ s

$\therefore \omega = 2\pi/T = 2\pi/7200 = 8.7 \times 10^{-4}$ rad s^{-1}.

(b) Speed, $\upsilon = \omega r = 8.7 \times 10^{-4} \times 8000 \times 10^3 = 7.0 \times 10^3$ m s^{-1}.

(c) Centripetal acceleration, $a = \frac{\upsilon^2}{r} = \frac{(7.0 \times 10^3)^2}{8 \times 10^6} = 6.1$ m s^{-2}.

26.2. (a) Angular speed, $\omega = 2\pi/T = 2\pi f = 2\pi \times 50 = 314$ rad s^{-1}.

(b) Speed, $\upsilon = \omega r = 314 \times 0.04 = 12.6$ m s^{-1}.

(c) Centripetal acceleration, $a = \upsilon^2/r = 12.6^2/0.04 = 3.9 \times 10^3$ m s^{-2}.

26.3. (a) Centripetal acceleration, $a = \upsilon^2/r = 15^2/60 = 3.75$ m s^{-2}.

(b) At its maximum speed, υ_{max}, the centripetal force, $m\upsilon^2_{max}/r = 0.5mg$.

$\therefore \upsilon^2_{max} = 0.5gr = 0.5 \times 9.8 \times 60 = \sqrt{294}$. Hence $\upsilon_{max} = \sqrt{294}$

$= 17$ m s^{-1}.

26.4. (a) (i) Gain of kinetic energy = loss of potential energy, hence the speed at the lowest point, υ, is given by the equation $\frac{1}{2}m\upsilon^2 = mgh$, where h is the drop in height. $\therefore \upsilon = \sqrt{2gh} = \sqrt{(2 \times 9.8 \times 45)} = 30$ m s^{-1}.

(ii) Frictional and drag forces are negligible.

(b) (i) Centripetal acceleration, $a = \frac{\upsilon^2}{r} = \frac{(2 \times 9.8 \times 45)}{25} = 35$ m s^{-2}.

(ii) Ratio $= (m\upsilon^2/r)/mg = \upsilon^2/rg = 35/9.8 = 3.6$.

26.5. (a) The horizontal distance between the two ends of each cable $= 5 \sin 30$ $\therefore$ radius of rotation, $r = 7.5 + 5 \sin 30 = 10.0$ m.

(b) The vertical component of the tension, T, in each cable, $T\cos 30 = mg$, where m is the total mass of each aircraft and occupants.

The horizontal component of the tension, $T\sin 30$ = centripetal force $= m\upsilon^2/r$, where υ is the speed of each aircraft.

Combining these two equations to eliminate T gives $m\upsilon^2/r = mg\tan 30$

$\therefore \upsilon^2 = gr\tan 30 = 9.8 \times 10 \times \tan 30 = 56$ m^{2}s^{-2}.

Hence $\upsilon = \sqrt{56} = 7.5$ m s^{-1}.

(c) Time for one rotation = circumference/speed $= 2\pi \times 10/7.5 = 8.4$ s.

26.2. (a) Let L = the lift force, θ = the angle of banking, r = radius of curvature of the flight path.

The vertical component of the lift force, $L\cos\theta = mg$, the aircraft's weight, The horizontal component of the lift force, $L\sin\theta = m\upsilon^2/r$, the centripetal force. Combining these equations to eliminate L gives $\tan\theta = \upsilon^2/gr = 210^2/(9.8 \times 6400) = 0.703$ $\therefore \theta = 35°$.

(b) (i) The lift force is due to the shape of the wings and is always perpendicular to the plane of the wings, pushing on the 'underside' of the wing. When upside down, the 'underside' is uppermost so the lift force is downwards.

(ii) At the highest point, the combined effect of the lift force, L, and the weight, mg, provides the centripetal force. Hence $L + mg = m\upsilon^2/r$. At the top of the loop, for L = mg, $m\upsilon^2/r = 2mg$ $\therefore \upsilon^2 = 2gr = 2 \times 9.8 \times 250 = 4900$ m^2 s^{-2} $\therefore$ at the top, $\upsilon = \sqrt{4900} = 70$ m s^{-1} for L = mg.

Unit 27

In-text questions

27.1

1. (a) $F = \frac{Gm_1m_2}{r^2} = \frac{6.67 \times 10^{-11} \times 70 \times 5.98 \times 10^{24}}{(6380 \times 10^3)^2} = 686$ N.

(b) $F = \frac{Gm_1m_2}{r^2} = \frac{6.67 \times 10^{-11} \times 7.35 \times 10^{22} \times 5.98 \times 10^{24}}{(384\,000 \times 10^3)^2}$

$= 1.99 \times 10^{20}$ N.

(c) $F = \frac{Gm_1m_2}{r^2} = \frac{6.67 \times 10^{-11} \times 7.35 \times 10^{22} \times 70}{(384\,000 \times 10^3)^2} = 2.32 \times 10^{-3}$ N.

2. (a) Force due to the Sun, F_2 at a distance of 150×10^6 km

$= G\frac{m_1m_2}{r^2} = \frac{6.67 \times 10^{-11} \times 1.99 \times 10^{30} \times 1.0}{(1.50 \times 10^8 \times 10^3)^2} = 5.90 \times 10^{-3}$ N.

(b) Let d = the distance from the centre of the Earth to the point at which the force on a mass, m, due to the Earth is equal and opposite to the force due to the Sun. See Fig. A27.1.

Force on m due to Earth, $F_1 = \frac{GM_1m}{d^2}$

where M_1 is the mass of the Earth.

Force on m due to theSun, $F_2 = \frac{GM_2m}{(D-d)^2}$

where M_2 is the mass of the Sun and D is the distance between the Earth and the Sun.

$\therefore F_1 = F_2$ gives $\frac{GM_1m}{d^2} = \frac{GM_2m}{(D-d)^2}$.

Cancelling G and m therefore gives

$\frac{M_1}{d^2} = \frac{M_2}{(D-d)^2}$ $\therefore \frac{(D-d)^2}{d^2} = \frac{M_2}{M_1} = \frac{1.99 \times 10^{30}}{5.98 \times 10^{24}} = 3.33 \times 10^5$.

Sun
D
Mass m
d
Earth

Fig. A27.1.

Taking square roots of both sides gives $\frac{(D-d)}{d} = \sqrt{(3.33 \times 10^5)} = 577$

$\therefore D - d = 577\,d$, which gives $578\,d = D$

$\therefore d = \frac{D}{578} = \frac{150 \times 10^6}{578}$ km $= 260\,000$ km.

The distance from the Earth's surface is therefore d − 6380 km = 254 000 km.

27.2

1. (a) $g_S = \frac{GM}{r_S^2} = \frac{6.67 \times 10^{-11} \times 1.99 \times 10^{30}}{(696\,000 \times 10^3)^2} = 2.74 \times 10^2$ N kg^{-1}.

(b) $g = \frac{GM}{r^2} = \frac{6.67 \times 10^{-11} \times 1.99 \times 10^{30}}{(57.9 \times 10^6 \times 10^3)^2} = 0.0396\ \text{N kg}^{-1}$.

2. Let d represent the distance from the Sun to the point at which the gravitational field strengths are equal and opposite.

Hence $\frac{GM_1}{d^2} = \frac{GM_2}{(D-d)^2}$ where D is the distance between the Sun and Jupiter, M_1 is the mass of the Sun and M_2 is the mass of Jupiter.

Hence $\frac{(D-d)^2}{d^2} = \frac{M_2}{M_1}$.

Since $M_1 = 1.99 \times 10^{30}$ kg and $M_2 = 1.90 \times 10^{27}$ kg,

then $\frac{M_2}{M_1} = \frac{1.90 \times 10^{27}}{1.99 \times 10^{30}} = 9.55 \times 10^{-4}$

$\therefore \frac{D-d}{d} = \sqrt{(9.55 \times 10^{-4})} = 3.09 \times 10^{-2}$

$\therefore D - d = 0.0309\,d$ which gives $1.0309\,d = D$, thus $d = 0.970\,D$.

Since D = 778 million kilometres, then d = 755 million km.

27.3

1. (a) Work done $= \frac{GMm}{r_S} = \frac{6.67 \times 10^{-11} \times 7.35 \times 10^{22} \times 1.0}{1740 \times 10^3} = 2.82$ MJ.

(b) Work done $= \frac{GMm}{r} = \frac{6.67 \times 10^{-11} \times 1.99 \times 10^{30} \times 1.0}{150 \times 10^{11}} = 8.84$ MJ.

2. $R = \frac{2GM}{c^2} = \frac{2 \times 6.67 \times 10^{-11} \times 1.99 \times 10^{30}}{(3.0 \times 10^8)^2} = 2.95 \times 10^3$ m.

27.4

1. (a) $v^2 = \frac{GM}{r} = \frac{g_S r_S^2}{r} = \frac{9.8 \times (6380 \times 10^3)^2}{7380 \times 10^3} = 5.41 \times 10^7\ \text{m}^2\,\text{s}^{-2}$

$\therefore v = 7.35\ \text{km s}^{-1}$.

$T = \frac{2\pi r}{v} = \frac{2\pi(7380\,000)}{7350} = 6307\ \text{s} = 105$ minutes.

(b) $r^3 = \frac{GMT^2}{4\pi^2} = \frac{g_S r_S^2 T^2}{4\pi^2} = \frac{9.8 \times (6380 \times 10^3)^2 \times (4 \times 3600)^2}{4\pi^2}$

$= 2.10 \times 10^{21}\ \text{m}^3$.

$r = 12.8 \times 10^6$ m $\therefore$ its height, h = 12 800 − 6380 = 6420 m.

Its speed, $v = \frac{2\pi r}{T} = \frac{2\pi \times (12800 \times 10^3)}{4 \times 3600} = 5590\ \text{m s}^{-1}$.

2. (a) Rearrange $r^3 = \frac{GMT^2}{4\pi^2}$ to give

$T^2 = \frac{4\pi^2 r^3}{GM} = \frac{4\pi^2(1840 \times 10^3)^3}{6.67 \times 10^{-11} \times 7.35 \times 10^{22}} = 5.02 \times 10^7\ \text{s}^2$

$\therefore T = 7080$ s = 118 minutes.

(b) Rearrange $v^2 = \frac{GM}{r}$ to give $r = \frac{GM}{v^2} = \frac{6.67 \times 10^{-11} \times 7.35 \times 10^{22}}{(1\,000)^2}$

$= 4.90 \times 10^6$ m $\therefore$ h = 4900 − 1740 = 3160 km.

$T = \frac{2\pi r}{v} = \frac{2\pi\,(4.90 \times 10^6)}{1000} = 3.08 \times 10^4$ s = 513 minutes.

Revision questions

27.1. (a) (i) Force due to the Earth, F_1

$= \frac{G\,m_1 m_2}{r^2} = \frac{6.67 \times 10^{-11} \times 7.35 \times 10^{22} \times 5.98 \times 10^{24}}{(384\,000 \times 10^3)^2} = 1.99 \times 10^{20}$ N.

(ii) Force due to the Sun, F_2

$= \frac{G\,m_1 m_2}{r^2} = \frac{6.67 \times 10^{-11} \times 7.35 \times 10^{22} \times 1.99 \times 10^{30}}{(150 \times 10^9)^2} = 4.33 \times 10^{20}$ N.

(b) Overall force $= F_1 - F_2 = 2.34 \times 10^{20}$ N towards the Sun.

27.2. (a) r = 5.20 − 1.00 = 4.20 A.U;

$g = \frac{GM}{r^2} = \frac{6.67 \times 10^{-11} \times 318 \times 5.98 \times 10^{24}}{(4.20 \times 150 \times 10^9)^2} = 3.20 \times 10^{-7}\ \text{N kg}^{-1}$.

(b) (i) $\frac{GM_1}{d^2} = \frac{GM_2}{(D-d)^2}$ where M_1 = the mass of the Earth, M_2 = the mass of Jupiter = 318 M_1, d = distance to be calculated and D = distance from Earth to Jupiter at opposition = 5.20 − 1.00 = 4.20 A.U.

Hence $(D-d)^2 = \frac{M_2 d^2}{M_1} = 318\,d^2$ $\therefore (D-d) = 17.8\,d$ so $18.8\,d = D$

hence $d = \frac{4.20}{18.8} = 0.22$ A.U.

(ii) r = 1.00 + 0.22 = 1.22 A.U. = 1.83×10^{11} m.

$g_{SUN} = \frac{GM_{SUN}}{r^2} = \frac{6.67 \times 10^{-11} \times 1.99 \times 10^{30}}{(1.83 \times 10^{11})^2} = 3.96 \times 10^{-3}\ \text{N kg}^{-1}$.

27.3. (a) (i) $\frac{1}{2}mv^2 = 62.5$ MJ $\therefore$ for m = 1 kg, $v^2 = 2 \times 62.5 \times 10^6$, hence $v = 1.12 \times 10^4\ \text{m s}^{-1}$.

(ii) $\frac{1}{2}mv^2 = 2.8$ MJ $\therefore$ for m = 1 kg, $v^2 = 2 \times 2.8 \times 10^6$, hence $v = 2.37 \times 10^3\ \text{m s}^{-1}$.

(b) (i) Its initial kinetic energy of 2.8 MJ kg^{-1} is used to enable it to escape from the Moon. Its potential energy at the Earth is −62 500 MJ (= −62.5 MJ kg^{-1} × 1000 kg). Its potential energy due to the Earth when it is on the Moon is negligible. Therefore its kinetic energy just before reaching the Earth = 62 500 MJ.

(ii) $\frac{1}{2}mv^2 = 62\,500$ MJ $\therefore$ for m = 1000 kg, $v^2 = 2 \times 62.5 \times 10^6$, hence $v = 1.12 \times 10^4\ \text{m s}^{-1}$.

27.4. (a) (i) $g_S = \frac{GM}{r_S^2} = \frac{6.67 \times 10^{-11} \times 0.108 \times 5.98 \times 10^{24}}{(3400 \times 10^3)^2} = 3.73\ \text{N kg}^{-1}$.

(ii) $v_{esc} = \sqrt{(2g_S r_S)} = 5.04 \times 10^3\ \text{m s}^{-1}$.

(b) (i) $r^3 = \frac{GMT^2}{4\pi^2} = \frac{g_S r_S^2\,T^2}{4\pi^2} = \frac{3.73 \times (3400 \times 10^3)^2 \times (7.65 \times 3600)^2}{4\pi^2}$

$= 8.28 \times 10^{20}\ \text{m}^3$ $\therefore r = 9.39 \times 10^6$ m = 9390 km so h = 9390 − 3400 = 5990 km.

(ii) r = 20 100 + 3400 = 23 500 km $\therefore r^3 = \frac{GMT^2}{4\pi^2}$ to give

$T^2 = \frac{4\pi^2 r^3}{GM} = \frac{4\pi^2 r^3}{g_S r_S^2} = \frac{4\pi^2 \times (23\,500 \times 10^3)^3}{3.73 \times (3400 \times 10^3)^2} = 1.19 \times 10^{10}\ \text{s}^2$

$\therefore T = 1.09 \times 10^5$ s = 30.3 hours.

27.5. (a) (i) Rearrange $r^3 = \frac{GMT^2}{4\pi^2}$ to give

$M = \frac{4\pi^2 r^3}{GT^2} = \frac{4\pi^2 \times (354\,000 \times 10^3)^3}{6.67 \times 10^{-11} \times (5.88 \times 24 \times 3600)^2} = 1.02 \times 10^{26}$ kg.

(ii) $r^3 = \frac{GMT^2}{4\pi^2} = \frac{6.67 \times 10^{-11} \times 1.02 \times 10^{26} \times (359 \times 24 \times 3600)^2}{4\pi^2}$

$= 1.66 \times 10^{29}\ \text{m}^3$ $\therefore r = 5.49 \times 10^9$ m.

(b) (i) r = 6380 + 1680 = 8060 km.

$T^2 = \frac{4\pi^2 r^3}{GM} = \frac{4\pi^2 r^3}{g_S r_S^2} = \frac{4\pi^2 \times (8060 \times 10^3)^3}{9.80 \times (6380 \times 10^3)^2} = 5.14 \times 10^7\ \text{s}^2$

$\therefore T = 7170$ s.

(ii) The Earth turns through $\frac{7170}{24 \times 3600}$ of a full rotation in 7170 s.

Therefore successive transits of the equator are separated by distance, d

$= \frac{7170}{24 \times 3600} \times$ Earth's circumference ($2\pi \times 6380$ km) = 3330 km.

Unit 28

In-text questions

28.1

1. (a) (i) Northwards **(ii)** Southwards.

(b) (i) its displacement is reversing from northwards to southwards, its velocity is southwards.

2. (a) (i) Upwards **(ii)** Zero **(b)** Downwards **(ii)** Zero.

28.2

1. (a) (i) $T = \frac{12}{10} = 1.2$ s. **(ii)** $\omega = \frac{2\pi}{T} = 5.2\ \text{rad s}^{-1}$.

(b) (i) amplitude, r = 0.20 m. **(ii)** $v_{MAX} = \omega r = 5.2 \times 0.20 = 1.0(4)\ \text{m s}^{-1}$.

(ii) $a_{MAX} = (-)\omega^2 r = (-)5.2^2 \times 0.20 = 5.4\ \text{m s}^{-2}$.

2. (a) (i) $T = \frac{1}{40} = 0.025$ s. **(ii)** $\omega = \frac{2\pi}{T} = \frac{2\pi}{0.025} = 250\ \text{rad s}^{-1}$.

(b) $v_{MAX} = \omega r = 250 \times 8.0 = 2000\ \text{mm s}^{-1}$.

28.3

1. (a) $T = \frac{22.0\text{ s}}{20} = 1.1$ s. **(b)** $\omega = \frac{2\pi}{T} = 5.7$ rad s^{-1}.
(c) $v_{MAX} = \omega r = 5.7 \times 25 \times 10^{-3} = 0.14$ m s^{-1}.
(d) The maximum tension = weight of 0.20 kg + extra tension due to a displacement of 25 mm.
Weight of 0.20 kg = $mg = 0.20 \times 9.8 = 1.96$ N.
Extra tension = ks, where s = 25 mm = 0.025 m and k is calculated from
$T = 2\pi\sqrt{\frac{m}{k}}$
Hence $k = \frac{4\pi^2 m}{T^2} = 6.5$ N m^{-1}, hence the extra tension = 6.5×0.025
= 0.16 N ∴ the maximum tension = 1.96 + 0.16 = 2.1 N.
2. (a) $T = 2\pi\sqrt{\frac{L}{g}} = 2\pi\left(\frac{2.0}{9.8}\right)^{½} = 2.83$ s.
(b) (i) For Fig. 28.11, $h = L - L\cos\theta = 2.0 - 2.0\cos 5° = 7.6 \times 10^{-3}$ m = 7.6 mm.
(ii) Maximum displacement from equilibrium, $s = L\sin\theta$, provided θ is less than about 10°. Hence amplitude, $r = s_{MAX} = 2.0 \sin 5° = 0.17(4)$ m
$\therefore v_{MAX} = \omega r = \frac{2\pi r}{T} = \frac{2\pi}{2.83} \times 0.174 = 0.39$ m s^{-1}.

28.4

1. (a) Each panel of the machine is capable of vibrating at its own natural frequency of oscillation. If the motor frequency is equal to the natural frequency of a panel, that panel vibrates in resonance.
(b) Each panel needs to be fixed to the chassis at sufficient points to prevent resonance at any possible motor frequency.
2. (a) (i) Additional weight = $300g/4 = 735$ N. **(ii)** k = additional weight ÷ change of length of spring = 735 N/0.05 m = 1.47×10^4 N m^{-1}.
(b) (i) Time period, $T = 2\pi\left(\frac{m}{k}\right)^{½} = 2\pi\left(\frac{375}{1.47 \times 10^4}\right)^{½} = 1.0$ s.
(ii) At this speed, the car would hit a speed bump once every second, so it would resonate as its time period is 1.0 s.

Revision questions

28.1. (a) (i) $T = 15.0/20 = 0.75$ s. **(ii)** $\omega = 2\pi/T = 2\pi/0.75$ = 8.4 rad s^{-1}; $v_{MAX} = \omega r = 8.4 \times 0.100 = 0.84$ m s^{-1}.
(iii) $a_{MAX} = (-)\,\omega^2 r = 8.4^2 \times 0.100 = 7.1$ m s^{-2}.
(b) (i) See Fig. A28.1. **(ii)** X is any point where the displacement is zero.

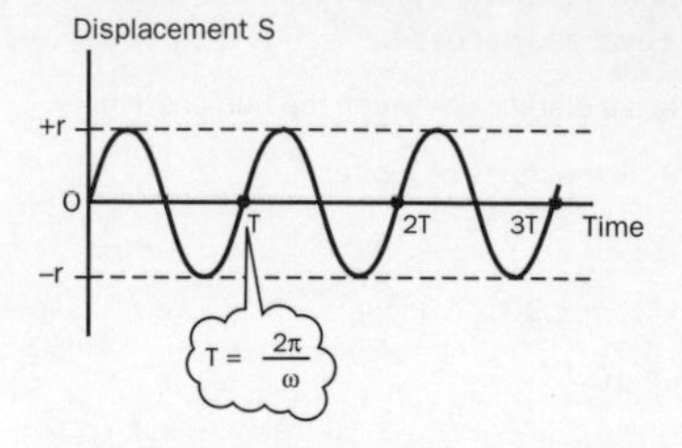

Fig. A28.1.

(iii) Y is any point where the displacement is a positive or a negative maximum.
28.2. (a) $k = 0.30 \times 9.8/0.090 = 33$ N m^{-1}.
(b) (i) $T = 2\pi(m/k)^{½} = 0.60$ s. **(ii)** $\omega = 2\pi/T = 2\pi/0.60$ = 10.5 rad s^{-1}; amplitude, $r = 0.060$ m ∴ $v_{MAX} = \omega r = 10.5 \times 0.060$ = 0.63 m s^{-1}.
28.3. (a) (i) $T = 3.0/2 = 1.50$ s. **(ii)** Rearrange $T = 2\pi(m/k)^{½}$ to give
$k = 4\pi^2 m/T^2 = \frac{4\pi^2 \times 900/4}{1.50^2} = 3950$ N m^{-1}.
(b) $m = 750/4$ kg, $k = 3950$ Nm^{-1}
$\therefore T = 2\pi(m/k)^{1/2} = 2\pi\left(\frac{750/4}{3950}\right)^{½} = 1.37$ s.
28.4. (a) Length of simple pendulum = 1.20 m, $T = 2\pi(L/g)^{½} = 2.2$ s.
(b) (i) For angles of no more than 10°, $s = L\sin\theta = 1.20\sin 10° = 0.21$ m.
(ii) $\omega = 2\pi/T = 2\pi/2.2 = 2.9$ rad s^{-1}; amplitude, $r = s_{MAX} = 0.21$ m
∴ $v_{MAX} = \omega r = 2.9 \times 0.21 = 0.60$ m s^{-1}.
28.5. (a) (i) Weight of sand = $120g$ = 1180 N, **(ii)** Additional weight on each spring = 1180/4 = 294 N ∴ $k = 294/0.020 = 1.47 \times 10^4$ N m^{-1}.
(iii) $T = 2\pi(m/k)^{½} = 2\pi(m/k)^{½} = 2\pi\left(\frac{130}{1.47} \times 10^4\right)^{½} = 0.59$ s.
(b) If the trailer is driven over speed bumps at a rate of 1 every 0.59 s, the trailer suspension system resonates.

Answers to Further Questions

Q1.1 (b) (i) 5.17×10^{14} Hz **(ii)** 3.18 m
Q1.2 (b) (i) 3.0 m **(ii)** $2\pi/3$ radians (= 120°)
Q2.1 (b) 86 dB
Q2.2 (b) (i) 0.40 ms **(ii)** 0.60 ms
Q2.3 (b) 3.6 m, 216 m s^{-1} **(ii)** 180 Hz
Q2.4 (b) (i) 340 m s^{-1} **(ii)** 360 Hz
Q3.1 (a) (i) 13° **(ii)** 31° **(b) (ii)** 42°
Q3.2 (a) (i) 1.5 mm **(ii)** 0.65 mm **(b)** 8.0 mm
Q3.3 (a)(i) v = −300 mm: virtual, upright, ×3 magnification
(ii) v = +300 mm: real, inverted, ×1 magnification
(b) v = −0.10 m
Q3.4 (b) 4.0° **(c)** 0.075 m
Q4.3 (a) 7.9° (33.4 − 25.5°) **(b)** 59.3° for 430 nm
Q5.1 (a) (i) 0.018 kg **(ii)** 0.044 kg
(b) (i) 3.0×10^{-26} kg
(ii) 0.31 nm
Q5.3 (b) 90% Ne 20, 10 % Ne 22
(ii) acts between the atoms
Q5.4 (a) (i) 1.0×10^{-5} m^3 **(b)** 6.0×10^{22}
(b) 0.26 nm
Q6.1 (a) 1.42 mm **(b)** 44 mm
Q6.2 (a) (i) 7.2 MJ **(ii)** 48°C **(b)** 383 s
Q6.3 (b) 128 W
Q6.4 (a) 5800 K **(b) (i)** 4.0×10^{26} W
(ii) 1.0
Q7.1 (b) (i) 325 mm **(ii)** 4.4 N
Q7.2 (b) (i) 1.0×10^9 Pa **(ii)** 102 N
Q7.3 (a) 3.2×10^8 Pa **(b)** 2.9 mm
Q7.4 (a) 1.26×10^{11} Pa **(b)** 0.6 J
Q8.1 (a) 40 kPa **(b)** 2.5 kPa
Q8.2 (a) 51.5 kPa
Q8.3 (b) (i) 500 kPa **(ii)** 3000 N
Q8.4 (a) (i) 490 Pa **(ii)** 36.7 m^2 **(b)** 72 kN
Q9.1 (a) (i) 6.93 N due North, 4.0 N **(ii)** 2.74 N due North, 7.52 N due West **(b) (i)** 2 N due North **(ii)** 12.8 N at 39° East of due North
(iii) 15.6 N at 26.3° East of due North
Q9.2 (a) (i) 1.0 N upward **(ii)** 7.81 N acting upwards at 40° to the vertical
Q9.3 (b) X: 48 N; Y: 32 N **(c)** 0.56 m
Q9.4 (a) 1.20 m **(b)** 70 N
Q10.1 (a) 15 m **(b)** 6.25 m s^{-2}
Q10.2 (b) (i) 0.2 m s^{-2}, 0, −0.25 m s^{-2}
(ii) 563 m, 3600 m, 450 m
(c) 4613 m, 12.3 m s^{-1}
Q10.3 (a) (i) 13.7 m s^{-1} **(ii)** 9.6 m
Q10.4 (a) (i) 1.6 s, **(ii)** 12.5 m
(iii) 15.7 m s^{-1} **(iv)** 15.7 m s^{-1}
Q11.1 (a) (i) −3.0 m s^{-2} **(ii)** 30 m s^{-1}
(iii) 150 kg m s^{-1}
(b) (i) 3.6 kg m s^{-1} **(ii)** 144 N
Q11.2 (a) (i) 12.8 m s^{-1} **(ii)** −1960 m s^{-2}
(iii) 7840 N
(b) (i) 2.6 m s^{-2} **(ii)** 98.7 kN
Q11.3 (a) (i) 1.87 kg m s^{-1} **(ii)** 1.87 N
(b) (i) 2030 m s^{-2}
(ii) 25.4 kN **(iii)** 2.7 m s^{-1}
Q11.4 (a) 1.03 m s^{-1} in the same direction as the first wagon
(b) 0.04 m s^{-1} in the opposite direction to the first wagon
Q12.1 (a) (i) 3.53 J, 3.53 J **(ii)** 0, 7.06 J
(b) (i) 1.18 J, 5.42 m s^{-1}
(ii) 0.94 J, 4.85 m s^{-1} **(iii)** 0.24 J
Q12.2 (a) 1.35 MJ **(b)** 0.78 MJ **(c)** 0.57 MJ
Q12.3 (a) 13 MW **(b)** 63 kW
Q12.4 (a) (i) 11 m s^{-1} **(ii)** 99 kJ ,
(b) (i) 80 kJ m^{-1} **(ii)** 80 kN
Q13.2 (a) (i) 9.0 Ω **(ii)** 2.0 Ω **(b)** 4.0 Ω: 1.0 A, 4.0 V, 4.0 W; 3.0 Ω: 0.67 A, 2.0 V, 1.33 W; 6.0 Ω: 0.33 A, 2.0 V, 0.67 W
Q13.3 (a) (i) 10.4 A **(ii)** 1870 C **(iii)** 450 kJ
Q13.4 (b) (iii) diode: 0.36 mW; resistor: 0.54 mW; cell: 0.90 mW
Q14.1 (a) 3.0 A **(b)** 7.5 V, 1.5 A
(c) (i) 36 W, 11.25 W **(ii)** 13.5 W dissipated due to internal resistance in the battery
Q14.2 (a) 6.1 Ω **(b)** 291 mm from the end at S
Q14.3 (a) (ii) 240 kΩ
Q14.4 (a) 4.81×10^{-7} Ω m **(b)** 592 mm
Q15.1 (a) 72 mC **(b)** 22 mF
Q15.2 (a) (i) 18 μF **(ii)** 4 μF **(b) (i)** 6 μF: 24 μC, 48 μJ; 12 μF: 24 μC, 24 μJ **(ii)** 24 μC, 72 μJ
Q15.3 (a) (i) 300 pF **(ii)** 3.6 nC, 21.6 nJ
(b) (i) 600 pF **(ii)** 6.0 V
Q15.4 (a) 42 μC, 190 μJ, **(b)** 6.0 μA,
(c) 21 μC, 46 μJ
Q16.2 (b) 3.0 V
Q16.3 (b) 2 kΩ
Q17.1 (a) 9.6×10^{10} V m^{-1}, 2.88 V
(b) 4.6×10^{-18} J
Q17.2 (a) 1.94×10^{-5} C
(b) 7.56 kV m^{-1}, 918 V
Q17.3 (a)(i) 125 kV m^{-1} **(b)** 1.35×10^{-11} J
Q17.4 (b) (i) 76.6 nC **(ii)** 600 V
Q18.1 (b) (i) 15.2 A **(ii)** 3.8 mT
Q18.2 (a) 55 mT **(b)** 0.96 mN
Q18.3 (a) 18 μT
Q19.1 (b) 0.19 V
Q19.2 (b) (i) 61 mWb **(ii)** 7.7 V
Q19.3 (b) (i) 11.4 V **(ii)** 1.72 A
Q19.4 (a) 12 V **(b)** 5.0 A
Q20.1 (a) 1.60 A, 50 Hz **(b) (i)** 1.13 A
(ii) 1.60 A **(iii)** −0.94 A
Q20.2 (a) (i) 1.7 A **(ii)** 17.0 V **(b) (i)** 1.2 A
(ii) 7.2 W
Q20.3 (a) (i) 2.43 mA **(ii)** 4.86 V across R, 3.52 V across C
(iii) 11.8 mW **(b) (i)** 2.95 mA **(ii)** 5.90 V across R, 1.09 V across C **(iii)** 17.4 mW
Q20.4 (a) 848 Hz **(b) (i)** 54 Ω **(ii)** 0.11 A
(iii) 8.3 V
Q21.1 (b) 7.2×10^{-16} J, 4.0×10^7 m s^{-1}
Q21.2 (b) (i) 2.2×10^7 m s^{-1}, 2.2×10^{-16} J
Q21.3 (a) 0.18 mm s^{-1} **(b)** 1.3×10^{-6} m, 8.2×10^{-15} kg
(c) 3.2×10^{-19} C
Q21.4 (a) 1.3×10^{19} Hz **(b)** 2.77×10^{-19} J, 4.2×10^{14} Hz
Q22.1 (b) (i) 0.16 MBq **(ii)** 270 Bq
Q22.2 (b) 1.37 mm
Q22.3 (b) 1770 years
Q22.4 (a) 360 kBq **(b)** 220 counts per second
Q23.1 (a) 2.0 fm
Q23.2 (b) (i) 7.1 MeV **(ii)** 7.4 MeV
Q23.4 (b) 2.82 MeV **(c)** 4.1×10^{13} Bq
Q24.1 (a) (i) 4.9 **(ii)** 22.7 kPa **(b)** 18.4 cm^3
Q24.2 (a) (i) 85 kPa **(ii)** 16 kPa
(b) 2.4×10^{24} m^{-3}
Q24.3 (b) (i) 12.4 moles **(ii)** 0.35 kg
(iii) 508 m s^{-1} **(iv)** 6.0×10^{-21} J
Q24.4 (b) (i) 474 m s^{-1} **(ii)** 461 m s^{-1}
(iii) 1845 m s^{-1}
Q25.1 (b) (i) 39°C **(ii)** 312 K
Q25.2 (a) (i) 2.4 moles **(ii)** 8.8 kJ
(b) (i) 129 kPa **(ii)** 2.4 kJ
Q25.3 (a) (i) 0.100 m^3 **(ii)** 0.116 m^3
(b) 1.63 kJ **(b)(ii)** 3.29 kJ
Q25.4 (a) 1720 kPa, **(b)** 0.55
Q26.1 (a) 0.10 s **(b)** 11.0 m s^{-1}, 690 m s^{-2}
Q26.2 (a) 46 ms **(b)** 22 Hz **(c)** 4270 m s^{-2}
Q26.3 (a) 9.31 kN **(b)** 4.5 m s^{-2}
(c) 5.03 kN
Q26.4 (a) 1.11 J **(b)** 4.31 m s^{-1}
(c) 19.6 m s^{-2} **(d)** 3.53 N
Q27.1 (a) 453 N **(b)** 6.52 km s^{-1}, 2 hours 31 minutes
Q27.2 (a) 278 km
Q27.3 (b) (ii) 2.37 km s^{-1}
Q27.4 (a) 5.9×10^{-3} N kg^{-1}
(b)(i) 3.7×10^9 m **(ii)** 0.025
Q28.1 (b) (i) 2.52 s **(ii)** 65 mm
(iii) 1.62 m s^{-1}
Q28.2 (a) 1.94 s **(b) (i)** 81 mm
(ii) 0.262 m s^{-1} **(iii)** 0.850 m s^{-2}
Q28.3 (a) (i) 51 N m^{-1} **(ii)** 0.44 s
(b) (i) 0.29 m s^{-1} **(ii)** 1.43 N

Glossary

absolute temperature The scientific scale of temperature, measured in kelvins (K). To convert from °C to K, add 273 to the temperature in °C. Absolute temperature in K = temperature in °C + 273.15.

absolute zero The lowest possible temperature (−273.16°C or 0 K).

acceleration The change of velocity of an object per second. The unit of acceleration is the metre per second^2 ($m\ s^{-2}$).

activity Number of radioactive nuclei that decay per second. The unit of activity is the becquerel (Bq).

adiabatic change A change in which no heat transfer occurs.

alpha particle A particle consisting of two protons and two neutrons emitted by a large unstable nucleus.

alternating current Current that repeatedly reverses its direction . A sinusoidal alternating current may be represented by the equation $I = I_o \sin(2\pi ft)$.

amorphous solid A solid in which the atoms are arranged in random order.

amplitude The maximum displacement of a wave from equilibrium (e.g. the height of a transverse wave from the middle).

amplitude modulation Variation of the amplitude of a carrier wave to carry information.

angular frequency 2π/time period.

angular speed Rate of rotation of an object.

anode Positive terminal in an electrolytic cell or an electron tube.

antimatter Made of antiparticles.

antinode A point of maximum displacement in a stationary wave pattern.

antiparticle An antiparticle and its particle annihilate each other to produce radiation.

atomic mass The mass of an atom on a scale where one atom of carbon-12 has a mass of exactly 12 units. The mass of a proton or a neutron is 1 on this scale.

atomic mass number (A) The number of protons and neutrons in the nucleus of an atom.

atomic number (Z) the number of protons in the nucleus of an atom.

avogadro constant, (N_A) the number of atoms in exactly 12 grams of carbon-12.

background radioactivity Radioactivity due to materials in buildings or due to cosmic radiation.

bandwidth The frequency width of a communication channel (e.g. a channel from 100 to 104 kHz has a bandwidth of 4 kHz).

barometer Instrument used to measure atmospheric pressure.

beta particle An electron created and emitted from an unstable nucleus.

binding energy The work done to separate a nucleus into its constituent nucleons.

bourdon gauge A pressure gauge used to measure the pressure of a gas or a liquid in a pipe.

Brownian motion The unpredictable motion of microscopic particles due to random and uneven collisions with fast-moving molecules.

byte Series of bits (1's and 0's) that are coded to carry information.

capacitor A device that can store charge.

cathode Negative terminal in an electrolytic cell or an electron tube.

cathode rays A beam of electrons.

centre of gravity The point where the weight of an object as a single force is considered to act.

centripetal acceleration Acceleration of an object on a circular path due to its circular motion.

centripetal force The force on an object that makes it go around on a circular path; the direction of the centripetal force is always towards the centre of the circle.

chain reaction Series of reactions in which each reaction causes one or more further reactions; a steady chain reaction is where each reaction causes exactly one more reaction.

change of state Change of physical state (e.g. solid changes to liquid).

chromatic aberration Splitting of white light into colours by a lens or a prism.

concave lens A lens that makes a parallel beam of light diverge.

condensation This occurs when a vapour turns to liquid.

conservation of energy The total energy is unchanged when energy changes from one form into other forms.

conservation of momentum When two or more objects collide, the total momentum is unchanged, provided no external forces act.

continuous spectrum A spectrum in which the colours are continuous from one end to the other.

convection Fluid flow due to hot fluid rising and cold fluid falling.

convex lens A lens that focuses a parallel beam of incident light.

covalent bond A bond in which electrons are shared.

critical angle The angle of incidence of a light ray that refracts along the boundary between two transparent substances.

damping Forces that cause an oscillating system to lose energy.

decay constant ln 2/half-life.

decibels A measure of the loudness level of a sound where the loudness increases tenfold for every extra ten decibels.

dielectric substance A substance that increases the capacitance of a capacitor.

diffraction The spreading of waves after passing through a gap or around an obstacle; the narrower a gap is, the greater the amount of diffraction there is.

diffraction grating A plate ruled with many regularly spaced slits.

diffusion Spreading of a substance into a fluid without stirring (e.g. a coloured crystal dissolves in water and the water gradually becomes evenly coloured).

digital circuit An electric circuit in which the voltage at any point can only take two values (high '1' or low '0').

diode A device that lets current through in one direction only.

dispersion The splitting of white light into colours by a prism; this happens because the speed of light in glass depends on the colour of the light.

displacement Distance from a given point in a certain direction; displacement is a vector quantity.

dynamo effect Generation of an e.m.f. due to changing magnetic flux.

echo Sound waves that reflect off a smooth surface.

eddy currents Unwanted currents in a metal caused by electromagnetic induction.

efficiency Fraction (or percentage) of the energy supplied to a machine that is used by the machine to do useful work.

elastic energy Energy stored in a body when its shape is changed.

elastic limit Limit beyond which a material is permanently extended when stretched.

elasticity Ability of a material to regain its shape after being distorted.

electric field Region where a force is exerted on an electric charge.

electric field strength (E) The force per unit charge on a test charge.

electric potential (V) The work done per unit charge to move a test charge from infinity.

electromagnetic induction A voltage induced in a coil as a result of a magnet moving near the coil or the magnetic field through a coil changing.

electromagnetic spectrum Radio waves, microwaves, infra-red radiation, light, ultraviolet radiation, X-rays, gamma radiation.

electromotive force (e.m.f.) The electrical energy produced per unit charge by a source of electrical energy; the unit of e.m.f. is the volt (V).

electron Negative particle in the atom.

electron volt The work done to move an electron through a p.d. of 1 volt.

element A substance that cannot be broken down by chemical methods into other substances.

enthalpy Heat transfer at constant pressure.

equipotential A surface of constant potential.

escape speed Mininum speed needed by an object to leave a planet.

evaporation Change of state from a liquid state to a gas.

excitation Process in which an electron in an atom moves to a higher energy level.

expansion Increase of volume of a substance when heated (e.g. increase in the length of a bar when it is heated).

exponential decay Process in which a quantity decreases by 50% every half-life.

extension Increase in length of a solid when it is stretched.

Faraday's law of electromagnetic induction The induced e.m.f. is proportional to the rate of change of magnetic flux linkage.

feedback Part of the output voltage from an amplifier is returned to the input.

ferromagnet A material that can be magnetised permanently (e.g. iron or steel).

fission When an unstable nucleus splits into approximately two equal fragments.

fluid Any liquid or gas.

focal length of a convex lens The distance from the lens to the point where parallel rays are brought to a focus.

force Anything that can change the state of motion of an object.

fossil fuel Material formed from the decay of long-dead organisms that releases energy when it undergoes chemical change.

free-body force diagram A diagram showing a body with the forces acting on it.

frequency The number of complete cycles per second made by a vibrating object. The unit of frequency is the hertz (Hz), equal to 1 cycle per second.

frequency modulation Variation of the frequency of a carrier wave to carry information.

fundamental frequency The lowest frequency of vibration of a string or pipe.

fuse A thin wire that melts when excess current passes through it.

fusion Thermal fusion is when a solid melts; nuclear fusion is when two nuclei are forced to join together.

gradient The gradient of a straight line on a graph is the increase of vertical distance/increase of horizontal distance.

gravitational potential The work done per unit mass to move a test mass from infinity.

gravity The force of attraction between any two objects due to their mass.

half-life The time taken for the number of atoms of a radioactive isotope to decrease to half.

hardness Resistance of a surface to scratches and dents.

heat Energy transfer due to temperature difference.

heat exchanger A device in which thermal energy is transferred from a hot fluid to a cooler fluid without the fluids mixing with each other.

Hooke's law The extension of a spring is proportional to the tension in the spring.

hydrometer An instrument used to measure the density of a liquid.

ideal gas A gas that obeys Boyle's law.

induced e.m.f. A voltage induced as a result of changing magnetic flux through a circuit.

inductance Flux linkage per unit current.

insulator A thermal insulator is a substance that will not conduct heat; an electrical insulator is a substance that will not conduct electricity.

interference When two or more sets of waves meet, points of cancellation and reinforcement are formed. Reinforcement occurs where a crest meets a crest or a trough meets a trough; cancellation occurs where a crest meets a trough.

internal energy The total kinetic and potential energy of the molecules due to their individual movements and positions.

internal resistance The resistance inside a source of electrical energy.

ion An atom that has become charged due to loss or gain of electrons.

ionic bond A chemical bond between two particles in which one or more electrons are transferred between the particles.

ionisation Formation of an ion by removal of an electron from an atom.

ionising radiation Radiation that causes ions to form; alpha radiation, beta radiation, X- and gamma radiation are all ionising.

Isothermal change A change at constant temperature.

Isotopes Atoms that have different numbers of neutrons but the same number of protons.

kilowatt hours The energy supplied to a 1 kW electrical appliance in 1 hour. One kilowatt hour (kW h) = 3.6 million joules (= 1000 W × 3600 s).

kinetic energy Energy of a moving object.

LDR A light-dependent resistor.

LED A light-emitting diode.

laser A light source that produces a coherent, monochromatic parallel beam of light.

latent heat Energy needed or released when a material changes its physical state.

Lenz's law The induced current always acts in a direction so as to oppose the change.

line emission spectrum The pattern of coloured lines seen when light from a slit illuminated by a suitable light source is passed through a prism.

linear magnification Image height ÷ object height.

logic gate A digital circuit which gives an output voltage that is determined by the state of the inputs.

longitudinal waves Waves in which the vibrations are parallel to the direction of travel of the wave.

loudness A measure of the amplitude of a sound wave.

magnetic field A region in which the pole of a magnet experiences a force.

magnetic field strength (B) See magnetic flux density.

magnetic flux Magnetic flux density, B × area, A.

magnetic flux density (B) The force per unit length per unit current on a current-carrying wire perpendicular to the lines of force of the field.

magnifying glass A convex lens used to make an object appear larger.

manometer An instrument used to measure the pressure of a gas or a liquid.

mass number The number of neutrons and protons in a nucleus.

metallic bonding A type of bonding which holds atoms together in a metal.

micrometer An instrument used to measure thicknesses to within 1/100th of a millimetre.

microwaves Electromagnetic waves of wavelength between about 0.1 mm and 10 mm.

moderator Used in a nuclear reactor to slow the neutrons down to enable further fission to occur.

modulation Where radio waves or light waves are used to carry information.

mole The amount of matter consisting of Avogadro's number of identical particles.

molecular bond Weak, attractive bond between two uncharged atoms or molecules.

molecule The smallest particle of a compound or element that can exist independently.

moment of a force about a point Force × perpendicular distance from the line of action of the force to the point.

momentum Mass × velocity.

motor effect The force on a current-carrying wire in a magnetic field. The wire mustn't be parallel to the lines of force of the field.

multimeter A meter that may be used as an ammeter or a voltmeter.

multivibrator A circuit which switches between two output states.

neutron An uncharged particle inside the nucleus that has approximately the same mass as the proton.

Newton's first law of motion An object remains at rest or in uniform motion unless acted on by a force.

Newton's second law of motion The force on an object is proportional to the change of momentum per second of the object. The equation for Newton's second law for constant mass is 'force = mass × acceleration'.

node A point of no displacement in a stationary wave pattern.

noise Sound produced at random.

nuclear energy Energy released when a heavy nucleus splits or light nuclei fuse together.

nuclear fission See fission.

nucleon A neutron or proton in the nucleus.

nucleus The nucleus of an atom is a tiny part of the atom where most of the atom's mass is located and which has a positive charge. It consists of neutrons and protons.

Ohm's law The current through a conductor that obeys Ohm's law is proportional to the potential difference across the conductor, provided the physical conditions do not change.

operational amplifier An amplifier designed to carry out mathematical operations.

orbit The path of a planet around the Sun or the path of a satellite around a planet.

oscilloscope An electronic instrument used to display and measure waveforms.

parallel components Components connected together in a circuit with the same potential difference across their terminals.

parallelogram of forces The rule used to add together two force vectors.

phase difference The fraction of a cycle between the motion of two objects vibrating at the same frequency.

photon Light consists of photons. Each photon is a wave packet of electromagnetic energy emitted by an atom; $E = hf$.

pitch The frequency of a note or sound.

plane mirror A flat mirror.

plastic behaviour Permanent change of shape of an object.

polarised waves Transverse waves that vibrate in one plane only.

polymer A solid composed of long molecules, each made from smaller, identical molecules.

positron The antimatter particle of the electron. If a positron and an electron collide, they annihilate each other to produce radiation.

potential difference (or voltage) The work done per unit charge when charge flows between two points in a circuit. The unit of potential difference is the volt, equal to 1 joule per coulomb.

potential divider Two resistors in series connected to a fixed voltage supply. The potential difference of the supply is shared or 'divided' between the two resistors.

potential energy Energy due to position.

potentiometer A potential divider used to compare cell e.m.f.s.

power Energy transferred per second; the unit of power is the watt (W), equal to 1 joule per second. Note that 1 kilowatt = 1000 watts.

prefixes Roman or Greek letters used to denote powers of ten. Prefixes to remember include milli- ($m = 10^{-3}$), micro- ($\mu = 10^{-6}$), kilo- ($k = 10^3$), and mega- ($M = 10^6$).

pressure Force per unit area applied at right angles to a surface. The unit of pressure is the pascal (Pa), equal to 1 newton per square metre.

primary colours of light Red, blue or green light.

principle of moments The sum of the clockwise moments about a fixed point is equal to the sum of the anticlockwise moments about the same point.

projectile An object in motion acted upon only by the force of gravity.

proton A positive particle inside the nucleus that has approximately the same mass as the neutron and a charge equal and opposite to that of the electron.

pulse code modulation Using a carrier wave to transmit pulses in a coded sequence.

quark Protons and neutrons are composed of quarks.

radio waves Electromagnetic waves of wavelengths longer than about a millimetre.

radioactive decay Disintegration of an unstable nucleus of an atom as a result of emitting an alpha or a beta particle or gamma radiation.

radioisotope An isotope that is radioactive.

random motion Unpredictable motion of a particle.

reactance $1/2\pi fC$ for a capacitor; $2\pi fL$ for an inductor.

real image An image that can be projected on a screen.

rectification Conversion of alternating current into direct current, achieved using one or more diodes.

reflection This occurs when a wave bounces off a smooth surface. For a light ray reflecting off a mirror, the angle of incidence is equal to the angle of reflection.

refraction This occurs when a wave changes direction on passing from one substance to another. A light ray passing from air to glass bends towards the normal.

refractive index of a substance The ratio of the wavelength in air to the wavelength in the substance.

relay An electromagnetic device in which a small current is used to switch a larger current on or off.

renewable energy Energy directly or indirectly from the Sun that does not need a chemical change to become available.

resistance The resistance of a resistor is defined as the potential difference ÷ current. The unit of resistance is the ohm (Ω), equal to 1 volt per ampere.

resistivity Resistance × area of cross section/length.

resistors in parallel The potential difference is the same; the current divides between the branches; the combined resistance of two resistors, R_1 and R_2, in parallel = $(1/R_1 + 1/R_2)^{-1}$.

resistors in series The current is the same; the potential difference divides across the components; the combined resistance of two resistors, R_1 and R_2, in series = $R_1 + R_2$.

resonance An oscillating system subjected to regular pushes at its natural frequency vibrates with a large amplitude.

root mean square current The value of direct current that gives the same power in a given resistor.

root mean square speed Square root of the mean of the squared speeds of the molecules in a gas.

scalar Any physical quantity that is not directional (e.g. speed, mass, energy).

semiconductors Substances that conduct better with increasing temperature.

series components Components in an electric circuit that are connected together so they pass the same current.

sound Vibrations in a substance that travel through the substance.

specific heat capacity The energy needed to raise the temperature of 1 kg of material by 1 K.

specific latent heat The energy lost or gained by 1 kg of material when it changes its physical state.

spectrometer An instrument used to measure accurately the wavelengths of a spectrum.

spectrum The colours of light seen when a parallel beam of white light is split into colours by a prism.

speed Distance ÷ time. The unit of speed is the metre per second.

spherical aberration A lens defect in which the outer rays of a parallel beam are brought to a different focus to that of the inner rays.

sphygmomanometer Used to measure blood pressure.

standard pressure The mean pressure of the atmosphere at sea level.

stationary waves Wave pattern formed in which there are fixed points of no displacement.

stiffness A measure of the force needed to change the shape of an object.

strain Change of length per unit length.

strength A measure of the force needed to break an object.

stress Force per unit area of cross-section acting perpendicular to an area.

sublimation Occurs when a solid changes directly to a vapour.

temperature A measure of the hotness of an object. Temperature is measured in degrees Celsius (°C) or in kelvins (K).

tension The force in an object that has been stretched.

terminal speed The speed that an object acted on by a constant force reaches when moving through a fluid.

thermal conduction Heat transfer through a substance due to individual movement of particles in the substance.

thermal expansion The increase of the length of an object due to increase of temperature.

thermal radiation Electromagnetic radiation emitted from a body due to its temperature.

thermionic emission The emission of electrons from a hot metal surface.

thermistor A resistor with a resistance designed to change with temperature.

thermocouple A junction formed by two metal wires; the potential difference across the junction depends on the temperature of the junction. A thermocouple connected to a suitable meter may be used as a thermometer.

total internal reflection This is seen when a light ray in a substance strikes the boundary of the substance and reflects totally. It occurs when the angle of incidence exceeds the critical angle.

toughness Ability of a material to withstand impacts.

transformer A device that steps an alternating voltage up or down.

transistor An electronic device in which a small current controls a much larger current.

transverse waves Waves in which the vibrations are at right angles to the direction of travel of the wave.

ultraviolet radiation Electromagnetic radiation between the violet end of the visible spectrum (wavelength about 400 nm) and X-rays (wavelengths less than about 1 nm).

ultrasonics Sound waves of frequency higher than 18 kHz which is the upper frequency range of the human ear.

upthrust The upward force on an object in a fluid due to the pressure of the fluid.

U-tube manometer A device used to measure the pressure of a gas or a liquid.

U-value The heat loss per second per square metre for a 1 K (or 1°C) temperature difference across a wall or window or roof or floor.

vaporisation The change of physical state when a liquid becomes a vapour.

vapour A gas that can be turned into a liquid by compression only.

vector Any physical quantity that has a direction as well as a magnitude (e.g. velocity, acceleration, force, weight, momentum).

velocity Speed in a given direction.

vernier calipers An instrument used to measure distances up to 200 mm or more to within 0.1 mm.

virtual image An image that is formed where light rays appear to come from after reflection by a mirror or refraction at a boundary.

viscosity A measure of the force needed to make a liquid flow.

voltage Another word for potential difference.

voltage gain Output voltage ÷ input voltage.

wavelength The distance between two adjacent wave-crests.

weight The force of gravity on an object. For a mass, m, its weight = mg.

wheatstone bridge A circuit used to compare resistances.

work Energy transferred by a force.

work function The work necessary for an electron to leave an uncharged surface.

X-rays Electromagnetic waves of wavelength less than about 1 nm.

Young modulus of elasticity Stress ÷ strain.

Index

A page number followed by *N* indicates that relevant information on that page is given only in a boxed note.